STEREOCHEMISTRY OF HETEROCYCLIC COMPOUNDS

GENERAL HETEROCYCLIC CHEMISTRY SERIES

Edward C. Taylor and Arnold Weissberger, Editors

MASS SPECTROMETRY OF HETEROCYCLIC COMPOUNDS
by Q. N. Porter and J. Baldas

NMR SPECTRA OF SIMPLE HETEROCYCLES
by T. J. Batterham

HETEROCYCLES IN ORGANIC SYNTHESIS
by A. I. Meyers

PHOTOCHEMISTRY OF HETEROCYCLIC COMPOUNDS
by Ole Buchardt

STEREOCHEMISTRY OF HETEROCYCLIC COMPOUNDS
Part I Nitrogen Heterocycles
by W. L. F. Armarego

Part II Oxygen; Sulfur; Mixed N, O, and S;
and Phosphorus Heterocycles
by W. L. F. Armarego

STEREOCHEMISTRY OF HETEROCYCLIC COMPOUNDS

Part II Oxygen; Sulfur; Mixed N, O, and S; and Phosphorus Heterocycles

W. L. F. ARMAREGO

The Australian National University
Canberra, Australia

With a chapter on phosphorus heterocycles by
M. J. GALLAGHER

The University of New South Wales
Sydney, Australia

A Wiley-Interscience Publication

JOHN WILEY & SONS New York • London • Sydney • Toronto

Library
I.U.P.
Indiana, Pa.

547.59 Ar54s
pt.2, c.1

Copyright © 1977 by John Wiley & Sons, Inc.

All rights reserved. Published simultaneously in Canada.

No part of this book may be reproduced by any means, nor transmitted, nor translated into a machine language without the written permission of the publisher.

Library of Congress Cataloging in Publication Data:

Armarego, W L F
Stereochemistry of heterocyclic compounds.

(General heterocyclic chemistry series)
Includes bibliographical references.
CONTENTS: pt. 1. Nitrogen heterocycles.—pt. 2. Oxygen, sulfur, mixed N, O, and S, and phosphorus heterocycles.
1. Heterocyclic compounds. 2. Stereochemistry.

I. Title.
QD400.3.A75 547′.59 76-26023
ISBN 0-471-03322-7

Printed in the United States of America

10 9 8 7 6 5 4 3 2 1

To the memory of

E. E. Turner, F.R.S.

INTRODUCTION TO THE SERIES

General Heterocyclic Chemistry

The series, "The Chemistry of Heterocyclic Compounds," published since 1950 by Wiley-Interscience, is organized according to classes of compounds. Each volume deals with syntheses, reactions, properties, structure, physical chemistry, etc., of compounds belonging to a specific class, such as pyridines, thiophenes, and pyrimidines, three-membered ring systems. This series has become the basic reference collection for information on heterocyclic compounds.

Many aspects of heterocyclic chemistry have been established as disciplines of *general* significance and application. Furthermore, many reactions, transformations, and uses of heterocyclic compounds have specific significance. We plan, therefore, to publish monographs that will treat such topics as nuclear magnetic resonance of heterocyclic compounds, mass spectra of heterocyclic compounds, photochemistry of heterocyclic compounds, X-Ray structure determination of heterocyclic compounds, UV and IR spectroscopy of heterocyclic compounds, and the utility of heterocyclic compounds in organic synthesis. These treatises should be of interest to *all* organic chemists as well as to those whose particular concern is heterocyclic chemistry. The new series, organized as described above, will survey under each title *the whole field of heterocyclic chemistry* and is entitled "General Heterocyclic Chemistry." The editors express their profound gratitude to Dr. D. J. Brown of Canberra for his invaluable help in establishing the new series.

Department of Chemistry
Princeton University
Princeton, New Jersey

Edward C. Taylor

Research Laboratories
Eastman Kodak Company
Rochester, New York

Arnold Weissberger

PREFACE

In 1974, the periodical *Tetrahedron* commemorated the centenary of van't Hoff and Le Bel's proposal of the tetrahedral carbon atom by devoting 500 pages to articles on stereochemistry. Over the years many aspects of the stereochemistry of organic compounds have been reviewed, but the stereochemistry of heterocyclic compounds has never been considered separately and described in its entirety. It is therefore timely that the stereochemistry of heterocyclic compounds should be collected in a systematic form in two volumes.

The availability of nuclear magnetic resonance spectrometers in the early sixties is mainly responsible for the explosion in the number of publications containing stereochemical data on heterocycles. Not only has nuclear magnetic resonance been useful in deducing relative configurations at chiral centers and conformational preferences with a high degree of certainty, but is has also provided thermodynamic and kinetic data for a variety of equilibria from which stereochemical data have been evaluated. For this reason a large proportion of the literature surveyed in these volumes is post-1960. The literature has been covered until the end of 1974 and incompletely for 1975. More than 4500 references on the stereochemistry of nitrogen, oxygen, sulfur, and phosphorus heterocycles are included in the two volumes, and to keep the books to a reasonable size it was necessary to limit the discussions for many references. Only a few leading references to the stereochemistry of natural products are given to identify them with the ring systems under discussion. These monographs take the form of "guides" into the literature and at the same time give a panoramic view of the stereochemistry of nitrogen heterocycles (in Part I) and oxygen; sulfur; mixed N, O, and S; and phosphorus heterocyclic compounds (in Part II).

No words can express my gratitude to Dr. D. J. Brown for his continued and inspiring guidance, and for his provision of every facility possible during the months of writing. The manuscript would have taken much longer to produce but for the efficient and accurate work of my research assistant, Mrs. Beverly A. Milloy, B. Sc, who carried out the painstaking job of checking the references and the typescript, and whose artistic talent has turned all the formulas into proper drawings that show the three-dimensional structures. My wife assisted immensely in reading and proofreading the whole manuscript. Finally, I am thankful to Mrs. A. Sirr for carrying out the arduous task of typing the manuscript.

W. L. F. Armarego

Canberra, Australia
The Australian National University, Canberra
June 1976

CONTENTS

Chapter 3 Sulfur Heterocycles

1 INTRODUCTION

One generally associates heterocyclic compounds with planar molecules because heterocycles are not usually regarded in three-dimensional terms. A glance at the reviews in the well established series *Advances in Heterocyclic Chemistry*[1] will confirm this statement. There are, however, a large number and variety of heterocyclic compounds that possess stereochemical properties. Many (but not all) of these heterocycles belong to the known classes of heterocyclic compounds, and in a large number of cases are related to or derived from them by reduction. It is the purpose of this monograph to describe systematically the stereochemical aspects of oxygen; sulfur; mixed N, O, and S; and phosphorus heterocycles (the nitrogen heterocycles are in Part I).

I. GENERAL STEREOCHEMICAL PROPERTIES OF HETEROCYCLIC COMPOUNDS

The stereochemical properties of heterocyclic compounds arise in a variety of ways. The first to be considered is the transition state in the electrophilic and nucleophilic substitution reactions of "aromatic" heterocycles. Here the stereochemistry of the approach of the reagent and the departure of the substituent involved, together with the nonplanar structure of the transition state, have to be accounted for in a full understanding of the processes. Cycloaddition reactions must also be considered in the above terms.

Reduced heterocyclic compounds possess many of the stereochemical features of their carbocyclic analogues. The presence of substituents on the carbon atoms in the ring can introduce chiral centers and display cis and trans isomerism. In addition to these properties, the heteroatom also alters the geometry of the ring with respect to the carbocyclic analogue by changing the bond distances and bond angles adjacent to the heteroatom. The alterations may be small or large depending on the heteroatom, the number of heteroatoms, and the relative positions of the heteroatoms in the ring. These changes are not relatively large when one heteroatom is involved, and the general structure of the reduced heterocycle can be extrapolated from the known structure of the analogous carbocycle. The differences are, however, subtle and show up in the nonbonded interactions and consequently in the conformational properties of the molecules.

The heteroatom in reduced heterocycles introduces properties which are characteristic of the heteroatom itself. Pyramidal atomic inversion[2] is a property which distinguishes nitrogen, oxygen, sulfur, and phosphorus atoms from tetravalent carbon. Trivalent nitrogen in a conformationally flexible ring, for example, inserts another conformational property to the ring. The hydrogen atom or substituent on the nitrogen atom in a ring can attain two equilibrating conformations by virtue of the inverting nitrogen atom. This points out the conformational property of the nitrogen lone pair of electrons, and considerable attention has been devoted in several laboratories to the "size" (space demand) of the nitrogen lone pair. The rate of atomic inversion is affected by the size of the ring in which the nitrogen atom is inserted and by the substituents. An oxygen atom directly attached to the ring nitrogen atom can decrease its inversion rate to the extent that the nitrogen atom is almost "locked" in a chiral configuration, and it introduces a source of optical activity. The inversion rate, and the effect of substituents on it, varies from one heteroatom to another. The oxygen heteroatom also undergoes inversion, but it is not as interesting as the nitrogen atom because it is divalent. Inversion of the oxygen atom does not alter the situation because it has two lone pairs of electrons which exchange place during the inversion without effecting a serious alteration. Oxonium ions derived from saturated oxygen heterocycles, on the other hand, will exhibit pyramidal inversion, but these present experimental difficulties because of their chemical reactivity. The properties of some oxonium ions, however, have been reported. The oxygen atom in reduced heterocycles produces strong dipolar effects with respect to polar substituents particularly on the adjacent carbon atoms. It tends to force the substituents into an axial conformation (anomeric effect) which affects the conformational properties of the molecule as a whole. Sulfur heterocycles are similar to oxygen heterocycles, but in addition the sulfur atom can expand its valency shell. Oxidation of the sulfur atom yields sulfoxides which have very high barriers to atomic inversion and are therefore possible centers of asymmetry in the molecule. Sulfur forms stable S–S bonds, and these adjacent sulfur atoms in a heterocyclic ring are a source of chirality in their own right by virtue of the helical nature of the C–S–S–C bonding arrangement. The energies involved in altering the torsion angles in this bonding arrangement are greater than in the corresponding carbocyclic systems and therefore have a large effect on the conformational properties of these sulfur containing molecules. Unlike the above heterocycles, many phosphorus heterocycles are known in which the valency state of the phosphorus atom is II, III, IV, V or VI. The phosphorus atom has a particular stereochemistry in each valency state. The heteroatom in phosphorus(III) heterocycles is similar to a nitrogen atom in its pyramidal atomic inversion but is generally much slower. The most interesting examples are found among the P(V) compounds. The trigonal bipyramid P(V) compounds undergo permutational isomerism (not observed in the other heteroatoms) in which the energy required for substituents to exchange places, for example, in inversion of configuration, without breaking and making bonds may be quite low, and the process may occur very readily. Not only have many of these P(V) structures been postulated as intermediates or transition states in reactions, but a host of them have been isolated and characterized.

Several dissymmetric heterocyclic molecules are known which exhibit optical activity. Biheteroaromatic compounds which owe their dissymmetry to restricted rotation about the bond which joins the two heteroaromatic rings have been resolved into their optical antipodes. Their optical stabilities, when compared with their carbocyclic counterparts, demonstrate the steric and other effects attributed to the heteroatoms. Rigid dissymmetric molecules such as heterotwistanes, heterohelicenes, and heteroadamantanes can also exhibit optical activity if the heteroatoms are placed in appropriate positions. A number of examples are known in which the heteroaromatic ring has a serious effect on the chiroptical properties of a chiral center in a side chain when there is enough interaction between the heteroaromatic ring and the asymmetric center.

The three-dimensional properties of many bi- and polycyclic reduced heterocycles, for example, heterobicyclo[*x.y.z*]alkanes, may not have special stereochemical features but are discussed in this monograph because a knowledge of the relative arrangement of all the atoms in space is necessary for a complete understanding of the properties and reactions of the systems. Transannular interactions due to the spatial proximity of a heteroatom and another distant atom in the same ring have been observed and have led to interesting polycyclic systems.

Short introductions to chapters and to several sections in chapters are provided throughout, and should be consulted for further general stereochemical properties of the relevant classes of compounds.

II. ORGANIZATION OF STEREOCHEMICAL DATA

The general arrangement of data for each class or group of compounds is in four basic parts. In the first part the stereochemical course of the syntheses of heterocyclic compounds is described. This is followed by a discussion of the configuration and then the conformation of heterocyclic molecules. The stereochemical reactions of heterocycles are reviewed in the fourth part.

A clear distinction between configuration and conformation is made in these sections, and these two terms have been defined in several excellent texts.[3–5] Although a detailed explanation of these terms is not warranted, a brief statement is necessary because they are major issues under discussion. The *configuration* of a molecule or a substituted atom is the relative arrangement of the atoms in a molecule, and this *relative arrangement is unaltered by bond rotation*. Except for pyramidal atomic inversion and permutational isomerism (see above) the configuration can only be altered by bond cleavage and the reformation of a new bond. Thus the configuration of an asymmetric carbon atom, and the cis and trans relationship of substituents in cyclic compounds can only be altered by breaking a bond(s) and reforming another bond(s) as in substitution and rearrangement reactions. The energy required to invert the configuration of the tetrahedral bonds of a carbon atom by way of a square planar arrangement of the bonds was calculated. It was found that the process required about 1046 kJ mol^{-1} and was about 2.5 times the C–C bond energy.[6] The *conformation* of a molecule is also a description of the relative arrangement of the atom and groups of atoms in space in a molecule. However, unlike the configuration, it is altered only by rotating one or more bonds. The essential difference between the two terms is that the configuration of a molecule describes the rigid arrangement of atoms or groups of atoms, whereas the conformation of a molecule describes the flexible arrangement of the atoms or groups of atoms. This is where the difference between the two terms becomes very fine. Thus the fixed arrangement of atoms in the crystal of 4-methylthiane (**1**) is in fact its configuration, but the same arrangement of atoms in solution is only the preferred conformation because it is in dynamic equilibrium with conformer **2**, among other conformers, and the hydrogen atoms in the methyl group are also rotating about the bond between the methyl carbon atom and C–4 of the thiane ring.

For absolute configuration see the following section.

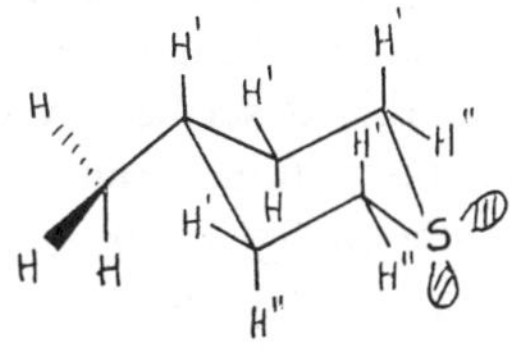

1

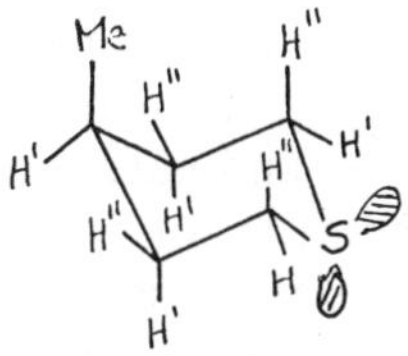

2

III. NOMENCLATURE

The substitutive naming system is adopted as much as possible in accordance with the International Union of Pure and Applied Chemistry (IUPAC) nomenclature for organic chemistry[7]; the monograph by Fletcher, Dermer, and Fox[8]; and the *Handbook for Chemical Society Authors.*[9] The *Ring Index* nomenclature[10] is used for the basic aromatic ring systems, and for the highly reduced bi- and polycyclic compounds the Baeyer system[7–9] is adopted throughout. In this system the heterocyclic compound is named after the parent alkane. The number of atoms between the bridgehead atoms, in decreasing number, is placed in square brackets before the name of the alkane. The numbering of the atoms begins at the bridgehead atom and proceeds by way of the longest chain to the second bridgehead atom, then through the second longest chain to the first bridgehead atom, and finally through the smallest chain to the second bridgehead atom. Whenever possible the double bond, the heteroatom, and the substituents (in this order) are given the smallest number. The most abundant examples in the present monograph are tricyclic systems, for example, bicyclo[*x*.*y*.*z*]alkanes (**3**). They are called *bicyclo* because the first two rings define the

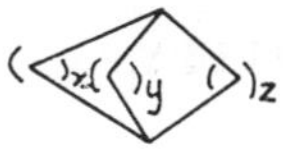

3

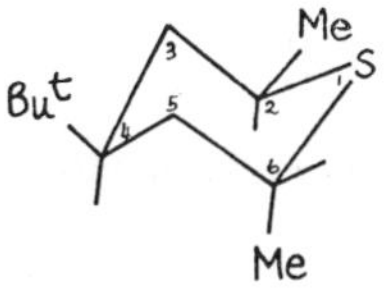

4

third ring. When *x*, *y*, or *z* is zero the molecule becomes a true bicyclic compound; for example, quinolizidine is 1-azabicyclo[4.4.0]decane. There are a few more complicated rings in the text, but these have been named and their formulas have been adequately numbered.

The rules for the notation of *absolute configuration* of asymmetric centers and for dissymmetric molecules proposed by Cahn, Ingold, and Prelog[11] have gained universal acceptance in the last decade. These rules and definitions of stereochemical properties were admirably described by Bentley[3] and Mislow[4] and have been used throughout this volume. Tentative rules for the nomenclature in fundamental stereochemistry proposed by the IUPAC[12] in 1969 and published in the Journal of Organic Chemistry in 1970[13] (see also ref. 8) for absolute and relative configurations, and for conformations are used as far as possible. The relative configurations of asymmetric centers are given in the usual *R* and *S* notation but are distinguished from absolute configurations by an asterisk, *R** and *S**. These are used in examples where the absolute configurations are not known. In a second system which is also used, one asymmetric center is taken as reference and denoted by *r* (reference) while the other centers are cis or trans with respect to it. This is particularly useful in rigid or cyclic molecules, for example, *r*-2-*trans*-6-dimethyl-*cis*-4-*t*-butylthiane (**4**).

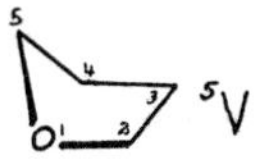

^5V **5** V_3 **5a** $^5T_4$ **6** $^5T_1$ **6a**

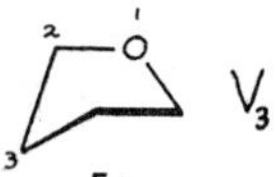

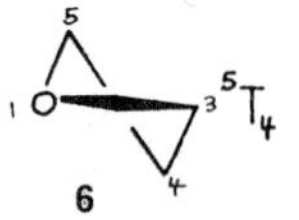

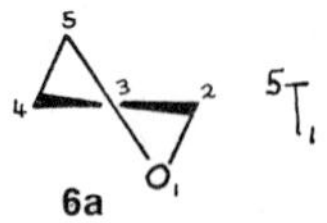

A proposal for the designation of conformations of small rings was made by Shaw,[14] but most of the generally accepted terminology in conformational analysis is adopted in this work. Also, for the pseudorotating systems, as in tetrahydrofuran (see Chapter 2, Scheme 5), the designations proposed by Shaw were not used; instead the conformers were denoted as V for "envelope" conformations **5** or **5a,** and T for the "twist" conformations **6** or **6a.** The superscript numbers denote atoms above the plane of at least three atoms of the ring, and the subscript numbers denote the atom below the plane of at least three atoms of the ring. The numbering in the ring begins at the heteroatom and is counterclockwise.

IV. ENERGY TERMS AND OPTICAL ACTIVITY

The main energy terms ΔG, ΔH, and E_a are expressed in kilojoules per mole, and for conversion into kilocalories per mole they should be divided by 4.185. In most cases the solvents used for these measurements are not indicated in the text because the intention was only to give some indication of the energies involved, and usually the solvent did not alter the magnitude of these values.

A rule of thumb relating optical stability with the free energy in the interconversion of enantiomers can be proposed. If the free energy of interconversion of enantiomers is approximately greater than 80 kJ mol^{-1}, then they can be separated into their optical antipodes by the usual methods of resolution.[17] When the free energy is between approximately 25 and 80 kJ mol^{-1} the molecules are optically labile and racemize very readily, but the enantiomers can be shown to possess optical activity by dynamic methods requiring kinetic resolution or asymmetric transformations.[18,19] For enantiomers with interconversion energies of approximately less than 25 kJ mol^{-1} it is not possible by the present methods to demonstrate optical activity, and the enantiomers are truly conformational isomers. However, they can be observed by nmr methods from which the energy parameters can be evaluated.

Although signs of rotation have been indicated in the text, sometimes together with the absolute configuration, they are truly meaningless because the solvent in which they were measured is not stated. They have, however, been included because they indicate that the compound mentioned is optically active and, in examples where the absolute configuration is provided, they demonstrate that the optical activity was correlated with the absolute configuration.

V. REFERENCES TO OTHER WORKS

A conscious effort has been made in this monograph to avoid repetition of many discussions that have been treated adequately in reviews or texts, and many of the relevant references are cited below. A basic knowledge of both heterocyclic chemistry and stereochemistry is assumed. Excellent texts and reviews are available on the basic principles of heterocyclic chemistry,[20–24] on the chemistry of separate heterocyclic systems,[25,26] and on advances in heterocyclic chemistry.[1] Specialist reports on the chemistry of saturated heterocyclic and heteroaromatic compounds by the London Chemical Society[27] and the *M.T.P. International Review of Science*[28] are continuously updating the field and contain much stereochemical data. Similarly, reviews and texts on optical rotatory power,[29] parity and optical rotation,[30] theory of optical activity,[31] optical rotatory dispersion[32–34] and circular dichroism,[33,34] general stereochemistry,[3–5,35,36] and conformational analysis[15,16,37–39] have been written by prominent authors. The continuing series entitled *Progress in Stereochemistry*[40] and *Topics in Stereochemistry*[41] are providing and updating stereochemical knowledge, and they are written by authorities on the various aspects of the subject. Fifty papers and reviews were published in *Tetrahedron*[42] in 1974 in commemoration of the centenary of the proposal of the tetrahedral carbon atom by van't Hoff[43] and Le Bel.[44] Three of these articles are particularly interesting to read because they are of historical interest,[45–47] and a short review on principal developments in stereochemistry during the last hundred years was also published in 1974.[48]

The relative and absolute configurations have been collated in an *Atlas of Stereochemistry*.[49] Compilations of interatomic distances and configurations of molecules and ions,[50] and molecular structures and dimensions of organic molecules in the crystalline state[51] contain useful reference material for a very large number of compounds, and they hopefully will be kept up to date.

VI. REFERENCES

1. A. R. Katritzky and A. J. Boulton, Eds., *Advances in Heterocyclic Chemistry*, Vol. 1 (1963) to Vol. 17 (1974), continuing series, Academic, New York.
2. J. B. Lambert, *Topics Stereochem.*, **6,** 19 (1971); A. Rauk, L. C. Allen, and K. Mislow, *Angew. Chem. Internat. Edn.*, **9,** 400 (1970).
3. R. Bentley, *Molecular Asymmetry in Biology*, Vol. 1, Academic, New York, 1969.
4. K. Mislow, *Introduction to Stereochemistry*, Benjamin, New York, 1966.
5. E. L. Eliel, *Stereochemistry of Carbon Compounds*, McGraw-Hill, New York, 1962.
6. H. J. Monkhorst, *Chem. Comm.*, 1111 (1968).
7. IUPAC, *Nomenclature for Organic Chemistry*, Sects. A, B, and C, Butterworths, London, 1965 and 1966.
8. J. H. Fletcher, O. C. Dermer, and R. B. Fox, *Nomenclature of Organic Chemistry*, Advances in Chemistry Series, Vol. 126, American Chemical Society, Washington, 1974.
9. The Chemical Society, *Handbook for Chemical Society Authors*, Burlington House, W.1., London, 1960.
10. A. M. Patterson, L. T. Capell, and D. F. Walker, *The Ring Index*, 2nd ed., American Chemical Society, Washington, 1960; Suppl. I, pp. 7728–9734; Suppl. II, pp. 9735–11524 (1964); and Suppl. III, p. 11525 (1965).
11. R. S. Cahn, C. K. Ingold, and V. Prelog, *Experientia*, **12,** 81 (1956) and *Angew. Chem. Internat. Edn.*, **4,** 385 (1966); R. S. Cahn and C. K. Ingold, *J. Chem. Soc.*, 612 (1951).
12. *IUPAC Information Bulletin*, **35,** 36 (1969).
13. *J. Org. Chem.*, **35,** 2849 (1970).
14. D. F. Shaw, *Tetrahedron Letters*, 1 (1965).
15. D. H. R. Barton and R. C. Cookson, *Quart. Rev.*, **10,** 44 (1956).
16. E. L. Eliel, N. L. Allinger, S. J. Angyal, and G. A. Morrison, *Conformational Analysis*, Interscience, New York, 1965.
17. S. H. Wilen, *Topics Stereochem.*, **6,** 107 (1971); P. H. Boyle, *Quart. Rev.*, **25,** 323 (1971).

18. M. M. Harris, *Progr. Stereochem.*, **2,** 157 (1958).
19. E. E. Turner and M. M. Harris, *Quart. Rev.*, **1,** 299 (1947).
20. A. Albert, *Heterocyclic Chemistry*, 2nd ed., Melbourne University Press, 1968.
21. A. R. Katritzky and J. M. Lagowski, *The Principles of Heterocyclic Chemistry*, Academic, New York, 1968.
22. J. A. Joule and G. F. Smith, *Heterocyclic Chemistry*, Van Nostrand Reinhold, London, 1972.
23. L. A. Paquette, *Principles of Modern Heterocyclic Chemistry*, Benjamin, New York, 1968.
24. R. M. Acheson, *An Introduction to the Chemistry of Heterocyclic Compounds*, 2nd ed., Interscience, New York, 1967.
25. A. Weissberger and E. C. Taylor, Eds., *The Chemistry of Heterocyclic Compounds*, Vol. 1 (1950) to Vol. 39 (1974), continuing series, Interscience, New York.
26. R. C. Elderfield, Ed., *Heterocyclic Compounds*, Vol. 1 (1950) to Vol. 9 (1967), series terminated, Wiley, New York.
27. W. Parker, Senior Reporter, *Saturated Heterocyclic Chemistry*, Vols. 1, 2, and 3, Specialist Periodical Reports, The Chemical Society, 1970–1975; C. W. Bird and G. W. H. Cheeseman, Senior Reporters, *Aromatic and Heteroaromatic Chemistry*, Vols. 1, 2, and 3, Specialist Periodical Reports, 1973–1975.
28. K. Schofield, Ed., *Heterocyclic Compounds*, *M.T.P. International Review of Science*, Vol. 4, Butterworths, London, 1973 and 1975.
29. S. F. Mason, *Quart. Rev.*, **17,** 20 (1963).
30. T. L. V. Ulbricht, *Quart. Rev.*, **13,** 48 (1959).
31. Symposium on the Theory of Optical Activity in memory of J. G. Kirkwood and W. Moffit, *Tetrahedron*, **13,** 1 (1961).
32. C. Djerassi, *Optical Rotatory Dispersion*, McGraw-Hill, New York, 1960.
33. P. Crabbé, *Optical Rotatory Dispersion and Circular Dichroism in Organic Chemistry*, Holden-Day, San Francisco, 1965; P. Crabbé, *An Introduction to the Chiroptical Methods in Chemistry*, Syntex, S.A. and Universidad Nacional Autonoma de Mexico, Universidad ibero americana, Mexico, 1971.
34. J. A. Schellman, *Chem. Rev.*, **75,** 323 (1975).
35. E. L. Eliel, *Elements of Stereochemistry*, Verlag Birkhaüser, Basle-Stuttgart, 1972.
36. M. S. Newman, Ed., *Steric Effects in Organic Chemistry*, Wiley, New York, 1956; J. D. Morrison and H. S. Mosher, *Asymmetric Organic Reactions*, Prentice-Hall, New Jersey, 1971.
37. E. L. Eliel, *Angew. Chem. Internat. Edn.*, **11,** 739 (1972).
38. F. G. Riddell, *Quart. Rev.*, **21,** 364 (1967).
39. R. Rahman, S. Safe, and A. Taylor, *Quart. Rev.*, **24,** 208 (1970).
40. W. Klyne, P. B. D. de la Mare, B. J. Aylett, and M. M. Harris, Eds., *Progress in Stereochemistry*, Vol. 1 (1954) to Vol. 4 (1968), Butterworths, London.
41. N. L. Allinger and E. L. Eliel, Eds., *Topics in Stereochemistry*, Vol. 1 (1967) to Vol. 8 (1974), continuing series, Interscience, New York.
42. *Tetrahedron*, **30,** 1477–2007 (1974).
43. J. H. van't Hoff, *Arch. neer.*, **9,** 445 (1874); *Bull. Soc. chim. France*, **23,** 295 (1875).

44. J. A. Le Bel, *Bull. Soc. chim. France*, **22,** 337 (1874).
45. R. Robinson, *Tetrahedron*, **30,** 1477 (1974).
46. M. J. T. Robinson, *Tetrahedron*, **30,** 1499 (1974).
47. F. G. Riddell and M. J. T. Robinson, *Tetrahedron*, **30,** 2001 (1974).
48. J. Weyer, *Angew. Chem. Internat. Edn.*, **13,** 591 (1974).
49. W. Klyne, *Atlas of Stereochemistry*, Chapman and Hall, London, 1974.
50. L. E. Sutton, Ed., *Tables of Interatomic Distances, Chem. Soc. Special Publ.*, The Chemical Society, Burlington House, W.1., London, No. 11 (1958) and No. 18 (1959).
51. O. Kennard, D. G. Watson, F. H. Allen, N. W. Isaacs, W. D. S. Motherwell, R. C. Petterson, and W. G. Town, Eds., *Molecular Structures and Dimensions*, Vol. 1 to 4 (1935–1972) and Vol. 1A (1960–1965, 1972), Crystallographic Data Centre, Cambridge and Interatomic Union of Crystallography.

2 OXYGEN HETEROCYCLES

I. INTRODUCTION

Oxygen heterocycles that possess features of stereochemical interest consist mainly of reduced systems, and are essentially cyclic ethers or enol ethers. These are generally simpler than the corresponding nitrogen analogues because they lack the complicating consequences that arise from pyramidal atomic inversion. This does not mean that pyramidal atomic inversion is not occurring at the oxygen atom; indeed it is, but inversion does not alter the stereochemical situation. The bivalency of oxygen, and its poor ability to expand its valency shell, allows the lone pair of electrons to exchange place as a result of inversion, but since these are equivalent no net change occurs. Dipolar effects, however, are strong in oxygen heterocycles and do influence conformational preferences as in the anomeric effect. The ability of saturated oxygen heterocycles to form intramolecular hydrogen bonds is quite marked and can influence the stereochemistry.

All the oxygen heterocycles are described in the present chapter, which, although not exhaustive, does cover most of the basic systems. Excellent reviews have been written on the various ring systems and references are provided in the respective sections. Most of the stereochemistry of naturally occurring oxygen heterocycles is omitted, for example, carbohydrates, flavanoids, and lactone antibiotics, because these have been adequately treated in major works. The oxygen heterocycles which are fundamentally cyclic lactones are very briefly discussed.

II. OXIRANES (ETHYLENE OXIDES)

The literature on the chemistry of oxiranes is very extensive. They are reactive substances, their reactions are usually predictable, and they have been used in countless syntheses. It is sometimes difficult to reconcile that they are heterocyclic compounds, partly because they are rarely classified as such in undergraduate textbooks. However, most detailed reviews on oxiranes have appeared in monographs on heterocyclic chemistry, for example, by Winstein and Henderson,[1] and by Rosowsky.[2] Other comprehensive reviews have appeared on general aspects of oxirane chemistry by Parker and Isaacs,[3] and Swern,[4] and on specialized aspects such as stereochemistry of epoxide synthesis by Berti,[5] stereochemistry of electrophilic additions to olefins by Fahey,[6] mechanism of epoxide formation by Henbest,[7] Darzens synthesis by Newman and Magerlein[8] and by Ballester,[9] rearrangements of α-haloepoxides and related α-substituted epoxides by McDonald,[10] and sugar epoxides by Newth.[11] The structure and stability of three-membered rings from *ab initio* MO theory were reviewed by Pople and coauthors.[11a] Clearly another review is not warranted, and only a few of the highlights of the stereochemistry of oxiranes are discussed in this section, including more recent references, for completeness.

A. Syntheses of Oxiranes

1. By Oxidation of Olefins

The most frequently used oxidizing agents for converting olefins into the corresponding oxiranes are peracids, for example, peracetic, perbenzoic, and monoperphthalic acids. The stereospecificity of the oxidation is almost complete in that the configuration of the olefin (i.e., cis or trans) is identical with that of the epoxide formed. The most acceptable mechanism is still the one proposed by Bartlett.[12] It involves a nonionic transition state (**1**) which has a structure quite similar to that of the olefin (Equation 1). Slight modifications of the transition state **1**, by introducing polarity to it, have been proposed to explain the finer details of some kinetic results.[13–15] Although cis addition of oxygen always occurs, a mixture of epoxides is possible if the olefin is not planar because the oxidant can attack from either side of the double bond. For example the oxidation of *cis*-4,5-dimethylcyclohexene gives an 87:13 ratio of anti and syn epoxides (Equation 2).[16] Numerous other examples are known: in some, one of two pos-

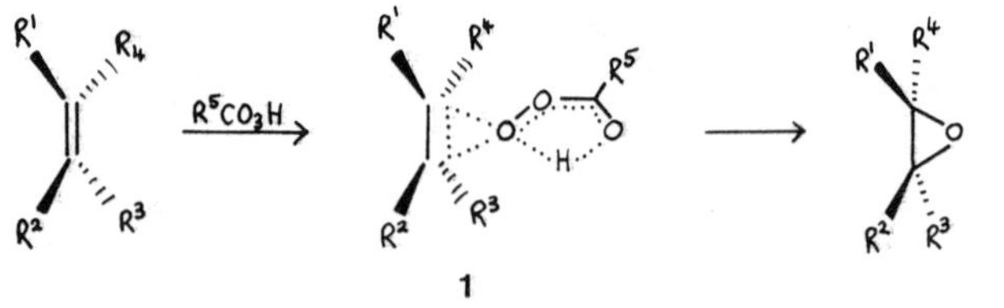

$$R^5CO_3H \qquad (1)$$

1

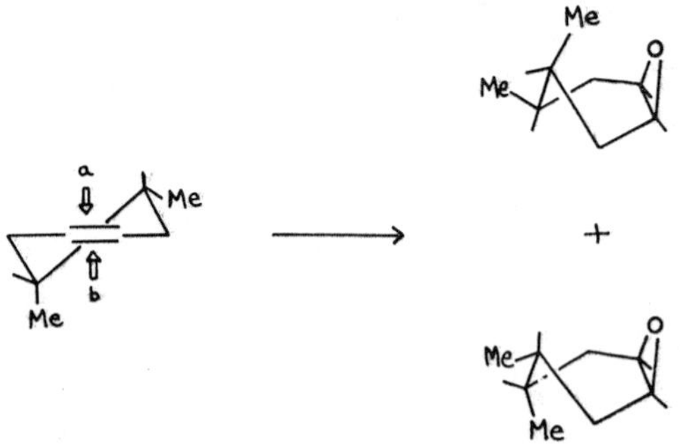

(2)

sible isomers predominates, and in others, equal, or almost equal, amounts of isomers are formed. The most important factor affecting the direction of attack is steric and this depends not only on the stereochemistry of the olefin but also on the peracid and the solvent used.[17] Monoperphthalic acid in this respect requires more space than performic acid. A few examples which demonstrate the steric effects in the olefins are the oxidation of 1,4,4*a*,5,6,7,8,8*a*-octahydro-4*a*-methylnaphthalene (**2**),[18] 1-*t*-butyl-4-methylenecyclohexane (**3**),[17] norbornene (**4**),[19] 7,7-dimethylnorbornene (**5**),[19] *endo*-8,8-diethoxytricyclo[3.2.1.0^{2,4}]oct-6-ene (**6**),[20] and 2-alkoxy-5,6-dihydro-2*H*-pyrans (**7**)[21] in which the direction of attack of the oxidant is indicated in the respective formulas.

Strong directive influence by allylic hydroxyl groups was observed by Henbest and Wilson[22] and by Albrecht and Tamm.[23] The hydroxyl group assists syn epoxidation; that is, the oxirane oxygen is cis with respect to the hydroxy group. This epoxidation of cyclohex-1-en-3-ol with perbenzoic acid in benzene gives a 91:9 ratio of **8** (R = H) to **9** (R = H), which is not seriously altered by varying the solvent (e.g., chloroform) or oxidant (e.g.,

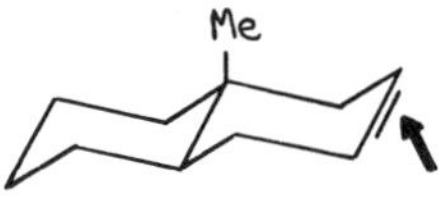

2

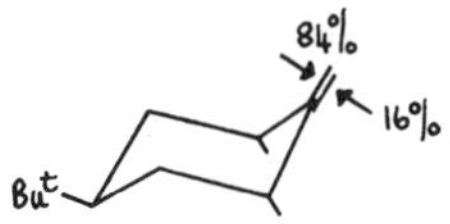

3

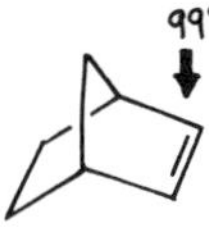

4

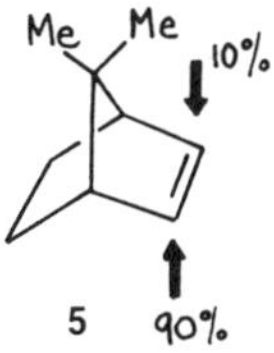

5

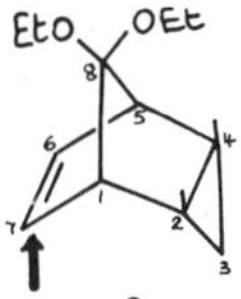

6

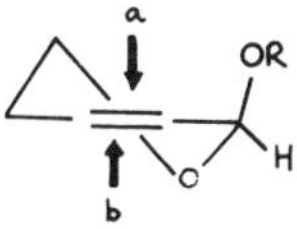

R = Me, a = 25%, b = 75%
R = Bu^t, a = 10%, b = 90%

7

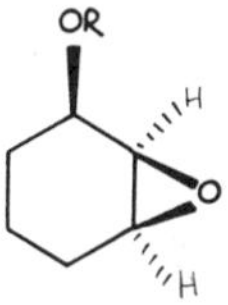

8

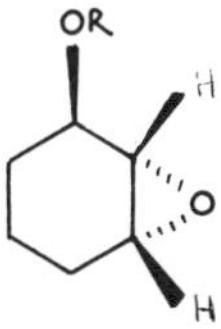

9

to peracetic or monoperphthalic acid).[24] Oxidation of 3-acetoxycyclohex-1-ene, on the other hand, is much slower and is only slightly stereoselective, giving a 43:57 ratio of **8** (R=OAc) to **9** (R=OAc).[24] A transition state similar to **10** has been proposed to account for such stereoselectivity.[22–24] Similar selectivity was also observed with allylic acylamino groups, and 3-benzamidocyclohex-1-ene provides almost exclusively the *cis*-oxirane.[25] A directing influence was noted in the intramolecular cis-stereospecific epoxidation of cyclohex-1-ene-3-hydroperoxide and its 1-methyl derivative. These rearranged, in the presence of vanadyl acetylacetone as catalyst, into 1,2-epoxycyclohexan-*cis*-3-ol and its 1-methyl derivative respectively.[26]

The epoxidation of suitably substituted olefins always results in a racemic mixture of oxiranes. If, however, an optically active peracid is used, for example, (+)-peroxycamphoric acid, an asymmetric synthesis occurs and an optically active oxirane is formed, for example, *S*(−)-phenyloxirane (**11**), from styrene.[27] Several studies of such asymmetric syntheses of oxiranes were reported,[7,28–31] but the optical yields were low—the highest was of the order of 10%.[32] Montanari, Moretti, and Torre[31,33] determined the absolute configuration of their chiral oxiranes, and from a knowledge of the absolute configuration of the chiral peroxy acids they proposed a transition state (**12**) in which the bulkiness of the substituents is accommodated with least steric interaction.

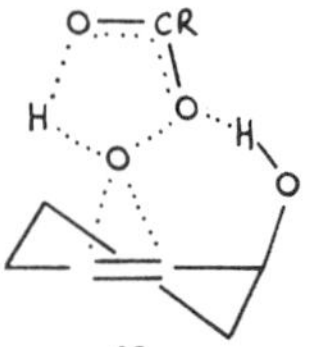

10

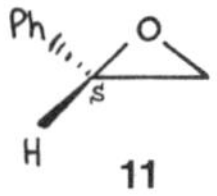

11

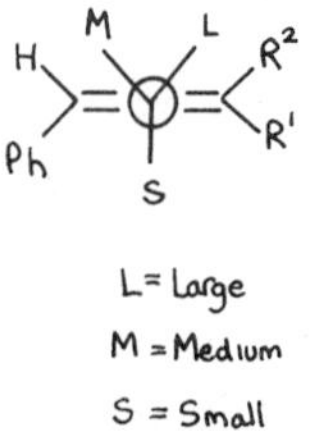

L = Large
M = Medium
S = Small

12

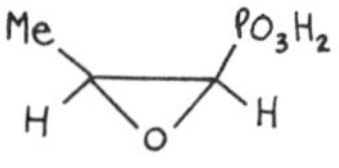

13

Epoxidation of olefins to oxiranes with oxidants other than peroxy acids has been successful. Alkaline hydrogen peroxide, in the presence of sodium tungstate, converted maleic and fumaric acids into *cis*- and *trans*-epoxysuccinic acids respectively. In contrast, these two acids were resistant to epoxidation by peracetic and perbenzoic acids.[34] An interesting application is the oxidation of prop-1-ene phosphonic acid with hydrogen peroxide–sodium tungstate in the presence of (+)-1-phenylethylamine, whereby epoxidation and optical resolution were performed in one operation and the (+) and (−) antipodes of the antibiotic phosphomycin, *cis*-2-methyloxirane phosphonic acid (**13**), were obtained.[35] Alkaline hydrogen peroxide epoxidations are, however, not always stereospecific.[35a] α,β-Unsaturated sulfones, such as *cis*- and *trans*-1-phenyl-2-*p*-tosylethylene, yield a mixture of *cis*- and *trans*-oxiranes. This is because the intermediate carbonium ion

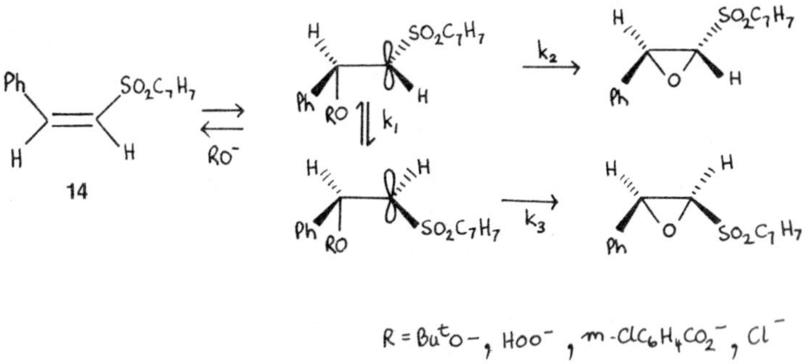

14

$R = Bu^tO^-, HOO^-, m\text{-}ClC_6H_4CO_2^-, Cl^-$

Scheme 1

isomerizes and cyclization to the epoxide can take place in two ways (Scheme 1). The stereospecificity of this synthesis depends on the relative rates of isomerization (K_1) of the intermediate carbonium ion and of the cyclization (K_2 or K_3). The better the leaving ability of the group R is, the higher the specificity. The order of leaving ability is t-BuO < OH < m-$ClC_6H_4CO_2$ < Cl. In the case where K_1 is large the relative rates of K_2 and K_3 will dictate the ratios of diastereomeric oxiranes formed. The oxidation of the *cis*-olefin (**14**) with *t*-butyl hydroperoxide and with alkaline hypochlorite yields approximately a 1:9 and 10:0 ratio of *cis*- to *trans*-2-phenyl-3-*p*-tosyloxiranes respectively.[36] Epoxidations with alkaline hypochlorite are therefore highly stereospecific with retention of configuration.[36,37] Chromium trioxide epoxidation is satisfactory only with highly substituted olefins.[38]

2. *By 1,3-Elimination Reactions*

A second very useful synthesis of oxiranes involves a 1,3-elimination of 2-substituted alcohols in which the 2-substituent is a good leaving group. The elimination proceeds by way of a transition state in which the 2-substituent and hydroxy functions are in an antiperiplanar conformation (Equation 3). The synthesis is thus highly stereospecific, and the configuration of oxirane formed is dictated by the stereochemistry of the alcohol. For example *S*-2-chloropropan-1-ol is cyclized with KOH into *R*-2-methyloxirane without loss of optical purity.[39] Other leaving groups that have proved satisfactory include bromide,[40,41] iodide,[42] sulfonate esters,[11,43,44] carboxylate esters,[45–47] trimethylammonium,[48–50] and diazonium.[51–53] The last mentioned group is derived from nitrous acid treatment of 2-amino alcohols, and although the reaction was successful when the groups were axial, it was found that ring contraction occurred when they were equatorially disposed.[51,52] These syntheses are highly stereospecific, but sometimes a racemic product may be formed, particularly when an intermediate carbonium ion can be stabilized as in the case of a benzylic carbon carrying the OH or other leaving group[43] (Equation 3).

B^- H–O ... R^1, R^2, R^3, R^4, X $\xrightarrow{-X^-}$ oxirane (R^1, R^2, R^3, R^4) (3)

X = Cl, Br, I, OSO_2R, OCOR, NMe_3^+, N_2^+

3. *From Carbonyl Compounds*

The Darzens reaction for the preparation of oxiranes possessing electron withdrawing substituents is the most widely investigated method in this section. The reaction has been reviewed in detail,[2,8,9] and only a short summary of it is reported here. It requires the condensation of an aldehyde or ketone with α-haloketones,[54,56]-esters,[57,58]-nitriles,[59]-sulfones,[60] or suitable activating groups in the presence of a base. The accepted mechanism is displayed in Scheme 2. The stereospecificity of the synthesis varies considerably with the substituents[57,60,61] and with the solvent.[58,59] Consideration of steric factors in predicting the stereochemistry of the products is important, but sometimes the thermodynamically less stable isomer is formed predominantly as in the case of *cis*-2-ethoxycarbonyl-2,3-diphenyl-

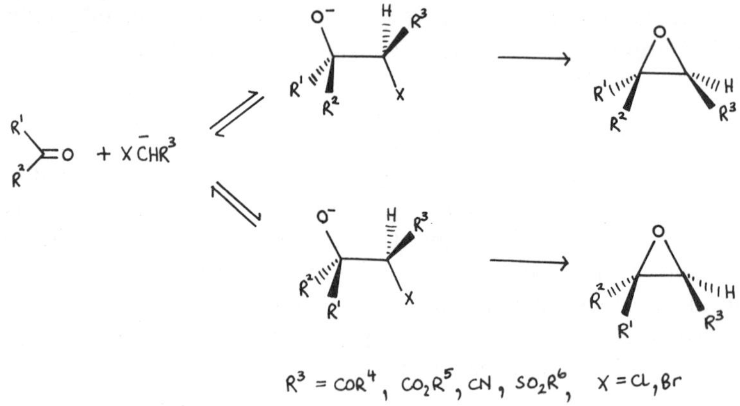

$R^3 = COR^4, CO_2R^5, CN, SO_2R^6, \quad X = Cl, Br$

Scheme 2

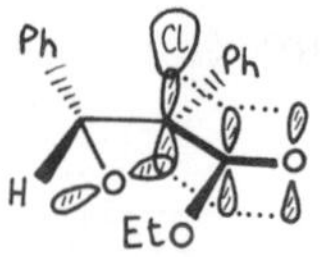

oxirane, which is obtained from benzaldehyde and ethyl α-chlorophenylacetate in the presence of potassium *t*-butoxide. Zimmermann and Ahramjian[62] proposed a transition state (**15**) in which the maximum overlap of orbitals is the stabilizing influence. A similar argument was adopted for the preferred formation of *trans*-2-arylsulfonyloxiranes.[60]

15

The stereochemistry of the formation of 3-aryl-2-methoxyoxiranes from α-bromoaldehydes and methoxide ions was investigated for a series of aldehydes. The preferential formation of the *trans*-oxiranes was attributed to attack of methoxide ions from the least hindered side (Equation 4).[63] When methoxide ions are replaced by cyanide ions, α-bromoaldehydes or ketones yield the corresponding cyanooxiranes.[64,65]

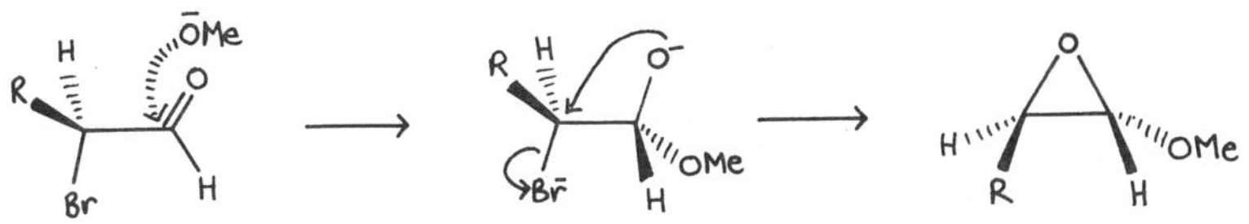

(4)

Diazomethane adds across carbonyl double bonds with the elimination of nitrogen and production of oxiranes. Only one chiral center is formed with diazomethane, but higher homologues such as diazoethane yield a diastereomeric mixture.[66] Conformationally immobile cyclohexanones, for example, *trans*-2-decalone, react with diazomethane preferentially by way of an equatorial attack,[67] for example, to form a 4:1 mixture of the oxiranes **16** and **17.** The main course of the reaction, however, is ring expansion of the ketone.

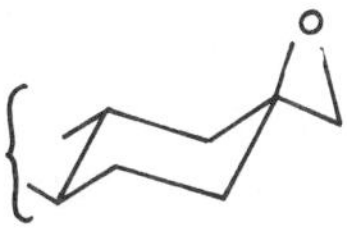

16  **17**

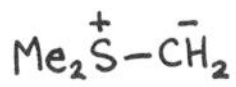

$Me_2\overset{+}{S}-\overset{-}{C}H_2$

18

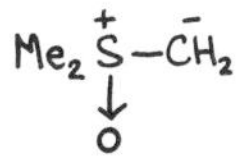

19

Sulfonium and oxysulfonium ylides are more effective for the preparation of oxiranes from carbonyl compounds than are diazoalkanes. When two oxiranes are possible, dimethylsulfonium methylide (**18**) attacks the carbonyl group preferentially from the more hindered side (e.g., axial), whereas the oxysulfonium derivative (**19**) attacks from the less hindered side (e.g., equatorial).[68] High stereoselectivity is also observed with diphenylsulfonium benzylide,[69] but not with diphenylsulfonium alkylylides.[70] The ylide from methylphenyl-*N*-*p*-tosylsulfoximine (**20**) yields the oxirane **21** exclusively by reaction with 4-*t*-butylcyclohexanone.[71]

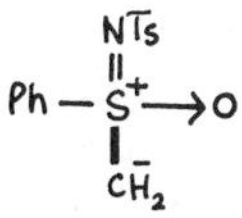

20 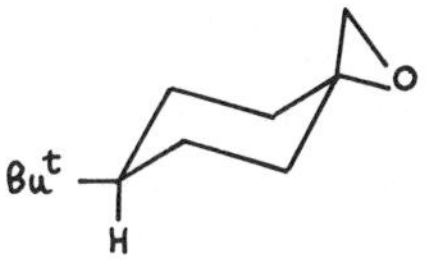 **21**

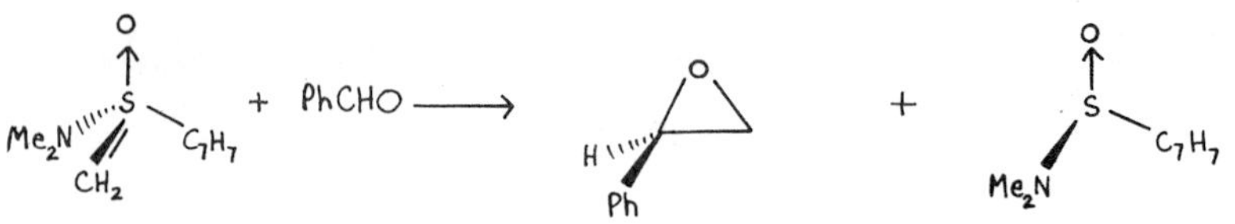

Me_2N–S(→O)(CH_2)–C_7H_7 + PhCHO ⟶ H, Ph oxirane + Me_2N–S(→O)–C_7H_7 (5)

Asymmetric synthesis has been achieved with optically active dimethylamino-*p*-tolyloxysulfonium methylide. It reacts with benzaldehyde to form optically active phenyloxirane and *N*,*N*-dimethyl-*p*-tolylsulfilimide which can be converted to the original oxysulfonium ylide with only 20% loss of optical activity. The optical yield of *R*(+)-phenyloxirane was unfortunately no better than 5% (Equation 5). Reaction with *p*-chlorobenzaldehyde and acetophenone similarly gave 2-*p*-chlorophenyl- and 2-methyl-2-phenyloxiranes enriched in the (+) antipode.[72]

B. Configuration and Conformation of Oxiranes

The configurations of oxiranes are similar to those of aziridines except that the complications from atomic inversion are absent in the oxiranes (Part I, Chapter 2, Section I.A.2 and 3). In pre-nmr days the configurations of *cis*- and *trans*-2,3-disubstituted oxiranes were deduced from the method of synthesis, for example, *erythro*-chlorohydrins yield *trans*-oxiranes and *threo*-chlorohydrins yield *cis*-oxiranes. The reverse reactions, that is, conversion of oxiranes into *erythro*- or *threo*-bromohydrins or 1,2-dibromides, were also useful in configurational assignments.[73] When the substituents in 2,3-disubstituted oxiranes were identical (e.g., phenyl[74] or carboxyl[75]) the distinction was made by obtaining the trans isomers in optically active forms. Optically active cis isomers, being meso, could not be obtained. In the case of 2-halomethyl-3-benzoyl oxiranes, Cromwell[76,77] and co-workers found uv and ir differences between the isomers. These cis and trans isomers react in different ways with aniline and yield a halogen-containing and a halogen-free *N*-phenylpyrrole respectively.[55,78] The products from reaction with hydroxylamine also are different,[79] and in conjunction with uv and ir data, Stevens and co-workers[80] arrived at the correct configurations. Differences in the Raman spectra were also noted.[81]

Bottini and Van Etten[82] prepared several 2,3-dialkyloxiranes and -aziridines and found that the *cis*-oxiranes had higher boiling points and refractive indices than the trans isomers, but the reverse was true for the aziridines. They had to use the reformulated Auwers-Skita rule of cis-trans isomers[83] which is based not on boiling points and refractive indices, but on heat content. This new rule states that in configurationally identical ring systems the isomers with the higher densities and refractive indices are those with the higher heat content. Since *cis*-2,3-dialkyloxiranes have a larger degree of nonbonded interactions than the trans isomers, it was concluded that the cis isomers should have a higher heat content.

The relative configurations of substituents in oxiranes can now be assigned from both the chemical shifts and the coupling constants of their ^{1}H nmr spectra.[84] If the absolute configuration of one of the chiral ring carbon atoms of a diastereomeric oxirane molecule is known, then the configuration of the second asymmetric ring carbon can be deduced from ^{1}H nmr spectra. The ^{1}H nmr spectra of enantiomeric chiral oxiranes are usually identical, but the use of optically active 2,2,2-trifluorophenyl ethanol as nmr solvent causes the spectra of the enantiomers to be different because of differences in the structures of the diastereomeric solvated molecules. This method is useful not only for assigning the absolute configuration (several of which have been deduced in this way) but also assessing the optical purity of the oxirane.[85] Optically active shift reagents should be very useful in this respect. ^{13}C nmr, like ^{1}H nmr, spectroscopy of oxiranes provides valuable information about their stereochemistry.[86]

Optically active oxiranes have been prepared by asymmetric epoxidation of olefins with optically active peroxy acids, but the optical yields are generally extremely low (see Section A.1). Oxiranes of high optical purity are best prepared from optically active intermediates by highly stereospecific syntheses, for example, (+)- and (−)-, *cis*- and *trans*-2-methyl-3-phenyl-;[49,87] (+)-*trans*-2,3-diphenyl-;[50] (−)-2-ethyl-2-methyl-;[88] and (−)-2-methyloxiranes;[41] or, if possible, by optical resolution, for example, *trans*-2-carboxy-3-methyl-,[89] *cis*- and *trans*-2-carboxy-3-phenyl,[90,91] and *cis*- and *trans*-2-carboxy-3-(*o*, *m*, and *p*)-chlorophenyl[92] oxiranes. These are obtained by recrystallization of the salts with optically active bases, but in the case of one acid the very rarely observed spontaneous resolution was reported.[92]

The absolute configuration of chiral oxiranes can be deduced from a knowledge of the absolute stereochemistry of the intermediates and a knowledge of the stereochemistry of the conversion to oxiranes (e.g., 2*R*,3*R*(+)-2,3-diphenyl- and *R*(+)-methyloxiranes[93]). More commonly, the absolute configuration is obtained by stereospecific cleavage of the ring (see Section C) and correlation of the products with compounds of known absolute configuration, for example, 1*S*,2*R*(−)-1,2-epoxy-1,2,3,4-tetrahydronaphthalene,[32] 2*R*,3*R*(+)-2-phenyl-2,3-tetramethylene-,[48] 2*S*,3*R*(−)-2-carboxy-3-methyl-,[89] 2*S*,3*S*(+)-2-carboxy-3-phenyl-,[90] 2*R*,3*S*(−)-2-carboxy-3-phenyl-,[91] 2*S*,3*S*(−)- and 2*S*,3*R*(+)-2-carboxy-3-(*o*, *m*, and *p*)-chlorophenyloxiranes.[92] The ord and cd curves of R(+)-phenyl-, 2*R*,3*S*(+)- and 2*R*,3*R*(+)-2-methyl-3-phenyl-, and 2*R*,3*R*(+)-2,3-diphenyloxiranes,[94] were measured and it was suggested that the phenyl rings are perpendicular to the oxirane ring. Analysis of the cd curves of *trans*-(+)-2,3-diphenyloxirane lead to the absolute configuration 2*R*,3*R*—in agreement with results from chemical correlation.[95]

Electron diffraction studies of *cis*-2,3-tetramethyleneoxirane (1,2-epoxycyclohexane) revealed that it had a deformed staircase structure with the four carbon atoms closest to the oxygen atom in one plane.[96] *Trans*-2,3-tetramethylene- and 2,3-pentamethyleneoxiranes would be too strained to be stable enough for isolation, but *cis*- and *trans*-2,3-hexamethylene oxiranes (1,2-epoxycyclooctanes) are known,[97] and their reactions have been studied in some detail.

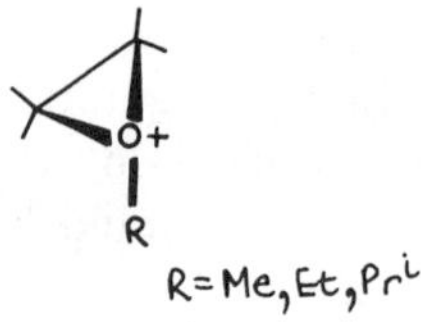

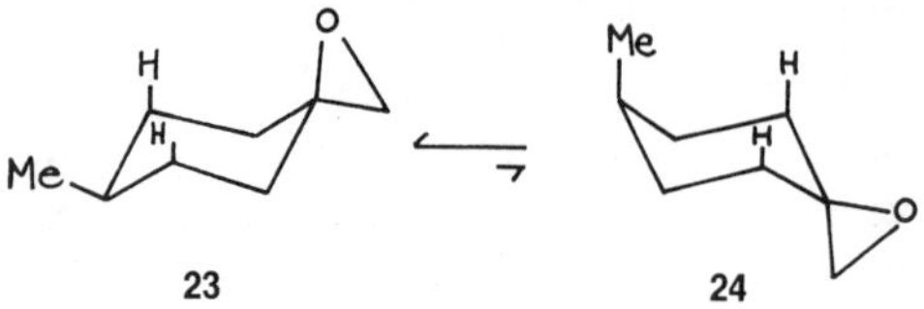

The oxirane nucleus is rigid and does not exhibit conformational changes. The oxygen atom does undergo inversion, but because it possesses two pairs of electron lone pairs no net change is observed (see above). However, if the oxygen atom is alkylated, that is, converted into an oxonium ion, then oxygen inversion should be observable. Lambert and Johnson[98] demonstrated that this is indeed true by studying the ^{1}H nmr spectra of *O*-methyl-, *O*-ethyl-, and *O*-isopropyloxiranium cations (**22**) in liquid SO_2 at various temperatures. The spectra were rather complicated, but inversion in the *O*-isopropyl salt was slow below $-70°C$ (coalescence temperature $\sim -50°C$) suggesting an activation energy of 41.8 ± 8 kJ mol^{-1}. They found that these oxiranes were relatively stable in comparison with Meerwein oxonium salts because no transfer of an alkyl group onto tetrahydro-2*H*-pyran was observed.

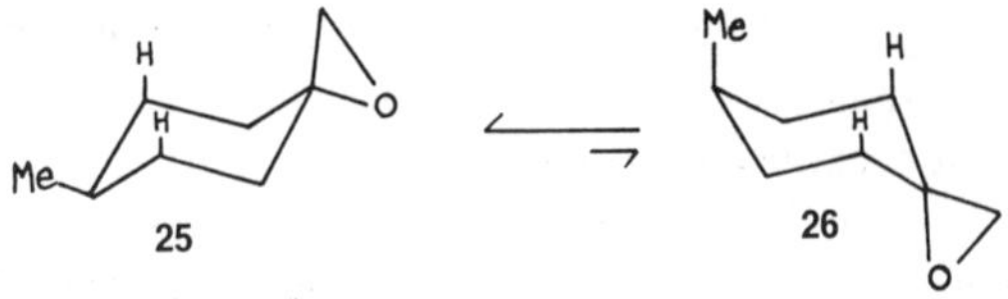

A comparison of the effect of a methyl group and an oxirane substituent on the conformation of cyclohexane was made by Uebel.[99] He examined the isomeric 4-methyl substituted 2,2-pentamethyleneoxiranes **23** and **25** by ^{1}H nmr spectroscopy and found that the ratios of **23**:**24** and **25**:**26** were

96:4 and 93:7 respectively. This demonstrated the larger space demanding properties of a CH_3 group compared with a $-CH_2-$ and an –O– group, and hence the larger "anchoring" (ananchomeric) ability of the methyl group. Anet and co-workers[100] studied the conformational changes of 2,3-pentamethyleneoxirane (cyclohepteneoxide) by ^1H and ^{13}C nmr spectroscopy at various temperatures. The assignments were facilitated by using the 1,3,3-trideuterated derivative, and they found that an equilibrium existed between two different chair conformations **27** and **28** which have slightly different energies. The ratios in vinyl chloride and in $CFHCl_2$–$CHClF_2$ were 71:29 ($\Delta G^{\ddagger} = 33.1$ kJ mol^{-1}) and 60:40 ($\Delta G^{\ddagger} = 31.4$ kJ mol^{-1}) respectively.

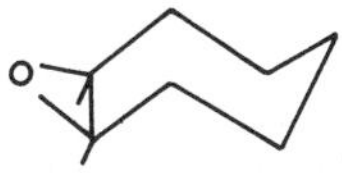

27 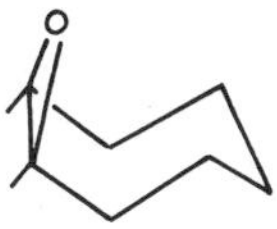 **28**

The bond distances and bond angles of 2-(*p*-bromophenyl)-*cis*-2,3-tetramethyleneoxirane [1-(*p*-bromophenyl)-1,2-epoxycyclohexane][101] and *p*-nitrophenyloxirane[102] were determined by X-ray analyses. The dihedral angles between the phenyl ring and the oxirane ring, 83 and 80.2° respectively, were explained by pseudoconjugation between the aryl and oxirane rings.

C. Reactions of Oxiranes

The reactions of oxiranes were reviewed in detail by Parker and Isaacs (1959)[3] and by Rosowsky (1964),[2] and only the important stereochemical features are outlined here together with more recent data.

1. *C–O Bond Cleavage Reactions*

The reactions involving C–O bond cleavage are by far the most numerous reactions of oxiranes. The cleavage is brought about in various ways: by protonation of the oxygen atom followed by nucleophilic attack at a carbon atom, by direct attack at the carbon atom by a nucleophile or hydride ion, and by catalytic reduction. In the first two the attack of the nucleophile occurs most commonly from the side opposite the oxygen atom, that is, with a Walden inversion at the carbon atom in which the C–O bond is broken (Equation 6). M is not necessarily a proton and could be a metal, as in $AlCl_3$. The extent of inversion depends on the nature of the transition state. In the case of the transition state **29** the C–C bond could rotate, that is, if the carbonium ion is long lived, and the product from retention of configuration is obtained. Generally, however, a fully

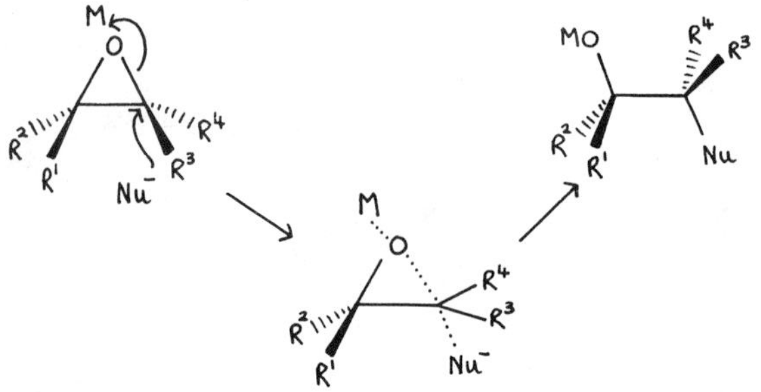

(6)

29

developed carbonium ion is not formed.[3] Lucas and co-workers[101–104] found that the reactions with acids (i.e., M=H^+) proceed as indicated in Equation 6, for example, *cis*- and *trans*-2,3-dimethyloxiranes yield *threo*- and *erythro*-2-halo-3-hydroxybutanes respectively. In slightly acidified water the ethylene glycol is obtained by way of the same mechanism[105–107]: *cis*- and *trans*-2,3-dimethyloxiranes give *d,l*- and *meso*-2,3-dihydroxybutanes,[105] but in the presence of trichloroacetic acid the hydroxy ester is formed, for example, *R*(−)-*erythro*-1-hydroxy-2-trichloroacetoxy-1,2-diphenylethane (**30**) from *trans*-*R*(+)-2,3-diphenyloxirane.[50] Similarly, in acidified methanol *trans*-(+)-2,3-dimethyloxirane cleaves to *erythro*-3-methoxybutan-2-ol because on methylation it furnishes the optically inactive *meso*-2,3-dimethoxybutane.[107]

Two sites of attack are available in unsymmetrical oxiranes, and the regiospecificity is demonstrated in the acid- and base-catalyzed addition of water to (+)-2-methyloxirane. In acidic medium (+)-1,2-dihydroxypropane is formed (Walden inversion), and in alkaline medium the (−) enan-

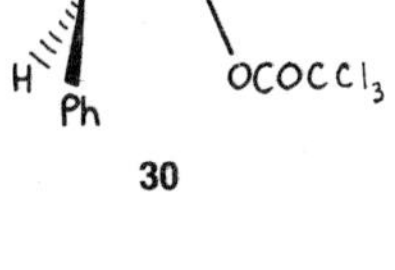

30

tiomer is formed.[108,109] Detailed studies by Pritchard and Long[110–112] using $H_2{}^{18}O$ showed that attack of water was at C–3 in basic medium and C–2

Library
I.U.P.
Indiana, Pa.

547.59 Ar54s
pt.2, c.1

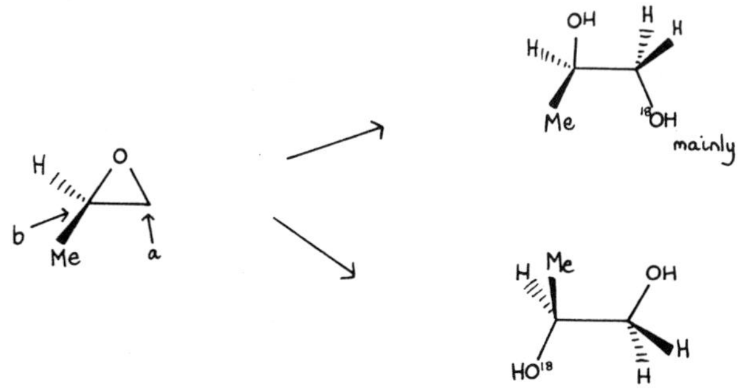

mainly

Scheme 3

in acidic medium (Scheme 3). Kinetic studies revealed that the base-catalyzed reactions are mainly S_N2, whereas in the acid-catalyzed reaction the carbonium ion character of the transition state may be large and leads to some racemization. Substituents which can stabilize a carbonium ion, for example, a benzene ring or alkyl substituents, will facilitate C–O bond cleavage mediated by the acid, that is, protonation of the oxygen atom. In the case of aryl substituents, as in 2-phenyl-2,3-tetramethylene-[113] and 2-(1-naphthyl)-2,3-tetramethyleneoxiranes,[114] not only is the carbonium ion more stabilized by the substituent and the C–O bond of the carbon atom α to the aryl group preferentially cleaved, but a transition state **31** different from **29** must be involved to account for the high, and occasionally complete, cis stereospecificity with retention of configuration (Equation 7). Generally when an electron donating substituent is present on C–2, or a substituent that can stabilize a positive charge on C–2 of the oxirane ring, ring cleavage is more likely to occur with retention rather than with inversion of configuration at the carbon atom involved.[115] Cleavage with retention of configuration could also result from a double inversion. Kuhn and Ebel[75] found that whereas *cis*-2,3-dicarboxyoxirane gave exclusively *erythro*-1,2-dihydroxysuccinic acid (*d,l*-tartaric acid), the trans isomer gave a 6:4 mixture of *meso*- and *erythro*-1,2-dihydroxysuccinic acids respectively. The erythro acid from the *trans*-oxirane is formed by a double inversion as shown in Equation 8.

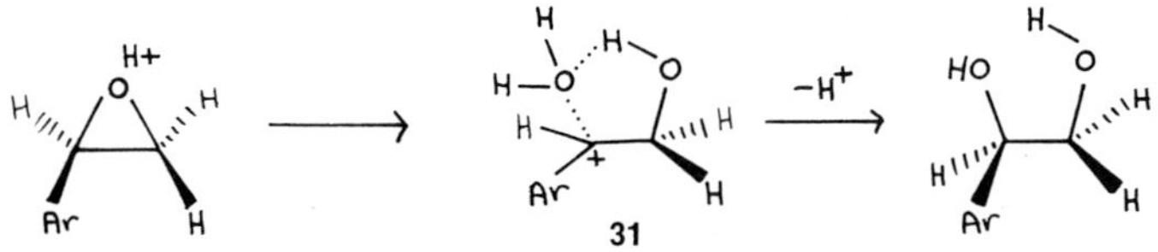

(7)

31

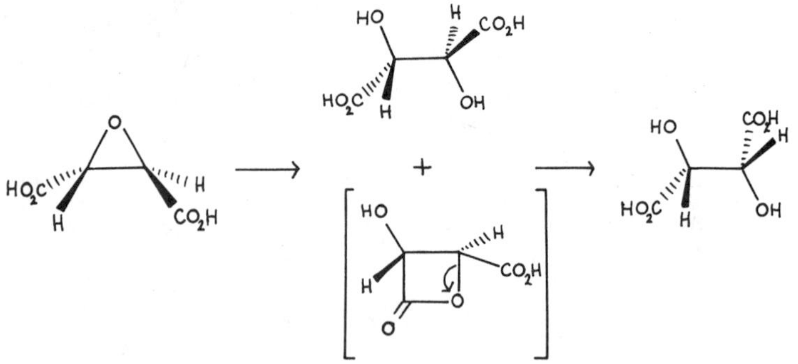

(8)

The reactions of oxiranes directly with nucleophiles have a larger S_N2 character, they follow the mechanism in Equation 9, and inversion of configuration is almost always the case. Reaction with hydroxyl ions yields the glycols,[49,116] and with ammonia[90,91,117] and amines,[48,118,119,120] the α-amino alcohols are formed with known stereochemistry. In unsymmetrical oxiranes the regiospecificity is not always complete, and the two products from nucleophilic attack at each of the carbon atoms of the oxirane are obtained (Equation 9). The ratio of the products depends on electronic and steric factors and on the nature and size of the nucleophile.[3] Other nucleophiles have been used and found very useful for synthetic purposes. The isocyanate ion is used for inverting the configuration of the epoxide (Equation 10),[121] and isothiocyanate,[122,123] thiourea,[124,125] methyl xanthate,[126] and thioacetic acid[127] behave in the same way but provide the thiirane (ethylene episulfide), also with inversion of configuration at both carbon atoms. Methyl- and phenylnitriles in sulfuric acid react with oxiranes and form the respective 2-oxazolines in which inversion of configuration at one carbon atom occurs with complete stereospecificity (Equation 11).[128]

The conversion of substituted oxiranes into cyclopropanes, with Walden inversion at one carbon atom by reaction with the triethyl phosphonoacetate anion, follow a similar mechanism.[129] Other nucleophiles which cleave oxiranes stereospecifically are phenylthioborane (to mercaptoethanols),[130] diphenylmethyl sodium (to β-diphenylmethyl alcohol),[93] and *N*-ethoxycarbonylhydrazine.[131] The last mentioned was used to convert *cis*-2,3-diphenyloxirane, obtained from *cis*-stilbene, into *trans*-stilbene. Extrusion of oxygen can take place stereospecifically when the oxirane reacts with triphenylphosphine. The *cis*-oxirane yields mainly the *trans*-olefin. The phosphorus compounds formed as intermediates are similar to those in the Wittig reactions between phosphorus ylides and carbonyl compounds.[132] Triphenyl phosphine selenide deoxygenates oxiranes in the presence of trifluoroacetic acid at room temperature and produces the

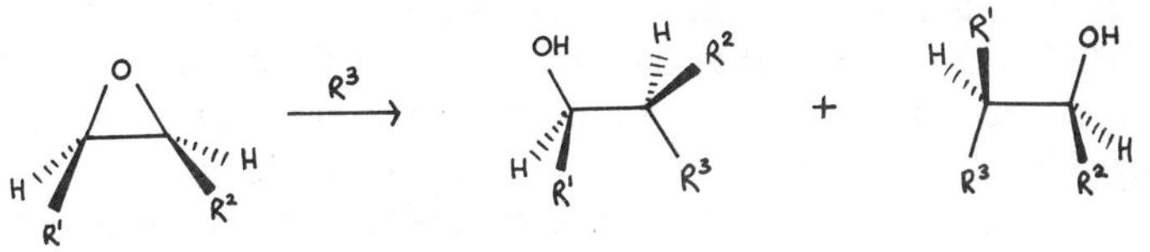

(9)

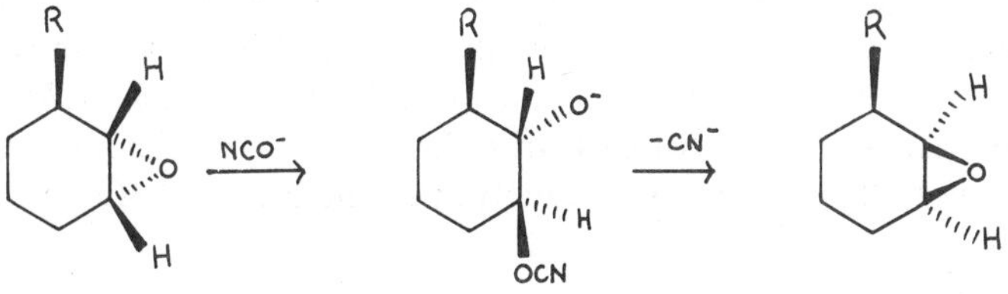

(10)

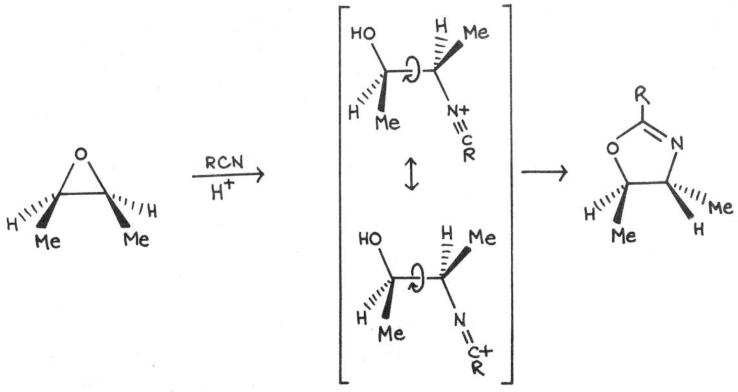

(11)

olefins with retention of relative configuration of substituents (Equation 12). The intermediate episelenide must lose selenium stereospecifically.[133] The mechanism is probably similar to the conversion of *cis*-oxiranes into *cis*-thiiranes with triphenyl phosphine sulfides in which an inversion at each carbon atom leads to relative retention of configuration of the substituents at C–2 and C–3 of the oxirane.[134]

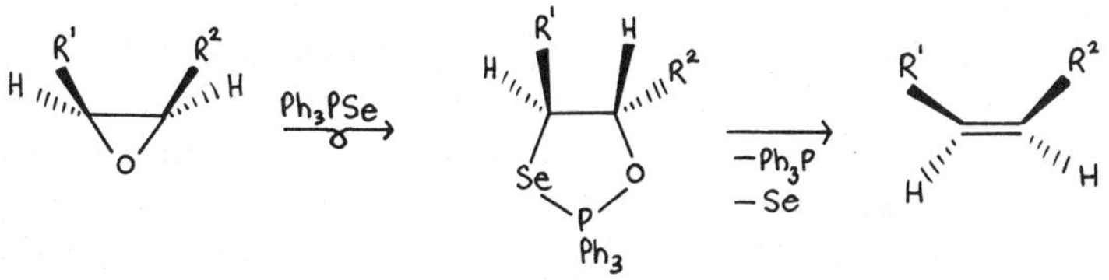

(12)

Reductive cleavage of oxiranes with LAH yields the corresponding alcohol in which the carbon atom of the C–O bond that is broken bears the least number of substituents; that is, hydride attack occurs at the least hindered carbon atom.[3,134] Eliel and co-workers[135–137] studied these reactions in detail and observed that the preference for C–O bond cleavage can be altered by addition of varying amounts of aluminum chloride. With careful choice of reagents considerable control of the products can be achieved.[138,139] The finer points of the stereochemistry of this reduction were elucidated by Leroux and Lucas.[140] They showed that the asymmetric carbon atom of the oxirane which is not involved in the cleavage retains its configuration completely; and Helmkemp and Schnautz[141] showed, by using lithium aluminum deuteride and optically active *trans*-*R*,*R*(+)-2,3-dimethyloxirane, that the hydride is transferred with complete inversion of configuration at the carbon atom involved in the cleavage (Equation 13).

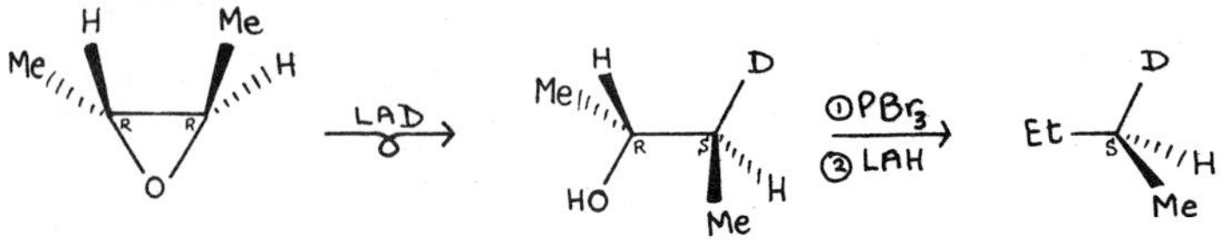

(13)

Lithium borohydride is also very effective in carrying out this reaction and its regiospecificity has been examined in detail by Fuchs and co-worker[142,143] and was found to alter on addition of magnesium bromide.[3,144] In the reduction of *trans*-2,3-diphenyl-*cis*-2-methoxyoxirane (**32**) with $LiBH_4$, the hydride ion is transferred to the carbon atom bearing the methoxy group almost exclusively and *d*,*l*-hydrobenzoinmonomethyl ether is formed. Lithium aluminum hydride, on the other hand, reacts with the ether **32** in a different way giving benzylphenyl carbinol, but $NaBH_4$ did not react at all.[145,146] The stereochemistry of the reduction of oxiranes with $LiBD_4$, lithium borotritide,[147–149] and of trideuteroborane[150] is the same as that of LAH, that is, with ring opening and a Walden inversion at the carbon atom that is reduced.

Oxiranes are reduced to alcohols by catalytic hydrogenation. The stereochemistry at the reduced center is not known, but the regiospecificity (i.e., which C–O bond is reduced) varies with the substituents and the condition of reduction.[3,151–153]

The cleavage of oxiranes with metal organic compounds is well known.[3] Recently the stereochemistry of this cleavage using dialkyl magnesium and zinc compounds was shown to be stereospecific and occurred with complete inversion of configuration at the benzylic carbon atom in phenyloxiranes.[154] The regioselectivity and stereospecificity of dialkyl magnesium reagents with *R*(+)-phenyloxirane were examined by Morrison, Atkins, and Tomaszewski.[155] They have determined that diethyl and diisopropyl magnesium react at the benzylic carbon atom with about 20% retention and 80% inversion of configuration, indicating that the reaction proceeded preferably by the intermediate **33.** Di-*t*-butyl magnesium is less stereoselective and the addition of magnesium bromide alters the selectivity slightly.

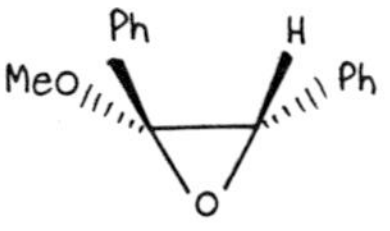

The alkylation of benzene with *R*(+)-2-methyloxirane in the presence of Lewis acids (e.g., $AlCl_3$, $SnCl_4$, and $TiCl_4$) was investigated under a variety of conditions. The products were (+)-2-phenylpropan-1-ol and a mixture of (−)-1-chloropropan-2-ol and (+)-2-chloropropan-1-ol. The optical rotations demonstrated that phenylation and ring cleavage occurred with complete inversion of configuration (Equation 14). Considerable racemization takes place, however, with aluminum bromide in nitromethane.[156]

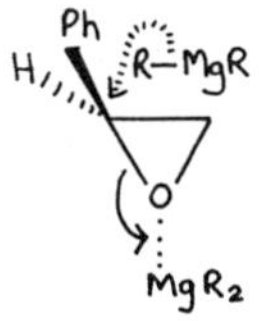

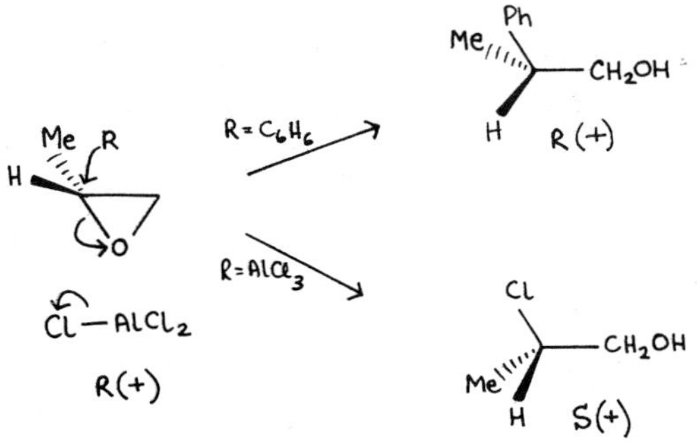

2. *C–C Bond Cleavage Reactions*

Cleavage of the C–C bond in oxirane was first suggested by Petrellis and Griffin[157] in 1967 in the photochemical fragmentation of 2-methoxycarbonyl-2,3,3-triphenyloxirane. Huisgen and co-workers[158,159] argued that oxirane should undergo electrocyclic C–C bond cleavage in much the same way as in the aziridines (Part I, Chapter 2, Section I.A.4.c) and should follow the same Woodward-Hoffmann orbital symmetry rules. They demonstrated that *cis*- and *trans*-2,3-dicyano-2,3-diphenyloxiranes can be equilibrated at 100–120°C. The equilibrium involves carbonylylides which react as 1,3-dipoles and can condense with a variety of acetylenic and olefinic compounds to yield 2,5-dihydro- and tetrahydrofurans respectively. The yields are high, and the stereochemistry of the products from reaction with dimethyl fumarate and *cis*- and *trans*-2-cyano-2,3-diphenyloxiranes establishes that the process is conrotatory as predicted (Scheme 4). The kinetics of thermal isomerization of optically active *cis*- and *trans*-2-phenyl-3-*p*-tolyloxirane were studied, and the slower racemization of the cis isomers compared to the trans isomers was caused by conrotatory ring opening of the oxirane ring.[160] The facile racemization of (+)-2,3-divinyloxirane at 150°C requires an equilibrium by way of an electrocyclic ring opening.[161] A slower, but competing, reaction is the conversion into racemic 2-vinyl-1,2-dihydrofuran. Thermal isomerization of *cis*- and *trans*-vinyloxiranes into 1,2-dihydrofurans may be partly or completely stereoselective depending on the substituents which influence the rates of electrocyclic equilibration (Equation 15).[162] Photolysis of *cis*- and *trans*-2,3-diphenyloxirane at low temperature produces deep red colors which have been attributed to oxygen ylides formed by disrotatory C–C ring cleavage.[163] The thermal isomerization of *trans*-2-cyano-2,3-diaryloxirane to the cis isomer probably involves an electrocyclic process.[164]

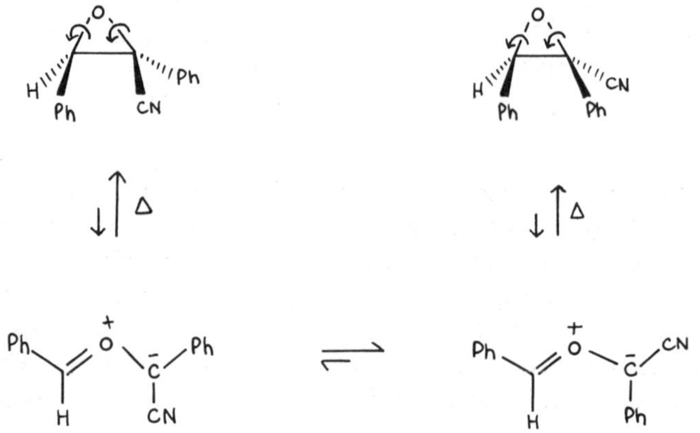

Scheme 4

3. *Rearrangements*

The acid- and BF_3-catalyzed rearrangements of oxiranes into carbonyl compounds with migration of one of the substituents is a synthetically very useful reaction. The effect of substituents on the rate of isomerization has been discussed.[3,165] The reaction is intramolecular and stereoselective but the selectivity is not necessarily high. In the unsymmetrical 2-methyl-2-*t*-butyloxirane it is the hydrogen atom (H_a) cis to the methyl group that

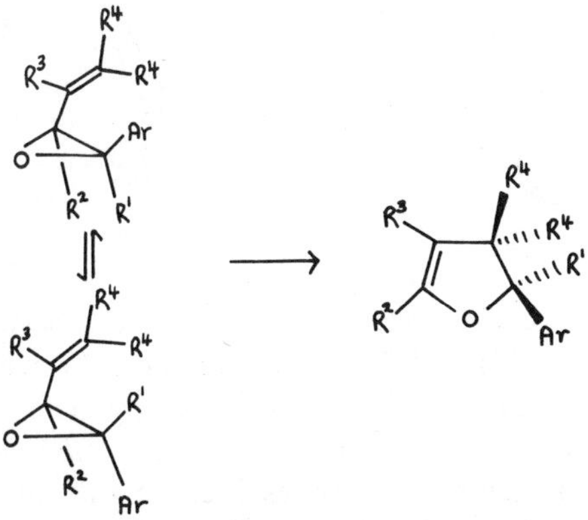

(15)

migrates preferentially to the trans hydrogen atom (H_b) (Equation 16). The selectivity (1.9 to 1) was explained by the relative stabilities of two conformers of the intermediate carbonium ion, and it was suggested that inversion occurred at the Me- and *t*-Bu-substituted carbon atom when H_a migrated but not when H_b migrated.[166]

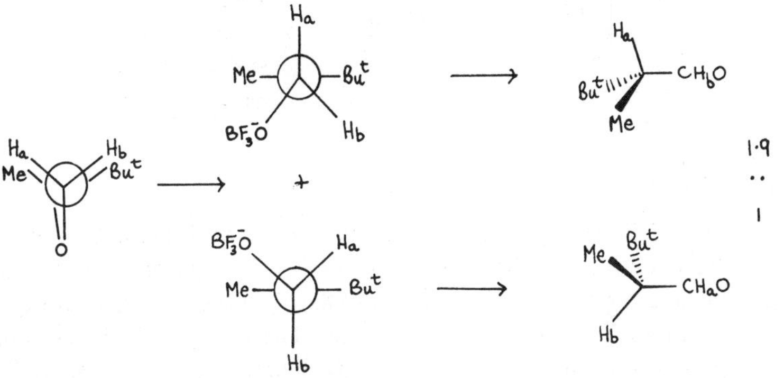

(16)

The epoxide function in 2-hydroxyalkyloxiranes migrates from one oxygen atom to the other, and an equilibrium is set up in 0.5 *N* aqueous sodium hydroxide. The equilibrium is in favor of the primary alcohol (e.g., **34** ⇌ **35**) or the lower substituted alcohol because it is strongly affected by steric factors.[167]

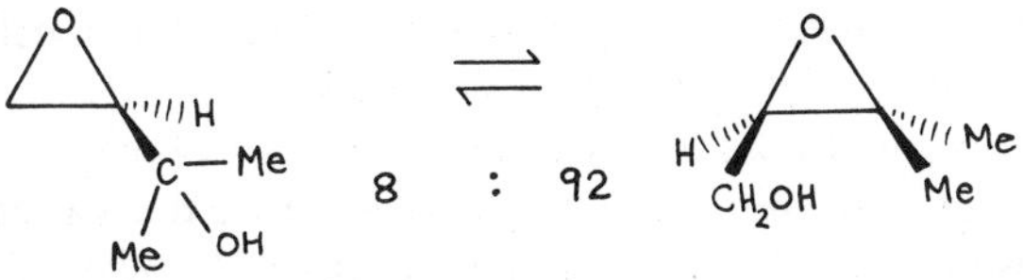

8 : 92

34 **35**

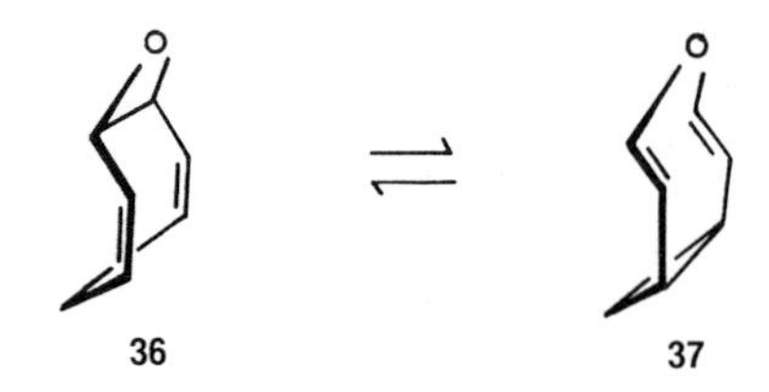

A rapid and reversible Cope rearrangement occurs between 8-oxabicyclo[5.1.0]octa-2,5-diene (**36**) and 4,5-homooxepine (**37**), which can be observed by variable temperature ^{1}H nmr spectroscopy. The ratio of **37** to **36** is 6.45 at 7°C and 4.12 at 30°C with an estimated $\Delta F^{\ddagger}$ value of 71.1 kJ mol^{-1}. Other rearrangements of oxiranes include the interconversion of benzene oxide (**38**) and oxepin (**39**),[168] the orbital symmetry allowed photochemical isomerization of cyclooctatetraene oxide to oxonine,[169–171] and the "NIH" rearrangement, for example, of 4-deuterotoluene 3,4-oxide into 3-deutero-4-hydroxytoluene. The stereochemistry of the mechanism of the latter reaction is still not completely clear.[172,173]

38 ⇌ 39

4. *Miscellaneous*

Cis- and *trans*-2,3-Dimethyloxiranes are chlorinated with *t*-butyl hypochlorite in the presence of uv light at 0°C with high retention of configuration. The retention of configuration is as high as 98.5% in the case of neat liquids but only around 80% in carbon tetrachloride. The halogen in the 2-chloro-2,3-dimethyloxiranes formed is replaced by hydrogen with triphenyl tin hydride, but whereas this reduction occurs essentially with retention of configuration in the chloro-*trans*-2,3-dimethyloxirane, it is only about 55% in the cis isomer.[174] The high stereospecificities observed are interesting because radicals are involved in these substitution reactions.

Bogert and Roblin[175] prepared cyclic acetals from the reaction of aldehydes and ketones with oxiranes and recognized that stereoisomers could be formed but were unable to separate the isomers. The mechanism of this reaction was investigated and is discussed in Section VIII.A (Equation 33).

The pyrolytic and Lewis acid-catalyzed ring cleavage of *cis*- and *trans*-2,3-hexamethyleneoxiranes (cyclooctene oxides) yield a variety of products including those formed by transannular reactions by way of intermediates related to hydroxycyclooctenes (Section XIV.B).[97]

III. DI- AND TRIOXIRANES (BIS- AND TRISETHYLENE OXIDES)

The epoxidation of di- and trienes should, if it is complete, provide di- and trioxiranes respectively. If the oxidation is random, stereoisomeric dioxiranes are possible. Generally, however, these oxidations proceed with some stereospecificity, particularly if the double bonds are not very far apart. Oxidation of 1,4-bismethylenecyclohexane with *m*-chloroperbenzoic acid produces the *trans*-1,4-bismethylenecyclohexane diepoxide (**40**) in over 97% yield. The cis isomer can also be obtained from 1,4-bismethylenecyclohexane, but by successive treatment with aqueous *N*-bromosuccinimide, to form the *cis*-1,4-bisbromomethyl-1,4-dihydroxycyclohexane, and then with aqueous potassium hydroxide.[176] Similar oxidation of 7,7-diethoxynorborna-2,5-diene gives the *endo,exo*-bisepoxide in high yield which is hydrolyzed in 80% aqueous acetic acid to 2,3-*endo*-5,6-*exo*-bisepoxynorbornan-7-one (**41**). This ketone is particularly interesting because on pyrolysis at 420°C it loses carbon monoxide and is converted into 4,8-dioxabicyclo-[5.1.0]octa-2,5-diene (**42,** $x=y=O$).[177] Unlike its monooxa analogue (**42,** $X=CH_2$, $Y=O$)[168] and the parent carbocycle, homotropilidene (**42,** $X=Y=CH_2$)[178] which undergo the valence tautomerism **42** ⇌ **43** ⇌ **44** with free enthalpies of activation of 71.1 and 60.3 kJ mol^{-1} respectively, it shows no signs of isomerization on heating even to 90°C.

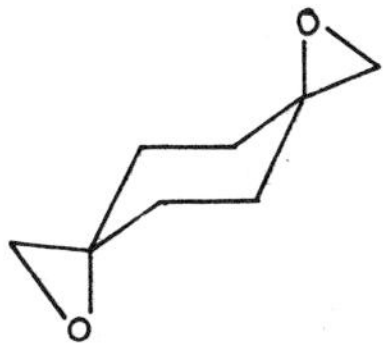

40 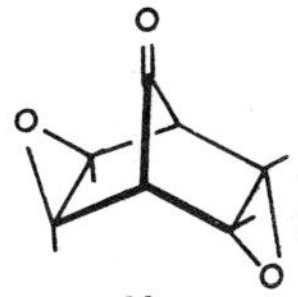 **41**

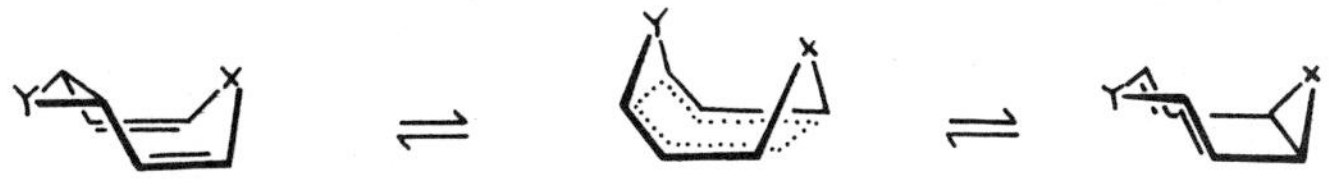

42 ⇌ **43** ⇌ **44**

With less rigid olefins, for example, butadiene or vinyloxirane, oxidation with peracids furnishes a mixture of dioxiranes.[179] 2,2′-Dioxirane (1,2,3,4-dioxobutane) exists in the meso and *d,l* forms. The former is obtained by dehydrobromination of 1,4-dibromo-2,3-dihydroxybutane, and the optically active forms of the latter are formed from a similar reaction on (+)- or (−)-2,3-dibromo(or dichloro)-1,4-dihydroxybutane. A change in the sign of rotation occurs on conversion into the dioxiranes together with Walden inversions at C–2 and C–3.[180]

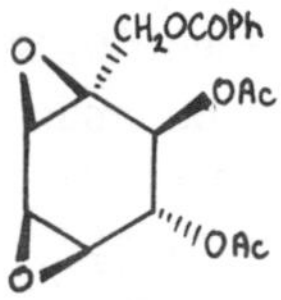

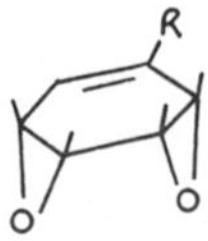

R = Br or H

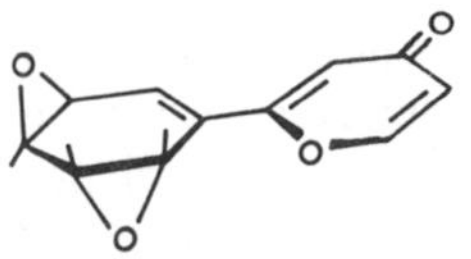

45 **46** **47** **48**

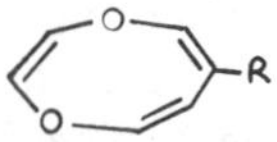

The discovery of benzene dioxides such as crotepoxide (**45**)[181] and LL-Z1220 (**46**),[182] and the possibility of valence tautomerism (e.g., into oxepines **38** ⇌ **39**, has aroused interest in the chemistry of benzene oxides. *syn*-Benzene-3,4:5,6-dioxides (*cis*-5,8-dioxatricyclo[5.1.0.$0^{4,6}$]oct-2-ene) (**47,** R = H, Br) were prepared by Vogel and co-workers[183,184] by starting from 4-epoxycyclohex-1-ene and by judiciously brominating, dehydrobrominating, debrominating, cleaving, and reforming oxirane rings. Crystalline *syn*-benzene dioxide (**47,** R = H) is stable at room temperature, but in solution at 50°C it rearranges rapidly into an equilibrium mixture with its valence tautomer 1,4-dioxocin (**48,** R = H). At 60°C in benzene a 5:95 ratio of **48** (R = H) to **47** (R = H) is obtained. The double bond in *syn*-benzene dioxide can be selectively reduced with Pd–C to give 1,2:3,4-diepoxycyclohexane.[184] From the same starting material, but with a modified sequence of reactions, Vogel and co-workers[185] prepared *anti*-benzene-1,2:3,4-dioxide (**49**) which, unlike the syn isomer, does not isomerize when heated even up to 150°C. The oxidation of 1,4-dihydrobenzene with perbenzoic acid or monoperphthalic acid gives only *anti*-cyclohexane-1,2:4,5-diepoxide (*trans*-4,8-dioxatricyclo[5.1.0.$0^{3,5}$]octane, **50**). Oxidation with excess *m*-chloroperbenzoic acid, on the other hand, gives a 2:1 mixture of *anti*- and *syn*-cyclohexane-1,2:4,5-diepoxides from which the syn isomer could not be isolated in a pure state because it formed a 1:1 complex with the anti isomer. The syn isomer can be obtained in a pure state from *trans*-4-bromo-5-hydroxycyclohex-1-ene by epoxidation, separation of isomers, and dehydrobromination. The structures of the *syn*- and *anti*-diepoxides were confirmed by their dipole moments which were 3.38 D and 0.0 D respectively.[186] *trans*-1,4-Dibromo-*syn*-2,3:5,6-diepoxycyclohexane (**51**) was an intermediate in the preparation of the 2,3:4,5-diepoxide (**47,** R = Br). It was obtained by oxidation, with peroxytrifluoroacetic acid, of *trans*-3,6-dibromo-4,5-epoxycyclohex-1-ene which in turn was formed together with the cis isomer from the reaction of 4,5-epoxycyclohex-1-ene and *N*-bromosuccinimide.[183]

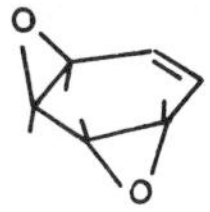

49 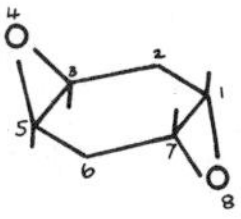 **50** 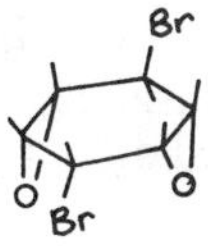 **51**

Vogel and co-workers[187] and Schwesinger and Prinzbach[188] prepared *syn*-benzene trioxide (3,6,9-trioxatetracyclo[6.1.0.0.^{2,4}0^{5,7}]nonane, **52**) independently, but starting from 4,5-epoxycyclohex-1-ene. The *syn*-trioxide undergoes an orbital symmetry allowed $[\sigma 2_s^2 + \sigma 2_s^2 + \sigma 2_s^2]$ cycloreversion to *cis,cis,cis*-1,4,7-trioxacyclononatriene (**53**) when heated to 200° for an extended period.[187] *anti*-Benzene trioxide (**54**) was obtained either by the direct oxidation of *syn*-benzene-3.4:5,6-dioxide (**47,** R=H) with peroxytrifluoroacetic acid or by thermolysis of 4,5-epoxy-3,6-*endo*-peroxycyclohex-1-ene (**55**), which was prepared by the addition of singlet oxygen to benzene oxide (**38**).[189] In a similar manner Foote and collaborators[190] obtained the *anti*-benzene trioxide derivative of 1-methyl-2-phenylindene (**56**). Although a $[\sigma 2_s^2 + \sigma 2_a^2 + \sigma 2_a^2]$ cycloreversion is also symmetry allowed for *anti*-benzene trioxide, the energy required to transform it into a trioxacyclononatriene would be very high for steric reasons. Vogel and co-

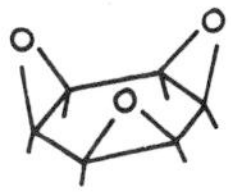

52 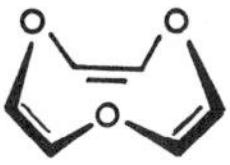 **53** 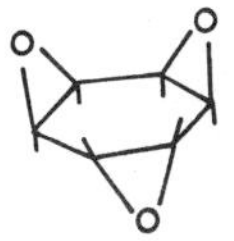 **54**

55 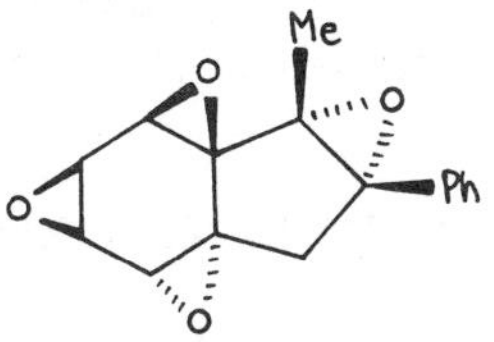 **56** 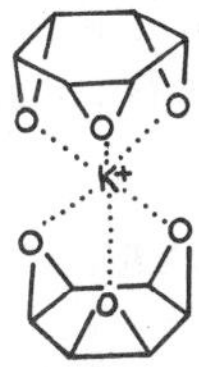 **57**

workers[187] were unable to isomerize it, and even when it was subjected to gas phase pyrolysis at 400 to 500°C it was unaltered. Interestingly, *syn*-benzene trioxide behaves as a crown ether and forms stable complexes with metal salts (Section XIV.D). Thus with KI in acetone it forms a 2:1 complex, probably **57**, which decomposes on melting at 180°C.[188]

The nucleophilic cleavage of the oxirane rings in benzene trioxide, as in simple oxiranes, occurs with inversion of configuration at the carbon atom that is involved in the nucleophilic attack (Section II.C.1).[191]

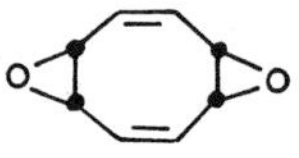

58 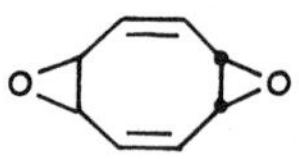 59

cis,cis-Cycloocta-1,5-diene yields a mixture of *syn*- and *anti*-1,2:5,6-diepoxycyclooctanes, on oxidation with perbenzoic acid, in which the syn isomer is the major product.[192] 1,2-Epoxycycloocta-3,5,7-triene, on the other hand, is oxidized by *m*-chloroperbenzoic acid to three of the four diepoxides: *syn*- (**58**) and *anti*-1,2:5,6-diepoxycycloocta-3,7-dienes (**59**) and *anti*-1,2:3,4-diepoxy-5,7-diene (**60**) in the ratio 24:67:9. Peroxytrifluoroacetic acid yields, on the other hand, 15% of 1,2(*r*):3,4(*anti*):5,6(*syn*)-triepoxycyclooct-7-ene (**61**) in addition to the above three diepoxides. The stereochemistries followed from their ^{1}H nmr spectra. When the diepoxide **58** is heated to 200°C appreciable amounts of the valence tautomer **62** are observed in equilibrium with it, and the vicinal diepoxide (**60**) rearranges

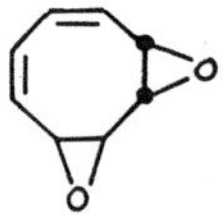

60 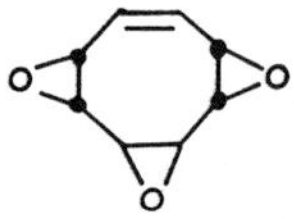 61 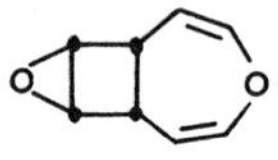 62 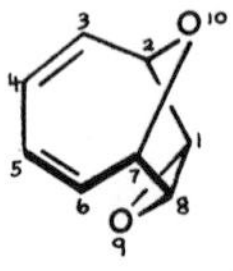 63

completely to 9,10-dioxatricyclo[6.1.1.^{2,7}0]deca-3,5-diene (**63**). The trioxirane however, remains unchanged on heating at 255°C for about 20 hr.[193] ^{13}C nmr spectroscopy of *cis,cis*-1,2:3,4-diepoxycyclooctane at low temperature is consistent with a twist-boat-chair conformation.[193a]

IV. OXETANES (TRIMETHYLENE OXIDES)

The stereochemistry of substituted oxetanes is similar to that observed in C-substituted azetidines (Part I, Chapter 2, Section II.A.2), and substituents can exhibit cis and trans relationships to one another. They differ from azetidines in that they are considerably less basic and in that it is more difficult to place a substituent on the oxygen atom than on a nitrogen atom. In general, the four-membered ring of oxetanes is much less puckered than that in azetidines and thietanes (Chapter 3, Section III.B), but this varies with substitution. In the parent substance it is essentially planar with a small energy barrier separating the two puckered conformations [compare oxetane (0.17–0.42 kJ mol^{-1}), thietane (4.48 kJ mol^{-1}), and azetidine (5.27 kJ mol^{-1})].[194] Here we are concerned only with stereochemical aspects; for the general chemistry of oxetanes reference should be made to the comprehensive review by Searles Jr.[195]

A. Syntheses

The most extensively studied synthesis of oxetanes is the Paterno-Büchi reaction in which carbonyl compounds add across electron rich olefins under the influence of uv light.[196] Irradiation of a mixture of benzophenone and isobutene produces a 9:1 mixture of 3,3-dimethyl-2,2-diphenyl- and 2,2-dimethyl-4,4-diphenyloxetanes. The regioselectivity has been attributed to the greater thermodynamic stability of the diradical **64** compared with the diradical **65** (Equation 17).[197] The stereospecificity of the reaction with acetaldehyde was tested with *cis*- and *trans*-but-2-ene. Three 2,3,4-trimethyloxetanes, **66–68**, were formed, but the *cis*-2,3-dimethyl isomers **66** and **68** predominated in the reaction with *cis*-but-2-ene, and the *trans*-2,3-dimethyl isomers **66** and **67** predominated in the reaction with *trans*-but-2-ene. The

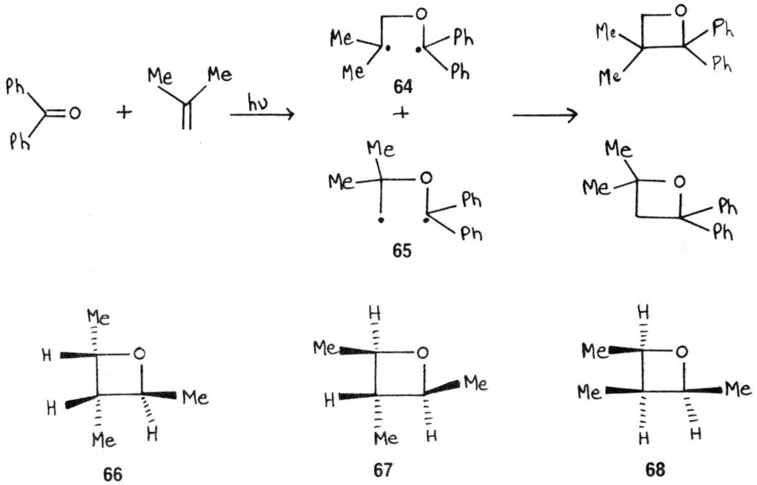

(17)

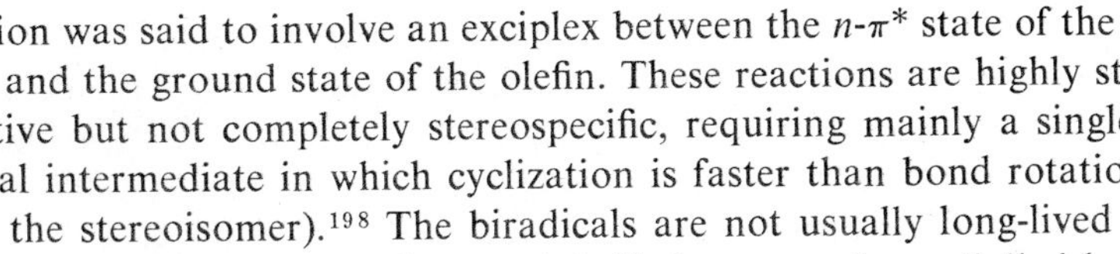

reaction was said to involve an exciplex between the n-π^* state of the aldehyde and the ground state of the olefin. These reactions are highly stereoselective but not completely stereospecific, requiring mainly a singlet biradical intermediate in which cyclization is faster than bond rotation (to yield the stereoisomer).[198] The biradicals are not usually long-lived as in the formation of *trans*-2,3-dicyano-4,4-alkyloxetanes from dialkyl ketones and *trans*-fumaronitrile. The rate of this reaction is faster than the photochemical isomerization of the dinitrile.[199,200] Several causes are responsible for the regiospecificities, one of which may be the reversible nature of the formation of the intermediate biradicals (e.g., **64** and **65**) thus making steric factors important in the direction of the reaction.[196] The complex formed from triplet benzophenone and simple alkenes, on the other hand, was shown to be irreversible.[201] Steric effects are observed in the preferential formation of the *exo*-oxetanes as in *exo*-4,4-diphenyl-3-oxatricyclo-[4.2.1.0^{2,5}]nonane (**69**, $R^1 = R^2 =$ Ph) from benzophenone and norbornene,[196,202] and the 1:1 mixture of the *exo*-oxetanes (**69**, $R^1 =$ Ph, $R^2 =$ 2-, 3-, or 4-pyridyl) and *endo*-oxetanes (**69**, $R^1 =$ 2-, 3-, or 4-pyridyl, $R^2 =$ Ph) from 2-, 3-, or 4-benzoylpyridine and norbornene when irradiation was performed in benzene solution. In acetic acid solution, however, 4-benzoylpyridine and norbornene yield a 4:1 mixture of **69** ($R^1 =$ 4-pyridyl, $R^2 =$ Ph) and **69** ($R^1 =$ Ph, $R^2 =$ 4-pyridyl); presumably the increased solvation of the pyridine cation forces the 4-pyridyl group into the exo configuration.[196] Other examples in which the orientation is probably influenced by the stability of the radicals, as in **64** and **65**, are the condensations of hexa-1,3-diene,[203] 2-methylbut-2-ene,[204] and styrene[205] with acetaldehyde to yield oxetanes regioselectively.

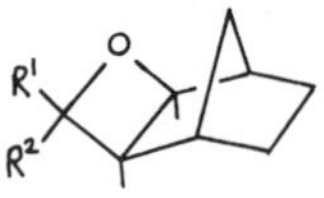

69 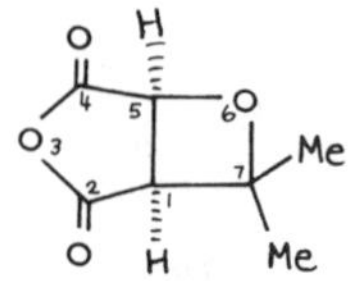 **70** 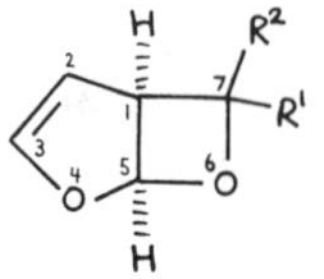 **71**

Generally the stereochemistry of the olefin is preferentially retained in the oxetane, and in certain instances, as in the condensation of acetone with maleic anhydride, only one oxetane is formed, for example, *cis*-7,7-dimethyl-3,6-dioxabicyclo[3.2.0]heptane-2,4-dione (**70**).[204] The latter is a useful intermediate for making *cis*-2,2,3,4-tetramethyloxetane unambiguously. Enol ethers such as *cis*- or *trans*-1-methoxybut-1-ene and acetone give 2-methoxyoxetanes together with 3-methoxyoxetanes; the stereochemistry of the original butene is retained with acetone singlets but not with acetone triplets, both of which are involved in the photochemical reaction.[206] Only one orientation is obtained with furans which yield the monooxetanes; 7-substituted *cis*-4,6-dioxabicyclo[3.2.0]hept-2-ene (**71**) is formed almost exclusively with aldehydes or ketones.[207–209] The 1:1 adducts **71** are basically bicyclic acetals, and unlike the isomeric 2,6-dioxabicyclo-[3.2.0]heptanes, their reduction products are acid labile.[208] Irradiation of furan with excess of a carbonyl compound forms 1:2 adducts (e.g., dioxetanes **72** and **73**). The completely acid labile isomers **72** are formed predominantly over the partially labile isomers **73.** There is an added stereochemical complication in these 1:2 adducts because of the possibility of syn and anti isomers. This was sorted out by reductive or acid degradation and by dipole moment measurements.[210–212] The oxetanes obtained by photocycloaddition of acetone or benzophenone and 1,4-diox-2-ene, 2,5,7-trioxabicyclo[4.2.0]octanes (**74**), are also cis-fused.[213]

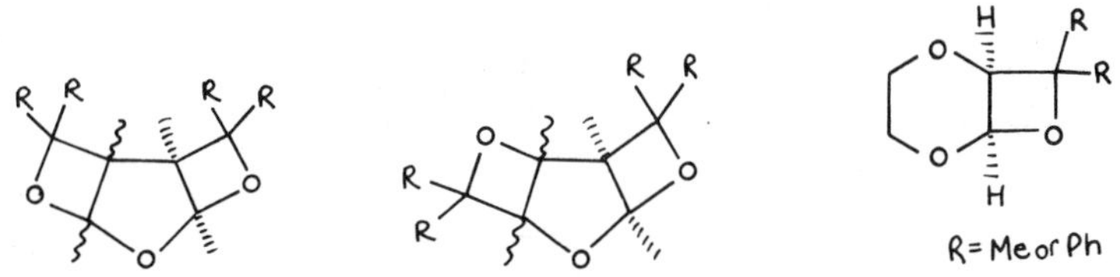

R = Me or Ph

72 **73** **74**

A direct chemical synthesis of oxetanes which proceeds stereospecifically is the base promoted cyclization of the monotosylate of propane-1,3-diols. If the tosyl group is originally placed on a primary hydroxyl group, then any chiral centers on the diol are unaffected by the reaction. Thus *erythro*- and *threo*-2-methyl-1-phenyl-3-tosyloxypropan-1-ol cyclize in the presence of *t*-BuOK in *t*-BuOH, at room temperature, into *cis*- and *trans*-3-methyl-2-phenyloxetanes (Equation 18).[214] Similarly *cis*-2-benzoyloxymethyl-[215] and *cis*-2-*p*-bromobenzenesulfonyloxycyclohexan-1-ol[216] yield the same *cis*-2,3-tetramethyleneoxetane (**75**, $R^1 = R^2 = H$), whereas the trans isomer of the latter alcohol solvolyzed much faster and gave a polymeric material. The photochemical condensation of methyl acetoacetate and cyclohexene produced a 1:1.5:2.4:2.2 mixture of methyl 3-cyclohexyl-3-hydroxybutyrate, methyl 3-(cyclohex-1-en-3-yl)butyrate, 8-*exo*-methyl-8-*endo*-methoxycarbonylmethyl- (**75**, $R^1 = CH_2CO_2Me$, $R^2 = Me$) and 8-*endo*-methyl-8-*exo*-methoxycarbonylmethyl-(**75**, $R^1 = Me$, $R^2 = CH_2CO_2Me$)*cis*-7-oxabicyclo-[4.2.0]octanes, and demonstrated the ease with which a cis-fused oxetane can be formed.[217]

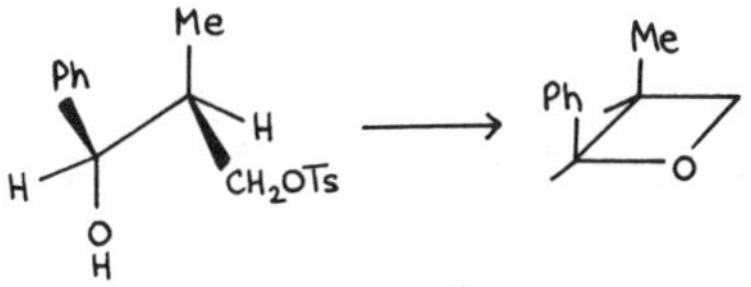

(18)

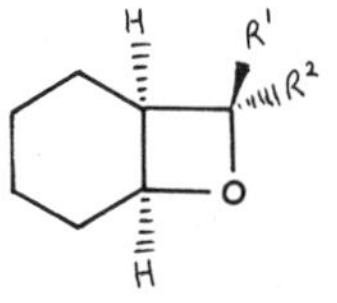

75

The ring-chain tautomers of 3,3-diethyl-2-ketoglutaric acid, described by Thorpe and co-workers,[218] were shown by Winberg and Holmquist to be the cis and trans isomers of 2,4-dicarboxy-3,3-diethyloxetane.[219]

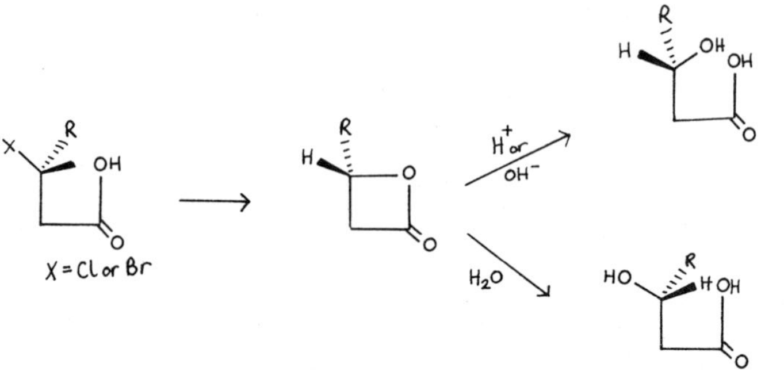

X = Cl or Br

(19)

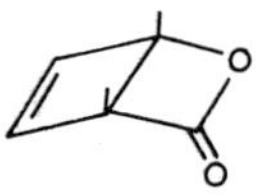

76

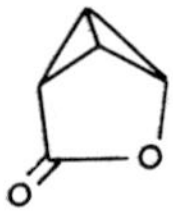

77

β-Lactones, which are indeed oxetan-2-ones, behave very much like cyclic esters and have been reviewed by Etienne and Fischer.[220] The most interesting stereochemical feature in these compounds is the nature of the asymmetric carbon atom at C–4 in 4-substituted oxetan-2-ones. The formation of these by intramolecular cyclization of optically active β-halo acids occurs with a Walden inversion at the β-carbon atom.[220–223] The hydrolysis of oxetan-2-ones takes place with inversion of configuration at C–4 in neutral medium (alkyl-oxygen fission), but with retention of configuration at C–4 in acidic or alkaline medium (acyl-oxygen fission) (Equation 19).[220,223–225] Very little other stereochemical work on oxetan-2-ones has been reported except for the valence bond isomerization of 2-pyrone, on irradiation, to the V-shaped 6-oxabicyclo[2.2.0]hex-2-en-5-one (**76**) which is relatively stable at −78°C, polymerizes readily at room temperature,[226] and rearranges into tricyclo[$2^{3,6}$.1.1.0]pyran-2-one (**77**) in polar aprotic solvents.[226]

B. Configuration and Conformation of Oxetanes

The relative configurations of substituents (cis or trans) in oxetanes sometimes follow from the method of synthesis,[215–217] but in cases where mixtures of isomers are formed the stereochemistry can be deduced from ^{1}H nmr spectroscopy.[227–230] The stereochemical relationship of substituents is deduced mainly from the coupling constants of vicinal hydrogen atoms; however, occasionally these are marginally different and incorrect assignments were reported.[231,232]

The oxetane ring is effectively planar[194,233,234] but does have a very small inversion barrier (see above). This barrier is completely removed, that is, the system becomes rigid, if fluorine atoms are introduced into the molecule as in 2,3-difluorooxetane.[235] Substituents may also render the four-membered ring puckered, and Macchia and co-workers[214] showed, by evaluating J_{trans}/J_{cis} values for *cis*- and *trans*-3-methyl-2-phenyloxetanes, that the cis isomer consists of equal amounts of the two conformers with pseudo-axial and pseudo-equatorial phenyl groups, but that the trans isomer exists mainly in the conformer **78** with dipseudo-equatorial substituents in equilibrium with the conformer **79.**

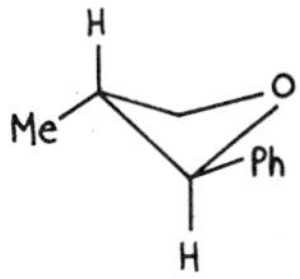

H Me O Ph H

78

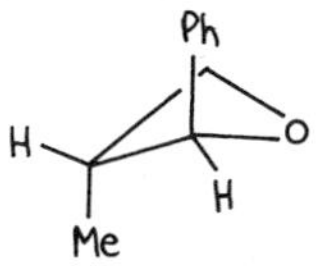

Ph H O H Me

79

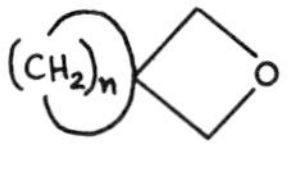

$(CH_2)_n$ O

80

The Thorpe-Ingold effect,[236] which states that "when the bond angle is distorted between one pair of substituents on a saturated carbon atom, the bond angle between the other pair of substituents will be distorted in the opposite manner," was tested by Searles and co-workers.[195,237] They prepared 3,3-spirooxetanes, **80** with $n=3$, 4, 5, and 6, and compared their physical and chemical properties with those of oxetane and 3,3-dimethyloxetane. They demonstrated that the shape of the oxetane ring is altered by the size of the carbocyclic ring. Thus the ir bands for oxetane, 3,3-dimethyloxetane, **80** with $n=2$, $n=3$, $n=4$, and $n=5$, were at 10.15, 10.20, 10.35, 10.23, 10.17, and 10.14 μ, there being a maximum bathochromic shift with **80** ($n=2$) in which the C–2, C–3, C–4 bond angle is at a maximum. The basicity of these oxetanes, as measured by the OD ir spectral shifts in MeOD and their reactivity towards LAH, confirmed that this effect is real.

C. Reactions of Oxetanes

The only reaction of oxetanes of interest here is their thermal decomposition into olefins. Pyrolysis of *trans*-3-*n*-propyl-2-phenyloxetane at 250–300°C proceeded stereospecifically and produced almost exclusively *trans*-1-

phenylpent-1-ene. Thermolysis of the cis isomer was slower, less stereospecific, and gave a mixture of *cis*- and *trans*-olefins with the *cis* to *trans* ratio varying from 8.1 at 242°C to 1.9 at 290°C.[238] Carless[239] examined the thermolysis of several di- and trimethyloxetanes at 430–450°C and provided evidence that intermediate biradicals are formed, and that the stereochemistry of the olefins formed depended on the relative half-lives of the biradicals (Equation 20). In all cases he obtained a mixture of *cis*- and *trans*-but-2-enes, in contrast with the decomposition of 2-methyl-3-phenyloxetanes[205] (and above) where high stereospecificity was observed. This reflected the difference between a radical on a benzilic carbon atom and on an aliphatic carbon atom. Adams, Bibby, and Grigg[240] found that specific cleavage of oxetanes occurred on thermolysis in the presence of a rhodium catalyst, $[Rh(CO)_2Cl]_2$. Carless[231] reexamined the pyrolysis of his series of alkyloxetanes in the presence of $[Rh(CO)_2Cl]_2$ or $AgBF_4$ and

Me Me Me Me Me Me Me Me Me Me Me Me Me Me + + + CH_3CHO (20)

demonstrated that the decomposition was highly stereoselective, and that it provided a useful method for the determination of the configurations of oxetanes in cases where the ^1H nmr data were not conclusive.

The stereochemistry of the hydrolysis of oxetan-2-ones is described in Section A.

V. 1,2-DIOXETANES (ETHYLENE PEROXIDES)

The addition of a singlet oxygen to a double bond is stereospecific and yields 1,2-dioxetanes by the cis addition of oxygen.[241] Because these are essentially ethylene peroxides, they are best formed and kept at low temperatures. Irradiation of *cis*- and *trans*-1-ethoxy-2-phenoxyethylene in the presence of oxygen and a sensitizer at −78°C gave *cis*- and *trans*-3-ethoxy-4-phenoxy-1,2-dioxetanes respectively.[242] Although they are stable at −78°C, they decompose into ethyl and phenyl formate on heating to

50°C. Similarly *cis*- and *trans*-1,2-diethoxyethylene gave *cis*- and *trans*-3,4-diethoxy-1,2-dioxetane[243] respectively (Equation 21), and 1,2-dimethoxy-*cis*-stilbene and 1,2-diphenyl-1,4-diox-2-ene gave the corresponding oxetanes.[244] Schaap and Bartlett[245] used triphenylphosphite ozonide to mimic the reaction of singlet oxygen and indeed obtained stereoselectively 1,2-dioxetanes from *cis*- and *trans*-1,2-diethoxyethylene. The stereospecificities, although good, were not higher than 83%. 1,4-Diox-2-ene yields the cis-fused 2,5,7,8-tetraoxabicyclo[4.2.0]octane (**81**) with triphenylphosphite ozonide, but at a slower rate than the diethoxyethylenes. Decomposition of 1,2-dioxetanes causes considerable breakdown of the molecule, and sometimes the fragments may be difficult to isolate. However, when diphenyl sulfide is added to the reaction, the 1,2-dioxetane formed from singlet oxygen addition can be trapped by the formation of intermediates such as *cis*-2,3-diphenyl-1,4-dioxane (**82**) which lose diphenyl sulfoxide (i.e., monodeoxygenation) without serious fragmentation of the original molecule.[244]

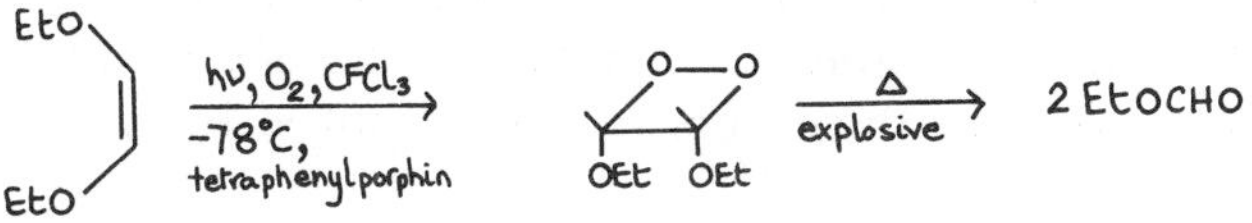

EtO, EtO —— hν, O_2, $CFCl_3$, −78°C, tetraphenylporphin ——→ [dioxetane: O—O, OEt OEt] —— Δ, explosive ——→ 2 EtOCHO (21)

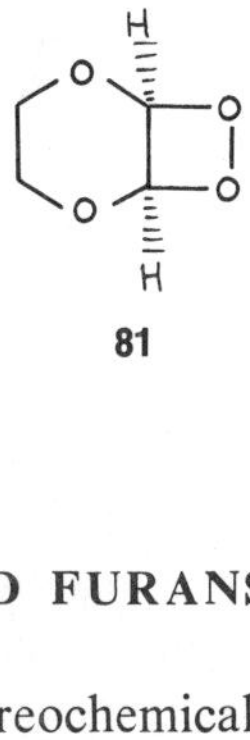

81 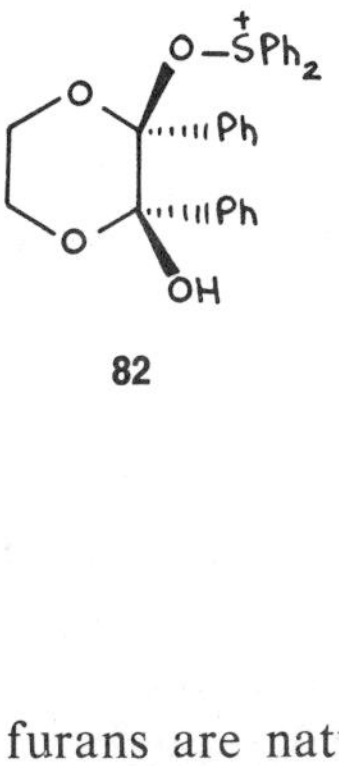 **82**

VI. REDUCED FURANS

Many of the stereochemically interesting furans are naturally occurring, and several of these are furan-2-ones, that is, cyclic γ-lactones. Only a few are briefly reported in this section, and other reviews should be consulted for further information.[246–249] The stereochemistry of reduced furans is treated here systematically, and reference to the comprehensive monograph by Dunlop and Peters[250] should be made for the general chemistry of furans prior to 1953. This was updated to 1963 in a review by Bosshard and Eugster.[246] The stereochemistry of furanoid sugars is beyond the scope of this monograph.

A. Syntheses of Reduced Furans

1. Reduction of Furans

The reduction of substituted furans yields stereochemically interesting dihydro- and tetrahydrofurans, but overreduction which results in ring cleavage is a serious drawback. If, however, adequate care is taken, substituted di- and tetrahydrofurans can be obtained. The general method of electrolytic alkoxylation of furans, developed by Clauson-Kaas and coworkers,[251–255] yields mixtures of *cis*- and *trans*-2,5-dialkoxy-2,5-dihydrofurans. The *cis*- and *trans*-2,5-dihydro-2,5-dimethoxyfurans are separable by distillation and are reduced catalytically with Raney nickel to the corresponding tetrahydrofurans.[256] Reduction with deuterium instead of hydrogen in the presence of the soluble catalyst triphenylphosphine rhodium chloride revealed that deuteration of the trans isomer gave one 3,4-dideutero-2,5-dihydro-2,5-dimethoxyfuran (**83**) as expected, but the cis isomer furnished one of two possible isomers stereospecifically, namely, 3-*trans*-deutero-4-*trans*-deutero-2-*r*-methoxy-5-*cis*-methoxytetrahydrofuran (**84**). An equilibrium mixture of the all-cis isomer **85,** with **83** and **84** is formed when the last two are allowed to stand in the presence of catalytic amounts of hydrochloric acid. The three isomers were identified by ^{1}H nmr spectroscopy.[257]

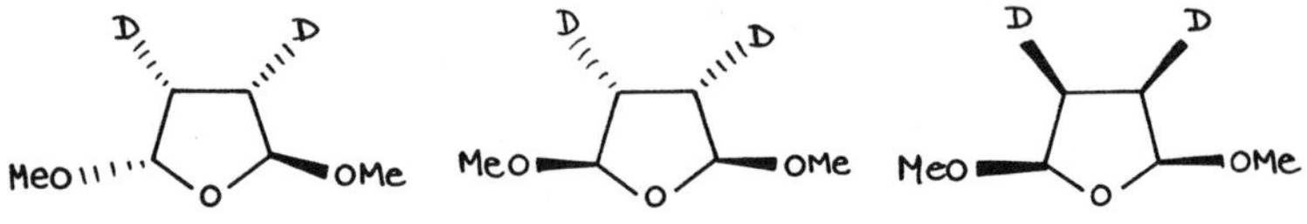

83 **84** **85**

Sodium amalgam reduces 2,5-dicarboxyfurans into a mixture of *cis*- and *trans*-2,5-dicarboxy-2,5-dihydrofurans together with 2,5-dicarboxy-2,3-dihydrofuran in varying amounts depending on the reaction time at 100°C. The amount of *cis*-2,5-dicarboxy-2,5-dihydrofuran decreases as the proportion of 2,5-dicarboxy-2,3-dihydrofuran increases, with *trans*-2,5-dicarboxy-2,5-dihydrofuran reaching an optimum of about 43% after 17 min. The latter is almost absent at the onset of reduction and after 60 min. The methyl esters of *cis*- and *trans*-2,5-dicarboxy-2,5-dihydrofurans were reduced catalytically (Pd/H_2) to the respective tetrahydrofurans and then reduced further with LAH to the corresponding 2,5-bishydroxymethyltetrahydrofurans.[258] They were identified by the ability of the *cis*-bistosyloxymethyltetrahydrofuran to cyclize into 8-oxa-3-azabicyclo[3.2.1]octane (**86**)[259] on treatment with methanolic ammonia. Reduction of the *cis*-bis-tosyl derivative with LAH gave *cis*-2,5-dimethyltetrahydrofuran identical with the only product obtained from catalytic reduction (Raney nickel) of 2,5-dimethylfuran.[258]

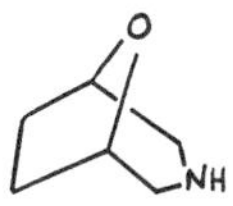

86 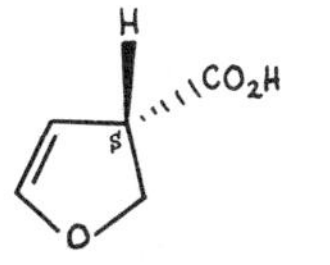 87 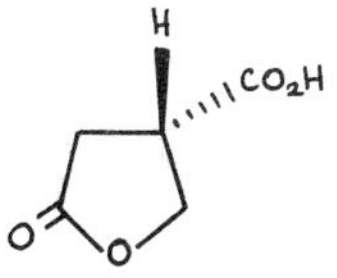 88

Asymmetric Birch reduction of 2- and 3-carboxyfurans was achieved in the presence of 1,2:5,6-di-*O*-isopropylidene-α-D-glucofuranose. 2-Carboxyfuran gave the 2,5-dihydro derivative whose methyl ester had $[\alpha]_D + 3.5°$. 3-Carboxyfuran, on the other hand, was reduced to 3-carboxy-2,3-dihydrofuran (**87**), and its methyl ester had $[\alpha]_D -8.7°$. The absolute configuration of this acid is *S* because on standing overnight in the presence of acid it was hydrated to 3-carboxy-5-hydroxytetrahydrofuran, which on oxidation with silver oxide, gave 4-carboxytetrahydrofuran-2-one (**88**) with $[\alpha]_D +$ 1.9°. The enantiomer, paraconic acid, $[\alpha]_D -60.4°$ is known and has an *R* configuration.[260]

2. *By Addition and Cycloaddition Reactions*

The electrolytic alkoxylation of furans mentioned in the previous subsection yields 2,5-dialkoxy-2,5-dihydrofurans and is satisfactory for 2- and 3-substituted furans.[251,253–255] Anodic oxidation of 2-acetoxy-5-methoxycarbonylfuran, however, leads to ring opening because the intermediate *cis*- and *trans*-2,5-dialkoxy adducts are unstable under the conditions used.[261] A cis and trans mixture of 2,5-diacetoxy-2,5-dihydrofuran is formed by the reaction of furan with lead tetraacetate.[262] The 3,4 double bond of *trans*-2,5-dihydro-2,5-dimethoxyfuran is hydroxylated to the 3,4-*cis*-diol with potassium permanganate, but attempts to resolve the *d,l* pair of 3-*cis*-hydroxy-4-*cis*-hydroxy-2-*r*-methoxy-5-*trans*-methoxytetrahydrofuran (**89**) by glc or recrystallization of *l*-menthoxyacetates failed.[256] Although *cis*-hydroxylation of *cis*-2,5-dihydro-2,5-dimethoxyfuran should yield two possible dihydroxy derivatives, only 3-*trans*-hydroxy-4-*trans*-hydroxy-2-*r*-methoxy-5-*cis*-methoxytetrahydrofuran (**90**) is formed. The methoxy groups obviously have a strong steric influence on the direction of hydroxylation.[256,263] The bromination of *cis*- and *trans*-2,5-dihydro-2,5-dimethoxyfuran, which takes place by a trans addition of bromine, should yield one and two bromides respectively. Because of epimerization, however, all three isomeric dibromo derivatives were formed. These were separated by chromatography and identified by ^1H nmr spectroscopy.[263] The chemistry of dihydroxy and diacetoxydihydrofurans was reviewed by Elming.[255]

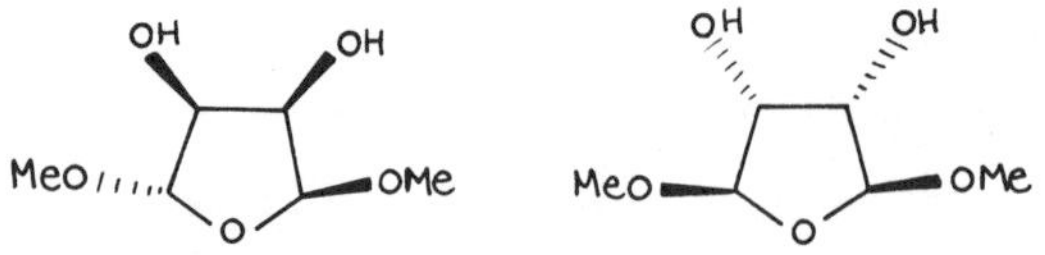

89 90

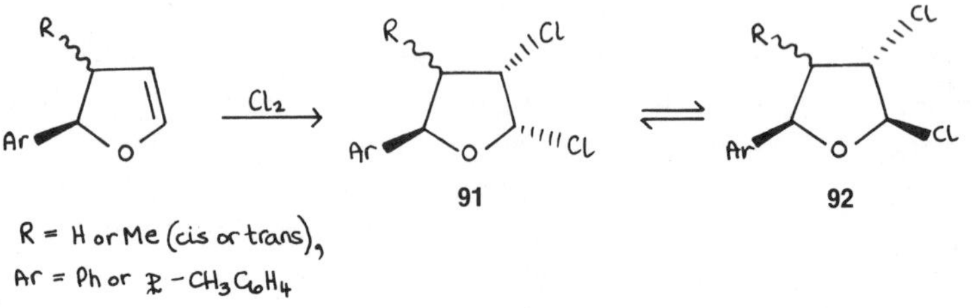

91 92 (22)

R = H or Me (cis or trans), Ar = Ph or *p*-$CH_3C_6H_4$

2,3-Dihydrofurans are basically cyclic enol ethers derived hypothetically (and experimentally) from compounds related to γ-hydroxybutyraldehyde by cyclization and dehydration of the cyclic acetals formed. The 3,4 double bond of 2,3-dihydrofurans is therefore subject to nucleophilic attack, and aniline readily adds across it to form 2-anilinotetrahydrofurans. Two aniline adducts are formed with 2,3-disubstituted 2,3-dihydrofurans.[264,265] Oxidation of 2,3-dihydrofuran with perbenzoic or *m*-chloroperbenzoic acids in the presence of alcohols produces *trans*-2-alkoxy-3-hydroxytetrahydrofurans. Acids isomerize these hydroxy alcohols into mixtures of cis and trans isomers.[266] The addition of chlorine to 2-aryl-2,3-dihydrofurans yields initially a mixture of 4,5-dichlorotetrahydrofurans in which the ratio **91**:**92** is high due to steric control. On standing, however, the ratio alters considerably, and the anomer **92** predominates at equilibrium (Equation 22). 2,3-Dihydrofuran is chlorinated to a 50:50 mixture of *cis*- and *trans*-2,3-dichlorotetrahydrofurans which changes almost completely into the trans isomer on standing.[267] Only the 2-halogen atom is epimerized in each case. Hydroboration followed by oxidation of 2-methyl-4,5-dihydrofuran is also stereospecific and furnishes *trans*-3-hydroxy-2-methyltetrahydrofuran.[268]

Furan undergoes cycloaddition reactions rather readily across the 2,3- and 2,5-positions. Irradiation of a mixture of furan and diazoacetic ester or amide with uv light produces 6-ethoxycarbonyl- or 6-amido-4-oxabicyclo[3.1.0]hex-2-enes (**93**) which are stabilized by reducing the double bond catalytically.[269]

6,6-Dibromo-2-oxabicyclo[3.1.0]hexane is demonobrominated stereoselectively with zinc and acetic acid, and gives a 62:38 mixture of *endo*- and *exo*-monobromo derivatives. This ratio is 9:1 if the 6,6-dibromo compound is treated with butyl lithium and hydrolyzed. The intermediate lithium derivative is converted into the carboxylic acid, by reaction with CO_2, with more than 99% retention of configuration.[269a] Stilbene, 1,1-diphenyl ethylene,[270] and 1,2-diphenyl acetylene[271] add onto the double bond of 4,5-dihydro-2-methylfuran to form the cis-fused 1-methyl-2-oxabicyclo[3.2.0]-heptane derivatives. Ethylene and 2-acetylfuran yield a similar $[2 + 2]\pi$ adduct with an acetyl group instead of a methyl group at C–1.[272] The photochemical addition of carbonyl compounds to the 2,3 and/or 3,4 double bond of furans to form oxetane derivatives is discussed in Section IV.A. The most common cycloaddition reactions of furans are those in which they behave as dienes and in which 2,5 addition occurs. These cis additions are facile and reversible. The cycloaddition of maleic anhydride to furan yields 5,6-endoxo-1,2,3,6-tetrahydrophthalic anhydride.[246,273,274] Two products are possible: the endo (**94**) and exo (**94a**) adducts. The exo isomer is formed twice as fast as the endo isomer in acetonitrile, but on standing at room temperature the mixture is converted completely into the exo isomer. The endo isomer can be isolated by careful manipulation and is stable in the crystalline form, but it dissociates in acetone solution. In aqueous solution, maleic acid adds onto furan to form a mixture of endo and exo products; with the endo adduct being formed four times as rapidly as the exo isomer. After standing for ten days an equal mixture of the two isomers is obtained. The evidence is that the isomerization takes place by dissociation and recombination.[274–276] Fumaric acid derivatives yield only one cis adduct with furan. A variety of substituted furans and maleic anhydrides or derivatives add in this manner[277–282]; acrylic acid derivatives,[283,284] acetylene dicarboxylic esters,[273,285–289] and related compounds, and even arynes,[290] butadiene,[291] and 1,2-diphenylcyclopropanone[292] act as dienophiles. The addition becomes more difficult when a substituent is present on the double bond of the furan involved in the reaction, and the adducts

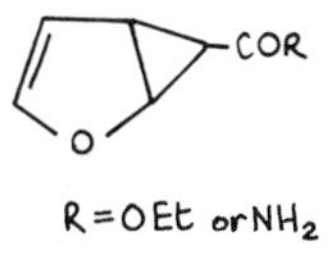

R = OEt or NH_2

93 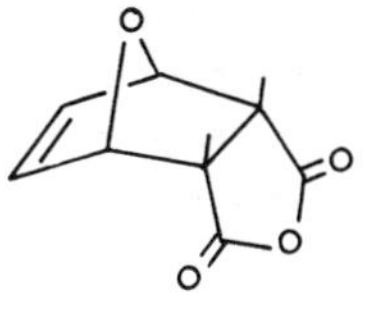 **94** 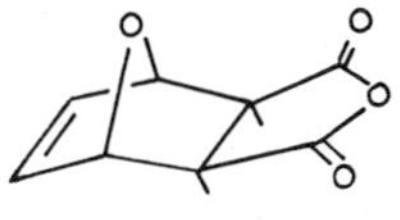 **94a**

are stabilized when the remaining double bond(s) is saturated. The reactions of these adducts have been studied extensively[274,275,277,279,282,284,292–301] because the products are generally more stable and because they are stereochemically interesting. The adducts from the condensation of acetylene derivatives (**95**) are particularly interesting because they react further

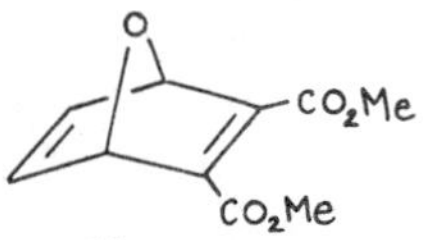

95

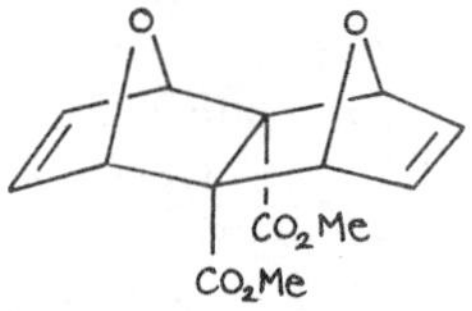

96

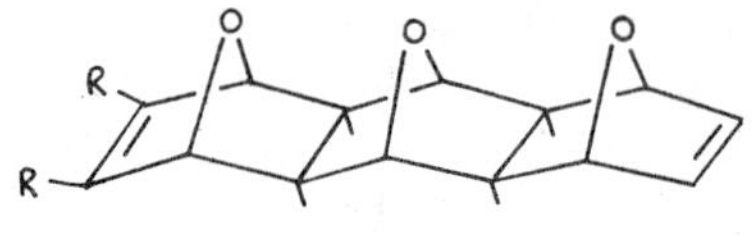

97

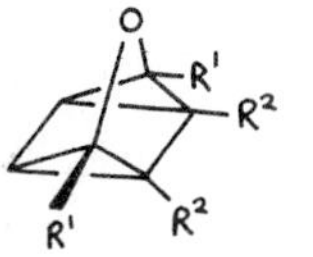

98

99

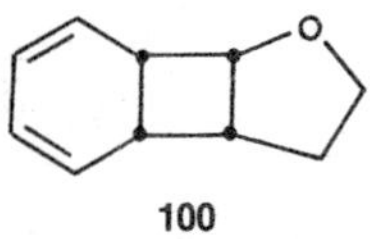

100

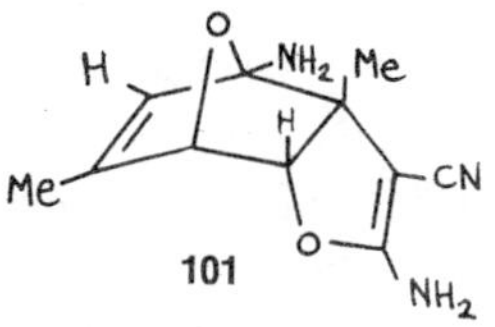

101

with furan to form naphthalene (**96**)[273,285–288] and anthracene (**97**) *endo*-oxides,[285] and form 3-oxaquadricyclanes (**98**) upon uv irradiation.[289] Uv irradiation of furan in benzene at 254 nm results in five 1 : 1, adducts of which the 2,5, 1′, 4′ cycloadduct **99** is predominant. This adduct undergoes an irreversible Cope rearrangement, on heating at 60°C, to the cyclobutane derivative (**100**) which is the 2,3,1′,2′ adduct of furan and benzene.[302] Only one furan derivative, 2-amino-3-cyano-4-methylfuran, is known to form a dimer, 2,4-diamino-3,5-dicyano-3*a*,6-dimethyl-*cis*-3*a*,4,7,7*a*-tetrahydro-*endo*-4,7-epoxybenzofuran (**101**), by a self Diels-Alder reaction. This was previously believed to be the monomer because it dissociated readily prior to reaction and behaved as a monomer.[303]

3. *By Intramolecular Cyclization*

The intramolecular cyclization of butane-1,4-diols with an asymmetric center at C–2 or C–3 by removal of water proceeds without upsetting the asymmetric center. Thus 2-*S*(+)-1,4-dihydroxy-2-methylbutane is dehydrated by the acid-catalyzed-azeotropic removal of water and yields *S*(+)-2-methyltetrahydrofuran.[304] When C–1 and C–4 in butane-1,4-diols are chiral centers, the stereochemistry of the tetrahydrofurans formed depends on the mechanism of the reaction. Mihailovic and co-workers[305] examined the cyclization of 1,4-disubstituted butane-1,4-diols with sulfuric and phosphoric acids, and of the corresponding 1,4-dimesylates in the presence of alkalies. All these reactions proceeded stereoselectively by S_N2 intramolecular substitution reactions. The *meso*- and *d,l*-1,4-diols give *trans*- and *cis*-2,5-disubstituted tetrahydrofurans in which one Walden inversion takes place (Equation 23). The *meso*- and *d,l*-1,4-bismesylates give the corresponding *cis*- and *trans*-2,5-disubstituted tetrahydrofurans by two Walden inversions—the first inversion occurs in the solvolysis of one mesylate group and the second in the intramolecular cyclization (Equation 23a).[306]

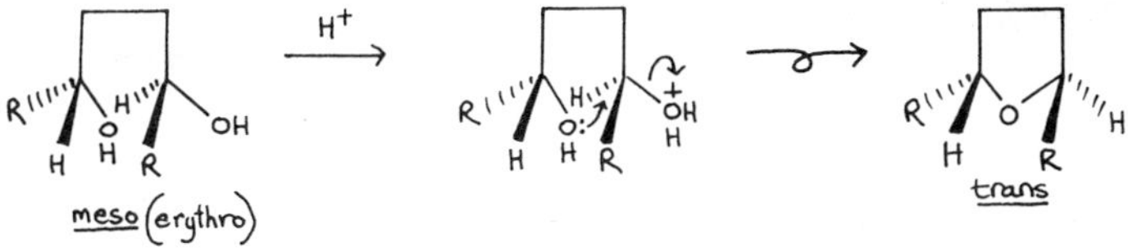

(23)

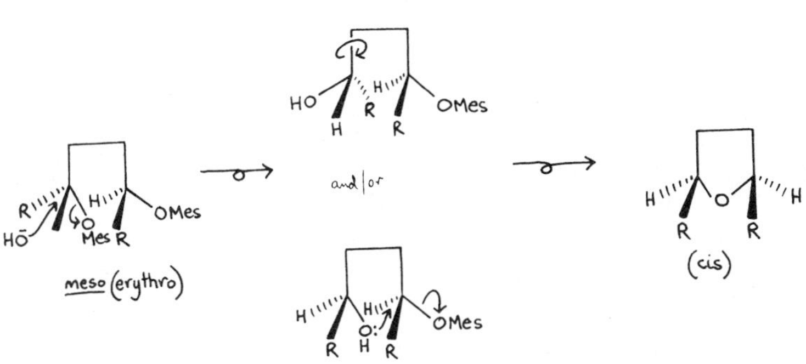

23a

In connection with the studies of naturally occurring tetrahydrofurans (−)-linalool was converted into 6,7-epoxy-6,7-dihydrolinalool which was cyclized into diastereoisomeric tetrahydrofurans.[306,307] Jacobus studied the mechanism and stereochemistry of a similar cyclization of *S*(−)-4-hydroxy-4-methylhexan-1-ol (**102**) into 2-ethyl-2-methyltetrahydrofuran (**103**) by acid-catalyzed azeotropic removal of water, heating with dimethylsulfoxide, heating with aluminum oxide, and by reaction with tosylchloride in pyridine. The yield of **103** in each case was high, but the first two reagents gave a racemic product and the other two gave the (+) isomers. Tosylchloride in pyridine furnished the optically purer tetrahydrofuran (**103**) whose absolute configuration, *S*, as determined unequivocally from *S*(−)-linalool, indicated that the configuration at the chiral center was unaltered.[308] It was concluded that the primary hydroxyl group of the diol **102** was tosylated and displaced by the tertiary hydroxyl group with inversion at C–1. The formation of 2,5-diphenyltetrahydrofurans from *d,l*- and *meso*-1,4-diphenylbutane-1,4-diols by acid-catalyzed dehydration is only slightly stereoselective, whereas tosyl chloride (1 mol equiv.) and sodium hydroxide in chloroform gave cis to trans ratios of diphenyltetrahydrofurans of 73:27 and 0:100 respectively.[309] By careful choice of reagents it should therefore be possible to determine whether butane-1,4-diols are the racemic or meso isomers or to determine the ratio of these in the diol mixtures.[310] Pyrolysis of 1-hydroxy-4-trimethylammoniobutane iodides, for example, α- and β-methadol methiodides, also yields tetrahydrofurans with elimination of trimethylamine and inversion of configuration at C–4.[311] The elimination of hydrogen chloride and formation of tetrahydrofurans by heating 4-chlorobutan-1-ols most probably proceeds with inversion of configuration at C–4.[312]

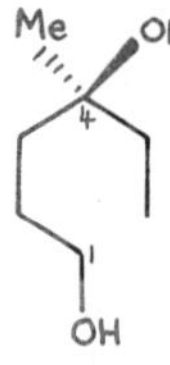

Me OH 4 1 OH **102**

Me Et O **103**

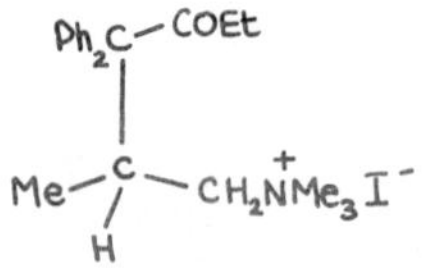

Ph_2C–COEt Me–C–$CH_2\overset{+}{N}Me_3$ I^- H **104**

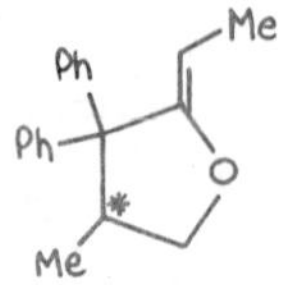

Me Ph Ph O * Me **105**

The intramolecular cyclization of 4-hydroxybutan-1-als to 1-hydroxytetrahydrofurans is reversible, but the latter dehydrate to 2,3-dihydrofurans without affecting a chiral center at C–3.[313] Isomethadone methiodide (**104**) eliminates trimethylamine irreversibly to form the 2-ethylidene-4-methyl-3,3-diphenyltetrahydrofuran (**105**), also without upsetting the asymmetric center at C–4.[314]

2,5-Dialkyltetrahydrofurans are formed in moderate yields by the reaction of secondary alcohols with the general formula **106** with lead tetraacetate. The ratio of *cis*- and *trans*-tetrahydrofurans is not far from 1. The synthesis, however, is very useful because when $R^1 = R^2$ the cis isomer is meso and optically inactive, but the trans isomer is optically active and the chiral center attached to the oxygen atom remains unaltered in the reaction (Equation 24). Thus when *R*(−)-hexan-2-ol is treated in this way, and the *cis*- and *trans*-2,5-dimethyltetrahydrofuran are separated by glc, 2*R*,5*R*(−)-2,5-dimethyltetrahydrofuran (the trans isomer) can be obtained in a pure form. Similarly 2*R*,5*R*- and 2*S*,5*S*-2,5-diethyltetrahydrofurans are obtainable from *R*(−)-octan-3-ol and *S*(+)-octan-3-ol respectively. If the relative stereochemistry of cis and trans isomers can be determined (e.g., by ^1H nmr) then this method could be adopted to a variety of chiral alcohols (**106**) with different R^1 and R^2 alkyl groups.[315] This reaction was studied further in some detail using secondary aliphatic alcohols with

R^2CH_2–CH_2–CH_2–CHR^1(OH) $\xrightarrow{Pb(OAc)_4}$ cis-2-R^2-5-R^1-tetrahydrofuran + trans-2-R^2-5-R^1-tetrahydrofuran (24)

106

$Pb(OAc)_4$, ceric ammonium nitrate, silver, lead, and mercury oxides, and salts together with chlorine, bromine, or iodine. It was shown that in the reaction of silver oxide and bromine with 2-hexanol to form tetrahydrofurans, 1,5-hydrogen abstraction by the alkoxy radical occurs with the formation of δ-bromohydrins which is followed by cyclization with elimination of HBr. This offered a new general method for preparing tetrahydrofurans (almost equal mixtures of isomers are formed) from alcohols, and it is best carried out with silver or mercury salts and chlorine.[316]

4. *By Rearrangement*

Three examples are worthy of mention under this heading. The first is the rearrangement of 3-vinyl-1,3-dioxalanes to 3-formyl-3-methylfurans by heating in the presence of activated alumina—which was studied by Mousset and co-workers.[317,318] Under moderate temperature conditions the rearrangement is highly stereoselective according to Equation 25. The mechanism was interpreted in terms of ring opening followed by ring closure to a pyran carbonium ion and ring contraction to the furan, with one of the oxygen atoms in the dioxalane becoming the oxygen atom of the formyl group. Grignard reagents also bring about this rearrangement but react further with the formyl group to yield the corresponding secondary alcohol.[319]

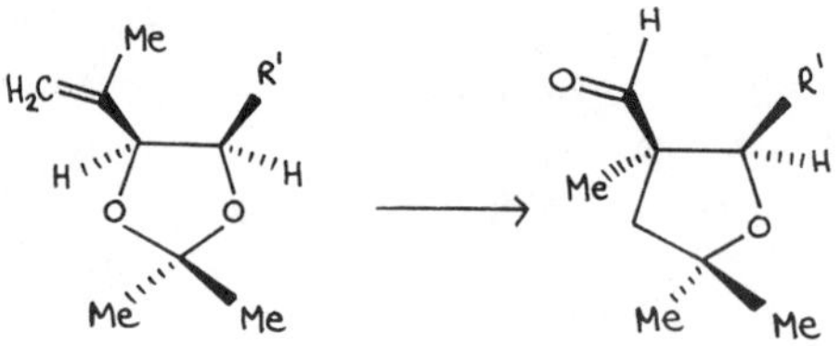

(25)

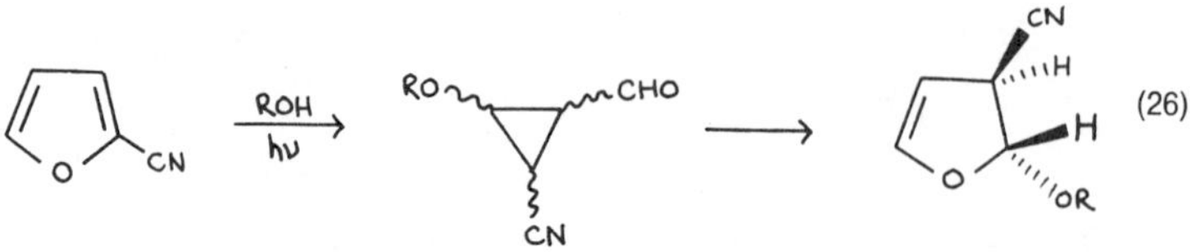

(26)

Photolysis of 2-cyanofuran, but not 3-cyanofuran, in alcoholic solution produces a mixture of 1-alkoxy-2-cyano-3-formylcyclopropanes which rearrange to *trans*-2-alkoxy-3-cyano-2,3-dihydrofurans predominantly on heating (Equation 26).[320] Uv irradiation of *cis*- or *trans*-2-ethyl-3-methoxycyclobutanone in methanolic solution yield *cis*- or *trans*-2-ethyl-3-methoxytetrahydrofurans, respectively, possessing another methoxy group at C–5 (anomeric) originating from methanol. Oxidation of these furnishes the corresponding 5-ethyl-4-methoxytetrahydrofuran-2-ones. These furan-2-ones are also obtained stereospecifically by direct oxidation (H_2O_2–AcOH, Baeyer-Villiger) of the butanones with retention of configuration at the chiral centers.[321]

5. *Miscellaneous*

Reductive condensation of furfural, benzaldehyde, or anisaldehyde, with crotonaldehyde by magnesium in acetic acid gave a mixture of 2-aryl-3-methyl-5-hydroxytetrahydrofurans which were dehydrated into a mixture

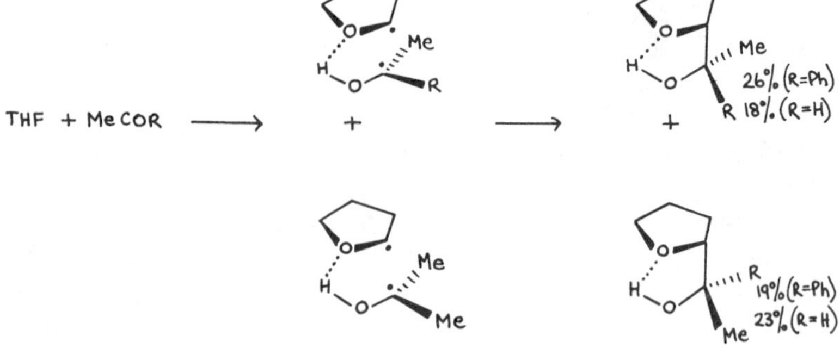

THF + MeCOR ⟶ + ⟶ +

26% (R=Ph) 18% (R=H)

19% (R=Ph) 23% (R=H)

(27)

of *cis*- and *trans*-2-aryl-3-methyl-2,3-dihydrofurans in which the trans isomers predominated. The corresponding tetrahydrofurans were obtained by catalytic reduction with Raney nickel or Pd–C.[322]

Irradiation of acetophenone or acetaldehyde in tetrahydrofuran produced a mixture of diastereoisomeric 1-phenyl-1-(tetrahydrofuran-2-yl)- and 1-(tetrahydrofuran-2-yl)-ethanols. The ratio of diastereoisomers formed suggests that the radicals may align themselves with the C–O bonds in the same direction, perhaps assisted by hydrogen bonding, with the proportion of the less sterically hindered transition state predominating (Equation 27).[323]

In a recent asymmetric synthesis using 2-oxazolines Meyers and Mihelich[324] reacted 4*R*,5*S*(−)-4-methoxymethyl-2-methyl-5-phenyl-2-oxazoline with butyl lithium and ethylene oxide at −78°C to form the 2-ω-hydroxypropyl-2-oxazoline. The trimethylsilyl derivative (**107**) of this was alkylated, with methyl, ethyl, *n*-propyl, and allylhalides, after treatment with lithium diisopropylamide, and hydrolyzed to give 3-*R*-3-methyl-, ethyl-, *n*-propyl-, and allyltetrahydrofuran-2-ones in over 64% optical yields (Equation 28). The amino alcohol **108** was optically pure and could be recycled. 4*S*,5*R*-2-Oxazolines would presumably yield the enantiomeric *S*-3-alkylfuran-2-ones.

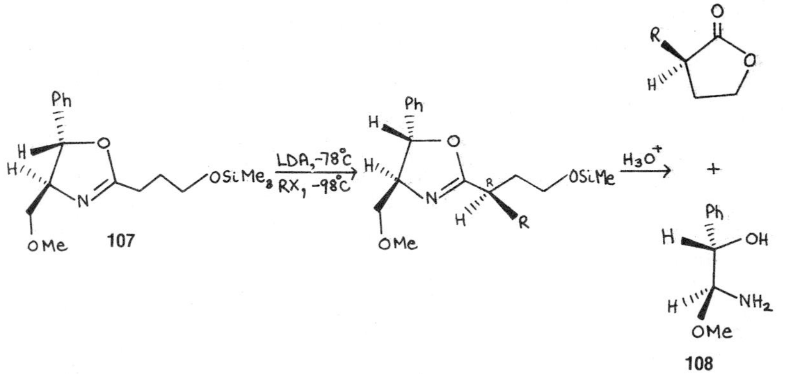

(28)

B. Configuration and Conformation of Reduced Furans

The relative configurations of substituents in di- and tetrahydrofurans can be deduced by nmr satisfactorily.[257,263,322,325,326] When reactive vicinal substituents as in 3,4-diaminotetrahydrofuran-2-ylvaleric acid are cis to each other, cyclic products (e.g., ureas) are formed readily. The trans isomers are either slower in reacting or polymerize.[327] A difference in glc behavior was observed between *meso*- and *d,l*-3,4-disubstituted tetrahydrofurans and provides data about the configuration of these compounds.[328] The absolute configurations in tetrahydrofurans can be deduced from the mechanism of synthesis as in the intramolecular cyclization of diols and secondary alcohols (see Section A.3) and by chemical correlation methods. (+)- and (−)-3-Carboxytetrahydrofurans, for example, which were obtained by optical resolution of the quinine salt, were correlated chemically with (−)- and (+)-methylsuccinic acid.[304,329] The configuration of the (+)-acid was correlated with that of malic acid by the quasiracemate method of Fredga.[330] The conversion of chiral open chain compounds such as isocitric acid into tetrahydrofurans turns them into rigid systems in which the relative orientations of substituents can be deduced more readily by ^{1}H nmr and these can be transformed into simpler chiral furans for stereochemical identification.[331,332] Ord and cd have been used to deduce the stereochemistry of hydrofurans.[324,333–335] Gerlach and Prelog used optical rotation, *p*K*a*, ir, and ^{1}H nmr data to determine the absolute configuration of the four chiral centers in the 2,5-disubstituted tetrahydrofuran-2*R*,3*R*,-6*S*,8*S*-nonactinic acid (**109**) obtained from Makrotetrolide I.[336] The configurations of a large number of naturally occurring furans have been elucidated including muscarine (**110**) and related furans,[337–340] monocro-

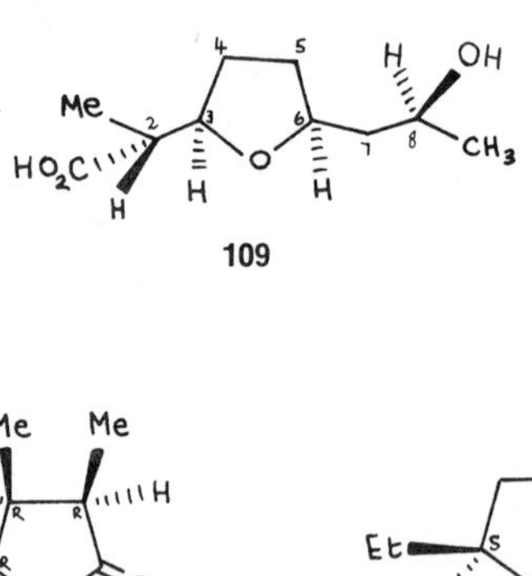

109 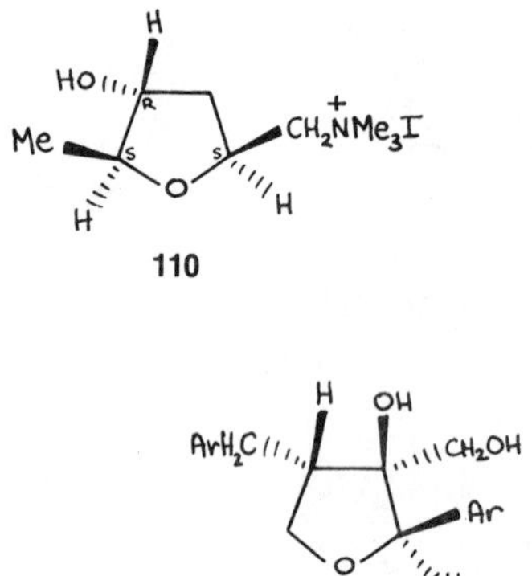 **110**

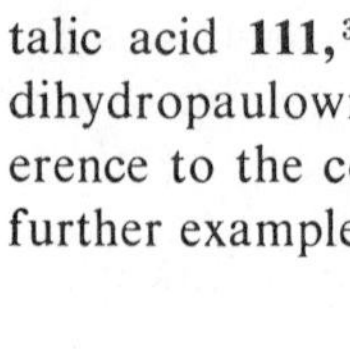

111 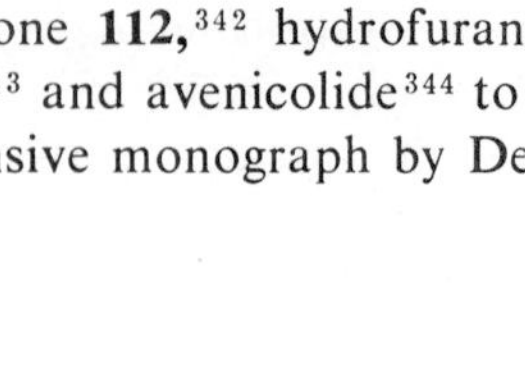 **112** **113**

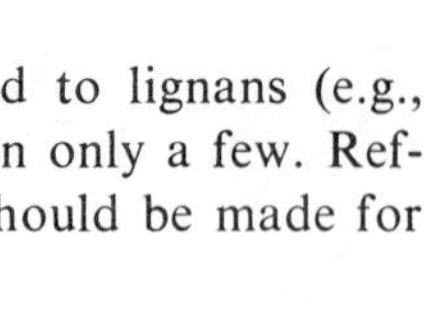

Ar = 3,4-$CH_2O_2C_6H_3$–

talic acid **111**,[341] oudenone **112**,[342] hydrofurans related to lignans (e.g., dihydropaulownin **113**,[343] and avenicolide[344] to mention only a few. Reference to the comprehensive monograph by Dean[247] should be made for further examples.

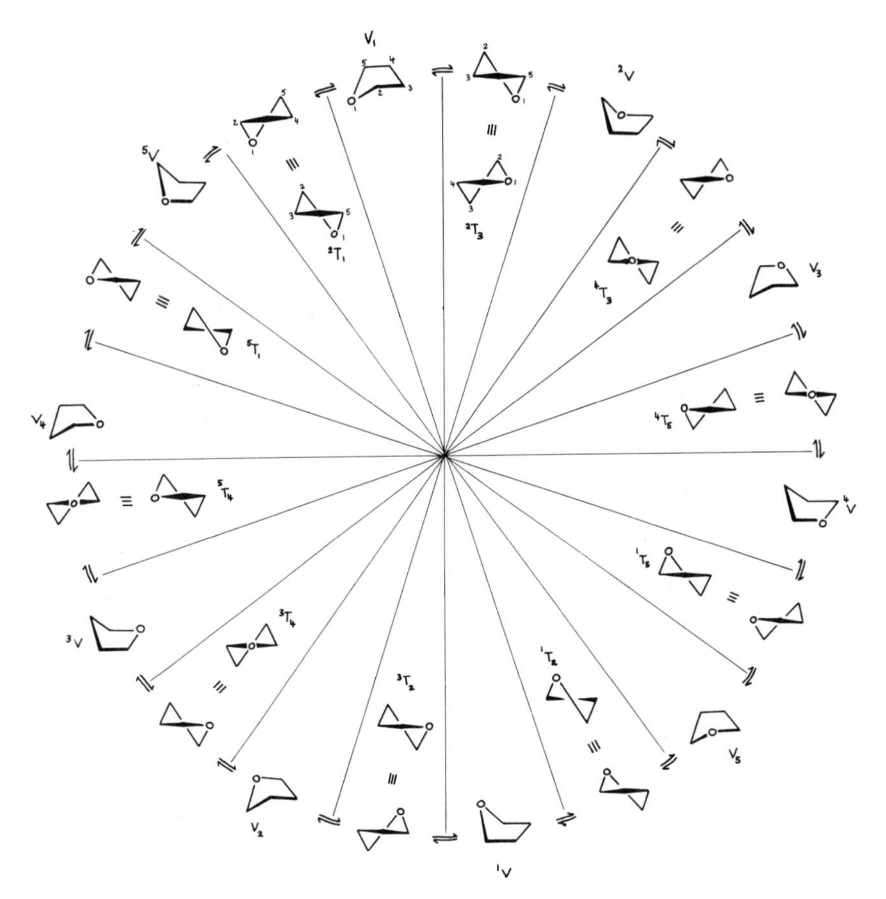

Scheme 5. Cycle of pseudorotation of tetrahydrofuran.

Tetrahydrofurans are conformationally labile like cyclopentane.[345] The ir,[346] dipole moment, and microwave data[347] were explained in terms of a freely pseudorotating system with very small energy barriers between the conformers. Hall, Steiner, and Pedersen[348] examined the ^1H nmr spectra of several furanose sugar derivatives and interpreted the spectra in terms of pseudorotation. They considered 10 twist and 10 envelope conformations which are related in a "free cycle of pseudorotation" which they termed a *cyclops*. Each atom in the ring goes out of the plane of the other four atoms in turn in the envelope conformation (V) through a twist conformation (T), in which only three adjacent atoms are in one plane, as it goes around the cyclops. The most populated conformations are represented by a segment in the cycle (Scheme 5). Theoretical treatment by Cremer and Pople[348a] of the relationship of the potential energy surface to ring puckering in tetrahydrofuran revealed that interconversion of conformers favors a path of pseudorotation with the half-chair conformation (T) marginally more stable than the envelope conformation, and consistent with microwave data.[347] The nmr spectra of 3,3,4,4-tetramethyltetrahydrofurans[349] and 2,5-diaryl-3,4-dimethyltetrahydrofurans[350] are adequately accommodated in this system. Ir and Raman spectra of 3-halogeno- and *trans*-3,4-dihalogenotetrahydrofuran indicate that the halogen atoms prefer the axial conformation with their carbon atom occupying the most puckered part of the ring. The structure of the 3-halogenotetrahydrofurans was confirmed by dipole moment measurements.[351]

The conformation of hydroxytetrahydrofurans may be dictated by the ability of the hydroxy group to form an intramolecular hydrogen bond with the furan oxygen atom.[352] Ir studies of *trans*-2-alkoxy-3-hydroxytetrahydrofurans and 3-hydroxytetrahydrofurans showed the presence of free and hydrogen bonded OH groups, whereas in the corresponding *trans*-1-alkoxy-2-hydroxy- and *trans*-1,2-dihydroxycyclopentanes only free OH stretching frequences were observed.[353] Conformers such as **114** account for the intramolecularly hydrogen bonded species.

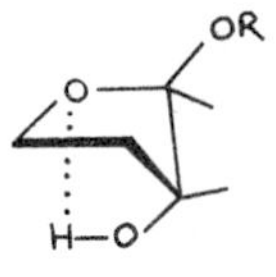

OR
O
H—O

114

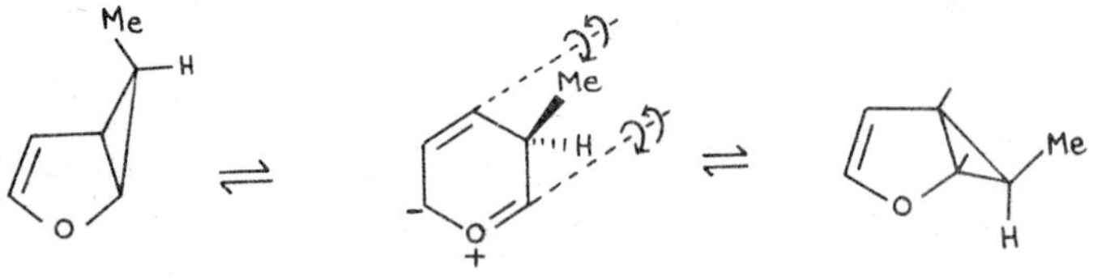

Me H ⇌ Me H ⇌ Me H (29)

The esr spectrum of the tetrahydrofuran radical was measured in 1964 and should be reappraised in terms of pseudorotation.[354]

C. Reactions of Reduced Furans

Several stereochemical reactions of reduced dihydro- and tetrahydrofurans were described in the previous sections and only two need to be mentioned here. *cis*-2,5-Dimethyltetrahydrofuran gives a stable oxonium ion in magic acid ($F \cdot SO_3H–SbF_5$) at −60°C. However, when the solution is heated to −20°C it alters rapidly to an equilibrium mixture of 58:42 cis and trans protonated species.[355] *Exo*- and *endo*-6-methyl-4-oxabicyclo[3.1.0]hex-2-enes are in equilibrium at temperatures above 380°C and are interconverted by way of electrocyclic reactions which also lead to ring cleavage products[356] (Equation 29).

VII. CONDENSED FURANS

The stereochemical data of condensed furans has come from studies on the synthesis and reactions of reduced systems,[357] and from structural studies of naturally occurring compounds rather than from systematic stereochemical investigations.[247,357] Only salient stereochemical points will be made here because the two references cited have a very comprehensive coverage of the literature.

A. 2,3-Dihydrobenzofurans

2,3-Disubstituted 2,3-dihydrobenzofurans exist as cis and trans isomers and several of these have been synthesized. Catalytic reduction, if allowed to stop at the dihydro stage, will yield the *cis*-2,3-dihydrobenzofuran, for example, *cis*-2,3-dihydro-7-methoxy-2,3-dimethylbenzofuran.[358] Dibenzofuran similarly gives cis-2,3-dihydro-2,3-tetramethylenebenzofuran,[359] a ring system found in Lunarine alkaloids.[360] The trans isomers, together with cis isomers, are obtained by chemical reduction, for example, of 2-methylcoumaran-3-one to a cis and trans mixture of 2,3-dihydro-3-hydroxy-2-methylbenzofuran with $NaBH_4$, and by intramolecular cyclization.[361,362] In connection with studies on fumigallin, 2,4-dichloro-5-methyl-6-(α-methylallyl)phenol was cyclized into a mixture of *cis*- and *trans*-5,7-dichloro-2,3-dihydro-2,3,4-trimethylbenzofurans on treatment with sulfuric acid (Equation 30). The *cis* to *trans* ratio varies from 6.4:0.33 after 1 min to 3.9:3.6 after 40 min.[258] Clearly it is C–2 that is epimerized under these conditions. Pappas and co-workers obtained a cis and trans mixture of 2,3-dihydro-3-hydroxy-3-methoxycarbonyl-2-phenylbenzofurans, **115** and **116,** by photolysis of methyl *o*-benzoyloxyphenylglyoxalate which proceeded predominantly by way of a triplet state. The ratio **115**:**116** varied considerably with the solvent used, for example, from 1:1 in *t*-butanol to greater than 20:1 in benzene.[363] Similarly *o*-benzyloxybenzaldehyde provides a mixture of *cis*- and *trans*-2,3-dihydro-3-hydroxy-2-phenylbenzofurans.[364] The cis and trans structures were assigned essentially by ^1H

nmr on the basis that the coupling constants for the C–2 and C–3 protons are larger for the cis than for the trans isomers in accordance with the Karplus equation.[358,364,365]

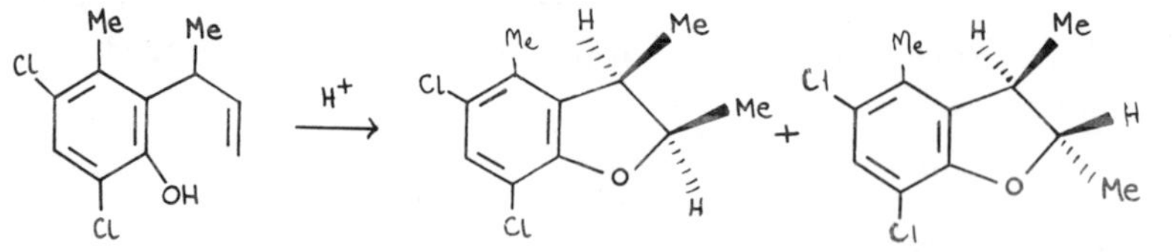

(30)

Zalkow and Ghosal[366,367] prepared racemic *cis*- (**117,** R=Br) and *trans*-5-bromo-2,3-dihydro-2-isopropyl-3-hydroxybenzofurans by treatment of 5-bromo-2-hydroxy-(α-bromo-β-methyl)butyrophenone with $NaBH_4$ in the absence and in the presence of potassium hydroxide respectively.

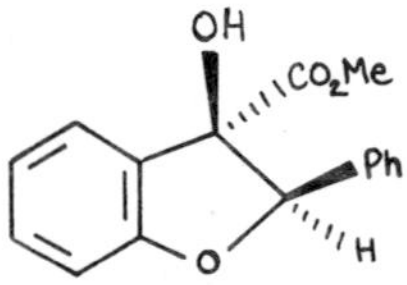

115 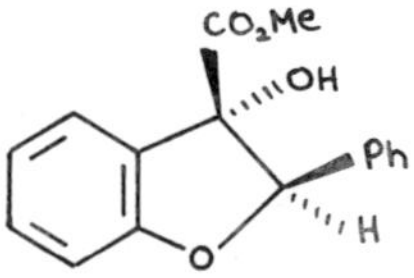 **116** 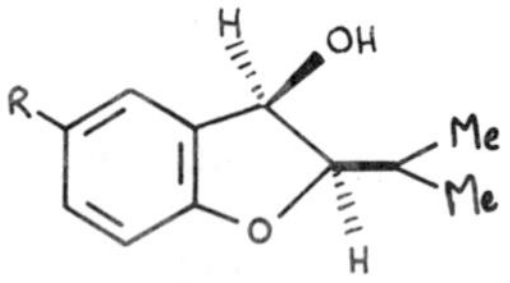 **117**

From these they also prepared several derivatives by reactions which did not affect the chiral centers and found that, contrary to previous assumptions, in all these the $J_{trans\text{-}2,3}$ values were larger than $J_{cis\text{-}2,3}$ values. They confirmed the stereochemistry by showing that optically active toxol (which was previously known to yield optically pure (+)-tartaric acid on ozonolysis)[368] gives dihydrotoxol on reduction, which is identical with **117** (R=Ac) prepared by the above method. Mertes and Powers[369] prepared cis and trans stereoisomers of 5-nitro- and 7-nitro-3-acetoxy-2,3-dihydro-2-methylbenzofurans from a cis and trans mixture of 3-acetoxy-2,3-dihydro-2-methylbenzofurans, separated the isomers, and confirmed that $J_{trans\text{-}2,3}$ (2Hz) $<$ $J_{cis\text{-}2,3}$ (6Hz), in contrast with the above. They converted one isomer into *cis*-7-dimethylamino-2,3-dihydro-3-hydroxy-2-methylbenzofuran methiodide and determined its structure by X-ray analysis. Clearly the Karplus equation generally holds for 2,3-dihydrobenzofurans. Zalkow and Ghosal's exception is caused by steric interaction of the bulky 2-isopropyl group with the 3-substituent, which must alter the geometry about C–2 and C–3 to the extent that the J_{cis} values are decreased considerably.[369] The kinetics of dehydration of 2-aroyl-2,3-dihydro-3-hydroxybenzofurans were studied by Ghelardoni, Pestellini, and Musante,[370] and although the trans isomer eliminated water at faster rates than the cis isomers, the relative rates were less than a factor of 10.

The oxidation of 1,1′-benzylidene-bis-2-naphthol with hypobromite gives the spiro compound 2*R**,3*S**-3-phenyl-4,5:2′,3′-dibenzogrisa-2′,4′-dien-6′-one (**118**, R^1=Ph, R^2=H), whereas oxidation with plumbic acetate furnishes the diastereomeric 2*R**,3*R** isomer **118** (R^1=H, R^2=Ph). There is considerable steric compression in both these spirans and the allylic alcohols derived from them by reduction of the carbonyl group. This compression is relieved by the acid-catalyzed conversion of the alcohols into 1,2:5,6-dibenzo-9-phenylxanthen, and occurs much earlier in the alcohol from the isomer **118** (R^1=H, R^2=Ph) than in its diastereoisomer because the transformation is 70 times faster.[371] An intramolecular Claisen rearrangement occurs with a number of 1,4-diaryloxy-2-butynes to yield hydrobenzofuro[3.2-*c*]benzopyrans (**119**) which must be relatively strain free and hence cis-fused.[372]

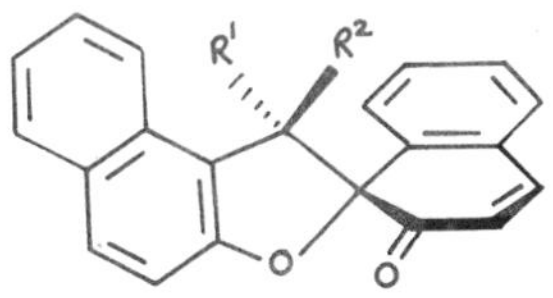

118

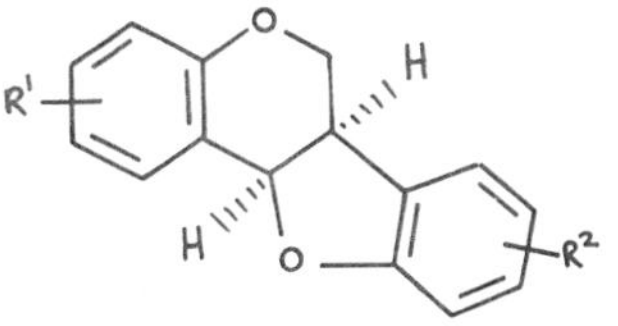

119

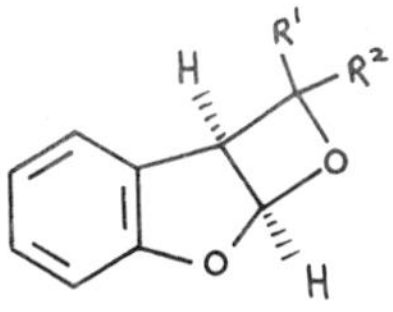

120

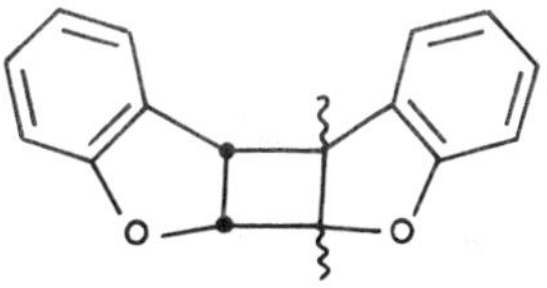

121

[2 + 2]π Cycloaddition between benzofuran and carbonyl compounds under the influence of uv light provides condensed *cis*-2,3-dihydrobenzofurans (**120**) in which the addition is highly regiospecific.[373,374] Photolysis of benzofuran in the presence of a sensitizer is regio- and stereospecific but yields a 1:3 mixture of syn and anti tetrahydrobenzofuran dimers (**121**).[374] Isobenzofuran undergoes cycloaddition reactions much more readily than benzofuran with dipolarophiles, uv light is not necessary, and 1,3 adducts are formed by a cis addition.[375] The electrocyclic photolytic ring contraction of 5-acetoxy- and 3,5-diacetoxy-4-phenyl-1-benzoxepin in

* Asterisk indicates only the relative, and not the absolute, configurations.

dioxane gives 5-acetoxy- and 5,7-diacetoxy-*cis*-3,4-benzo-6-phenyl-2-oxabicyclo[3.2.0]hept-6-enes respectively (**122**).[376]

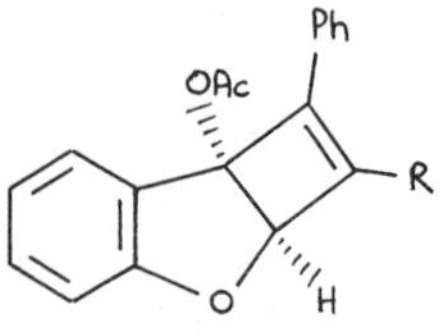

R = H or OAc

122

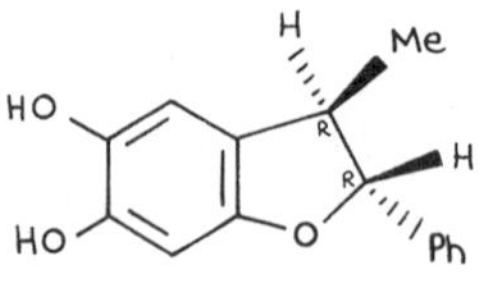

123

The stereochemistry and absolute stereochemistry of (−)-2-carboxy-2,3-dihydrobenzofuran,[377] (+)-obtusafuran (**123**),[377a] aflatoxin,[378] griseofulvin,[378a] and many naturally occurring 2,3-dihydrobenzofurans[247] have been reported.

B. Oxabicyclo[3.3.0]octanes

The ring strains involved in bicyclo[3.3.0]octanes have been mentioned previously, and it was stated that cis-fused isomers are more stable than the trans-fused isomers (Part I, Chapter 2, Section III.E.3). Very few derivatives of 2-oxabicyclo[3.3.0]octane are known. The parent *cis*-2-oxabicyclo[3.3.0]octane was obtained in 20% yield by the $Pb(OAc)_4$ oxidation of cyclopentylethanol and by the mercuric acetate oxidation of 2-(cyclopent-2-en-1-yl)-ethanol followed by $NaBH_4$ reduction.[379] The 2,6-dioxabicyclo-[3.3.0]octanes, on the other hand, are derived from monosaccharides and are 1,4,3,6-anhydrohexitols. Their structures and reactivities have been investigated in detail in connection with studies of carbohydrates.[380,381] The cis-fused structure **124** is consistent with the physical data and with X-ray analyses.[382] The parent substance **124** ($R^1 = R^2 = R^3 = R^4 = H$) was identified as one of the products (in 15% yield) from the oxidation of hexane-1,6-diol with lead tetraacetate in boiling benzene. No trans isomer was formed in this reaction.[383] Weyerstahl and co-workers[384] reduced 2,2,5,5-tetrasubstituted 1,6-ditosyloxyhexan-3,4-dione with $NaBH_4$ and obtained a variety of tetrahydrofurans depending on the C-2 and C-5 substituents. When these were $-(CH_2)_n-$ with $n=4$ (cyclopentane), $n=5$ (cyclohexane), and $n=6$ (cycloheptane), the *cis*-4,4,8,8-bisspiro-2,6-dioxabicyclo[3.3.0]octanes were formed predominantly (Equation 31). However, when $R^1,R^1 = -(CH_2)_5-$ and $R^2 = R^2 = Me$, the cis to trans ratio of bicyclo compounds was 66:34, but this was altered to 85:15 on increasing the reduction time. The trans isomers were obtained in a purer form by treating the intermediate monotosylates (**125**) with methanolic sodium hydroxide. The ^{1}H nmr spectra of the two series have been compared.

Wood and co-workers[385] obtained a trimer (m.p. 73.5°C) by passing biacetyl through an ion exchange resin (Amberlite CG400, OH^- form) and showed that it was *cis*-3,7-diacetyl-5-hydroxy-1,3,7-trimethyl-2,4,8-trioxabicyclo[3.3.0]octane (**126**).

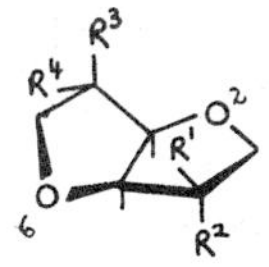

124

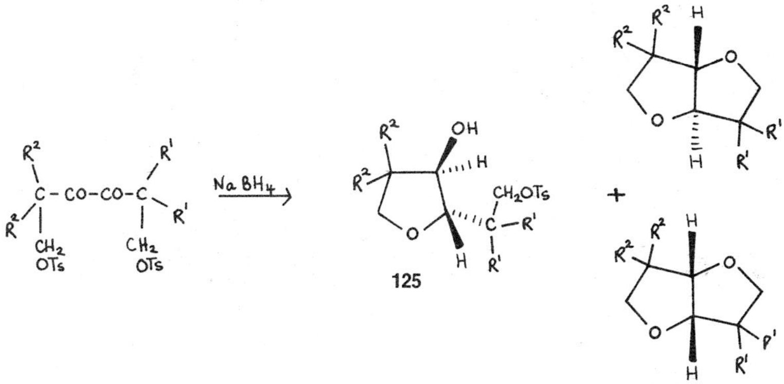

$NaBH_4$

125

(31)

Owen and Peto[386] prepared *cis*- and *trans*-3-oxabicyclo[3.3.0]octanes from authentic *cis*- and *trans*-1,2-bishydroxymethylcyclopentanes by way of the sulfonate esters. When cyclopentene was treated with paraformaldehyde and hydrogen chloride at low temperatures, 32% of a 2:1 mixture of *trans*-6-chloro-*cis*-3-oxabicyclo[3.3.0]octane and *anti*-8-chloro-3-oxabicyclo[3.2.1]octane was formed. The cis configuration of the former was identified by its ^{1}H nmr spectrum and by dechlorination to the known parent substance.[387] *trans*-3-Oxabicyclo[3.3.0]octane has a higher boiling point and refractive index than the cis isomer[386] in accordance with the Auwers-Skita rule,[388] which states that higher boiling points and refractive indices in strained systems are attributed to the less stable configuration.

A considerable number of 3,7-dioxabicyclo[3.3.0]octanes are known and most are naturally occurring lignans. They are all cis-fused (e.g., **127**) and have aryl substituents at C–2 and C–6 with varying relative stereochemistry.[389–393] Some of them have been synthesized[394] and degraded for determination of absolute configuration,[395,396] and some ^1H nmr spectral[397,398] and X-ray data have been recorded.[399,400] The parent *cis*-3,7-dioxabicyclo[3.3.0]octane (**127**, $R^1=R^2=R^3=R^4=H$) was prepared by Weinges and Spänig[401] by dehydration of the tetrahydric alcohol 2,3-bishydroxymethylbutane-1,4-diol. This leads initially to a 1:1 mixture of *cis*- and *trans*-3,4-bishydroxymethyltetrahydrofurans of which only the cis isomer cyclizes further. The chemistry of lignans was summarized by Dean,[247] and a proposal for the systematic naming of lignans and related compounds has been proposed by Freudenberg and Weinges.[402]

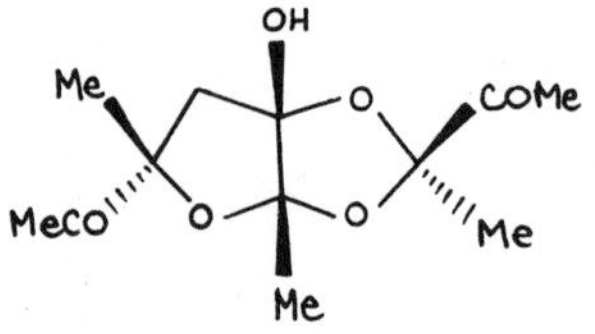

126

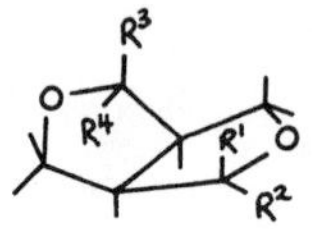

127

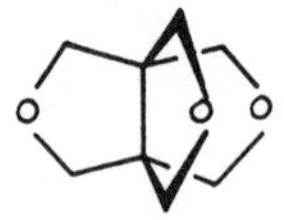

128

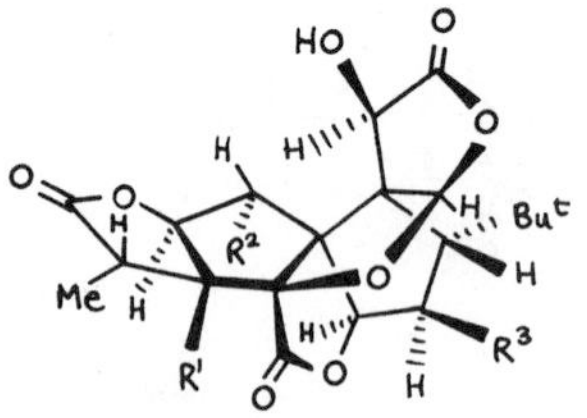

129

Condensed furans related to propellanes[403,404] (e.g., **128**) have attracted much attention recently (see Part I, Chapter 4, Section IX), and mixed oxygen and sulfur heterocycles of this nature were prepared.[405,406] The most complicated system of cis-fused five-membered rings containing only oxygen heteroatoms are the ginkgolides (**129**), which were isolated from the ginkgo tree, *Ginko biloba*, which is a "living-fossil tree" that has an origin dating back to Paleozoic times. The Ginkgolides were reviewed by Nakanishi[407] in 1967.

C. Oxabicyclo[4.3.0]nonanes

7-Oxabicyclo[4.3.0]nonanes are reduced benzofurans, and cis and trans isomers are possible when the bridgehead carbon atoms are saturated. Catalytic reduction of benzofurans or partially reduced benzofurans always yields the cis-fused system,[379,408–412] but the stereochemistry of further substituents in the 7-oxabicyclo[4.3.0]nonanes formed is not always, though it is predominantly, trans to the bridgehead protons; that is, all the hydrogens are cis as in the 4,9-dimethyl derivative **130.**[413] The trans isomers are obtained by intramolecular cyclization,[408,411] and various examples of these are known. Dehydration of *trans*-2-(2-hydroxy-2-methylpropyl)cyclohexan-1-ol in dimethyl sulfoxide at 165–180°C gives 78% of pure *trans*-8,8-dimethyl-7-oxabicyclo[4.3.0]nonane. This dehydration most probably occurs by elimination of the tertiary hydroxyl group because it is stereospecific.[411] Lead tetraacetate oxidation of 2-hydroxyethylcyclohexane furnishes *trans*- and *cis*-7-oxabicyclo[4.3.0]nonanes in 38 and 10% yields, respectively, together with 2-acetoxyethylcyclohexane.[414,379] In contrast, the mercuric oxide oxidation of 1-(2-hydroxyethyl)cyclohex-2-ene followed by $NaBH_4$ reduction is completely stereospecific and produces only the *cis*-bicyclo compound.[379] *cis*- and *trans*-2-Methoxycyclohexylethanol, and some of its homologues, were shown to form *cis*- and *trans*-7-oxabicyclo-[4.3.0]nonanes with tosyl chloride in pyridine. Studies with ^{18}O labeled compounds showed that the hydroxyl OH is eliminated with retention of configuration at the chiral centers.[408] The intramolecular cyclization of *trans*-2-(2-aminoethyl)cyclohexan-1-ol, which is promoted by nitrous acid, proceeds with inversion of configuration at the oxygen atom and yields *cis*-7-oxabicyclo[4.3.0]nonane.[415] On the other hand, considerable steric control is observed when the *N,N,N*-trimethyl iodides of compounds related to 2-(2-aminoethyl)cyclohexan-1-ols are treated with thallium hydroxide, affording the respective 7-oxabicyclononanes with retention of configuration at the chiral centers; that is, the hydroxyl oxygen is retained. These cyclizations have been useful in the preparations of 9-oxabenzomorphans.[416] A stereospecific intramolecular cyclization occurs when 1-(2-carboxymethyl)cyclohex-2-ene in sodium bicarbonate is treated with iodine and 5-iodo-*cis*-7-oxabicyclo[4.3.0]nonan-8-one is formed exclusively.[417,418] 1-(2-Cyanomethyl)cyclohex-1-ene also cyclizes to *cis*-7-oxabicyclo[4.3.0]nonan-8-one in hydrochloric or hydrobromic acids.[419] The greater stability of the cis-fused system in these bicyclononanes is clearly demonstrated by the relative ease with which *cis*- and *trans*-2-(2-carboxymethyl)cyclohexan-1-ols form the lactones *cis*- and *trans*-7-oxabicyclo[4.3.0]nonan-8-ones, with retention of configuration. The cis hydroxy acid cyclizes spontaneously whereas the trans isomer is stable.[419] This is also reflected in the faster rates of hydrolysis of *trans*-7-oxabicyclo[4.3.0]nonan-8-ones compared with the corresponding cis isomers,[420,421] and in the isomerization of the *cis*- to the *trans*-bicyclic compound in 50% sulfuric acid.[418]

The intramolecular cycloaddition of *cis,cis,cis*-oxonin into its geometrical isomer 7-oxabicyclo[4.3.0]nona-2,4,8-triene at 30°C was shown by Anastassiou and Cellura[169,170] to be in accord with orbital symmetry rules.

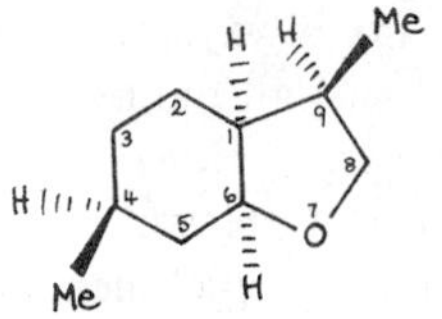

130 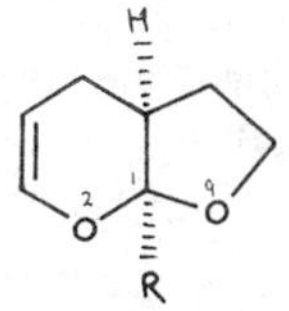 131 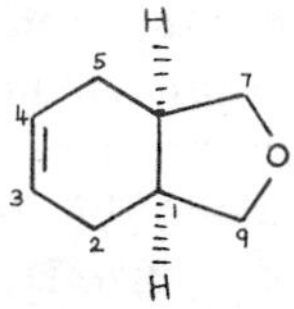 132

This furan was previously thought to be the trans isomer. The trans isomer, however, can be formed by low temperature photolysis of cyclooctatetraene oxide through the cyclization of *trans,cis,cis*-oxonin, which is formed by isomerization of the cis,cis,cis isomer (Scheme 6)[422]

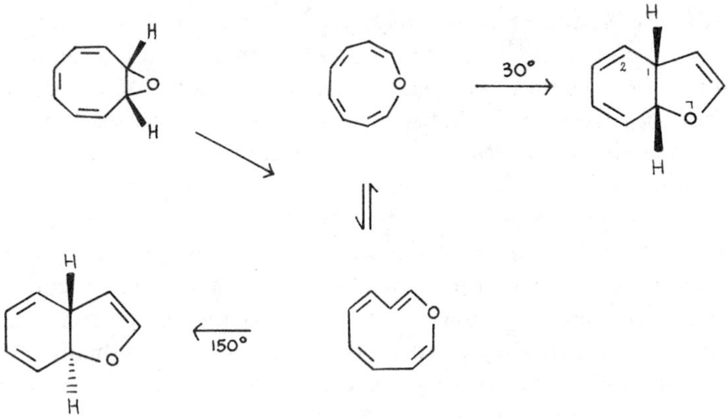

30° 150°

Scheme 6

The Diels-Alder reaction is very useful for preparing cis-fused 2,9-dioxabicyclo[4.3.0]non-3-enes (**131**) from 2,3-dihydrofurans and acrolein, and is particularly useful for preparing 9-oxabicyclo[4.3.0]nonanes.[423] Butadiene condenses with 2,5-dihydrofuran to yield *cis*-8-oxabicyclo[4.3.0]non-3-ene (**132**),[424] and several 7,9-diketo derivatives have been obtained by using maleic anhydride instead of the dihydrofuran with butadienes.[425–427] Reduction of these diketo compounds, which are in reality reduced phthalic anhydrides (**133**), gives the respective diols which can be dehydrated in various ways to *cis*-8-oxabicyclo[4.3.0]nonanes.[425–432] If fumaric ester is used in place of maleic anhydride, then the condensation product is a reduced ester of *trans*-phthalic acid, and this can be converted into derivatives of *trans*-8-oxabicyclo[4.3.0]nonane by the appropriate reactions.[433,434]

A Prins reaction between cyclohexane and formaldehyde in the presence of acid,[426] or 1-hydroxymethycyclohex-2-ene and formaldehyde in the pres-

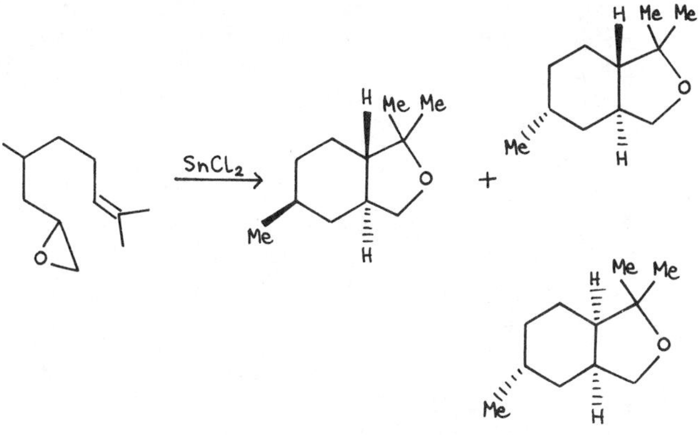

(32)

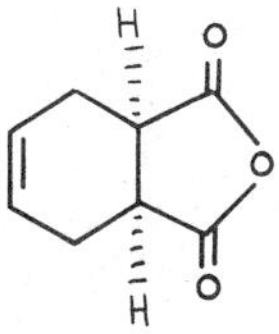

133 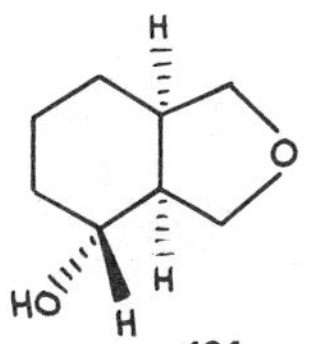 134 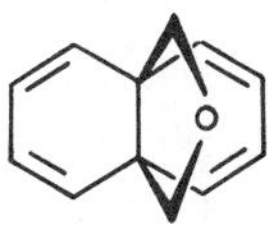 135

ence of acid, yields *trans*-2-hydroxy-*cis*-8-oxabicyclo[4.3.0]nonane (**134**).[435] An open chain compound such as the epoxide derived from citronellal cyclizes, with $SnCl_2$ in benzene, into a mixture of 8-oxabicyclo[4.3.0]nonanes in which the trans-fused isomers predominate (Equation 32).[436] Propellanes (e.g., **135**) related to *cis*-8-oxabicyclo[4.3.0]nonane were prepared by Ginsburg and co-workers,[437,438] and their properties including their cycloaddition reactions were investigated (see Section B and Part I, Chapter 4, Section IX).

The photochemical acetoxylation,[439] hydroxylation,[440,441] epoxidation,[442] and hydroboration[443] of *cis*-8-oxabicyclo[4.3.0]non-3-enes (**132**), and in the first case also the trans isomers show some stereoselectivity. The selectivity was attributed to a ring oxygen effect. A comparison of the relative rates of oxymercuration of cyclohexene, **132,** the *trans*-2-methyl derivative of **132,** and *cis*-bicyclo[4.3.0]non-3-ene (1:0.93:0.35:0.11) revealed that the ring oxygen atom enhances the rate in comparison with a CH_2. The slower rate for the *trans*-2-methyl derivative of **132** is attributed to a steric effect.[430] The data are generally consistent with the reactivity of the exo (**136**) rather than the endo (**137**) conformer.[442] Some selectivity was also observed in the epoxidation of *trans*-7-oxabicyclo[4.3.0]non-3-en-8-one.[444] The **ir** spectra of several *cis*- and *trans*-8-oxabicyclo[4.3.0]nonanes were compared, and the C–O–C stretching frequencies, around 1000 cm^{-1} for the cis isomers, were at higher frequencies, by about 20–40 cm^{-1}, than the trans isomers, and are a reflection of the differences in the ring strain between the two systems.[445]

136

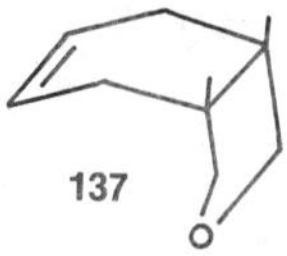

137

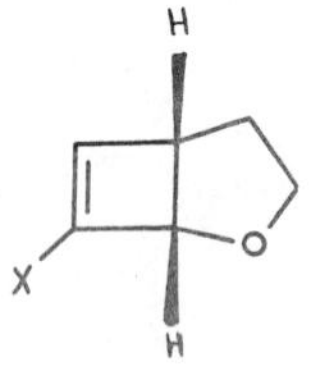

138

D. Miscellaneous

Oxabicyclo[3.2.0]heptanes obtained by $[2 + 2]\pi$ cycloaddition reactions of olefins, or carbonyl compounds, with furans are described in Section VI.A.2, but two more examples are mentioned here because the methods of preparation are unique. The first is the photolytic rearrangement of 2,3-dihydrooxepin and its 6-chloro derivative into *cis*-2-oxabicyclo[3.2.0]-hept-6-ene (**138,** X=H) and its 7-chloro derivative (**138,** X=Cl) respectively, described by Paquette and his collaborators.[446] The second example is the oxidation of *syn*-octamethyltricyclo[4.2.0.0.^{2,5}]octa-3,7-diene with peroxytrifluoroacetic acid, in the presence of boron trifluoride, to octamethyl-2-oxatricyclo[5.2.0.0^{3,6}]nona-4,8-diene (**139**)[447]—not the isomeric furan **140** as previously proposed.[448] The reaction is slightly in favor of the syn configuration, but this is not conclusive.

Tetrahydrofurans with a three-carbon bridge between C–2 and C–4 are formed by intramolecular cyclization of hydroxymethylcyclohexenes. Thus 5-methoxy,[449] 2-hydroxymethyl-4-chloromercuri-,[450] 4-bromo-2-hydroxymethyl-,[451] 6-oxabicyclo[3.2.1]octanes, and the parent compound (**141**)[452] are obtained from 3-hydroxymethyl-1-methoxycyclohex-1-ene (by distillation), 4,4-bis(hydroxymethyl)cyclohex-1-ene [$Hg(OAc)_2$], *cis*-4,5-bis(hydroxymethyl)cyclohex-1-ene (*N*-bromosuccinimide), and 3-tosyloxymethylcyclohexanol (LAH) respectively.

Although tetrahydrofurans with a four-carbon bridge between C–3 and C–4 were discussed in Section VI.A.2, an important point must be made here about *exo*- (**142**) and *endo*-2-oxa-1,2-dihydrodicyclopentadiene (**143**). By comparison with the carbocyclic analogues and the exo isomer **142,** the endo isomer **143** is reluctant to undergo rearrangement reactions. This property has been attributed to the stabilization of the system by forming

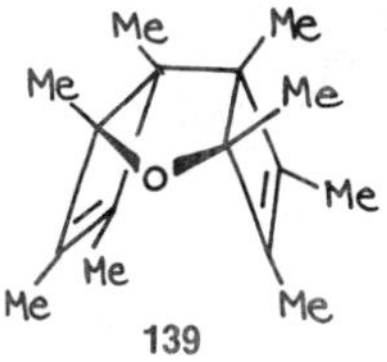

139

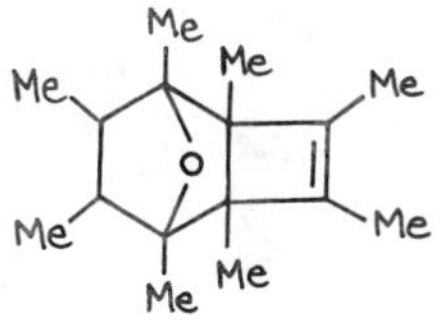

140

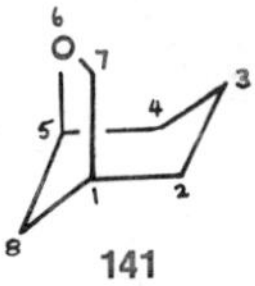

141

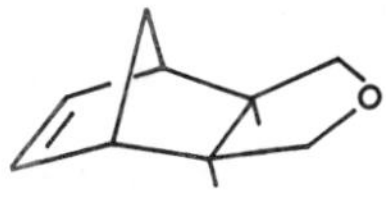

142

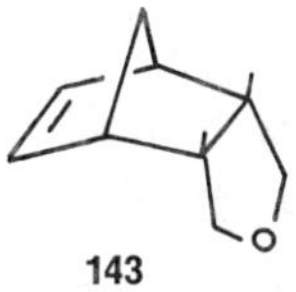

143

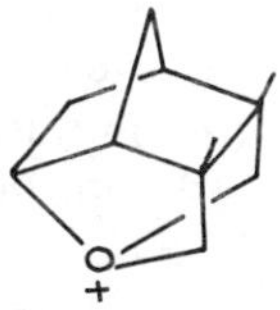

144

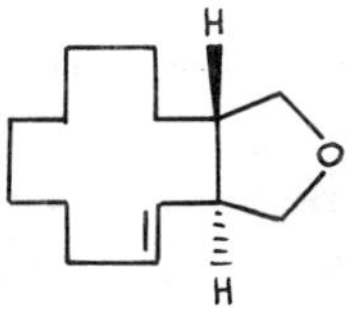

145

the oxonium ion **144** which is in resonance with carbonium ions related to it.[453] The Prins reaction between cyclododecene and paraformaldehyde is highly stereoselective, yielding *trans*-14-oxabicyclo[10.3.0]pentadec-2-ene (**145**) in which C–3 and C–4 are bridged by a 10-carbon chain.[454]

146

147

148

Tetrahydrofurans with a bridge between C–2 and C–5 containing two, three, four, and six carbon atoms are now described. Alkaline hydrolysis of *trans*-4-chlorocyclohexanol causes the formation of a 1,4-oxygen bridge giving 7-oxabicyclo[2.2.1]heptane (**146**). This cyclization is rather slow, being 10^{-3} times as rapid as the formation of tetrahydrofuran from 4-chlorobutan-1-ol.[455] The formation of 8-oxabicyclo[3.2.1]octane (5%) from treatment of cycloheptanol with $Pb(OAc)_4$[456] and the solvolysis of such bicycles[457,458] demonstrate the proximity of C–1 to C–4 in cycloheptanes. Similarly oxidation of cyclooctanol with $Pb(OAc)_4$ produces 9-oxabicyclo[4.2.1]nonane (26%) but in higher yield than with cycloheptanol.[456] Oxymercuration of *cis*-4,5-epoxycyclooct-1-ene produces *exo*-2-hydroxy-9-oxabicyclo[4.2.1]nonane (**147**) and *exo*-2-hydroxybicyclo[3.3.1]nonane (**148**) by intramolecular neighboring-group participation.[459] Cyclodecanol is also oxidized in the same way to form 11-oxabicyclo[6.2.1]undecane[460] in 2.5% yield. For further discussion see Section XIV.B.

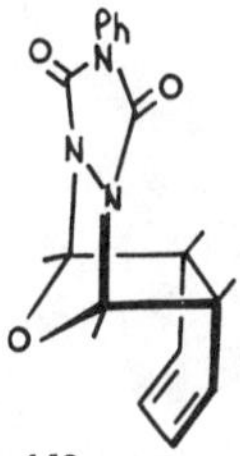

149

4-Phenyl-1,2,4-triazolin-3,5-dione reacts with oxonin at −78°C quantitatively and stereospecifically to give the cycloadduct **149** by a 1,8-addition. A double Diels-Alder reaction occurred in this case and is explained either by a $[2 + 2 + 2]\pi$ cycloaddition or by an initial $[2 + 8]\pi$ addition followed by a $[2 + 2]\pi$ addition.[461]

VIII. 1,3-DIOXALANES

The stereochemistry of 1,3-dioxalanes has received considerable attention in recent years. To a large extent this is because the five-membered oxygen-containing ring presents interesting stereochemical problems which can be most conveniently investigated by nmr spectroscopy. Other physical techniques have been used but to a much lesser extent. 1,3-Dioxalanes are essentially cyclic acetals or ketals of ethylene glycols.

A. Syntheses of 1,3-Dioxalanes

2-Unsubstituted 1,3-dioxalanes are made from the respective racemic (trans) or meso (cis) ethylene glycols by reaction with methylene bromide or chloride in dimethyl formamide containing sodium hydride.[462,463] The condensation proceeds with retention of configuration at the carbon atoms bonded to the oxygen atoms. Lucas and co-workers[464] showed that (−)-2,3-butanediol gave (−)-*trans*-4,5-dimethyl-1,3-dioxalane with methylene chloride or iodide, or chloromethylacetate, and is the same as the acid-catalyzed condensation with formaldehyde. Thus neither an S_N2 displacement nor a carbonium ion was involved; otherwise the resulting dioxalane would have been optically inactive or highly racemized due to the formation of the meso, cis isomer. The condensation of 1,2-diols with aldehydes and ketones, or their acetals and ketals, furnishes 2-mono and 2,2-disubstituted 1,3-dioxalanes respectively. Benzaldehyde reacts with *d,l*-1,2-diols such as butanediol to form one 4,5-dimethyl-2-phenyl-1,3-dioxalane (**150**), but with the *meso*-diol a 3:1 mixture of the syn (**151**) and anti (**152**) isomers is formed.[465] Various methods for preparing 2-methyl-1,3-dioxalanes from *trans*- and *cis*-1,2-dihydroxycyclohexanes were compared, and whereas the trans isomer gave only 2-methyl-*trans*-4,5-tetramethylene-1,3-dioxalane, the *cis*-diol gave mixtures of *syn*- and *anti*-2-methyl-*cis*-4,5-tetramethylene-1,3-dioxalanes in varying proportions depending on the method of preparation. Although the dioxalane from *trans*-1,2-dihydroxycyclohexane was quite stable, it was more difficult to prepare than the dioxalanes from the cis isomer.[466] Several 4- and/or 5- and/or 2-substituted phenyl-,[467] methyl-,[468] methoxycarbonyl-,[469] and 4,5-tetramethylene-

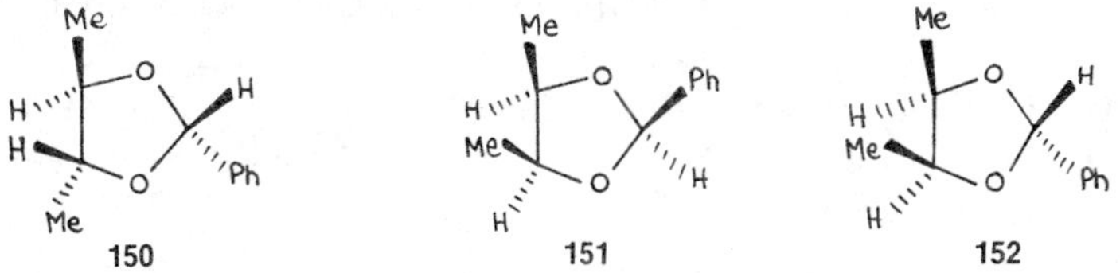

150 **151** **152**

1,3-dioxalanes[470] were prepared, and the isomers were separated and identified by nmr spectroscopy. These are consistent with a mechanism in which the stereochemistry of the diol is retained in the dioxalane. The condensation of racemic and *meso*-1,2-diphenylethane-1,2-diol with dimethyl formamide dimethylacetal gave *r*-2-dimethylamino-*cis*-4-phenyl-*trans*-5-phenyl-1,3-dioxalane and a 1:2 mixture of *r*-2-dimethylamino-*cis*-4-phenyl-*cis*-5-phenyl- and *r*-2-dimethylamino-*trans*-4-phenyl-*trans*-5-phenyl-1,3-dioxalanes, which reacted with methyl iodide in ethanol to form the corresponding 2-ethoxydioxalanes, presumably by forming the 2-trimethylammonium salt followed by S_N2 displacement of this group with ethyl alcohol.[471]

4-Hydroxymethyl-1,3-dioxalanes from glycerol have been prepared by condensation with aldehydes and ketones, and the stereoisomers have been separated and identified mainly by ^{1}H nmr spectroscopy.[472–476] 4-Hydroxy-1,3-dioxanes are also formed in these reactions if the hydroxy groups are not adequately protected. Assignment of configuration from nmr data is not infallible and can be misleading,[477,478] and chemical methods have to be used when there are some doubts. For example the two isomeric 1,3-dioxalanes, *cis*-2-bromomethyl-*r*-4-hydroxymethyl-*trans*-2-phenyl- (**153**) and *trans*-2-bromomethyl-*r*-4-hydroxymethyl-*cis*-2-phenyl-1,3-dioxalanes (**154**), which were separated after condensing glycerol with phenacyl bromide, were identified by the fact that only the former, that is, **153,** cyclizes to a 1,4-dioxane (**155**) on treatment with sodium hydride. The CH_2Br or CH_2OH groups can then be converted into methyl groups with known configuration.[477]

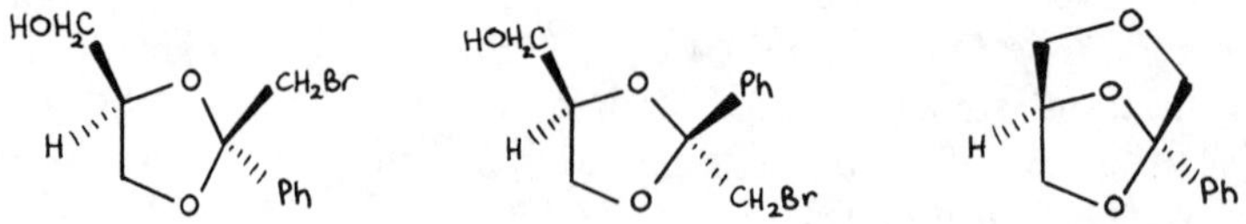

HOH$_2$C, H, O, CH_2Br, Ph, O — **153**; HOH$_2$C, H, O, Ph, CH_2Br, O — **154**; H, O, O, Ph — **155**

The proposed mechanism[479] for the conversion of *cis*- and *trans*-2,3-dimethyloxiranes into *trans*- and *cis*-2,2,4,5-tetramethyl-1,3-dioxalanes by reaction with acetone in the presence of BF_3 was confirmed, and the details clarified. By using ^{18}O-enriched acetone, a mechanism in which attack of the oxirane carbon by the acetone carbonyl oxygen and retention of this oxygen in the 1,3-dioxalane was established. This mechanism proceeds even in the presence of water; that is, the free diol is not an intermediate[480] (Equation 33).

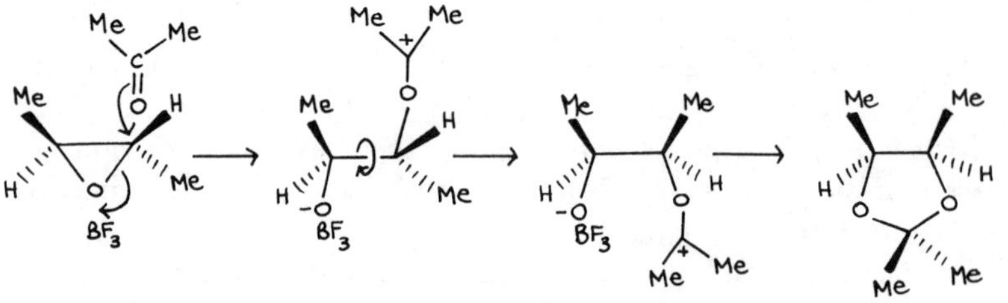

(33)

Some stereoselectivity was observed in the reaction of diphenyl diazomethane with pyruvic nitrile. The major products were 3-cyano-3-methyl-2,2-diphenyloxirane (43%), *trans*-4,5-dicyano-*trans*-4,5-dimethyl- (33–37%), and *cis*-4,5-dicyano-*cis*-4,5-dimethyl-2,2-diphenyl-1,3-dioxalanes (8–9%).[481]

A new chiral catalyst, (−)-2,3-*O*-isopropylidene-2,3-dihydroxy-1,4-bis-(diphenylphosphine)butane rhodium(I), was prepared by condensing the (−)-*trans*-2,2-dimethyl-4,5-diphenylphosphinomethyl-1,3-dioxalane (**156**) (prepared from ethyl (−)-tartrate and acetone followed by the necessary elaboration of the ethoxycarbonyl groups) with $[RhCl(cyclooctene)_2]_2$. This catalyst will reduce α-acylaminocinnamic acids to *N*-acyl-*R*-phenylalanines in 55 to 72% optical yields.[482]

The double bond in vinylene carbonate (1,3-dioxal-4-en-2-one) is reactive towards carbenes, and when vinylene carbonate or its 4-methyl derivative is heated with diazoacetic ester in the presence of cuprous cyanide, addition occurs and *exo*-ethoxycarbonyl-*cis*-2,4-dioxabicyclo[3.1.0]hexan-3-ones (**157**, R = H or Me) are formed stereospecifically.[483] A cycloaddition reaction was observed with vinylene thiocarbonates which added across the 9,10-positions of anthracene to yield 9,10-dihydro-9,10-ethanoanthracene-11,12-diol thionocarbonates (**158**).[484]

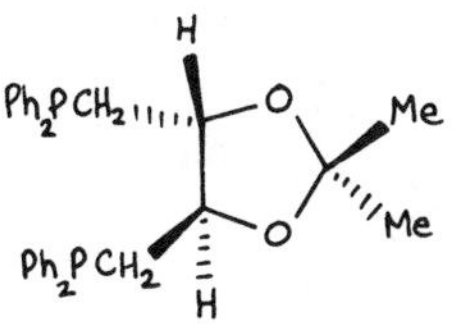

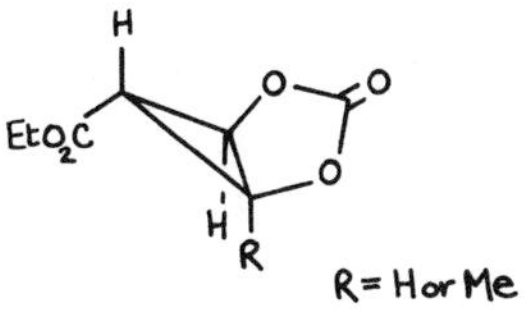

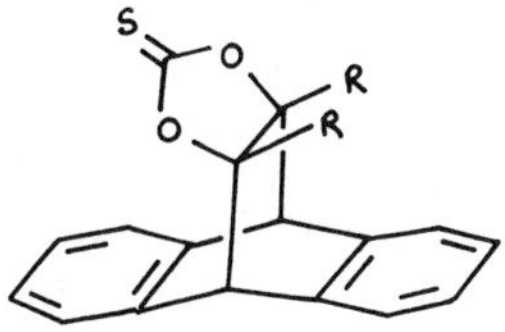

B. Configuration and Conformation of 1,3-Dioxalanes

Ir studies of 1,3-dioxalane revealed that, like cyclopentane and tetrahydrofuran (see Section VI.C), it exhibits pseudorotation with very small energy barriers between the envelope and twist conformers.[346] Five-membered rings generally form pseudorotating systems in the gas and liquid states, and their structures alter by way of a continuous set of conformations (cyclops) which consists of ten envelope (C_s) and ten twist (C_2, half-chair) structures as the phase angle varies from 0 to 720° (see Scheme 5).[485,486] Abraham[487] made a detailed analysis of the A_2B_2 system in 2-methyl-1,3-dioxalane and found that the $\cos^2 \phi$ law did not accurately account for the magnitude of the coupling constants, most probably because of pseudorotation. MO studies of five-membered rings by Cremer and Pople[348a] showed that a pseudorotation path is favored for conformational changes in 1,3-dioxalane with the twist conformation slightly more stable than the envelope conformation. The molecular polarizability of 1,3-dioxalane and 1,3-dioxalan-2-one have been measured, by Le Fèvre and co-workers,[488] and discussed in terms of puckered conformations.

It is not always easy to decide which conformation is present in solution because of time-averaged coupling constants and chemical shifts which provide ^1H nmr evidence for pseudorotation.[486,489] Also incorrect assignments of configuration, for example, of *cis*- and *trans*-2,4-dimethyl-1,3-dioxalanes,[490] have been made from considerations of shielding effects of protons by methyl groups in a 1,3-cis arrangement, when in fact the result is an overall deshielding effect.[491] Eliel and co-workers[492,493] studied the acid-catalyzed equilibration of 2,4-dialkyl- and *r*-2-*cis*-4,5-trialkyl-substituted 1,3-dioxalanes, in which case C–2 is epimerized, and found that the free energy differences between the diastereoisomers are small (~0.8 to 2.8 kJ mol^{-1}). Only the very bulky substituents (e.g., *t*-butyl) showed any signs of steric interactions, and the data suggested a very flexible system. Flexibility of conformation was also observed in the cyclooctane derivative 2,2-dimethyl-*trans*-4,5-hexamethylene-1,3-dioxalane, and the dynamic nmr data indicate that it cannot even exist predominantly in one conformation.[494] The J_{gem} values for the C–2 protons in *cis*-4,5-hexamethylene-1,3-dioxalane were shown to have atypically positive values.[495] [For the prediction of $^2J(XCH_2Y)$ values see the paper by Anteunis and coworkers.[496]] Substituents do tend to have a conformationally anchoring effect (ananchomeric), and although this may be small, nonetheless, in several cases it is such that correct configurations have been assigned on nmr evidence[486,497] (see also examples in Section A). Foster and collaborators[498] were able to distinguish between the cis and trans isomers of 2-phenyl-4-substituted 1,3-dioxalanes from the high and low field chemical shifts, respectively, of the benzylic C–2 protons (see above, however). Anisotropic substituents at C–2, for example, *p*-nitrophenyl, greatly influence the chemical shifts of protons at C–4 and C–5, and provide information which assists in deciding the stereochemistry of C–4 and C–5 substituents.[498a] The preparation and identification of these 1,3-dioxalanes are useful for determining the stereochemistry of the original 1,2-diols.

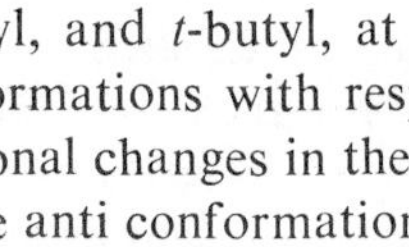

159 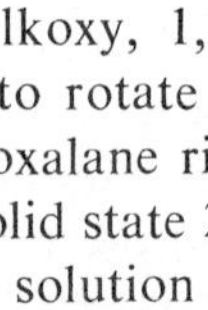 **160** 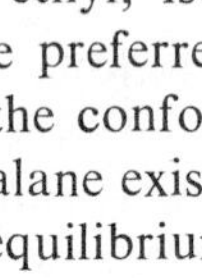 **161**

Large substituents, for example, alkoxy, 1,3-dioxalan-2-yl, ethyl, isopropyl, and *t*-butyl, at C–2 are free to rotate and can assume preferred conformations with respect to the dioxalane ring apart from the conformational changes in the ring. In the solid state 2,2′-bis-1,3-dioxalane exists in the anti conformation **159**,[499] but in solution an anti-gauche equilibrium is observed with the gauche form predominating.[485,491] The dipole moments of 2-methoxy-, 2-methoxy-2-methyl-, and 2-*t*-butoxy-1,3-dioxalanes were explained in terms of gauche-anti (with respect to H–2) equilibria. In the case of 2-methoxy-1,3-dioxalane the anti structure **160** is preferred, whereas in the 2-*t*-butoxy derivative the gauche conformer **161** is present exclusively in solution. The effect of these conformations on the mean phase angle of the ring has also been considered.[485,491] The preferred orientations of 2-alkyl groups were examined by nmr[500] and evidence of hindered rotation in 2-*t*-butyl-2-methyl-1,3-dioxalane ($\Delta G^{\ddagger} = 31.4$ kJ mol^{-1} at -124.7°C) was obtained from spectra at low temperatures.[501] Katritzky and co-workers[502] studied the nmr spectra at various temperatures of cyclohexanespiro-4′-(1,3-dioxalane) and observed that the conformation with the axial oxygen (**162**) was slightly preferred, but this preference was decreased by the presence of two methyl groups on C–2 of the dioxalane ring.

Intramolecular hydrogen bonding in 4-hydroxymethyl-1,3-dioxalanes (derived from glycerol) gives some idea of the structure of the system—the hydrogen bonding, however, is not always complete.[503]

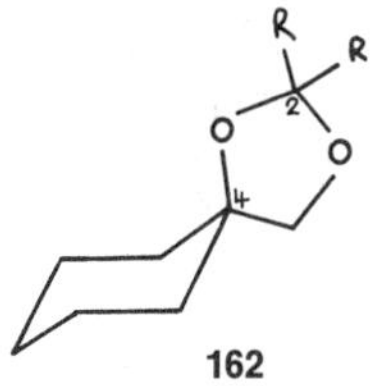

162

1,3-Dioxalan-2-one has been studied by microwave spectroscopy. In the solid state it is planar, but in the gas phase the evidence is that it is not planar.[504] The cyclic carbonate structure is a chromophore, as is that of 1,3-dioxalane-2-thione, and the chiroptical properties of such systems in which C–4 and/or C–5 are chiral centers have been reported.[505–507] The ^{1}H nmr spectra of *cis*- and *trans*-2,5-disubstituted 1,3-dioxalan-4-one were examined and showed long range coupling between H–2 and H–5, but could not be used to decide between the cis and trans isomers.[508] Asabe and co-workers,[509] however, have analyzed the spectra of a large number of *cis*- and *trans*-2,5-disubstituted 1,3-dioxalan-4-ones and found that $J_{2,5(\mathrm{trans})}$ were always larger than $J_{2,5(\mathrm{cis})}$ and that they had a positive sign. The nmr spectra of 1,3-dioxalanes were discussed by Batterham.[325]

C. Reactions of 1,3-Dioxalanes

One of the most actively studied reactions of stereochemical interest is the conversion of 1,3-dioxalanes into olefins in which C–4 and C–5 of the ring are retained. Treatment of *r*-2-dimethylamino-4-*cis*-phenyl-5-*trans*-phenyl-1,3-dioxalane with acetic anhydride furnishes *t*-1,2-diphenylethylene in high yield. Similarly the mixture of *r*-2-dimethylamino-4-*cis*-phenyl-5-*cis*-phenyl- and 4-*trans*-phenyl-5-*trans*-phenyl-1,3-dioxalanes yields *cis*-1,2-diphenylethylene stereospecifically.[471] Corey and collaborators[510] heated the 1,3-dioxalane-2-thiones derived from meso and racemic 1,4-diphenylbutane-2,3-diols in boiling triphenylphosphite and showed that a concerted cis elimination occurred giving *cis*- and *trans*-1,4-diphenylbut-2-enes. Ana-

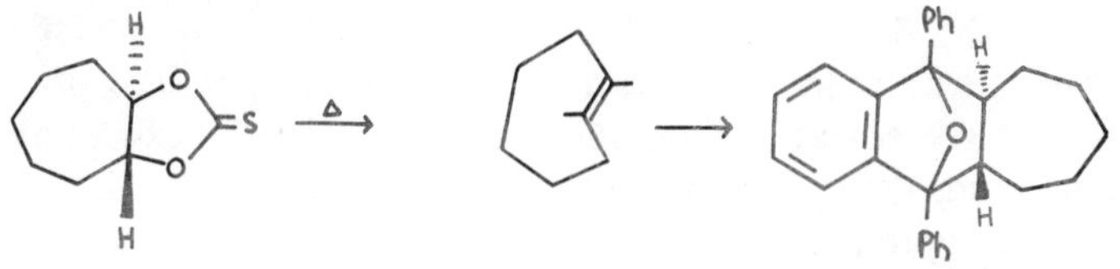

(34)

logously, *trans*-4,5-hexamethylene-1,3-dioxalane-2-thione (but in triisooctylphosphite) stereospecifically gave *trans*-cyclooctene in quantitative yield. The reaction was then extended to *t*-4,5-pentamethylene-1,3-dioxalane-2-thione, and they obtained the highly strained *trans*-cycloheptene for the first time. It isomerizes to *cis*-cycloheptene readily, but if 2,5-diphenylbenzoisofuran is added to the mixture then the trans adduct is isolable (Equation 34). By using (−)-*trans*-cyclooctane-1,2-diol, Corey and Shulman[511] prepared [via (−)-*trans*-4,5-hexamethylene-1,3-dioxalane-2-thione] *trans*-cyclooctene in over 99% purity with $[\alpha]_D$ −423° (in CH_2Cl_2) in a similar manner. This synthesis has been applied by others,[512] and the photofragmentation of 1,3-dioxalan-2-ones also proceeds with some stereoselectivity.[513] The cis elimination in 4,5-dialkyl-1,3-dioxalane-2-thiones can be carried out at 25°C if bis(cycloocta-1,5-diene)Ni° is added to the solution, and occurs with complete stereospecificity.[514] In a variant of this method *cis*- and *trans*-2-phenyl-4,5-hexamethylene-1,3-dioxalane furnished *cis*- and *trans*-cyclooctenes respectively, together with lithium benzoate, when treated with *n*-butyl lithium. The reaction is highly cis stereospecific and obviously concerted.[515]

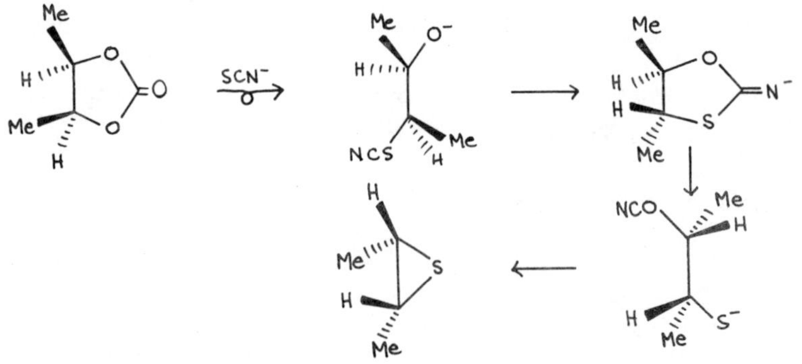

(35)

In a stereospecific conversion involving two Walden inversions *cis*-4,5-dimethyl-1,3-dioxalan-2-one reacts with thiocyanate ions to produce *cis*-2,3-dimethylthiirane (Equation 35).[516] On the other hand, the conversion of *cis*-4,5-dimethyl-2-phenyl-1,3-dioxalane into *cis*-2,3-dimethyloxirane, with *N*-bromosuccinimide in CCl_4 followed by alkali, proceeds with retention of configuration, whereas the same reaction in water followed by tosylation and treatment with alkali involves one inversion of configuration, and *trans*-2,3-dimethyloxirane is formed. The mechanism is depicted in Scheme 7 and provides an example in which two isomers can be obtained from the same compound.[517]

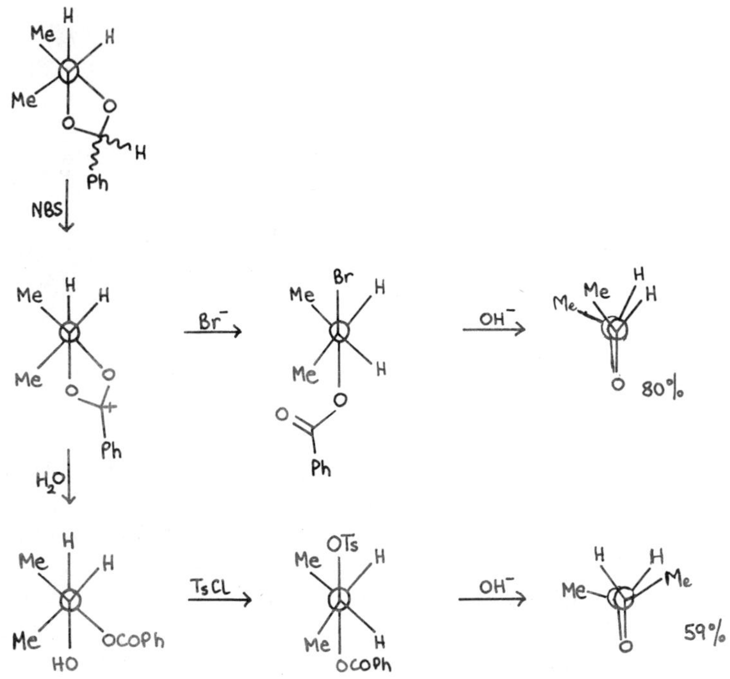

Scheme 7

The photolysis of 2-methyl-1,3-dioxalanes in trifluoromethyl chloride results in ring cleavage and formation of 2-chloroethyl acetates. The reaction is highly specific and apparently proceeds by substitution of the hydrogen on C–2 by chlorine (from the solvent) to form the unstable 2-chloro derivative which decomposes to yield a carbonium ion. The carbonium ion is attacked by a halide ion with inversion of configuration at C–4. The stereospecificity was exemplified by the photolysis of 2*R*,*S*,4*R*(−)-2-methyl-4-phenyl-1,3-dioxalane which gave *S*(+)-2-chloro-2-phenylethyl acetate (Scheme 8).[518] A not too dissimilar reaction is the treatment of 2-carboxy-2-methyl-*trans*-4,5-dimethyl-1,3-dioxalane with phosphorus pentachloride which yields *trans*-2,3-dimethyloxirane stereospecifically, by way of decarboxylation and chlorination at C–2.[519] Allyl cations are thermally generated from 2-dimethylamino-4-methylene-1,3-dioxalane by elimination of dimethylamine and can be used stereospecifically in cycloaddition reactions.[520]

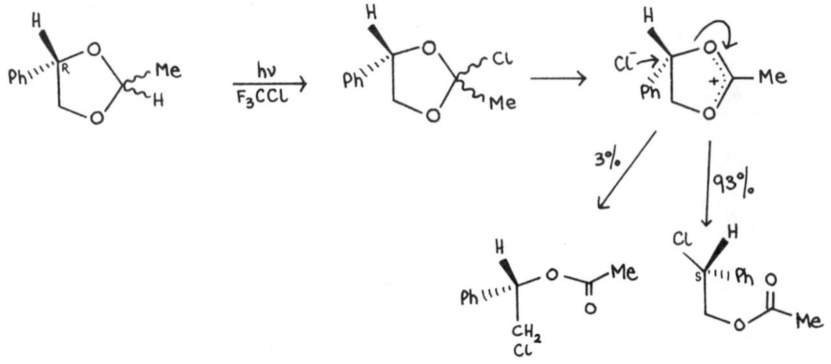

Scheme 8

IX. TRIOXALANES

There are two isomers of trioxalane, 1,2,3-trioxalane [molozonide, **163**) and 1,2,4-trioxalane (ozonide, **164**], both obtained by ozonization of olefins. They are unstable and explosive at high temperatures, but by working at low temperatures considerable spectroscopic information has been obtained. Cis and trans isomers are possible when they are substituted at the carbon atoms. Although the chemistry of cyclic peroxides was admirably reviewed by Schulz and Kirschke[521] in 1967, a few pertinent stereochemical papers have appeared since then.

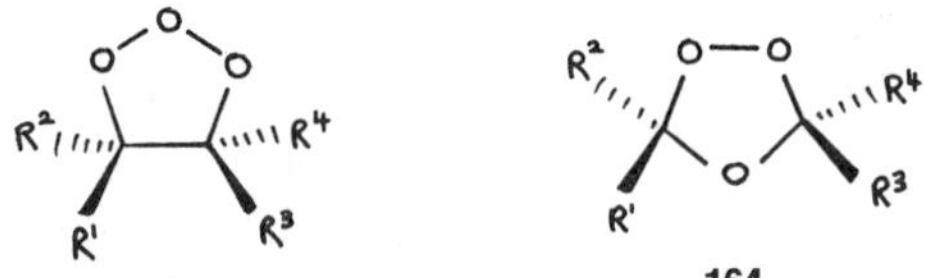

163 **164**

The mechanism of ozonization of olefins originally proposed by Criegee[522] still holds, although a few refinements have been made (Scheme 9). The first step is the formation of the respective 1,2,3-trioxalane which is highly cis stereospecific. Trans-substituted 1,2,3-trioxalanes, prepared by ozonization of *trans*-olefins have been known for some time and are stable at −100°C. The *cis*-1,2,3-trioxalanes, obtainable by ozonization of *cis*-olefins, were elusive until it was discovered that they were less stable than the trans isomers and could only be prepared, and examined by nmr spectroscopy, below −130°C.[523,524] At temperatures above −100°C the products of

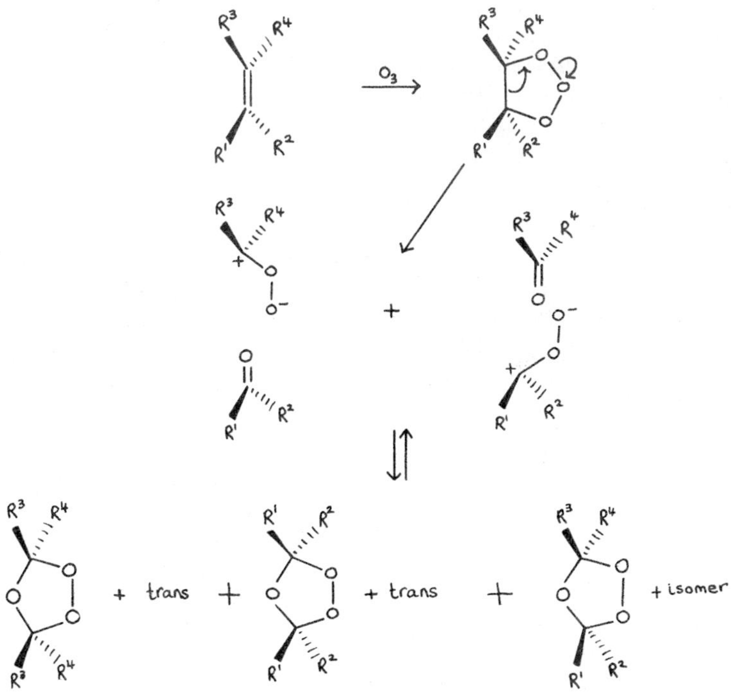

165 166 167

Scheme 9

ozonolysis are formed, namely, a carbonyl compound and a peroxidic zwitterion. This mechanism was upheld by demonstrating that if benzaldehyde-^{18}O was added to the mixture from the ozonolysis of *cis*- and *trans*-stilbene in CCl_4, then ^{18}O was incorporated into the ozonide. When starting with *cis*-stilbene a 63.3:36.4 *cis* to *trans* ratio of stilbene ozonide was produced, but when starting with *trans*-stilbene a 59.7:40.3 ratio of ozonide was formed.[525] Criegee and Korber[526] went so far as to separate *cis*- and *trans*-3,5-diphenyl-, diisopropyl-, and bisbromomethylozonides and partially resolved the trans isomers into their optical antipodes. The *cis*-ozonides are, naturally, meso forms.

According to the mechanism in Scheme 9 six theoretically possible ozonides can be formed from 1,2-disubstituted olefins, or molozonides, in which all the substituents are different. Indeed, this is the case; however, some regio- and stereoselectivity is observed. The proportion of crossed ozonides, that is, **165** and **166,** is generally less than the normal ozonides, **167,** in cases of dialkyl substituted olefins, and the cis-trans stereoselectivity is not normally very large (i.e., near a 1:1 mixture and no more than a 2:1 mixture).[527,528] The stereospecificity of the conversion of molozonide into a mixture of aldehyde and zwitterionic peroxide, which is affected by solvents and nucleophiles,[529] and their recombination to form ozonides were studied in great detail by Bailey and collaborators.[530] They formulated three rules regarding the stereochemical transformations of molozonide into ozonides and considered the cleavage of the molozonide as a disrotatory concerted process. The first rule states that "equatorial substituents in the molozonide are preferentially converted into *anti*-, and axial substituents into *syn*-zwitterions." The second states that "equatorial substituents in the molozonide are transformed into the zwitterion in preference to axial substituents," and the third states that "aldehydes prefer to react with *anti*-zwitterions so as to orient bulky substituents diequatorially (cis), and with *syn*-zwitterions to orient bulky substituents in an axial-equatorial (trans) configuration." These rules are depicted in Scheme 10.

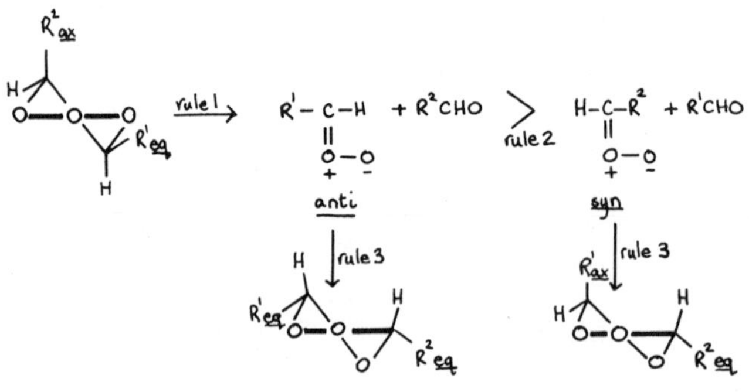

Scheme 10

X. REDUCED PYRANS

The stereochemical properties of tetrahydropyrans* are quite similar to those found in piperidines. The geometries of the two systems are similar and the conformational changes are therefore analogous, that is, the chair

* Tetrahydropyran is generally used to mean tetrahydro-2-*H*- or tetrahydro-4-*H*-pyran.

conformation being more stable than the boat conformation.[531] The essential differences arise from (*a*) the weaker basicity of tetrahydropyran (a cyclic ether) compared with piperidine (a cyclic aliphatic amine), (*b*) the presence of two lone pairs of electrons on the oxygen atom compared with one pair on the nitrogen atom, (*c*) the longer C–O bond and different C–O–C angles, and (*d*) the higher electronegativity of oxygen which introduces serious dipolar effects with electronegative substituents and is a major cause of the anomeric effect.[485] Reduced pyrans occur abundantly in nature, and the pyranose sugars, derivatives of 2-hydroxytetrahydropyran, form a very large class of stereochemically interesting carbohydrates. These are not included here, and texts on carbohydrate chemistry[532] should be consulted for further reading.

A. Syntheses of Reduced Pyrans

2,3-Dihydro-4*H*-pyrans are basically cyclic enol ethers and reagents add across the double bond with the nucleophile attacking C–2. The tetrahydropyran-2-yl group is a useful base-stable acid-labile protecting group for alcohols and has been used for this purpose extensively.[533] 2,3-Dihydro-4-*H*-pyrans are obtained by cis-stereospecific $[2 + 2]\pi$ cycloaddition reactions between acrolein and an olefin such as propenyl ethyl ether (Equation 36),[534] 2-methyl-dihydrofuran (to form 1-methyl-*cis*-2,9-dioxabicyclo-[4.3.0]non-3-ene), and 2,3-dihydro-4*H*-pyran (to form *cis*-2,10-dioxabicyclo[4.4.0]dec-3-ene]stereospecifically.[423] Further, addition reactions can be carried out on the dihydro-4*H*-pyrans. Addition of bromine to 2,3-dihydro-3-methyl-4*H*-pyran, for example, provides *r*-2-bromo-*trans*-3-bromo-*cis*-3-methyltetrahydropyran,[534] and the addition of ethanesulfinyl chloride to 2,3-dihydro-4*H*-pyran provides *trans*-2-chloro-3-ethylthiotetrahydropyran regio- and stereospecifically,[535] but *m*-chloroperbenzoic acid in aqueous ether yields a 3:7 mixture of *cis*- and *trans*-2,3-dihydroxytetrahydropyrans. The latter are a good source of *cis*- and *trans*-1,4,8-trioxadecalins.[536] However, if the peracid oxidation is carried out in the presence of an alcohol the reaction is more regio- and stereospecific, and 65 to 80% yields of *trans*-2-alkoxy-3-hydroxytetrahydropyrans are obtained.[266] Bromination of 2-pyrone in the presence of uv light involves a trans addition of bromine and gives *trans*-5,6-dibromo-5,6-dihydro-2-pyrone, which is a good source of related compounds.[537] The addition of hypochlorous acid to 2,3-dihydro-4*H*-pyran is highly regiospecific, but partly stereospecific, yielding a 37:63 mixture of *cis*- and *trans*-3-chloro-2-hydroxytetrahydropyran. The cis to trans ratio for the 2,2-dimethyl derivative is 20:80.[537a] Hydroboration, followed by oxidation, of 2-ethoxy-2,3-dihydro-4*H*-pyran is regio- but not stereospecific and yields 2-alkoxy-5-hydroxytetrahydro-4*H*-pyran. Diborane, however, cleaves the ring of 2,3-dihydro-4*H*-pyran.[268]

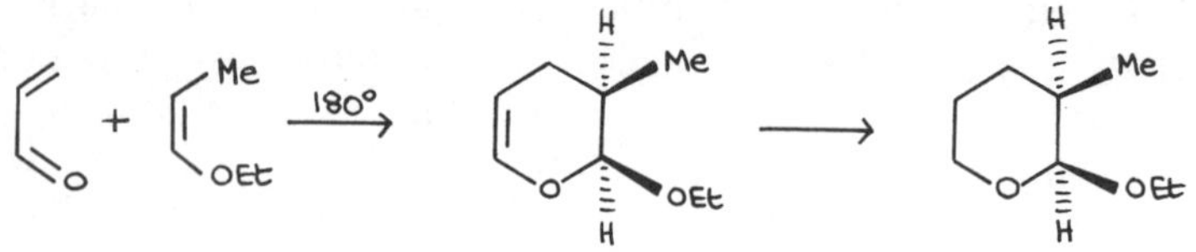

(36)

2,3-Dihydro-4*H*-pyran undergoes [2 + 2]π cycloaddition reactions with *cis*- and *trans*-1,2-diphenylethylene,[538] acrolein,[539] diphenylacetylene,[540] and bismethoxycarbonylacetylene[541] under uv light to form *cis*-2-oxabicyclo-[4.2.0]octane derivatives. The last named reagents yield the exo (**168**) rather than the endo adducts. The uniparticulate electrophile arylsulfonyl isocyanate also adds on to 2,3-dihydro-4*H*-pyran in a cis fashion, and the adduct reacts with alcohols to form *trans*-2-alkoxy-3-arylsulfonylcarbamoyltetrahydropyran.[542] Acetylenedicarboxylic ester adds 3,6 to 2-pyrone by a [2 + 4]π cycloaddition,[543] and the [2 + 2]π photochemical cycloaddition of olefins to 2,3-dihydro-2,2-dimethyl-4-pyrone produces a mixture of *cis*- and *trans*-3,3-dimethyl-2-oxabicyclo[4.2.0]octan-5-ones (**169**) in which the ratio of isomers is dependent on the temperature used.[544] Carbenes also add across the double bond of 2,3-dihydro-4*H*-pyrans to form *cis*-2-oxabicyclo[4.1.0]heptanes, and some selectivity is obtained when the substituents on the carbene are different.[545–549] Thus, chlorofluorocarbene yields a 1.5 to 1 mixture of the *endo*-chloro-*exo*-fluoro (**170,** R^1=Cl, R^2=F) and *endo*-fluoro-*exo*-chloro (**170,** R^1=F, R^2=Cl) isomers.[545,546] 5,6-Dihydro-2*H*-pyran behaves similarly towards carbenes and provides *cis*-3-oxabicyclo[4.1.0]heptanes.[549]

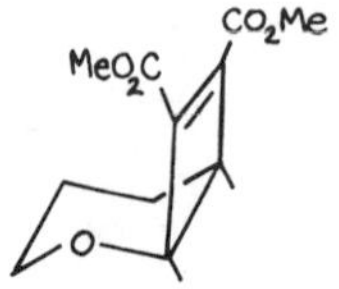

168 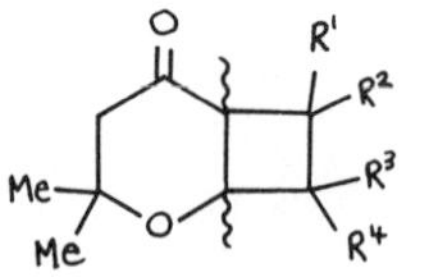 **169** 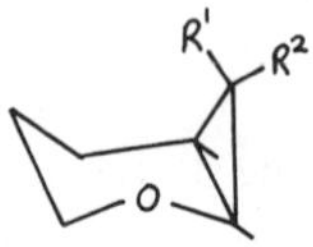 **170**

The photodimerization of pyrones is particularly interesting because of the multiplicity of products that are formed. Irradiation of α-pyrone in ether yields the valence tautomer 6-oxabicyclo[2.2.0]hex-2-en-5-one (**76**)[550] (see Section IV.A), but in the presence of a sensitizer a nonequilibrating mixture of [2 + 4]π cyclodimers **171** and **171a** are formed in which the exo-endo structure is not established.[551] 3,5-Dimethyl-2-pyrone, on the other hand, gives a mixture of cis-anti (**172**) and trans-anti (**173**) [4 + 4]π dimers together with the [2 + 2]π cyclobutane dimer **174.**[551] 4,6-Diphenyl-

2-pyrone forms similar dimers,[552] but 2,3-dihydro-2,6-dimethyl-4-pyrone produces a mixture of three *cis,anti,cis*-cyclobutane photodimers, **175, 175a,** and **175b,** in a 9:15:1 ratio; interconversion of these occurs on irradiation.[553] Intramolecular photodimerization takes place with 4-methyl-2-pyrone in which two molecules are joined by a polymethylene chain between the six-positions. When there are three or four methylene groups between the heterocycles, addition yields the syn and anti intramolecular dimers **176** and **177,** but when there are five or six methylene groups, the cyclobutane derivative **178** is formed.[553a] 2,6-Dimethyl-4-pyrone can also be converted into the Yates[554] cage compound, 2,4,8,10-tetramethyl-3,9-dioxapentacyclo-[6.4.0.0.^{2,7}0.^{4,11}0^{5,10}]dodecan-6,12-dione (**179**), which can be opened up into *trans,syn*-cyclobutane dimers by heating[555] or the action of bromine.[556] 2,3-Dihydro-3,3-dimethyl-2,4-dioxopyran similarly yields cyclobutane dimers on photolysis.[557] An alternative synthesis is the Baeyer-Villiger oxidation of the cis- or trans- anti-photo dimer, for example, **180,** of cyclopentenone.[558] For further examples of the stereochemistry of cyclobutane dimers see the recent review by Moriarty.[559]

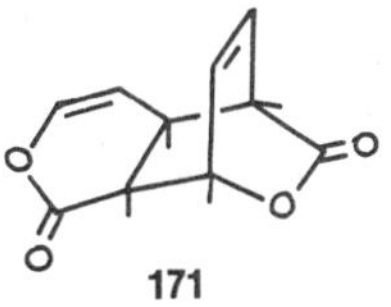

171 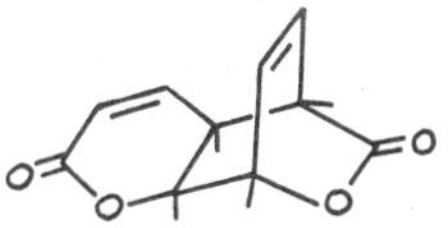 **171a**

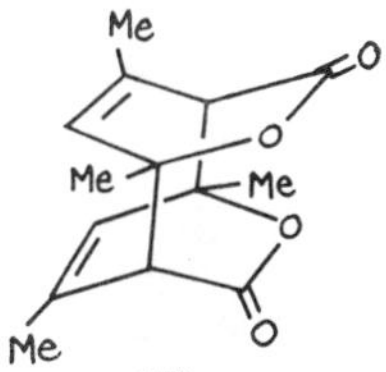

172 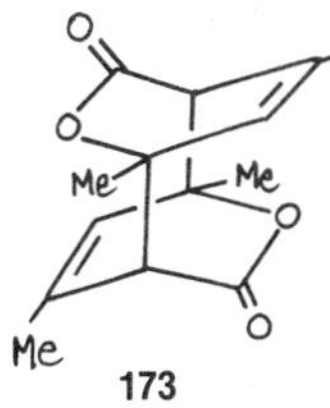 **173** 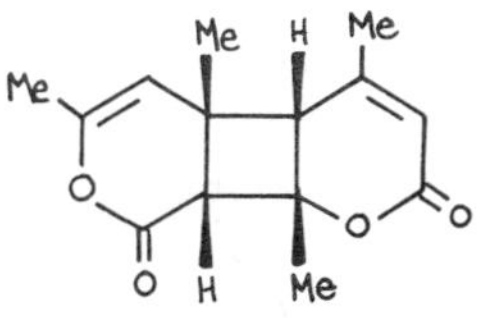 **174**

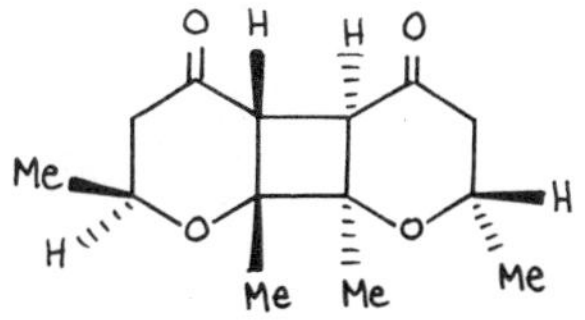

175 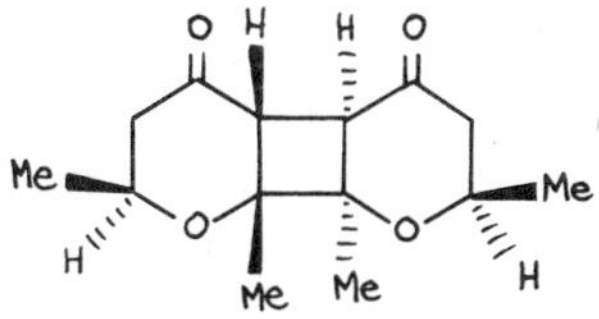 **175a**

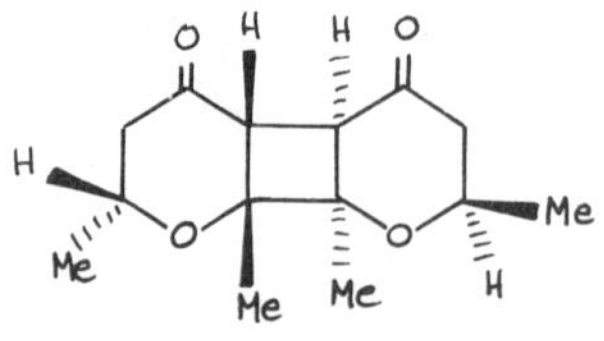

175b 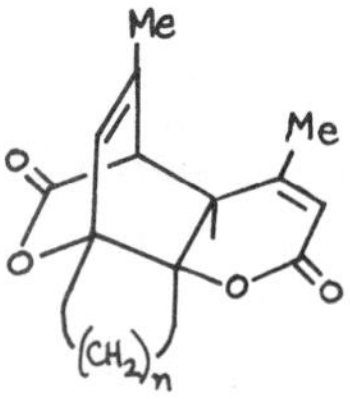 176 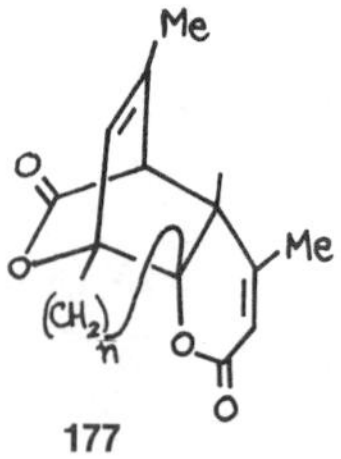 177

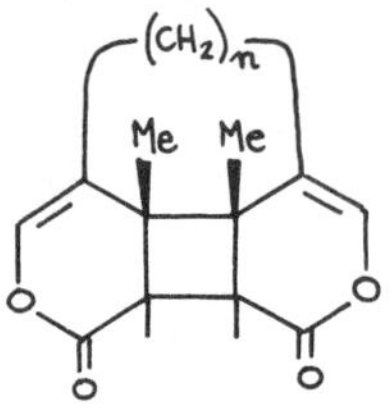

178 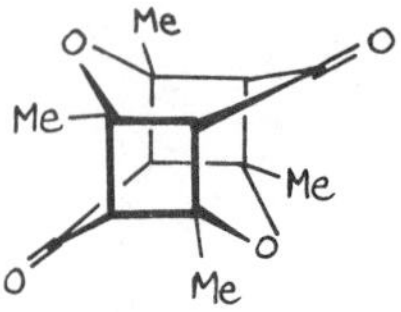 179 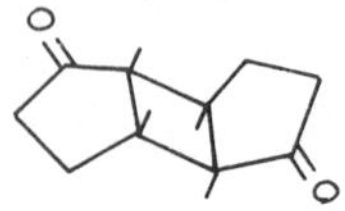 180

5-Hydroxyhexanoic acids cyclize to tetrahydropyrones, that is, δ-lactones, and as long as the substituents are not on C–5 their configuration remains unaltered.[560–563] This lactonization is very useful for determining the relative configuration of substituents on the carbon chain because the cyclic structure is conformationally more rigid than in the δ-hydroxy acids, and the coupling constants can be used to determine the relative (i.e., cis or trans) configurations of substituents.[562,563]

The cyclization of pentane-1,5-diols to tetrahydropyrans occurs without affecting the three carbon atoms which are not bonded to the oxygen atom. The configuration of substituents at C–1 and C–5, on the other hand, are affected, and the ratio of cis and trans isomers varies with the reagents used in the cyclization and with the configuration of the original diol. If clean inversion takes place at one carbon atom then, in theory, the *meso*-diol and racemic diol should yield the *trans*- and *cis*-2,6-disubstituted tetrahydropyrans respectively. When the substituents are phenyl groups the stereoselectivity is not very good. Better selectivity in favor of the cis isomer was obtained by catalytic reduction of 2,6-diphenyl-4*H*-pyran.[309] 2,6-Diallyltetrahydropyran was obtained as a cis-trans mixture from cyclization of nona-1,8-dien-3,7-diol, and the isomers were separated by glc.[564] Intramolecular cyclization of δ-hydroxyolefins (e.g., *trans*-2,6-dimethylocta-1,3-dien-8-ol) yields cis:trans mixtures of tetrahydropyrans which can usually be separated.[306,565]

B. Configuration of Reduced Pyrans

The relative configurations of substituents in several reduced pyrans were mentioned in Section A, and the relative and absolute configurations of several naturally occurring reduced pyrans, for example, cinenic acid,[566] loganine (**181**), and related compounds[567–569]; the acetylenic hydropyrans (**182**) from *Compositae*[570,571]; and reduced pyran-2-ones, for example, *R*- and *S*-mevalonolactones (**183**),[572] necic acids[573,574] (clivonecic acid, **184**), and the *R*,*R*-tetrahydropyran-2-one (**185**) from the hydrazide antibiotic Negamycin,[375] to name only a few examples, have been elucidated by physical, chemical, and synthetic methods. Further examples can be found in Dean's monograph.[247]

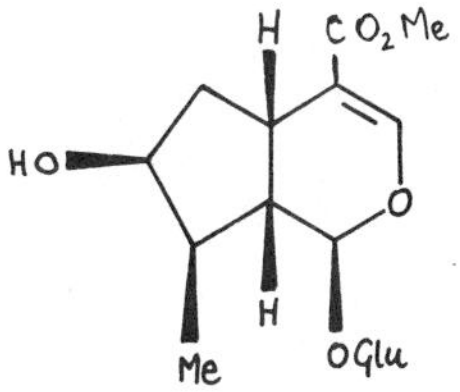

H CO$_2$Me HO H Me OGlu O
181

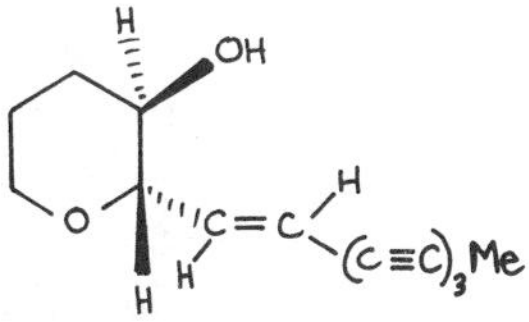

H OH O H H H (C≡C)$_3$Me
182

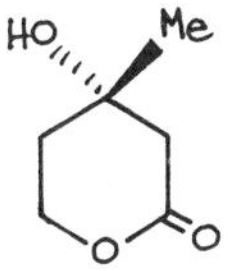

HO Me O O
183

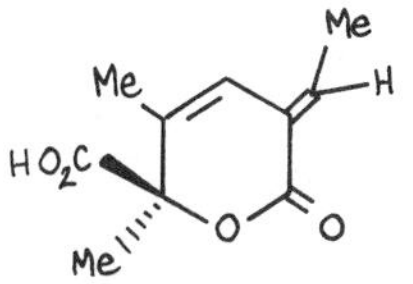

Me H Me HO$_2$C Me O O
184

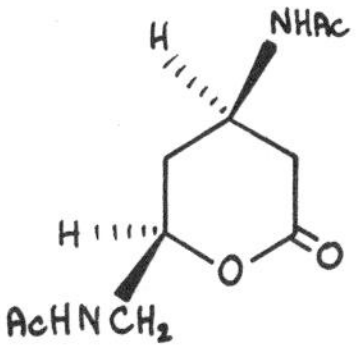

H NHAc H AcHNCH$_2$ O O
185

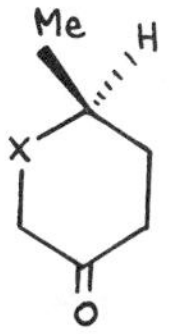

Me H X O
186

The carbonyl group in reduced 2- and 3-pyranones is a useful chromophore in ord and cd measurements for establishing absolute configuration. However, the sector rule is not always satisfactory because conformational changes may not allow a fixed model,[576,577] and because lactone ring chirality rules, and not the sector rule, should be applied.[577] Nmr correlations in such examples can be very useful in assisting to obtain the correct structures.[577,578] Cook and Djerassi found that the rotatory effects obtained by altering the ring heteroatom in *S*-6-methyl-2,4,5,6-tetrahydropyran-3-one (**186,** R=O) from O to S to NMe are very large although the configuration of the chiral center is unaltered and can point to completely misleading conclusions about the nature of the chiral center. Obviously the effects of the heteroatoms on the carbonyl chromophore are vastly different in these compounds.[579] 2-Carboxymethyltetrahydropyran was resolved with quinine into its optical antipodes, and the (−) acid converted into the (+) amide and then, by way of a Hofmann reaction, into (+)-2-aminomethyltetrahydropyran.[580] The ord curves of a number of pairs of anomeric tetrahydropyran-2-yl ethers were measured and the absolute configuration of the C–2 carbon atom of the tetrahydropyran ring was obtained by comparison with data from 2-desoxyglycosides of known configuration.[581]

The structures of *cis*- and *trans*-2-alkyl-3-chlorotetrahydropyrans, prepared from the corresponding 2,3-dichloropyrans and alkyl magnesium halide, were established by applying the Auwers-Skita rule (see Section VII.B). Thus the cis isomers had higher boiling points, refractive indices, and densities than the trans isomers and were consistent with the rates of reaction with sodium ethoxide.[582] 2,3-Dihydro-4*H*-pyran reacted with hydroquinone to give two isomeric 1,4-bis-(pyran-2-yloxy)benzenes. The higher melting compound has been tentatively assigned a meso configuration.[583] The configurations of *cis*- (meso) and *trans*-3,5-diphenyltetrahydropyran (racemic) were deduced from their differences in behavior on glc.[328]

Nmr spectroscopy is useful in assigning relative configurations (see above), and the application of shift reagents[584] has greatly improved the utility of the method. By using Eu(fod)$_3$, *cis*-7,8-diphenyl- and 7,7-diphenyl-2-oxabicyclo[4.2.0]octanes could be distinguished,[585] and with the chiral shift reagent Eu(facam)$_3$ (europium tristrifluoroacetyl-*d*-camphorate) the configuration at C–2 of several optically active 2-methyl-5,6-dihydro-2*H*-pyran-6,6-dicarboxylic acid derivatives was evaluated.[586]

Reference should be made to recent texts on carbohydrate chemistry for the configuration of pyranose sugars.[532]

C. Conformation of Reduced Pyrans (Including the Anomeric Effect)

Tetrahydro-4-*H*-pyran is like cyclohexane and piperidine (Part I, Chapter 3, Section II.C) in that the most stable conformation of the ring is the chair form. The kinetic parameters of ring inversion of tetrahydropyran (Equation 37) were determined by Gatti, Segre, and Morandi[587] who obtained values of $\Delta G^{\ddagger}$ and $\Delta H^{\ddagger}$ equal to 41.4 and 42.3 kJ mol^{-1} at 208.16°K respectively. These results indicate that the conformational mobility of the ring in comparison with cyclohexane ($\Delta G^{\ddagger}$ and $\Delta H^{\ddagger}$ are 42.3 and 38.1 kJ mol^{-1},[588] is not seriously affected by the oxygen atom. On the other hand, the barrier of isomerization of dihydro-4*H*-pyran ($\Delta G^{\ddagger}=27.6$ kJ mol^{-1}) is higher than that for cyclohexene ($\Delta G^{\ddagger}=21.8$ kJ mol^{-1}), suggesting an increased stabilization of the half-chair conformations by electronic delocalization between the two sp^2 carbon atoms and the oxygen atom (Equation 38).[589]

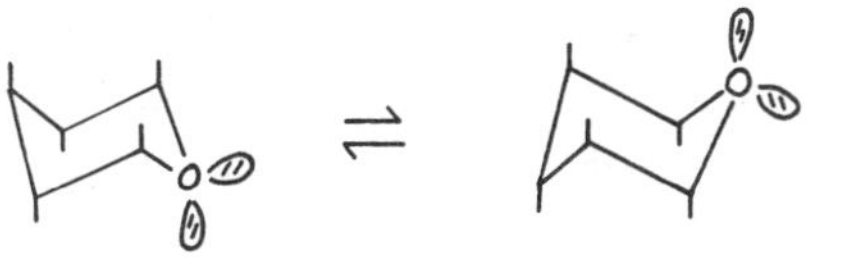

(37)

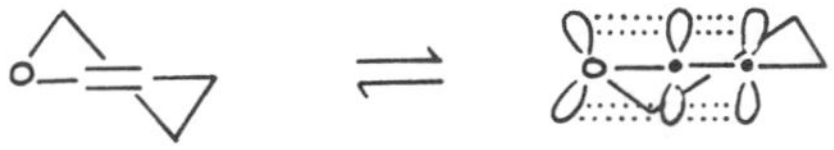

(38)

Substituents acquire axial or equatorial orientations, and any conformational equilibria invariably involve two conformations in which the tetrahydropyran ring is in the chair form. Substituents in positions other than on C–2 are generally subject to the usual interactions, except that axial substituents on one side of the ring have an interaction with the oxygen axial lone pair instead of an axial hydrogen atom (Equation 37). Substituents on the carbon atoms adjacent to the ring oxygen, however, are strongly influenced by the oxygen atom because of its polar effect and the presence of two electron lone pairs. Electronegative groups such as OH, OR, and Cl are influenced to such a large extent that, of the two possible orientations, they prefer the (more sterically encumbered) axial conformation. This was clearly observed in sugar pyranose ethers, and an electrostatic effect was pointed out by Edward.[590] It was called the "anomeric effect" by Lemieux[591] and has been discussed by Riddell,[592] Angyal,[593] Romers and coauthors,[485] Eliel and coauthors,[531] Zefirov and coauthors,[594] and Eliel.[595] Martin[596] presented theoretical as well as practical aspects of the effect, and another thorough review is not warranted. A brief discussion including a few references, however, is given here for the sake of completeness.

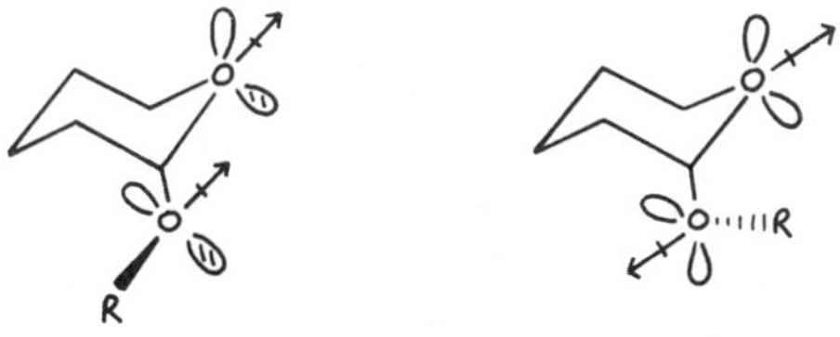

187 A_1 **188** A_2

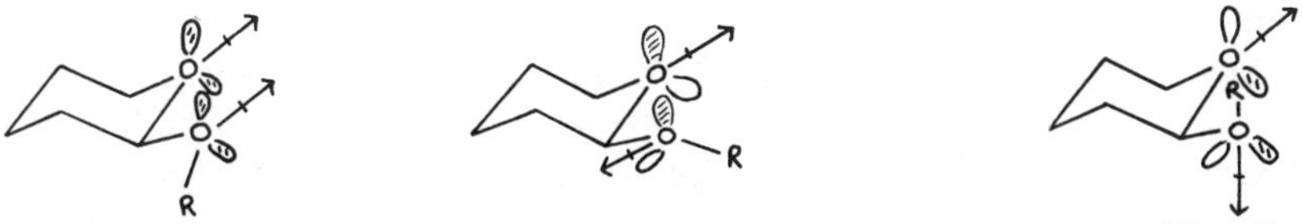

189 E_1 **190** E_2 **191** E_3

The anomeric effect in 2-alkoxytetrahydropyrans is very pronounced, and several studies of such systems were made. Both dipole moment and nmr methods were used to elucidate structures, and the most important axial and equatorial conformations being considered are A_1 (**187**), A_2 (**188**) and E_1 (**189**), E_2 (**190**) and E_3 (**191**) respectively. By examining *cis*- and *trans*-2-methoxy-, 2-ethoxy-, and 2-*t*-butoxy-4-methyltetrahydropyran, de Hoog and Buys[597] concluded that the axial conformer A_2 (**188**) and equatorial conformer E_2 (**190**) and/or E_3 (**191**) were the most populated. The unfavorable arrangement of the syn diaxial lone pair, indicated as shaded lone pairs in **187** and **189–191**, has been called the *rabbit ear effect* by Eliel,[598] and because it is a general anomeric phenomenon it was renamed the *generalized anomeric effect*[599–601] (see Part I, Chapter 3, Section IV.C). In nonpolar solvents the energy involved in the anomeric effect for alkoxytetrahydropyrans is 5.4–11.8 kJ mol^{-1}.[602,603] The proportion of 2-axial alkoxy group is affected by the position and relative stereochemistry of other substituents in the ring, for example, the percentage of 2-axial conformer in 2-methoxy-, *cis*-2-methoxy-3-methyl-, *trans*-2-methoxy-3-methyl-, *cis*-2-methoxy-4-methyl-, and *trans*-2-methoxy-4-methyltetrahydropyrans are respectively 73, 96, 35, 2, and 98. This demonstrates that ananchomeric substituents (anchoring substituents) can enhance or suppress the anomeric effect depending on whether the preferred conformation of the substituent is such that the alkoxy group has to be axially oriented or not.[604,605] The solvent has a large effect on the position of equilibrium in 2-alkoxytetrahydropyrans, and this was elegantly demonstrated by Lemieux and co-workers,[606] who determined the percentage of axial conformer in 4,4,5,5-tetradeutero-2-methoxytetrahydropyran and *S*(+)-2-methoxytetrahydropyran by measuring the nmr spectra and optical rotations in various solvents. They found that the proportion of axial conformer decreases as the dielectric constant decreases, but that strong hydrogen bonding solvents (e.g., H_2O) have the ability to stabilize an equatorial *O*-alkyl group considerably, that is, a counter anomeric effect.

The acid-catalyzed addition of alcohols to 2-alkoxy-2,3-dihydro-4*H*-pyrans yields more than 77% of *trans*-2,6-dialkoxytetrahydropyrans; that is, one alkoxy group is axial and the other equatorial, and the group with the bulkiest alkyl substituent prefers the equatorial orientation.[607] The acid-catalyzed addition of methanol to 2-methoxy-5,6-dihydro-2*H*-pyran provides a 4:1 mixture of *trans*- and *cis*-2,4-dimethoxytetrahydropyrans from which the magnitude of the anomeric effect of the 2-methoxy group was calculated (5.8 kJ mol^{-1}).[608] This ratio is unaltered when the mixture is allowed to stand in the presence of *p*-toluenesulfonic acid and demonstrates that a highly selective reaction occurred originally between methanol and the dihydromethoxypyran.

Booth and Ouellette[609] used nmr spectroscopy to show that the anomeric effect was operating in 2-chloro- and 2-bromotetrahydropyrans which existed predominantly with the halogen atom in the axial orientation. The results were confirmed by Zefirov and Shekhtman,[610,611] and Anderson and Sepp,[612] in an investigation of the conformational equilibria of 2-bromo- and 2-chloro-4-methyltetrahydropyrans, estimated the free energy involved as 13 and 11.3 kJ mol^{-1} respectively. Dipole moment and molar Kerr constants of 2-chlorotetrahydropyran, by Eckert and Le Fèvre,[613] in CCl_4 are consistent with a chair conformation possessing an axial chlorine atom. The effect of solvent polarity is negligible on the conformational equilibria in 2-halotetrahydropyrans.[609] The conformational equilibria in 3-chloro- and 3-bromotetrahydropyrans were also measured by nmr and ir, and the appreciable proportions of axial halogen were explained by dipole, electronic, and van der Waals effects consistent with a *homoanomeric effect*.[614] As in the addition of methanol to 2-methoxy-3,4-dihydro-2*H*-pyran, the addition of ethylsulfinyl chloride to 2-methoxy-5,6-dihydro-2*H*-pyran is also highly stereoselective yielding the all-equatorial 4-chloro-3-ethylthio-2-methoxytetrahydropyran by way of the equivalent of a trans addition to a dihydropyran with an axial methoxy group (**192**).[615] The nmr spectra and dipole moment measurements on six 2-alkylthiotetrahydropyrans revealed that, in general, the 2-alkylthio group has only a slight preference for the axial position,[616] in contrast with 2-azido and 2-isocyanato groups which are mainly axial.[616a]

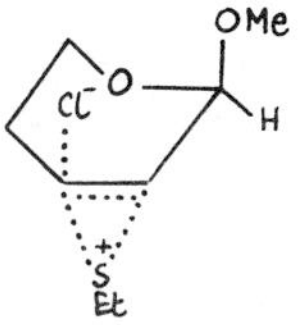

OMe H Cl O S Et

192

CH₂OAc O N Me OAc AcO OAc

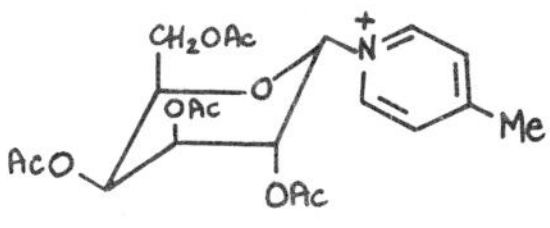

193

The 2-acetoxy group in 2-acetoxytetrahydropyrans is also subject to the anomeric effect, and a high preference for the axial conformation is observed.[603,610] A methoxycarbonyl group in the 2-position, however, causes an inverse or reverse anomeric effect. Examination of 4-, 5-, and 6-methyl-2-methoxycarbonyltetrahydropyrans demonstrated that the CO_2Me group prefers an equatorial orientation even if it has to place a methyl substituent in the ring into an axial conformation.[617] This effect is of the order of 6.8 kJ mol^{-1} (compare 4.6 kJ mol^{-1} for cyclohexanecarboxylic ester). Lemieux and Morgan[618] observed this reverse effect in pyranose sugars. They found that the *N*-pyridinium substituent prefers the equatorial position in the

α-anomer of tetraacetoxyglucopyranose (**193**) even if it means that the other substituents have to be axially oriented. The groups $-CH_2OAc$, $-CH_2NHAc$, $-CH(SO_2Et)_2$, and perhaps $-CH_2OH$, have a weak reverse anomeric effect.[659]

The half-chair conformation of 2,3-dihydro-4*H*-pyran is mentioned above, and a 2-alkoxy group in this system prefers the axial conformation (**194**).[610,619] The epimers of 6,6-diethoxycarbonyl-2,5-dimethyl-5,6-dihydro-2*H*-pyran also appear to exist almost exclusively in the half-chair conformation, and for these a modified Karplus equation was proposed.[620] The preferred conformation of tetrahydropyran-2-ones, δ-lactones, is a slightly flattened half-chair, deduced from nmr studies, and has some support from X-ray data.[621]

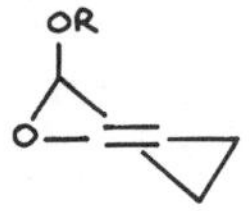

194 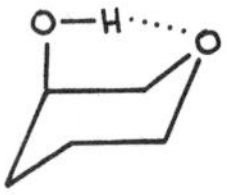 **195** 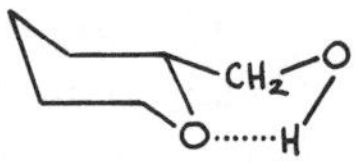 **196**

Intramolecular hydrogen bonding was observed between an hydroxyl substituent and the tetrahydropyran oxygen atom in some, but not all, hydroxytetrahydropyrans. Hydrogen bonding occurs in 3-hydroxy- (**195**), 2-hydroxymethyl- (**196**), but not in 2- and 4-hydroxytetrahydropyrans.[352] In certain cases the preferred conformation of the tetrahydropyran ring (e.g., **195**) is fixed by intramolecular hydrogen bonding.[622]

The base-catalyzed equilibria between *cis*- and *trans*-3,5-dimethyltetrahydropyran-4-ones were compared, by Katritzky and co-workers,[622a] with those of the corresponding 3,5-dimethyl-*N*-*t*-butylpiperidin-4-ones, 3,5-dimethyltetrahydrothiapyran-4-one (and *S*,*S*-dioxide) and 2,6-dimethylcyclohexanone. They concluded from the thermodynamic data that the steric interaction of the 2-axial methyl groups in the trans isomers, was in the order: 1-*S* (due to larger bond distances), 1-SO_2 (little steric repulsion between axial *O* and axial Me) $<$ 1-*O*, 1-*N*-Bu^t $<$ CH_2; this clearly demonstrates the smaller steric requirements of the oxygen lone pair compared with an axial CH when a methyl group was used as a probe.

D. Reactions of Reduced Pyrans

The stereochemistry of addition reactions across the double bond of dihydro-2-*H*- and 4*H*-pyrans is influenced by a substituent at C–2 or C–6, and several examples are cited in Sections A and C. A few more reactions are mentioned here and include addition of borane, followed by oxidation with alkaline peroxide, to 2-methoxy-3,4-dihydro-2*H*-pyran which is highly regiospecific but only partly stereospecific (Equation 39). The 6-methyl derivative behaves similarly.[623] The acid-catalyzed addition of methanol to 2-methoxymethyl-3,4-dihydro-2*H*-pyran is also regiospecific, but only stereoselective, with the anomeric effect causing the resulting stereoselectivity. The product is a 7:3 mixture of *trans*-(2-OMe-*ax*)- and *cis*-(2-OMe-*eq*)2-methoxy-6-methoxymethyl(*eq*)tetrahydropyrans.[624] The addition of HCl to 2-alkoxy-5,6-dihydro-2*H*-pyran (**197**) yields *trans*-2-alkoxy-4-chlorotetrahydropyran in a highly regiospecific manner by attack of chloride ion from the *b* side (**197**), and the addition of HOCl in acetic anhydride produces a 55:45 mixture of *r*-2-alkoxy(*eq*)-*trans*-3-chloro-*cis*-4-acetoxy- and *r*-2-alkoxy(*ax*)-*cis*-3-chloro-*trans*-4-acetoxytetrahydropyrans. Epoxidation is also stereoselective, and 80% of the attack takes place from *b* side (**197**), but cis hydroxylation with permanganate occurs mainly from the *a* side.[625] Epoxidation of 2-methyl-5,6-dihydro-2-*H*-pyran is not highly stereospecific because, unlike the alkoxy analogues (**197**), the methyl group is pseudo-equatorial and a 55:45 mixture of epoxides is formed. The cleavage of these epoxides with LAH or hydrogen halide, however, is highly regiospecific (in contrast with the cyclohexane analogues) with the nucleophile attacking C–4; that is, the 3-hydroxy derivative is formed.[626]

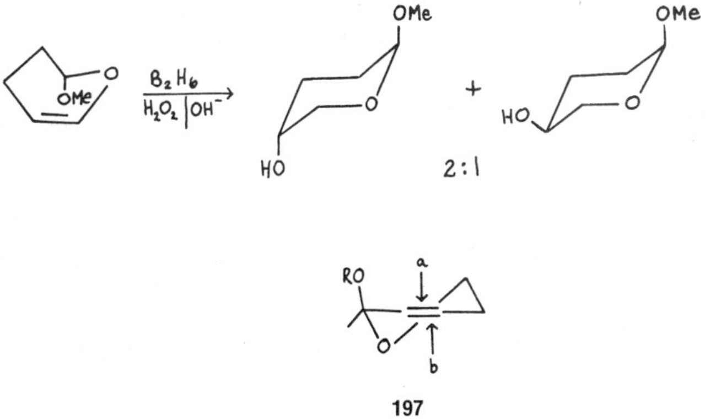

OMe, O, B_2H_6, H_2O_2 | OH^-, OMe, O, HO, +, OMe, HO, O, 2:1 (39)

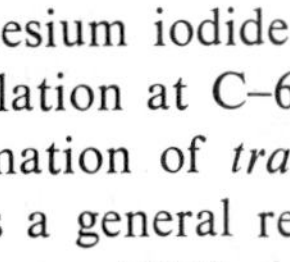

RO, a, O, b

197

The reaction of methyl magnesium iodide with 2-ethoxy-6-methyl-2,3-dihydro-4*H*-pyran causes methylation at C–6 with stereospecific cleavage of the ether bond and the formation of *trans*-1-ethoxy-2-(2′-hydroxy-2′-methylethyl)cyclobutane. This is a general reaction of Grignard reagents with 2-alkoxy-2,3-dihydro-4*H*-pyrans.[627] Hydrolysis and transetherification of 2,2-dialkoxy-5,6-dihydro-2*H*-pyrans takes place by way of a ring-opened ω-hydroxy hemiorthoester which decomposes in a highly stereoselective manner to form either the ω-alkoxycarbonylbutyl alcohol or the trans-etherified 2-alkoxydihydropyran. The selective decomposition was explained in terms of the stereochemistry of the lone pair orbitals of the oxygen atom which control the reaction.[628,628a]

Lutz and Roberts[629] demonstrated that the acrolein dimer 2-formyl-2,6-dimethyl-2,3-dihydro-4*H*-pyran isomerizes thermally without apparent alteration in structure. The optically active dimer is not racemized, indicating retention of configuration. By heating the deuterium labeled (at C*H*O) dimer they clearly showed, using ^{1}H nmr spectroscopy, that the deuterium atom from C*D*O exchanges place with H–6 (Equation 40).

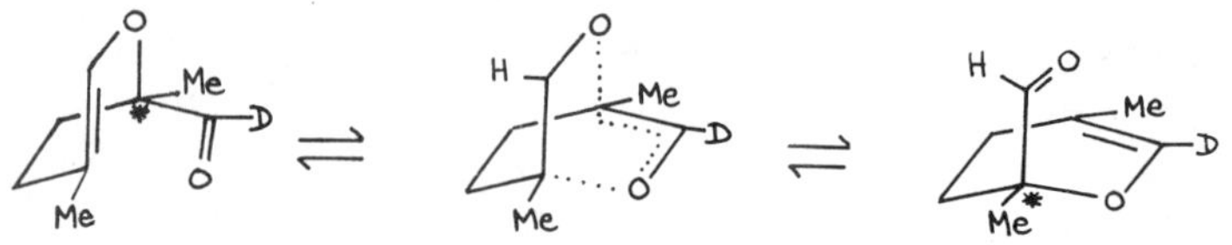

(40)

XI. CONDENSED PYRANS

A. Benzopyrans and Isobenzopyrans

The benzopyrans included in this subsection are reduced in the oxygen ring and in both rings. The stereochemical interest is in the relationship of substituents in the oxygen ring and the bridgehead carbon atoms. The data come mainly from the syntheses and reactions of benzopyrans and benzoisopyrans rather than from deliberate stereochemical studies. Benzopyrans form a large group of naturally occurring heterocycles which are not discussed here because they have been adequately treated by Dean,[247] and the flavanoids were recently reviewed by Harborne, Mabry, and Mabry.[630] Chiral biflavanoids and flavanoid glycosides were admirably discussed by Locksley[631] and Wagner respectively.[632]

1. *Benzopyrans and Isobenzopyrans Reduced in the Oxygen Ring (Chromans and Isochromans)*

The double bond of coumarins is reactive towards olefins in the presence of uv light. Irradiation of coumarin and tetramethylethylene in the presence of benzophenone probably proceeds through an eximer-exciplex exchange and forms a cis-fused tricyclic system.[633] Similar [2 + 2]π cycloadditions with cyclopentene and 1,1-diethoxyethylene are reported, and in the latter case regioselectivity is observed (Equation 41).[634] In the absence of olefins and sensitizers, coumarin dimerizes in a cis head-to-head fashion to yield the syn (**198**) and anti (**199**) dimers.[635–638] In the presence of a sensitizer (e.g., benzophenone) the trans head-to-tail anti dimer **200** is also

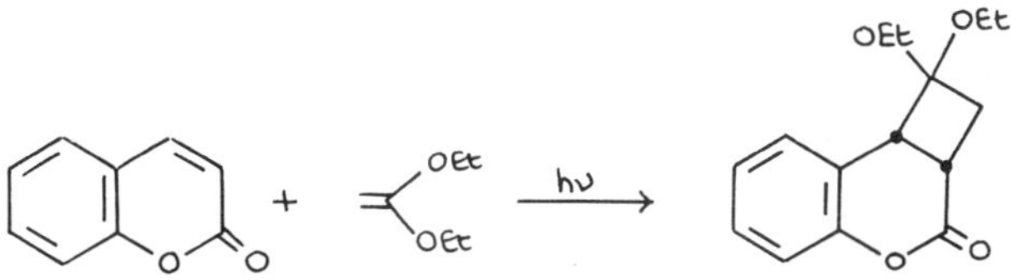

(41)

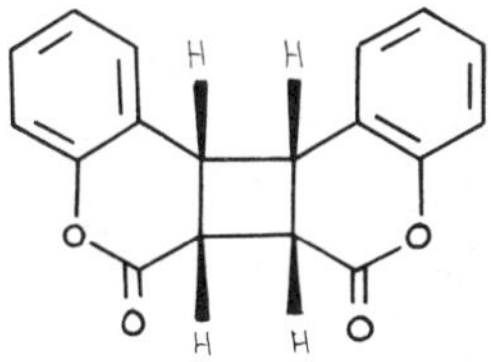

198

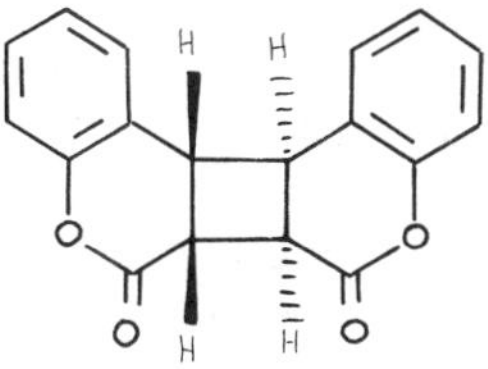

199

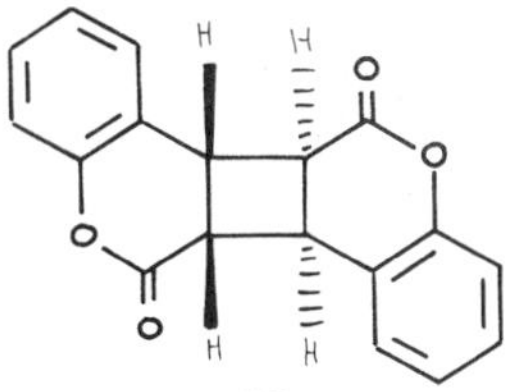

200

formed,[639,640] and the effects of solvent polarity on these dimerizations were investigated. An interesting application of this dimerization is the irradiation of 7,7′-polymethylene-ω-dioxy coumarins in which more than four methylene units are present. Intramolecular dimerization occurs to yield predominantly the syn head-to-tail isomer when there are no substituents in the oxygen ring, and predominantly the syn head-to-head isomer when ester groups are present in the 3,4- and 3′,4′-positions (Equation 42)[641] (compare formulas **176–178**).

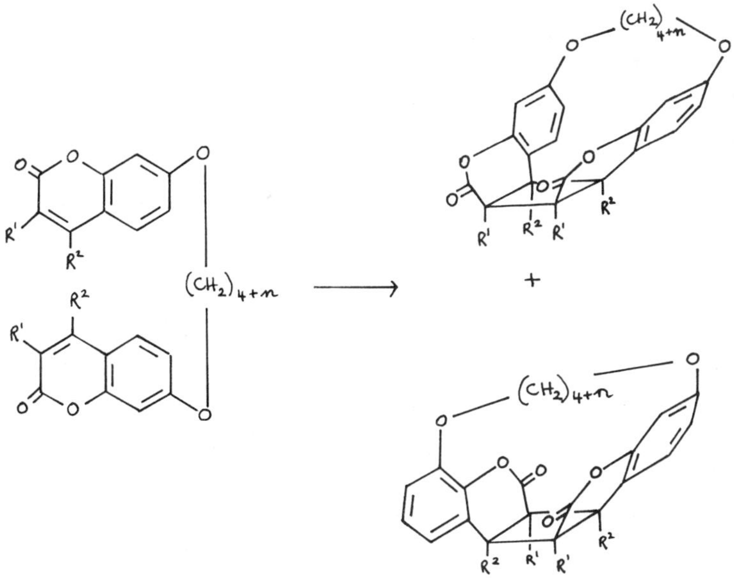

(42)

In a similar manner chromones undergo $[2 + 2]\pi$ photocycloaddition reactions with olefins[642] and diphenylacetylene[643] to form the cis-fused cyclobutane derivatives. The cycloaddition of β,γ-unsaturated aldehydes with indene to yield 8,9-benzo-*cis*-2-oxabicyclo[4.3.0]nona-3,8-dienes does not require uv light.[643a]

The addition reactions to the isolated double bond of 4*H*-chromenes occur readily with the nucleophilic portion of the reagent attacking the 2-position. Thus HOCl, HOBr, and BrN_3 add across the double bond with the formation of *trans*-2-hydroxy- and 2-azido-3-halochromans. The substituent at C–2 can be readily displaced by a nucleophile and may afford the *cis*-2,3-substituted chroman by a Walden inversion. ^1H nmr studies of these suggest that the anomeric effect may be operating because in both *cis*- and *trans*-2-methoxy- and 2-acetoxychromans the 2-substituent prefers the axial orientation (Equation 43).[644,645] Similar studies were made with *cis*- and *trans*-3-amino-2-methylchromans, and it was found that in the cis isomer (free base and salt) the conformer with the axial amino group is favored, whereas in the trans isomer the conformer with the axial amino group is favored only in the cation but not in the free base. The free bases of *cis*- and *trans*-(?) 2-methyl-3-dimethylaminochromans have the dimethylamino group equatorial, but in the cation of the latter the protonated dimethylamino group prefers the axial conformation.[646] The kinetics of solvolysis of *trans*-3,4-dichloro-2,2-dimethylchroman were measured and found to involve a benzylic carbonium ion which is subject to both steric and neighboring group effects.[647]

Cyclopropa-[*b*]- or [*c*]-chromans are formed by the reaction of carbene, dichloro-, and dibromocarbenes with the respective 4-*H*- and 2*H*-chromenes. Two conformations are possible in which a substituent on the cyclopropane ring can have an endo or exo configuration. One of the halogen atoms can be replaced by hydrogen (with LAH) stereospecifically, or not, depending on the preferred conformation **201** or **202.**[648]

(43)

201

202

Reduction of 3-bromochroman-4-one and 3-bromo-2,6-dimethylchroman-4-one with $NaBH_4$ is stereospecific, and the *cis*-3-bromo-4-hydroxy derivatives are formed. These *cis*-4-hydroxy compounds were converted into the *trans*-3,4-dihalo compounds, with $SOCl_2$, PCl_3, or PBr_3, which solvolyze in the presence of alcohols to give the corresponding *trans*-4-alkoxy-3-bromochromans with retention of configuration at C–4.[649] Here again a benzylic cation may be involved. Clark-Lewis and co-workers[650] reduced 4-oximino-2-phenylchroman stereospecifically and obtained *cis*-4-amino-2-phenylchroman (4-aminoflavan) which they resolved into its enantiomers. The cis configuration was unequivocally established by ^1H nmr spectroscopy, and the heterocyclic ring was shown to be in the half-chair conformation. Controlled catalytic reduction of 3-oximino-2-phenylchroman-4-one gave *trans*-3-amino-2-phenylchroman-4-one.

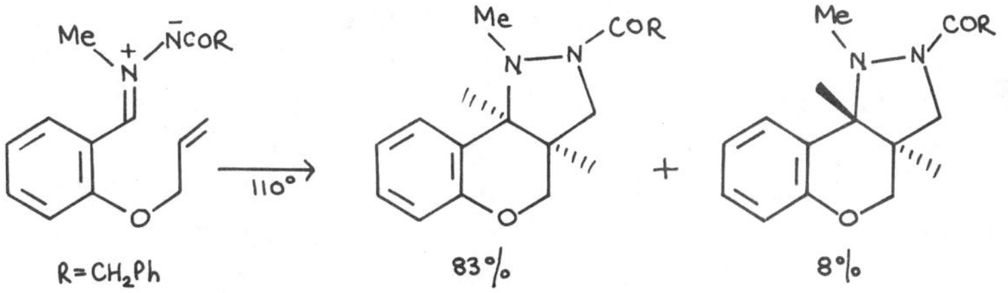

Me, $\bar{N}COR$, 110°, R=CH_2Ph → Me–N–N–COR 83% + Me–N–N–COR 8% (44)

Oppolzer[651] extended his intramolecular cyclization reactions to the synthesis of 3,4-substituted chromans from azomethinium ylides derived from *N*-acyl-*N'*-methylhydrazine and *o*-allyloxybenzaldehyde. The *cis*-3,4-disubstituted cycloadducts predominate (Equation 44). Phenylhydroxylamine reacts similarly and yields the *cis*-isoxazolo[*c.c*]chroman (**203**). The nitrogen containing heterocyclic rings of these adducts can be cleaved to furnish *cis*-2,3-disubstituted chromans.

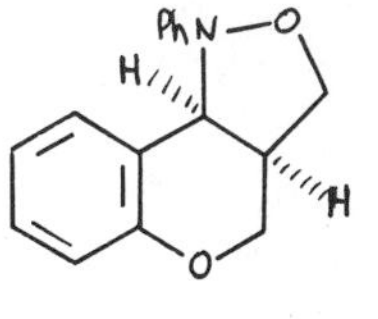

203  **204** 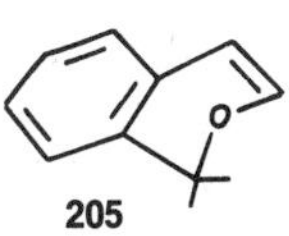 **205**

^1H nmr studies of isochromenes and homoisochromenes revealed that these compounds are conformationally mobile and involve structures such as **204** and **205**,[652] in which the CH_2 group undergoes a flip-flap motion.

2. *Benzopyrans and Isobenzopyrans Reduced at the Bridgehead Carbon Atoms (Oxabicyclo[4.4.0]decanes)*

Benzopyrans (and isobenzopyrans), which are reduced at the bridgehead carbon atoms, exist in cis and trans geometrical forms and have been obtained by intramolecular cyclization reactions, by cycloaddition and condensation reactions, and by reduction or addition reactions in which specificity of the addition across the double bond common to the two rings dictates the proportion of cis- and trans-fused products.

Lead tetraacetate oxidation of 3′-hydroxypropylcyclohexane gives a low yield of comparable amounts of *cis*- and *trans*-2-oxabicyclo[4.4.0]decanes,[414] among other cyclohexanes, but the intramolecular cyclization of 2′-carboxyethylcyclohex-2-ene in sodium bicarbonate containing iodine affords *trans*-10-iodo-*cis*-2-oxabicyclo[4.4.0]decan-3-one stereospecifically in 95% yield (Equation 45).[418] Meerwein-Ponndorf reduction of 6-[2′-(dioxalan-2-yl)-1′-methylethyl]-6-ethoxycarbonyl-3-methylcyclohex-2-enone furnished 6-ethoxycarbonyl-5,9-dimethyl-*cis*-2-oxabicyclo[4.4.0]deca-3,9-diene (**206**), a key intermediate in the total synthesis of (±)-12,13-epoxytrichothec-9-ene.[653] Intramolecular cyclization of the open chain 3,6,10-trimethylundeca-5,9-dien-2-one is catalyzed by BF_3 and affords 1,3,7,7-tetramethyl-*trans*-2-oxabicyclo[4.4.0]dec-3-ene (**207**).[654] Carvone and benzaldehyde in the presence of dry HCl, on the other hand, produce 6-chloro-7,7,9-trimethyl-2,4-diphenyl-*cis*-3-oxabicyclo[4.4.0]decan-10-one (**208**).[655]

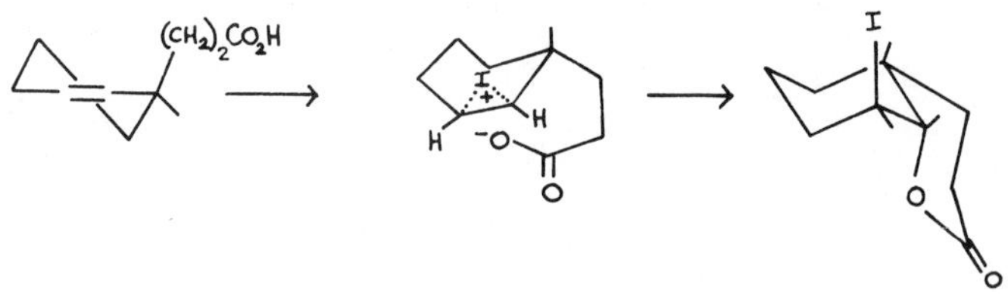

$(CH_2)_2CO_2H$ (45)

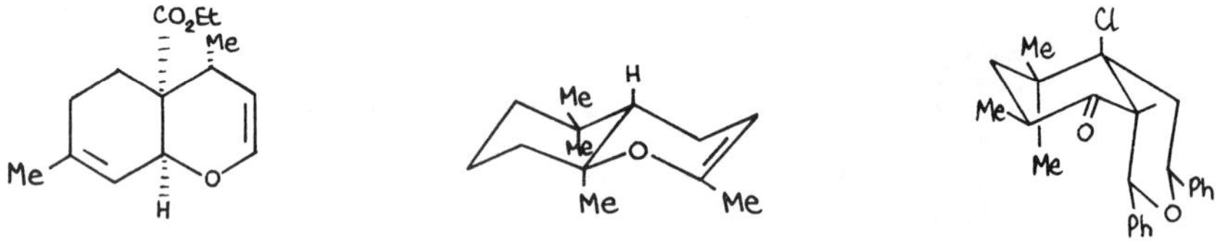

206 **207** **208**

2-Hydroxy-5-oxo-5,6-dihydro-2*H*-pyran undergoes cycloaddition reactions with olefins to form 2-hydroxy-*cis*-3-oxabicyclo[4.4.0]dec-8-en-5-ones stereospecifically.[656] The crystalline product isolated from the wastes of furfural purification, by using butadiene, was shown to be 4-hydroxy-*cis*-2-oxabicyclo[4.4.0]dec-8-en-3-one (**209**) and must involve a preliminary [2 + 2]π cycloaddition between butadiene and furfural.[657]

Catalytic reduction of 1-oxa-1,2,3,4,5,6,7,8-octalins usually furnishes *cis*-2-oxabicyclo[4.4.0]decanes,[658,659] but the catalytic hydrogenation of 6- and 7-methoxycarbonyl or 6-methylchromones with 5% ruthenium on charcoal at 100°C causes reduction of the benzene ring with formation of a mixture of the corresponding *cis*- and *trans*-2-oxabicyclo[4.4.0]decanes, which probably have a high proportion of cis isomers.[660] Base-catalyzed equilibration of *cis*- and *trans*-2-oxabicyclo[4.4.0]decanes and their 5-keto derivatives was studied by ^{1}H nmr spectroscopy, and it was demonstrated that the thermodynamically favored isomers are the trans-fused decalins in all cases.[661] Peroxy acid oxidation of 1-oxa-1,2,3,4,5,6,7,8-octalin yields the 4*a*,8*a*-epoxide which hydrolyzes to 1,6-dihydroxy-*cis*-2-oxabicyclo[4.4.0]-decane (**210**).[662,663] The cis isomer can be obtained by OsO_4 hydroxylation of the oxaoctalin so that it can be converted into the trans isomer by aqueous acid.[664]

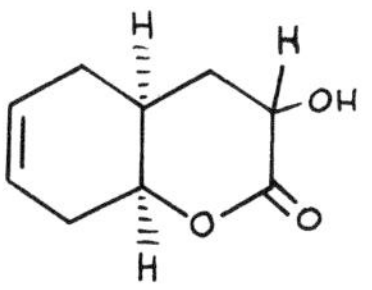

209 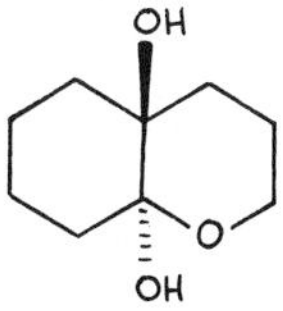 **210** 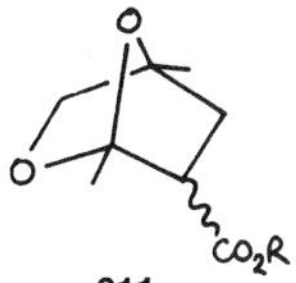 **211**

B. Miscellaneous Oxabicycloalkanes

m-Chloroperbenzoic acid oxidation of hex-5-en-2-ones was shown, by Gaoni,[665–667] to proceed in a specific manner and form 1-methyl-2,7-dioxabicyclo[2.2.1]heptanes (e.g., **211**) rather than the epoxides. Exo- and endo-substituted derivatives are produced, and several were obtained from substituted γ,δ-unsaturated ketones, demonstrating the generality of the reaction.

Although *trans*-bicyclo[4.2.0]octanes are known it is, all the same, interesting that the trans-fused 1*R**,6*S**,7*R**,8*S**-7,8-diphenylbicyclo[4.2.0]-octane (**212**) should be formed in a photocycloaddition reaction between *cis*-stilbene and 2,3-dihydro-4*H*-pyran. Its relative configuration was deduced by ^{1}H nmr spectroscopy.[668] Photolysis of nitrite esters of 4-hydroxymethylcyclohex-1-enes caused intramolecular addition across the double bond to form 6-oxabicyclo[3.2.1]octane derivatives (e.g., **213**) in about 30% yields.[669] Intramolecular addition of hydroxyl groups was observed in the reaction of 2-formyl-[670] and 2-acetyl-2,3-dihydro-4-*H*-pyran[671] with methyl magnesium iodide, followed by acid treatment, to yield 7-methyl- and 7,7-dimethyl-6,8-dioxabicyclo[3.2.1]octanes (**214**) respectively. Other Grignard reagents produced the corresponding derivatives, and the cyclization was most probably brought about the acid on the alcohol formed because 2-hydroxymethyl-2,3-dihydro-4-*H*-pyran cyclizes in acid to 6,8-dioxabicyclo[3.2.1]octane.[672,673] 8-Oxabicyclo[3.2.1]oct-6-en-3-ones (**215**) were prepared by stereospecific cycloaddition of a three-carbon unit to the 2,5-positions of furans. Thus they can be obtained by heating cyclopropanones with furans[292,674] or by the addition of $Fe_2(CO)_9$ to a mixture of 1,3-dibromoacetones and furans and keeping at 40°C for 24 hr.[675]

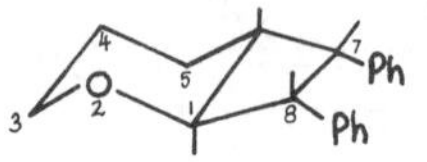

212 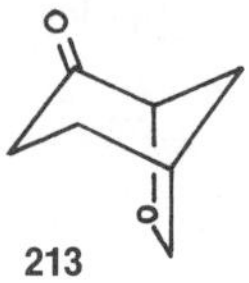 **213** 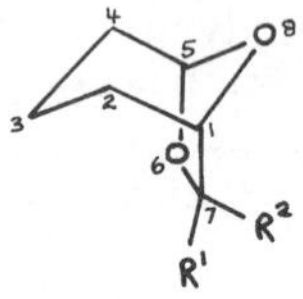 **214**

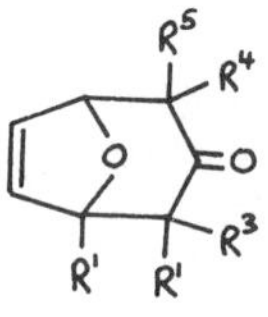

215 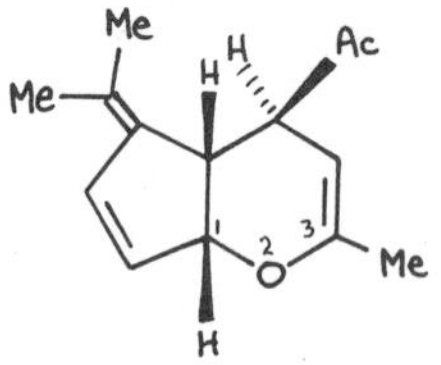 **216**

Catalytic reduction of 4-oxa-1,2,3,5,6,7-4*H*-hexahydroindene with Pd–$BaSO_4$ gave exclusively *cis*-2-oxabicyclo[4.3.0]nonane. The 2-(3′-bromopropyl)cyclopentanol, prepared by LAH reduction of the corresponding cyclopentanone, cyclized, with 20% potassium hydroxide, into a 9:1 mixture of *trans*- and *cis*-2-oxabicyclo[4.3.0]nonanes.[676] The physical properties of these two isomers were consistent with the Auwers-Skita rule (see Sections VII.B and X.B). In refluxing ether, dimethylfulvene condenses with *cis*-hex-3-ene-2,5-dione stereospecifically to yield only one of four possible isomeric compounds—*trans*-5-acetyl-7-isopropylidene-3-methyl-*cis*-2-oxabicyclo[4.3.0]nona-3,8-diene (**216**). The relative configuration of substituents in this product was established by reduction of the acetyl group to an 2′-hydroxyethyl group followed by intramolecular addition of the hydroxyl group onto C–3 to form the tricyclo compound **217**.[677,678]

The two isomers of 9-phenyl-9-hydroxy-3-oxabicyclo[3.3.1]nonane (**218**) are formed by the reaction of 1-phenylcyclohexene with formaldehyde and acetic acid in the presence of H_2SO_4. The ratio of these and the by-products, *cis*-4-phenyl-4,5-tetramethylene-1,3-dioxane and 3-acetoxymethyl-2-phenylcyclohex-1-ene, vary with the time of reaction and provide some indication of its mechanism.[679,680] The oxa derivative 1,6-bishydroxymethyl-3,7-dioxabicyclo[3.3.1]nonan-9-one is formed by intramolecular cyclization of 3,3,5,5-tetrahydroxymethylpyran-4-one, which is formed in the base-catalyzed condensation of acetone and formaldehyde.[681] 1,4-Bishydroxymethylcyclohexane also cyclizes into 3-oxabicyclo[3.2.2]nonane (**219**) by dehydrating the vapors over aluminum oxide at 325°C.[682]

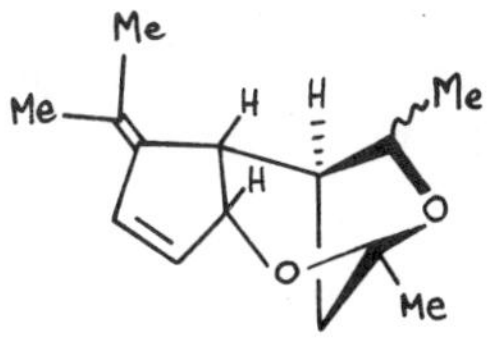

217

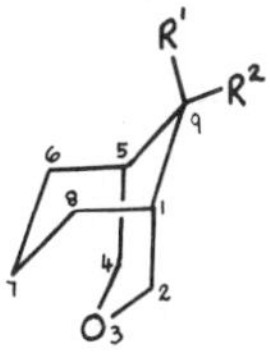

218

219

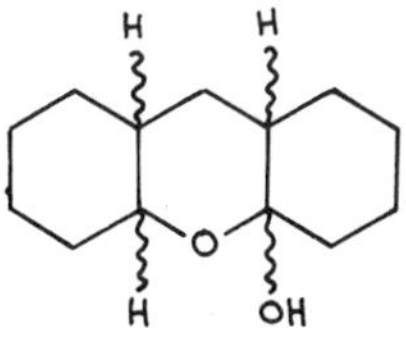

220

C. Reduced Xanthenes

2-(*o*-Hydroxybenzyl) cyclohexanone is in tautomeric equilibrium with 10*a*-hydroxy-1,2,3,4,4*a*,10*a*-hexahydroxanthene which can be reduced to *cis*-1,2,3,4,4*a*,10*a*-hexahydroxanthene.[683] *Trans*-10*a*-hydroxy-1,2,3,4,4*a*,10*a*-hexahydroxanthene is dehydrated to 1,2,3,4-tetrahydroxanthene which disproportionates into *cis*- and *trans*-1,2,3,4,4*a*,10*a*-hexahydroxanthene and xanthene in varying amounts depending on the catalyst used. The ratios and preferred conformations were deduced by ^1H nmr spectroscopy.[684]

Eight 4*a*-hydroxydodecahydroxanthenes (**220**) were identified by Mounet, Huet, and Dreux[685,686] in the reduction of racemic and *meso*-2-(2-oxocyclohexylmethyl)cyclohexanones, and their preferred conformations were discussed.

D. Oxatwistanes, Oxaadamantanes, and Related Compounds

Oxatwistanes are oxygen analogues of azatwistanes, which are described in Part I, Chapter 3, Section X.C. 2,7-Dioxatwistanes and isotwistanes were prepared by Ganter and co-workers[687] starting from the monoepoxide of *cis,cis*-cycloocta-1,5-diene which was converted into 2-iodo-6-hydroxy-(and acetoxy)-1,5-*endo*-oxocyclooctane. The latter 6-hydroxy compound undergoes intramolecular cyclization in pyridine to yield 2,7-dioxatwistane (**221**).[687] *trans*-6-Hydroxy-1,5-*endo*-oxocyclooct-2-ene (**222**) reacts with mercuric acetate to yield the key intermediate **223** which furnishes two iodides (about 4:1), the precursors of 2,7-dioxatwistane (**221,** by cyclization and reduction) and isotwistane (**224,** by reduction) (Scheme 11).[688] Treatment of the iodide **225** with silver acetate gives the hypothetical oxonium salt **226,** which is the source of the acetoxy derivative **225** (R^1=OAc, R^2=H) and 4-acetoxy-2,8-dioxahomotwistbrendane (**227**) by attack of the nucleophile (according to the arrows shown in the oxonium ion, **226**).[689] *trans*-3-

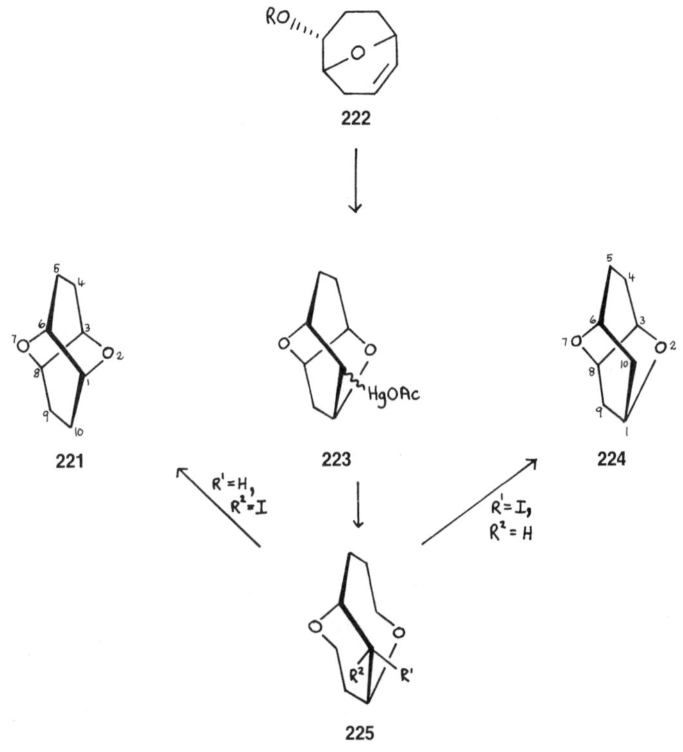

RO
222
HgOAc
221
223
224
R'=H, R²=I
R¹=I, R²=H
R² R'
225

Scheme 11

Hydroxy-6-tosyloxy-1,5-*endo*-oxocyclooctane solvolyzes in N-methanolic KOH to 2,7-dioxaisotwistane.[690] By judicious substitution and elimination both 2,7-dioxatwist-4-ene and 2,7-dioxatwist-4,9-diene were prepared by Ackermann and Ganter.[691] 1*S*,3*R*,6*R*,8*R*-(−)-2,7-Dioxaisotwistane and 1*R*,3*R*,6*R*,8*R*-(−)-2,7-dioxatwistane were prepared by Ackermann, Tobler, and Ganter[692] from (−)-*endo*-2-hydroxy-9-oxabicyclo[3.3.1]non-6-ene (**222,** R = H). The absolute configuration of the latter was correlated with that of *S*-malic acid, and the absolute configurations of the twistane and isotwistane were deduced from a knowledge of the reaction mechanisms. Sulfur analogues of twistanes are described in Chapter 3, Section XIII.

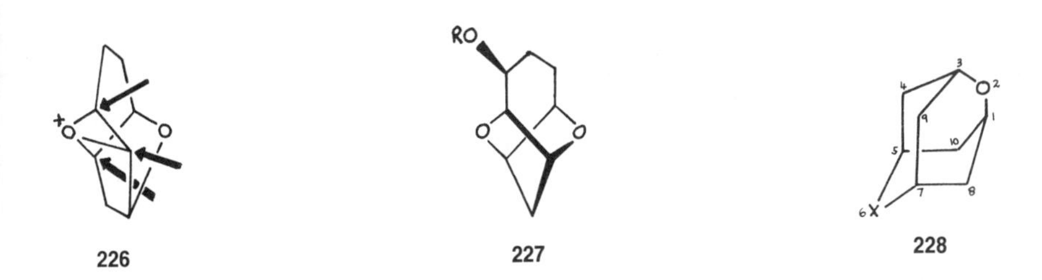

226 **227** **228**

Adamantanes in which a CH_2 group is replaced by an oxygen atom have been reported. 3,6-Dihydroxybicyclo[3.3.1]nonane can be dehydrated to 2-oxaadamantane (**228,** X = $H_2C\lessdot$),[693] and *cis*-3,7-dibromocyclooct-1,5-diene cyclizes to 4,8-dibromo-2,6-dioxaadamantane which, after reduction with LAH, yields 2,6-dioxaadamantane (**228,** X = O).[694] 1,3,5-Trishalomethyl-2,4,9-trioxaadamantane is formed by intramolecular cyclization of trisdiazoacetonyl methane in the presence of halogen acids.[695] Oxa-modified adamantanes, such as 5-oxahomobrendane (**229**) and 5-oxaprotoadamantane (**230**), are obtained from intramolecular cyclization of 5-*endo*-hydroxymethylnorborn-2-ene followed by catalytic reduction, and from intramolecular cyclization of *endo*-6-hydroxymethylbicyclo[3.2.1]oct-2-ene followed by $NaBH_4$ reduction, respectively.[696] Finally, photolysis of cyclohex-2-en-1-ylacetaldehyde in degassed pentane solution gives 7-oxatricyclo[4.3.0.0^{5,8}]nonane (**231**) which is cleaved to bicyclo[3.2.1]octan-*exo*-8-ol with LAH.[697] All the reactions in this subsection clearly demonstrate that an oxygen atom in close proximity to a reactive center in the same molecule results in a facile intramolceular cyclization with the formation of a cyclic ether (see further examples in Section XIV.B).

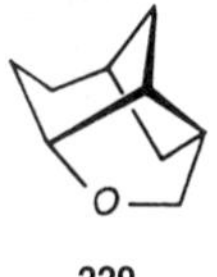

229

230

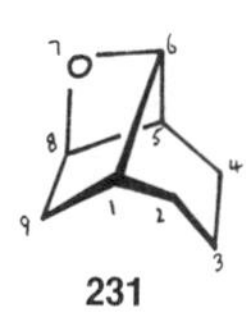

231

XII. DIOXANES

A. 1,2-Dioxanes

1,2-Dioxanes are, essentially, cyclic peroxides.[521] The parent compound, 1,2-dioxane, was prepared by Criegee and Müller[698] from butane-1,4-diol bismethanesulfonate and hydrogen peroxide under carefully controlled conditions. It is most probably in the chair conformation like 3,3,6,6-tetramethyl-1,2-dioxane, which was shown by Calvin and co-workers[699] to exist in the interconverting chair conformations **232** and **233.** The energy barrier for the conversion is unusually high ($\Delta G^{\ddagger}=60.7$ kJ mol^{-1}, $\Delta H^{\ddagger}=$ 77.4 kJ mol^{-1}). 3,6-Bridged 1,2-dioxanes (e.g., **234**) are readily obtainable

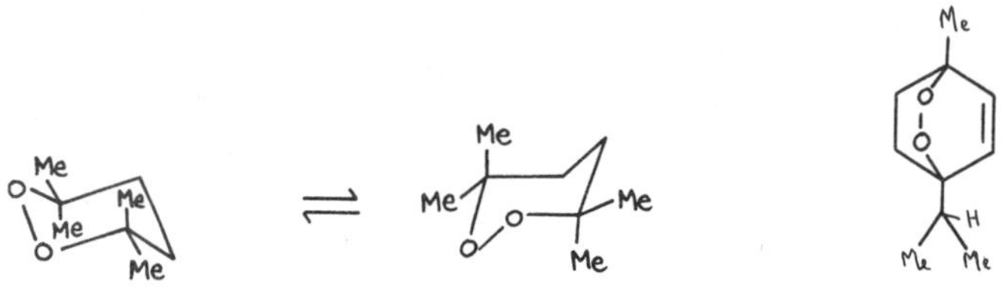

232 ⇌ 233 234

by photochemical oxidation of hexa-1,3-dienes in the presence of a sensitizer.[700] These are rigid structures and their stereochemistry is clear from the known cis addition of oxygen. Some derivatives (e.g., ascaridol, **234**) are naturally occurring but can be prepared by photochemical oxidation (e.g., α-terpinine yields ascaridol).[701]

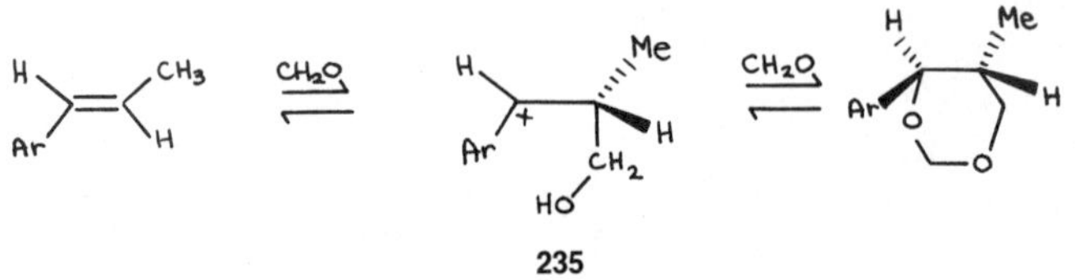

235 (46)

B. 1,3-Dioxanes

In the past decade considerable interest in 1,3-dioxanes arose because these compounds were amenable to extensive studies by nmr spectroscopy. These studies have provided valuable thermodynamic data about substituent groups, intramolecular interactions, interaction with solvents, and the effects of oxygen atoms in conformationally mobile six-membered rings.

1. *Syntheses of 1,3-Dioxanes*

1,3-Dioxanes are cyclic acetals or ketals of propane-1,3-diols and are prepared readily by standard methods from the diol and an aldehyde or ketone, or their respective methyl or ethyl acetals or ketals. With aldehydes or unsymmetrical ketones two isomers are formed, not necessarily in equal amounts, which can be separated by standard methods. Many such syntheses are found in the references used in the discussion of the configuration and conformation of 1,3-dioxanes in this section. A synthesis that is not generally used for preparing 1,3-dioxanes but which is useful for making 4-aryl-substituted 1,3-dioxanes is the Prins reaction. The reaction involves the condensation of a styrene with formaldehyde in the presence of acid. Smissman, Schnettler, and Portoghese[702] condensed *trans*-1-*p*-methoxyphenylprop-1-ene with formaldehyde in the presence of acetic acid and obtained *trans*-4-methoxyphenyl-5-methyl-1,3-dioxane. They suggested that the stereochemistry of the product was the result of thermodynamic control. Similarly, safrole and *trans*-cinnamic acid provided *trans*-4-(3,4-methylenedioxyphenyl)-5-methyl- and *trans*-5-carboxy-4-phenyl-1,3-dioxanes respectively. It was proposed that the two molecules of formaldehyde added in a cis fashion,[702,703] and that a benzyl carbonium ion (**235**) was involved (Equation 46). Detailed investigations with *cis*- and *trans*-1-phenylprop-1-ene, β-bromo- and β-chlorostyrenes,[704] and *trans*-β-deuterostyrene[705] indicated that there was very little stereoselectivity in the reaction. A study of this reaction with 1-(4-hydroxy-3,5-dimethylphenyl)-prop-1-ene in alkaline medium revealed that the steric course can be explained in terms of the conformation of the transition states.[706]

2. *Configuration and Conformation of 1,3-Dioxanes*

1,3-Dioxane is akin to cyclohexane in that physical data are consistent with a preferred chair conformation for the ring.[485,531,592,595] The ring, however, as evidenced by X-ray measurements,[707–709] is slightly puckered in the O–C–O region and flattened in the C–C–C region because the C–O bonds are shorter than the C–C bonds. This "pinching" of the molecule shows up in the J values of the ^{1}H nmr spectra[710–712] and its conformational properties (see below). A very large number of substituted isomeric 1,3-dioxanes have been prepared and separated, and invariably the relative configurations of substituents were deduced from chemical shift and coupling constant data obtained from ^{1}H nmr spectra[713] (see most references below). Occasionally X-ray measurements,[707–709] ultrasonic relaxation,[714] dipole moment (see references below), and, more recently, ^{13}C nmr methods[715,716] were used to determine the configuration and conformation of 1,3-dioxanes. The ease with which 1,3-diols can be converted into the comparatively more rigid 1,3-dioxanes affords a convenient and reliable method for determining the relative configuration of substituents in the original diols by nmr spectroscopy—coupling constants provide more meaningful information in the dioxanes than in the diols.[717,718,718a] Calorimetric studies have also provided valuable information about the configuration of methyl substituted 1,3-dioxanes.[711,719]

Conformational analysis of 1,3-dioxanes has been a fruitful field of study and a considerable amount of data have accumulated. Space does not permit a full discussion of the topic, particularly since excellent reviews which include discussions on 1,3-dioxanes have been published.[531,592,594,596,599,600,720,721] References to studies of conformational equilibria of various groups in the different positions in the ring, however, are provided for completeness and to give leads into the literature.

Anderson, Riddell, and Robinson[722] questioned the value of 9.2 kJ mol^{-1} for the enthalpy difference between the chair and boat (chair → 1,4-twist) forms of 1,3-dioxane, and they showed from a study of the coupling constants of 4-phenyl-, 2,4-diphenyl-, 2,4,6-triphenyl-, 2-*t*-butyl-, and 2-*t*-butyl-2-methyl-1,3-dioxanes and related compounds that it is rather low compared with cyclohexane. Pihlaja[723] obtained $\Delta G°$ (24.7 kJ mol^{-1}) and $\Delta H°$ (28.5 kJ mol^{-1}) values for the chair-boat interconversion in 1,3-dioxane, from heat-of-combustion measurements, which are higher and consistent with previous prediction. Eliel and co-workers[724] studied the acid-catalyzed equilibria in several four-component systems (e.g., Equation 47) and concluded that the chair-twist enthalpy difference in 1,3-dioxane is larger than the value for cyclohexane (24.7 kJ mol^{-1}). The higher value (e.g., 31 kJ mol^{-1} for 2,2-*trans*-4,6-tetramethyl-1,3-dioxane) is supported by values for 1,3-dioxane from microcalorimetric methods (37.3 kJ mol^{-1})[725] and mass spectrometric measurements (35.6 kJ mol^{-1})[726] (see twist conformations below). Wood and collaborators[727] tabulated the barriers to ring inversion of oxa-, thiaoxa-, and thiacyclohexanes and the following data for 1,3-dioxanes were included: 1,3-dioxane and 2,2-dimethyl-, 4,4-dimethyl-5,5-dimethyl-, and 2,2,4,4-tetramethyl-1,3-dioxanes, and cyclohexane have $\Delta G^{\ddagger}$ values of 41.4, 33.5, 36.0, 46.5, 37.2, and 43.9 kJ mol^{-1} respectively.

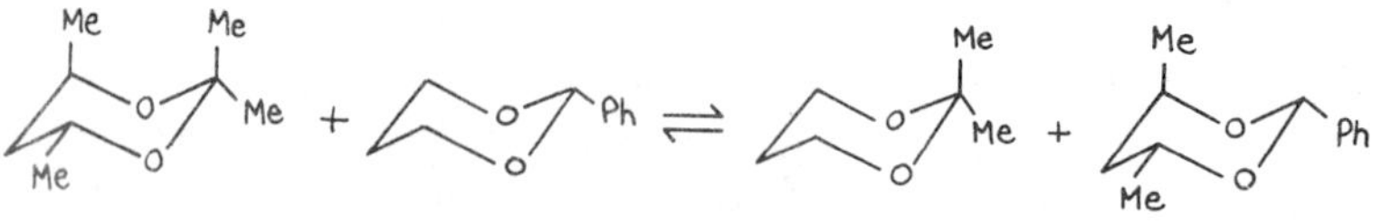

(47)

Interconversion of chair conformers alters the axial and equatorial substituents, but their relative configurations remain unchanged. In presence of an acid, however, the 2-carbon atom of 1,3-dioxanes can undergo inversion irrespective of the rest of the molecule. One of the oxygen atoms is protonated, a C–O bond is broken, and C–2 can undergo inversion and cyclize, or cyclize without inversion, to furnish the thermodynamically more stable diastereomer. Such acid-catalyzed equilibrations were used extensively to determine conformational preferences and obtain thermodynamic parameters for substituents. The shorter C–O bond, compared with the C–C bond, causes an axial substituent at C–2 in 1,3-dioxane to be closer to a syn axial proton, or substituent, than similar substituents in cyclohexane. This is an important factor when the 2-substituents in 1,3-dioxanes are compared with those in cyclohexane analogues.

A large number of 2-substituted 1,3-dioxanes were examined.[472,500,728–731] An equatorial 2-phenyl group[710,732,733] (and 2,2′-pyridyl)[734] acts as an anchoring group (ananchomeric) and orients itself in a "parallel" conformation in which the phenyl group bisects the dioxane ring (**236**).[709,733] An axial 2-phenyl group, on the other hand, orients itself "perpendicular" to the C_2–C_5 axis (**237**).[708] The effect of C–2 substituents and the free energies involved were discussed by Eliel recently.[595] Equilibration studies (BF_3 catalyzed) of 2-alkyl-5-*t*-butyl-1,3-dioxanes gave free energy values which for the first time revealed an example in which a *t*-butyl group is axial in a six-membered ring.[735,736] This information provided further evidence for the smaller space requirements of a lone pair of electrons compared with a hydrogen atom. The anomeric effect in 1,3-dioxanes is operative and an axial alkoxy group at C–2 is highly favored.[737,738,739] 2-Ethenyl substituents also prefer the axial conformation, but 2-CH_2Cl, 2-CH_2Br, and 2-carboxy groups show little equatorial-axial preference and provide evidence for a reverse anomeric effect.[740] By cooling a solution of 2-*t*-butyl-2-methyl-1,3-dioxane it was possible to demonstrate restricted rotation of the *t*-butyl group and to deduce the free energy involved ($\Delta G^{\ddagger} = 36.4$ kJ mol^{-1}).[501] The anomeric effect and the generalized anomeric effect (rabbit ear effect) in 1,3-dioxanes has been adequately reviewed.[485,594–596,600,720,721]

A large number of papers has been published on the effect of substituents at C–3, C–4, and C–5 on equilibria, on the conformational preference of substituents, and on the acid-catalyzed equilibration of diastereomers in which C–2 is substituted. The anomeric effect is, no doubt, operating for substituents at C–4, and is exemplified by 4-CF_3 and 4-CHF_2 groups which prefer an axial orientation,[741] unlike similar substituents at C–5.[742] Halogen atoms at C–5 are unusual, and in the acid-catalyzed equilibrium of 2-isopropyl-1,3-dioxanes a 5-F substituent prefers to orient itself axially in contrast to a 5-Cl or 5-Br atom; 5,5-difluoro substituents lower the ring inversion barrier in comparison with the 5-*gem*-dimethyl derivative.[743] In 4-vinyl-5-methyl- and 4-vinyl-6-methyl-1,3-dioxanes the vinyl group favors the axial conformation.[744] Considerable data on 1,3-dioxanes with alkyl and aryl groups at C–4, C–5, and/or C–6 have accumulated, and thermodynamic data were reported.[710,745–751] Conformational equilibria (BF_3) of 2-isopropyl (as ananchomeric substituent) 1,3-dioxanes, in which two different substituents are present at C–5, were measured in order to compare the relative conformational characteristics of the substituents, for example, alkyl, alkoxy, nitro, methoxymethyl, and carboxymethyl. The free energies of these substituents were determined but were not additive.[752] 5,5-Disubstituted and 5,5-spiro-1,3-dioxanes were examined, and the spiro substituent was found to cause a marked lowering of the ring inversion barrier.[753–756] The origin for the lowering of the ring inversion barriers was discussed by Anteunis and co-workers,[757] who examined 5,5-diisopropyl-1,3-dioxane and showed that the topology of the isopropyl groups plays an important role.[757] 5-Ethenyl-, 5-ethynyl- and 5-isopropyl-5-methyl-2,2-disubstituted 1,3-dioxanes were also studied, and these groups were found to prefer an equatorial conformation with the 5-methyl groups axial. Solvent effects were investigated, but the results were difficult to interpret.[758]

The free energy difference for equatorial conformational preference between a methyl and phenyl group in *trans*-6-methyl-4-phenyl-1,3-dioxane is rather small.[759] Generally the benzene ring in the axial 5-phenyl derivative bisects the dioxane ring (**238**). The equatorial 5-phenyl group, on the other hand, assumes a parallel orientation (**239**) or a bisecting orientation (**240**) depending on the size of the other (gem) alkyl substituent. In 5,5-diphenyl-1,3-dioxane the two benzene rings orient themselves in positions of least nonbonded interaction (**241**),[760] and the phenyl group in a series of 5-nitro-5-phenyl-1,3-dioxanes prefers the equatorial conformation.[761]

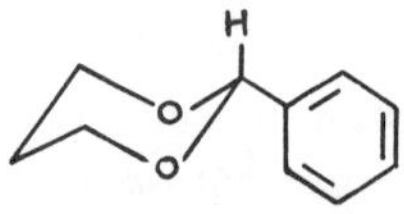

H O O

236

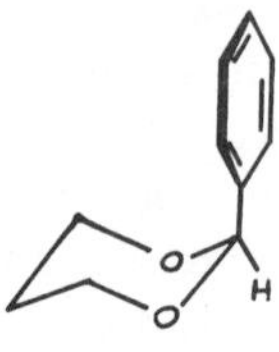

O H O

237

The base-catalyzed decarboxylation of 5,5-dicarboxy-2-isopropyl-1,3-dioxane yields a mixture of *cis*- and *trans*-5-carboxy-2-isopropyl(*eq*)-1,3-dioxanes in which the ratio is not far from one, in 2,6-dimethylpyridine, but varies with the solvent used.[762] It is interesting that the 5-substituent in 5-trimethylammonio-1,3-dioxane is almost exclusively axial, and the $C–N^+$ bond prefers to orient itself parallel to the lobes of the electron pairs on the oxygen atoms (**242**).[763] A dipolar effect is also operating in 5-methyl-

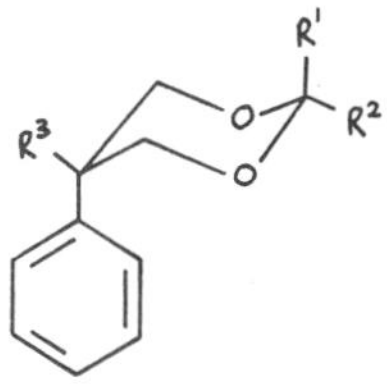

238

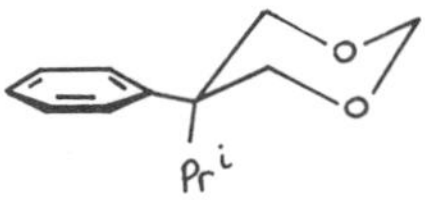

239

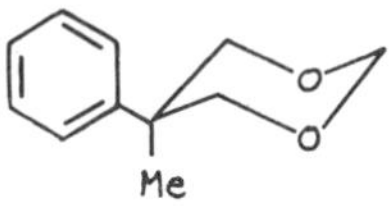

240

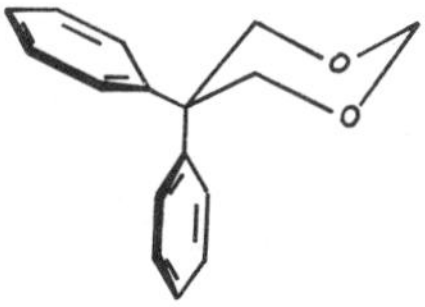

241

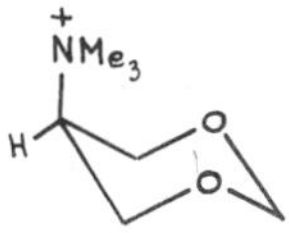

242

sulfinyl- and 5-methylsulfonyl-2-isopropyl-1,3-dioxanes because the groups prefer to be axial (e.g., **243**) compared with the 5-methylthio derivative in which the SMe group prefers to be equatorial.[764]

The effect of solvent on conformational equilibria can be strong or negligible and can provide valuable information about the properties of substituents.[765–768] These effects have been discussed previously in some detail.[592,595]

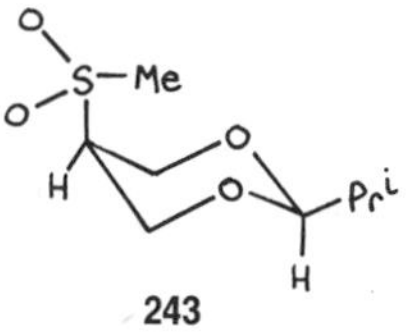

243

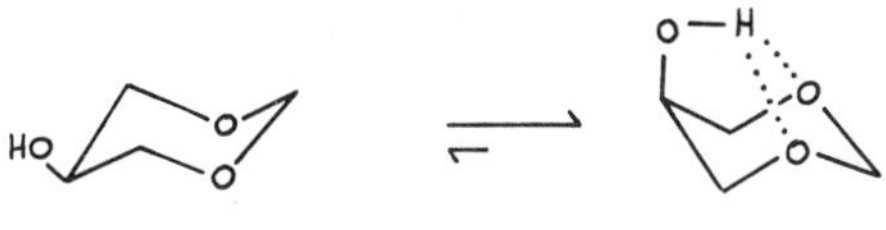

244

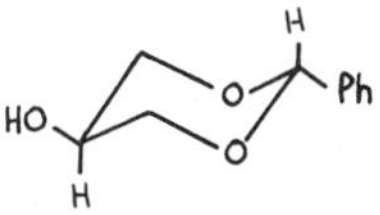

245

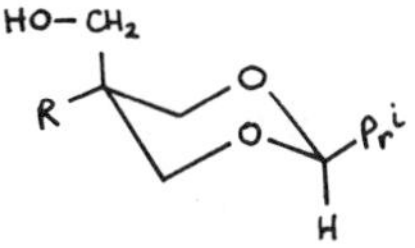

246

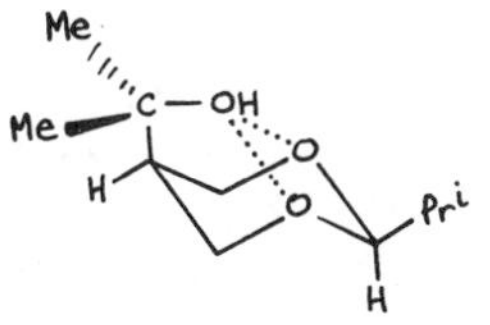

247

A hydroxyl group on the carbon skeleton of 1,3-dioxanes exhibits considerable hydrogen bonding when in the appropriate position. A hydroxyl group at C-5 is strongly hydrogen bonded with the ring oxygen atoms when it is allowed to assume an axial conformation (**244**), but not otherwise (e.g., **245**).[352,475,476,503,769,770] The pattern of intra- versus intermolecular hydrogen bonding in variously substituted 5-hydroxy-1,3-dioxanes has been examined and has provided valuable information regarding the preferred conformations.[771,772] Intramolecular hydrogen bonding was not observed in *cis*-2-isopropyl-5-hydroxymethyl-1,3-dioxane (**246,** R = H) or its 5-methyl derivative (**246,** R = Me), but weak hydrogen bonding was found in *cis*-2-isopropyl-5-(1-hydroxy-1-methylethyl)-1,3-dioxane (**247**). Hydrogen bonding in this case is caused by some eclipsing dipolar process.[773–776] 2- and 5-Hydroxymethyl-1,3-dioxanes were examined by Gelas and Peterquin, and while confirming the previous results with the 5-hydroxymethyl derivatives, they also showed that almost complete hydrogen bonding occurred in the 2-hydroxyalkyl group.[777] Intramolecular hydrogen bonding in 1,3-dioxanes has been discussed previously.[531,592,595]

Mention was made earlier of the enthalpy involved in the conversion chair → twist in 1,3-dioxanes and a few words on non-chair, twisted 1,3-dioxanes are warranted here because most of the studies on the non-chair oxygen heterocycles have been in this series. The presence of syn diaxial alkyl groups in 1,3-dioxanes cause the chair conformation to alter to either (or both) of two twist-boat conformations, the 1,4-twist C_1 (**248**) or the 2,5-twist C_2 (**249**) depending on the substituents[728,778] (compare with twist-boat piperidines in Part I, Chapter 3, Section II.C). Twist-boat conformations are therefore common among polysubstituted 1,3-dioxanes in which 1,3-syn diaxial interactions would be otherwise large if the ring had a chair conformation. There are ^{1}H nmr,[723,768,778–782] ^{13}C nmr,[783] thermochemical,[723,784,785] molecular rotation,[786] and ultrasonic relaxation[787] data to support these twist-boat conformations. A few pertinent examples which exist in twist-boat conformations are 2,2-*t*-4,6-tetramethyl-,[723] *r*-2-*trans*-6-dimethyl-*trans*-4-*t*-butyl-, *r*-2-*cis*-6-dimethyl-*trans*-4-*t*-butyl- (Equation 48, the two twist-boat forms are observed),[779] 2,2,4,4,6-pentamethyl-,[785] and 2,2,4-trimethyl-*trans*-6-*t*-butyl-1,3-dioxanes.[781] An X-ray study of 2-(4-bromophenyl)-*r*-2,4,4-*cis*-6-tetramethyl-1,3-dioxane revealed that, in the crystal, the molecule is in a deformed chair with the phenyl ring axial and perpendicular to the dioxane ring, and not in a twist-boat conformation.[788] The stereochemistry of non-chair conformations in six-membered rings was reviewed recently by Kellie and Riddell.[789]

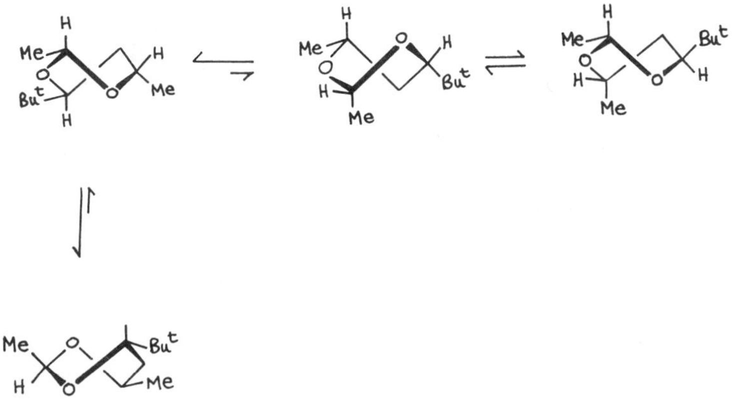

(48)

Detailed nmr spectral and kinetic data on several substituted 1,3-dioxan-2-ones demonstrated that the ring is in a chair conformation and that it is slightly flattened at the O–C–O end.[790] The ^{1}H nmr spectra of a large number of 1,3-dioxane-4,6-diones were examined and were shown to exist in rapid interconverting conformers which are quite flattened in the –O–C(=O)–C–C(=O)–O–region **250** and **251.** As a solvent, benzene orients itself over the acylal ring and is similar to a benzene ring attached to C–5 by a polymethylene bridge (**252**).[791]

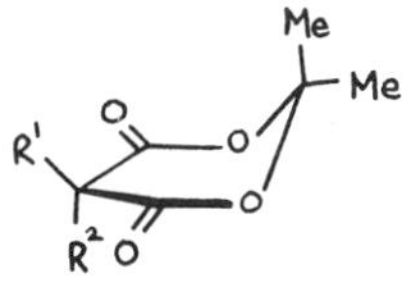

250 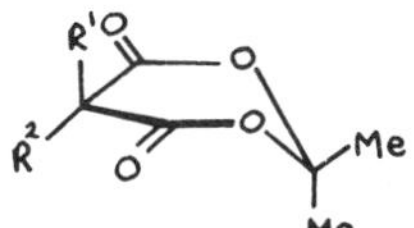 ⇌ 251 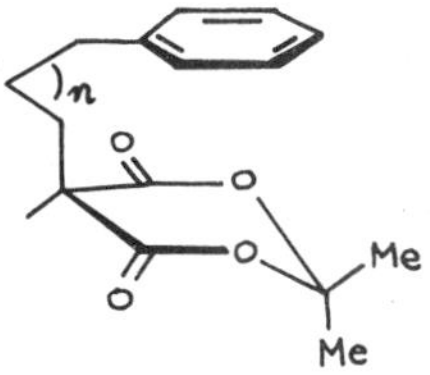 252

3. Reactions of 1,3-Dioxanes

The reduction of 2-phenyl-1,3-dioxan-5-one with LAH and the addition of Grignard reagents to the 5-carbonyl group proceed with high stereoselectivity to yield the *cis*-5-hydroxy-2-phenyl-1,3-dioxane derivatives. These alcohols epimerize almost completely in the presence of acid, but they also undergo a ring contraction to an unequal mixture of epimeric 4-hydroxymethyl-2-phenyl-1,3-dioxalanes (Equation 49).[792,793] The reduction of *cis*-

and *trans*-2-methoxy-4,4,6-trimethyl- and *r*-2-methoxy-*cis*-4,6-dimethyl-1,3-dioxanes with LAD or AlD_3 provided predominantly the 2-deuterio (replacing 2-OMe) with retention of configuration. Reduction with $AlCl_2D$, on the other hand, gave predominantly the compound with an axial 2-D in all cases. Several other 2-methoxy-1,3-dioxanes gave similar results.[794] The displacement of an axial 2-OMe group in 1,3-dioxane by Grignard reagents proceeds with high retention of configuration, whereas an equatorial 2-OMe group is barely reactive. The stereospecificity is undoubtedly controlled by the overlap of the orbitals from the ring oxygen atoms (Equation 50).[795] Catalytic reduction of 2-methyl-, *t*-butyl-, and phenyl-5-methylene-1,3-dioxanes is highly stereoselective and provides more than 93% of the *cis*-2-methyl-, *t*-butyl-, and phenyl-5-methyl-1,3-dioxanes respectively.[796]

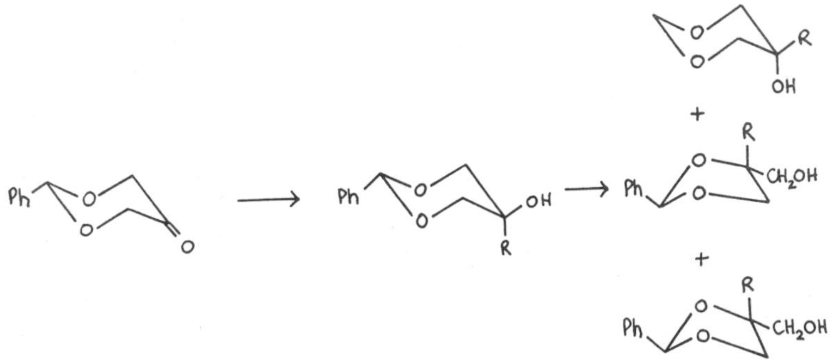

(49)

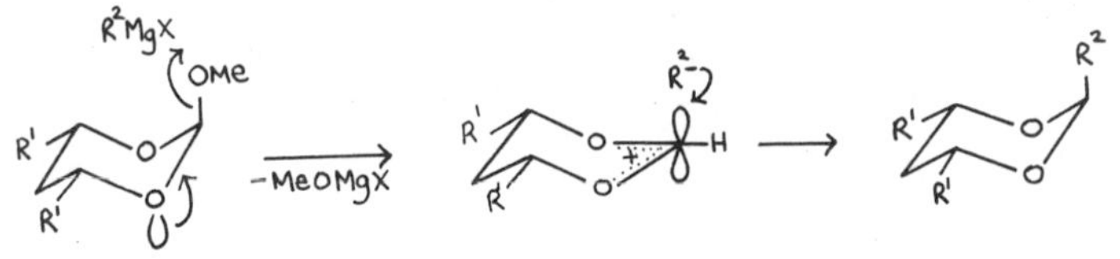

(50)

4*S*,5*S*(+)-5-Amino-2,2-dimethyl-4-phenyl-1,3-dioxane (**253**, available in quantities as an intermediate in the synthesis of chloramphenicol) is a valuable reagent for the asymmetric synthesis of *S*-α-methylamino acids. It undergoes a Strecker synthesis with benzylmethyl ketones in the presence of cyanide to yield the α-aminonitrile with almost complete steric induction (Equation 51). The nitrile is then treated with acid, whereby it rearranges to a 6-hydroxymethyl-5-phenyl-1,4-oxazin-3-one which furnishes the *S*-α-benzyl-α-methyl-α-amino acid on hydrogenolysis.[797] When alkylmethyl ketones were used an interesting result was obtained. By using the 4*S*,5*S*-dioxane **253** and a methyl ketone with an even number of carbon atoms in the alkyl chain, the α-methyl-α-amino acids formed have *S* configurations, but when there are an even number of carbon atoms the *R*-amino acid is formed. In branched alkyl chains the number of carbon atoms in the longest chain dictates the stereochemistry. The stereochemical course in these examples is still not clear.[798]

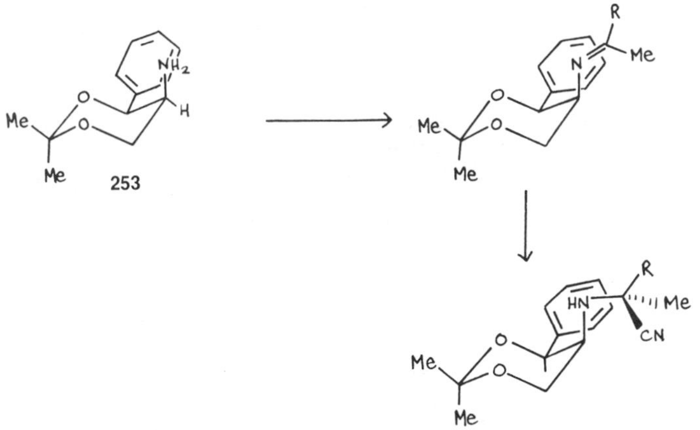

253

(51)

4. *Condensed 1,3-Dioxanes*

The Prins reaction, mentioned in Section B.1, has been extended to the synthesis of condensed 1,3-dioxanes. 1,2-Dihydro-7-methoxynaphthalene and formaldehyde in the presence of acid yield 8-methoxy-1,2,3,4,4*a*,5,6,-10*b*-octahydro-*cis*-1,3-dioxaphenanthrene (**254**) stereospecifically.[702] Indene, 1,2-dihydro-1-thianaphthalene,[702] and *N*-methyl-, *N*-benzyl-, and *N*-*t*-butyl-4-phenyl-1,2,3,6-tetrahydropyridines[799] similarly provide the corresponding cis-fused 1,3-dioxanes. The products from the last three reactions have, in solution, the general structure **255** with the oxygen near the bridgehead carbon on the "inside" of the molecule. Blomquist and Wolinsky[799a] isolated several products including *trans*-4,5-tetramethylene-1,3-dioxane, but not the cis isomer, from a Prins reaction with cyclohexene. The cis isomer was formed from *cis*-2-hydroxymethylcyclohexanol and formaldehyde. Coryn and Anteunis[800] used the Prins reaction with cyclic olefins and formaldehyde to prepare a number of condensed 1,3-dioxanes and analyzed their products carefully. They demonstrated that the primary reaction occurred with various specificities and that all the condensed 1,3-dioxanes formed were not necessarily cis-fused; the regiospecificity varied with the olefin. Thus cyclopentene gave *cis*- and *trans*-4,5-trimethylene-1,3-dioxane (14:86); 4-*t*-butylcyclohexene gave the two *trans*-4,5-tetramethylene-1,3-dioxanes **256** and **257**; 1,3-cycloheptadiene gave only *trans*-8,10-dioxabicyclo[5.4.0]undecene; and cyclooctene gave exclusively *cis*-4,5-hexamethylene-1,3-dioxane among several other reactions that were described. General rules for the stereospecificity of this reaction cannot,

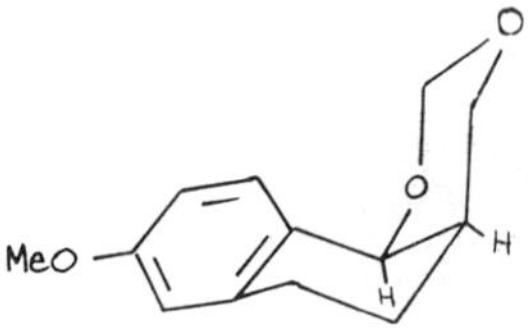

254 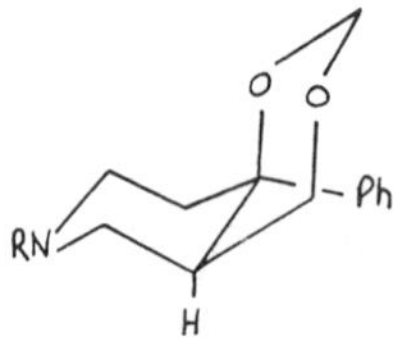 **255**

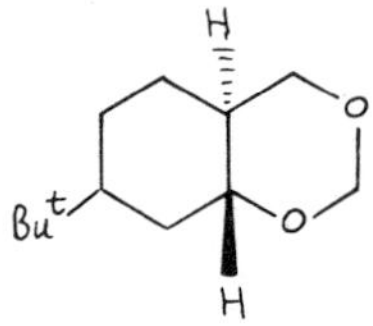

256 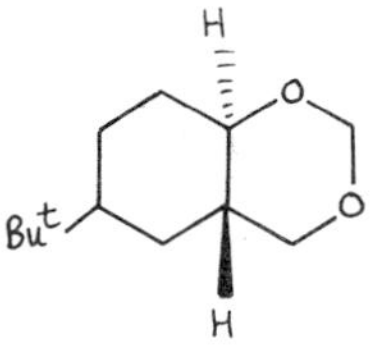 **257**

therefore, be laid down. The obvious method for the preparation of authentic condensed 1,3-dioxanes is from 1,3-diols of known configuration. Authentic *cis*- and *trans*-2,2-dimethyl-4,5-tetramethylene-1,3-dioxanes are formed from *cis*- and *trans*-2-hydroxymethylcyclohexanols by condensation with acetone (see above).[801] An interesting cyclization was observed when *cis*-1-acetoxy-2-tosyloxymethylcyclohexane was treated with potassium acetate in ethanol because the acetoxy carbonyl group was attacked, intramolecularly, to form 2-ethoxy-2-methyl-*cis*-4,5-tetramethylene-1,3-dioxane. This gave 2-methyl-*cis*-4,5-tetramethylene-1,3-dioxenium tetrafluoroborate on treatment with $BF_3 \cdot Et_2O$, which was also formed when 2-methyl-*cis*-4,5-tetramethylene-1,3-dioxane was reacted with trityl tetrafluoroborate. The O–C(Me)–O portion of this dioxenium ion is most probably planar (Equation 52).[802]

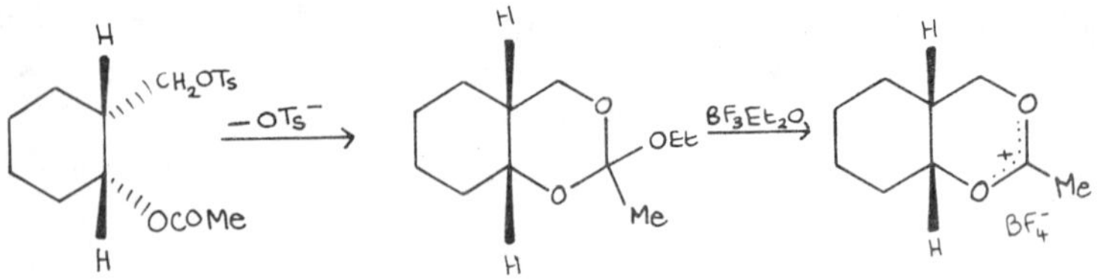

(52)

Hydroformylation of cyclohexene oxide under stoichiometric conditions gave a dimer, **258,** of the expected 1-formyl-2-hydroxymethylcyclohexane in which the stereochemistry was all-trans. The dimer exhibits ring-chain tautomerism with the open-chain hemiacetal aldehyde **259.**[803] Propionaldehyde and butyraldehyde reacted with 1,3,5-trioxane (formaldehyde trimer) in the presence of BF_3 to form 4*a*-methyl- and 4*a*-ethyl-1,3,6,8-tetraoxanaphthalene (**260**) respectively. The stereochemistry at the bridgehead carbon atoms, however, was not established.[804] In a study on the synthesis of brevicomin and related bicyclo[3.2.1] systems, Wasserman and Barber[805] heated 1-(acetoxy)-1-(2,3-epoxypropyl)cyclopropane to 100°C and obtained 3-spirodimethylene-2,7,8-trioxa[3.2.1]octane (**261,** R^1, R^2=$-CH_2-CH_2-$). The analogous 2-acetoxy-4,5-epoxypentane and 2-acetoxy-4,5-epoxy-2-methylpentane behaved similarly and provided **261** (R^1=H, R^2=Me or R^1=Me, R^2=H) and **261** (R^1=R^2=Me) respectively.

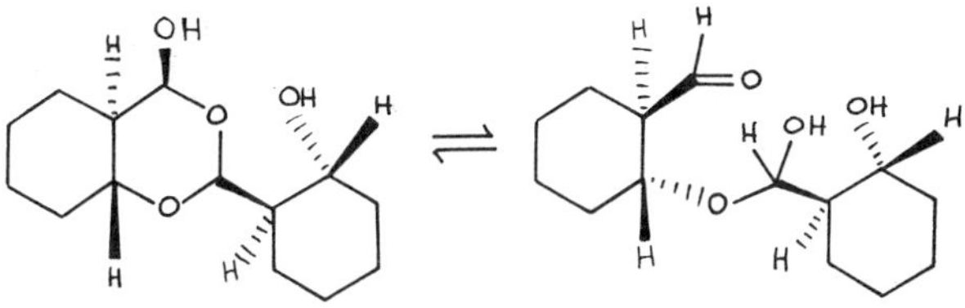

258

259

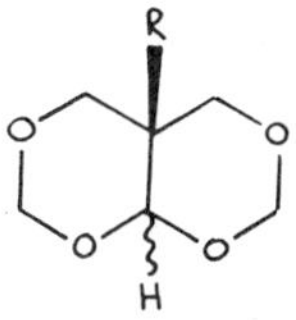

260

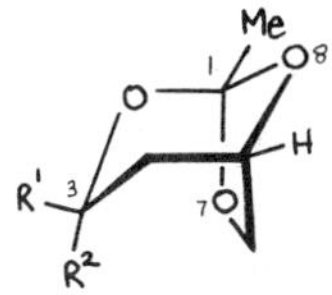

261

Anteunis and his school have devoted much effort in investigating the conformational equilibria of 4,5-polymethylene-1,3-dioxanes. *cis*-4,5-Tetramethylene-1,3-dioxane exists in two stable conformations: one has O–1 in the axial position and is termed O-inside (**262**), and the other has the oxygen atom in the equatorial position and is termed the O-outside conformer (**263**). The conformers can be readily distinguished by ^1H nmr spectroscopy. In the parent compound and several substituted derivatives the O-inside conformer (**262**) predominates and is probably a consequence of fewer 1,4-syn axial interactions in conformer **262** compared with conformer **263** because of the small space demanding properties of the oxygen lone pair of electrons compared with the axial hydrogen atoms. The free energy difference in the parent compound is of the order of 10 kJ mol^{-1}.[806,807] This value varies with the position and stereochemical configuration of substituents relative to the bridgehead carbon atoms. Alkyl substituents, which in general prefer an equatorial conformation in a labile system, will stabilize or destabilize the conformer **262** depending on whether they assume an equatorial or axial orientation, respectively, in this conformer.[807] All four 4-*t*-butyl-5,6-tetramethylene-1,3-dioxanes have been prepared and examined. The trans isomers are in their similar stable chair-chair conformations, but the cis epimers are in the two different conformations shown in formulas **264** and **265** and are such that the *t*-butyl

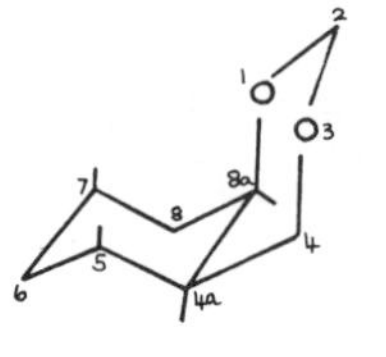

262

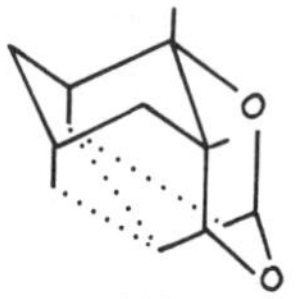

263

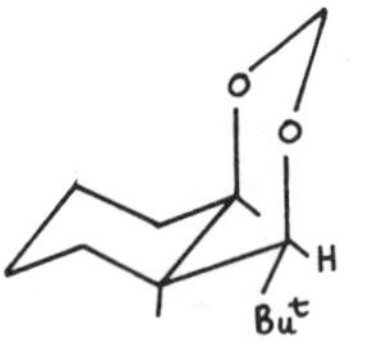

264

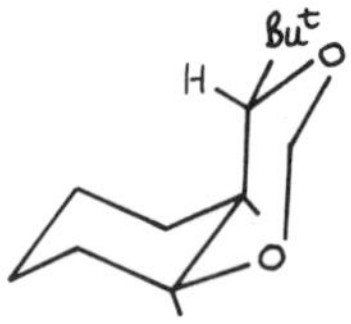

265

group is equatorial in each case. ¹H nmr data on **265,** however, indicate that the chairs are distorted.[808] The ratios of O-inside to O-outside for 5,6-trimethylene-, 5,6-tetramethylene-, 5,6-pentamethylene-, and 5,6-hexamethylene-*cis*-2,4-dimethyl-1,3-dioxanes, among others studied, are 2.15, 3.27, 0.17, and 0.46 respectively, and are a clear reflection of the non-bonded interactions involved.[809] *cis*-10,12-Dioxabicyclo[6.4.0]dodec-2-ene favors the conformation **266,** with the axial vinyl group, over the conformation **267,** in contrast to the isomeric *cis*-dodec-3-ene in which the conformation **268** is preferred.[810]

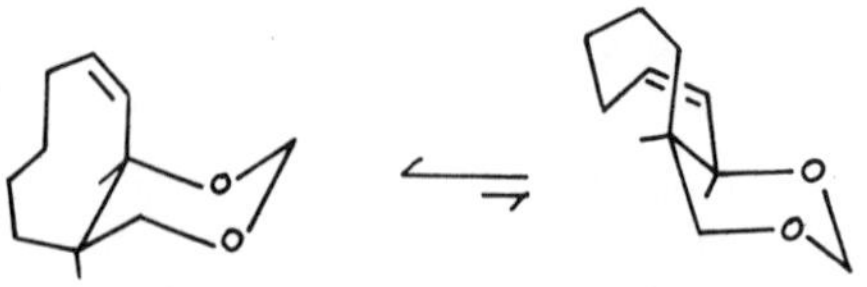

266

267

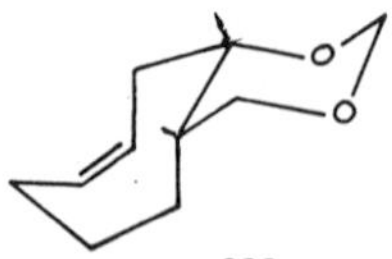

268

Anteunis and co-workers investigated the conformation of a variety of condensed 1,3-dioxanes, including 4-substituted 1,2,3,4-tetrahydro-1,3-dioxanaphthalenes, and assessed the preferred form from $^2J(O–CH_2–O)$ values. The latter compounds were thus shown to exist in a half-chair conformation (**269**).[808]

Bishop, Sutton, Robinson, and Pumphrey[811] measured the dipole moments of the 1,3-dioxanes **270** (R=*p*-Cl–C_6H_4– and *p*-NO_2–C_6H_4–) and concluded that the 1,3-dioxane ring is in a preferred boat conformation; and by inference, using ^1H nmr data, they concluded that 2,4-dioxabicyclo-[3.3.1]nonane is in the boat-chair form. Swaelens and Anteunis[812] examined the ^1H nmr spectra of 2,4-dioxabicyclo[3.3.1]nonane by multiple radiation and put forward arguments in favor of two flattened chair conformations (**271**). The spectra, however, were unchanged on cooling, inferring that the system is either very rigid (unlikely) or very mobile. The spectra of 3-methyl-2,4-dioxaadamantane could provide valuable information in this connection.[813]

269

270

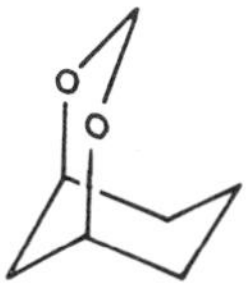

271

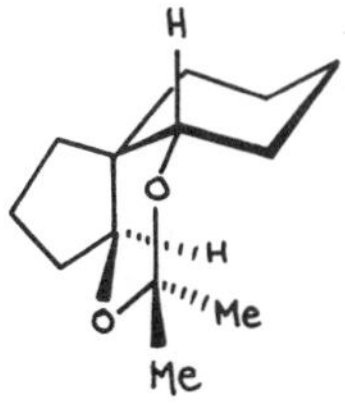

272

The stereochemical relationships of the three diastereomeric spiro-compounds, 2,2-dimethyl-4,5-trimethylene-5,6-tetramethylene-1,3-dioxanes (**272**, only one shown), were determined by Hardegger, Maeder, Semarne, and Cram.[814]

C. 1,4-Dioxanes

1,4-Dioxanes are more stable chemically than the isomeric 1,2- and 1,3-dioxanes because they are true cyclic ethers. Although the 1,4-disposition of the oxygen atoms introduces elements of symmetry in this nonplanar ring system, two substituents in the ring carbon atoms are a source of geometrical isomerism. Identical substituents at the 2,3- and 2,5-positions give two possible 1,4-dioxanes in each case. The *cis*-2,3-disubstituted 1,4-dioxanes are meso forms, and the trans isomers are racemic. On the other hand, the *cis*-2,5-disubstituted 1,4-dioxanes are racemic, and the trans isomers are meso (compare with disubstituted piperazines in Part I, Chapter 3, Section V.B). Several of these were synthesized and their configurations and conformational mobility were the subjects of several investigations.

1. *Syntheses of 1,4-Dioxanes*

Catalytic reduction of 2,3-dehydro-2,3-diphenyl-1,4-dioxane (Pt_2O–AcOH–Ac_2O) stereospecifically gave a 2,3-diphenyl-1,4-dioxane (m.p. 136°C) which was assigned the cis configuration. Reduction with sodium and pentan-2-ol, on the other hand, gave this dioxane (18%) together with its isomer (34%, m.p. 48°C).[815] Dipole moment measurements, however, later showed that the higher melting isomer indeed had the trans configuration.[816] Complete trans stereospecificity in catalytic reduction is unusual, and clearly the *cis*-2,3-diphenyl-1,4-dioxane was originally formed and isomerized in the presence of acid. 2,3-Dehydro-1,4-dioxane reacts with iodobenzene dichloride to yield almost exclusively *trans*-2,3-dichloro-1,4-dioxane, but direct chlorination provided a cis-trans mixture of 2,3-dichlorodioxanes in which the cis isomer was predominant. The trans isomer was apparently resolved with brucine, whereas the cis isomer had epimerized slightly under the reaction conditions and gave a little of the optically active *trans*-2,3-dichloro-1,4-dioxane. The cis isomer hydrolyzed 14 times faster than the trans isomer in 50% aqueous dioxane, and this may be a reflection of the 2,3-diaxial configuration of the halogens in the trans isomer (see Section XII.C.2).[817]

Phenylchlorocarbene, generated from benzaldichloride and *t*-BuOK, added across the double bond of 2,3-dehydro-1,4-dioxane in a regio- and stereospecific reaction yielding 45% of *endo*-7-chloro-*exo*-7-phenyl-2,5-dioxabicyclo[4.1.0]heptane (**273**). The stereochemistry was confirmed by an X-ray analysis.[818]

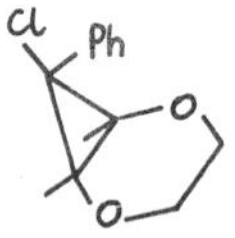

273 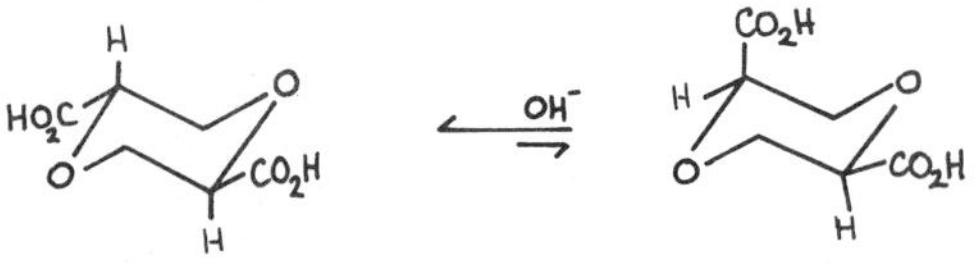 **274** **275**

Diepiiodohydrin, prepared by the reaction of mercuric oxide with allyl alcohol followed by potassium iodide, was shown to be *trans*-2,5-bisiodomethyl-1,4-dioxane. It was converted in several steps into *trans*-2,5-dicarboxy-1,4-dioxane (**274**) which was epimerized into a mixture of cis (**275**) and trans (**274**) isomers. The trans isomer was more stable than the cis isomer to the epimerization conditions and, by analogy with the corresponding cyclohexanedicarboxylic acids, was assigned this configuration. This was further confirmed by partial optical resolution of *cis*-2,5-dicarboxy-1,4-dioxane by way of the brucine salts and by the generally higher melting points of the trans derivatives compared with the cis isomers. Diglycerol diacetate obtained from glycerol was shown to be *trans*-2,5-bisacetoxymethyl-1,4-dioxane by correlation with the above series.[819] The reaction of morpholine (and other secondary amines) with epichlorohydrin, followed by treatment with alkali, gave 2,5-bismorpholinomethyl(and bis-dialkylaminomethyl)-1,4-dioxane, which was identical with the product from the condensation of *trans*-2,5-bisiodomethyl-1,4-dioxane and thus assigned the stereochemistry to these amines.[820] In contrast with the 2,5-dicarboxylic acids, the *cis*-2,6-dicarboxylic acid is more stable than the trans acid.[821] This was confirmed in the synthesis of *cis*- and *trans*-2,6-bis-acetoxymercurimethyl-1,4-dioxanes from the reaction of diallyl ether and aqueous $Hg(OAc)_2$. Analysis of the products after 2.5 hr indicated a cis to trans ratio of 1.8, but on completion of reaction (48 hr) the ratio was 16 (Equation 53).[822] This reaction was extended to monoalkyl polyol ethers

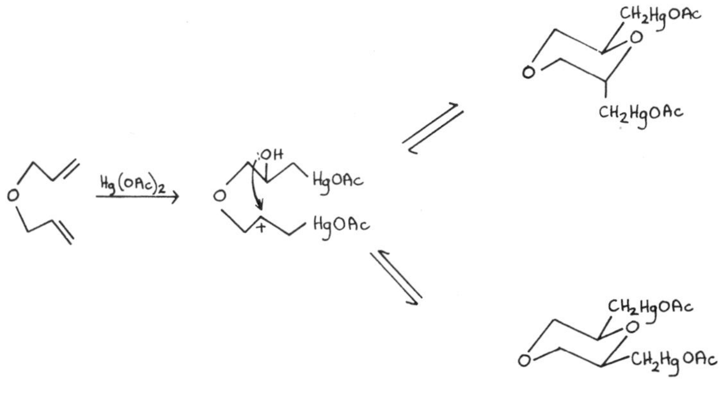

(53)

derived from carbohydrates, for example, 3-allyloxy-1,2,4,5,6-pentahydroxyhexane (3-allyl-D-mannitol), and gave optically active 1,4-dioxanes directly, for example, 2,3-bis-(1,2-dihydroxyethyl)-6-iodomethyl-1,4-dioxane, $[\alpha]_D + 10.8°$.[823] These reactions are essentially intramolecular but have also been carried out intermolecularly. Thus dienes and glycol condense in the presence of mercuric salts to form a mixture of *cis*- and *trans*-2,3-bismercurimethyl-1,4-dioxanes in which the latter isomer predominates.

Butadiene and ethylene glycol in aqueous mercuric nitrate gave a mixture of mercurimethyldioxanes which were converted into the respective 2,3-bis-iodomethyl-1,4-dioxanes and separated. The stereochemistries were deduced by oxidation to the 2,3-dicarboxylic acids. The cis isomer readily gave an anhydride and could not be resolved into its enantiomers, whereas the *trans*-diacid was stable and was partially resolved into its enantiomers.[824]

The chlorination of dioxane is catalyzed by stannous chloride and presumably forms a mixture of *cis*- and *trans*-2,3-dichloro-1,4-dioxanes of unknown composition. If zinc chloride is used as catalyst the reaction is highly stereospecific and only the cis isomer is formed. This is because zinc chloride coordinates with the two oxygen atoms of 1,4-dioxane (e.g., **276**) enabling attack to occur only from one side of the molecule.[825]

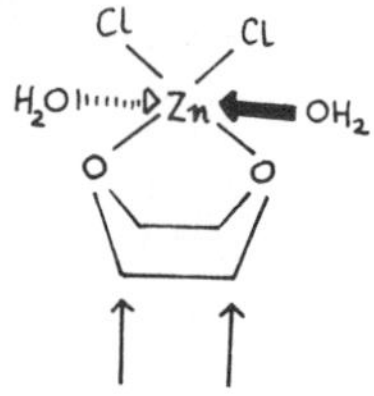

276

2,3,5,6-Tetradeutero-1,4-dioxane is a useful derivative for ^1H nmr studies. A mixture of two isomers was formed by reacting racemic 1,2-dideutero-ethyleneglycol (from OsO_4 oxidation of *trans*-1,2-dideuteroethylene) and its di-*O*-tosyl derivative in the presence of 2,6-di-*t*-butylpyridine (Equation 54).[826]

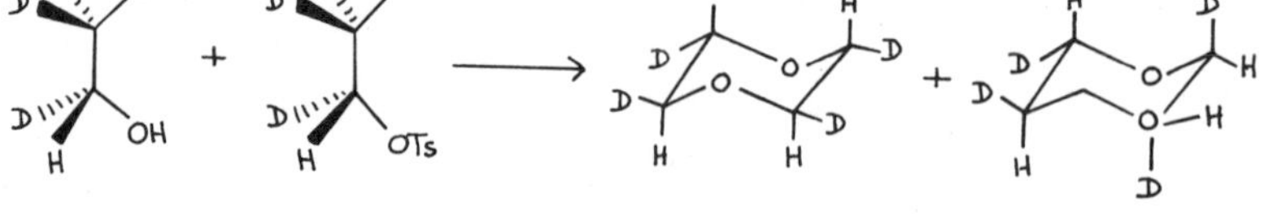

(54)

2. *Configuration and Conformation of 1,4-Dioxanes*

1,4-Dioxane has a center of symmetry in the chair conformation. That the chair conformation is the stable form is demonstrated by Raman and ir spectroscopy,[827,828] variation of total polarization with temperature,[829] dipole moment and molecular polarizability,[488] and electron diffraction.[830,831] ^1H nmr spectral evidence revealed that 1,4-dioxane is a conformationally mobile system of two interconverting chair conformations by way of a twist-chair intermediate. Variable temperature studies on 2,3,5,6-

tetradeutero-1,4-dioxane[832] and examination of the ^{13}C side bands[833] gave a $\Delta G^\ddagger$ value for the inversion barrier of 40.6 kJ mol^{-1} at $-93.6°C$. Coupling constants[496] and Lambert R values of 1,4-dioxanes are also consistent with a preferred chair conformation.[712,834]

The relative configurations of the substituents in *cis*- and *trans*-2,3-dimethyl-1,4-dioxane,[835] two of the five isomers of 2,3,5,6-tetramethyl-1,4-dioxanes,[836] and *cis*- and *trans*-2,3-diphenyl-[837] and 2,3-dichloro-1,4-dioxanes[837,838] were assigned by ^1H nmr spectroscopy. The structures of the last four compounds were also confirmed by dipole moment measurements.[816,839] The rather low chair-chair inversion barriers in *cis*-2,3-,[837] 2,5-diphenyl-,[840] and 2,3-dichloro-1,4-dioxanes[841] were deduced from nmr spectroscopy.

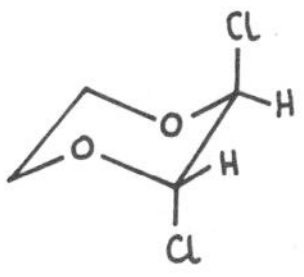

277

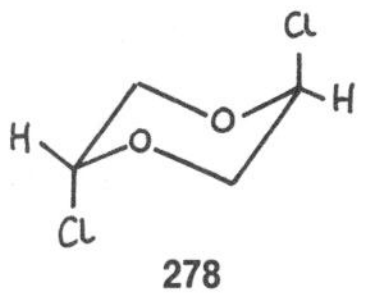

278

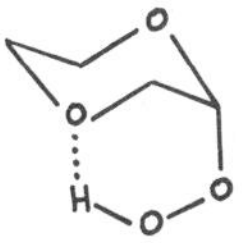

279

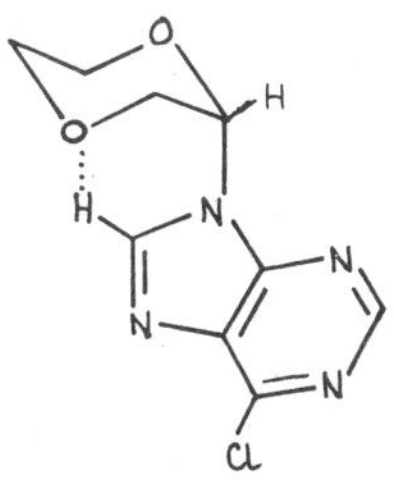

280

Dipole moment, ir, and Raman spectra of *trans*-2,3- and *trans*-2,5-dihalo-1,4-dioxanes demonstrated that the halogen atoms prefer the axial conformation (e.g., **277** and **278**) in contrast to the corresponding dihalocyclohexanes.[842] It is clearly a dipolar effect similar to the one observed in the anomeric effect (see Section X.C). These structures were confirmed by X-ray analyses.[843–845] Strong intramolecular hydrogen bonding was observed in 2-hydroperoxy-1,4-dioxane which is consistent with an axial orientation of the substituent **279.** 2-Hydroxy-1,4-dioxane shows no signs of intramolecular hydrogen bonding with the OH group equatorial, in contrast with 2-acetoxy-1,4-dioxane whose ^1H nmr spectrum is similar to that of 2-hydroperoxy-1,4-dioxane, and clearly demonstrates the axial conformation of the OAc and OOH groups.[846] A high proportion (44% in

$CDCl_3$) of the molecules in 2-(6-chloropurin-9-yl)-1,4-dioxane have the heterocyclic substituent in the axial conformation (**280**). The stabilization of this conformer was attributed to hydrogen bonding between H–8 of the purine ring and O–4 of the dioxane ring. The percentage of axial conformer is less (35% in $CDCl_3$) in the analogous 2-(6-chloropurine-9-yl)-tetrahydropyran.[847]

The assignment of configuration of substituents by optical resolution is mentioned in Section XII.C.1.

3. *Reactions of 1,4-Dioxanes*

The halogen atoms in *trans*-2,3-dichloro-1,4-dioxane undergo substitution reactions. With phenyl magnesium bromide[815] and *p*-bromophenyl magnesium bromide,[848] the respective *trans*-2,3-diaryl-1,4-dioxanes are formed. The latter, 2,3-bis-*p*-bromophenyl-1,4-dioxane, was converted into a bis-*p*-carboxyphenyl derivative, through a bis-Grignard compound, and resolved with brucine into its optical antipodes. The bis-Grignard intermediate was also hydrolyzed to the known *trans*-2,3-diphenyl-1,4-dioxane. This is evidence that the products had the trans configuration but is not proof that two Walden inversions had occurred.[848] Condensation of *trans*-2,3-dichloro-1,4-dioxane with dimethyl (+)-tartrate provided dimethyl (−)-1,4,5,8-tetraoxa-*cis*-decalin-*trans*-2,3-dicarboxylate (**281**), which was confirmed by nmr spectroscopy.[849] Similarly, reaction with catechol provided *cis*-2,3,4*aα*-10*aα*-tetrahydrobenzo[*b*]*p*-dioxino[2,3-*e*]dioxin.[850] *O,O*-Diethyl phosphorodithioate also displaces the chlorine atoms from *trans*-2,3-dichloro-1,4-dioxane and forms only *trans*-2,3-bis-(*O,O*-diethylphosphorodithioato)-1,4-dioxane. *cis*-2,3-Dichloro-1,4-dioxane, on the other hand, provided a 2:3 mixture of *cis*- and *trans*-2,3-phosphorodithioate esters.[851] Further examples of such substitutions can be found in Section XII.C.4.

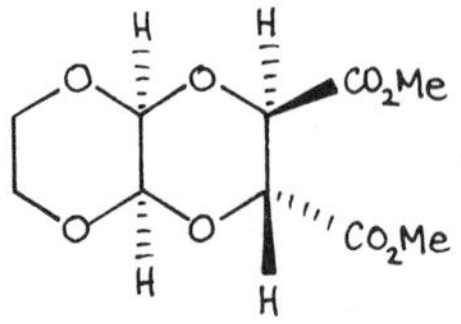

281

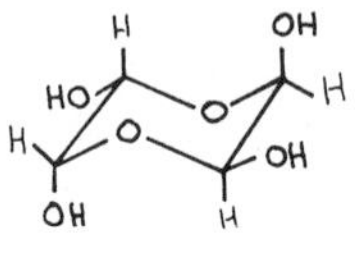

282

In solution, glycol aldehyde is in equilibrium with its hydrated form, 4-hydroxy-2-hydroxymethyl-1,3-dioxalane and *trans*-2,5-dihydroxy-1,4-dioxane.[852] The proportion of these varies with the solvent, and the kinetic details for these equilibria were reported.[853] Spectral and cryoscopic data for glyoxal monohydrate revealed that it consisted of dimeric species, and the structure **282** was proposed for it.[854]

4. *Condensed 1,4-Dioxanes*

3,4,5,6-Tetrachloro-*o*-quinone condenses with stilbene (at 128°C in the dark) almost stereospecifically to yield 4,5-(tetrachlorobenzo)-*trans*-2,3-diphenyl-1,4-dioxene (**283**). When this reaction was carried out in the presence of uv light lower specificity was observed, and a 88:12 mixture of **283** and its cis isomer was formed.[855] Tetrachloro-*o*-quinone also condenses with norbornadiene in a nonphotolytic reaction with cis stereospecificity, and a 21:2.5 ratio of *exo*- and *endo*-1,4-dioxene adducts was formed among other products.[856] Phenanthra-5,6-quinone undergoes similar reactions with olefins, but in the presence of uv light, and the stereoselectivity varies with the olefin but is reduced at higher temperatures.[857,858] The cis and trans configurations of these 1,4-dioxenes were deduced by ^{1}H nmr spectroscopy,[857,859] and, together with studies of 2-substituted 5,6-benzo-1,4-dioxenes,[860] it was concluded that the 1,4-dioxene ring existed as an equilibrium between two half-chair conformations, **284** and **285.**

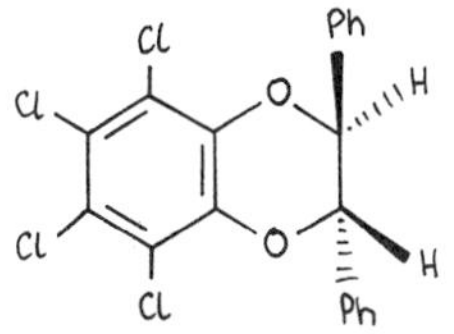

283 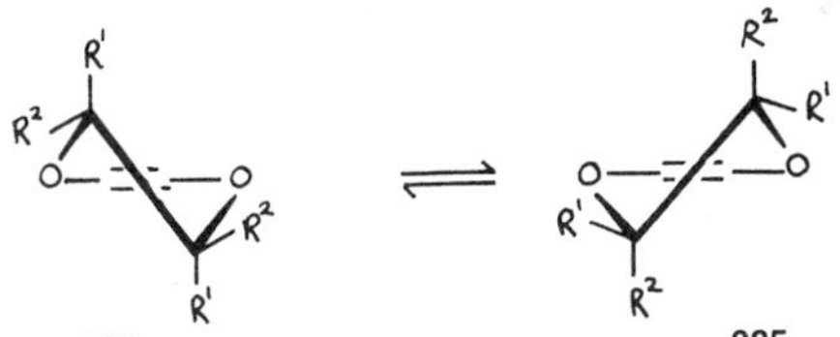 **284** **285**

Bottini and co-workers[861,862] cyclized 1-bromo-6-(2-hydroxyethoxy)-cyclohex-1-ene with *t*-BuOK in dimethyl sulfoxide and obtained equal amounts of *cis*-2,5-dioxabicyclo[4.4.0]dec-7-ene and cyclohex-2-enone ethylene ketal. Interestingly, 1-chloro-6-(2-hydroxyethoxy)cyclohex-1-ene, under similar conditions, gave a 92:8 ratio of dioxabicyclodecene to the ethylene ketal (Equation 55). Reduction of the decene provided *cis*-2,5-dioxabicyclo-[4.4.0]decane. The electrochemical reduction of benzo-1,4-dioxene in methanol, according to Belleau and Weinberg's method,[863] caused the addition of methoxy groups at the bridgehead carbon atoms and formed 1,6-dimethoxy-2,5-dioxabicyclo[4.4.0]deca-7,9-diene (**286**). The trans stereochemistry at the bridgeheads was deduced from its ^{1}H nmr spectrum as was that of its reduction product, 1,6-dimethoxy-2,5-dioxabicyclo[4.4.0]-decane.[864]

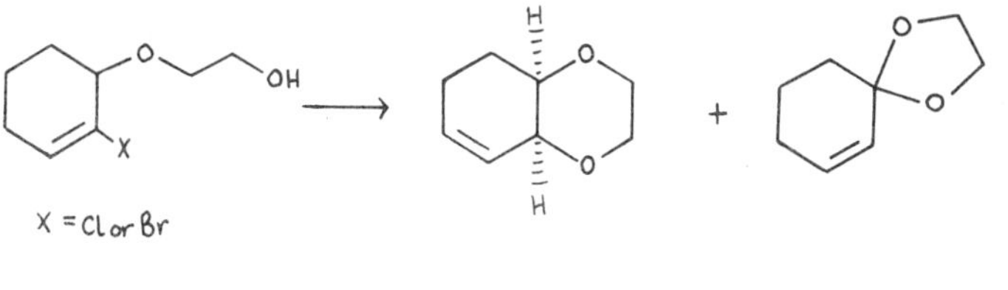

X = Cl or Br (55)

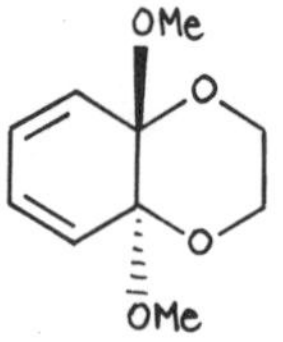

286

Trans-2,3-Dichloro-1,4-dioxane, in concentrated H_2SO_4, is converted into 4,4′,5,5′-tetrachlorodi-(1,3-dioxan-2-yl) and not into tetrachloro-1,4,5,8-tetraoxadecalin,[865–867] but it reacts with dimethyl 2,3-bistrimethylsilyloxy succinic acid, in the presence of silver perchlorate, to form 2,3-bismethoxycarbonyl-1,4,5,8-tetraoxadecalin. *cis*-2,3-Dichloro-1,4-dioxane reacts similarly, and all stereoisomers of 2,3-bismethoxycarbonyl-1,4,5,8-tetraoxadecalin were prepared in this way. The conformational free energy for the CO_2Me groups ($\Delta G^{\ddagger} = 1.6$ kJ mol^{-1}) was deduced by ^1H nmr spectra. Variable temperature ^1H nmr spectra of *cis*-2,3-bismethoxycarbonyl-*anti,cis*-1,4,5,8-tetraoxadecalin (**287**) gave a ring inversion barrier of $\Delta G^{\ddagger} = 41.8$ kJ mol^{-1} at $-93°C$. The equilibrium proportions (methoxide catalyzed) of the three cis conformers **287, 288,** and **289** are 34, 59, and 7 respectively. Chemical elaboration of the CO_2Me groups into Me groups provided 2,3-dimethyl-1,4,5,8-tetraoxadecalins of known configuration.[868,869]

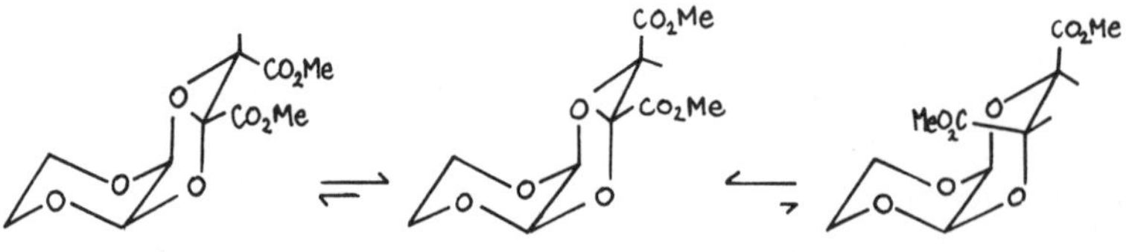

287 ⇌ **288** ⇌ **289**

1,4-Dibromobutane-2,3-dione condenses with two molecular equivalents of ethylene glycol and produces a mixture of 2,2′-bisbromomethyl-di(1,3-dioxan-2-yl) and 4*a*,8*a*-bisbromomethyl-1,4,5,8-tetraoxadecalin.[870,871] The ^1H nmr spectrum of the latter is consistent with a cis configuration (**290,** R = Br) in which the ring inversion barrier is lower (48.1 kJ mol^{-1}) compared with that of *cis*-4*a*,8*a*-bisbromomethyldecalin ($\Delta G^{\ddagger} = 61.5$ kJ mol^{-1}). The cis structure was confirmed by an X-ray study.[872] The bisbromomethyl compound **290** (R = Br) was reduced to the corresponding 4*a*,8*a*-dimethyl-1,4,5,8-tetraoxadecalin (**290,** R = H),[871,872] and related compounds have been described.[873] The ^1H nmr spectrum of *trans*-1,4,5,8-tetraoxadecalin,[837] together with an earlier X-ray determination of the crystal structure of its mercuric chloride complex, were strong evidence for the trans chair-chair arrangement of the atoms.[874]

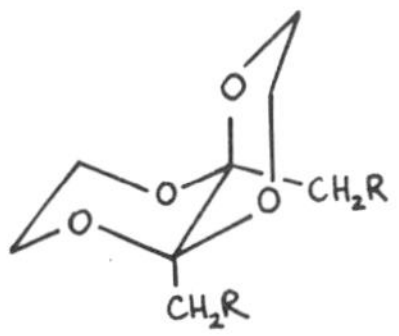

290

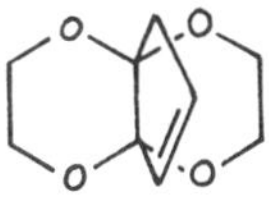

291

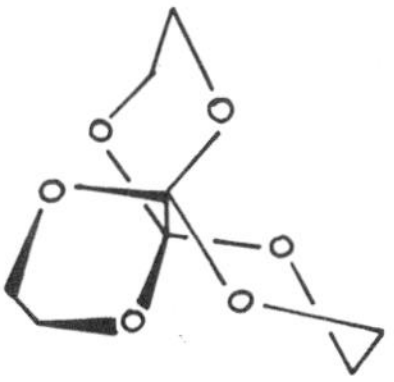

292

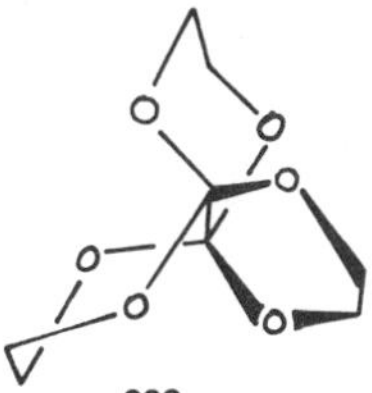

293

The acid-catalyzed dehydration of equimolar amounts of cyclopent-1-en-3,4-dione and ethylene glycol provided a 65% yield of 2,5,7,10-tetraoxa-[4.4.3]propell-11-ene (**291**) which could be catalytically reduced to the propellane without degradation.[875] 2,2,3,3-Tetrachloro-1,4-dioxane or 1,4-dioxane-2,3-dione condensed with ethylene glycol to yield 2,5,7,10,11,14-hexaoxa[4.4.4]propellane (**292**). The conformational mobility, **292** ⇌ **293**, of this propellane was revealed by the coalescence of its ^{1}H nmr signals at $-24°C$.[876] (See Part I, Chapter 4, Section IX for azapropellanes, and Chapter 3, Section XIV.E for thiapropellanes.)

Several 3,6,8-trioxabicyclo[3.2.1]octanes (**294**) were synthesized by intramolecular cyclization of *cis*-2-chloromethyl-4-hydroxymethyl-1,3-dioxalanes with a base, and their physical properties were examined.[877–879] This class of compounds provided a further example of the anomeric effect. In the acid-catalyzed equilibration of the 4,5-dimethyl derivative the isomer with the equatorial 4-methyl group (**294,** $R^1=R^3=Me$, $R^2=H$) was more abundant, whereas the thermal rearrangement (as yet unexplained) of the chloromethyl compound (**294,** $R^1=CH_2Cl$, $R^2=R^3=H$) into the 4-chloro-5-methyl derivative gave the isomer with the axial 4-chloro substituent (**294,** $R^1=Me$, $R^2=Cl$, $R^3=H$) predominantly.[880] The trimethylammonio group of the $Me_3N^+–CH_2–$ substituent at C–7 of **294** ($R^1=Me$, $R^2,R^3=H$) orients itself in an antiperiplanar conformation to O–6 and is another consequence of dipolar effects.[881]

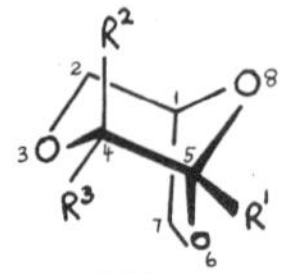

294 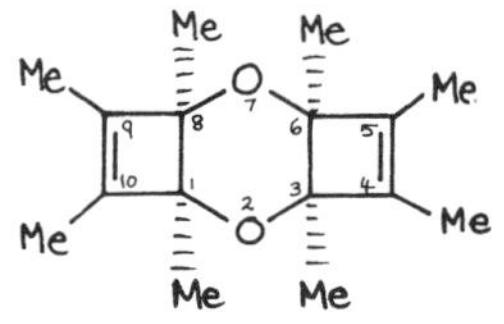 295 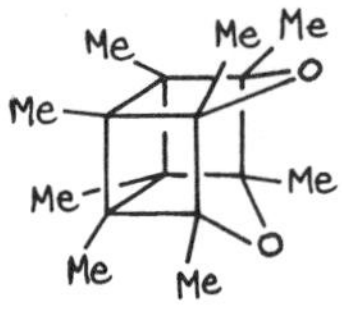 296

Acid-catalyzed dehydration of *cis*-3,4-dihydroxy-1,2,3,4-tetramethylcyclobut-1-ene gave a 22:18 mixture of *syn*- (**295**) and *anti*-1,3,4,5,6,8,9,10-octamethyl-2,7-dioxatricyclo[6.2.0.0^{3,6}]deca-4,9-dienes. The *syn*-diene **295** was identified by its conversion into the cubane structure, 1,2,3,4,5,6,8,9-octamethyl-7,10-dioxapentacyclo[4.4.0.0.^{2,5}0.^{3,9}0^{4,8}]decane (**296**), upon irradiation with uv light.[882]

XIII. 1,3,5-TRIOXANES AND 1,2,4,5-TETRAOXANES

Of the three possible trioxanes the 1,3,5 isomer is the more stable; the other two are cyclic peroxides. 1,3,5-Trioxanes are cyclic trimers of aldehydes or ketones and are consequently prepared by the trimerization of these. 1,3,5-Trioxane (formaldehyde trimer) is like cyclohexane in preferring a chair conformation, and evidence from crystal studies,[883,884] dipole moment,[488] and ^{1}H nmr[496] measurements confirm this. Similarly 2,4,6-trimethyl-1,3,5-trioxane (paraldehyde, acetaldehyde trimer) was shown by Raman,[885] ir,[886] microwave,[887] and ^{1}H nmr[888] spectra, and X-ray[889] and dipole moment measurements[890] to be in a chair conformation with the three methyl groups oriented equatorially (**297,** R = Me). Barón and Hollis[888] prepared several 2,4,6-trialkyl-1,3,5-trioxanes and found that in most instances a single product was formed whose structure, as determined by ^{1}H nmr spectroscopy, was in the all-cis chair conformation **297,** as in 2,4,6-trimethyl-1,3,5-trioxane. In a few cases, however, as in the trimerization of 2,2,3-trichlorobutyraldehyde, two trimers were formed: the *cis,cis,-cis*-2,4,6-tris-2′,2′,3′-trichloropropyl-1,3,5-trioxane (**297,** R = CH_3CHCl–CCl_2–, β-parabutylchloral) and the trans,cis,cis isomer **298** (R = $CH_3 \cdot CHCl \cdot CCl_2$, α-parabutylchloral). The ^{1}H nmr spectrum of the *cis,cis,cis*-trioxane was rather simple whereas that of the trans,cis,cis isomer was, as expected, more complex. Dipole moment evidence supported these particular structures. 2,4,6-Trisepoxyethyl-1,3,5-trioxane (glycidaldehyde cyclic trimer) was also obtained as two isomers with simple (**297,** R = $\overset{O}{\widehat{CH_2\text{–}CH}}$–) and complex (**298,** R = $\overset{O}{\widehat{CH_2\text{–}CH}}$–) nmr spectra, and were respectively noted as the *d,d,l* and *l,l,d* racemate, and *d,d,l* and *l,l,d* racemate.[891]

297 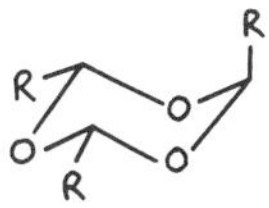 **298** 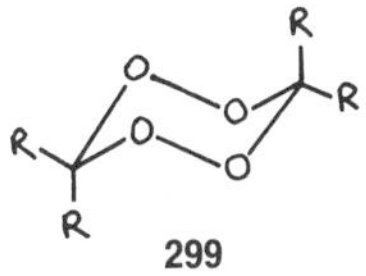 **299**

3,3,6,6-Tetraalkyl-1,2,4,5-tetraoxanes have attracted some stereochemical interest because they are conformationally mobile. X-Ray data on the spiro compounds 3,3,6,6-bispentamethylene-[892] and bishexamethylene-1,2,4,5-tetraoxanes[893] showed that the tetraoxane ring is in the chair conformation but is slightly distorted because of intramolecular repulsive forces. Variable temperature ^{1}H nmr data of several alkyl substituted 1,2,4,5-tetraoxanes, that is, **299** with R = Me, Et, Pr, CH_2Ph, CH_2Br, and CH_2Cl, provided the thermodynamic parameters for ring inversion.[727,894,895] The $\Delta G^{\ddagger}$ values fall within the range of 52.7 and 64.0 kJ mol^{-1}. These are rather high compared with cyclohexane (43.9 kJ mol^{-1}) and 1,1,4,4-tetramethylcyclohexane (49 kJ mol^{-1}), and they are dependent on the volume and inductive effects, but not weights, of the substituents.[894] The study of inversion barriers in solvents with acceptor properties (e.g., nitrobenzene) demonstrated that the repulsive forces of the oxygen lone pairs were responsible for much of the enthalpy of activation of the conformational changes.[895] For further reading on the chemistry of 1,2,4,5-tetraoxanes, see the review by Schulz and Kirschke.[521]

XIV. MISCELLANEOUS OXYGEN COMPOUNDS

A. Oxepins and Related Compounds

The chemistry of seven-membered oxygen heterocycles was discussed by Rosowsky,[896] and details about reduced derivatives can be found in the reference cited. The stereochemistry of a few derivatives are described in this subsection.

The preferred suggested conformations for benzo-3-oxacyclohepta-1,4-diene and the more rigid benzo-*cis*-6,7-dihydroxy-3-oxacyclohept-1-ene are shown in formulas **300** and **301,** respectively, and are based on ^{1}H nmr spectra.[897] Variable temperature ^{1}H nmr spectra of benzo-5-oxacyclohept-1-ene revealed that it is a mobile system ($\Delta G^{\ddagger} = 39.7$ kJ mol^{-1}) which is very sensitive to substitution of the hydrogen atoms at C–3 and C–4.[898] The conformational properties of several 10-substituted 10,11-dihydrobenzo[*b.f*]oxepins were examined. Although the preferred conformation has the 10-substituent in the equatorial orientation, it was not possible to decide between the two possible forms.[899] The tribenzo compound, tribenzo[*b.d.f*]oxepin, undergoes ring inversion so slowly that the derivative with a methyl and carboxyl group in the [*d*]-benzo ring (**302**) could be resolved into its optically active forms, but it racemized readily at room temperature with a $\Delta F^{\ddagger}$ value of 87.0 kJ mol^{-1} at 20.5°C.[900] Overberger and Kaye[901] synthesized 4*R*-4-methyl- and 6*R*-6-methyl-oxepan-2-ones and examined their chiroptical properties. The data suggested that the oxepanones existed as slightly deformed chair conformations with the lactone function in the flat portion of the molecule (**303**), and this was consistent with predictions from the application of the quadrant rule.[902]

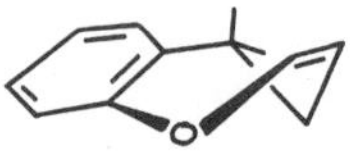

300

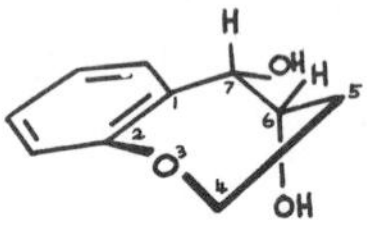

301

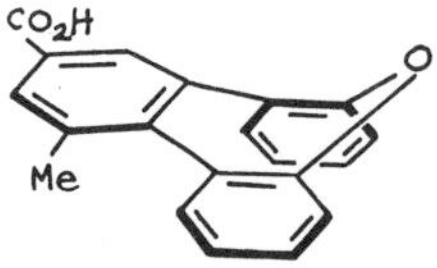

302

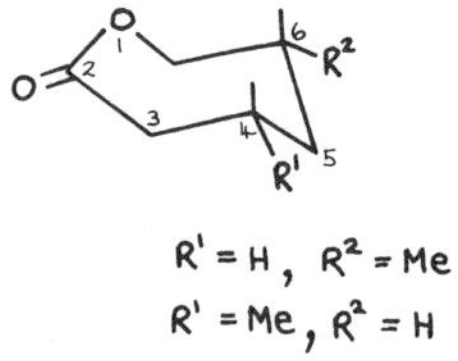

303

R' = H, R² = Me
R' = Me, R² = H

Dihydrooxepins derived from 2,2′-bridged biphenyl were the subject of several investigations. Turner and co-workers[903] observed that, although the benzene rings in 2,7-dihydrodibenz[*c.d*]oxepin (**304**) were not in the same plane, there was substantial conjugation between them. Several of these were obtained in optically active forms from the respective 2,2′-diphenic acids by way of 2,2′-bishydroxymethyl derivatives followed by cyclization to the required dihydrooxepins. These included derivatives similar to **304** with various R groups,[904–906] dinaphtho (**305**)[905,906] and dianthraceno (**306**)[907] derivatives, and even dihydrooxepins in which the R^1,R^2-positions in compound **304** are bridged by CH_2OCH_2 and CH_2SCH_2[908,909] links. Their chiroptical properties were examined in detail,[904–906,908,909] and the optical stabilities, that is, racemization characteristics, were determined.[908,909] Absolute configurations were also deduced from ord and cd spectra.[904–906] Mislow and Glass[908] deduced, from the activation energies of racemization of 2,2′,6,6′-doubly bridged biphenyls (e.g., **304,** R^1, $R^2$$=$$CH_2OCH_2$ or CH_2SCH_2), a model for calculating the angle strain. Sutherland and Ramsay[910] obtained the energy barrier for the conformational inversion of 2,7-dihydro-2,2,7,7-tetramethyldibenzo-

[*c.d*]oxepin ($\triangle E_a = 55.7$ kJ mol^{-1}) from ^{1}H nmr data and found that, unlike in the sulfur or dimethylammonium analogues (i.e., O replaced by S or $N^+Me_2 \cdot Br^-$), the value did alter from $CDCl_3$ to pyridine. Kessler[911] prepared two isomeric 2,7-dihydro-2,7-dimethylbenzo[*c.d*]oxepins, a trans-(*d,l*) and a cis-(meso) form. The trans form was more stable with the methyl groups in pseudo-equatorial conformations; that is, methyl groups are almost in the same plane as the vicinal benzene rings (**307**). Ōki and collaborators measured the barriers of inversion of several dihydrooxepins, **304** in which R^1,R^2 = H,H; Me,Me; $-CH_2OCH_2-$; $-CH_2CH_2-$; H,NO_2, and obtained respectively 40.2, 83.7, 84.5, 43.9, 69.0 kJ mol^{-1}, which were in fair agreement with energies of racemization of these compounds.[912,913] For an excellent account of 2,2′-bridged biphenyls, see the review by Hall.[914]

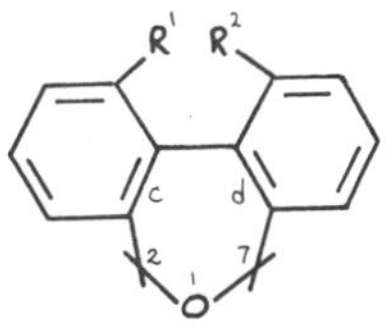

304

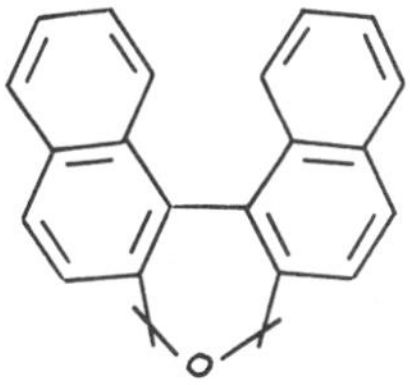

305

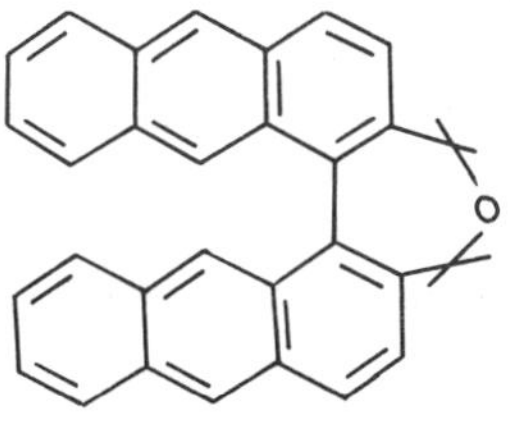

306

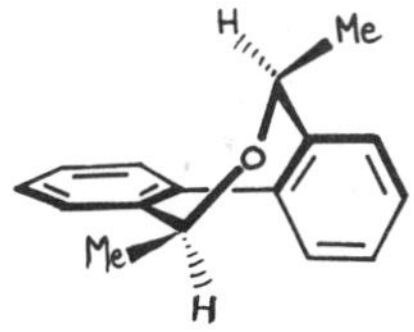

307

The valence-bond isomerization of epoxides to oxepins was mentioned earlier (Section II.C.3), and Vogel and co-workers[915,916] have extended this study to bisepoxides, for example, **308** $\rightleftharpoons$ **309** which occurs at 80–100°C. Related transformations have also been described,[917,918] and the molecular structure of 1,6-epoxy[10]annulene (**310**) was determined by an X-ray investigation.[919]

The conformational properties of *cis*- and *trans*-4,7-dimethyl-, 2-*t*-butyl-4-methyl-, and 2-*t*-butyl-5-methyl-1,3-dioxacycloheptanes were examined by ^{1}H and ^{13}C nmr spectroscopy. The data suggested the presence of two barriers, a rotational and a pseudorotational barrier. The $\triangle G°$ values

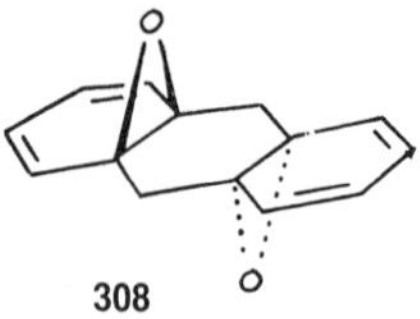

308

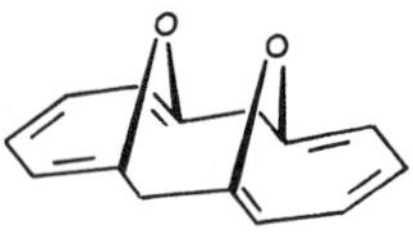

309

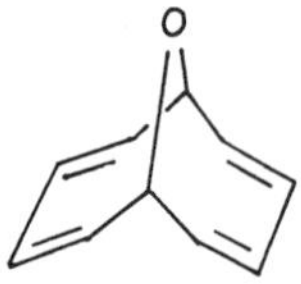

310

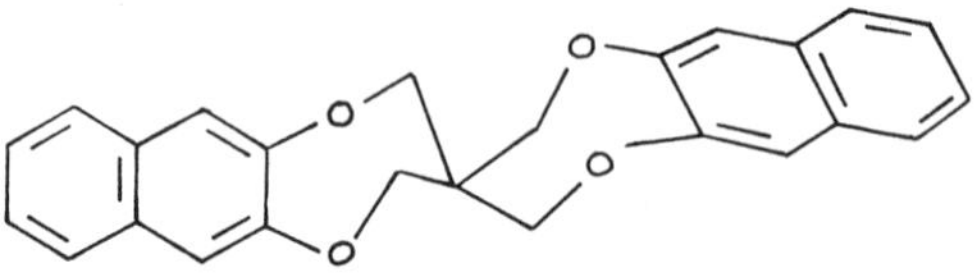

311

deduced indicated that there were several conformations with low energy barriers for each isomer.[920] In 3,3′-spirobis-[naphtho(ββ)-[1,5]dioxepen-6], on the other hand, the seven-membered rings are anchored in a preferred chair conformation which has a twofold axis of symmetry (**311**).[921]

B. Transannular Reactions Involving the Ring Oxygen Atom

Several examples of intramolecular attack of an oxygen atom on to a carbon atom are mentioned in Sections VII.D and XI.D. A few more references are described here which demonstrate that an apparently unreactive oxygen atom can participate in intramolecular reactions if the steric situation is favorable.

Lead tetraacetate oxidation of cycloheptanol, cyclooctanol, 1-methylcyclooctanol,[456] and cyclodecanol,[460] but not cyclohexanol, provided 8-oxabicyclo[3.2.1]octane (5%), 9-oxabicyclo[4.2.1]nonane (26%), a 1:3 mixture of 1-methyl-9-oxabicyclo[4.2.1]- and 1-methylbicyclo[3.3.1]nonanes (23%), and 11-oxabicyclo[6.2.1]undecane (2.5%) respectively. This data shows that the reacting atoms in the cyclooctanes are sterically more suited for intramolecular cyclization. This system was tested with *cis*- and *trans*-, 4- and 5-phenylcyclooctanols, and it was found that the phenyl group also had a directive influence. Although all these isomers gave oxabicyclononanes, the most favorable arrangement was found with *trans*-5-phenylcyclooctan-1-ol which gave a 72% yield of 1-phenyl-9-oxabicyclo[3.3.1]nonane.[922] A comprehensive study of proximity effects was reported by Cope and his collaborators[923] who examined the reactions of strong bases (e.g., PhLi or Et_2NLi) on *cis*- and *trans*-cyclooctene oxides. The cis isomer furnished *cis*-bicyclo[3.3.0]octan-*endo*-2-ol, and the trans isomer produced *cis*-bicyclo[3.3.0]octan-*exo*-2-ol indicating the mechanisms in Equations 56 and 57 respectively. Similarly *cis*- and *trans*-cyclodecene oxide gave the respective all-*cis*-1-hydroxydecalin and *trans*-1-hydroxy-*cis*-decalin **(312)** among other products.[924] That transannular interactions proceed in the acid-catalyzed reactions of *cis*-cyclooctene oxide was shown by the formation of 1,4-dihydroxycyclooctane, 4-hydroxy-, and 5-hydroxycyclooct-1-enes. However, the course of the reactions is influenced by the acid used because, although CF_3CO_2H provides the above products from transannular reactions, CH_3CO_2H yields equal amounts of this mixture and the normal product *trans*-1,2-dihydroxycyclooctane.[925]

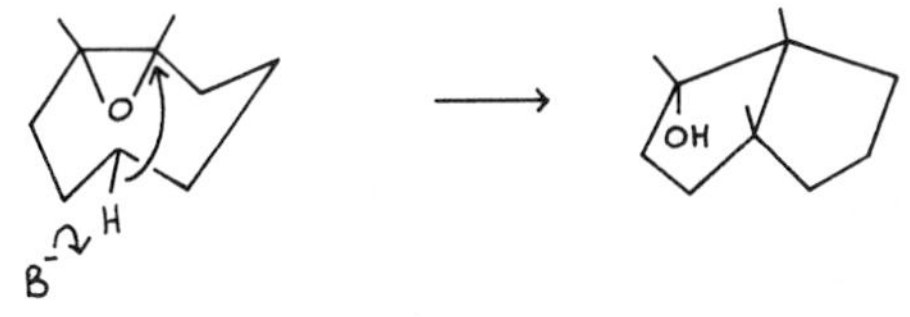

(56)

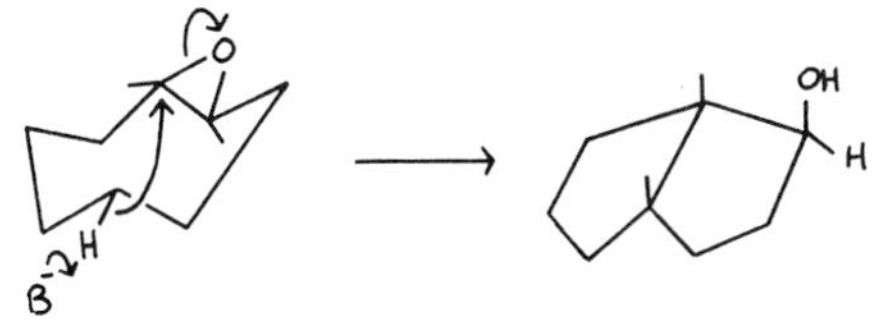

(57)

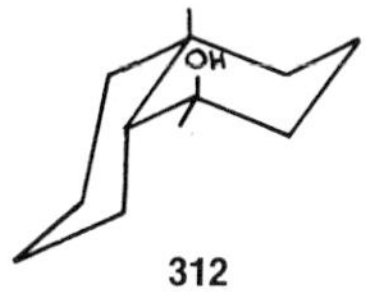

312

The transannular formation of a cyclic ether is demonstrated by the oxymercuration of 5-hydroxycyclooctene with mercuric acetate.[926] When this is carried out a second time on the same molecule a dioxatwistane can be formed (Scheme 11). The cyclization was studied in some detail by Bordwell and Douglass[927] who found that the intermediate formed rapidly produces 9-oxabicyclo[4.2.1]nonane by kinetic control, but this slowly equilibrates, by thermodynamic control, to the isomeric 9-oxabicyclo[3.3.1]-nonane derivative (Equation 58). Similar neighboring group participation was observed in the oxymercuration of 5,6-epoxy-*cis*-cyclooct-1-ene and *trans*-5,6-dihydroxy-*cis*-cyclooct-1-ene which provide the same two products. In a comparison of the rates of oxymercuration, the following relative rates were obtained: 5,6-epoxy-*cis*-cyclooct-1-ene (1.0); 5-hydroxy-*cis*-cyclooct-1-ene (~4); 5,6-cyclopropa-*cis*-cyclooct-1-ene, that is, bicyclo[6.1.0]-non-4-ene (0.02); and *cis*-cyclooctene (0.0005),[459] and clearly demonstrate the effect of the participation of the oxygen atom already present in the ring.

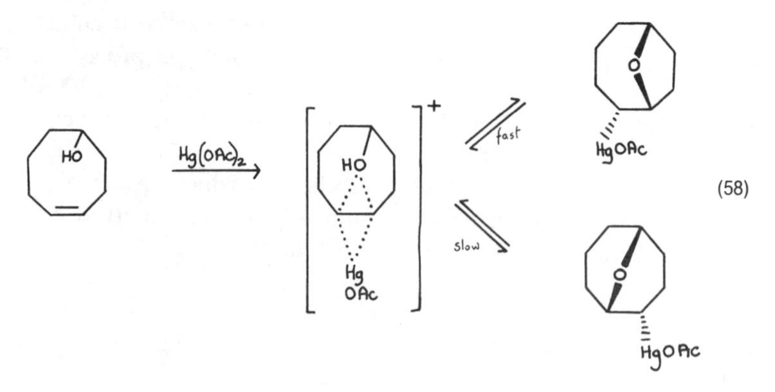

(58)

The above mercuric acetates are readily converted into the corresponding iodides by reaction with iodine, and the iodine can be removed with LAH without disrupting the ring system. 2-Iodo-9-oxabicyclo[4.2.1]nonane thus exclusively gave 9-oxabicyclo[4.2.1]nonane. However, when 2-iodo-9-oxabicyclo[3.3.1]nonane was reduced in the same way, a 1:1 mixture of 9-oxabicyclo[3.3.1]- and 9-oxabicyclo[4.2.1]nonanes was formed.[928] This led Paquette and co-workers to postulate the existence of an oxonium ion intermediate such as **313.** They extended their work to a study of the acetolysis of *endo*- and *exo*-2-*p*-bromobenzenesulfonyloxy-9-oxabicyclo-[4.2.1]nonanes and found that a mixture of *endo*-2-acetoxy-9-oxabicyclo-[3.3.1]nonane and *endo*-2-acetoxy-9-oxabicyclo[4.2.1]nonane was formed which provided further evidence for the intermediate cation **313.**[929] Final confirmation came from the acetolysis of optically active (+)-*endo*-2-*p*-bromobenzenesulfonyloxy-8-oxabicyclo[3.2.1]octane which gave racemic *endo*-2-acetoxy-8-oxabicyclo[3.2.1]octane. Racemization was not caused by epimerization at C–2, but occurred because the intermediate oxonium ion produced two similar acetoxy compounds which are mirror images (Equation 59). Comparison of kinetic data and deuterium isotope effects of the 2-deutero derivatives further strengthened the proposed mechanism. The exo derivative, on the other hand, behaved differently by producing 4-formylcyclohexanols and their acetates.[930]

313

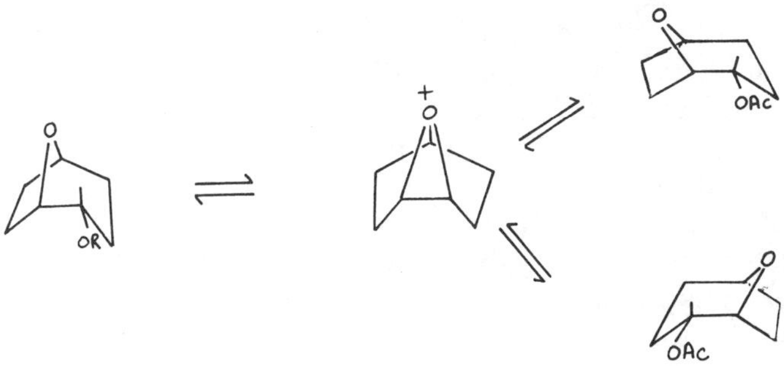

(59)

Despite the poor nucleophilic nature of the ethereal oxygen atom, Paquette and co-workers demonstrated further that it can be involved in transannular reactions, and again the cyclooctane series provided the examples. Acid treatment (e.g., HCl) of 3,4,7,8-tetrahydro-2-*H*-oxocin-3-one (**314**) caused a deep seated reaction which gave 2,3′-chloropropyltetrahydrofuran-4-one, and the mechanism in Equation 60 was postulated.[931] Hexahydro-2*H*-oxocin-3-one is unreactive under these conditions, and the movement of the double bond in the molecule was indicated by the acid stability of the 4,4-dimethyl derivative of **314.** The carbonyl group is not necessary for the reaction if it is possible to stabilize a carbonium ion at C–5, as shown by the conversion of 5-phenyl-3,6,7,8-tetrahydro-2*H*-oxocin into 2-chloropropyl-2-phenyltetrahydrofuran.[932] The 3,6,7,8-tetrahydro-2*H*-oxocin ring occurs in nature, as in the stereochemically interesting bromo compound laurencin from *Laurencia glandulifera*.[933]

From a study of the autolysis of 3-*p*-bromobenzenesulfonyloxy-2*H*-

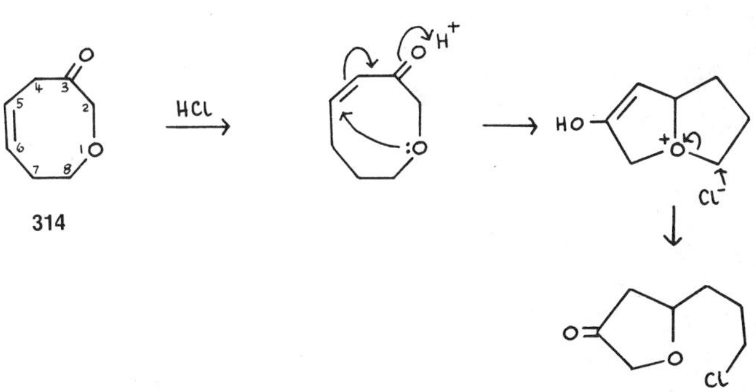

314

(60)

oxocan, which proceeds exclusively through a bicyclic oxonium intermediate, and 3-*p*-bromobenzenesulfonyloxy-3,4,7,8-tetrahydro-2*H*-oxocin, which uses a homoallylic mechanism, it was concluded that participation of the latter mechanism overrules the oxygen neighboring group effect.[934] A comparison of the solvolysis of 3- and 4-arenesulfonates of tetrahydropyran with open chain models suggests that the ring-oxygen atom assists in the reaction of the 3- (**315**) but not the 4-isomer. Lack of ring-oxygen assistance in the hydrolysis of 3-*p*-bromobenzenesulfonyloxytetrahydrofuran is probably due to the less favorable geometry of the system.[935] The solvolyses of *syn*- and *anti*-3-*p*-bromobenzenesulfonyloxy-*cis*-cyclooctene 1,2-oxides are also effected by the ring-oxygen atom—although several products are formed. The predominant product in the former case is suberaldehyde (52%) (Equation 61), whereas in the latter case substitution occurs with retention of configuration (32%), probably by steric obstruction.[936]

In a thorough investigation of the ring cleavages of *cis*-3-oxabicyclo-[5.1.0]octane (**316**), *cis*-4-oxabicyclo[5.1.0]octane (**316a**), and *cis*-4-oxabicyclo[6.1.0]nonane (**317**) promoted by formic acid, Paquette and collaborators[937] demonstrated that only in the last named compound does the oxygen participate in a transannular reaction. The product is the formic ester of 2,3′-hydroxypropyltetrahydro-2*H*-pyran, and the proposed mechanism is summarized in Equation 62.

OR

315

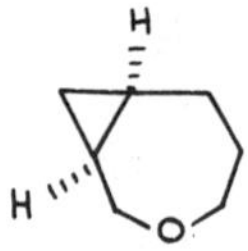

316

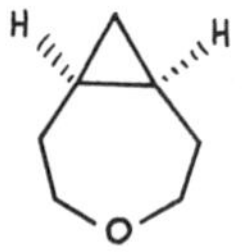

316a

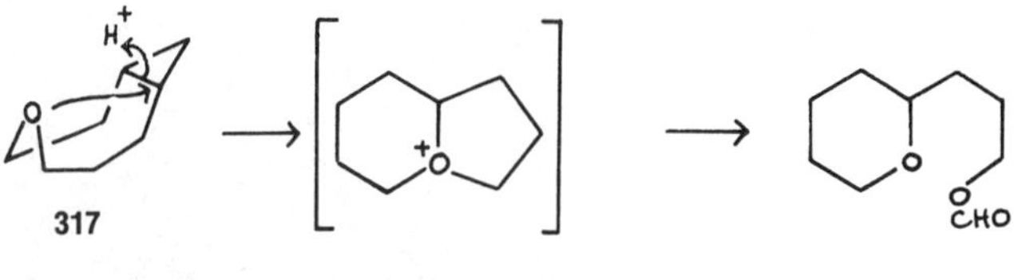

OBs → → CHO $(CH_2)_6$ CHO (61)

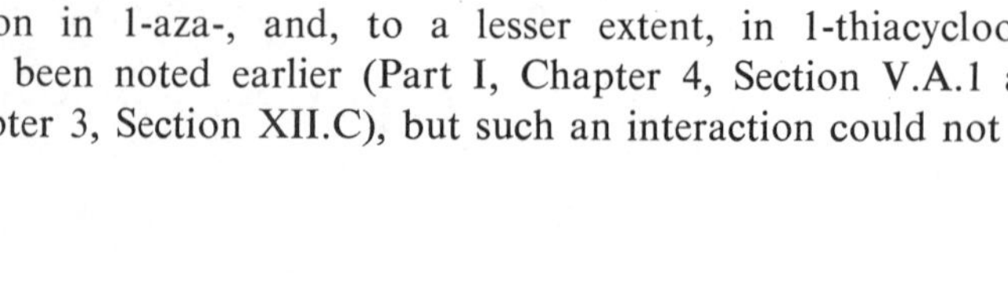

317 → → CHO (62)

Strong transannular interactions between the heteroatom and carbonyl carbon in 1-aza-, and, to a lesser extent, in 1-thiacyclooctan-5-ones, have been noted earlier (Part I, Chapter 4, Section V.A.1 and Part II, Chapter 3, Section XII.C), but such an interaction could not be observed in 1-oxacyclooctan-5-one (oxocan-5-one). However, the hydrolysis of the latter with hydrochloric acid to yield 1,7-dichloroheptan-4-one is suggestive of a transannular interaction involving an oxonium intermediate.[938] Dipole moment measurements of this ketone indicated that the preferred conformation is boat-boat, but more recent ^1H and ^{13}C nmr spectroscopy could not support this structure, but the data agreed with a preferred boat-chair conformation.[939] For further details on the conformations of oxocanes see Section XIV.D.

Metalation of 9,9-dibromo-3,5-dioxabicyclo[5.1.0]octane with MeLi is stereospecific and provides the *exo*-lithium-*endo*-bromo derivative **318.** This lithium compound is stable below −20°C, but above this temperature it cyclizes by intramolecular C_4–H insertion and yields essentially 2,8-dioxatricyclo[3.3.0.^{1,5}0^{4,6}]octane (**319**). The corresponding *endo*-lithium-*exo*-bromo isomer, which is obtained from the reaction of the *exo*-chloro-*endo*-bromo derivative with MeLi, is less stable and reacts intermolecularly. The difference in stability between these two isomers may well be due to chelation of the metal with the proximal oxygen atoms.[940]

C. Oxygen Containing Ansa Compounds

In 1937, Lüttringhaus[941] used a high dilution technique[942] to effect the intramolecular cyclization of *m*- and *p*-ω-bromoalkoxyphenols with potassium in amyl alcohol. Several interesting results arose from this study because intra- and intermolecular reactions were competing with each other. Intramolecular cyclization of the *p*-hydroquinol derivatives with $n=10$ and $n=8$ proceeds in 79 and 18% yields respectively, but form intermolecular products with shorter chains (Equation 63). In the case of resorcinol derivatives cyclization to **320** with $n=10$ and $n=7$ takes place in 62 and 10% yields. The limit here is 7, whereas in the 1,5-disubstituted naphthalene (**321**) the lower limit is $n=10$—the yield being only 22%. Lüttringhaus and Gralheer[943,945] adopted the general term *ansa* (latin, handle) compounds to name these and other compounds in which a bridge connection is present between two nonadjacent positions on an aromatic nucleus, for example, meta or para in benzene. The azacyclophanes and some catenanes, which are discussed in Part I, Chapter 4, Section V.A.2 and B, have been called ansa compounds. Some antibiotics with similar rings have been named Ansamycins.[944]

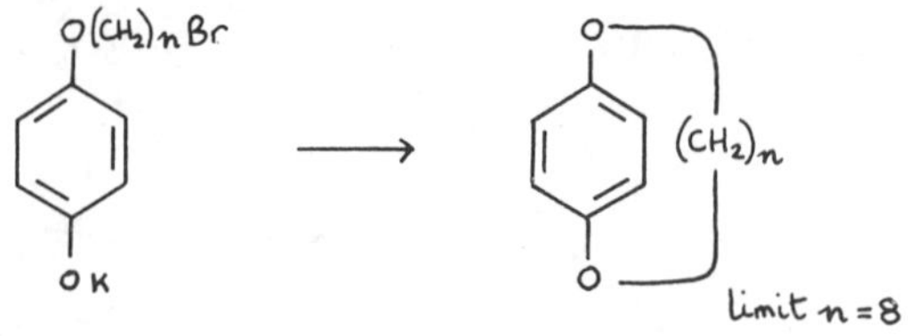

$O(CH_2)_nBr$ … OK → $(CH_2)_n$ limit $n=8$ (63)

If the aromatic ring in the above ethers is not free to rotate, the molecule as a whole becomes dissymetric. Lüttringhaus and Gralheer[943] recognized this property and confirmed it experimentally by resolving 4-bromo-2,5-(ω-decamethylenedioxy)benzoic acid (**322,** $R^1 = CO_2H$, $R^2 = Br$, $n = 10$) into its optically active enantiomers and showed that they did not racemize easily. The (+) acid was converted into (−)-1,4-dibromo-2,5-(ω-decamethylenedioxy)benzene (**322,** $R^1 = R^2 = Br$, $n = 10$), by way of a Curtius reaction followed by diazotization and reaction with HBr, without apparent racemization.[946] When the (−)-hydrazinium salt was shaken with Raney nickel, debromination occurred; the optical rotation fell to zero in a few minutes; and inactive 2,5-(ω-decamethylenedioxy)benzoic acid (**322,** $R^1 = H$, $R^2 = CO_2H$, $n = 10$) was isolated. The lower homologous optically active acid 2,5-(ω-octamethylenedioxy)benzoic acid (**322,** $R^1 = H$, $R^2 = CO_2H$, $n = 8$) gave a methyl ester which could not be racemized at 200°C in diphenyl ether, but hydrobromic acid destroyed the molecule. An interesting feature of this acid is its rotatory dispersion in chloroform: $[\alpha]_{656}$ − 7.2°, $[\alpha]_D$ 0° and $[\alpha]_{546}$ + 5.7°.[947] In demonstrating new variations of asymmetry, Helmchen and Prelog[948] prepared 1,12-dioxa[12]paracyclophanes; for example, **323** in which the substituents R^1 and R^2 contained chiral centers (e.g., CONH–C*HPhMe). When R^1 and R^2 have the same configuration the system is chiral, but if they are of opposite configuration the system is pseudo-asymmetric (meso).

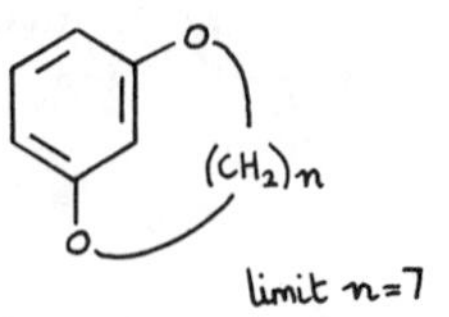

$(CH_2)_n$ limit $n=7$

320

$(CH_2)_n$ limit $n = 10$

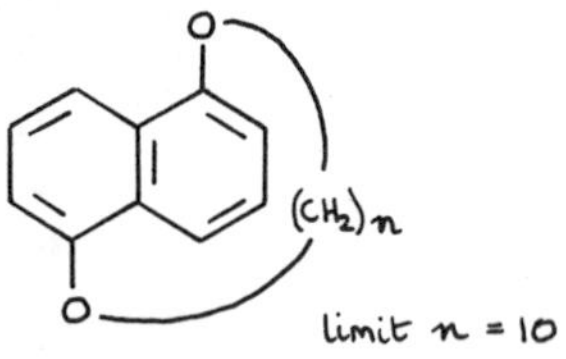

321

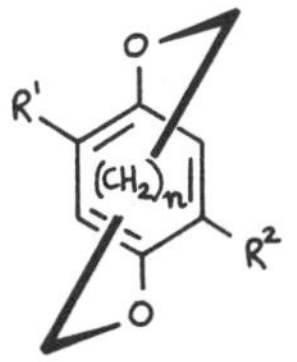

322

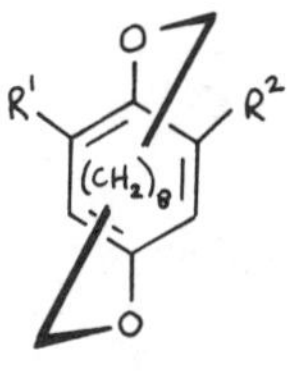

323

The furanophanes can be legitimately classified as ansa compounds, and a few interesting reports on their stereochemistry are worthy of mention. Winberg and collaborators[949] pyrolyzed 5-methyl-2-furfuryltrimethylammonium hydroxide and obtained the reactive 2,5-bismethylene-2,5-dihydrofuran (stable at $-78°C$) which dimerized in the presence of a stabilizer to [2.2](2,5)furanophane (Equation 64). A structure with D_{2h} symmetry for the furanophane was excluded on Raman spectral evidence which indicated that the molecule was centrosymmetric, and by analogy with the X-ray crystal structure of di-*m*-xylxylene(metacyclophane)[950] the step structure **324** with C_{2h} symmetry was proposed. This synthesis was used to prepare mixed furanophanes[951] and layered mixed furanophanes

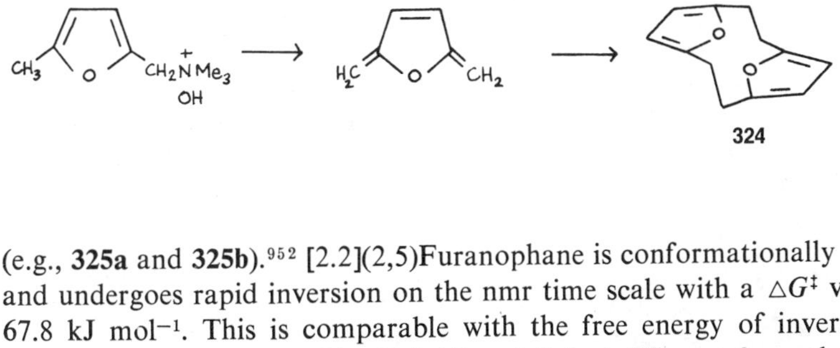

(64)

324

(e.g., **325a** and **325b**).[952] [2.2](2,5)Furanophane is conformationally flexible and undergoes rapid inversion on the nmr time scale with a $\Delta G^{\ddagger}$ value of 67.8 kJ mol^{-1}. This is comparable with the free energy of inversion of [2.2](2.6)pyridophane ($\Delta G^{\ddagger} = 61.9$ kJ mol^{-1}) but different from the stable metacyclophane (>113.0 kJ mol^{-1})[953] (See Part I, Chapter 4, Section V.A.2). In order to compare the nonbonding interactions of a lone pair of electrons with an N–H group, Keehn and co-workers[954,955] prepared [2.2](2,5)pyrrolofuranophane by a Paal-Knorr reaction (see Equation 65) and examined the ^{1}H nmr spectral properties. These did not give clear results and the hexadeutero derivative **326** was prepared. They showed that it was conformationally quite rigid with an inversion barrier greater than 113.0 kJ mol^{-1}.[956] These energy barriers should be compared with those for the equilibrium **325a** $\rightleftharpoons$ **325b** which is 42.7 kJ mol^{-1}.[952]

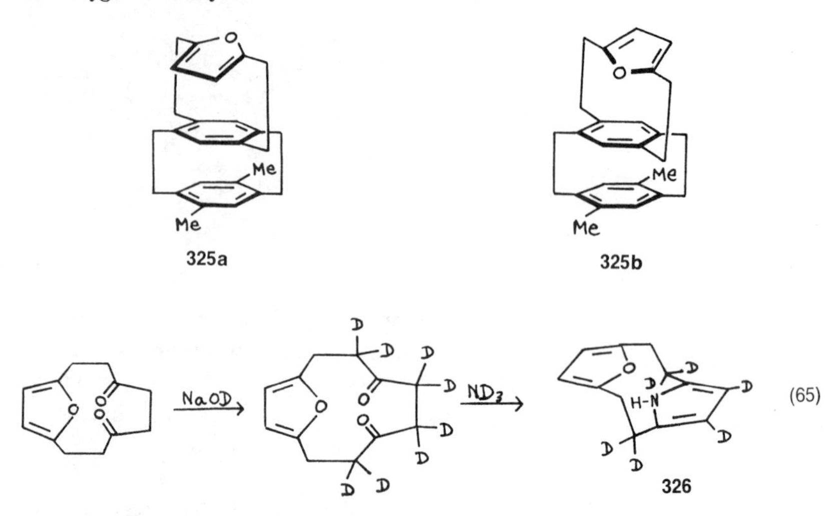

325a

325b

326

[2.2](2.5)Furanophane forms a 1:1 adduct with tetrachlorocyclopropene, and the structure of the adduct was shown by X-ray diffraction to be 5,6,7,7-tetrachloro-16,17-*syn*-dioxahexacyclo$[9.2.2.1.1^{4,8}0.^{4,14}0^{8,15}]$heptadeca-5,12-diene (**327**).[957] Helder and Wynberg[958] found that [8](2.5)furanophane also yields a 1:1 adduct with dicyanoacetylene in 80% yield (Equation 66). Addition occurs across C–2 and C–5 of the furan ring, and this adduct is as close as any workers have been in making a simple "paddlane" (**328**). The singlet signal at +30°C for the ethylenic protons broadens at −80°C and becomes a clear doublet at −100°C. These are consistent with the equilibrium in Equation 66 and support the proposed structure. Attempts to make paddlanes by cyclization of compounds like 1,4-bis-hydroxymethylbicyclo[2.2.2]octanes and sebacoyl chloride lead to dimeric structures (e.g., **329**) in which two paddlane units are joined together.[959]

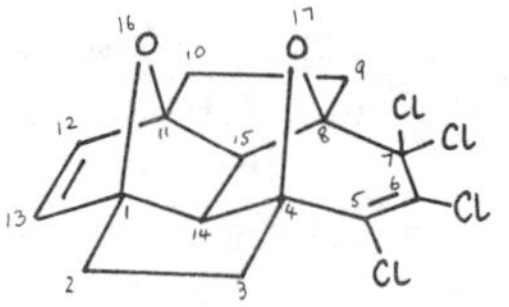

327

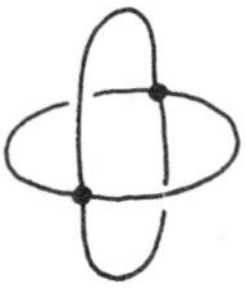

328

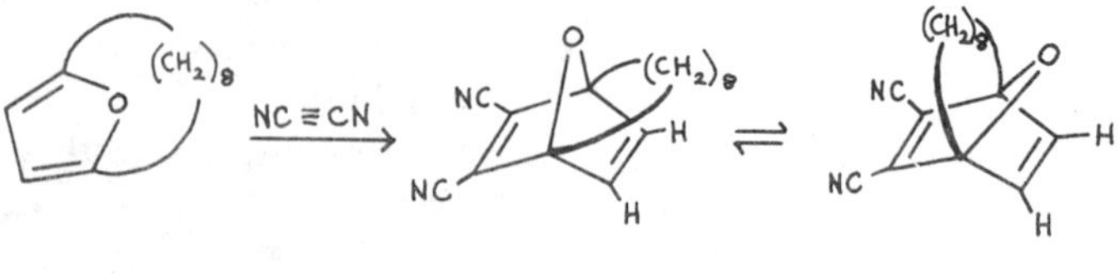

(66)

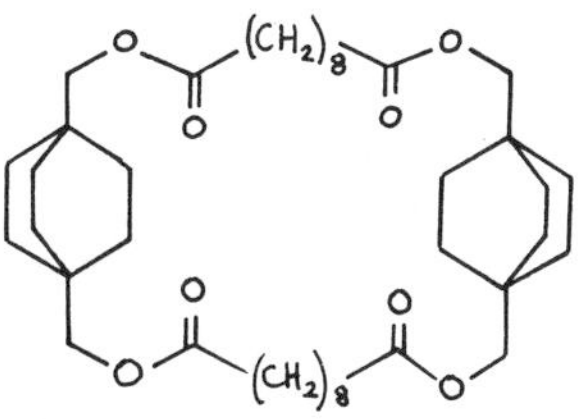

329

D. Oxygen Heterocycles with Eight-Membered and Larger Rings—Crown Ethers

Cyclooctane is a conformationally mobile molecule in which the more stable conformation is the boat-chair form. The ring inversion barrier, $\Delta G^{\ddagger}$, is 33.9 kJ mol^{-1}. Replacement of CH_2 by O in cyclooctane, as in oxocane, 1,3-di-, 1,3,6-, and 1,3,7-trioxocanes, decreases the nonbonded 1,5-H interactions and lowers the ring inversion barrier.[960] ^{1}H nmr, ir, and Raman spectra indicate that changes in conformations from boat-chair to chair-boat, for example, **330, 331,** and similar conformers, occur by pseudorotation as in all members of the *crown family.*[960,961] Large substituents can anchor the conformation of oxocane in a different manner as in 2,4,6,8-tetrakis(trischloromethyl)-1,3,5,7-tetraoxocane, metachoral. This compound was shown by nmr, ir and dipole moment studies to exist in a crown conformation (**332**) with the alkyl groups equatorially displayed.[962]

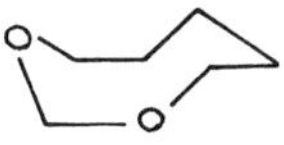

330 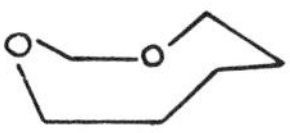 331 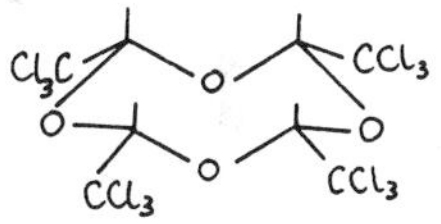 332

A few nine-membered ring oxygen heterocycles were investigated and conformational changes examined. 10,15-Dihydrotribenzo[*b.e.h*]oxonin (**333**) was found to undergo ring inversion between two crown conformers whose equilibrium can be frozen at −90°C, unlike the sulfur and carbocyclic analogues which were more rigid.[963] 3-Oxa- (**334,** X=O, Y=Z=CH_2), 3,7-dioxa- (**334,** X=Y=O, Z=CH_2), and 3,7,9-trioxabicyclo[3.3.1]nonanes (**334,** X=Y=Z=O) exist in double-chair conformations, albeit with the wings slightly flattened because of lone pair repulsion (compare with the sulfur analogues of the last named compound, in which the boat-chair conformation is preferred; Chapter 3, Section XIII, formula **200**).[964]

Boat-chair and chair-chair conformations for 3-methyl-2,4-dioxabicyclo-[3.3.1]nonane were considered, but the data was not conclusive.[965] 3,3′-Spirobis-[dinaphtho[1.2-*f*.1.′2.′-*h*]-1,5-dioxacyclononane][966] and its dibenzo analogue[967] (**335**) were prepared, and their physical data are consistent with a conformation that is chiral with a three twofold axis of symmetry. The uv spectra of the naphtho compound, when compared with the "monomer" suggest that the chromophoric groups are apparently conjugated in some way through the spiro system. Conformational studies of 1,4,7-trioxacyclononane (unsymmetrical conformation)[968] and the tetra-oxanonane (**336**) derived from carbohydrates[969] were also reported.

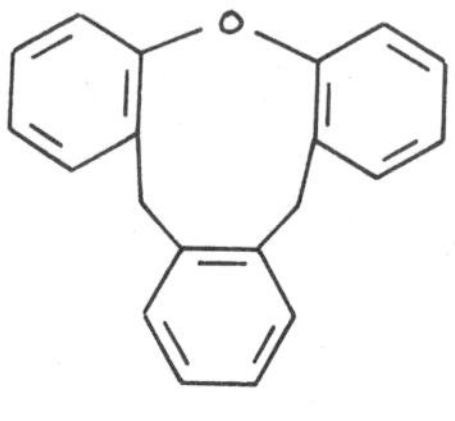

333

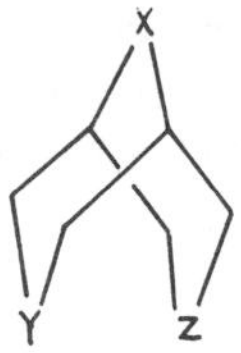

334

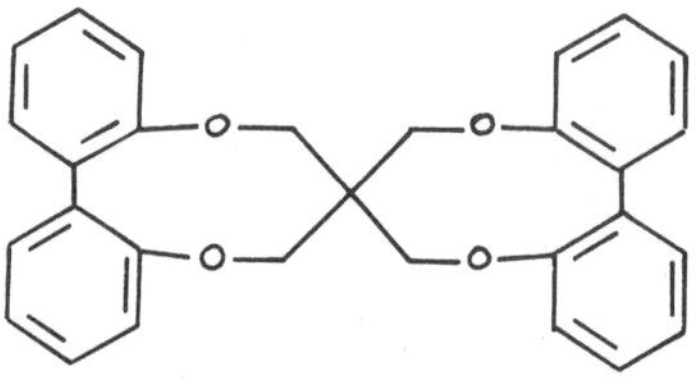

335

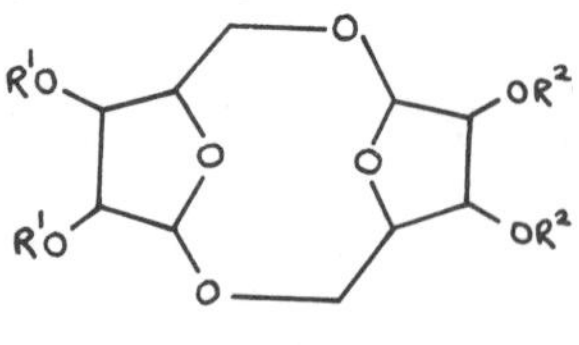

336

The three 4,5:9,10-bistetramethylene-1,3,6,8-tetraoxacyclodecanes known were assigned the structures trans,syn,trans; trans,anti,trans; and cis,anti,-cis by ^{1}H nmr spectroscopy. The boat-chair-boat conformation (e.g., **337**) was stable and was favored by the trans,anti,trans isomer.[462,463,970] The spectral data for the diastereotopic methylene protons of these compounds were discussed.

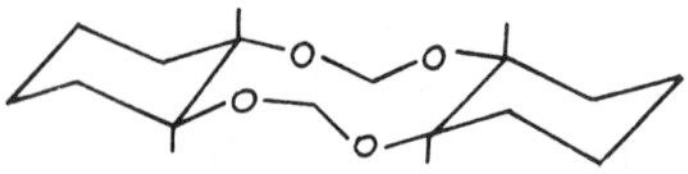

337

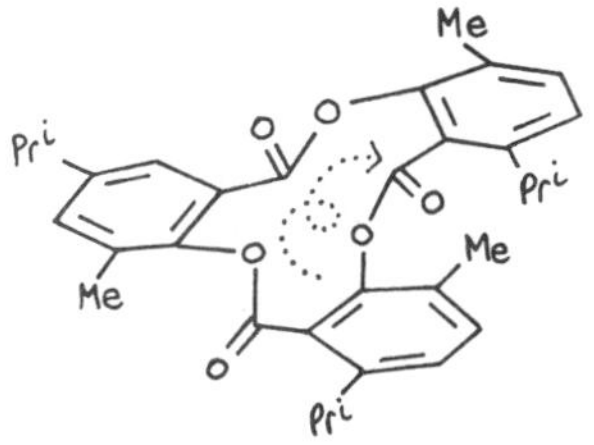

338

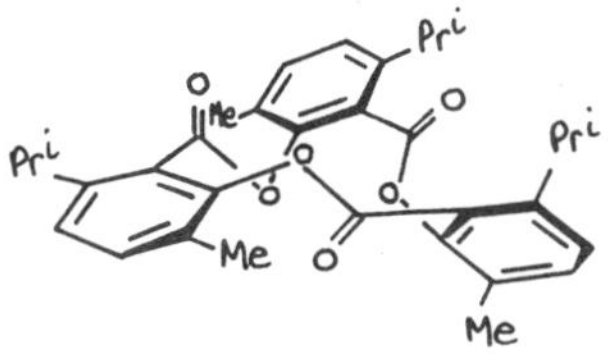

339

1,3-Dioxane dimerizes to 1,3,7,9-tetraoxacyclododecane whose preferred conformation is such that the oxygen atoms form the corners of an imaginary square. This confers high conformational stability to this ring ($\Delta G^{\ddagger} =$ 46 kJ mol^{-1}, compare with oxocanes above).[971] The 1,5,9-trioxadodecane ring system in the "trimeric" lactones of salicylic acid and derivatives has interested stereochemists for over two decades.[972] Tri-*o*-thymotide has been obtained in optically active forms by spontaneous resolution on recrystallization of its clathrate complexes with *n*-hexane, benzene, or chloroform.[973] The solid state conformation and the absolute configuration of (+)-tri-*o*-thymotide was determined by cd and shown to have the *M*-propeller configuration.[974] Detailed ^{1}H nmr studies of trisalicylides by Downing, Ollis, and Sutherland[975] revealed that these compounds are in chiral nonplanar conformations of the helical (e.g., **338**) and propeller (e.g., **339**) types in equilibrium. Activation parameters obtained from ^{1}H nmr spectra compared favorably with those obtained from racemization measurements of the optically active isomers of tri-*o*-thymotide, and the differences were discussed.

Macrocyclic polyethers have attracted considerable interest in recent years.[976–978] The particularly interesting ones which are being studied extensively are those in which the oxygens in the macrocycle are placed 1,4 to each other. Basically they are cyclic ethers derived from ethylene glycol.[976–980] Variations of these with a nitrogen atom in place of oxygen; that is, mixed heteromacrocycles were prepared,[981,982] and also macrocyclic amines derived from ethylenediamines were reported (see Part I, Chapter 4, Section V.B). The interest in the macrocyclic oxygen ethers, now commonly called crown ethers, is because of their ability to coordinate with weakly chelating metal ions (e.g., Na, K, Ca, Rb, Li, Mg, etc.). The metal is enveloped by the ethereal oxygen atoms, and, because the sizes of metal ions are different, crown ethers which chelate selectively can be made simply by altering the size of the ring.[976–978,983] Several antibiotics (e.g., nonactin and dinactin) possess macrocyclic ether rings and act in a similar way.[984] Macrocyclic ethers that possess two nitrogen atoms were studied by Lehn and co-workers.[984a] These have ball-like structures (e.g., **340**) and can hold a metal ion within them. Because of this property, these ethers have been called *cryptates.*

Crown ether complexes with metals alter the solubility of the metal, which can be used to advantage. Potassium superoxide, for example, will dissolve in benzene solution if coordinated with a crown ether (e.g., **341,** with the oxygen atoms arranged octahedrally about the K ion), and provides a new oxidizing agent.[985] The stereochemistry of the complex is dictated both by the ether and by the metal ion, and the three-dimensional structures of several of these complexes have been determined by X-ray analysis.[986–989] Several chiral crown ethers were prepared and separated by Parsons,[990] in which he had methyl groups on the carbon chain, whereby he introduced asymmetric centers into the macromolecule. The X-ray structures of the Cs complexes of some of these were determined by Mallinson.[991] A functional group (e.g., CO_2H) was also introduced in the carbon skeleton of crown ethers to alter the complexing properties.[992]

The most ingeneous application of crown ethers was provided by Cram and his co-workers. They synthesized a crown ether containing a 2,2′-disubstituted 1,1′-binaphthyl unit (e.g., **342**) which can be prepared in its optically active forms,[993] and they observed that the complexing properties of the enantiomers towards certain optically active amino acids were different. Thus *S*-valine combined much more strongly with *S*-**342** than the *R* isomer of the ligands. A chromatographic procedure was developed whereby *S*-valine was absorbed on a Celite column, and the *S* and *R* isomers of the complex were separated by passage through the column in the order stated. Used in reverse, this technique can be adapted to resolve *R,S*-valine.[994] Cram and co-workers[995,995a] also prepared optically active cyclic ethers containing two 2,2′-substituted 1,1′-binaphthyl units which were capable of resolving the hexafluorophosphonate salts of *R,S*-methyl phenylglycinate and methyl valinate into their enantiomers. They succeeded in immobilizing a complex on to silica gel which they used to separate the antipodes of hexafluorophosphonate salts of α-phenyl ethylamine, α-phenylglycinate esters, and a phenylalaninate ester.[996] Macrocyclic ethers containing nine oxygen atoms which form complexes with guanidinium tetraphenyl borate or hexafluorophosphate were also described.[996a]

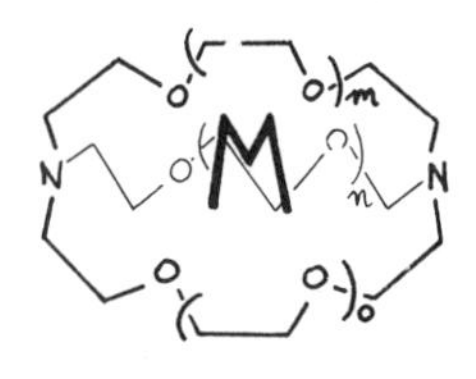

340

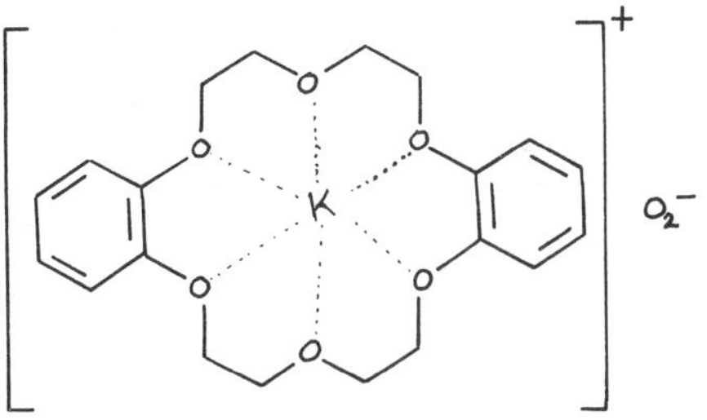

341

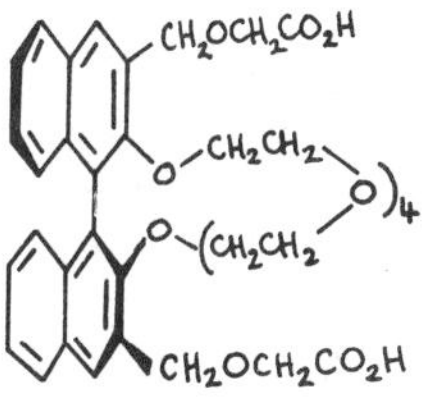

342

E. Miscellaneous

Several examples of aromatic nitrogen (Part I, Chapter 2, Section III.D; Part I, Chapter 3, Section XIV.C) and sulfur (Chapter 3, Section XIV.B and C) compounds which are dissymmetric because of restricted rotation about a single bond were reported previously. The furans also exhibit similar behavior and 3-(2-methyl-6-nitrophenyl)- (**343,** $R^1 = Me$, $R^2 = H$), 3-(2,4-dimethyl-6-nitrophenyl)- (**343,** $R^1 = R^2 = Me$), and 3-(2-nitrophenyl)-2,5-dimethyl-4-carboxyfurans (**343,** $R^1 = R^2 = H$) were resolved through their quinine salts. The enantiomers of **343** ($R^1 = Me$, $R^2 = H$) and **343** ($R^1 = R^2 = Me$) were stable in $CHCl_3$ at 25°C for one week but racemized in boiling EtOH, whereas the quinine salt of **343** ($R^1 = R^2 = H$) was quite unstable and racemized at 25°C in 3 hr.[997] It is not surprising then that the 2,5-dimesityl-3,4-dimethylfurans (**344,** $R^1 = Me$, Br, or NO_2; $R^2 = H$, Br, or NO_2) did not exhibit diastereoisomerism because if these were separated they would be easily interconvertible.[998] Diastereoisomerism could perhaps be demonstrated by nmr spectroscopy at low temperature.

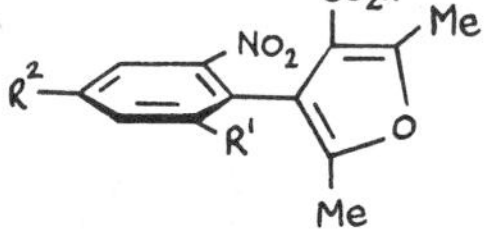

343

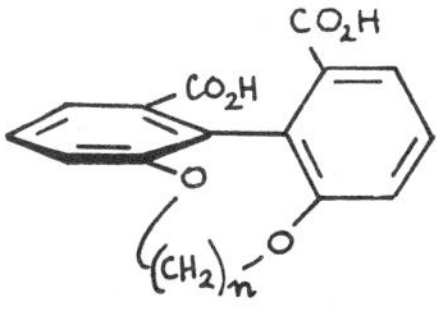

345

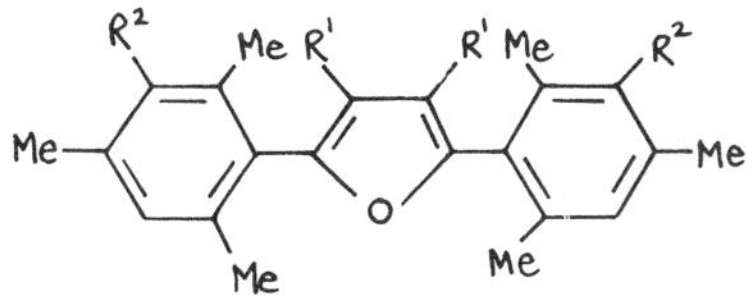

344

Diphenic acid could not be resolved into its optical forms, but Adams and Kornblum[999] succeeded in resolving the diphenic acids **345,** $n=8$, and **345,** $n=10$, which have a diether chain linking the 6,6′-positions. They found that the latter compound with $n=10$ racemized faster than the former compound with the shorter 6,6′-bridge. Since racemization is the result of the two carboxyl and two ether groups crossing each other, the greater flexibility of the longer 6,6′-chain would allow more facile crossing of these groups, and hence more rapid racemization. Kataoka, Ando, and Nakagawa[1000] demonstrated that the strain in a molecule such as the cyclic acetylene ether **346** can be observed in the uv spectra. The spectral shifts for $n=7$, 8, 9, 10, 11, and 12 varied in a zig-zag fashion being large at $n=8$, 10, and 12 and relatively smaller at $n=7$, 9, and 11.

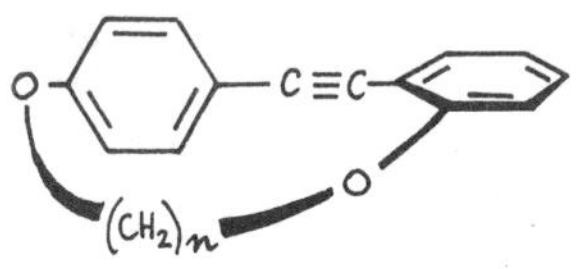

O C≡C $(CH_2)_n$ O

346

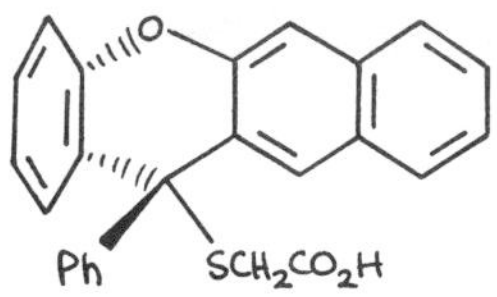

O Ph SCH_2CO_2H

347

Me H HO C_5H_{11} O H Me Me

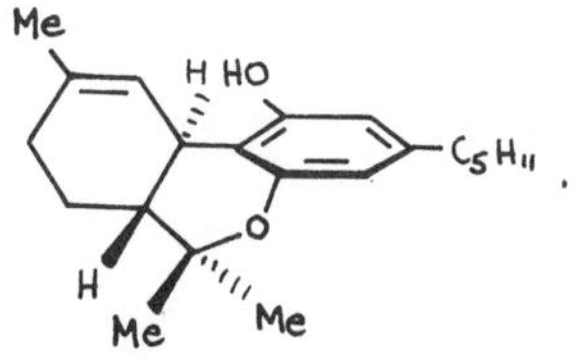

348

The nonplanar structure of 12-carboxymethylthio-12-phenyl benzoxanthene **(347)** was confirmed by resolution of its brucine salt into the optical antipodes.[1001]

The stereochemistry of cannabinoids, for example, (−)-Δ^9-tetrahydrocannabinol **(348)**, was investigated by Adams and co-workers,[1002–1005] and more recently, ^{1}H nmr spectroscopy has provided information regarding the conformation of the pyran ring.[1006] The absolute configuration of the cannabinol **348,** a major constituent of hashish, was reinvestigated and the configuration confirmed by conversion into (−)-menthanecarboxylic acid whose structure was related to (−)-menthol of known configuration.[1007] The chemistry (including absolute configurations), biogenesis, and biological activity of cannabinoids were reviewed by Mechoulam and Gaoni.[1008]

349

The synthesis and properties of 9-oxabicyclo[3.3.1]non-1-ene (**349**) were described by Quinn and Wiseman.[1009] It is another example that violates the Bredt rule (see also the 9-thia analogue, formula **198,** in Chapter 3, Section XIII).

XV. REFERENCES

1. S. Winstein and R. B. Henderson in *Heterocyclic Compounds*, Vol. 1, R. C. Elderfield, Ed., Wiley, New York, 1950, p. 8.
2. A. Rosowsky in *Heterocyclic Compounds with Three- and Four-Membered Rings*, Vol. 19, Part 1, A. Weissberger, Ed., Interscience, New York, 1964, p. 1.
3. R. E. Parker and N. S. Isaacs, *Chem. Rev.*, **59,** 737 (1959).
4. D. Swern, *Org. Reactions*, **7,** 378 (1953).
5. G. Berti, *Topics Stereochem.*, **7,** 93 (1972).
6. R. C. Fahey, *Topics Stereochem.*, **3,** 294 (1968).
7. H. B. Henbest in "Organic Reaction Mechanisms," *Chem. Soc. Special Publ.*, No. 19, p. 83 (1965).
8. M. S. Newman and B. J. Magerlein, *Org. Reactions*, **5,** 413 (1949).
9. M. Ballester, *Chem. Rev.*, **55,** 283 (1955).
10. R. N. McDonald in *Mechanisms of Molecular Migrations*, Vol. 3, B. S. Thyagarajan, Ed., Wiley-Interscience, New York, 1971, p. 67.
11. F. H. Newth, *Quart. Rev.*, **13,** 30 (1959).

11a. W. A. Latham, L. Radom, P. C. Hariharan, W. J. Hehre, and J. A. Pople, *Fortschr. Chem. Forsch.*, **40,** 1 (1973).

12. P. D. Bartlett, *Rec. Chem. Progr.*, **11,** 47 (1950).
13. A. Ažman, B. Borštnik, and B. Plesničar, *J. Org. Chem.*, **34,** 971 (1969).
14. R. Kavčič and B. Plesničar, *J. Org. Chem.*, **35,** 2033 (1970).
15. N. N. Schwartz and J. H. Blumbergs, *J. Org. Chem.*, **29,** 1976 (1964).
16. B. Rickborn and S-Y. Lwo, *J. Org. Chem.*, **30,** 2212 (1965).
17. R. G. Carlson and N. S. Behn, *J. Org. Chem.*, **32,** 1363 (1967).
18. A. Casadevall, E. Casadevall, and M. Lasperas, *Bull. Soc. chim. France*, 4506 (1968).
19. H. C. Brown, J. H. Kawakami, and S. Ikegami, *J. Amer. Chem. Soc.*, **92,** 6914 (1970).
20. H. Klein, W. Kursawa, and W. Grimme, *Angew. Chem. Internat. Edn.*, **12,** 580 (1973).
21. F. Sweet and R. K. Brown, *Canad. J. Chem.*, **46,** 707 (1968).
22. H. B. Henbest and R. A. L. Wilson, *J. Chem. Soc.*, 1958 (1957).
23. R. Albrecht and Ch. Tamm, *Helv. Chim. Acta*, **40,** 2216 (1957).

24. P. Chamberlain, M. L. Roberts, and G. H. Whitham, *J. Chem. Soc.* (*B*), 1374 (1970).
25. L. Goodman, S. Winstein, and R. Boschan, *J. Amer. Chem. Soc.*, **80,** 4312 (1958).
26. T. Itoh, K. Kaneda, and S. Teranishi, *Bull. Chem. Soc. Japan*, **48,** 1337 (1975).
27. R. C. Ewins, H. B. Henbest, and M. A. McKervey, *Chem. Comm.*, 1085 (1967).
28. D. R. Boyd and M. A. McKervey, *Quart. Rev.*, **22,** 95 (1968).
29. G. V. Pigulevskiĭ and G. V. Markina, *Doklady. Akad. Nauk S.S.S.R.*, **63,** 677 (1948), *Chem. Abs.*, **43,** 4628 (1949).
30. R. M. Bowman and M. F. Grundon, *J. Chem. Soc.* (*C*), 2368 (1967); R. M. Bowman, J. F. Collins, and M. F. Grundon, *Chem. Comm.*, 1131 (1967).
31. F. Montanari, I. Moretti, and G. Torre, *Chem. Comm.*, 135 (1969).
32. D. R. Boyd, D. M. Jerina, and J. W. Daly, *J. Org. Chem.*, **35,** 3170 (1970).
33. F. Montanari, I. Moretti, and G. Torre, *Gazzetta*, **104,** 7 (1974).
34. G. B. Payne and P. H. Williams, *J. Org. Chem.*, **24,** 54 (1959).
35. E. J. Glamkowski, G. Gal, R. Purick, A. J. Davidson, and M. Sletzinger, *J. Org. Chem.*, **35,** 3510 (1970).
35a. H. O. House and R. S. Ro, *J. Amer. Chem. Soc.*, **80,** 2428 (1958); H. H. Wasserman and N. E. Aubrey, *J. Amer. Chem. Soc.*, **77,** 590 (1955).
36. R. Curci and F. DiFuria, *Tetrahedron Letters*, 4085 (1974).
37. A. Robert and A. Foucaud, *Bull. Soc. chim. France*, 2531 (1969).
38. G. E. M. Moussa and N. F. Eweiss, *J. Appl. Chem.*, **19,** 313 (1969); W. J. Hickinbottom, D. Peters, and D. G. M. Wood, *J. Chem. Soc.*, 1360 (1955).
39. W. Fickett, H. K. Garner, and H. J. Lucas, *J. Amer. Chem. Soc.*, **73,** 5063 (1951).
40. E. E. van Tamelen and K. B. Sharpless, *Tetrahedron Letters*, 2655 (1967).
41. B. Franzus and J. H. Surridge, *J. Org. Chem.*, **31,** 4286 (1966).
42. J. W. Cornforth and D. T. Green, *J. Chem. Soc.* (*C*), 846 (1970).
43. E. L. Eliel and D. W. Delmonte, *J. Org. Chem.*, **21,** 596 (1956).
44. S. Mitsui and S. Imaizumi, *J. Chem. Soc. Japan* (*Pure Chemistry*), **86,** 219 (1965), *Chem. Abs.*, **68,** 4133 (1965).
45. A. S. Hallsworth and H. B. Henbest, *J. Chem. Soc.*, 4604 (1957).
46. M. Davis and V. Petrow, *J. Chem. Soc.*, 2536 (1949).
47. G. Berti, F. Bottari, and A. Marsili, *Tetrahedron*, **25,** 2939 (1969).
48. G. Berti, B. Macchia, F. Macchia, and L. Monti, *J. Org. Chem.*, **33,** 4045 (1968).
48a. M. Rosenberger, A. J. Duggan, R. Borer, R. Müller, and G. Saucy, *Helv. Chim. Acta*, **55,** 2663 (1972).
49. F. Fischer, *Chem. Ber.*, **94,** 893 (1961).
50. G. Berti and F. Bottari, *J. Org. Chem.*, **25,** 1286 (1960).
51. R. J. W. Cremyln, D. L. Garmaise, and C. W. Shoppee, *J. Chem. Soc.*, 1847 (1953).
52. R. Wylde and F. Forissier, *Bull. Soc. chim. France*, 4508 (1969).
53. M. Chérest, H. Felkin, J. Sicher, F. Šipoš, and M. Tichý, *J. Chem. Soc.*, 2513 (1965).
54. H. Kwart and L. G. Kirk, *J. Org. Chem.*, **22,** 116 (1957).
55. H. H. Wasserman and J. P. Brous, *J. Org. Chem.*, **19,** 515 (1954).
56. N. H. Cromwell and J. L. Martin, *J. Org. Chem.*, **33,** 1890 (1968).
57. F. W. Bachelor and R. K. Bansal, *J. Org. Chem.*, **34,** 3600 (1969).

58. J. Seyden-Penne, M. C. Roux-Schmitt, and A. Roux, *Tetrahedron*, **26,** 2649 (1970).
59. G. Kyriakakou and J. Seyden-Penne, *Tetrahedron Letters*, 1737 (1974).
60. P. F. Vogt and D. F. Tavares, *Canad. J. Chem.*, **47,** 2875 (1969).
61. M. Charpentier-Morize, R. Doukhan, and J. Sansoulet, *Bull. Soc. chim. France*, 685 (1968).
62. H. E. Zimmermann and L. Ahramjian, *J. Amer. Chem. Soc.*, **82,** 5459 (1960).
63. J. J. Riehl and L. Thil, *Tetrahedron Letters*, 1913 (1969).
64. J. Cantacuzène and R. Jantzen, *Tetrahedron*, **26,** 2429 (1970).
65. J. Cantacuzène, M. Atlani, and J. Anibié, *Tetrahedron Letters*, 2335 (1968); J. Cantacuzène and M. Atlani, *Tetrahedron*, **26,** 2447 (1970).
66. J. A. Marshall and J. J. Partridge, *J. Org. Chem.*, **33,** 4090 (1968).
67. R. G. Carlson and N. S. Behn, *J. Org. Chem.*, **33,** 2069 (1968); P. A. Hart and R. A. Sandmann, *Tetrahedron Letters*, 305 (1969).
68. E. J. Corey and M. Chaykovsky, *J. Amer. Chem. Soc.*, **87,** 1353 (1965) and **84,** 867 (1962).
69. A. W. Johnson, V. J. Hruby, and J. L. Williams, *J. Amer. Chem. Soc.*, **86,** 918 (1964).
70. E. J. Corey and W. Oppolzer, *J. Amer. Chem. Soc.*, **86,** 1899 (1964).
71. C. R. Johnson and G. F. Katekar, *J. Amer. Chem. Soc.*, **92,** 5753 (1970).
72. C. R. Johnson and C. W. Schroeck, *J. Amer. Chem. Soc.*, **90,** 6852 (1968).
73. H. J. Lucas, M. J. Schlatter, and R. C. Jones, *J. Amer. Chem. Soc.*, **63,** 22 (1941).
74. J. Read and I. C. M. Campbell, *J. Chem. Soc.*, 2377 (1930).
75. R. Kuhn and F. Ebel, *Ber.*, **58,** 919 (1925) and **59,** 2514 (1926).
76. N. H. Cromwell and R. A. Setterquist, *J. Amer. Chem. Soc.*, **76,** 5752 (1954).
77. N. H. Cromwell, F. H. Schumacher, and J. L. Adelfang, *J. Amer. Chem. Soc.*, **83,** 974 (1961).
78. H. H. Wasserman and J. B. Brous, *J. Amer. Chem. Soc.*, **76,** 5811 (1954).
79. C. L. Stevens and V. J. Traynelis, *J. Org. Chem.*, **19,** 533 (1954).
80. C. L. Stevens, R. J. Church, and V. J. Traynelis, *J. Org. Chem.*, **19,** 522 (1954).
81. M. R. Lespieau and B. Gredy, *Bull. Soc. chim. France*, 769 (1933).
82. A. T. Bottini and R. L. VanEtten, *J. Org. Chem.*, **30,** 2994 (1965).
83. E. L. Eliel, *Stereochemistry of Carbon Compounds*, McGraw-Hill, New York, 1962, pp. 216, 217.
84. T. J. Batterham, *N.M.R. Spectra of Simple Heterocycles*, E. C. Taylor and A. Weissberger, Eds,. Wiley-Interscience, New York, 1973, pp. 365–368.
85. I. Moretti, F. Taddei, G. Torre, and N. Spassky, *J.C.S. Chem. Comm.*, 25 (1973).
86. D. R. Paulson, F. Y. N. Tang, G. F. Noran, A. S. Murray, B. P. Pelka, E. M. Vasquez, *J. Org. Chem.*, **40,** 184 (1975).
87. F. Fischer and H. Rönsch, *Chem. Ber.*, **94,** 901 (1961).
88. W. Kirmse, H. Arold, and B. Kornrumpf, *Chem. Ber.*, **104,** 1783 (1971).
89. K. Harada and J-i. Oh-hashi, *Bull. Chem. Soc. Japan*, **39,** 2311 (1966); Y. Liwschitz, Y. Rabinsohn, and D. Perera, *J. Chem. Soc.*, 1116 (1962).
90. K. Harada and Y. Nakajima, *Bull. Chem. Soc. Japan*, **47,** 2911 (1974).
91. K. Harada, J. *Org. Chem.*, **31,** 1407 (1966).
92. A. Collet, *Bull. Soc. chim. France*, 215 (1975).

93. H. M. Walborsky and C. G. Pitt, *J. Amer. Chem. Soc.*, **84,** 4831 (1962).
94. I. Moretti and G. Torre, *Tetrahedron Letters*, 2717 (1969).
95. G. Gottarelli, S. F. Mason, and G. Torre, *J. Chem. Soc. (B)*, 1349 (1970).
96. B. Ottar, *Acta Chem. Scand.*, **1,** 283 (1947).
97. A. C. Cope and J. H. Kecht, *J. Amer. Chem. Soc.*, **84,** 4872 (1962).
98. J. B. Lambert and D. H. Johnson, *J. Amer. Chem. Soc.*, **90,** 1349 (1968).
99. J. J. Uebel, *Tetrahedron Letters*, 4751 (1967).
100. K. L. Servis, E. A. Noe, N. R. Easton, Jr., and F. A. L. Anet, *J. Amer. Chem. Soc.*, **96,** 4185 (1974).
101. S. Merlino, G. Lami, B. Macchia, F. Macchia, and L. Monti, *J. Org. Chem.*, **37,** 703 (1972).
102. D. J. Williams, P. Crotti, B. Macchia, and F. Macchia, *Tetrahedron*, **31,** 993 (1975).
103. H. J. Lucas and H. K. Garner, *J. Amer. Chem. Soc.*, **70,** 990 (1948) and **72,** 2145 (1950); H. J. Lucas and C. W. Gould, Jr., *J. Amer. Chem. Soc.*, **63,** 2541 (1941).
104. S. Winstein and H. J. Lucas, *J. Amer. Chem. Soc.*, **61,** 1576, 2845 (1939).
105. C. E. Wilson and H. J. Lucas, *J. Amer. Chem. Soc.*, **58,** 2396 (1936).
106. W. Hückel and U. Wörffel, *Chem. Ber.*, **88,** 338 (1955).
107. H. B. Henbest, M. Smith, and A. Thomas, *J. Chem. Soc.*, 3293 (1958); G. K. Helmkamp and H. J. Lucas, *J. Amer. Chem. Soc.*, **74,** 951 (1952).
108. P. A. Levene and A. Walti, *J. Biol. Chem.*, **68,** 415 (1926).
109. P. A. Levene and A. Walti, *J. Biol. Chem.*, **73,** 263 (1927).
110. F. A. Long and J. G. Pritchard, *J. Amer. Chem. Soc.*, **78,** 2663 (1956).
111. J. G. Pritchard and F. A. Long, *J. Amer. Chem. Soc.*, **78,** 2667 (1956).
112. J. G. Pritchard and F. A. Long, *J. Amer. Chem. Soc.*, **78,** 6008 (1956).
113. G. Berti, B. Macchia, and F. Macchia, *Tetrahedron*, **28,** 1299 (1972).
114. P. Crotti, B. Macchia, and F. Macchia, *Tetrahedron*, **29,** 155 (1973).
115. A. A. Akhrem, A. M. Moiseenkov, and V. N. Dobrynin, *Russ. Chem. Rev.*, **37,** 448 (1968).
116. H. E. Audier, J. F. Dupin, and J. Jullien, *Bull. Soc. chim. France*, 3844 (1968).
117. F. H. Dickey, W. Fickett, and H. J. Lucas, *J. Amer. Chem. Soc.*, **74,** 944 (1952).
118. W. Stühmer and G. Messwarb, *Arch. Pharm.*, **286,** 19 (1953).
119. R. M. Laird and R. E. Parker, *J. Amer. Chem. Soc.*, **83,** 4277 (1961).
120. A. J. Castro, D. K. Brain, H. D. Fischer, and R. K. Fuller, *J. Org. Chem.*, **19,** 1444 (1954).
121. K. Jankowski and J-Y. Daigle, *Canad. J. Chem.*, **49,** 2594 (1971).
122. C. C. Price and P. J. Kirk, *J. Amer. Chem. Soc.*, **75,** 2396 (1953).
123. P. L. Nichols, Jr., and J. D. Ingham, *J. Amer. Chem. Soc.*, **77,** 6547 (1955).
124. R. Ketcham and V. P. Shah, *J. Org. Chem.*, **28,** 229 (1963).
125. F. G. Bordwell and H. M. Andersen, *J. Amer. Chem. Soc.*, **75,** 4959 (1953).
126. C. G. Overberger and A. Drucker, *J. Org. Chem.*, **29,** 360 (1964).
127. L. Goodman, A. Benitez, and B. R. Baker, *J. Amer. Chem. Soc.*, **80,** 1680 (1958).
128. R. A. Wohl and J. Cannie, *J. Org. Chem.*, **38,** 1787 (1973).
129. R. A. Izydore and R. G. Ghirardelli, *J. Org. Chem.*, **38,** 1790 (1973).

130. D. J. Pastro, C. C. Cumbo, and J. Frazer, *J. Amer. Chem. Soc.*, **88,** 2194 (1966).
131. M. Rosenblum, A. Longroy, M. Neveu, and C. Steel, *J. Amer. Chem. Soc.*, **87,** 5716 (1965).
132. D. E. Bissing and A. J. Speziale, *J. Amer. Chem. Soc.*, **87,** 2683 (1965).
133. D. L. J. Clive and C. V. Denyer, *J.C.S. Chem. Comm.*, 253 (1973).
134. L. A. Paquette, A. A. Youssef, and M. L. Wise, *J. Amer. Chem. Soc.*, **89,** 5246 (1967).
135. E. L. Eliel and D. W. Delmonte, *J. Amer. Chem. Soc.*, **78,** 3226 (1956).
136. E. L. Eliel and D. W. Delmonte, *J. Amer. Chem. Soc.*, **80,** 1744 (1958).
137. M. N. Rerick and E. L. Eliel, *J. Amer. Chem. Soc.*, **84,** 2356 (1962).
138. D. K. Murphy, R. L. Alumbaugh, and B. Rickborn, *J. Amer. Chem. Soc.*, **91,** 2649 (1969).
139. R. Gyuon and P. Villa, *Bull. Sci. chim. France*, 1375 (1972).
140. P. J. Lerouz and H. J. Lucas, *J. Amer. Chem. Soc.*, **73,** 41 (1951).
141. G. K. Helmkamp and N. Schnautz, *Tetrahedron*, **2,** 304 (1958).
142. R. Fuchs, *J. Amer. Chem. Soc.*, **78,** 5612 (1956).
143. R. Fuchs and C. A. VanderWerf, *J. Amer. Chem. Soc.*, **76,** 1631 (1954).
144. G. J. Park and R. Fuchs, *J. Org. Chem.*, **21,** 1513 (1956).
145. C. L. Stevens and T. H. Coffield, *J. Org. Chem.*, **23,** 336 (1958).
146. A. Rashid and G. Read, *J. Chem. Soc.*, (*C*), 2053 (1969).
147, J. W. Cornforth, R. H. Cornforth, C. Donninger, and G. Popják, *Proc. Roy. Soc., Ser. B.*, **163,** 492 (1965).
148. J. W. Cornforth, R. H. Cornforth, G. Popják, and L. Yengoyan, *J. Biol. Chem.*, **241,** 3970 (1966); J. W. Cornforth and G. Ryback, *Ann. Reports, Chem. Soc.*, **52,** 428 (1965).
149. J. W. Cornforth, J. W. Redmond, H. Eggerer, W. Buckel, and C. Gutschow, *European J. Biochem.*, **14,** 1 (1970), *Nature*, **221,** 1212 (1969).
150. D. J. Pasto, C. C. Cumbo, and J. Hickman, *J. Amer. Chem. Soc.*, **88,** 2201 (1966).
151. G. J. Park and R. Fuchs, *J. Org. Chem.*, **22,** 93 (1957).
152. M. S. Newman, G. Underwood, and M. Renoll, *J. Amer. Chem. Soc.*, **71,** 3362 (1949).
153. A. R. Graham, A. F. Millidge, and D. P. Young, *J. Chem. Soc.*, 2180 (1954).
154. D. Abenhaim, J-L. Namy, and G. Boireau, *Bull. Soc. chim. France*, 3254 (1971).
155. J. D. Morrison, R. L. Atkins, and J. E. Tomaszewski, *Tetrahedron Letters*, 4635 (1970).
156. T. Nakajima, S. Suga, T. Sugita, and K. Ichikawa, *Tetrahedron*, **25,** 1807 (1969).
157. P. C. Petrellis and G. W. Griffin, *Chem. Comm.*, 691 (1967).
158. H. Hamberger and R. Huisgen, *Chem. Comm.*, 1190 (1971).
159. A. Dahmen, H. Hamberger, R. Huisgen, and V. Markowski, *Chem. Comm.*, 1192 (1971).
160. H. H. J. MacDonald and R. J. Crawford, *Canad. J. Chem.*, **50,** 428 (1972).
161. R. J. Crawford, V. Vukov, and N. Tokunaga, *Canad. J. Chem.*, **51,** 3718 (1973); V. Vukov and R. J. Crawford, *Canad. J. Chem.*, **53,** 1367 (1975).
162. J-C. Paladini and J. Chuche, *Bull. Soc. chim. France*, 197 (1974).

163. T. Do-Minh, A. M. Trozzolo, and G. W. Griffin, *J. Amer. Chem. Soc.*, **92,** 1402 (1970).

164. R. N. McDonald and D. G. Hill, *J. Org. Chem.*, **35,** 2942 (1970).

165. H. O. House and G. D. Ryerson, *J. Amer. Chem. Soc.*, **83,** 979 (1961).

166. B. N. Blackett, J. M. Coxon, M. P. Hartshorn, and K. E. Richards, *Austral. J. Chem.*, **23,** 2077 (1970).

167. G. B. Payne, *J. Org. Chem.*, **27,** 3819 (1962).

168. E. Vogel and H. Günther, *Angew. Chem. Internat. Edn.*, **6,** 385 (1967); D. M. Jerina, H. Yagi, and J. W. Daly, *Heterocycles*, **1,** 267 (1973).

169. A. G. Anastassiou and R. P. Cellura, *Chem. Comm.*, 903 (1969).

170. A. G. Anastassiou and R. P. Cellura, *Chem. Comm.*, 1521 (1969).

171. J. M. Holovka, R. R. Grabbe, P. D. Gardner, C. B. Strow, M. L. Hill, and T. V. Van Auken, *Chem. Comm.*, 1522 (1969).

172. D. M. Jerina, J. W. Daly, and B. Witkop, *J. Amer. Chem. Soc.*, **90,** 6523, 6525 (1968).

173. G. J. Kasperek and T. C. Bruice, *J. Amer. Chem. Soc.*, **94,** 198 (1972).

174. L. J. Altman and R. C. Baldwin, *Tetrahedron Letters*, 981 (1972).

175. M. T. Bogert and R. O. Roblin, *J. Amer. Chem. Soc.*, **55,** 3741 (1933); K. Bodenbenner, *Annalen*, **623,** 183 (1959).

176. A. G. Causa, H. Y. Chen, S. Y. Tark, and H. J. Harwood, *J. Org. Chem.*, **38,** 1385 (1973).

177. H. Klein and W. Grimme, *Angew. Chem. Internat. Edn.*, **13,** 672 (1974).

178. H. Günther, J. B. Pawliczek, J. Ulmen, and W. Grimme, *Angew. Chem. Internat. Edn.*, **11,** 517 (1972).

179. J. L. Everett and G. A. R. Kon, *J. Chem. Soc.*, 3131 (1950).

180. P. W. Feit, *Chem. Ber.*, **93,** 116 (1960).

181. S. M. Kupchan, R. J. Hemingway, and R. M. Smith, *J. Org. Chem.*, **34,** 3898 (1969).

182. D. B. Borders, P. Shu, and J. E. Lancaster, *J. Amer. Chem. Soc.*, **94,** 2540 (1972).

183. E. Vogel, H-J. Altenbach, and D. Cremer, *Angew. Chem. Internat. Edn.*, **11,** 935 (1972).

184. H-J. Altenbach and E. Vogel, *Angew. Chem. Internat. Edn.*, **11,** 937 (1972).

185. E. Vogel, H-J. Altenbach, and E. Schmidbauer, *Angew. Chem. Internat. Edn.*, **12,** 838 (1973).

186. T. W. Craig, G. R. Harvey, and G. A. Berchtold, *J. Org. Chem.*, **32,** 3743 (1967); R. W. Gleason and J. T. Snow, *J. Org. Chem.*, **34,** 1963 (1969).

187. E. Vogel, H-J. Altenbach, and C-D. Sommerfeld, *Angew. Chem. Internat. Edn.*, **11,** 939 (1972).

188. R. Schwesinger and H. Prinzbach, *Angew. Chem. Internat. Edn.*, **11,** 942 (1972).

189. C. H. Foster and G. A. Berchtold, *J. Amer. Chem. Soc.*, **94,** 7939 (1972).

190. C. S. Foote, S. Mazur, P. A. Burns, and D. Lerdal, *J. Amer. Chem. Soc.*, **97,** 586 (1973).

191. L. Knothe and H. Prinzbach, *Tetrahedron Letters*, 1319 (1975).

192. A. C. Cope, B. S. Fisher, W. Funke, J. M. McIntosh, and M. A. McKervey, *J. Org. Chem.*, **34,** 2231 (1969).

193. A. G. Anastassiou and E. Reichmanis, *J. Org. Chem.*, **38,** 2421 (1973).

193a. F. A. L. Anet and I. Yavari, *Tetrahedron Letters*, 1567 (1975).

194. R. M. Moriarty, *Topics Stereochem.*, **8,** 271 (1974).

195. S. Searles, Jr. in *Heterocyclic Compounds with Three- and Four-Membered Rings*, Part 2, A. Weissberger, Ed., Interscience, New York, 1964, Ch. 9, p. 983.

196. D. R. Arnold, *Adv. Photochem.*, **6,** 301 (1968).

197. D. R. Arnold, R. L. Hinman, and A. H. Glick, *Tetrahedron Letters*, 1425 (1964).

198. N. C. Yang and W. Eisenhardt, *J. Amer. Chem. Soc.*, **93,** 1277 (1971).

199. N. J. Turro, P. Wriede, J. C. Dalton, D. R. Arnold, and A. H. Glick, *J. Amer. Chem. Soc.*, **89,** 3950 (1967).

200. J. J. Beereboom and M. S. von Wittenau, *J. Org. Chem.*, **30,** 1231 (1965).

201. R. A. Caldwell, G. W. Sovocool, and R. P. Gajewski, *J. Amer. Chem. Soc.*, **95,** 2549 (1973).

202. D. Scharf and F. Korte, *Tetrahedron Letters*, 821 (1963).

203. T. Kubota, K. Shima, S. Toki, and H. Sakurai, *Chem. Comm.*, 1462 (1969).

204. L. E. Friedrich and G. B. Schuster, *J. Amer. Chem. Soc.*, **91,** 7204 (1969).

205. G. Jones II and J. C. Staires, *Tetrahedron Letters*, 2099 (1974).

206. N. J. Turro and P. A. Wriede, *J. Amer. Chem. Soc.*, **92,** 320 (1970).

207. K. Shima and H. Sakurai, *Bull. Chem. Soc. Japan*, **39,** 1806 (1966).

208. G. O. Schenck, W. Hartmann, and R. Steinmetz, *Chem. Ber.*, **96,** 498 (1963); D. Gagnaire and E. Payo-Subiza, *Bull. Soc. chim. France*, 2623 (1963).

209. S. Toki and H. Sakurai, *Bull. Chem. Soc. Japan*, **40,** 2885 (1967).

210. J. Leitich, *Tetrahedron Letters*, 1937 (1967).

211. M. Ogata, H. Watanabe, and H. Kano, *Tetrahedron Letters*, 533 (1967).

212. G. R. Evanega and E. B. Whipple, *Tetrahedron Letters*, 2163 (1967).

213. N. R. Lazear and J. H. Schauble, *J. Org. Chem.*, **39,** 2069 (1974).

214. A. Balsamo, G. Ceccarelli, P. Crotti, and F. Macchia, *J. Org. Chem.*, **40,** 473 (1975).

215. A. Rosowsky and D. S. Tarbell, *J. Org. Chem.*, **26,** 2255 (1961).

216. H. B. Henbest and B. B. Millward, *J. Chem. Soc.*, 3575 (1960).

217. M. Tada, T. Kokubo, and T. Sato, *Bull. Chem. Soc. Japan*, **43,** 2162 (1970).

218. E. W. Lanfear and J. F. Thorpe, *J. Chem. Soc.*, 1683 (1923); S. S. Deshapande and J. F. Thorpe, *J. Chem. Soc.*, 1430 (1922).

219. K. B. Wiberg and H. W. Holmquist, *J. Org. Chem.*, **24,** 578 (1959).

220. Y. Etienne and N. Fischer in *Heterocyclic Compounds with Three- and Four-Membered Rings*, Part 2, A. Weissberger, Ed., Interscience, New York, 1964, Ch. 6, p. 729.

221. H. E. Zuagg, *Org. Reactions*, **8,** 305 (1954).

222. W. A. Cowdrey, E. D. Hughes, C. K. Ingold, S. Masterman, and A. D. Scott, *J. Chem. Soc.*, 1252 (1937).

223. A. R. Olson and R. J. Miller, *J. Amer. Chem. Soc.*, **60,** 2687 (1938).

224. B. Holmberg, *J. prakt. Chem.*, **87,** 456 (1913).

225. B. Holmberg, *Finska Kemistsamfundets Medd.*, **54,** No. 3/4, 116 (1945), *Chem. Abs.*, **44,** 9380 (1950).

226. E. J. Corey and W. H. Pirkle, *Tetrahedron Letters*, 5255 (1967).

227. T. J. Batterham, *N.M.R. Spectra of Simple Heterocycles*, E. C. Taylor and A. Weissberger, Eds., Wiley-Interscience, New York, 1973, p. 369.

228. K. Pihlaja, K. Polviander, R. Keskinen, and J. Jalonen, *Acta Chem. Scand.*, **25,** 765 (1971).
229. J. Jokisaari, *Z. Naturforsch*, **26A,** 136 (1971); J. Jokisaari, E. Rahkamaa, and H. Malo, *Z. Naturforsch*, **26A,** 973 (1971); J. Jokisaari, E. Rahkamaa, and P. O. I. Virtanen, *Soumen Kemistilehti*, B, **43,** 14 (1970), *Chem. Abs.*, **72,** 72956 (1970).
230. A. V. Bogat-skii, G. A. Filip, Y. Y. Samitov, N. G. Izotova, and G. V. P'yankova, *J. Gen. Chem. (U.S.S.R.)*, **39,** 909 (1969).
231. H. A. J. Carless, *J.C.S. Chem. Comm.*, 982 (1974).
232. Y. Y. Samitov, A. V. Bogatskii, and G. A. Filip, *Zhur. org. Khim.*, **7,** 585 (1971).
233. S. I. Chann, J. Zinn, J. Fernandez, and W. D. Gwinn, *J. Chem. Phys.*, **33,** 1643 (1960); S. I. Chann, J. Zinn, and W. D. Gwinn, *J. Chem. Phys.*, **34,** 1319 (1961).
234. R. F. Zürcher and H. H. Günthard, *Helv. Chim. Acta*, **40,** 89 (1959).
235. G. L. McKnown and R. A. Beaudet, *J. Chem. Phys.*, **55,** 3105 (1971).
236. R. M. Beesley, C. K. Ingold, and J. F. Thorpe, *J. Chem. Soc.*, 1080 (1915).
237. S. Searles, E. F. Lutz, and M. Tamres, *J. Amer. Chem. Soc.*, **82,** 2932 (1960).
238. N. Shimizu and S. Nishida, *J.C.S. Chem. Comm.*, 734 (1974).
239. J. A. H. Carless, *Tetrahedron Letters*, 3425 (1974).
240. G. Adams, C. Bibby, and R. Grigg, *J.C.S. Chem. Comm.*, 491 (1972).
241. D. J. Bogan, R. S. Sheinson, R. G. Gann, and F. W. Williams, *J. Amer. Chem. Soc.*, **97,** 2560 (1975).
242. A. P. Schaap and N. Tontapanish, *J.C.S. Chem. Comm.*, 490 (1972).
243. P. D. Bartlett and A. P. Schaap, *J. Amer. Chem. Soc.*, **92,** 3223 (1970).
244. H. H. Wasserman and I. Saito, *J. Amer. Chem. Soc.*, **97,** 905 (1975).
245. A. P. Schaap and P. D. Bartlett, *J. Amer. Chem. Soc.*, **92,** 6055 (1970).
246. P. Bosshard and C. H. Eugster, *Adv. Heterocyclic Chem.*, **7,** 377 (1966).
247. F. M. Dean, *Naturally Occurring Oxygen Ring Compounds*, *Butterworths*, London, 1963.
248. R. Tschesche, *Fortschr. Chem. org. Naturstoffe*, **4,** 1 (1945).
249. A. R. Pinder in *The Chemistry of Carbon Compounds*, Vol. IV.C, E. H. Rodd, Ed., Elsevier, Amsterdam, 1960, pp. 2019–2057.
250. A. P. Dunlop and F. N. Peters, *The Furans*, Reinhold, New York, 1953.
251. N. Clauson-Kaas, *Acta Chem. Scand.*, **6,** 569 (1952); N. Elming, *Acta Chem. Scand.*, **6,** 572 (1952); N. Clauson-Kaas, F. Limborg, and K. Glens, *Acta Chem. Scand.*, **6,** 531 (1952).
252. N. Clauson-Kaas and Z. Tyle, *Acta Chem. Scand.*, **6,** 667 (1952).
253. N. Clauson-Kaas and F. Limborg, *Acta Chem. Scand.*, **6,** 551 (1952).
254. N. Clauson-Kaas, *Acta Chem. Scand.*, **6,** 556 (1952); N. Clauson-Kaas, F. Limborg, and P. Dietrich, *Acta Chem. Scand.*, **6,** 545 (1952).
255. N. Elming, *Adv. Org. Chem.*, **2,** 67 (1960).
256. J. T. Nielson, N. Elming, and N. Clauson-Kaas, *Acta Chem. Scand.*, **12,** 63 (1958).
257. D. Gagnaire and P. Vottero, *Bull. Soc. chim. France*, 164 (1970).
258. D. Gagnaire and P. Monzeglio, *Bull. Soc. chim. France*, 474 (1965).
259. L. F. Wiggins and D. J. C. Wood, *J. Chem. Soc.*, 1566 (1950).
260. T. Kinoshita and T. Miwa, *J.C.S. Chem. Comm.*, 181 (1974).

261. N. Elming and N. Clauson-Kaas, *Acta Chem. Scand.*, **6,** 535 (1952).

262. D. Gagnaire and P. Vottero, *Bull. Soc. chim. France*, 2779 (1963).

264. C. Glacet, *Bull. Soc. chim. France*, 990 (1952).

265. C. Glacet, *Bull. Soc. chim. France*, 575 (1954).

266. F. Sweet and R. K. Brown, *Canad. J. Chem.*, **44,** 1571 (1966).

267. G. Dana and C. Roos, *Bull. Soc. chim. France*, 371 (1973).

268. G. Zweifel and J. Plamondon, *J. Org. Chem.*, **35,** 898 (1970).

269. G. O. Schenck and R. Steinmetz, *Annalen*, **668,** 19 (1963).

269a. H. Maskill, *J.C.S. Perkin II*, 197 (1975).

270. H. M. Rosenberg and M. P. Servé, *J. Org. Chem.*, **36,** 3015 (1971).

271. M. P. Servé and H. M. Rosenberg, *J. Org. Chem.*, **35,** 1237 (1970).

272. T. S. Cantrell, *J. Org. Chem.*, **39,** 2242 (1974).

273. O. Diels and K. Adler, *Annalen*, **489,** 243 (1931).

274. R. B. Woodward and H. Baer, *J. Amer. Chem. Soc.*, **70,** 1161 (1948).

275. J. A. Berson and R. Swidler, *J. Amer. Chem. Soc.*, **76,** 4060 (1954).

276. F. A. L. Anet, *Tetrahedron Letters*, 1219 (1962).

277. R. Daniels and J. L. Fischer, *J. Org. Chem.*, **28,** 320 (1963).

278. G. F. D'Alelio, C. J. Williams, Jr., and C. L. Wilson, *J. Org. Chem.*, **25,** 1025 (1960).

279. N. S. Zefirov, R. A. Ivanova, and Yu. K. Yur'ev, *J. Gen. Chem.* (*U.S.S.R.*), **35,** 54 (1965).

280. R. B. Woodward and R. B. Loftfield, *J. Amer. Chem. Soc.*, **63,** 3167 (1941); K. Ziegler, G. Schenck, E. W. Krockow, A. Siebert, A. Wenz, and H. Weber, *Annalen*, **551,** 1 (1942).

281. G. M. Brown and P. Dubreuil, *Canad. J. Chem.*, **46,** 1577 (1968).

282. A. W. McCulloch, B. Stanovnik, D. G. Smith, and A. G. McInnes, *Canad. J. Chem.*, **49,** 241 (1971).

283. W. L. Nelson and D. R. Allen, *J. Heterocyclic Chem.*, **9,** 561 (1972).

284. M. P. Kuntsmann, D. S. Tarbell, and R. L. Autrey, *J. Amer. Chem. Soc.*, **84,** 4115 (1962).

285. J. D. Slee and E. LeGoff, *J. Org. Chem.*, **35,** 3897 (1970).

286. R. P. Gandhi and V. K. Chadha, *Chem. Comm.*, 552 (1968); P. Vogel and A. Florey, *Helv. Chim. Acta*, **57,** 200 (1974).

287. J. Kallos and P. Deslongchamps, *Canad. J. Chem.*, **44,** 1239 (1966).

288. C. D. Weis, *J. Org. Chem.*, **28,** 74 (1963).

289. J. Laing, A. W. McCullloch, D. G. Smith, and A. G. McInnes, *Canad. J. Chem.*, **49,** 574 (1971); H. Prinzbach, M. Argüelles, and E. Druckrey, *Angew. Chem. Internat. Edn.*, **5,** 1039 (1966); W. Eberbach, M. Perroud-Argüelles, H. Achenbach, E. Druckery, and H. Prinzbach, *Helv. Chim. Acta*, **54,** 2579 (1971).

290. T. Kauffmann and F-P. Boettcher, *Chem. Ber.*, **95,** 949 (1962); G. Wittig and L. Pohmer, *Chem. Ber.*, **89,** 1334 (1956).

291. N. S. Zefirov, R. A. Ivanova, R. J. Filatova, and Yu. K. Yur'ev, *J. Gen. Chem.*, (U.S.S.R.), **36,** 1893 (1966).

292. A. W. Fort, *J. Amer. Chem. Soc.*, **84,** 4979 (1962).

293. N. S. Zefirov, A. F. Davydova, V. F. Bystrov, A. U. Stepanyants, and Yu. K. Yur'ev, *J. Gen. Chem.* (*U.S.S.R.*), **36,** 1734 (1966).
294. W. Eberbach and M. Perroud-Argüelles, *Chem. Ber.*, **105,** 3078 (1972).
295. J. Jolivet, *Ann. Chim.* (*France*), **5,** 1165 (1960).
296. N. S. Zefirov, L. P. Prikazchikov, and Yu. K. Yur'ev, *J. Gen. Chem.* (*U.S.S.R.*), **35,** 825 (1965).
297. N. S. Zefirov, A. F. Davydova, F. A. Abdulvaleeva, and Yu. K. Yur'ev, *J. Gen. Chem.* (*U.S.S.R.*), **36,** 206 (1966).
298. N. S. Zefirov, A. F. Davydova, and Yu. K. Yur'ev, *J. Gen. Chem.* (*U.S.S.R.*), **35,** 817 (1965).
299. N. S. Zefirov and R. S. Filatova, *J. Gen. Chem.* (*U.S.S.R.*), **37,** 2322 (1967).
300. N. S. Zefirov, R. A. Ivanova, R. S. Filatova, and Yu. K. Yur'ev, *J. Gen. Chem.* (*U.S.S.R.*), **35,** 1794 (1965).
301. K. Adler and K. H. Backendorf, *Annalen*, **535,** 113 (1937).
302. J. Berridge, D. Bryce-Smith, and A. Gilbert, *J.C.S. Chem. Comm.*, 964 (1974).
303. J. L. Isidor, M. S. Brookhart, and R. L. McKee, *J. Org. Chem.*, **38,** 612 (1973).
304. R. K. Hill and W. R. Schearer, *J. Org. Chem.*, **27,** 921 (1962).
305. M. Lj. Mihailović, S. Gojković, and Ž. Čekovic, *J.C.S. Perkin I*, 2460 (1972).
306. D. Felix, A. Melera, J. Seibl, and E. sz. Kováts, *Helv. Chim. Acta*, **46,** 1513 (1963).
307. A. F. Thomas and M. Ozainne, *Helv. Chim. Acta*, **57,** 2062 (1974), and earlier parts in the series.
308. J. Jacobus, *J. Org. Chem.*, **38,** 402 (1973).
309. H. Neudeck and K. Schlögl, *Monatsh.*, **106,** 229 (1975).
310. A. R. Jones, *Chem. Comm.*, 1042 (1971).
311. P. S. Portoghese and D. A. Williams, *J. Heterocyclic Chem.*, **6,** 307 (1969).
312. E. G. E. Hawkins, *J. Chem. Soc.*, 248 (1959).
313. C. Botteghi, G. Consiglio, G. Ceccarelli, and A. Stefani, *J. Org. Chem.*, **37,** 1835 (1972).
314. N. R. Easton and V. B. Fish, *J. Amer. Chem. Soc.*, **76,** 2836 (1954).
315. M. Lj. Mihailović, R. I. Mamuzić, Lj. Žigić-Mamuzić, J. Bošnjak, and Ž. Čeković, *Tetrahedron*, **23,** 215 (1967).
316. M. Lj. Mihailović, S. Gojković, and S. Konstantinović, *Tetrahedron*, **29,** 3675 (1973).
317. D. Chambenois and G. Mousset, *Bull. Soc. chim. France*, 2969 (1974).
318. P. Martinet and G. Mousset, *Bull. Soc. chim. France*, 4093 (9171).
319. G. Mousset, *Bull. Soc. chim. France*, 4097 (1971).
320. H. Hiraoka, *Tetrahedron*, **29,** 2955 (1973).
321. N. J. Turro and D. M. McDaniel, *J. Amer. Chem. Soc.*, **92,** 5727 (1970).
322. G. Dana and A. Zysman, *Bull. Soc. chim. France*, 1951 (1970).
323. M. Tokuda, M. Hasegawa, A. Suzuki, and M. Itoh, *Bull. Chem. Soc. Japan*, **47,** 2619 (1974).
324. A. I. Meyers and E. D. Mihelich, *J. Org. Chem.*, **40,** 1186 (1975).
325. T. J. Batterham, *N.M.R. Spectra of Simple Heterocycles*, E. C. Taylor and A. Weissberger, Eds., Wiley-Interscience, New York, 1973, pp. 370–382.
326. L. E. Erickson, *J. Amer. Chem. Soc.*, **87,** 1867 (1965).

327. K. Hofmann, *J. Amer. Chem. Soc.*, **66,** 157 (1944) and **71,** 164 (1949).
328. B. Feibush and L. Spialter, *J. Chem. Soc.* (*B*), 106 (1971).
329. T. Kaneko, H. Katsura, H. Asano, and K. Wakabayashi, *Chem. and Ind.*, 1187 (1960).
330. J. A. Mills and W. Klyne, *Progr. Stereochem.*, **1,** 202–203 (1954).
331. T. C. Bruice and S. J. Benkovic, *Bioorganic Mechanisms*, Vol. 1, Benjamin, New York, 1966, p. 298.
332. O. Gawron, A. J. Glaid III, and T. P. Fondy, *J. Amer. Chem. Soc.*, **83,** 3634 (1961).
333. O. Červinka and L. Hub., *Coll. Czech. Chem. Comm.*, **33,** 2927 (1968).
334. R. S. Airs, M. P. Balfe, M. Irwin, and J. Kenyon, *J. Chem. Soc.*, 531 (1942).
335. S. F. Mason and G. W. Vane, *Chem. Comm.*, 598 (1967).
336. H. Gerlach and V. Prelog, *Annalen*, **669,** 121 (1963).
337. S. Wilkinson, *Quart. Rev.*, **15,** 153 (1961).
338. C. H. Eugster, *Adv. Org. Chem.*, **2,** 427 (1960).
339. P. Matzinger, Ph. Catalfomo, and C. H. Eugster, *Helv. Chem. Acta*, **55,** 1478 (1972).
340. J. Whiting, Y-K. Au-Young, and B. Belleau, *Canad. J. Chem.*, **50,** 3322 (1972).
341. D. J. Robins and D. H. G. Crout, *J. Chem. Soc.* (*C*), 1334 (1970).
342. M. Ohno, M. Okamoto, N. Kawabe, H. Umezawa T. Takeuchi, M. Iinuma, and S. Takahashi, *J. Amer. Chem. Soc.*, **93,** 1285 (1971).
343. K. Takahashi, Y. Hayashi, and M. Takani, *Chem. and Pharm. Bull.* (*Japan*), **18,** 421 (1970).
344. W. L. Parker and F. Johnson, *J. Amer. Chem. Soc.*, **91,** 7208 (1969).
345. E. L. Eliel, N. L. Allinger, S. J. Angyal, and G. A. Morrison, *Conformational Analysis*, Interscience, New York, 1966, pp. 200–206 and p. 244.
346. J. A. Greenhouse and H. L. Strauss, *J. Chem. Phys.*, **50,** 124 (1969).
347. G. G. Engerholm, A. C. Lutz, W. D. Gwinn, and D. O. Harris, *J. Chem. Phys.*, **50,** 2446 (1969).
348. L. D. Hall, P. R. Steiner, and D. Pedersen, *Canad. J. Chem.*, **48,** 1155 (1970).
348a. D. Cremer and J. A. Pople, *J. Amer. Chem. Soc.*, **97,** 1358 (1975).
349. J. Bogner, J-C. Duplan, Y. Infarnet, J. Delmau, and J. Huet, *Bull. Soc. chim. France*, 3616 (1972).
350. K. V. Sarkanen and A. F. A. Wallis, *J. Heterocyclic Chem.*, **10,** 1025 (1973).
351. H. R. Buys, C. Altona, and E. Havinga, *Tetrahedron*, **24,** 3019 (1968).
352. S. A. Barker, J. S. Brimacombe, A. B. Foster, D. H. Whiffin, and G. Zweifel, *Tetrahedron*, **7,** 10 (1959).
353 W. W. Zajac, Jr., F. Sweet, and R. K. Brown, *Canad. J. Chem.*, **46,** 21 (1968).
354. W. T. Dixon and R. O. C. Norman, *J. Chem. Soc.*, 4850 (1964).
355. G. A. Olah and P. J. Szilagyi, *J. Org. Chem.*, **36,** 1121 (1971).
356. J. Wolfhugel, A. Maujean, and J. Chuche, *Tetrahedron Letters*, 1635 (1973).
357. A. Mustafa, *Benzofurans*, Vol. 29, A. Weissberger and E. C. Taylor, Eds., Wiley, New York, 1974.
358. D. P. Brust, D. S. Tarbell, S. M. Hecht, E. C. Hayward, and L. D. Colebrook, *J. Org. Chem.*, **31,** 2192 (1966).
359. J. I. Jones and A. S. Lindsey, *J. Chem. Soc.*, 1836 (1950).

360. M. M. Badawi, K. Bernauer, P. van den Broek, D. Groger, A. Guggisberg, S. Johne, I. Kompis, F. Schnieder, H-J. Veith, M. Hesse, and H. Scmid, *Pure Appl. Chem.*, **33,** 81 (1973).

361. L. J. Powers and M. P. Mertes, *J. Medicin. Chem.*, **13,** 1102 (1970).

362. M. P. Mertes, L. J. Powers, and M. M. Hava, *J. Medicin. Chem.*, **14,** 361 (1971).

363. S. P. Pappas, B. C. Pappas, and J. E. Blackwell, Jr , *J. Org. Chem.*, **32,** 3066 (1967).

364 S. P. Pappas and J. E. Blackwell, Jr., *Tetrahedron Letters*, 1171 (1966).

365. E. C. Hayward, D. S. Tarbell, and L. D. Colebrook, *J. Org. Chem.*, **33,** 399 (1968).

366. L. H. Zalkow and M. Ghosal, *Chem. Comm.*, 922 (1967).

367. L. H. Zalkow and M. Ghosal, *J. Org. Chem.*, **34,** 1646 (1969).

368. W. A. Bonner, N. I. Burke, W. E. Fleck, R. K. Hill, J. A. Joule, B. Sjöberg, and J. H. Zalkow, *Tetrahedron*, **20,** 1419 (1964).

369. M. P. Mertes, L. J. Powers, and E. Shefter, *J. Org. Chem.*, **36,** 1805 (1971).

370. M. Ghelardoni, V. Pestellini, and C. Musante, *Gazzetta*, **99,** 1273 (1969).

371. D. J. Bennett, F. M. Dean, and A. W. Price, *J. Chem. Soc.* (*C*), 1557 (1970).

372. K. C. Majumdar, B. S. Thyagarajan, and K. K. Balasubramanian, *J. Heterocyclic Chem.*, **90,** 159 (1973).

373. Y. Kawase, S. Yamaguchi, H. Ociai, and H. Horita, *Bull. Chem. Soc. Japan*, **47,** 2660 (1974).

374. C. H. Krauch, W. Metzner, and G. O. Schenck, *Chem. Ber.*, **99,** 1723 (1966).

375. R. N. Warrener, *J. Amer. Chem. Soc.*, **93,** 2346 (1971).

376. H. Hofmann and P. Hofmann, *Tetrahedron Letters*, 4055 (1971).

377. B. Sjoberg, *Arkiv Kemi*, **15,** 451 (1960).

377a. M. Gregson, W. D. Ollis, B. T. Redman, I. O. Sutherland, and H. H. Dietrichs, *Chem. Comm.*, 1394 (1968).

378. G. Büchi and S. M. Weinreb, *J. Amer. Chem. Soc.*, **91,** 5408 (1969).

378a. J. F. Grove, *Fortschr. Chem. org. Naturstoffe*, **22,** 203 (1964).

379. S. Moon and B. H. Waxman, *J. Org. Chem.*, **34,** 288 (1969).

380. M. Jackson and L. D. Hayward, *Canad. J. Chem.*, **37,** 1048 (1959); A. C. Cope and T. Y. Shen, *J. Amer. Chem. Soc.*, **78,** 3177 5916 (1956).

381. K. W. Buck, A. B. Foster, A. R. Perry, and J. M. Webber, *J. Chem. Soc.*, 4171 (1963).

382. L. D. Hayward, D. J. Livingstone, M. Jackson, and V. M. Csizmadia, *Canad. J.* Chem., **45,** 2191 (1967).

383. V. M. Mićović, S. Stojčić, M. Bralović, S. Mladenović, D. Jeremić, and M. Stefanović, *Tetrahedron*, **25,** 985 (1969).

384. H-R. Krüger, H. Marschall, P. Weyerstahl, and F. Nerdel, *Chem. Ber.*, **106,** 91 (1973).

385. R. M. Cresswell, W. R. D. Smith, and H. C. S. Wood, *J. Chem. Soc.*, 4882 (1961).

386. L. N. Owen and A. G. Peto, *J. Chem. Soc.*, 2383 (1955).

387. P. R. Stapp and J. C. Randall, *J. Org. Chem.*, **35,** 2948 (1970).

388. N. L. Allinger, *Experientia*, **10,** 328 (1954).

389. M. Beroza, *J. Amer. Chem. Soc.*, **77,** 3332 (1955).

390. M. Beroza, *J. Amer. Chem. Soc.*, **78,** 5082 (1956).

391. L. Crombie and T. L. Tayler, *J. Chem. Soc.*, 2760 (1957); E. E. Dickey, *J. Org. Chem.*, **23,** 179 (1958).
392. A. J. Birch, B. Moore, E. Smith, and M. Smith, *J. Chem. Soc.*, 2709 (1964).
393. B. Carnmalm, H. Erdtman, and Z. Pelchowicz, *Acta Chem. Scand.*, **9,** 1111 (1955).
394. M. Beroza and M. S. Schechter, *J. Amer. Chem. Soc.*, **78,** 1242 (1956).
395. A. W. Schrecker and J. L. Hartwell, *J. Amer. Chem. Soc.*, **79,** 3827 (1957).
396. L. R. Row, P. Satyanarayana, and G. S. R. Subba Rao, *Tetrahedron*, **23,** 1915 (1967).
397. W. A. Jones, M. Beroza, and E. D. Becker, *J. Org. Chem.*, **27,** 3232 (1962).
398. E. D. Becker and M. Beroza, *Tetrahedron Letters*, 157 (1962).
399. E. W. Lund, *Acta Chem. Scand.*, **14,** 496 (1960).
400. K. Freudenberg and G. S. Sidhu, *Chem. Ber.*, **94,** 851 (1961).
401. K. Weinges and R. Spänig, *Chem. Comm.*, 9 (1966).
402. K. Freudenberg and K. Weinges, *Tetrahedron*, **15,** 115 (1961).
403. D. Ginsburg, *Accounts Chem. Res.*, **2,** 121 (1959).
404. J. Altman, E. Babad, J. Pucknat, N. Reshef, and D. Ginsburg, *Tetrahedron*, **24,** 975 (1968).
405. K. Weinges and A. Weissenhütter, *Annalen*, **746,** 70 (1971).
406. K. Weinges and K. Klessing, *Chem. Ber.*, **107,** 1925 (1974).
407. K. Nakanishi, *Pure Appl. Chem.*, **14,** Sect. 1, 89 (1967).
408. S. E. Cantor and D. S. Tarbell, *J. Amer. Chem. Soc.*, **86,** 2902 (1964).
409. W. E. Harvey and D. S. Tarbell, *J. Org. Chem.*, **32,** 1679 (1967).
410. N. I. Shuikin, I. I. Demitriev, and T. P. Dobrynina, *J. Gen. Chem.* (*U.S.S.R.*), **10,** 967 (1940.)
411. A. I. Meyers and K. Baburao, *J. Heterocyclic Chem.*, **1,** 203 (1964).
412. D. P. Brust and D. S. Tarbell, *J. Org. Chem.*, **31,** 1251 (1966).
413. G. Ohloff, K. H. Schulte-Elte, and B. Willhalm, *Helv. Chim. Acta*, **49,** 2135 (1966).
414. M. Lj. Mihailović, S. Konstantinović, A. Milovanović, J. Janković, Ž. Ceković, and D. Jeremić, *Chem. Comm.*, 236 (1969).
415. M. Mousseron and M. Canet, *Bull. Soc. chim. France*, 190 (1952).
416. E. L. May and H. Kugita, *J. Org. Chem.*, **26,** 188 and 1954 (1961).
417. J. Klein, *J. Amer. Chem. Soc.*, **81,** 3611 (1959).
418. H. O. House, R. G. Carlson, and H. Babad, *J. Org. Chem.*, **28,** 3359 (1963).
419. J. Klein, *J. Org. Chem.*, **23,** 1209 (1958).
420. E. H. Charlesworth, H. J. Campbell, and D. L. Stachiw, *Canad. J. Chem.*, **37,** 877 (1959).
421. W. Herz and L. A. Glick, *J. Org. Chem.*, **28,** 2970 (1963).
422. S. Masamune, S. Takeda, and R. T. Seidner, *J. Amer. Chem. Soc.*, **91,** 7769 (1969).
423. R. Paul and S. Tchelitcheff, *Bull. Soc. chim. France*, 672 (1954).
424. N. O. Brace, *J. Amer. Chem. Soc.*, **77,** 4157 (1955).
425. A. P. Krapcho and B. P. Munday, *J. Heterocyclic Chem.*, **2,** 355 (1965).
426. L. J. Dolby, *J. Org. Chem.*, **27,** 2971 (1962).
427. S. F. Birch, N. J. Hunter, and D. T. McAllan, *J. Org. Chem.*, **21,** 970 (1956).

428. E. Buchta and G. Scheuerer, *Chem. Ber.*, **89,** 1002 (1956).
429. J. E. Ladbury and E. E. Turner, *J. Chem. Soc.*, 3885 (1954).
430. B. P. Munday and R. D. Otzenberger, *J. Org. Chem.*, **37,** 677 (1972).
431. B. T. Gillis and P. E. Beck, *J. Org. Chem.*, **28,** 1388 (1963).
432. D. L. Garin, *J. Org. Chem.*, **34,** 2355 (1969).
433. Y. Fujiwara, S. Kimoto, and M. Okamoto, *Chem. and Pharm. Bull.* (*Japan*), **21,** 1166 (1973).
434. S. F. Birch, R. A. Dean, and E. V Whitehead, *J. Org. Chem.*, **19,** 1449 (1954).
435. L. J. Dolby and M. J. Schwarz, *J. Org. Chem.*, **28,** 1456 (1963).
436. D. J. Goldsmith, B. C. Clark, Jr., and R. C. Joines, *Tetrahedron Letters*, 1211 (1967).
437. C. Amith, M. Cais, D. Fraenkel, and D. Ginsburg, *Heterocycles*, **3,** 25 (1975).
438. M. Korat and D. Ginsburg, *Tetrahedron*, **29,** 2373 (1973).
439. T-Y. Leong, T. Imagawa, K. Kimoto, and M. Kawanisi, *Bull. Chem. Soc. Japan*, **46,** 596 (1973).
440. K. Kimoto, T-Y. Leong, T. Imagawa, and M. Kawanisi, *Canad. J. Chem.*, **50,** 3805 (1972).
441. A. P. Krapcho and B. P. Mundy, *J. Org. Chem.*, **32,** 2041 (1967).
442. B. P. Mundy, K-R. Sun, and R. D. Otzenberger, *J. Org. Chem.*, **37,** 2793 (1972).
443. B. P. Mundy, A. R. De Bernardis, and R. D. Otzenberger, *J. Org. Chem.*, **36,** 3830 (1971).
444. P. E. Manni, G. A. Howie, B. Katz, and J. M. Cassady, *J. Org. Chem.*, **37,** 2769 (1972).
445. M. Mousseron-Canet, D. Lerner, and J-C. Mani, *Bull. Soc. chim. France*, 4639 (1968).
446. L. A. Paquette, J. H. Barrett, R. P. Spitz, and R. Pitcher, *J. Amer. Chem. Soc.*, **87,** 3417 (1965).
447. H. Hart and L. R. Lerner, *Tetrahedron*, **25,** 813 (1969).
448. G. Maier, *Chem. Ber.*, **96,** 2238 (1963).
449. C. A. Matuszak and L. Dickson, *J. Org. Chem.*, **37,** 1864 (1972).
450. E. J. Grubbs, R. A. Froehlich, and H. Lathrop, *J. Org. Chem.*, **36,** 504 (1971).
451. N. L. Wendler, D. Taub, and C. H. Kuo, *J. Org. Chem.*, **34,** 1510 (1969).
452. H. L. Goering and C. Serres, Jr., *J. Amer. Chem. Soc.*, **74,** 5908 (1952).
453. C. F. Culberson, H. J. Seward, and P. Wilder, Jr., *J. Amer. Chem. Soc.*, **82,** 2541 (1960).
454. E. Klein, F. Thömel, A. Roth, and H. Struwe, *Annalen*, 1797 (1973).
455. H. W. Heine, *J. Amer. Chem. Soc.*, **79,** 6268 (1957).
456. A. C. Cope, M. Gordon, S. Moon, and C. H. Park, *J. Amer. Chem. Soc.*, **87,** 3119 (1965).
457. A. C. Cope and B. C. Anderson, *J. Amer. Chem. Soc.*, **79,** 3892 (1957).
458. A. C. Cope, T. A. Liss, and G. W. Wood, *J. Amer. Chem. Soc.*, **79,** 6287 (1957)
459. J. L. Jernow, D. Gray, and W. D. Closson, *J. Org. Chem.*, **36,** 3511 (1971).
460. M. Lj. Mihailović, V. Andrejević, M. Jakovljević, D. Jeremić, A. Stojiljković, and R. E. Partch, *Chem. Comm.*, 854 (1970).
461. A. G. Anastassiou and R. P. Cellura, *Chem. Comm.*, 484 (1970).

462. J. G. Brimacombe, A. B. Foster, B. D. Jones, and J. J. Willard, *Chem. Comm.*, 174 (1965).

463. J. S. Brimacombe, A. B. Foster, B. D. Jones, and J. J. Willard, *J. Chem. Soc.*, (*C*) 2404 (1967).

464. P. Schlichta, J. K. Inman, and H. J. Lucas, *J. Amer. Chem. Soc.*, **77,** 3784 (1955).

465. D Gagnaire and J-B. Robert. *Bull. Soc. chim. France*, 3646 (1965).

466. R. I. T. Cromartie and Y. K. Hamied, *J. Chem. Soc.*, 3622 (1961).

467. M. Farines, J. Soulier, and R. Soulier, *Bull. Soc. chim. France*, 1066 (1972).

468. F. Kametani and Y. Sumi, *Chem. and Pharm. Bull.* (*Japan*), **20,** 1479 (1972).

469. F. I. Carroll, *J. Org. Chem.*, **31,** 366 (1966).

470. F. S. H. Head, *J. Chem. Soc.*, 1778 (1960).

471. F. W. Eastwood, K. L. Harrington, J. S. Josan, and J. L. Pura, *Tetrahedron Letters*, 5223 (1970).

472. W. J. Baumann, *J. Org. Chem.*, **36,** 2743 (1971).

473. N. Baggett, J. M. Duxbury, A. B. Foster, and J. M. Webber, *J. Chem. Soc.* (*C*), 208 (1966).

474. D. J. Triggle and B. Belleau, *Canad. J. Chem.*, **40,** 1201 (1962).

475. J. Gelas, *Bull. Soc. chim. France*, 2341 (1970).

476. J. Gelas, *Bull. Soc. chim. France*, 2349 (1970).

477. T. D. Inch and N. Williams, *J. Chem. Soc.* (*C*), 263 (1970).

478. T. D. Inch, *Ann. Rev. N.M.R. Spectroscopy*, **2,** 35 (1969).

479. Y. N. Yandowskii and T. I. Temnikova, *J. Org. Chem.*, (*U.S.S.R.*), **4,** 1695 (1968).

480. B. N. Blackett, J. M. Coxon, M. P. Hartshorn, A. J. Lewis, G. R. Little, and G. J. Wright, *Tetrahedron*, **26,** 1311 (1970).

481. G. F. Bettinetti and A. Donetti, *Gazzetta*, **97,** 730 (1967).

482. H. B. Kagan and T-P. Dang, *J. Amer. Chem. Soc.*, **94,** 6429 (1972)

483. F W. Breitbeil III, J. J. McDonnell, T. A. Marolewski, and D. T. Dennerlein, *Tetrahderon Letters*, 4627 (1965).

484. H-M. Fischler and W. Hartmann, *Chem. Ber.*, **105,** 2769 (1972).

485. C. Romers, C. Altona, H. R. Buys, and E. Havinga, *Topics Stereochem.*, **4,** 39 (1969).

486. F. Alderweireldt and H. Anteunis, *Bull. Soc. chim. belges*, **74,** 488 (1965).

487. R J. Abraham, *J. Chem. Soc.*, 256 (1965).

488. R. J. W. Le Fèvre, A. Sundaram, and P. K. Pierens, *J. Chem. Soc.*, 479 (1963).

489. R. U. Lemieux, J. D. Stevens, and R. R. Fraser, *Canad. J. Chem.*, **40,** 1962 (1955).

490. M. Anteunis and F. A. Alderweireldt, *Bull. Soc. chim. belges*, **73,** 889 (1964).

491. C. Altona and A P. M. van der Veek, *Tetrahedron*, **24,** 4377 (1968).

492. E. L. Eliel and W. E. Willy, *Tetrahedron Letters*, 1775 (1969).

493. W. E. Willy, G. Binsch, and E. L. Eliel, *J. Amer. Chem. Soc.*, **92,** 5394 (1970).

494. F. A L. Anet and M. St. Jacques, *J. Amer. Chem. Soc.*, **88,** 2586 (1966).

495. M Anteunis, F. Anteunis-De Ketelaere, and F. Borremans, *Bull. Soc. chim. belges*, **80,** 701 (1971).

496. M. Anteunis, G. Swaelens, and J. Genal, *Tetrahedron*, **27,** 1917 (1971).

497. M. Anteunis, R. van Cauwenberge, and C. Becu, *Bull. Soc. chim. belges*, **83,** 591 (1973).

498. N. Baggett, K. W. Buck, A. B. Foster, M. H. Randall, and J. M. Webber, *J. Chem. Soc.*, 3394 (1965).

498a. G. J. Karabatsos, J. S. Fleming, N. Hsi, and R. H. Abeles, *J. Amer. Chem. Soc.*, **88,** 849 (1966).

499. S. Furberg and O. Hassel, *Acta Chem. Scand.*, **4,** 1584 (1950).

500. F. G. Riddell and M. J. T. Robinson, *Tetrahedron*, **27,** 4163 (1971).

501. C. Hackett Bushweller, G. U. Rao, W. G. Anderson, and P. E. Stevenson, *J. Amer. Chem. Soc.*, **94,** 4743 (1972).

502. R. A. Y. Jones, A. R. Katritzky, D. L. Nicol, and R. Scattergood, *J.C.S. Perkin II*, 337 (1973).

503. J. S. Brimacombe, A. B. Foster, and A. H. Haines, *J. Chem. Soc.*, 2582 (1960).

504. I. Wang, C. O. Britt, and J. E. Boggs, *J. Amer. Chem. Soc.*, **87,** 4950 (1965).

505. V. Usieli, A. Pilerdorf, S. Shor, J. Katzhendler, and S. Sarel, *J. Org. Chem.*, **38,** 2073 (1974).

506. A. H. Haines and C. S. P. Jenkins, *Chem. Comm.*, 350 (1969).

507. A. H. Haines and C. S. P. Jenkins, *J. Chem. Soc.*, (*C*) 1438 (1971).

508. R. Brettle and I. D. Logan, *J.C.S. Perkin II*, 687 (1973); M. Farines and J. Soulier, *Bull. Soc. chim. France*, 332 (1970).

509. Y. Asabe, S. Takitani, and Y. Tsukuki, *Bull. Chem. Soc. Japan*, **48,** 966 (1975).

510. E. J. Corey, F. A. Carey, and R. A. E. Winter, *J. Amer. Chem. Soc.*, **87**, 934 (1965).

511. E. J. Corey and J. I. Shulman, *Tetrahedron Letters*, 3655 (1968).

512. W. Hartmann, L. Schrader, and D. Wendisch, *Chem. Ber.*, **106,** 1076 (1973).

513. R. L. Smith, A. Manmade, and G. W. Griffin, *Tetrahedron Letters*, 663 (1970).

514. M. F. Semmelhack and R. D. Stauffer, *Tetrahedron Letters*, 2667 (1973).

515. J. N. Hines, M. J. Peagram, E. J. Thomas, and G. H. Whitham, *J.C.S. Perkin I*, 2332 (1973).

516. S. Searles, Jr., H. R. Hays, and E. F. Lutz, *J. Org. Chem.*, **27,** 2832 (1962).

517. D. A. Seeley and J. McElwee, *J. Org. Chem.*, **38,** 1691 (1973).

518. J. W. Hartgerink, L. C. J. van der Laan, J. B. F. N. Engberts, and T. J. de Boer, *Tetrahedron*, **27,** 4323 (1971).

519. M. S. Newman and C. H. Chen, *J. Org. Chem.*, **38,** 1173 (1973).

520. H. M. R. Hoffmann, K. E. Clemens, and R. H. Smithers, *J. Amer. Chem. Soc.*, **94,** 3940 (1972).

521. M. Schulz and K. Kirschke, *Adv. Heterocyclic Chem.*, **8,** 165 (1967); P. S. Bailey, *Chem. Rev.*, **58,** 925 (1958).

522. R. Criegee, *Annalen*, **583,** 1 (1953).

523. L. J. Durham and F. L. Greenwood, *Chem. Comm.*, 24 (1968).

524. L. J. Durham and F. L. Greenwood, *J. Org. Chem.*, **33,** 1629 (1968).

525. S. Fliszár and J. Carles, *J. Amer. Chem. Soc.*, **91,** 2637 (1969).

526. R. Criegee and H. Korber, *Chem. Ber.*, **104,** 1807 (1971).

527. R. W. Murray, R. D. Youssefyeh, and P. R. Story, *J. Amer. Chem. Soc.*, **89,** 2429 (1967).

528. L. D. Loan, R. W. Murray, and P. R. Story, *J. Amer. Chem. Soc.*, **87,** 737 (1965).

529. F. L. Greenwood, *J. Org. Chem.*, **30,** 3108 (1965).

530. N. L. Bauld, J. A. Thompson, C. E. Hudson, and P. S. Bailey, *J. Amer. Chem. Soc.*, **90,** 1822 (1968).

531. E. L. Eliel, N. L. Allinger, S. J. Angyal, and G. A. Morrison, *Conformational Analysis*, Interscience, New York, 1966, pp. 362–403.

523. J. F. Stoddart, *Stereochemistry of Carbohydrates*, Wiley-Interscience, New York, 1971; W. Pigman and D. Horton, *The Carbohydrates*. Vol. 1A, Academic, New York, 1972; *Adv. Carbohydrate Chem.*, **31** (1975), *Carbohydrate Res.*, **4** (1975) and earlier vols, in the series.

533. C. B. Reese in *Protective Groups in Organic Chemistry*, J. F. W. McOmie, Ed., Plenum Press, London, 1973, p. 104.

534. M. F. Ansell and B. Gadsby, *J. Chem. Soc.*, 3388 (1958).

535. M. J. Baldwin and R. K. Brown, *Canad. J. Chem.*, **46,** 1093 (1968).

536. F. Sweet and R. K. Brown, *Canad. J. Chem.*, **45,** 1007 (1967).

537. W. H. Pirkle and M. Dines, *J. Org. Chem.*, **34,** 2239 (1969).

537a. R. Aguilera and G. Descotes, *Bull. Soc. chim. France*, 3318 (1966).

538. H. M. Rosenberg, R. Rondeau, and P. Servé, *J, Org. Chem.*, **34,** 471 (1969).

539. L. Ghosez, R. Montaigne, A. Roussel, H. Vanlierde, and P. Mollet, *Tetrahedron*, **27,** 615 (1971).

540. H. M. Rosenberg and P. Servé, *J. Org. Chem.*, **33,** 1653 (1968).

541. R. P. Gandhi and V. K. Chadha, *Indian J. Chem.*, **6,** 402 (1968).

542. F. Effenberger and R. Glieter, *Chem. Ber.*, **97,** 1576 (1964).

543. K. Alder and H. F. Rickert, *Ber.*, **70,** 1354 (1937).

544. P. Margaretha, *Annalen*, 727 (1973).

545. T. Ando, H. Yamanaka, and W. Funasaka, *Tetrahedron Letters*, 2587 (1967).

546. T. Ando, H. Hosaka, and H. Yamanaka, and W. Funasaka, *Bull. Chem. Soc. Japan*, **42,** 2013 (1969).

547. T. Ando, H. Yamanaka, F. Namigata, and W. Funasaka, *J. Org. Chem.*, **35,** 33 (1970).

548. D. B. Ledlie and W. H. Hearne, *Tetrahedron Letters*, 4837 (1969).

549. K. G. Taylor, W. E. Hobbs, and M. Saquet, *J. Org. Chem.*, **36,** 369 (1971).

550. E. J. Corey and J. Streith, *J. Amer. Chem. Soc.*, **86,** 950 (1964).

551 W H. Prikle and L. H. McKendry, *Tetrahedron Letters*, 5279 (1968).

552. R. D. Rieke and R. A. Copenhafer, *Tetrahedron Letters*, 879 (1971).

553. P. Yates and D. J. MacGregor, *Canad. J. Chem.*, **51,** 1267 (1973).

553a. M. Van Meerbeck, S. Toppet, and F. C. De Schryver, *Tetrahedron Letters*, 2247 (1972).

554. P. Yates, M. J. Jorgenson, and P. Singh, *J. Amer. Chem. Soc.*, **91,** 4739 (1969); P. Yates and E. S. Hand, *J. Amer. Chem. Soc.*, **91,** 4749 (1969).

555. P. Yates and D. J. MacGregor, *Chem. Comm.*, 1209 (1967).

556. P. Yates and P. Singh, *J. Org. Chem.*, **34,** 4052 (1969).

557. P. Margaretha, *Tetrahedron*, **27,** 6209 (1971).

558. P. E. Eaton, *J. Amer. Chem. Soc.*, **84,** 2344 (1962).

559. R. M. Moriarty, *Topics Stereochem.*, **8,** 271 (1974).

560. E. Honkanen, T. Moisio, and P. Karvonen, *Acta Chem. Scand.*, **23,** 531 (1969).

561. S. Dev and C. Rai, *J. Indian Chem. Soc.*, **34,** 266 (1957).
562. E. Honkanen, T. Moisio, P. Karvonen, and A. I. Virtanen, *Acta Chem. Scand.*, **22,** 2041 (1968).
563. Y. Suhara, F. Sasaki, G. Koyama, K. Meada, H. Umezawa, and M. Ohno, *J. Amer. Chem. Soc.*, **94,** 6501 (1972).
564. G. C. Corfield, A. Crawshaw, S. J. Thompson, and A. G. Jones, *J.C.S. Perkin II*, 1549 (1973).
565. E. H. Eschinasi, *J. Org. Chem.*, **35,** 1097 (1970).
566. H. Strickler and G. Ohloff, *Helv. Chim. Acta*, **49,** 2157 (1966).
567. P. J. Lentz, Jr., and M. G. Rossmann, *Chem. Comm.*, 1269 (1969).
568. A. Horeau and A. Nouaille, *Tetrahedron Letters*, 1939 (1971).
569. P. Eigtved, S. R. Jensen, and B. J. Nielsen, *Acta Chem. Scand.*, **28B,** 85 (1974).
570. C. Chin, M. C. Cutler, J. Lee, and V. Thaller, *Chem. Comm.*, 202 (1966).
571. C. Chin, E. R. H. Jones, V. Thaller, R. T. Aplin, L. J. Durham, C. S. Cascon, W. B. Mors, and B. M. Tursch, *Chem. Comm.*, 152 (1965).
572. R. H. Cornforth, J. W. Cornforth, and G. Popják, *Tetrahedron*, **18,** 1351 (1962).
573. O. Červinka, L. Hub, A. Klásek, and F. Šantavý, *Chem. Comm.*, 261 (1968).
574. O. Červinka and L. Hub., *Coll. Czech. Chem. Comm.*, **33,** 2933 (1968).
575. S. Kondo, S. Shibahara, S. Takahashi, K. Maeda, H. Umezawa, and M. Ohno, *J. Amer. Chem. Soc.*, **93,** 6305 (1971).
576. O. Korver, *Tetrahedron*, **26,** 2391 (1970).
577. F. I. Carroll, A. Sobti, and R. Meck, *Tetrahedron Letters*, 405 (1971).
578. F. I. Carroll, G. N. Mitchell, J. T. Blackwell, A. Sobti, and R. Meck, *J. Org. Chem.*, **39,** 3890 (1974).
579. M. M. Cook and C. Djerassi, *J. Amer. Chem. Soc.*, **95,** 3678 (1973).
580. R. P. Zelinski, N. G. Peterson, and H. R. Wallner, *J. Amer. Chem. Soc.*, **74,** 1504 (1952).
581. W. Klyne, W. P. Mose, P. M. Scopes, G. M. Holder, and W. B. Whalley, *J. Chem. Soc. (C)*, 1273 (1967)
582. O. Riobé, *Ann. Chim. (France)*, **4,** 593, 597 (1949).
583. R. Stern, J. English, Jr., and H. G. Cassidy, *J. Amer. Chem. Soc.*, **79,** 5797 (1957).
584. R. E. Sievers, *N.M.R. Shift Reagents*, Academic, New York, 1973.
585. P. Servé, R. E. Rondeau, and H. M. Rosenberg, *J. Heterocyclic Chem.*, **9,** 721 (1972).
586. K. Jankowski and A. Rabczenko, *J. Org. Chem.*, **40,** 960 (1975).
587. G. Gatti, A. L. Segre, and C. Morandi, *J. Chem. Soc.*, *(B)* 1203 (1967).
588. E. L. Eliel, N. L. Allinger, S. J. Angyal, and G. A. Morrison, *Conformational Analysis*, Interscience, New York, 1965, pp. 41 and 42.
589. C. H. Bushweller and J. W. O'Neill, *Tetrahedron Letters*, 4713 (1969).
590. J. T. Edward, *Chem. and Ind.*, 1102 (1955).
591. R. U. Lemieux in *Molecular Rearrangements*, P. de Mayo, Ed., Interscience, New York, 1964, p. 723, *Adv. Carbohydrate Chem.*, **9,** 1 (1955).
592. F. G. Riddell, *Quart. Rev.*, **21,** 364 (1967).
593. S. J. Angyal, *Angew. Chem. Internat. Edn.*, **8,** 157 (1969).
594. N. S. Zefirov and N. M. Shekhtman, *Russ. Chem. Rev.*, **40,** 315 (1971).

595. E. L. Eliel, *Angew. Chem. Internat. Edn.*, **11,** 739 (1972).

596. J-C. Martin, *Ann. Chim. (France)*, **6,** 205 (1971).

597. A. J. de Hoog and H. R. Buys, *Tetrahedron Letters*, 4175 (1969); A. J. de Hoog, H. R. Buys, C. Altona, and E. Havinga, *Tetrahedron*, **25,** 3365 (1969).

598. R. O. Hutchins, L. D. Kopp, and E. L. Eliel, *J. Amer. Chem. Soc.*, **90,** 7174 (1968).

599. E. L. Eliel, *Accounts Chem. Res.*, **3,** 1 (1970).

600. R. U. Lemieux, *Pure Appl. Chem.*, **22,** 527 (1971).

601. H. Booth and R. U. Lemieux, *Canad. J. Chem.*, **49,** 777 (1971).

602. G. O. Pierson and O. A. Runquist, *J. Org. Chem.*, **33,** 2572 (1968).

603. C. B. Anderson and D. T. Sepp, *Chem. and Ind.*, 2054 (1964).

604. M. Gelin, Y. Bahurel, and G. Descotes, *Bull. Soc. chim. France*, 3723 (1970).

605. D. T. Sepp and C. B. Anderson, *Tetrahedron*, **24,** 6873 (1968).

606. R. U. Lemieux, A. A. Pavia, J. C. Martin, and K. A. Watanabe, *Canad. J. Chem.*, **47,** 4427 (1969).

607. F. Mutterer, J-M. Morgen, J-M. Biedermann, J-P. Fluery, and F. Weiss, *Bull Soc. chim. France*, 4778 (1969).

608. F. Sweet and R. K. Brown, *Canad. J. Chem.*, **46,** 1543 (1968).

609. G. E. Booth and R. J. Ouellette, *J. Org. Chem.*, **31,** 544 (1966).

610. N. S. Zefirov and N. M. Shekhtman, *Doklady Akad. Nauk S.S.S.R.*, **180,** 1363 (1968).

611. N. S. Zerfirov and N. M. Shekhtman, *Doklady Akad. Nauk S.S.S.R.*, **177,** 842 (1967).

612. C. B. Anderson and D. T. Sepp, *J. Org. Chem.*, **32,** 607 (1967).

613. J. M. Eckert and R. J. W. Le Fèvre, *J. Chem. Soc. (B)*, 855 (1969).

614. C. B. Anderson and M. P. Geis, *Tetrahedron*, **31,** 1149 (1975).

615. M. J. Baldwin and R. K. Brown, *Canad. J. Chem.*, **47,** 3553 (1969).

616. A. J. de Hoog and E. Havinga, *Rec. Trav. chim.*, **89,** 972 (1970).

616a. N. S. Zefirov and N. M. Shekhtman, *Zhur. org. Khim.*, **6,** 863 (1970).

617. C. B. Anderson and D. T. Sepp, *J. Org. Chem.*, **33,** 3272 (1968).

618. R. U. Lemieux and A. R. Morgan, *Canad. J. Chem.*, **43,** 2205 (1965).

619. N. M. Shekhtman, E. A. Viktorova, A. E. Karakhanov, N. M. Khvorostukhina, and N. S. Zefirov, *Doklady Akad Nauk S.S.S.R.*, **196,** 367 (1971).

620. K. Janowski and J. Couturier, *J. Org. Chem.*, **37,** 3997 (1972).

621. R. C. Sheppard and S. Turner, *Chem. Comm.*, 77 (1968).

622. J. S. Brimacombe, A. B. Foster, M. Stacey, and D. H. Whiffen, *Tetrahedron*, **4,** 351 (1958).

622a. M. D. Brown, M. J. Cook, and A. R. Katritzky, *J. Chem. Soc. (B)*, 2358 (1971).

623. R. M. Srivastava and R. K. Brown, *Canad. J. Chem.*, **48,** 2334 (1970).

624. U. E. Diner and R. K. Brown, *Canad. J. Chem.*, **45,** 2547 (1967).

625. M. Cahu and G. Descotes, *Bull. Soc. chim. France*, 2975 (1968).

626. G. Berti, G. Catelani, M. Ferretti, and L. Monti, *Tetrahedron*, **30,** 4013 (1974).

627. J. d'Angelo, *Bull. Soc. chim. France*, 181 (1969).

628. P. Deslongchamps, P. Atlani, D. Fréhel, and A. Malaval, *Canad. J. Chem.*, **50,** 3405 (1972).

628a. P. Deslongchamps, R. Chênevert, R. J. Taillefer, C. Moreau, and J. K. Saunders, *Canad. J. Chem.*, **53,** 1601 (1975).
629. R. P. Lutz and J. D. Roberts, *J. Amer. Chem. Soc.*, **83,** 2198 (1961).
630. J. B. Harborne, T. J. Mabry, and H. Mabry, *The Flavanoids*, Chapman and Hall, London, 1975.
631. H. D. Locksley, *Fortschr. Chem. org. Naturstoffe*, **30,** 207 (1973).
632. H. Wagner, *Fortschr. Chem. org. Naturstoffe*, **31,** 153 (1974).
633. P. P. Wells and H. Morrison, *J. Amer. Chem. Soc.*, **97,** 154 (1975).
634. J. W. Hanifin and E. Cohen, *Tetrahedron Letters*, 1419 (1966).
635. H. Morrison and R. Hoffman, *Chem. Comm.*, 1453 (1968).
636. R. Hoffman, P. Wells, and H. Morrison, *J. Org. Chem.*, **36,** 102 (1971).
637. L. Paolillo, H. Ziffer, and O. Buchardt, *J. Org. Chem.*, **35,** 38 (1970).
638. D. G. Rao, H. Ulrich, F. A. Stuber, and A. A. R. Sayigh, *Chem. Ber.*, **106,** 388 (1973).
639. H. Morrison, H. Curtis, and T. McDowell, *J. Amer. Chem. Soc.*, **88,** 5415 (1966).
640. G. O. Schenck, I. von Wilucki, and C. H. Krauch, *Chem. Ber.*, **95,** 1409 (1962).
641. L. H. Leenders, E. Schouteden, and F. C. De Schryver, *J. Org. Chem.*, **38,** 957 (1973).
642. J. W. Hanifin and E. Cohen, *Tetrahedron Letters*, 5421 (1966).
643. A. Schönberg and G. D. Khandelwal, *Chem. Ber.*, **103,** 2780 (1970).
643a. G. Descotes and A. Jullien, *Tetrahedron Letters*, 3395 (1969).
644. J. Badin and G. Descotes, *Bull. Soc. chim. France*, 1949 (1970).
645. G. Descotes and D. Missos, *Bull. Soc. chim. France*, 696 (1972).
646. H. Booth, D. Huckle, and I. M. Lockhart, *J.C.S. Perkin II*, 227 (1973).
647. R. Binns, W. D. Cottrill, and R. Livingstone, *J. Chem. Soc.*, 5049 (1965).
648. B. Graffe, M-C. Sacquet, and P. Maitte, *Bull. Soc. chim. France*, 4016 (1971).
649. H. Hofmann and G. Salbeck, *Chem. Ber.*, **103,** 2768 (1970).
650. R. Bognár, J. W. Clark-Lewis, A. Liptákné-Tökés, and M. Rákosi, *Austral. J. Chem.*, **23,** 2015 (1970).
651. W. Oppolzer, *Tetrahedron Letters*, 3091 (1970).
652. C. Normant-Chefnay, *Bull. Soc. chim. France*, 1351 and 1362 (1971).
653. Y. Fujimoto, S. Yokura, T. Nakamura, T. Morikawa, and T. Tatsuno, *Tetrahedron Letters*, 2523 (1974).
654. G. Ohloff and G. Schade, *Chem. Ber.*, **91,** 2017 (1958).
655. C. A. R. Baxter, G. C. Forward, and D. A. Whiting, *J. Chem. Soc. (C)*, 1162 (1968).
656. G. Jones, *Tetrahedron Letters*, 2231 (1974).
657. J. C. Hillyer and J. T. Edmonds, Jr.. *J. Org. Chem.*, **17,** 600 (1952).
658. C. Normant-Chefnay and P. Maitte, *Bull. Soc. chim. France*, 1090 (1974).
659. E. N. Marvell, T. Gosink, P. Churchley, and T. H. Li, *J. Org. Chem.*, **37,** 2989 (1972).
660. J. A. Hirsch and G. Schwartzkopf, *J. Org. Chem.*, **38,** 3534 (1973).
661. J. A. Hirsch and G. Schwartzkopf, *J. Org. Chem.*, **39,** 2040 (1974).
662. I. J. Borowitz, G. J. Williams, L. Gross, and R. Rapp, *J. Org. Chem.*, **33,** 2013 (1968).
663. I. J. Borowitz, G. Gonis, R. Kelsey, R. Rapp, and G. J. Williams, *J. Org. Chem.*, **31,** 3032 (1966).

664. I. J. Borowitz and R. D. Rapp, *J. Org. Chem.*, **34,** 1370 (1969).

665. Y. Gaoni, *J. Chem. Soc.* (*B*), 382 (1968).

666. Y. Gaoni, *J. Chem. Soc.* (*C*), 2925 (1968).

667. Y. Gaoni, *J. Chem. Soc.* (*C*), 2934 (1968).

668. M. P. Servé, *Canad. J. Chem.*, **50,** 3744 (1972).

669. R. Nouguier and J-M. Surzur, *Bull. Soc. chim. France*, 2399 (1973).

670. J. Cologne, J. Buendia, and H. Guingard, *Bull. Soc. chim. France*, 956 (1969).

671. Y. Naya and M. Kotake, *Tetrahedron Letters*, 2459 (1967).

672. T. P. Murray, C. S. Williams, and R. K. Brown, *J. Org. Chem.*, **36,** 1311 (1971).

673. P. Clasper and R. K. Brown, *J. Org. Chem.*, **37,** 3346 (1972).

674. N. J. Turro, S. S. Edelson, J. R. Williams, and T. R. Darling, *J. Amer. Chem. Soc.*, **90,** 1926 (1968).

675. R. Noyori, Y. Baba, S. Makino, and H. Takaya, *Tetrahedron Letters*, 1741 (1973).

676. M. Procházka and J. V. Černý, *Tetrahedron*, **16,** 25 (1961).

677. M. T. Hughes and R. O. Williams, *Chem. Comm.*, 559 (1967).

678. M. T. Hughes and R. O. Williams, *Chem. Comm.*, 587 (1968).

679. G. Lippi, B. Macchia, and M. Pannocchia, *Gazzetta*, **100,** 14 (1970).

680. P. Bucci, G. Lippi, and B. Macchia, *J. Org. Chem.*, **35,** 913 (1970).

681. S. Olsen, C. Schönheyder, A. Henriksen, and B. Alstad, *Chem. Ber.*, **92,** 1072 (1959).

682. V. L. Brown and W. H. Seaton, *J. Heterocyclic Chem.*, **5,** 575 (1968).

683. M. Moreau, R. Quagliero, R. Longeray, and J. Dreux, *Bull. Soc. chim. France*, 1362 (1969).

684. M. Moreau, R. Longeray, and J. Dreux, *Bull. Soc. chim. France*, 2490 (1969).

685. J. Mounet, J. Huet, and J. Dreux, *Bull. Soc. chim. France*, 3006 (1971).

686. J. Mounet, J. Huet, and J. Dreux, *Bull. Soc. chim. France*, 3010 (1971).

687. C. Ganter and W. Zwahlen, *Helv. Chim. Acta*, **54,** 2628 (1971).

688. C. Ganter and K. Wicker, *Helv. Chim. Acta*, **53,** 1693 (1970).

689. K. Wicker, P. Ackermann, and C. Ganter, *Helv. Chim. Acta*, **55,** 1842, 2744 (1972).

690. R. O. Duthaler, K. Wicker, P. Ackermann, and C. Ganter, *Helv. Chim. Acta*, **55,** 1809 (1972).

691. P. Ackermann and C. Ganter, *Helv. Chim. Acta*, **56,** 3054 (1973).

692. P. Ackermann, H. Tobler, and C. Ganter, *Helv. Chim. Acta*, **55,** 2731 (1972).

693. H. Stetter and J. Mayer, *Chem. Ber.*, **92,** 2664 (1959).

694. E. Cuthbertson and D. D. MacNicol, *Tetrahedron Letters*, 2367 (1974).

695. H. Stetter and H. Stark, *Chem. Ber.*, **92,** 732 (1959).

696. R. Sasaki, S. Eguchi, and T. Kiriyama, *J. Org. Chem.*, **38,** 2230 (1973).

697. J. Meinwald and A. T. Hamner, *Chem. Comm.*, 1302 (1969).

698. R. Griegee and G. Müller, *Chem. Ber.*, **89,** 238 (1956).

699. G. Claeson, G. Androes, and M. Calvin, *J. Amer. Chem. Soc.*, **83,** 4357 (1961).

700. A. Schönberg, G. O. Schenck, and O-A. Neumuller, *Preparative Organic Photochemistry*, Springer-Verlag, Berlin, 1968, pp. 46, 47, 383, and 384.

701. G. O. Schenck and K. Ziegler, *Naturwiss.*, **32,** 157 (1944).

702. E. E. Smissman, R. A. Schnettler, and P. S. Portoghese, *J. Org. Chem.*, **30,** 797 (1965).

703. K. B. Schowen, E. E. Smissman, and R. L. Schowen, *J. Org. Chem.*, **33,** 1873 (1968).

704. L. J. Dolby, C. Wilkins, and T. G. Frey, *J. Org. Chem.*, **31,** 1110 (1966).

705. C. L. Wilkins and R. S. Marianelli, *Tetrahedron*, **26,** 4131 (1970).

706. H. Griengl and W. Sieber, *Monatsh.*, **104,** 1008 (1973).

707. A. J. de Kok and C. Romers, *Rec. Trav. chim.*, **89,** 313 (1970).

708. F. W. Nader, *Tetrahedron Letters*, 1591 (1975).

709. F. W. Nader, *Tetrahedron Letters*, 1207 (1975).

710. M. Anteunis, D. Tavernier, and F. Borremans, *Bull. Soc. chim. belges*, **75,** 396 (1966).

711. K. Pihlaja and J. Heikkilä, *Acta Chem. Scand.*, **21,** 2430 (1967); C. Barbier, J. Delmau, and J. Ranft, *Tetrahedron Letters*, 3339 (1964).

712. T. P. Forrest, *J. Amer. Chem. Soc.*, **97,** 2628 (1975); J. B. Lambert, *Accounts Chem. Res.*, **4,** 87 (1971).

713. T. J. Batterham, *N.M.R. Spectra of Simple Heterocycles*, E. C. Taylor and A. Weissberger, Eds., Wiley-Interscience, 1973, pp. 404–406.

714. G. Eccleston and E. Wyn-Jones, *Chem. Comm.*, 1511 (1969).

715. F. G. Riddell, *J. Chem. Soc.* (*B*), 331 (1970).

716. A. J. Jones, E. L. Eliel, D. M. Grant, M. C. Knoeber, and W. F. Bailey, *J. Amer Chem. Soc.*, **93,** 4772 (1971).

717. J-P. Maffrand and P. Maroni, *Bull. Soc. chim. France*, 1408 (1970).

718. Y. Fujiwara and S. Fujiwara, *Bull. Chem. Soc. Japan*, **37,** 1010 (1964).

718a. A. W. Bogatskij, J. J. Samitow, A. I. Gren, and S. G. Sobolewa, *Tetrahedron*, **31,** 489 (1975).

719. K. Pihlaja and J. Heikkilä, *Acta Chem. Scand.*, **21,** 2390 (1967).

720. E. L. Eliel, *Pure Appl. Chem.*, **25,** 509 (1971).

721. R. U. Lemieux, *Pure. Appl. Chem.*, **25,** 527 (1971).

722. J. E. Anderson, F. G. Riddell, and M. J. T. Robinson, *Tetrahedron Letters*, 2017 (1967).

723. K. Pihlaja, *Acta Chem. Scand.*, **22,** 716 (1968).

724. E. L. Eliel, J. R. Powers, Jr., and F. W. Nader, *Tetrahedron*, **30,** 515 (1974).

725. R. M. Clay, G. M. Kellie, and F. G. Riddell, *J. Amer. Chem. Soc.*, **95,** 4632 (1973).

726. K. Pihlaja and J. Jalonen, *Org. Mass Spectrometry*, **5,** 1363 (1971).

727. G. Wood, R. M. Srivastava, and B. Adlam, *Canad. J. Chem.*, **51,** 1200 (1973).

728. F. W. Nader and E. L. Eliel, *J. Amer. Chem. Soc.*, **92,** 3050 (1970).

729. H. R. Buys and E. L. Eliel, *Tetrahedron Letters*, 2779 (1970).

730. V. I. P. Jones and J. A. Ladd, *J. Chem. Soc.* (*B*), 567 (1971).

731. J. E. Anderson, *J. Chem. Soc.* (*B*), 712 (1967).

732. J. K. Porter, L. C. Martinelli, and J. P. LaRocca, *J. Heterocyclic Chem.*, **8,** 867 (1971).

733. B. J. Hutchinson, R. A. Y. Jones, A. R. Katritzky, K. A. F. Record, and P. J. Brignell, *J. Chem. Soc.* (*B*), 1224 (1970).

734. P. Aeberli and W. J. Houliman, *J. Org. Chem.*, **32,** 3211 (1967).

735. E. L. Eliel and M. C. Knoeber, *J. Amer. Chem. Soc.*, **88,** 5347 (1966).
736. E. L. Eliel and M. C. Knoeber, *J. Amer. Chem. Soc.*, **90,** 3444 (1968).
737. G. Eccleston, E. Wyn-Jones, and W. J. Orville-Thomas, *J. Chem. Soc. (B)*, 1551 (1971).
738. E. L. Eliel and C. A. Giza, *J. Org. Chem.*, **33,** 3754 (1968).
739. E. L. Eliel and F. W. Nader, *J. Amer. Chem. Soc.*, **92,** 584 (1970).
740. W. F. Bailey and E. L. Eliel, *J. Amer. Chem. Soc.*, **96,** 1798 (1974).
741. P. Dirinck and M. Anteunis, *Canad. J. Chem.*, **50,** 412 (1972).
742. M. Anteunis and P. Dirinck, *Canad. J. Chem.*, **50,** 423 (1972).
743. G. Binsch, E. L. Eliel, and S. Mager, *J. Org. Chem.*, **38,** 4079 (1973).
744. M. Anteunis and M. Coryn, *Bull. Soc. chim. belges*, **82,** 413 (1973).
745. J. Delmau, J-C. Duplan, and M. Davidson, *Tetrahedron*, **23,** 4371 (1967).
746. J. Delmau, J-C. Duplan, and M. Davidson, *Tetrahedron*, **24,** 3939 (1968).
747. M. Anteunis, D. Tavernier, and G. Swaelens, *Rec. Trav. chim.*, **92,** 531 (1973).
748. D. Tavernier and M. Anteunis, *Bull. Soc. chim. belges*, **82,** 405 (1973).
749. F. G. Riddell and M. J. T. Robinson, *Tetrahedron*, **23,** 3417 (1967).
750. M. Anteunis, *Bull. Soc. chim. belges*, **80,** 3 (1971).
751. D. Tavernier and M. Anteunis, *Bull. Soc. chim. belges*, **80,** 219 (1971).
752. E. L. Eliel and R. M. Enanoza, *J. Amer. Chem. Soc.*, **94,** 8072 (1972).
753. J. E. Anderson, *Chem. Comm.*, 417 (1970).
754. A. Greenberg and P. Laszlo, *Tetrahedron Letters*, 2641 (1970).
755. E. Coene and M. Anteunis, *Bull. Soc. chim. belges*, **79,** 37 (1970).
756. E. Coene and M. Anteunis, *Bull. Soc. chim. belges*, **79,** 25 (1970).
757. D. Tavernier, M. Anteunis, and E. Bernaert, *Bull. Soc. chim. belges*, **83,** 497 (1974).
758. M. Anteunis, R. Camerlynck, and R. De Waele, *Bull. Soc. chim. belges*, **83,** 483 (1974).
759. J. Feeney, M. Anteunis, and G. Swaelens, *Bull. Soc. chim. belges*, **77,** 121 (1968).
760. E. Bernaert, M. Anteunis, and D. Tavernier, *Bull. Soc. chim. belges*, **83,** 357 (1974).
761. J-M. Kamenka P. Herrmann, *Bull. Soc. chim. France*, 3432 (1972).
762. H. D. Banks, *J. Org. Chem.*, **38,** 4084 (1973).
763. R. Van Cauwenberghe, M. Anteunis, and L. Valckz, *Bull. Soc. chim. belges*, **83,** 285 (1974).
764. E. L. Eliel and S. A. Evans, Jr., *J. Amer. Chem. Soc.*, **94,** 8587 (1972).
765. E. L. Eliel and O. Hofer, *J. Amer. Chem. Soc.*, **95,** 8041 (1973).
766. E. L. Eliel and D. I. C. Railenau, *Chem. Comm.*, 291 (1970).
767. E. L. Eliel and M. K. Kaloustian, *Chem. Comm.*, 290 (1970).
768. K. Pihlaja and P. Äyräs, *Acta Chem. Scand.*, **24,** 531 (1970).
769. B. Dobinson and A. B. Foster, *J. Chem. Soc.*, 2338 (1961).
770. N. Baggett, J. S. Brimacombe, A. B. Foster, M. Stacey, and D. H. Whiffen, *J. Chem. Soc.*, 2574 (1960).
771. S. A. Barker, A. B. Foster, A. H. Haines, J. Lehmann, J. M. Webber, and G. Zweifel, *J. Chem. Soc.*, 4161 (1963).

772. N. Baggett, M. A. Bukhari, A. B. Foster, J. Lehmann, and J. M. Webber, *J. Chem. Soc.*, 4157 (1963).
773. E. L. Eliel and H. D. Banks, *J. Amer. Chem. Soc.*, **94,** 171 (1972).
774. E. L. Eliel and H. D. Banks, *J. Amer. Chem. Soc.*, **92,** 4730 (1970).
775. T. A. Crabb and R. F. Newton, *Tetrahedron*, **26,** 693 (1970).
776. R. Dratler and P. Laszlo, *Tetrahedron Letters*, 2607 (1970).
777. J. Gelas and D. Petrequin, *Bull. Soc. chim. France*, 3471 (1972).
778. K. Pihlaja, G. M. Kellie, and F. G. Riddell, *J.C.S. Perkin II*, 252 (1972).
779. D. Tavernier and M. Anteunis, *Tetrahedron*, **27,** 1677 (1971).
780. D. Tavernier and M. Anteunis, *Bull. Soc. chim. belges*, **76,** 157 (1967).
781. P. Maroni and J-P. Gorrichon, *Bull. Soc. chim. France*, 785 (1972).
782. H. P. Maffrand and P. Maroni, *Tetrahedron Letters*, 4201 (1969).
783. G. M. Kellie and F. G. Riddell, *J. Chem. Soc. (B)*, 1030 (1971).
784. K. Pihlaja and S. Luoma, *Acta Chem. Scand.*, **22,** 2401 (1968).
786. G. M. Kellie and F. G. Riddell, *J.C.S. Chem. Comm.*, 42 (1972); G. M. Kellie and F. G. Riddell, *J.C.S. Perkin II*, 740 (1975).
786. J-F. Tocanne, *Bull. Soc. chim. France*, 750 (1970).
787. G. Eccleston and E. Wyn-Jones, *J. Chem. Soc. (B)*, 2469 (1971).
788. G. M. Kellie, P. Murray-Rust, and F. G. Riddell, *J.C.S. Perkin II*, 2384 (1972).
789. G. M. Kellie and F. G. Riddell, *Topics in Stereochem.*, **8,** 225 (1974).
790. J. Katzhendler, L .A. Poles, and S. Sarel, *Israel J. Chem.*, **10,** 111 (1972).
791. I. Schuster and P. Schuster, *Tetrahedron*, **25,** 199 (1969).
792. J. C. Jochims and Y. Kobayashi, *Tetrahedron Letters*, 575 (1974).
793. J. C. Jochims, Y .Kobayashi, and E. Skrzelewski, *Tetrahedron Letters*, 571 (1974).
794. E. L. Eliel and F. W. Nadar, *J. Amer. Chem. Soc.*, **92,** 3045 (1970).
795. E. L. Eliel and F. W. Nadar, *J. Amer. Chem. Soc.*, **91,** 536 (1969).
796. J. E. Anderson, F. G. Riddell, J. P. Fluery, and J. Morgen, *Chem. Comm.*, 128 (1966).
797. K. Weinges, G. Graab, D. Nagel, and B. Stemmle, *Chem. Ber.*, **104,** 3594 (1971).
798. K. Weinges and B. Stemmle, *Chem. Ber.*, **106,** 2291 (1973).
799. A. F. Casy, A. B. Simmonds, and D. Staniforth, *J. Org. Chem.*, **37,** 3189 (1972).
799a. A. T. Blomquist and J. Wolinsky, *J. Amer. Chem. Soc.*, **79,** 6025 (1957).
800. M. Coryn and M. Anteunis, *Bull. Soc. chim. belges*, **83,** 83 (1974).
801. D. Fârcaşiu, C. Kascheres, and L. H. Schwartz, *J. Amer. Chem. Soc.*, **94,** 180 (1972).
802. G. Schneider and O. K. J. Kovács, *Chem. Comm.*, 202 (1965).
803. L. Roos, R. W. Goetz, and M. Orchin, *J. Org. Chem.*, **30,** 3023 (1965).
804. B. Wesslén and L. O. Ryrfors, *Acta Chem. Scand.*, **20,** 1731 (1966).
805. H. H. Wasserman and E. H. Barber, *J. Amer. Chem. Soc.*, **91,** 3674 (1969).
806. G. Swaelens and M. Anteunis, *Tetrahedron Letters*, 561 (1970).
807. G. Swaelens and M. Anteunis, *Bull. Soc. chim. belges*, **78,** 321 (1969).
808. M. Anteunis, G. Swaelens, F. Anteunis-De Ketelaere, and P. Dirinck, *Bull. Soc. chim. belges*, **80,** 409 (1971).
809. A. K. Bhatti and M. Anteunis, *Tetrahedron Letters*, 71 (1973).

810. M. Anteunis and M. Coryn, *Bull. Soc. chim. belges*, **83**, 133 (1974).

811. R. J. Bishop, L. E. Sutton, M. J. T. Robinson, and N. W. J. Pumphrey, *Tetrahedron*, **25**, 1417 (1969).

812. G. Swaelens and M. Anteunis, *Bull. Soc. chim. belges*, **78**, 471 (1969).

813. H. Stetter, and R. Hesse, *Monatsh.*, **98**, 755 (1967).

814. E. Hardegger, E. Maeder, H. M. Semarne, and D. J. Cram, *J. Amer. Chem. Soc.*, **81**, 2729 (1959).

815. R. K. Summerbell and D. R. Berger, *J. Amer. Chem. Soc.*, **81**, 633 (1959).

816. C-Y. Chen and R. J. W. Le Fèvre, *J. Chem. Soc.*, 558 (1965).

817. R. K. Summerbell and H. E. Lunk, *J. Amer. Chem. Soc.*, **79**, 4802 (1957).

818. J. D. Oliver, J. D. Woodyard, P. E. Rush, and J. R. Curtis, *J. Heterocyclic Chem.*, **11**, 1125 (1974).

819. R. K. Summerbell and J. R. Stephens, *J. Amer. Chem. Soc.*, **76**, 6401 (1954).

820. D. L. Heywood and B. Phillips, *J. Amer. Chem. Soc.*, **80**, 1257 (1958).

821. R. K. Summerbell and J. R. Stephens, *J. Amer. Chem. Soc.*, **76**, 731 (1954).

822. R. K. Summerbell, G. Lestina, and H. Waite, *J. Amer. Chem. Soc.*, **79**, 234 (1957).

823. L. H. Werner and C. R. Scholz, *J. Amer. Chem. Soc.*, **76**, 2701 (1954).

824. R. K. Summerbell and G. J. Lestina, *J. Amer. Chem. Soc.*, **79**, 3838 (1957).

825. L. F. Hatch and G. D. Everett, *J. Org. Chem.*, **33**, 2551 (1968).

826. F. R. Jensen and R. A. Neese, *J. Org. Chem.*, **37**, 3037 (1972).

827. F. E. Malherbe and H. J. Bernstein, *J. Amer. Chem. Soc.*, **74**, 4408 (1952).

828. D. A. Ramsay, *Proc. Roy. Soc.*, **190A**, 562 (1947).

829. J. H. Gibbs, *Discuss. Faraday Soc.*, **10**, 122 (1951).

830. L. E. Sutton and L. O. Brockway, *J. Amer. Chem. Soc.*, **57**, 473 (1935).

831. O. Hassel and H. Viervoll, *Acta Chem. Scand.*, **1**, 149 (1947); M. Davis and O. Hassel, *Acta Chem. Scand.*, **17**, 1181 (1963).

832. F. A. L. Anet and J. Sandstrom, *Chem. Comm.*, 1558 (1971).

833. F. R. Jensen and R. A. Neese, *J. Amer. Chem. Soc.*, **93**, 6329 (1971), and **97**, 4345 (1975).

834. H. R. Buys, *Rec. Trav. chim.*, **88**, 1003 (1969).

835. G. Gatti, A. L. Segre, and C. Morandi, *Tetrahedron*, **23**, 4385 (1967).

836. Y. Sumi and F. Kametani, *Chem. and Pharm. Bull. (Japan)*, **21**, 1103 (1973).

837. E. Caspi, T. A. Wittstruck, and D. M. Piatak, *J. Org. Chem.*, **27**, 3183 (1962).

838. D. Jung, *Chem. Ber.*, **89**, 566 (1966).

839. J. Böeseken, F. Tellegen, and P. C. Henriquez, *Rec. Trav. chim.*, **54**, 733 (1935).

840. C. Altona and E. Havinga, *Tetrahedron*, **22**, 2275 (1966).

841. R. R. Frazer and C. Reyes-Zamora, *Canad. J. Chem.*, **43**, 3445 (1965).

842. C. Altona, C. Romers and E. Havinga, *Tetrahedron Letters*, No. **10**, 16 (1959).

843. C. Altona, C. Knobler, and C. Romers, *Rec. Trav. chim.*, **82**, 1089 (1963).

844. C. Altona and C. Romers, *Rec. Trav. chim.*, **81**, 1080 (1963).

845. C. Altona, C. Knobler, and C. Romers, *Acta Cryst*,. **16**, 1217 (1963).

846. J. Gierer and I. Petersson, *Acta Chem. Scand.*, **22**, 3183 (1968).

847. W. A. Szarek, D. M. Vyas, and B. Achmatowicz, *Tetrahedron Letters*, 1553 (1975).

848. R. K. Summerbell, B. S. Sokolski, J. P. Bays, D. J. Godfrey, and A. S. Hussey, *J. Org. Chem.*, **32,** 946 (1967).

849. R. Bramley, L. A. Cort, and R. G. Pearson, *J. Chem. Soc.* (*C*), 1213 (1968); J. Böeseken, F. Tellegen, and P. Cohen Henriques, *J. Amer. Chem. Soc.*, **55,** 1284 (1933).

850. R. E. Ardrey and L. A. Cort, *J. Chem. Soc.* (*C*), 2457 (1970).

851. W. R. Diveley, A. H. Haubein, A. D. Lohr, and P. B. Moseley, *J. Amer. Chem. Soc.*, **81,** 139 (1959).

852. G. C. S. Collins and W. O. George, *J. Chem. Soc.* (*B*), 1352 (1971).

853. C. I. Stassinopoulou and C. Zioudrou, *Tetrahedron*, **28,** 1257 (1972).

854. G. C. S. Collins and W. O. George, *Chem. Comm.*, 501 (1971).

855. D. Bryce-Smith and A. Gilbert, *Chem. Comm.*, 1318, 1701, 1702 (1968).

856. W. Friedrichsen and R. Epbinder, *Tetrahedron Letters*, 2059 (1973).

857. S. Farid, *Chem. Comm.*, 1268 (1967).

858. S. Farid and K-H. Scholz, *Chem. Comm.*, 412 (1968).

859. G. Pfundt and S. Farid, *Tetrahedron*, **22,** 2237 (1966).

860. M. J. Cook, A. R. Katritzky, and M. J. Sewell, *J. Chem. Soc.* (*B*), 1207 (1970).

861. A. T. Bottini and W. Schear, *J. Amer. Chem. Soc.*, **87,** 5802 (1965).

862. A. T. Bottini, F. P. Corson, K. A. Frost, and W. Schear, *Tetrahedron*, **28,** 4701 (1972).

863. B. Belleau and N. L. Weinberg, *J. Amer. Chem. Soc.*, **85,** 2525 (1963).

864. R. R. Fraser and C. Reyes-Zamora, *Canad. J. Chem.*, **45,** 929 (1967).

865. L. A. Cort, *J. Chem. Soc.*, 3167 (1960).

866. L. A. Cort, B. C. Stace, and D. P. C. Thackeray, *J.C.S. Perkin I*, 177 (1972).

867. S. Furberg and O. Hassel, *Acta Chem. Scand.*, **4,** 1584 (1950).

868. Y. Auerbach, M. Sprecher, and B. Fuchs, *Tetrahedron Letters*, 5207 (1970).

869. B. Fuchs, Y. Auerbach, and M. Sprecher, *Tetrahedron*, **30,** 437 (1974).

870. B. Fuchs, *Tetrahedron Letters*, 3571 (1969).

871. B. Fuchs, *Tetrahedron Letters*, 1747 (1970).

872. B. Fuchs, I. Goldberg, and U. Shmueli, *J.C.S. Perkin II*, 357 (1972).

873. M. Anteunis, M. Vandewalle, and L. Van Wijinsberghe, *Bull. Soc. chim. belges*, **80,** 423 (1971).

874. O. Hassel and J. Hvozlef, *Acta Chem. Scand.*, **10,** 138 (1956).

875. C. Maignan and F. Rouessac, *Bull. Soc. chim. France*, 3041 (1971).

876. B. Fuchs, Y. Auerbach, and M. Sprecher, *Tetrahedron Letters*, 2267 (1972).

877. P. Calinaud, J. Gelas, and S. Veyssieres-Rambaud, *Bull. Soc. chim. France*, 2769 (1973).

878. P. Calinaud and J. Gelas, *Bull. Soc. chim. France*, 1155 (1974).

879. J. Gelas, *Bull. Soc. chim. France*, 3721, 4041, 4046 (1970).

880. J. Gelas, *Bull. Soc. chim. France*, 4465 (1970).

881. R. van Cauwenberghe, M. Anteunis, and C. Becu *Heterocycles*, **3,** 101 (1975).

882. R. Griegee and R. Ruchtäschel, *Chem. Ber.*, **103,** 50 (1970).

883. N. F. Moerman, *Rec. Trav. chim.*, **56,** 161 (1937).

884. M. Kimura and K. Aoki, *J. Chem. Soc. Japan*, **72,** 169 (1951).
885. H. Gerding and G. W. A. Rijnders, *Rec. Trav. chim.*, **58,** 603 (1939).
886. H. Gerding and J. Lecomte, *Rec. Trav. chim.*, **58,** 614 (1939).
887. R. Kewley, *Canad. J. Chem.*, **48,** 852 (1970).
888. M. Barón and D. P. Hollis, *Rec. Trav. chim.*, **83,** 391 (1964).
889. D. C. Carpenter and L. O. Brockway, *J. Amer. Chem. Soc.*, **58,** 1270 (1936).
890. R. J. W. Le Fèvre and P. Russell, *J. Chem. Soc.*, 496 (1936).
891. J. L. Jungnickel and C. A. Reilly, *Rec. Trav. chim.*, **84,** 1526 (1965).
892. P. Groth, *Acta Chem. Scand.*, **21,** 2608 (1967).
893. P. Groth, *Acta Chem. Scand.*, **21,** 2631 (1967).
894. H. A. Brune, K. Wulz, and W. Hetz, *Tetrahedron*, **27,** 3629 (1971).
895. K. Wulz, H. A. Brune, and W. Hetz, *Tetrahedron*, **26,** 3 (1970).
896. A. Rosowsky, *Seven-Membered Heterocyclic Compounds Containing Oxygen and Sulfur*, Vol. 26, A. Weissberger and E. C. Taylor, Eds., Wiley-Interscience, 1972.
897. G. Fontaine, *Ann. Chim. (France)*, **3,** 469 (1968).
898. L. Canuel and M. St. Jacques, *Canad. J. Chem.*, **52,** 3581 (1974).
899. T. Kametani, S. Shibuya, and W. D. Ollis, *J. Chem. Soc. (C)*, 2877 (1968).
900. W. Tochtermann, D. Schäfer, and C. Rohr, *Chem. Ber.*. **104,** 2923 (1971).
901. C. G. Overberger and H. Kaye, *J. Amer. Chem. Soc.*, **89,** 5640 (1967).
902. C. G. Overberger and H. Kaye, *J. Amer. Chem. Soc.*, **89,** 5646 (1967).
903. G. H. Beaven, D. M. Hall, M. S. Lesslie, and E. E. Turner, *J. Chem. Soc.*, 854 (1952).
904. E. Bunnenberg, C. Djerassi, K. Mislow, and A. Moscowitz, *J. Amer. Chem. Soc.*, **84,** 2823 (1962).
905. K. Mislow, M. A. W. Glass, R. E. O'Brien, P. Rutkin, D. H. Steinberg, J. Weiss, and C. Djerassi, *J. Amer. Chem. Soc.*, **84,** 1455 (1962).
906. K. Mislow, M. A. W. Glass, R. E. O'Brien, P. Rutkin, D. Steinberg, and C. Djerassi, *J. Amer. Chem. Soc.*, **82,** 4740 (1960).
907. G. M. Badger, P. R. Jefferies, and R. W. L. Kimber, *J. Chem. Soc.*, 1837 (1957).
908. K. Mislow and M. A. W. Glass, *J. Amer. Chem. Soc.*, **83,** 2780 (1961).
909. K. Mislow, M. A. W. Glass, H. B. Hopps, E. Simon, and G. H. Wahl, Jr., *J. Amer. Chem. Soc.*, **86,** 1710 (1964).
910. I. O. Sutherland and M. V. J. Ramsay, *Tetrahedron*, **21,** 3401 (1965).
911. H. Kessler, *Tetrahedron Letters*, 1461 (1968).
912. M. Ōki, H. Iwamura, and N. Hayakawa, *Bull. Chem. Soc. Japan*, **37,** 1865 (1964).
913. M. Ōki and H. Iwamura, *Tetrahedron*, **24,** 2377 (1968).
914. D. Muriel Hall, *Progr. Stereochem.*, **4,** 1 (1968).
915. E. Vogel, U. Haberland, and J. Ick, *Angew. Chem. Internat. Edn.*, **9,** 517 (1970).
916. E. Vogel and H. Günther, *Angew. Chem. Internat, Edn.*, **6,** 385 (1967).
917. F-G. Klärner and E. Vogel, *Angew. Chem. Internat. Edn.*, **12,** 840 (1973).
918. G. R. Ziegler, *J. Amer. Chem. Soc.*, **91,** 446 (1969).
919. N. A. Bailey and R. Mason, *Chem. Comm.*, 1039 (1967).
920. M. H. Gianni, J. Saavedra, and J. Savoy, *J. Org. Chem.*, **38,** 3971 (1973).

921. S. Smoliński and A. Malata, *Tetrahedron*, **25,** 5427 (1969).

922. A. C. Cope, M. A. McKervey, N. M. Weinshenker, and R. B. Kinnel, *J. Org. Chem.*, **35,** 2918 (1970).

923. A. C. Cope, H-H. Lee, and H. E. Petree, *J. Amer. Chem. Soc.*, **80,** 2849 (1958).

924. A. C. Cope, M. Brown, and H-H. Lee, *J. Amer. Chem. Soc.*, **80,** 2855 (1958).

925. A. C. Cope, J. M. Grisar, and P. E. Peterson, *J. Amer. Chem. Soc.*, **81,** 1640 (1959).

926. C. Ganter, R. O. Duthaler, and W. Zwahlen, *Helv. Chim. Acta*, **54,** 578 (1971).

927. F. G. Bordwell and M. L. Douglass, *J. Amer. Chem. Soc.*, **88,** 993 (1966).

928. L. A. Paquette and P. C. Storm, *J. Org. Chem.*, **35,** 3390 (1970).

929. L. A. Paquette and P. C. Storm, *J. Amer. Chem. Soc.*, **92,** 4295 (1970).

930. L. A. Paquette, I. R. Dunkin, J. P. Freeman, and P. C. Storm, *J. Amer. Chem. Soc.*, **94,** 8124 (1972).

931. L. A. Paquette and R. W. Begland, *J. Amer. Chem. Soc.*, **87,** 3784 (1965).

932. L. A. Paquette, R. W. Begland, and P. C. Storm, *J. Amer. Chem. Soc.*, **90,** 6148 (1968).

933. A. F. Cameron, K. K. Cheung, G. Ferguson, and J. M. Robertson, *Chem. Comm.*, 638 (1965).

934. L. A. Paquette, R. W. Begland, and P. C. Storm, *J. Amer. Chem. Soc.*, **92,** 1971 (1970).

935. D. S. Tarbell and J. R. Hazen, *J. Amer. Chem. Soc.*, **91,** 7657 (1969).

936. D. L. Whalen, *J. Amer. Chem. Soc.*, **92,** 7619 (1970).

937. L. A. Paquette and M. K. Scott, *J. Amer. Chem. Soc.*, **94,** 6751 (1972).

938. N. J. Leonard, T. W. Milligan, and T. L. Brown, *J. Amer. Chem. Soc.*, **82,** 4075 (1960).

939. F. A. L. Anet and P. J. Degen, *Tetrahedron Letters*, 3613 (1972).

940. K. G. Taylor and J. Chaney, *J. Amer. Chem. Soc.*, **94,** 8924 (1972).

941. A. Lüttringhaus, *Annalen*, **528,** 181 (1937).

942. K. Ziegler and A. Lüttringhaus, *Annalen*, **511,** 1 (1934).

943. A. Lüttringhaus and H. Gralheer, *Annalen*, **550,** 67 (1942).

944. K. L. Rinehart, Jr., *Accounts Chem. Res.*, **5,** 57 (1972).

945. A. Lüttringhaus and H. Gralheer, *Naturwiss*, **61,** 255 (1940).

946. A. Lüttringhaus and H. Gralheer, *Annalen*, **557,** 108 (1947).

947. A. Lüttringhaus and H. Gralheer, *Annalen*, **557,** 112 (1947).

948. G. Helmchen and V. Prelog, *Helv. Chim. Acta*, **55,** 2612 (1972).

949. H. E. Winberg, F. S. Fawcett, W. E. Mochel, and C. W. Theobald, *J. Amer. Chem. Soc.*, **82,** 1428 (1960).

950. C. J. Brown, *J. Chem. Soc.*, 3278 (1953).

951. H. H. Wasserman and P. M. Keehn, *Tetrahedron Letters*, 3227 (1969).

952. S. Mizogami, T. Otsubo, Y. Sakata, and S. Misumi, *Tetrahedron Letters*, 2791 (1971).

953. I. Gault, B. J. Price, and I. O. Sutherland, *Chem. Comm.*, 540 (1967).

954. J. F. Haley, Jr. and P. M. Keehn, *Tetrahedron Letters*, 4017 (1973).

955. S. Rosenfeld and P. M. Keehn, *Tetrahedron Letters*, 4021 (1973).

956. S. M. Rosenfeld and P. M. Keehn, *J.C.S. Chem. Comm.*, 119 (1974).

957. M. A. Battiste, L. A. Kapicak, M. Mathew, and G. J. Palenik, *Chem. Comm.*, 1536 (1971).

958. R. Helder and H. Wynberg, *Tetrahedron Letters*, 4321 (1973).

959. E. H. Hahn, M. Bohm, and D. Ginsburg, *Tetrahedron Letters*, 507 (1973).

960. F. A. L. Anet and P. J. Degen, *J. Amer. Chem. Soc.*, **94,** 1390 (1972).

961. J. Dale, T. Ekeland, and J. Krane, *J. Amer. Chem. Soc.*, **94,** 1389 (1972).

962. M. Barón, O. B. De Mandirola, and J. F. Westerkamp. *Canad. J. Chem.*, **41,** 1893 (1963).

963. T. Sato and K. Uno, *J.C.S. Perkin I*, 895 (1973).

964. N. S. Zefirov and S. V. Rogozina, *Tetrahedron*, **30,** 2345 (1974).

965. E. N. Marvell and S. Provant, *J. Org. Chem.*, **29,** 3084 (1964).

966. S. Smoliński and I. Deja, *Tetrahedron*, **27,** 1409 (1971).

967. S. Smoliński and B. Golabek, *Tetrahedron*, **25,** 5431 (1969).

968. G. Borgen, J. Dale, F. A. L. Anet, and J. Krane, *J.C.S. Chem. Comm.*, 243 (1974).

969. J. F. Stoddart and W. A. Szarek, *Canad. J. Chem.*, **46,** 3061 (1968).

970. T. B. Grindley, J. F. Stoddart, and W. A. Szarek, *J. Amer. Chem. Soc.*, **91,** 4722 (1969).

971. G. Borgen and J. Dale, *J.C.S. Chem. Comm.*, 484 (1974).

972. P. G. Edgerley and L. E. Sutton, *J. Chem. Soc.*, 1069 (1951).

973. A. C. D. Newman,and H. M. Powell, *J. Chem. Soc.*, 3747 (1952).

974. A. P. Downing, W. D. Ollis, I. O. Sutherland, J. Mason, and S. F. Mason, *Chem. Comm.*, 329 (1968).

975. A. P. Downing, W. D. Ollis, and I. O. Sutherland, *J. Chem. Soc. (B)*, 24 (1970); *Chem. Comm.*, 171 (1967).

976. C. J. Pedersen and H. K. Fernsdorff, *Angew. Chem. Internat. Edn.*, **11,** 16 (1972).

977. J. J. Christensen, J. O. Hill, and R. M. Izatt, *Science*, **174,** 459 (1971).

978. M. R. Truter and C. J. Petersen, *Endeavour*, **30,** 142 (1971).

979. C. J. Petersen, *J. Amer. Chem. Soc.*, **92,** 386, 391 (1970).

980. J. Cooper and P. W. Plesch, *J.C.S. Chem. Comm.*, 1017 (1974).

981. M. Cinquini, F. Montanari, and P. Tundo, *J.C.S. Chem. Comm.*, 393 (1975).

982. G. R. Newkome, G. L. McClure, J. B. Simpson, and F. Danesh-Khoshboo, *J. Amer. Chem. Soc.*, **97,** 3232 (1975).

983. R. M. Izatt, D. P. Nelson, J. H. Rytting, B. L. Haymore, and J. J. Christensen, *J. Amer. Chem. Soc.*, **93,** 1619 (1971).

984. S. N. Graven, H. A. Lardy, D. Johnson, and A. Rutter, *Biochemistry*, **5,** 1729 (1966).

984a. B. Dietrich, J. M. Lehn, and J. P. Sauvage, *Tetrahedron Letters*, 2885, 2889, (1969); J. M. Lehn and J. P. Sauvage, *Chem. Comm.*, 440 (1971); E. Graf and J. M. Lehn, *J. Amer. Chem. Soc.*, **97,** 5022 (1975).

985. R. A. Johnson and E. G. Nidy, *J. Org. Chem.*, **40,** 1680 (1975).

986. M. A. Bush and M. R. Truter, *Chem. Comm.*, 1439 (1970).

987. M. A. Bush and M. R. Truter, *J. Chem. Soc. (B)*, 1440 (1971).

988. D. Bright and M. R. Truter, *Nature*, **225,** 176 (1970).

989. D. Bright and M. R. Truter, *J. Chem. Soc. (B)*, 1544 (1970).
990. D. G. Parsons, *J.C.S. Perkin I*, 245 (1975).
991. P. R. Mallinson, *J.C.S. Perkin II*, 261 (1975).
992. M. Newcomb and D. J. Cram, *J. Amer. Chem. Soc.*, **97**, 1257 (1975).
993. R. C. Helgeson, J. M. Timko, and D. J. Cram, *J. Amer. Chem. Soc.*, **97**, 3023 (1973).
994. R. C. Helgeson, K. Koga, J. M. Timko, and D. J. Cram, *J. Amer. Chem. Soc.*, **97**, 3021 (1973).
995. G. W. Gokel, J. M. Timko, and D. J. Cram, *J.C.S. Chem. Comm.*, 394 (1975).
995a. G. W. Gokel, J. M. Timko, and D. J. Cram, *J.C.S. Chem. Comm.*, 444 (1975).
996. G. Dotsevi, Y. Sogah, and D. J. Cram, *J. Amer. Chem. Soc.*, **97**, 1259 (1975).
996a. K. Madan and D. J. Cram, *J.C.S. Chem. Comm.*, 427 (1975).
997. A. Khawam and E. V. Brown, *J. Amer. Chem. Soc.*, **74**, 5603 (1952).
998. R. E. Lutz and C. J. Kibler, *J. Amer. Chem. Soc.*, **62**, 1520 (1940).
999. R. Adams and N. Kornblum, *J. Amer. Chem. Soc.*, **63**, 188 (1941).
1000. M. Kataoka, T. Ando, and M. Nakagawa, *Bull. Chem. Soc., Japan*, **44**, 177 (1971).
1001. E. S. Wallis and F. H. Adams, *J. Amer. Chem. Soc.*, **55**, 3838 (1933).
1002. R. Adams, C. K. Cain, W. D. McPhee, and R. B. Wearn, *J. Amer. Chem. Soc.*, **63**, 2209 (1941).
1003. R. Adams, H. Wolff, C. K. Cain, and J. H. Clark, *J. Amer. Chem. Soc.*, **62**, 2215 (1940).
1004. R. Adams, C. M. Smith, and S. Loewe, *J. Amer. Chem. Soc.*, **64**, 2087 (1942).
1005. R. Adams, D. C. Pease, C. K. Cain, and J. H. Clark, *J. Amer. Chem. Soc.*, **62**, 2402 (1940).
1006. R. A. Archer, D. B. Boyd, P. V. Demarco, I. J. Tyminski, and N. L. Allinger, *J. Amer. Chem. Soc.*, **92**, 5200 (1970).
1007. R. Mechoulam and Y. Gaoni, *Tetrahedron Letters*, 1109 (1967).
1008. R. Mechoulam and Y. Gaoni, *Fortschr. Chem. org. Naturstoffe*, **25**, 175 (1967).
1009. C. B. Quinn and J. R. Wiseman, *J. Amer. Chem. Soc.*, **97**, 1342 (1973).

3 SULFUR HETEROCYCLES

I. INTRODUCTION

The stereochemistry of the carbon skeleton around the sulfur atom in saturated and partly saturated sulfur heterocycles is quite similar to the one found in the corresponding nitrogen and oxygen heterocycles. The longer C–S bond compared to the C–N and C–O bond lengths slightly modifies the geometry of the molecules, and although this makes the nonbonded interactions somewhat different, the final outcome is not generally unlike that observed in the other heterocycles. The ability of sulfur to expand its valency shell[1] gives the sulfur heterocycles another stereochemical property by creating an asymmetric center in the appropriately substituted heterocycle as in sulfoxides and related compounds. Pyramidal inversion[2] in tervalent sulfur (e.g., Me_2SO, 157 kJ mol^{-1}) is much more energy demanding than in a trivalent nitrogen atom (e.g., Et_3N, 24 kJ mol^{-1}—see sections on conformation in Part I, Chapters 2 and 3), and inversion of configuration at the trivalent sulfur atom in sulfoxides requires an S_N2 substitution reaction.[3] Thus conformational analyses of saturated sulfur heterocycles, as in oxygen heterocycles (Chapter 2), are not complicated by atomic inversion as in the nitrogen heterocycles.

Sulfur is capable of forming di- and polysulfides which, if they are not planar as in rigid systems (e.g., *trans*-1,2-dithiadecalin), are a source of chirality in their own right.[4] The sulfur atom in saturated and unsaturated heterocycles is more strongly chromophoric; that is, it absorbs uv light at longer wavelengths, than an oxygen or nitrogen atom[5] and consequently enhances the optical rotation of a chiral derivative if the atom is associated with the source of asymmetry.

The stereochemistry of sulfur heterocycles and their *S*-oxides are systematically treated, as were the nitrogen and oxygen heterocycles in previous chapters. A comparison between the heterocycles can be readily made by reading only the individual sections on three-, four-, or five-membered rings, *et cetera*, in the respective chapters. For further general information on sulfur heterocycles, including some stereochemistry, reference should be made to review articles[1,4,6–8] and texts[9–11] on the subject.

II. THIIRANES (ETHYLENE SULFIDES, EPISULFIDES, OR THIACYCLOPROPANES)

The stereochemistry of thiiranes is similar to that of oxiranes (Chapter 2, Section II) except that these can also exist as *S*-oxides and *S,S*-dioxides. One substituent on a carbon atom introduces an asymmetric center, and two substituents, one on each carbon atom, produce cis and trans isomerism. When these are similar the cis isomer is meso, and the trans isomer is racemic. The *S*-oxide creates another chiral center if one is already present in the molecule, and syn and anti isomers are possible.

The chemistry of thiiranes up to January 1961, was reviewed by Reynolds and Fields.[12]

A. Syntheses of Thiiranes

Thiocyanic acid reacts with oxiranes and yields thiiranes. *S*(−)-2-Methyloxirane provided *R*(+)-2-methylthiirane indicating that at least one Walden inversion had occurred. This thiirane was also formed by treating the cyclic carbonate, *S*(−)-4-methyl-1,3-dioxalan-2-one (obtained from *S*(+)-propane-1,2-diol), with thiocyanate ions showing that a Walden inversion

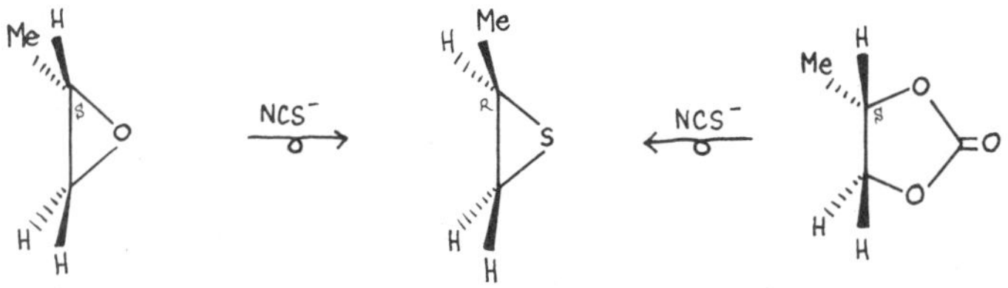

$$\xrightarrow[\Theta]{NCS^-} \qquad \xleftarrow[\Theta]{NCS^-} \qquad (1)$$

occurred here too (Equation 1).[13] The mechanism proposed for the thiocyanate conversion of oxiranes into thiiranes involved the intermediate formation of the cyclic 1,3-oxathiolane anion (**1**).[14] Evidence in support of this mechanism (Scheme 1) was provided by van Tamelen,[15] who demonstrated that whereas *cis*-2,3-tetramethyleneoxirane, $n=2$, gave 73% yield of *cis*-2,3-tetramethylenethiirane, *cis*-2,3-trimethyleneoxirane, $n=1$, was recovered unchanged when it was reacted with thiocyanate ions using identical procedure. The lack of reactivity in the latter case was attributed to the difficulty in forming the intermediate strained *trans*-bicyclo[3.3.0]octane system (**1**), when $n=1$. The isolation of 2-(4-nitrobenzimido)-5-methyl-1,3-oxathiolane from the reaction of 2-methyloxirane (propylene oxide) and KCNS in the presence of 4-nitrobenzoyl chloride further strengthens the case for the intermediacy of the anion (**1**).[16] The operation of this mechanism indicates that a double Walden inversion results in the transformation of oxiranes into thiiranes, and *R*,*R*-oxiranes should yield *S*,*S*-thiiranes. Stereoisomersm of thiiranes occurs with KCNS and lends further support to the double inversion mechanism. Jankowski and Harvey[16a] observed that *cis*- and *trans*-3-hydroxycyclohex-2-ene episulfides and their *O*-acetyl derivatives separately gave cis–trans mixtures on treatment with KCNS in DMF or aqueous ethanol. The conversion may also be brought about with thioureas and most probably involves two Walden inversions[12] (see Section

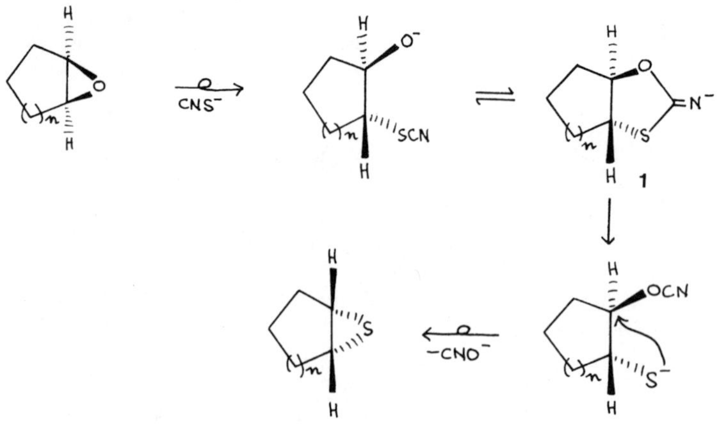

Scheme 1

II.B). Intramolecular cyclization of *trans*-2-acylthio-1-tosyloxycyclopentane and -cyclohexane, in the presence of a base, into *cis*-2,3-trimethylene- and *cis*-2,3-tetramethylenethiiranes requires an internal displacement of the tosyloxy group with inversion of configuration at C–1.[17]

Kellogg and co-workers[18] discovered a novel thiirane synthesis by pyrolysis of 2,5-diethyl-1,3,4-thiadiazoline which promotes the elimination of nitrogen. The reaction is highly stereospecific and apparently involves an intermediate sulfur ylide which cyclizes to the thiirane by a conrotatory ring closure. *trans*-2,5-Diethyl-,[18] *trans*-2,5-di-*t*-butyl-, and *cis*-2,5-di-*t*-butyl-1,3,4-thiadiazolines[19] furnish *cis*-2,3-dimethyl-, *cis*-2,3-di-*t*-butyl-, and *trans*-2,3-di-*t*-butylthiiranes respectively (Equation 2).

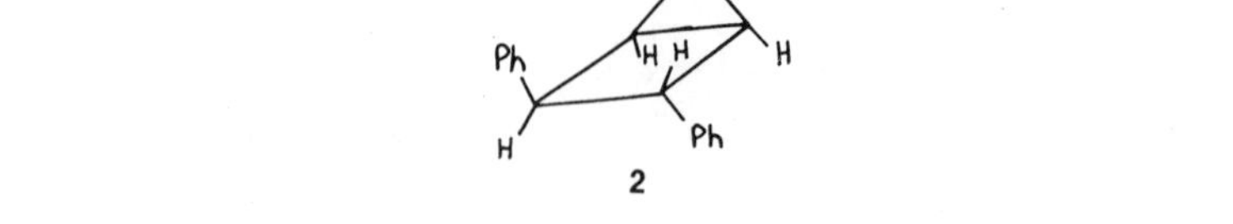

(2)

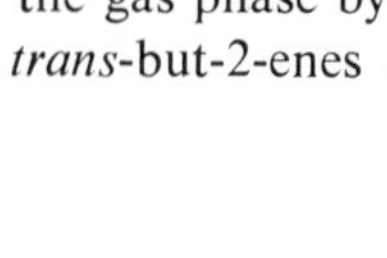

2

Sulfur atoms, generated in the gas phase by dissociation of carbon oxysulfide, react with *cis*- and *trans*-but-2-enes and produce the respective *cis*- and *trans*-2,3-dimethylthiiranes with retention of configuration. The reaction apparently can involve singlet and triplet sulfur, and the stereospecificity of the reaction cannot be used in this case to identify the spin states.[20] Irradiation of 2,2′-diphenyldivinyl sulfide in diethyl ether results in two intramolecular cyclizations with the formation of *trans*-2,3-diphenyl-5-thiabicyclo[2.1.0]pentane (**2**). This is the first example in which a thiirane ring is fused to a cyclobutane ring.[21]

The chemistry of thiirane *S*-oxides is better understood and this has made it possible to explore them more fully. They have been obtained by the oxidation of substituted thiiranes with hydrogen peroxide,[22] perbenzoic acid,[23] and *m*-chloroperbenzoic acid.[19] The oxidation of 2,3-dibenzoyl-2,3-diphenylthiiranes gave two *S*-oxides: a *d,l* pair and a meso form.[22] Perbenzoic acid oxidation of 2-mono- and *cis*-2,3-disubstituted thiiranes furnished the *anti-S*-oxides.[23] *anti-S*-Oxides are relatively stable, but *S*-oxides which have a syn substituent with an α-hydrogen atom decompose readily to thiosulfonates (see Section II.C). Diarylmethylene sulfines react with aryldiazomethanes with elimination of nitrogen and provide *syn*- and *anti*-2,2,3-triarylthiirane *S*-oxides in which the anti isomer predominates (Equation 3).[24] Diazopropane also reacts with diaryl-, arylchloro-, and arylarylthiomethylene sulfines, but the products, obtained with various specificities depending on the substituents, are *syn*- or *anti*-2,2-dimethyl-5-substituted 1,3,4-thiadiazoline *S*-oxides. This suggests that thiadiazoline *S*-oxides are intermediates in the previously mentioned reaction because under the same conditions mesitylphenyl sulfonyl sulfine and diazopropane give an equal mixture of diastereomeric 2,2-dimethyl-3-mesityl-3-phenylsulfonylthiirane *S*-oxides (Equation 4)[25] (compare with Equation 2).

Ar₂C=S=O → syn + anti (Ar, Ar, H, XC_6H_4 thiirane S-oxides) (3)

Mesityl–C(=S=O)–SO_2Ph + $Me_2C{=}N_2$ → syn + anti ($PhSO_2$, $Me_3C_6H_2$, Me, Me thiirane S-oxides) (4)

Thiirane *S,S*-dioxides are formed by the oxidation of thiiranes with hydrogen peroxide,[22] but they are subject to solvolytic decomposition under these conditions. 2,3-Diphenylthiirane *S,S*-dioxide attracted the most attention but was prepared in a different way. A 45:55 mixture of cis and trans isomers is obtained from phenyldiazomethane and sulfur dioxide at −20°C (Equation 5).[26] Tokura and co-workers[27] isolated only the cis isomer from this reaction, but the other products (e.g., *cis*- and *trans*-stilbene) formed suggest that perhaps the *trans*-dioxide might have been formed but decomposed readily. Caprino and co-workers[28,29] also isolated *cis*-2,3-diphenylthiirane *S,S*-dioxide from this reaction and obtained it by reduction of 2,3-diphenylthiirene 1,1-dioxide (**3**) with aluminum amalgam in the presence of wet ether. The thiirene dioxide **3**, the first example of a potentially aromatic three-membered ring heterocycle, was prepared in 70% yield from bis-α-bromobenzyl sulfone and Et_3N in chloroform. It decomposed into diphenylacetylene at its melting point.[28]

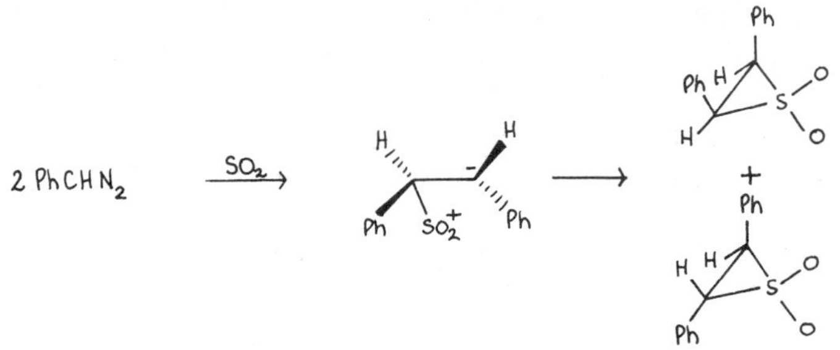

$$2\ PhCHN_2 \xrightarrow{SO_2} \qquad (5)$$

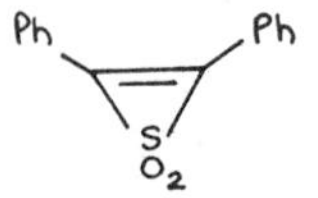

3

B. Configuration of Thiiranes

There are no conformational changes involved in thiiranes, but their configurations are discussed in this section. Theoretical treatment of thiirane, its *S*-oxide and *S,S*-dioxide, is consistent with bond distances and angles derived from microwave spectroscopy.[30,31]

The cis and trans configurations of thiiranes were deduced from the method of synthesis and from ¹H nmr data.[32] The ord spectra of *R*(+)-2-methyl-, *R*(+)-2-*t*-butyl-, and *R*(−)-2-phenylthiiranes were measured, and the chiral band at 260 nm was discussed in terms of a dynamic coupling mechanism.[33] Cd data of *S*(+)-2-phenyl-, 2*S*,3*S*(−)-(*trans*)-2,3-diphenyl-, 2*S*,3*S*(−)-(*trans*)-2-methyl-3-phenyl-, and 3*R*,2*S*(−)-(*cis*)-2-methyl-3-

phenylthiiranes clearly showed the characteristic bands for the aromatic ring and the thiirane chromophore. These thiiranes were prepared by the KNCS and thiourea methods from the respective oxiranes, and confirm that two Walden inversions occurred.[34]

The spin coupling constants in the nmr spectra of thiirane *S*-oxides are anomalous, and the syn and anti configurations of these are best deduced from a study of benzene-induced chemical shifts.[35] These make use of the anisotropic effects of the S–O group.[35,23]

C. Reactions of Thiiranes

Cleavage of the thiirane ring with nucleophiles proceeds, as in the oxiranes (Chapter 2, Section II.C.1), with inversion of configuration at the carbon atom involved in the reaction. This is exemplified in the reactions that follow. The transformation of thiiranes into Δ^2-thiazolines, by reaction with nitriles in the presence of strong acid, is accomplished with one inversion of configuration (Equation 6).[36] The more highly substituted carbon atom is attacked by the nitrile and is the one that undergoes inversion. The reaction is stereospecific, and *trans*-thiiranes yield *cis*-Δ^2-thiazolines; for example, 2*S*,3*S*(−)-*trans*-2,3-dimethylthiirane gave 4*R*,5*S*(−)-*cis*-2,4,5-trimethyl-Δ^2-thiazoline. The thiirane was prepared from 2*R*,3*R*(+)-*trans*-2,3-dimethyloxirane.[37]

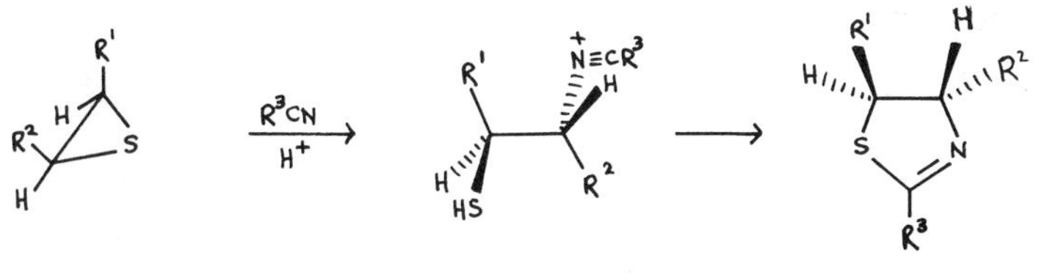

R^3CN, H^+ (6)

Unlike *cis*-2,3-trimethyleneoxirane (cyclopentene oxide), *cis*-2,3-trimethylenethiirane reacts with methyl xanthate to furnish the strained *trans*-2,3-trimethylene-1,3-dithiolan-2-thione (**4**). The difference is probably due to the longer C–S bond length compared with C–O bond length.[38]

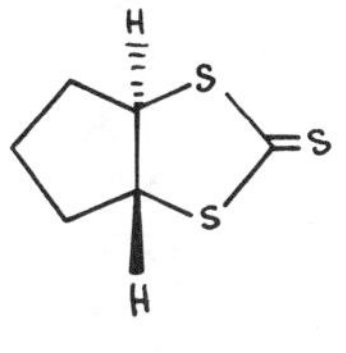

4 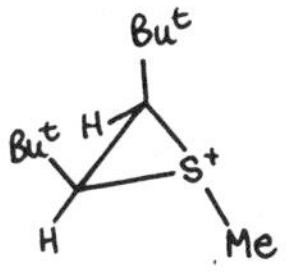 **5**

Attempts to *S*-methylate *cis*-2,3-dimethylthiirane with methyl halides leads to elimination of sulfur stereospecifically with formation of *cis*-butylene. A mechanism involving the unstable methylsulfonium ion in which C–2 is attacked by halide ion was proposed.[39] Butyl lithium converts *cis*- and *trans*-2,3-dimethylthiiranes into *cis*- and *trans*-butylene.[40] The *S*-methylation of the crowded *cis*- and *trans*-2,3-di-*t*-butylthiirane was examined by Kellogg and co-workers[19] using magic acid ($MeSO_3F$). The cis isomer gave one *S*-methyl derivative which must be the anti isomer **5**, but attempts to obtain the trans isomer (shielded from both sides) produced intractable products because higher temperatures were necessary for the reaction. Both *S*-oxides, on the other hand, were prepared.

Stereospecific sulfur extrusion, by way of a sulfonium ylide, occurs when 2,3-tetramethylene-*cis*-, *trans*-2-isopropyl-3-methyl-, and *cis*-2,3-diphenylthiiranes[41] are reacted with ethyl diazoacetate, in the presence of copper acetoacetate, and form cyclohexene, *cis*- and *trans*-4-methylpent-2-enes, and *cis*-stilbene, respectively, in high yields. Stereospecific desulfurization of thiiranes with triethyl phosphite,[42] triphenyl phosphine,[43,44] and tributyl phosphine[43] also yields the corresponding olefins. Kinetic studies provided evidence for initial attack of trivalent phosphorus on the sulfur atom followed by ring opening and release of the respective phosphine sulfide.[43] Only one observation was reported in which a C–C bond was cleaved, and it occurred in the reaction of triphenyl phosphine with the cyclobutane **2** which gave *trans,trans*-1,4-diphenylbutadiene and 2,3-dihydro-3,4-diphenylthiophene.[21]

(7)

Pyrolysis of *trans*-2,3-diphenylthiirane *S*-oxide is highly stereospecific and provides *trans*-stilbene, but pyrolysis of the cis isomer (*anti*-oxide) yields *cis*- and *trans*-stilbenes. At low temperatures the cis to trans ratio is more than one, but at higher temperatures it is less than one. This suggests a stepwise decomposition.[45] Thermolysis of *cis*- and *trans*-2,3-dimethylthiiranes at 150° is stereoselective in favor of retention of stereochemistry.[46] The loss of SO and retention of configuration in these reactions is suggestive of a nonlinear cheletropic reaction.[47] *cis*-2,3-Dialkylthiirane *anti-S*-oxides are relatively stable, whereas thiirane oxides with alkyl substituents that have an α-CH group syn to the oxide function decompose easily

(Equation 7).[23] A similar mechanism was proposed by Baldwin, Höfle, and Choi[48] for the decomposition of such compounds in which SO is not lost, and they isolated a new sulfur structure—the sulfoxylate ester R–S–S–OR. Thiirane *S*-oxides are cleaved by acids; the nucleophile attacks the less hindered carbon atom with inversion of configuration at that atom. The products are thiolsulfinates, and the stereochemistry was deduced by reduction with LAH followed by methylation (Equation 8).[49]

$$\xrightarrow[\text{MeOH}]{H^+} \quad \text{MeO–CH(Me)–CH(Me)–SO–S–CH(Me)–CH(Me)–OMe} \quad \xrightarrow[\text{(2) MeI/OH}]{\text{(1) LAH}} \quad \text{MeO–CH(Me)–CH(Me)–SMe} \qquad (8)$$

cis-2,3-Diphenylthiirane *S,S*-dioxide decomposes stereospecifically on melting and gives *cis*-stilbene.[28] Decomposition also occurs in alkaline solution, and it is subject to the dielectric constant of the medium. The base-calayzed decomposition in 2 *N* sodium hydroxide gave 92% of *trans*-stilbene,[50] but in weaker base (e.g., 0.2 *N* alkali) a 39:61 mixture of *cis*- and *trans*-stilbenes is obtained.[27] A much higher retention of configuration, however, was observed when BuLi was used in which a 70:6 ratio of cis to trans stilbenes was formed.[50] In a kinetic study of the solvolytic decomposition of *cis*- and *trans*-diphenyl *S,S*-dioxide, alkoxide ions accelerated the reaction and gave *cis*- and *trans*-stilbenes stereoselectively with preferred retention of configuration. Deuterium exchange experiments revealed that the methoxide-catalyzed exchange of the cis isomer with retention of configuration occurred much faster than isomerization to the *trans*-dioxide. A mechanism of decomposition involving biradicals was suggested.[26] *cis*-2,3-Diphenylthiirane *S,S*-dioxide was postulated as an intermediate in the reaction of *d,l*- and *meso*-bis-α-bromobenzylsulfones with triphenyl phosphine, which furnish *cis*- and *trans*-stilbenes in ratios depending on the reaction conditions.[51]

III. THIETANES (PROPYLENE SULFIDES)

The four-membered sulfur-containing ring heterocycles, the thietanes, were reviewed by Etienne, Soulas, and Lumbroso[52] and details about the physical properties including molecular dimensions and geometry, general chemical syntheses, and reactions were described. Stereochemical aspects of thietanes published after their review are discussed in the following sections.

A. Syntheses of Thietanes

The general method for preparing thietanes by reaction of 1,3-dibromopropane with sodium sulfide proceeds with inversion at C–1 and C–3. Paquette and Freeman[53,54] prepared *S*-2-methylthietane from *R*-2,4-dibromobutane, which is obtainable from *S*(+)-3-acetoxybutyric acid of known configuration (Equation 9). A mixture of stereoisomers of 1,3-diphenyl-3-hydroxypropan-1-thiol, prepared by LAH reduction of 1,3-diphenyl-3-acetylthiopropan-1-one, was converted to the chloro compound with HCl and cyclized into equal amounts of *cis*- and *trans*-2,4-diphenylthietanes which were separated by recrystallization.[55] In an application of this syn-

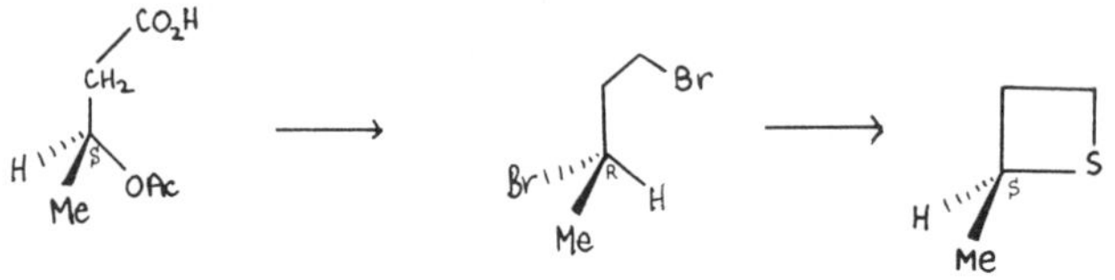

(9)

thesis 2-hydroxypentan-4-thiol was converted into the corresponding thiocyanato derivative with BrCN, which cyclized into a mixture of *cis*- and *trans*-2,4-dimethylthietanes.[56] A highly stereospecific synthesis is the reaction of 1,3-dioxan-2-ones (e.g., the 4*R*-4-methyl derivative) with KCNS to form the thietane (e.g., 2*S*-2-methylthietane). In the example cited inversion of configuration occurred at the substituted carbon atom C–4, but there is no doubt the reaction proceeded with two S_N2 displacements, and inversion of configuration must have also taken place at C–6 (Scheme 2).[54]

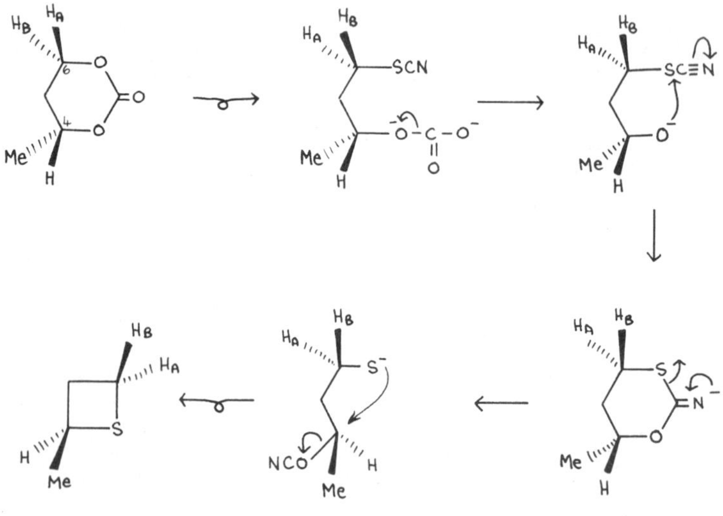

Scheme 2

2,2-Diphenylthietanes are formed in the photocycloaddition of thiobenzophenone to olefins. The stereospecificity of the addition depends on the substituents in the olefins and their stereochemistry. *trans*-1-Phenylprop-1-ene yields exclusively *trans*-4-methyl-2,2,3-triphenylthietane, whereas the cis isomer gives a cis and trans mixture of thietanes. Olefins with electron withdrawing substituents provide thietanes with retention of configuration; that is, *cis*- and *trans*-1,2-dichloroethylenes yield *cis*- and *trans*-3,4-dichloro-2,2-diphenylthietanes respectively. 1,2-Dicyanoethylenes react with the same high stereospecificity, as do *cis*- and *trans*-1-cyanoprop-1-enes, except that the regiospecificities of the latter compounds are not complete. The trans compound (crotonitrile), for example, gives a 60:30 mixture of *trans*-3-cyano-4-methyl- and *trans*-4-cyano-3-methyl-2,2-diphenylthietanes.[57]

In connection with studies on 1,2-dithiolanes, Padwa and Gruber[58] found that tris(dimethylamino)phosphine desulfurized *trans*-4-benzoyl-3-phenyl-1,2-dithiolane into *trans*-3-benzoyl-2-phenylthiirane with retention of configuration. This thiirane was also obtained by the stereospecific elimination of SO_2 from *trans*-4-benzoyl-3-phenyl-1,2-dithiolane 2,2-dioxide by thermolysis in benzene at 230°C. It is not known whether S–1 or S–2 is extruded in the former case (Equation 10).

COPh Ph (dithiolane) $\xrightarrow{(Me_2N)_3P}$ PhCO (thiirane, H, Ph, H, S) $\longleftarrow$ PhCO Ph (dithiolane, S, SO_2) (10)

Careful oxidation of thietanes with H_2O_2 in formic acid,[55] with $NaIO_4$[59] or *m*-chloroperbenzoic acid,[60] provides the 1-oxides of which there are two possibilities. The S–O bond can be cis or trans to a substituent already present which clearly influences the direction of oxidation. The specificity is not always complete, and mixtures are formed which can be separated[60–62] and identified by physical methods (see Section III.B). The ratio of cis to trans thietane 1-oxides varies considerably with the reagent used. Further oxidation furnishes the corresponding thietane 1,1-dioxides.[54–56,59,62]

A direct synthesis of thietane 1,1-dioxides, which is stereochemically interesting and has been explored by several investigators, involves the addition of a sulfene (generated from an alkanesulfonyl chloride and a base) to an activated double bond; generally it is an enamine double bond. Stork and Borowitz[64] treated 1-morpholinocyclohex-1-ene with methanesulfonyl chloride and Et_3N, and they obtained *cis*-2,3-tetramethylene-3-morpholinothietane 1,1-dioxide. 1-Morpholinocyclopent-1-ene and 1-propene reacted similarly and gave *cis*-2,3-trimethylene-3-morpholino- and *trans*-2-methyl-3-morpholinothietane 1,1-dioxides respectively. The reaction is a general one; the addition is always cis, and the reaction is regiospecific (Equation 11). A Hofmann type elimination of the amino group to yield a thiete 1,1-dioxide can be achieved by methylation and treatment with silver oxide,[65] or by oxidation to the intermediate *N*-oxide which undergoes elimination on heating.[63] 3,4-Tetramethylenethiete 1,1-dioxide (**8**, R^5, R^3=–$(CH_2)_4$–, R^4=H) was reduced catalytically, with LAH, to *cis*-2,3-tetramethylenethietane 1,1-dioxide and *cis*-2,3-tetramethylenethietane respectively.[65] The enamines derived from 2-formyl-5-norbornene reacted with sulfene to form the spirothietane 1,1-dioxide (**9**) stereospecifically,[66] and when the sulfene was derived from sulfonyl chlorides other than $MeSO_2Cl$, a further substituent α to the thietane sulfur atom was introduced.[67] This substituent which can be aryl, methyl,[68] or halide,[69,70] however, is not introduced with complete stereospecificity, and a mixture of isomers is formed, although the stereo- and regiospecificity towards the initial double bond is unaltered; that is, the amino group ends up at C–3 of the thietane 1,1-dioxide obtained. Paquette and co-workers[53,54,71] achieved a partial asymmetric synthesis by using an enamine derived from an optically active amine. The chiral enamine (**6**, R^1=R^4=Me, R^2=*R*–Ph–CHMe–, R^3=H) was condensed with sulfene to give the optically active thietane 1,1-dioxide (**7**) which, after elimination of the amino group by a Hofmann reaction, gave (+)-4-methylthiete 1,1-dioxide (**8**, R^3=R^5=H, R^4=Me). The absolute configuration *S* was deduced by comparison with optically active 2-methylthietane 1,1-dioxide, of known configuration, which was synthesized from *S*-2-acetoxybutyric acid. The asymmetric induction, however, was only 6%. Better asymmetric induction, 25%, was obtained with the 1-propeneenamine derived from 2*R*-2-methylpyrrolidine,

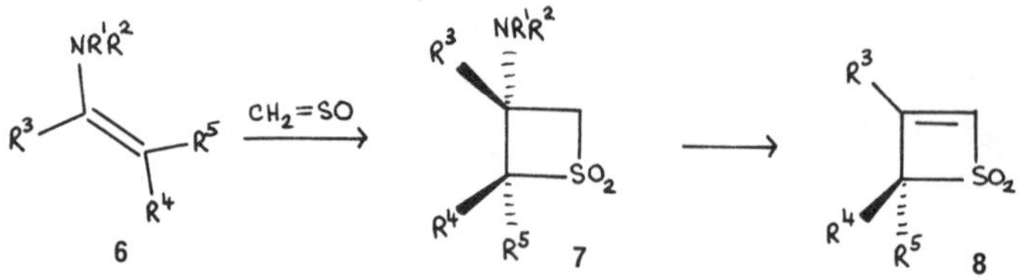

(11)

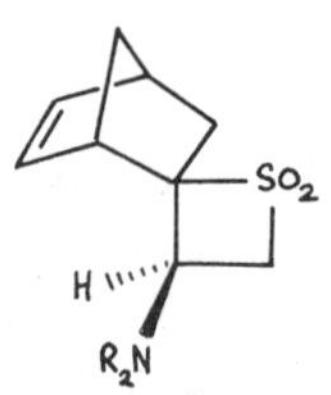

6 **7** **8**

9

but in this case the product was enriched in the opposite enantiomer *R*(–)-4-methylthiete 1,1-dioxide (**8**, R^3=R^4=H, R^5=Me). The sign of rotation is altered when these thiete 1,1-dioxides are reduced to the corresponding thietane 1,1-dioxides without, of course, any change in the configuration of the chiral centers.

The double bond in thiete 1,1-dioxide is highly reactive and undergoes cycloaddition reactions. Diazoalkanes provide the [2 + 2]π adducts, for example, 4-substituted *cis*-6-thia-2,3-diazabicyclo[3.2.0]hept-2-en 6,6-dioxide (**10**), but the regiospecificity varies with the diazo compound.[72] Enamines, dienamines, and ynamines react similarly,[73] and 1,3-diphenylbenzoisofuran gives the adduct **11** which most probably has the exo configuration.[74] *R*(−)-4-Methylthiete 1,1-dioxide (**8,** $R^3=R^4=H$, $R^5=Me$) provides the adduct **11** ($R^1=Me$, $R^2=H$) which is transformed into the *R*(+)-2-methyl-3,8-diphenyl-2*H*-naphtho[2.3-*b*]thiete 1,1-dioxide, **12,** on treatment with HBr in acetic acid.[75]

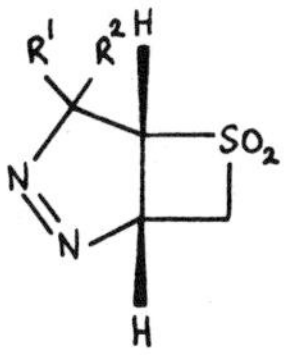

10 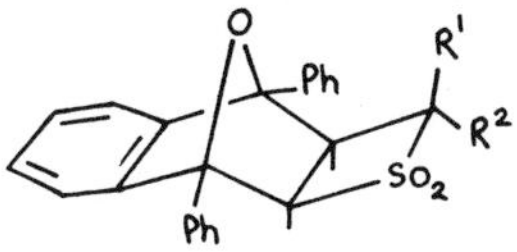 11 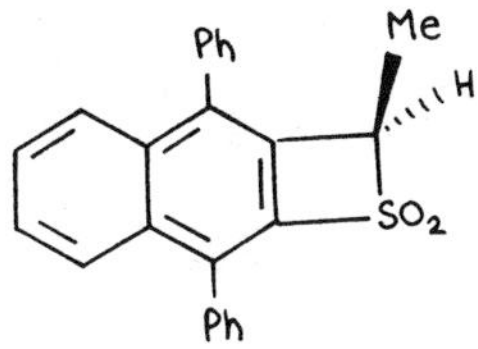 12

B. Configuration and Conformation of Thietanes

Microwave[76] data for thietane, unlike oxetane, strongly favors a puckered structure which is supported by dipole moment[62,59] and ^{1}H nmr[32,59,62,77,78] measurements and theoretical calculations.[79] Physical data, however, does not always show that this puckering exists.[80,81] The same situation is present in the 1-oxides, and the overall concensus is that the four-membered ring is puckered. X-Ray data of several thietane 1-oxides, for example, 3-carboxy-,[82] 3-*p*-bromophenyl-,[60] and *cis*-2,4-diphenylthietane 1-oxides[83] clearly demonstrate that a puckered ring is present in the crystalline forms.

Cis and trans configurations in thietanes,[32,55,80] thietane 1-oxides,[55,59,62] and thietane 1,1-dioxides[55,70,84,85] can be deduced by nmr spectroscopy or from dipole moment measurements. The configurations of thietane 1-oxides have attracted attention because of the conformational equilibria involved in the substituted derivatives. *cis*-3-Substituted thietane 1-oxides prefer the di-quasi-equatorial conformation **13** to the di-quasi-axial conformation **14** both in solution[62] and in the solid state.[60] *trans*-3-Substituted thietane 1-oxides prefer the conformation with the oxygen oriented axially (**15**).[62,86] X-Ray data on *trans*-*p*-bromophenylthietane 1-oxide, however, show that the conformation **16**, that is, equatorial oxygen, is present in the crystalline form, but because the nmr data of cis and trans isomers are not vastly

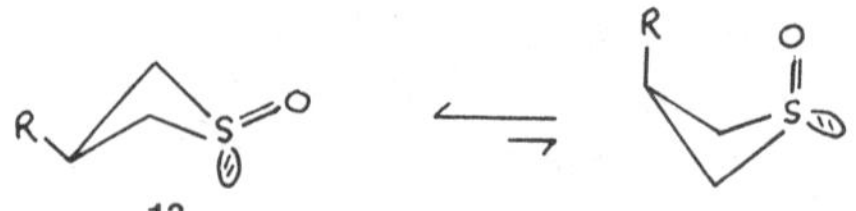

13 ⇌ 14

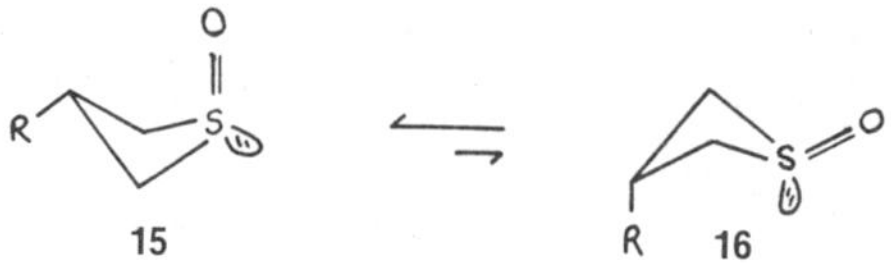

15 ⇌ 16

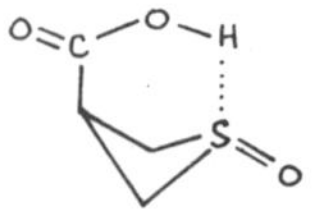

17

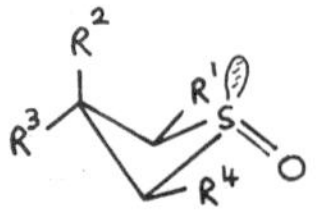

$R^2 = R^3 = Me$, $R^1 = R^4 = H$
$R^2 = OH$, $R^3 = Me$, $R^1 = R^4 = H$
$R^1 = R^4 = Ph$, $R^2 = R^3 = H$

18

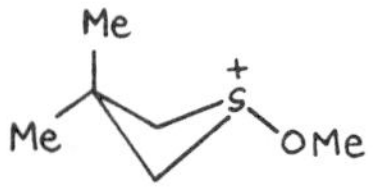

19

20

21

different, the authors feel that it should be reexamined.[60] The crystal structure of *trans*-3-carboxythietane 1-oxide also supports the axial orientation of the carboxyl group **17**,[82] but this may be assisted by hydrogen bonding, which is evident in the ir spectra.[61] The substituent on the sulfur atom generally prefers the quasi-equatorial conformation (e.g., **18**–**20**).[59,62,83] Benzene induced chemical shifts have been very useful in assigning configurations in thietane 1-oxides because of the anisotropic effect of the S–O group.[23,62]

Optically active thietanes,[54] their 1-oxides,[53,54,75] and thiete 1,1-dioxides[53,54,71] were mentioned in the previous section. A chiral thietane 1-oxide worthy of mention is 2,6-dithiaspiro[3.3]heptane 2,6-dioxide (**21**). This was resolved into its enantiomers by its cobalt complex and *d*-camphorsulfonic acid. It resisted racemization by N–HCl at 100°C or sodium hydroxide, and it clearly demonstrated nonplanarity at the sulfur atom. Oxidation, however, gave the disulfone which is now a symmetrical molecule.[87]

C. Reactions of Thietanes

Most of the reactions of stereochemical interest in the thietanes involve the *S*-oxides because they are rather reactive. *trans*-2,4-Diphenylthietane decomposes with *t*-BuOK in DMF into a number of products, including 2,3,5-triphenylthiophene and 1,2,4,5-tetraphenylbenzene, indicating that extrusion of sulfur had occurred.[88] *cis*- or *trans*-2,4-Diphenylthietane 1-oxides, under similar conditions, yield the same *cis*-1,2-diphenylcyclopropane-1-thiol and 1-sulfinic acid irrespective of the original configuration.[89] The *trans*-oxide isomerizes, however, into a 96:4 mixture of cis and trans isomers in methanolic sodium methoxide.[55] Treatment of these 1-oxides with methyl magnesium iodide gives a mixture of *cis*- and *trans*-1,2-diphenylcyclopropanes in which the trans isomer is predominant in each case.[90]

Thermolysis of *cis*- and *trans*-2,4-dimethylthietane 1,1-dioxides is nonstereospecific, SO_2 is lost, and mixtures of *cis*- and *trans*-1,2-dimethylcyclopropanes and propenes are formed in about the same proportions.[56] Although *cis*- and *trans*-2,4-dimethylthietanes are usually stable to desulfurization, the 1-methylthianonium salts lose sulfur, on treatment with *n*-BuLi at low temperatures, and give mixtures of *cis*- and *trans*-2,3-dimethylcyclopropanes stereoselectively in which inversion of configuration was better than 87%.[56] Similar pyrolysis of *cis*- and *trans*-2,4-diphenylthietane 1,1-dioxides gave a mixture of *cis*- and *trans*-1,2-diphenylcyclopropanes nonstereospecifically.[55] Both isomers rearranged into the same *trans*-1,2-diphenylcyclopropane sulfinic acid on treatment with ethyl magnesium bromide in a highly selective manner (contrast above).[91] It must be said that equilibration experiments showed that *trans*-2,4-diphenylthietane 1,1-dioxide epimerizes to the more stable cis isomer in the presence of NaOEt[67] (compare with 1-oxide above). Reaction with *t*-butoxy magnesium bromide, on the other hand, was stereospecific and provided the respective *cis*- and *trans*-3,5-diphenyl-1,2-oxathiolanes with retention of configuration (Equation 12).[92]

$$\xrightarrow{MgBr(OBu^t)} \qquad (12)$$

The elimination of hydrogen halide in a series of *cis*- and *trans*-2-halo-3-morpholinothietane 1,1-dioxides was investigated by Maiorana and coworkers.[84] They found that trans elimination of hydrogen halide in these compounds was favored when it was possible, as in the cis compounds, whereas the trans compounds, in which a cis elimination was necessary to promote reaction but was not possible, were stable. The rates of isomerization and hydrogen-deuterium exchange at C–2 were dictated by the halogen atom at C–2 and were in the order I > Br > Cl.[93]

Activation parameters for racemization and H-D exchange in *R*(−)-2-methylthietane 1,1-dioxide and *R*(+)-2-methyl-3,8-diphenyl-2*H*-naphtho-[2.3-*b*]thiete 1,1-dioxide (**12**) were compared. Hydrogen-deuterium exchange and racemization in the former were too slow to measure and were considerably slower than in the napththo compound **12**.[75]

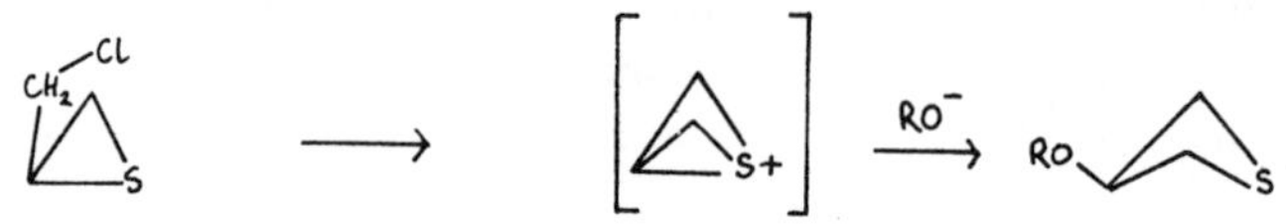

(13)

A molecular rearrangement which may well involve a bicyclothianium intermediate is the conversion of 2-chloromethylthiirane into 3-hydroxythietane by a base,[94] or into 3-aryloxythietanes with phenolates (Equation 13).[95] The reactions of these 3-substituted thietanes are worthy of further investigation in view of the reactions of related azetidines (see Part I, Chapter 2, Section I.A.4.d and II.A.3).

IV. THIOLANES AND THIOLENES

Thiolanes and Thiolenes are tetrahydro- and dihydrothiophenes. Although the name thiophane instead of thiolane has been used, it will not be adopted here because with this nomenclature the monodehydro derivative should be called thiophene. The stereochemical properties of these are similar to those of the corresponding nitrogen and oxygen heterocycles, but in addition thiolanes and thiolenes can be converted into sulfoxides, in which the sulfur atom may be chiral, and into sulfones, in which the sulfur atom is symmetrical again. The general chemistry of thiophenes was reviewed by Blicke,[96] Hartough (monograph),[97] and more recently by Gronowitz.[7] Although these reviews and the monograph do not contain up-to-date material, they nevertheless contain a wealth of information and the earlier references to thiolanes and thiolenes.

A. Syntheses of Thiolanes and Thiolenes

Catalytic hydrogenation (Pd/C) of thiophenes yields thiolanes, and Claeson and Jonsson[98] used this method to obtain 2-(ω-carboxybutyl)thiolane. C–2 of this compound is an asymmetric center, and the acid was resolved into its enantiomers with cinchonidine. The (−) acid **22** was correlated with (−)-2-carboxythiolane, by way of (−)-2-formylthiolane, by a Wittig reaction with γ-ethoxycarbonylpropyltriphenylphosphonium iodide. The *S*(−) absolute configuration of the valeric acid **22** was correlated by the Fredga quasi-racemate method with *R*(+)-lipoic acid of known absolute configuration (see Section IV.B).[99] Degradation of the naturally occurring (+)-biotin (**23**), of known absolute configuration,[100] was effected by conversion into the all-*cis*-3,4-diamino-2-(4-carboxybutyl)thiolane. Its (−)-di-*N*-tosyl derivative (**24**, $R^1=CO_2H$, $R^2=NHTs$) was reduced with LAH to the corresponding (−)-alcohol (**24**, $R^1=CH_2OH$, $R^2=NHTs$) and deaminated to *S*(−)-2-(5-hydroxypentyl)thiolane (**24**, $R^1=CH_2OH$, $R^2=H$) with H_2NOSO_3H. The latter was identical with the alcohol obtained by reduction of the above (−)-2-(4-carboxybutyl)thiolane, thus confirming and correlating the configurations of *S*(−)-2-carboxythiolanes, **22** and (+)-biotin.

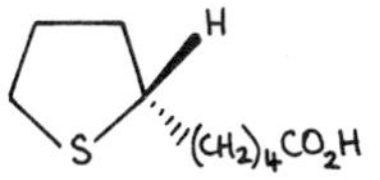

H
S
$(CH_2)_4CO_2H$
22

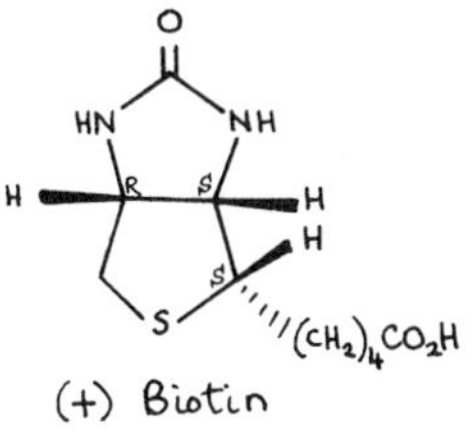

O
HN NH
H R S H
H
S S
$(CH_2)_4CO_2H$
(+) Biotin
23

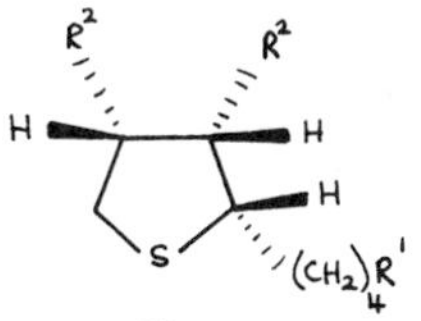

R^2 R^2
H H
H
S
$(CH_2)_4R^1$
24

In an excellent series of papers by Baker and collaborators[101,102] on the synthesis of biotins, several stereochemically interesting 2,3,4-trisubstituted thiolanes are described. The syntheses start from thioglycolic ester and β-alkyl substituted acrylic esters. *cis*- and *trans*-2-Alkyl-3,4-dicarboxythiolanes were separated, and, interestingly, the trans acid was readily converted into the *trans*-3,4-diamino derivative by a Curtius reaction, whereas the cis isomer failed, and the *cis*-diamine could only be prepared by stepwise conversions.[101,102] Both the *cis*- and *trans*-3,4-diamino-2-(4-carboxybutyl)thiolanes gave the corresponding *d,l*-biotin, *d,l*-allobiotin,[103] and *d,l*-epibiotin.[104] There are three asymmetric centers in biotin and eight stereoisomers (2^3) are possible: (±)-biotin (of which the *R,S,S*(+) isomer **23** is naturally occurring), (±)-epibiotin (**25**), (±)-allobiotin (trans-fused), and (±)-epiallobiotin (trans-fused). (+)-Biotin also yields two diastereomeric sulfoxides because of the introduction of yet another chiral center.[105,106] For further reading and for leading references on biotin see Bruice and Benkovic[107] and Bentley.[108]

Sulfur analogues of prostaglandins (e.g., **26** and their *S*-oxides) were synthesized recently from open chain compounds, as the biotins above, by condensation of thioglycolic ester with 4-*t*-butoxybut-2-enoate. The isomeric mixture of 2-*t*-butoxymethyl-3-methoxycarbonylthiolan-3-one was transformed into the thiaprostaglandin derivative (**26**), and isomers were separated and identified.[109] Substituted thiolanes have, however, been prepared by direct substitution. Bromination of thiolane in CH_2Cl_2 provides the *trans*-2,3-dibromo derivative which reacts with methanol in pyridine to give *trans*-3-bromo-2-methoxythiolane. Chlorination is similar in forming *trans*-3-chloro-2-methoxythiolane. Clearly substitution of the 2-halogen atom by a methoxide group occurred with retention of configuration because the trans configuration of substituents is thermodynamically more stable. Addition of halogen to thiol-2-ene gives the respective *trans*-2,3-dihalothiolanes.[110]

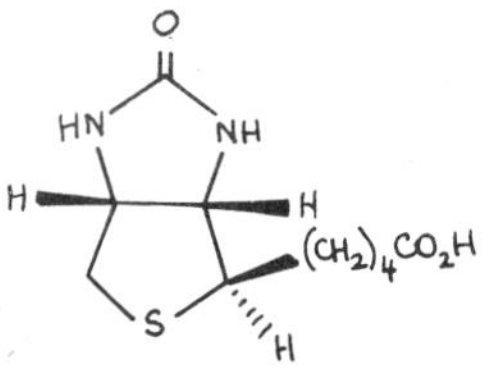

O HN NH H H $(CH_2)_4CO_2H$ S H

25

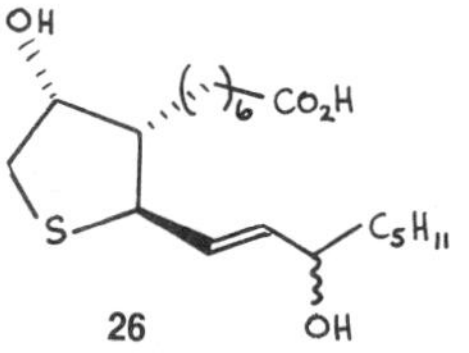

OH $(\cdot)_6 CO_2H$ S C_5H_{11} OH

26

MeO_2C CO_2Me R^1 R^1 R^2 S R^2

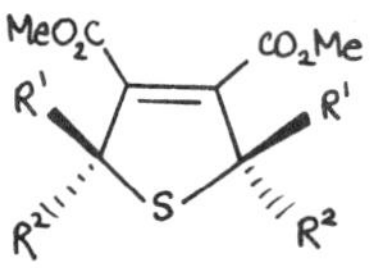

27

Polymerization of thiophene with phosphoric acid gave two main products, one of which turned out to be a mixture, *cis*- and *trans*-2,4-di-(2-thienyl)thiolanes, which were separated by tlc and identified by ^1H nmr spectroscopy.[111]

Thermolysis of 2,5-dialkyl-1,3,4-thiadiazol-3-enes causes elimination of molecular nitrogen with conrotatory ring opening to form an intermediate unstable thiocarbonyl ylide. The ylide may cyclize to a thiirane, with inversion of configuration (Equation 2),[18,112] or condense with polarophiles (e.g., dimethyl acetylene dicarboxylate) to give 2,5-dialkyl-3,4-dimethoxycarbonylthiol-3-enes (**27**), with retention of configuration of the alkyl substituents.[112]

Thiophene and 2-acetylthiophene undergo $[2 + 2]\pi$ photocycloaddition reactions with olefinic compounds, for example, tetramethylethylene[113] or maleic anhydride,[114] to give condensed 2-thiolenes, but unlike pyrroles and furans, some $[4 + 2]\pi$ cycloaddition also takes place to yield 7-thiabicyclo[2.2.1]hexanes.[113] Thermal $[2 + 2]\pi$ cycloadditions between thiophenes and acetylene do occur, and are catalyzed by $AlCl_3$.[115] The products 2-thiabicyclo[3.2.0]heptenes (e.g., **28**) are cis-fused and rearrange on further heating.[115,116] Stork and Stotter[117] effected a Diels-Alder reaction between butadienes and 3-thiolenes to obtain *cis*-8-thiabicyclo[4.3.0]non-3-enes (e.g., **29**) stereospecifically.

An example of the oxidation of a thiolane to two isomeric 1-oxides was mentioned above in the case of (+)-biotin. Similarly *cis*-3,4-dicarboxythiolane furnishes two diastereomeric 1-oxides, the syn (**30**) and anti (**31**) isomers, which of course are oxidized to the same 1,1-dioxide. These have been distinguished by their nmr spectra and by *p*K*a* measurements. The *syn*-oxide is a weaker acid than the *anti*-oxide. *trans*-3,4-Dicarboxythiolane, on the other hand, gives only one *trans*-3,4-dicarboxythiolane 1-oxide.[118] The oxidation of 2-methylthiolane was studied in detail with a variety of oxidizing agents and reaction conditions. The ratios of cis to trans 2-methylthiolane 1-oxides varied from 62:38 (N_2O_4 at 0°C) to 16:84 ($HCrO_4^-$–pyridine at 0–25°C). The oxides were separated by glc, and their acid-catalyzed equilibration was studied. Under the conditions examined the cis to trans ratio at equilibrium was always greater than one; that is, the cis isomer is thermodynamically more stable.[119] Pyrolysis of 5-*t*-butylsulfinylpent-1-ene produces a 74% yield of *cis*-2-methylthiolane 1-oxide free from the trans isomer in a stereospecific cyclization.[120] Preference for the cis configuration is also demonstrated in the chlorination (in pyridine) of thiolane 1-oxide which yields an 8:2 mixture of *cis*- and *trans*-2-chlorothiolane 1-oxides.[121]

Sulfur monoxide, generated by the thermolysis of thiirane 1-oxide, reacts with hexa-2,4-diene with most remarkable stereospecificity. The *trans,trans*-diene produces approximately 87% of *cis*-2,5-dimethyl-3-thiolene *anti*-1-oxide (**32**), and the *cis,trans*-diene furnishes around 95% of *trans*-2,5-dimethyl-3-thiolene 1-oxide (**33**). The *cis,cis*-diene, however, gives a 52:48 ratio of *cis*-2,5-dimethyl-3-thiolene *syn*- and *anti*-1-oxides.[122]

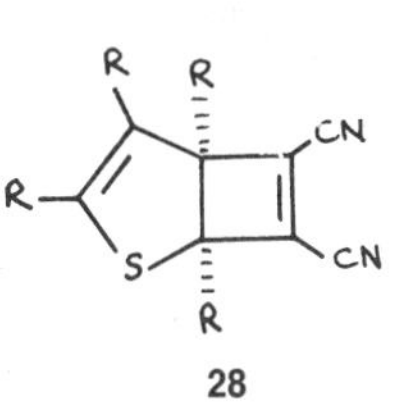

28 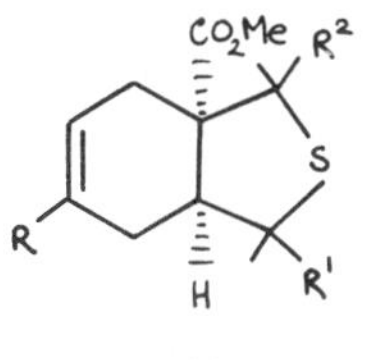 **29** 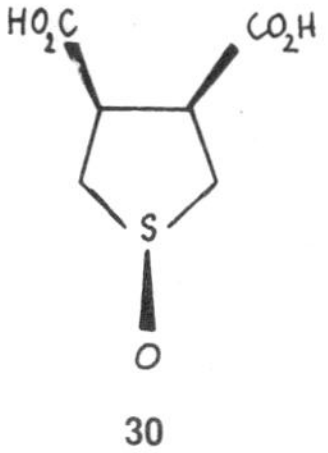 **30**

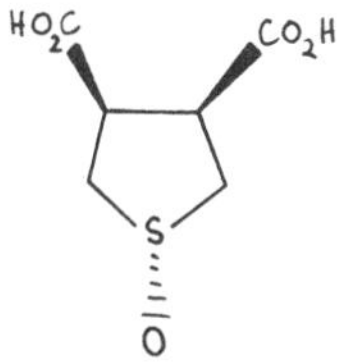

31 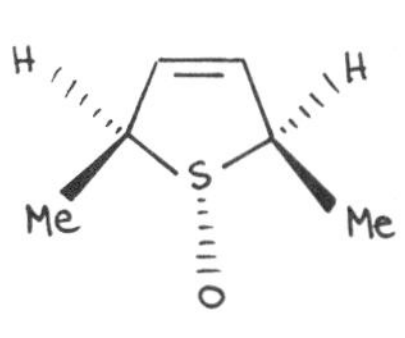 **32** 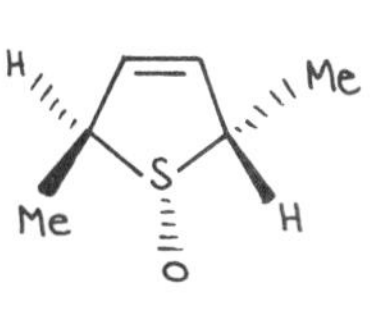 **33**

The double bond in 3-thiolene 1,1-dioxide (sulfolene, butadiene sulfone) is susceptible to the usual addition reactions, such as oxidation (H_2O_2/AcOH or HOCl),[123] bromination,[124,125] and epoxidation,[123] to yield 3,4-disubstituted thiolanes with the usual stereospecificities of these reagents. The reaction of sulfur dichloride was also a trans addition, but two thiolenes were involved for every molecule of reagent. A mixture of two *trans*-4,4′-dichloro-3,3′-bi-(1,1′-dioxythiolan-3-yl)sulfides was obtained which were obviously meso (**34**) and racemic (**35**). Both gave the same 4,4′-bi-(1,1′-dioxy-2-thiolen-3-yl)sulfide **36** on dehydrochlorination. When (−)-brucine, however, was used for this reaction, the meso isomer **34** gave the optically inactive bithiolene **36**, whereas the *d,l* isomer gave the (−)-bithiolene with $[\alpha]_D$ − 13.7°.[126]

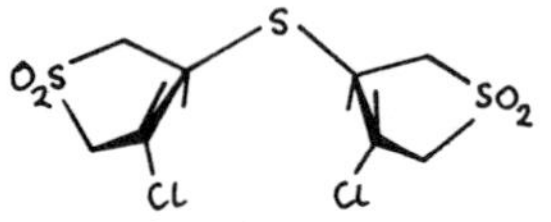

34

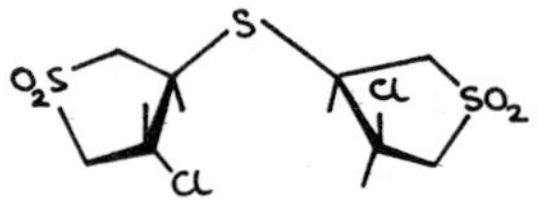

35

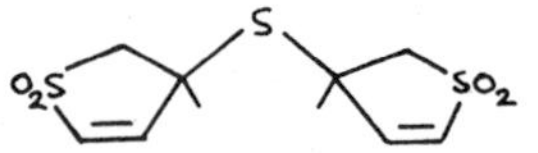

36

B. Configuration and Conformation of Thiolanes and Thiolenes

The configurations of several thiolanes and thiolenes and their oxides are described in the previous section, but a few more examples are noted in this section. 2-Carboxythiolane was resolved into its enantiomers with brucine and cinchonidine, and the absolute configuration of the (−) isomer was shown to be *R* by synthesis from *S*(+)-ornithine. The cd curves of *R*(−)-2-carboxythiolane had opposite ellipticities from those of (+)-2,5-dicarboxythiolane, inferring that the configuration of the latter was 2*S*,5*S* The melting point diagrams for various compositions of (+)- and (−)-2-carboxythiolane and (+)-3-carboxythiolane were obtained, but a definite conclusion for the absolute configuration of the latter could not be deduced.[127] The absolute configuration of *S*(−)-2-(4-carboxybutyl)thiolane (**22**) was deduced also by the Fredga quasi-racemate method with *R*(+)-lipoic acid (**37,** see Section IV.A). The crystals from a 1:1 mixture of these two gave an X-ray powder photograph different from that of *R*(+)-lipoic acid, which infers that a quasi racemate was formed and that the (−)-thiolane had the enantiomeric configuration.[98] This procedure is very useful when very small amounts of materials are available. The absolute configuration of *trans*(−)-3,4-dicarboxythiolane, prepared by optical resolution of the *d,l* acid with strychnine, was deduced as 3*R*,4*R* because desulfurization with Raney nickel provided 2*R*,3*R*(−)-2,3-dimethylsuccinic acid of known configuration.[128] The absolute configuration of 3-substituted thiolan-2-ones and their preferred conformation (**38,** R = $NH_3^+Cl^-$, NHAc, $NHSO_2Me$, Cl, and Br) were evaluated from cd, ord, and nmr measurements.[129]

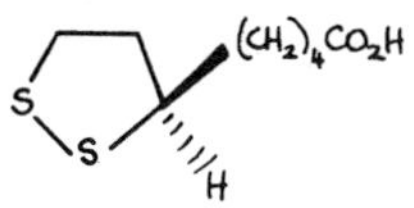

37 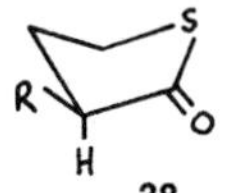 **38** 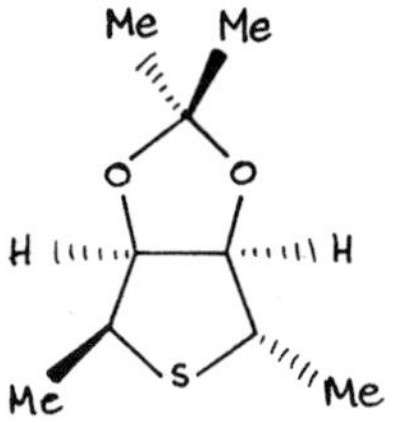 **39**

Thermodynamic data of thiolane from calorimetric, vapor pressure, and Raman spectral measurements support a pseudorotating system[130] in which the half-chair conformation is slightly more stable than the envelope conformation (compare with tetrahydrofuran, Scheme 5 in Chapter 2, Section VI.B). Nmr data not only supports such a system for substituted thiolanes (e.g., the 2,3-dihalo derivatives)[110] but also for the bicyclic isopropylidene derivative of *cis*-3,4-dihydroxy-*trans*-2,5-dimethylthiolane (**39**).[131,132] Lett and Marquet[133] have drawn the full cycle of pseudorotation for this compound. Although the ^1H nmr data of 2,3-dihalothiolanes are consistent with pseudorotation, the conformers with the two substituents quasi-axially oriented (e.g., **40**) are, however, preferred.[110] A dipolar effect similar to the anomeric effect may be operating here (compare with tetrahydropyrans in Chapter 2, Section X.C).

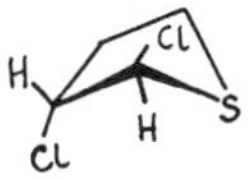

40  **41**

The rates of base-catalyzed hydrogen-deuterium exchange of the four α-hydrogen atoms in the 1-methylthiolanium cation have been measured. These demonstrated that there is an approximate 12:1 ratio in rates between the two sets of diastereotopic protons as compared with only a 1:1 ratio for the thiacyclohexane system.[134] Nuclear Overhauser Effect (NOE)

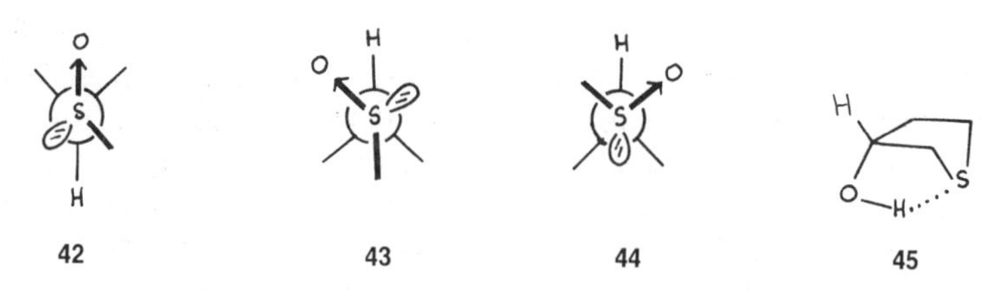

studies revealed that the α-protons cis to the *S*-methyl group are the ones that are exchanged more rapidly (e.g., **41**). Values of J_{gem} indicate that the preferred conformation for the cation is a half-chair.[135] Similar studies with derivatives of 3,4-dihydroxy-2,5-dimethylthiolane 1-oxides, and thietane 1-oxides, revealed that in this system α-hydrogen atoms antiperiplanar with the S–O substituent **42** undergo exchange faster than α-hydrogen atoms in other conformations (e.g., **43** or **44**).[136]

Some intramolecular hydrogen bonding has been observed in 3-hydroxythiolane which would fix it in the conformation **45,** but the bonding is not very strong.[137,138]

C. Reactions of Thiolanes and Thiolenes

The rates of acid-catalyzed reduction of cyclic sulfoxides with iodide ions were studied by Tamagaki, Mizuno, and Yoshida. The rates were in the order: thiolane 1-oxide (**46,** $n=2$) > thietane 1-oxide (**46,** $n=1$) > R.SO.R. (open chain) > thiepan 1-oxide (**46,** $n=4$) > thiane 1-oxide (**46,** $n=3$), and were explained in terms of steric effects than of basicity. The oxygen atom is protonated, and then the sulfur atom is attacked by the iodine from the rear. The least hindrance is in thiolane 1-oxide and the most in thiane 1-oxide. In the latter case the six-membered ring would need to react through the twist-boat conformation **47,** which is more energy demanding, whereas the half-chair conformation **48** is the preferred conformation for the thiolane 1-oxide.[139] The acid-catalyzed rates of equilibration of cis-trans isomers of 2-methylthiacycloalkane 1-oxides similarly follow the above pattern. The relative rates are 2-methylthiolane 1-oxide (*390*) > 2-methylthiane 1-oxide (*1*), and the slow step is the nucleophilic attack of the halide on the protonated sulfoxide.[140] That electrostatic effects are also important in the thiolane 1-oxide series is demonstrated in the nucleophilic displacement of chlorine in 2-chlorothiolane 1-oxides. *cis*-2-Chlorothiolane 1-oxide undergoes substitution with sodium methylmercaptide with inversion of configuration at C–2, whereas the trans isomer (chlorine quasi-axial) is unreactive because of the electrostatic repulsion of the nucleophile by the sulfinyl oxygen atom.[141] The displacement of the 2-halogen atom in *trans*-2,3-dihalothiolane 1-oxides by methoxide, which yields *trans*-3-halo-

2-methoxythiolane 1-oxide (see Section IV.A) with retention of configuration, most probably proceeds by way of an S_N1 mechanism.[110]

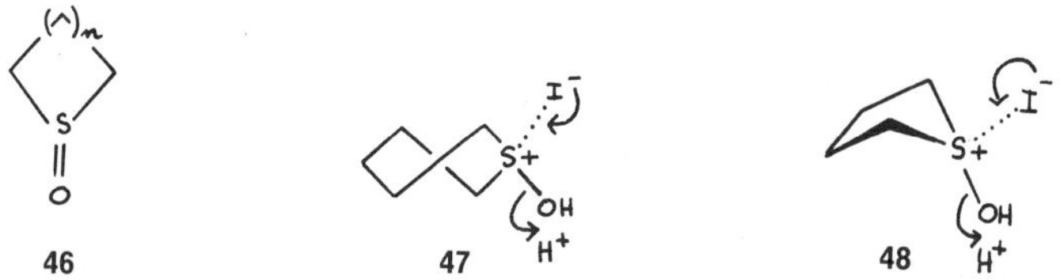

46 47 48

The sulfonyl group in 2-substituted thiolane 1,1-dioxides has a large steric effect on the reaction of the 2-substituent at the α-carbon atom. Thus reduction of 2-acyl derivatives, for example, 2-benzoylthiolane 1,1-dioxide (**49**), with $NaBH_4$ gives the *threo*-alcohol (e.g., **50**) in very high yields. The *threo*-alcohol **50,** together with the erythro isomer, is formed from the reaction of 1,1-dioxothiolan-2-yl magnesium bromide with benzaldehyde. The influence of the sulfonyl group is also observed in these alcohols; the *threo*-alcohol is converted into the corresponding chloro compound, with thionyl chloride and inversion of configuration, whereas similar substitution in the erythro alcohol occurred with retention of configuration.[142]

49 50

The thermolysis of 2,5-dialkyl-3-thiolene 1,1-dioxide into sulfur dioxide and the respective olefin follows the orbital symmetry rules; that is, they proceed by a disrotatory process. Thus *cis*- and *trans*-2,5-dimethyl-3-thiolene 1,1-dioxides decompose to *trans,trans*- and *cis,trans*-hexa-2,4-dienes in almost quantitative yields.[143,144] Kellogg and Prins[145,146] studied the thermolysis of several 2,5-dialkyl derivatives and confirmed the above results. The photolytic extrusion of SO_2 does not show the same high specificity observed in the thermolysis reaction. Thermolysis of 2,5-dialkyl-3-thiolenes yields intractable materials, but photolysis proceeds smoothly and yields a mixture of 2-vinylthiiranes (Equation 14). The stereoselectivity, however, is not always as good as shown with the di-*t*-butyl compounds. Thermolytic extrusion of SO from 2,5-dialkyl-3-thiolene 1-oxides also proceeds with disrotatory participation, but the stereoselectivity is not as good as in the corresponding sulfones. Photolysis of the 1-oxides, like the 1,1-dioxides, is not as stereoselective and obviously involves radical intermediates.

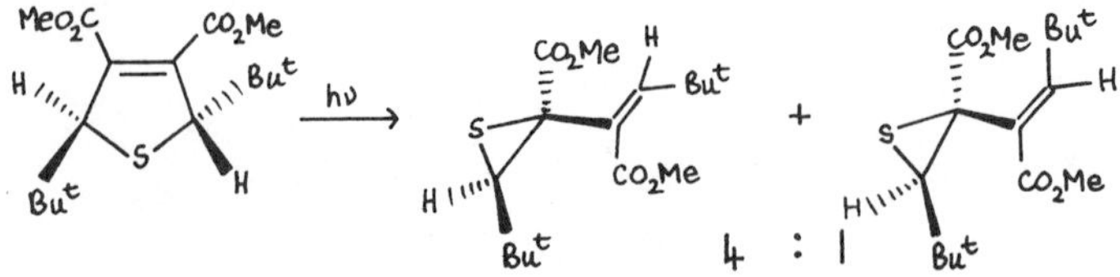

MeO2C CO2Me H Bu^t S H Bu^t hν → CO2Me H Bu^t S CO2Me H Bu^t + CO2Me Bu^t H S CO2Me H Bu^t 4 : 1 (14)

V. CONDENSED THIOLANES AND THIOLENES

In the present section the stereochemistry of 2,3-dihydrobenzothiophenes and other condensed thiolanes are treated together. For an excellent account of the chemistry of condensed thiophenes and reduced systems, albeit out-of-date, the monograph by Hartough and Meisel should be consulted.[147]

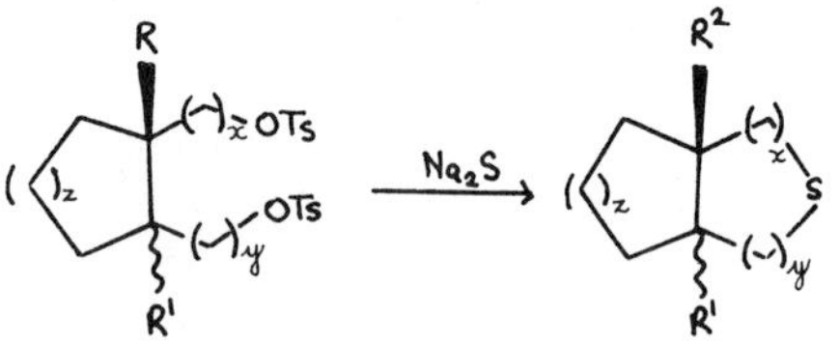

R ( )z ( )x OTs ( )y OTs R' Na2S → R2 ( )z ( )x S ( )y R' (15)

The reaction of bistosyloxy derivatives of cycloalkanes, of known configuration, with sodium sulfide provided a variety of thiabicycloalkanes (Equation 15). Thus *cis*-2-thiabicyclo[3.3.0]octane,[148,149] 2-thiabicyclo-[2.2.1]heptane,[149] 2-thiabicyclo[2.2.2]octane,[149] *cis*-1,6-dimethyl-8-thiabicyclo[4.3.0]nonane,[150] and *cis*- and *trans*-3-thiabicyclo[3.3.0]octane[151] of known configuration were prepared. This reaction was duplicated in the same molecule in 2,2,3,3-tetrakismethanesulfonyloxymethylbutane, and a 1:1 mixture of *cis*- (**51**) and *trans*-1,5-dimethyl-3,7-dithiabicyclo[3.3.0]octanes (**52**) was formed. The isomers were separated and identified by ^1H nmr spectroscopy.[152]

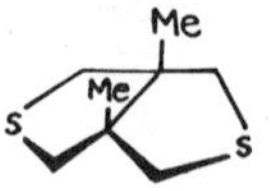

Me Me S S **51**

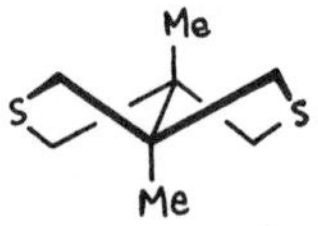

Me S S Me **52**

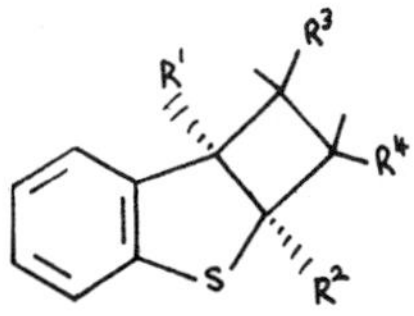

53

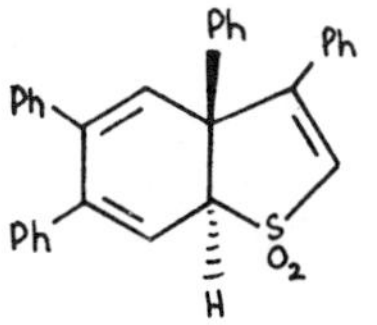

54

2,3-Dihydrobenzo[*b*]thiophenes, which are 2,3-cis-fused with a four-membered ring (e.g., **53**), are formed by photocycloaddition of benzo[*b*]-thiophenes with olefins.[153–155] The $[2 + 2]\pi$ cycloaddition can take place at room temperature in the absence of uv light if the thiophene is electron deficient, as in 3-nitrobenzo[*b*]thiophene, and if the olefin is electron rich, as in ynamines,[156] or if the electronic properties of these are reversed.[116]

Thiophene sulfones dimerize with loss of SO_2 on heating to yield cis-fused dihydrobenzo[*b*]thiophene sulfones,[147] for example, 3,4,6,7-tetraphenyl-*cis*-9-thiabicyclo[4.3.0]nona-2,4,7-triene 9,9-dioxide (**54**) is obtained from 3,4-diphenylthiophene 1,1-dioxide.[157] Benzo[*b*]thiophene 1,1-dioxide dimerizes on uv irradiation to a 73:27 mixture of the anti head-to-head and anti head-to-tail cyclobutane dimers. The configurations were deduced by desulfurization with Raney nickel (after reduction of the SO_2 to S with LAH) to *trans*-1,2-diphenyl- and *trans*-1,3-diphenylcyclobutanes respectively.[158] The double bond in 3-thiolene 1,1-dioxide is reactive towards diazoalkanes and forms cis-condensed 1-pyrazolenes, which lose molecular nitrogen on uv irradiation to give *cis*-3-thiabicyclo[3.1.0]hexane 3,3-dioxides (Equation 16).[159] 1-Vinylcyclopent-1-ene behaves like butadiene and forms *cis*-2-

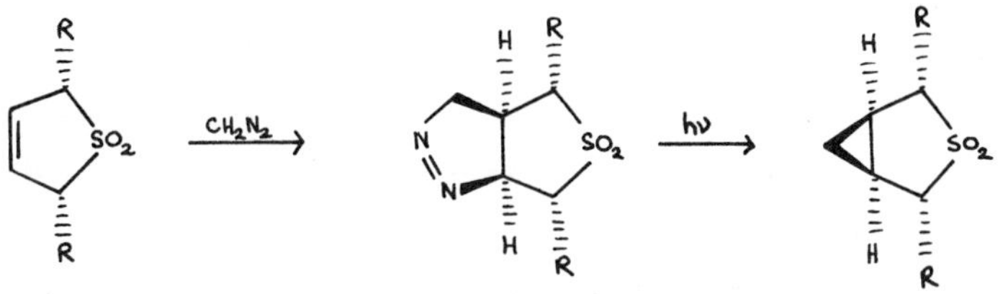

(16)

thiabicyclo[3.3.0]oct-3-ene 2,2-dioxide, which can be reduced to the parent *cis*-thiahydrocarbon. 2-Thiolene 1,1-dioxide condenses with butadiene to give *cis*-7-thiabicyclo[4.3.0]non-3-ene 7,7-dioxide, which was also reduced to *cis*-7-thiabicyclo[4.3.0]nonane. The trans isomer was obtained by isomerization of *cis*-7-thiabicyclo[4.3.0]nonane 7,7-dioxide with alkali, followed by reduction of the sulfone.[160]

The chiroptical properties of (−)-2,3-dihydro-2-carboxybenzo[*b*]thiophene (and benzofuran),[161] *trans*-8-thiabicyclo[4.3.0]nonanes,[162,163] biotin, and related compounds[164] were determined, and the sulfur chromophore was assessed. The absolute configurations were deduced from these properties and from unambiguous syntheses.

The stereospecificity of the oxidation of 2,3-dihydro-3-carboxybenzo[*b*]-thiophene depends on the reagent. Hydrogen peroxide or $NaIO_4$ provide the pure *trans*-1-oxide, whereas N_2O_4 furnishes 80% of the *cis*-1-oxide and 20% of trans isomer. These were identified by their ^1H nmr spectra and by their *pKa* values—the cis isomer being the weaker acid. These two acids are epimerized in the presence of acid.[165] Microbiological oxidation (*Aspergillus niger*) of racemic *trans*-8-thiabicyclo[4.3.0]non-3-ene produces the *S*-oxide in high yield and 14% optical enrichment, indicating partial biological specificity.[166] *cis*- and *trans*-2,3-Dihydro-2-methylbenzo[*b*]thiophene 1-oxides are *O*-alkylated with $Me_3O^+BF_4^-$. These products react with methyl magnesium bromide or Me_2Cd at −78°, with inversion of configuration, and yield *trans*- and *cis*-2,3-dihydro-1,2-dimethylbenzo[*b*]-thiophene (Equation 17). These isomers were identified by ^1H nmr spectroscopy and by making use of shifts induced by aromatic solvents and Lanthanide reagents. Loss of some stereospecificity in the reaction in Equation 17 was attributed to epimerization of the *O*-methylated salt.[167]

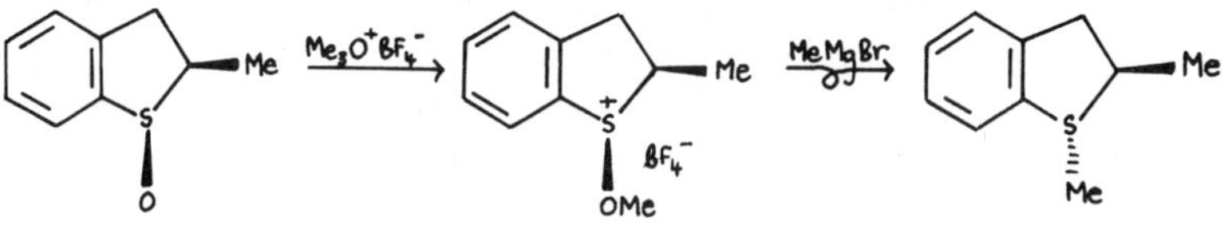

Me $\xrightarrow{Me_3O^+BF_4^-}$ Me, BF_4^-, OMe $\xrightarrow{MeMgBr}$ Me, Me (17)

A detailed study of the hydrogen-deuterium exchange of the four α-hydrogen atoms in 8-methyl-*trans*-8-thioniabicyclo[4.3.0]nonane was made by Fava and co-workers.[168] The rates of exchange observed were about 200:3:3:1, and NOE experiments revealed that the hydrogen atom which exchanged most rapidly was cis to the S^+–Me group **55.** Pyramidal inversion of the sulfur atom in this molecule does not alter its configuration, but by replacing a hydrogen atom by deuterium the rate of stereomutation ($K = 2 \times 10^{-5}$ sec^{-1} at 90°C) can be deduced.

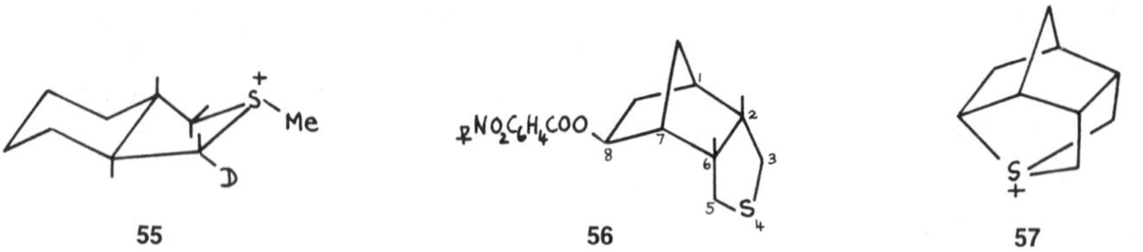

55 **56** **57**

Gratz and Wilder[169] examined the solvolysis of *endo*-4-thiatricyclo-[5.2.1.$0^{2,6}$]dec-8-yl *p*-nitrobenzoate (**56**), its exo isomer, and the respective endo and exo hydrocarbon esters (CH_2 in place of S), and they obtained the following relative rates: 752, 0.8, 1.0, and 0.7 respectively. The enhanced rate observed for the ester **56** was attributed to the formation of the sulfonium ion **57** (compare with formula **144** in Chapter 2, Section VII.D). Wilder and co-workers[170–172] obtained other evidence for the formation of such a cation and even isolated the bromide salt of **57.**

The chlorination of *trans*-8-thiabicyclo[4.3.0]nonane 8,8-dioxide is stereoselective and gives a mixture of α-chloro group which was equatorial (80%) and axial (20%). The *trans*-8-oxide, on the other hand, gave exclusively the equatorial α-chloro derivative.[173]

The stereochemical reactions of 2,3-dihydrobenzo[*b*]thiophenes have attracted much attention. Cram and co-workers[174] studied the stereochemistry of substitution reactions at the sulfur atom in 2,3-dihydrobenzo[*b*]-thiophene 1-oxide. One example in Equation 17 has already been mentioned. Treatment of the oxide with hydrazoic acid provided the corresponding 1-oxide-1-imine, which was resolved into its enantiomers by way of its *N*-*d*-10-camphorsulfonyl derivative. The optically active diastereoisomers were then hydrolyzed to yield optically pure (+)-2,3-dihydrobenzo[*b*]thiophene 1-oxide-1-imine, or treated with nitrosyl hexafluorophosphate to give (−)-2,3-dihydrobenzo[*b*]thiophene 1-oxide. They established the transformations shown in Scheme 3 (R=H and R=Br) and correlated the absolute configurations by comparison of the ord data of (+)-5-bromo-2,3-dihydrobenzo[*b*]thiophene 1-tosylimine with that of *R*(+)-methyltosylsulfide *S*-tosylimine.

Cram and collaborators also examined the base-catalyzed deuterium exchange in (+)-2,3-dihydro-2-methylbenzo[*b*]thiophene 1,1-dioxide,[175] and the decarboxylation of the (−)-2-carboxy derivative,[176] and in sorting out the kinetic parameters, including deuterium isotope effects, showed that decarboxylation occurs with inversion of configuration. They also showed that LAH reduction of (+)-2,3-dihydro-2-methylbenzo[*b*]thiophene 1,1-dioxide to the thiophene occurred with retention of configuration (but inversion in sign of rotation), demonstrating that the reduction does not proceed by ring opening. The optically active 2,3-dihydro-2-methylbenzo-[*b*]thiophene was oxidized to a mixture of optically active *cis*- and *trans*-1-oxides which were separated by tlc and identified by ^1H nmr spectroscopy.[177] Whitney and Pirkle[178] reexamined the above decarboxylation and confirmed that inversion had occurred. They also prepared the optically active (+)- and (−)-2,3-dihydro-2-methylbenzo[*b*]thiophene 1-oxides and deduced their absolute configuration as 1*R*,2*S* and 1*S*,2*S*, respectively, by studying the nmr spectra in the chiral solvent R(−)-2,2,2-trifluoro-2-

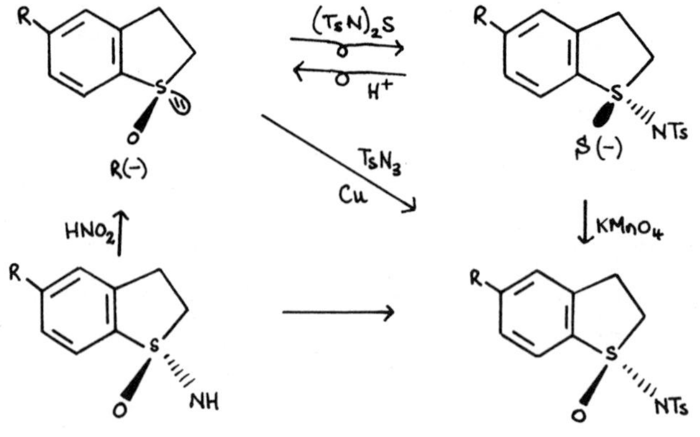

Scheme 3

phenylethanol. They had previously shown that this solvent could be used to determine the absolute configuration of chiral sulfoxides by producing characteristic ^1H nmr spectra for the diastereomeric interactions.[179]

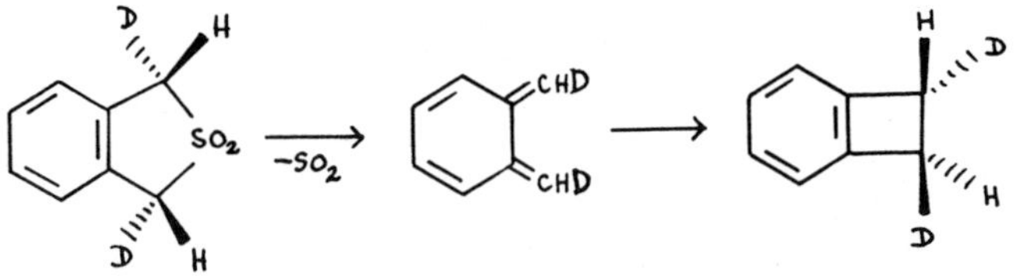

(18)

Extrusion of SO_2 occurs on thermolysis of *cis*-[1.3-$^2H_2$]-1,3-dihydrobenzo-[*c*]thiophene 2,2-dioxide. At 430–440°C the reaction proceeds by a disrotatory ring opening to give the intermediate *o*-quinodimethane, which is followed by conrotatory ring closure to yield *trans*-dideuterobenzocyclobutane (Equation 18). The reaction is highly stereospecific at this temperature, but the specificity decreases as the temperature is raised.[180]

VI. DITHIOLANES

A. 1,2-Dithiolanes

1,2-Dithiolanes are five-membered ring cyclic disulfides in which the disulfide group is a source of chirality in its own right. The two conformers **58** and **59** are indeed enantiomeric (nonsuperimposable mirror images). However, they would not be stable in their optical forms because the free energy of interconversion is relatively small ($\triangle G^{\ddagger}=42$–63 kJ mol^{-1}).[181] Most of the interest in 1,2-dithiolanes arose from the naturally occurring coenzyme *R*(+)-lipoic acid (**37**, also called thioctic acid; see Sections IV.A and B) which catalyzes the enzymic oxidative decarboxylation of pyruvic acid.[182] It is the dextro isomer of 3-(4-carboxybutyl)-1,2-dithiolane, and its oxidation and reductions,[183,184] synthesis,[99,185,186] biosynthe sis,[187] syntheses of the ^{35}S acid[188] and of analogues in which the carboxylic

58

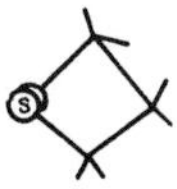

59

acid group was replaced by a sulfonic acid group[189] have been reported. The absolute configuration was deduced by Mislow and Meluch[99,190] by correlating it with 3-thioloctane-1,8-dioic acid (**60**) and with (+)-3-methyloctane-1,8-dioic acid (**61**), of known configuration, and using the Fredga method of melting point-composition diagrams (Equation 19).

There is, apparently, considerable steric strain in the 1,2-dithiolane derivatives, and several estimates (from 25 to 126 kJ mol^{-1}) based on uv, thermochemical, and conformational analysis data were compared.[191] A minimum value of 67 kJ mol^{-1} was estimated by Bergson and Schotte[192] from X-ray diffraction data on 4-carboxy-1,2-dithiolane. The torsion angle between the C–S bonds in C–S–S–C for 4-carboxy-1,2-dithiolane is 26°36′, which is much less than in unstrained disulfides (torsion angle = 90°). This strain is responsible for the ease with which lipoic acid is reduced and photolyzed—properties which Calvin[193] pointed out were necessary for its biological activity.

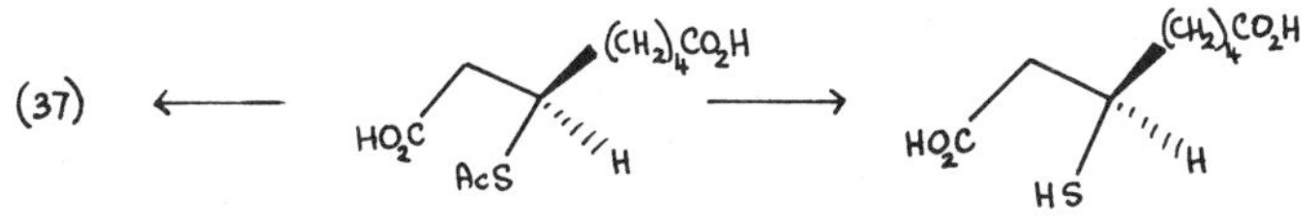

(37) ⟵ HO_2C–CH_2–C(AcS)(H)–$(CH_2)_4CO_2H$ ⟶ HO_2C–CH_2–C(HS)(H)–$(CH_2)_4CO_2H$ (19)

60

HO_2C–CH_2–C(Me)(H)–$(CH_2)_4CO_2H$

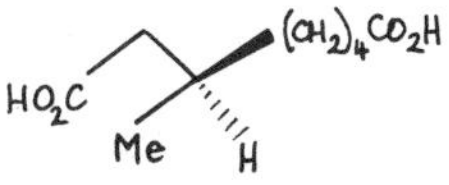

61

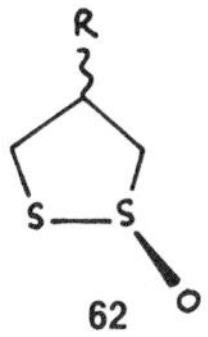

62

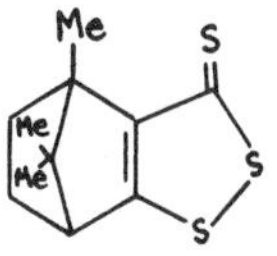

63

3-Carboxy-1,2-dithiolane was synthesized and resolved by Claeson,[194] and the (+) acid was shown to be *S* by the quasi-racemate method and by using *R*(−)-2-carboxythiolane. The configuration *R*,*R*(−) was deduced for 3,5-dicarboxy-1,2-dithiolane by similar comparisons with *R*,*R*(−)-2,5-dicarboxythiolane.[127]

The naturally occurring *anti*- and *syn*-4-carboxy-(**62,** R = CO_2H, asparagusic acid oxides)[195] and 4-hydroxy-1,2-dithiolane 1-oxides (**62,** R = OH, brugierol and isobrugierol)[196] were isolated and identified by ^1H nmr using the lanthanide shift reagents. The former are plant growth regulators.

The 1,2-dithiolene 3-thiones, derived from (+)-camphor (**63**), (+)-α-pinene, and sterols, exhibited anomalous ord curves caused by the heterocyclic chromophore.[197]

B. 1,3-Dithiolanes

1,3-Dithiolanes are the sulfur analogues of 1,3-dioxalanes and are cyclic acetals and ketals of alkane-1,2-dithiols. The formation of these compounds from 1,2-dithiols and aldehydes, ketones, phosgene, and potassium methyl xanthate proceeds with retention of configuration at the carbon atoms bearing the thiol groups.[198–201]

The exocyclic sulfur atom in *trans*-4,5-tetramethylene-1,3-dithiolane 2-thione was oxidized to a sulfinyl group (S–O) with $Pb(OAc)_4$[198] and can be replaced by two hydrogen atoms, that is, to *trans*-4,5-tetramethylene-1,3-dithiolane by reduction in the presence of molybdenum sulfide.[202] Whereas cyclohexane-1,2-dithiol and 2-hydroxycyclohexane-1-thiol reacted with Me_2CO, $COCl_2$, and $MeOCS_2K$ to yield the respective 1,2-dithiolanes, *trans*-2-hydroxycyclopentane 1-thiol gave the *trans*-4,5-trimethylene-1,3-oxathiolanes which must be highly strained.[201] 3-Hydroxypropane-1,2-dithiol condenses with benzaldehyde at the two sulfur atoms to give a mixture of *cis*- and *trans*-4-hydroxymethyl-2-phenyl-1,3-dithiolanes.[203] An unusual dimerization occurred when thiaisochroman-1-thione was treated with diazomethane and a 1:1 mixture of *cis*- (**64,** X = S) and *trans*-dispiro-1,3-dithiolanes (**65,** X = S) was formed. Isochroman-1-thiones behaved similarly, but in this case the cis isomer was always predominant.[204]

The cd curves of 21 1,3-dithiolanes, derived mainly from steroid ketones and ethylene-1,2-dithiol, were measured by Cookson and co-workers,[205] and the bands at 245 and 270 nm were ascribed to the 1,3-dithiolane chromophore. *cis*-4,5-Dicarboxy-1,3-dithiolane is a meso compound, but the trans isomer, which is racemic, was resolved by Hedblom[206] into its optically active forms. He determined the absolute configuration, 4*S*,5*S*(−), by the quasi-racemate method and by correlation with 2*S*,3*S*-2,3-dithiolsuccinic acid. *cis*- and *trans*-2,2-Diphenyl-1,3-dithiolane 1,3-dioxides were prepared, together with several 2,2-disubstituted 1,3-dithiolane oxides, dioxides, and tetroxides, and were identified by ir spectroscopy.[207]

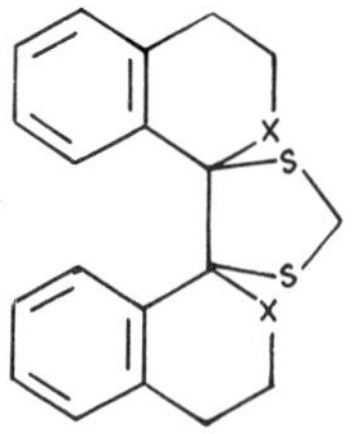

64

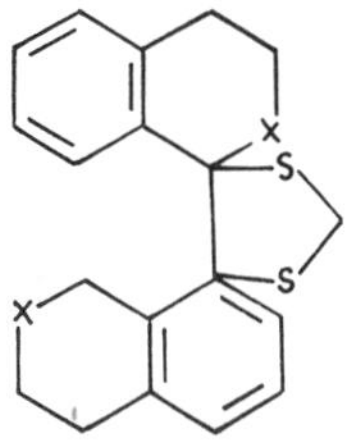

65

1,3-Dithiolane has a slightly larger ring than 1,3-dioxalane because of the longer C–S bonds and is more puckered, but it also undergoes pseudorotation as do most five-membered ring compounds.[208] The acid-catalyzed equilibration of several 2-alkyl-4-methyl- and 2-alkyl-2,4-dimethyl-1,3-dithiolanes were examined by ^{1}H nmr spectroscopy. The presence of substituents raises the free energy barriers between the half-chair conformations of enantiomeric 1,3-dithiolanes, nevertheless they are small. For the dialkyl series the free energy between diastereoisomers **66** and **67** are in the range 0.13–0.54 kJ mol^{-1}, and for the trialkyl substituents they are higher (0.71–1.0 kJ mol^{-1}). The longer C–S bonds in 1,3-dithiolanes, compared with the C–O bonds in dioxalanes, make the 1,3-diaxial nonbonded interactions small, and the strain involved in the ring due to the S–C–S bonds becomes more important than the conformational preference of the substituents.[209] The presence of oxygen atoms on sulfur, as in 2-substituted 1,3-dithiolane 1,1,3,3-tetroxides obtained by direct oxidation of dithiolanes, also introduces barriers into the pseudorotation cycle. The oxygen atoms increase the electronegativity of the sulfur atoms and shorten the C–S bond, which causes a flattening of the ring. The flattening is also enhanced by the nonbonded interactions between the 1,3-oxygen atoms.[210] A barrier to hindered rotation of the *t*-butyl groups was observed in 2-*t*-butyl- ($\triangle G^{\ddagger}=31.4$ kJ mol^{-1}) and 2-*t*-butyl-2-methyl-1,3-dithiolanes ($\triangle G^{\ddagger}=44.$ kJ mol^{-1}) by low temperature ^{1}H nmr spectroscopy, and the barrier for the latter compound was higher than that for the corresponding 1,3-dioxalane (31.4 kJ mol^{-1}).[211]

VII. TRITHIOLANES

1,2,3- and 1,2,4-Trithiolanes are the two possible isomers, and the substituents on the carbon atoms possess cis or trans configurations. One example of the former of stereochemical interest is *exo*-3,4,5-trithiatricyclo-[5.2.1.0^{2,6}]decane (**68**). It was prepared by direct reaction of bicyclo[2.2.1]-hept-2-ene (norbornene) with sulfur. The exo stereochemistry was deduced by ^{1}H nmr spectroscopy and by cleavage with Na/liq. NH_3 to bicyclo-[2.2.1]heptane-*exo,cis*-2,3-dithiol. 4-Isopropylidenebicyclo[2.2.1]hept-1-ene reacted similarly with sulfur, and interestingly the exocyclic double bond remained unaffected.[211a]

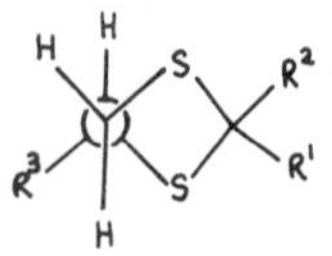

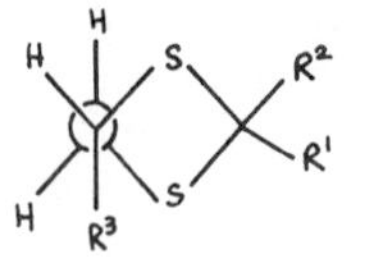

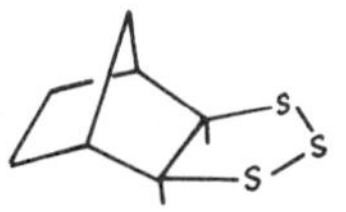

cis- and *trans*-3,5-Dialkyl-1,2,4-trithiolanes were formed by reacting di-α-chloroalkyl disulfides with sodium sulfide. The isomers were separated by glc, or by chromatography on an alumina column, and identified unequivocally by NOE measurements. ^{1}H nmr data showed that these compounds undergo pseudorotation with estimated energy barriers of less than 25 kJ mol^{-1}.[212] 1,2,4-Trithiolane and the 3,3,5,5-tetramethyl derivative were studied by photoelectron spectroscopy. The spectra could only be explained in terms of one conformation, and in the vapor phase 1,2,4-trithiolane exists in the half-chair conformation (C_2) (**69**) and not the envelope form (C_s) (**70**).[213]

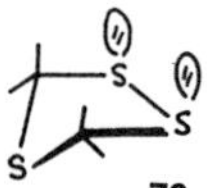

VIII. THIANES

Thianes are the sulfur analogues of tetrahydro-4*H*-pyrans, and, apart from the slightly longer C–S bonds compared with C–O bonds (and the consequences thereof), the systems are stereochemically quite similar. Thianes, however, can be oxidized to *S*-oxides and *S,S*-oxides. The oxygen atoms alter the C–S bond distances, the electronegativity of the sulfur atom, and introduce nonbonded interactions between them and the hydrogen atoms.

A. Syntheses of Thianes

The reaction of benzylidene acetone with H_2S in the presence of sodium acetate produced the less stable *trans*-2,6-diphenylthiane-4-one, which subsequently isomerized to the thermodynamically more stable cis isomers. These isomers were identified by ^1H nmr spectroscopy (Equation 20). The cis isomer is reduced with LAH, reacts with Grignard reagents, and forms the respective alcohols in which the 4-OH group is predominantly equatorial, that is, cis to the phenyl groups.[214] 3-Carboxythiane-4-one is a useful compound for preparing *cis*- and *trans*-3,4-disubstituted thianes by way of the corresponding dicarboxylic acids.[215] Methyl cyanodithioformate undergoes a Diels-Alder reaction with butadiene and forms 2-cyano-2-methylthio-4,5-dehydrothiane with the expected cis stereospecificity. 1-Methoxybutadiene reacts in a similar way, but it is also highly regiospecific giving 2-cyano-3-methoxy-2-methylthio-4,5-dehydrothiane (Equation 21). It was concluded from ^1H nmr studies that the half-chair conformer **71** was predominant and that the relative configuration of the methoxy group was cis with respect to the methylthio group.[216]

Ph Ph =O $\xrightarrow[NaOAc]{H_2S}$ (20)

NC SMe C=S, OMe (21)

71

The syntheses of several thianes are described in references in the following sections on configuration and conformation, and the syntheses of *S*-oxides and *S*,*S*-dioxides by direct oxidation of thianes are mentioned in Section VIII.C.

B. Configuration and Conformation of Thianes

The thiane ring is conformationally mobile and inverts between the two chair conformations **72** and **73.** This is consistent with data from coupling constants,[217,218] and variable temperature ^1H nmr studies on 3,3,5,5-tetradeuterothiane (in CH_2Cl_2) gave an E_a value of 48.5 kJ mol^{-1}. This value should be compared with the values for thiane 1-oxide, thiane 1,1-dioxide, tetrahydropyran, and piperidine, which are 59.4 (CH_2Cl_2), 62.3 (CH_2Cl_2), 44.8 (CD_3OD), and 60.7 (CD_3OD) kJ mol^{-1} respectively.[219] Lambert, Keske, and Weary[217] concluded from the ^{1}H nmr spectra of thiane (also selenane and tellurane) in strong acid solution that the hydrogen atom on the sulfur is exclusively axial. Despite syn-axial interactions, the protonated species **74** of 3,3-dimethylthiane has the S–H oriented axially.[220] The predominant equatorial approach in the alkylation of thianes was demonstrated by Katritzky and co-workers.[221] They showed that methyl bromoacetate reacted with 4-hydroxy-4-phenylthiane to yield exclusively the equatorial thionia compound **75** because the carboxy group lactonized readily with the 4-OH group, inferring cis orientation. Methylation and ethylation gave 79:21 and 95:5 ratios of equatorial to axial isomers, and methylation of 4-phenylthiane indicated that the 4-OH group had not interfered with the approach of the reagent. Methylation of the conformationally biased 4-*t*-butylthiane produced a 12:88 mixture of axial and equational *S*-methyl derivatives which were separated on an ion exchange column. When heated to 100°C in $CDCl_3$ the perchlorate salts equilibrate to a 1.4:1 mixture of equatorial and axial isomers, again favoring the equatorial orientation. The configurations of these isomers were confirmed by X-ray analysis,[222] and X-ray data on 1-methyl-1-thioniacyclohexane iodide (**76**) revealed that in the crystal the methyl group is equatorial.[223]

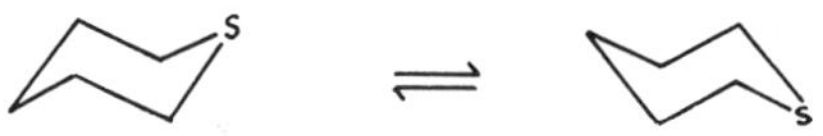

S ⇌ S

72 **73**

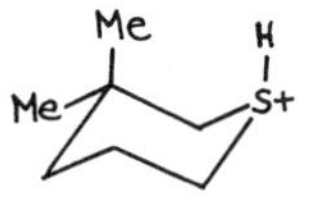

Me, Me, H, S+

74

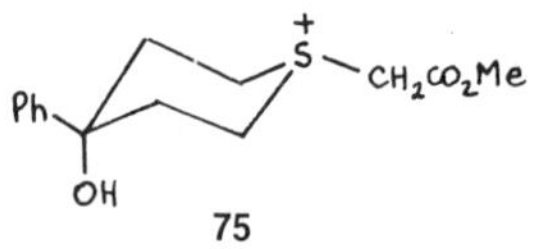

S+, CH_2CO_2Me, Ph, OH

75

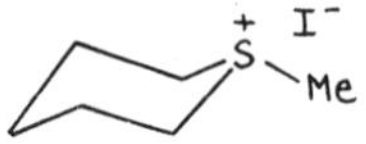

+S, I⁻, Me

76

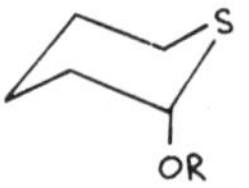

S, OR

77

O—H, R, S

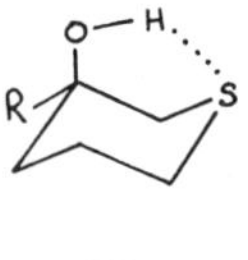

78

The anomeric effect is also operating in the thiane series, and ^{1}H nmr data of the 2-hydroxy derivative indicate that the substituent is axial (**77**, R = H)[224] (compare with tetrahydropyrans in Chapter 2, Section X.C).

Zefirov and co-workers[225,226] found that 2-alkoxythianes generally prefer a conformation with the alkoxy group oriented axially, and in the 2-methoxy and 2-*n*-butoxy derivatives over 90% are in this form. 2-Alkylthiothianes also exhibit a distinct anomeric effect[225] (see Section X). 2-Benzoyloxythiane, obtained from the reaction of *t*-butylperoxybenzoate–Cu_2Cl_2 with thiane, prefers the conformation with the axial substituent.[226a] The effects of solvents on the ratio of axial and equatorial conformers are similar to those observed in the tetrahydropyran series.[227,228] Unlike the tetrahydropyran, the substituent in 2-(6-chloropurin-9-yl)thiane shows a stronger preference for the equatorial conformation.[229]

There is no evidence from ir measurements to support intramolecular hydrogen bonding in 4-hydroxythiane. In contrast, 3-hydroxythiane (**78**, R = H) exhibits intramolecular hydrogen bonding, and it is stronger than in the five-membered ring compound 3-hydroxythiolane (**45**). Hydrogen bonding in 3-hydroxythiane is enhanced by the presence of a 3-phenyl group (**78**, R = Ph).[137]

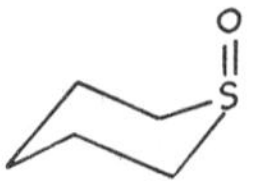

O=S (79)

Ts–N=S (80)

S=NH (81)

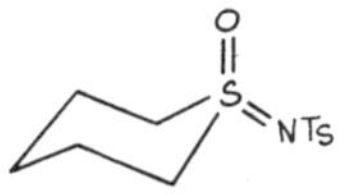

O=S=NTs (82)

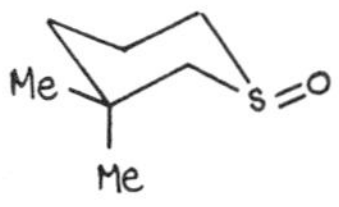

Me, Me, S=O (83)

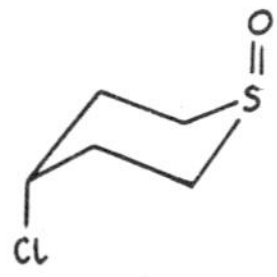

Cl, S=O (84)

Thiane 1-oxide is remarkable because the preferred conformation (1.62:1; by 0.73 kJ mol^{-1}) has the S–O oriented axially (**79**). Similarly the tosylimino group on sulfur is axial (**80**; by 0.63 kJ mol^{-1}), whereas the sulfimino group has a slight preference for the equatorial conformation (**81**; by 0.31 kJ mol^{-1}).[230] Oxidation of the sulfimide **80** yields the corresponding sulfoximide which shows a 2:1 ratio of two conformers where probably the one having the axial sulfoxide equatorial *N*-tosylimido groups predominates (**82**).[231] The conformation of thiane 1-oxide (**79**) (and the imide, **80**) is altered by the presence of *gem*-methyl groups at C–3. The axial-3-methyl group forces the oxygen (and nitrogen) atom into the equatorial orientation **83** (compare with **74**). Methyl groups at C–4, however, have little effect on the axial orientation of the oxygen atom.[220] *cis*- and *trans*-4-Chloro-thiane 1-oxides were studied by ^{1}H nmr spectroscopy at low temperatures, and the conformations with the oxygen axial were favored to the extent of 74:26 and 96:4 respectively. The proportion of the axial conformation **84** for the trans isomer is high, probably because of a strong 1,4-dipolar effect.[232]

The configuration of *cis-* and *trans*-4-*t*-butylthiane 1-oxides was confirmed by ^{13}C nmr spectroscopy.[233] The oxygen atom and *t*-butyl group in the trans isomer are both equatorial. *p*K*a* measurements in aqueous sulfuric acid showed that the thiane with the axial oxygen atom (i.e., cis isomer) is more basic than *trans*-4-*t*-butylthiane 1-oxide.[234] Attempts to determine the orientation of the 2-lithium salts of *cis-* and *trans*-4-*t*-butylthiane 1-oxide and deuterated derivatives were unsuccessful, and it was not possible to say whether C–2 had a pyramidal or a planar configuration. Methylation of these lithium salts gave the 2-methyl derivatives in which the methyl group was trans with respect to the oxygen atom.[235]

C. Reactions of Thianes

The attack of carbenes on 4-*t*-butylthiane is highly stereoselective in giving predominantly the equatorial sulfonium ylides (**85,** R = Ac, CO_2R). Nitrenes, on the other hand, are unselective and furnish about equal amounts of *cis-* and *trans*-iminosulfonium ylides.[236] The rate of reduction of thiane 1-oxide was compared with those of thietane, thiolane, and thiepane *S*-oxides,[139] and the acid-catalyzed isomerizations of these oxides[140] are discussed in Section IV.C.

The oxidation of 4-*t*-butyl-, 4-*p*-chlorophenyl-, and 4-methylthianes to the respective *cis-* and *trans*-1-oxides was examined using ten oxidants and twelve reaction conditions. The ratio of *cis-* and *trans*-oxides formed varied considerably with the oxidant, and hence the mechanism involved. For example, 4-*t*-Butylthiane gave a cis to trans ratio of 1:9 with ozone and 8:1 with N_2O_4. Similar specificities were found with the three 4-substituted thianes,[237] and the cis to trans ratio does alter slightly with temperature.[238] The oxidation of thianes (e.g., *cis*-2,6-diphenylthiane) with wet bromine occurs by equatorial electrophilic attack to yield the equatorial 1-oxide (e.g., *cis,cis*-2,6-diphenylthiane 1-oxide). The presence of a hydroxy or carbonyl group at C–4 alters the steric course and the axial 1-oxides are formed (e.g., *trans*-4-hydroxythiane 1-oxide).[239]

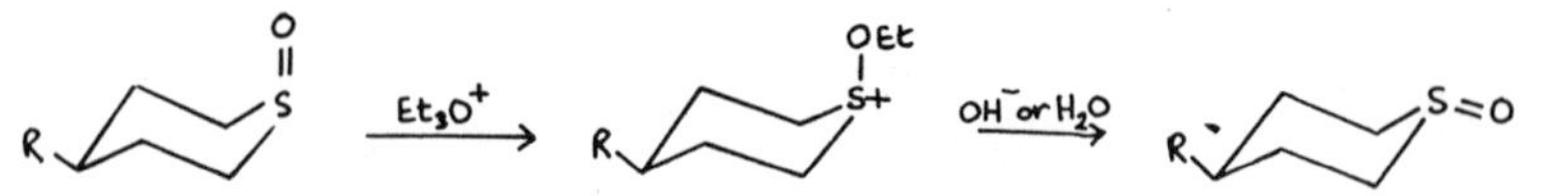

$\xrightarrow{Et_3O^+}$ $\xrightarrow{OH^- \text{ or } H_2O}$ (22)

Johnson and McCants[240,241] have shown that the configuration at the sulfur atom in thiane 1-oxides can be altered by *O*-alkylation (with retention of configuration) followed by inversion of configuration on treatment with alkali or water (Equation 22). Considerable selectivity in the methylation on the carbon atoms α to the SO group in thiane 1-oxides has been observed.[235,242,243] Similarly α-halogenation of thiane 1-oxides is highly stereoselective and is subject not only to the orientation of the oxygen atom but also to the configuration of other substituents in the ring.[243–248] Johnson and McCants[249] examined the equilibration of *cis*- and *trans*-4-*t*-butyl-, 4-*p*-chlorophenyl-, and 4-methylthiane 1-oxides in HCl (25°C), decalin (190°C), and N_2O_4 (0°C), and in each case the cis isomer (**86**), in which the oxygen atom is axial, predominated by factors of 2 to 9. The free energy changes were of the order of 2.5 to 5.4 kJ mol^{-1}. The solvolyses of *cis*- and *trans*-4-chloro- and 4-tosyloxythiane 1-oxides were compared by Martin and Uebel.[250] The K_{trans}/K_{cis} ratio for the chloro and tosyloxy compounds were 630 (at 140° in 50% ETOH) and 150 (at 100°C in 80% ETOH), and this suggested that the S–O group assisted in the solvolysis by forming the intermediate **87.**

85  **86**  **87**

The base-catalyzed α-hydrogen-deuterium exchange of rigid thiane 1-oxides was explained in terms of the relative stabilities of tetrahedral and planar α-sulfinyl carbanions.[251] Katritzky and collaborators[252] measured base-catalyzed H-D exchange rates in conformationally rigid thiane 1,1-dioxides (e.g., *cis*-4-oxo-3,5-diphenylthiane 1,1-dioxides, *cis*-3,5-diphenyl-*trans*-4-hydroxythiane 1,1-dioxides, and deuterated species. The rate ratio K_{eq}/K_{ax} was 1.6 and is consistent with the stability of α-sulfinyl carbanions from tetrahedral (or planar) species. A quantum mechanical study of proton exchange in sulfoxides with inversion and retention of configuration was made by Rauk, Wolfe, and Csizmadia.[253]

The reaction of 4-*t*-butylthiane with chloramine T produces *trans*-4-*t*-butyl-1-tosyliminothiane in which the sulfilimine is equatorial. This compound can be formed from *cis*-4-*t*-butylthiane 1-oxide (oxide axial) with inversion of configuration at the sulfur, and the many stereochemical transformations of this group were described by Johnson and co-workers.[254,255]

D. Condensed Thianes

Most of the condensed thianes described in this section are related to thiachromans, isothiachromans, and reduced derivatives.

The reduction of 3-bromothiachroman-4-one with $NaBH_4$ proved to be highly stereoselective in yielding *cis*-3-bromo-4-hydroxythiachroman.[256] The cis stereochemistry was deduced from the fact that this bromohydrin was unchanged when treated with methanolic KOH, and it is similar to the one observed in the $NaBH_4$ reduction of 3-halochroman-4-ones.[257] The acid-catalyzed disproportionation of 3,4-dialkyl-substituted 3-thiachromenes is stereoselective. Intramolecular hydride transfer occurs producing

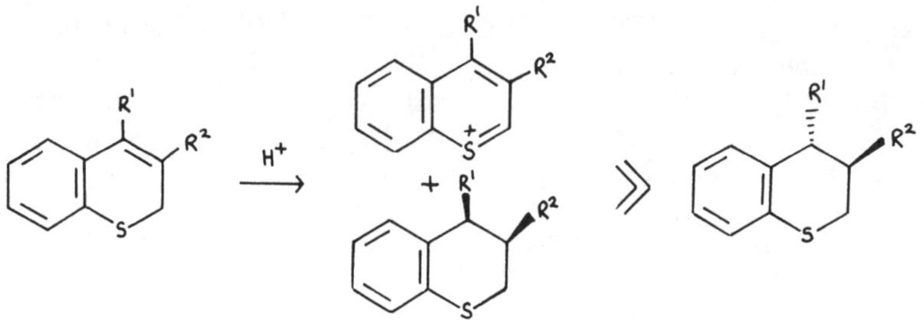

(23)

a much larger proportion of *cis*- than *trans*-3,4-disubstituted thiachromans (Equation 23). The corresponding 3-chromenes (oxygen analogue) also exhibit this disproportionation, but there is no selectivity in the chromans formed. Intermediate sulfonium ions (involving a C–4,S–1 bond) were invoked in the disproportionation of 3-thiachromenes.[258]

3-Thiachromene, like 1,2-dihydronaphthalene, undergoes the Prins reaction with formaldehyde in the presence of acid to yield condensed products such as the dioxanothiane **88** which is cis-fused.[259] An easy entry into the 3-ketothiapyrans is through the reaction of 2-chloroallylthiol with 1-methylcyclohexene oxide that forms the intermediate 1-(2-chloroallylthio)-*trans*-2-hydroxy-*cis*-2-methylcyclohexane, which cyclizes in the presence of acid into *trans*-6-methyl-2-thiabicyclo[4.4.0]decan-4-one (**89**) stereospecifically.[260] The stereospecificity of α-chlorination and α-H-D exchange in *cis*- and *trans*-2-thiabicyclo[4.4.0]decane 2,2-oxides was examined by de Waard and co-workers.[261] The chlorinations insert a chlorine atom preferentially in the equatorial direction and α-H-D exchange proceeds in protic solvents, preferentially, with retention of configuration. Young and Heitz[262] prepared *trans*-2,7-dithiabicyclo[4.4.0]decane in several steps from thian-3-one by way of the pyrrolidine enamine followed by reaction with ethyl acrylate.

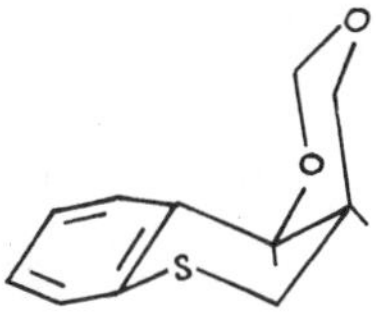

88

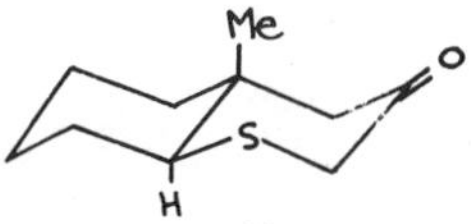

89

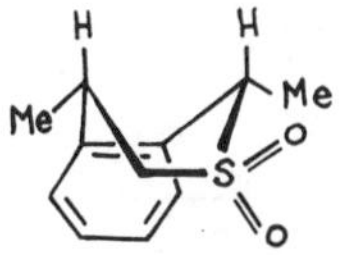

90

cis-1,4- and *cis*-1,3-Dimethylisothiachroman were obtained by stereoselective synthesis and oxidized to the 2,2-dioxides. *cis*-1,3-Dimethylthiachroman 2,2-dioxide readily isomerizes to the trans isomer, and interestingly, an X-ray crystallographic study of *cis*-1,4-dimethylisothiachroman 2,2-dioxide showed that it is in a boat conformation (**90**).[263,264]

Bi-3-isothiachromen-1,1′-yl (m.p. 230°C) was formed by reduction of 2-thianaphthalenium perchlorate with zinc dust or dithionite. A 1:1 mixture of this dimer and a second isomer (m.p. 200°C) was formed from the perchlorate by reaction with *t*-butyl magnesium chloride. One should be the *d,l* and the other the meso isomer, but these have not been identified.[265] Holliman and Mann[266] resolved the 2-*p*-chlorophenacylisothiachromanium cation **91,** by way of the bromocamphorsulfonate salt, which demonstrated that the sulfur atom was asymmetric. Although the salt was optically stable at room temperature, it racemized slowly in boiling ethanol. The selenium and tellurium analogues were also prepared, and the latter was found to racemize much faster than the sulfur analogue. Racemization was explained by either dissociation or inversion of the cation. 1,2,3,4-Tetrahydro-8-methoxy-6*H*-benzo[*b,d*]thiachroman (**92**), like the 3-isothiachromenes (Equation 23), disproportionates in aqueous perchloric acid into the *cis*-4*a*,10*a*-dihydro derivative, and an intermediate sulfonium ion **93** was proposed to account for the stereoselectivity.[267]

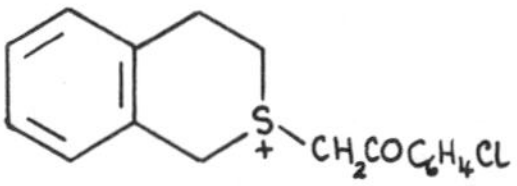

91

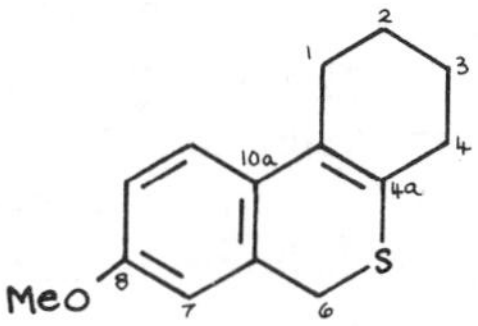

92

93

The configurations of 2- and 3-thiabicyclo[4.4.0]decan-4-ones, and deuterated derivatives, were deduced by ^{1}H nmr spectroscopy and by using the shift reagent Eu(dpm)$_3$.[268]

IX. DITHIANES

A. 1,2-Dithianes

The inherent dissymmetry of the C–S–S–C bond system noted in 1,2-dithiolanes (Section VI.A) is also found in 1,2-dithianes. The torsion angle in this case is around 60°. Calvin and co-workers[269] demonstrated the interconversion of enantiomers **94** and **95** by examining the ^1H nmr spectrum of 4,4,5,5-d_4-1,2-dithiane at low temperatures, and determined the free energy of inversion as 48.5 kJ mol^{-1}. The 3,3,6,6-tetramethyl derivative inverts more slowly, and the free energy for this inversion, 56.1 kJ mol^{-1}, was determined in a similar manner.[181,270] The system is therefore conformationally flexible. Dipole moment[271] and X-ray[272] data, and the evidence below, indicate that the chair conformation for 1,2-dithianes is the most stable form.

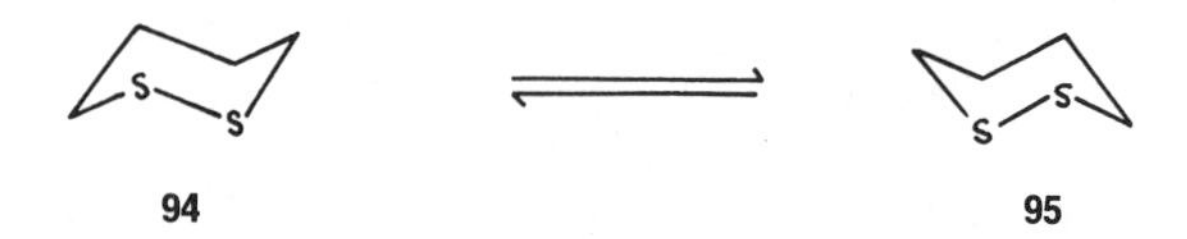

A mixture of meso and racemic 2,5-dibromohexane (8:2) gave a mixture of bisthiocyanates which provided an 8:2 mixture of *cis*- and *trans*-3,6-dimethyl-1,2-dithianes. The synthesis, by way of the bisisothiuronium salt, gave, on the other hand, a 1:1 mixture of these isomers. They were readily identified by ^1H nmr spectroscopy which supports a preferred chair conformation, and the uv band at 286 nm was more intense in the cis isomer ($\epsilon = 321$) than in the trans isomer ($\epsilon = 174$).[273] *cis*-3,6-Dimethyl-1,2-dithiane is meso, whereas the trans isomer is a racemic form. Dodson and Nelson[274] resolved *d,l*-2,5-dihydroxyhexane, by way of the brucine salt of the hydrogen phthalate ester, and reaction of the di-*O*-tosyl ester from the 2*R*,5*R*(−)-alcohol with Na_2S_2–DMF gave 3*S*,6*S*(+)-*trans*-3,6-dimethyl-1,2-dithiane with complete inversion at the two asymmetric centers. This antipode can undergo the conformational changes **96** ⇌ **97** ⇌ **98**, but the ^1H nmr spectra show that the diequatorial conformer **96** is predominant. The 3*R*,6*R*(−) isomer was also prepared and behaved similarly. A comparison of the uv and cd curves of these with those of the open-chain dithiols (formed by reduction with LAH) revealed that, "the positive cd band at long wavelength (e.g., 285 nm) and the cd band at short wavelength (e.g., 243 nm) are associated with *P* helicity about the restricted C–S–S–C bonds." For the nomenclature *P* ("plus," right handed screw) and *M* ("minus," left handed screw) see the specification of molecular chirality by Cahn, Ingold, and Prelog.[275] The cd curves of 1,2-dithianes are consistent with the nmr spectra (that is, the 3*S*,6*S* and 3*R*,6*R* isomers have *P* and *M* helicites respectively) and with X-ray data on gliotoxin (**99**), sporidesmin (**100**),[276] acetylaranotin,[277] and the dithio–diketopiperazines (formula **122** in Part I, Chapter 3, Section V.D). The piperazine rings in these

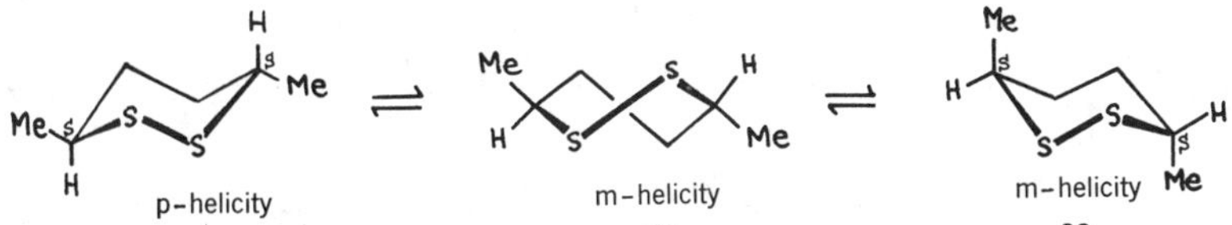

p-helicity
96

m-helicity
97

m-helicity
98

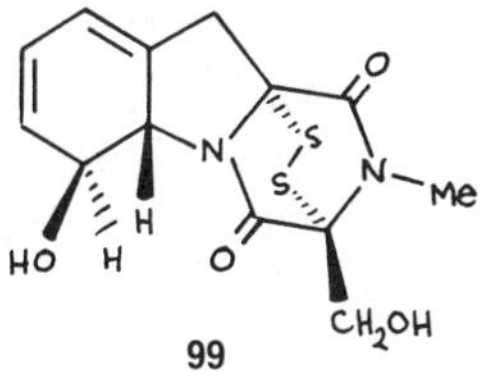

99

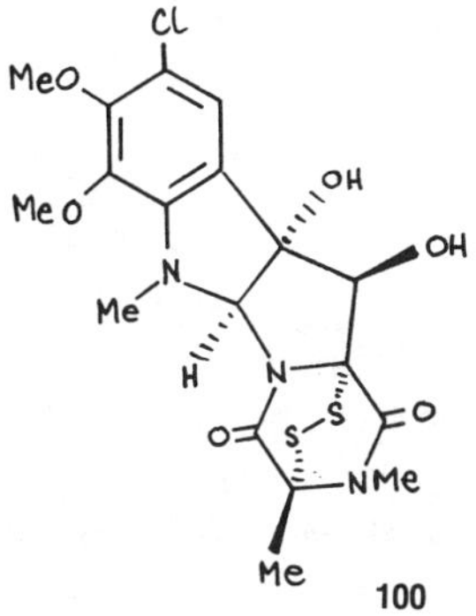

100

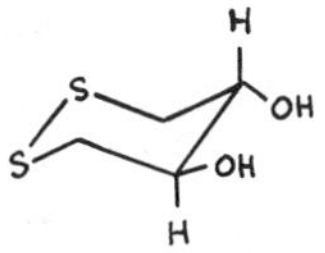

101

compounds have a boat structure. Derivatives of *cis*- and *trans*-4,5-dihydroxy-1,2-dithianes were studied,[278] and 4*R*,5*R*(+)-*trans*-4,5-dihydroxy-1,2-dithiane (**101,** the oxidized Cleland reagent dithiothreitol), prepared from (+)-tartaric acid, had nmr and cd spectra consistent with *P* helicity.[279] The more rigid isopropylidene derivative demonstrated similar spectral properties.[280]

Lüttringhaus and Brechlin[281] prepared *cis*- and *trans*-2,3-dithiadecalins from the intermediate *cis*- and *trans*-1,2-bishalomethylcyclohexanes and were aware that the trans-fused compound is very rigid. 9*S*,10*S*(−)-*trans*-2,3-Dithiadecalin (**102**) was synthesized,[280,282] starting from optically active 1*S*,2*S*(+)-*trans*-1,2-dicarboxycyclohexane. The cd curve for *M* chirality was consistent with the above rule, which was also applied to a study of the conformation of L-cystine derivatives.[282] The cis isomer of **102** is more flexible, and, from variable temperature nmr spectroscopy, free energy values of 60.3 and 52.3 kJ mol^{-1} were obtained for the inversion rates of *cis*-2,3-dithiadecalin and *cis*-2,3-dithia-1,2,3,4,4*a*,5,8,8*a*-octahydronaphthalene respectively.[4,283] These should be compared with 2,3-dithia-

1,2,3,4-tetrahydronaphthalene ($\Delta G^{\ddagger} = 35.6$ kJ mol^{-1}),[283] 4,5-d_2-4,5-dimethyl- (51.5 kJ mol^{-1}), 4,4-dimethyl- (52.7 kJ mol^{-1}), and 4,5-d_2-4,5-dihydroxy-1,2-dithianes (52.3 kJ mol^{-1}).[4]

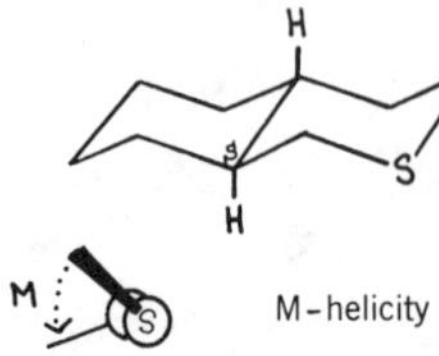

102

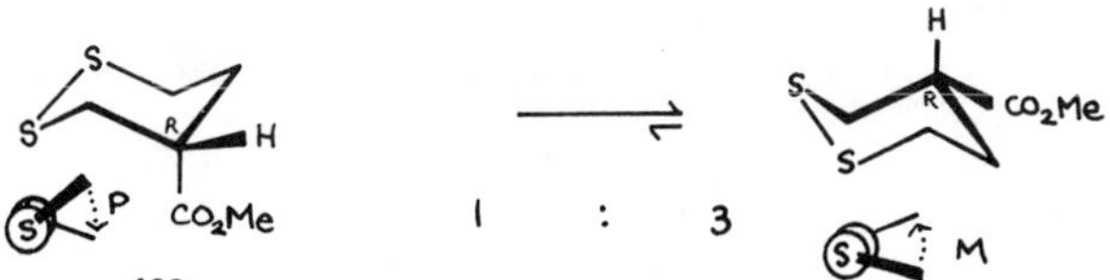

103 **104**

3- and 4-Carboxy-1,2-dithianes were resolved by Claeson,[284] who determined the absolute configuration 3*R*(+) for the former acid, by synthesis from L(+)-ornithine, and 4*S*(+) for the latter acid, by desulfurization to the known *S*(+)-2-methylbutanoic acid with Raney nickel. He also synthesized (+)-*trans*-3,6-dicarboxy-1,2-dithiane which has ord and cd curves similar to (+)-3-carboxy-1,2-dithiane, and hence has the 3*R*,6*R*(+) configuration. The conformational preference of a substituent in a flexible system has a direct influence on the helicity of the disulfide bond. This is exemplified in 4*R*(−)-4-methoxycarbonyl-1,2-dithiane which, in solution, is an equilibrium mixture of the conformer with the axial ester group **103** (33.3%) and the conformer with the equatorial ester group **104** (66.6%). The helicity of the disulfide bond is therefore dictated by the conformation of the substituent.[4,284] The X-ray structure of *trans*-3,6-dicarboxy-1,2-dithiane, determined by Foss and co-workers,[285] revealed that molecules with C–S–S–C bonds with *P* and *M* helicity are arranged alternately, that is, an equal number of two-chair conformers in the crystal. The rearrangement of *cis*-(*meso*)-3,6-dicarboxy-1,2-dithiane (**105**) into *trans*-2,5-dicarboxy-2-mercaptothiolane (**106**) in 0.1 *N* NaOH is 100 times faster than the rearrangement of the *trans*-(racemic)-dicarboxydithiane into *cis*-2,5-dicarboxy-2-mercaptothiolane. The anhydride or the diethyl ester of the *cis*-dithiane rearranges even faster, and these rates are most probably governed by the ease with which the proton at C–3 or C–6 is abstracted by the base.[286] A similar base-promoted rearrangement was observed with *trans*-3,6-dimethoxycarbonyl-4,5-benzo-4,5-dehydro-1,2-dithiane.[287] The desulfurization of *cis*- and *trans*-3,6-bismethoxycarbonyl-1,2-dithianes with [tris(dimethylamino)]phosphine proceeded quantitatively with loss of one sulfur atom and inversion of configuration at one carbon atom, and gave *trans*- and *cis*-2,5-bismethoxycarbonylthiolanes respectively.[287a]

Lüttringhaus and collaborators[137] observed two types of intramolecular hydrogen bonding in hydroxy-1,2-dithianes. They found that one type involved the ring sulfur atom and the other only the hydroxy group. In 4-hydroxy- (**107**), *cis*-4,5-dihydroxy- (**108**), and *trans*-4,5-dihydroxy-1,2-

105 → **106**

dithianes (**109**), the hydroxyl group was hydrogen bonded with the ring sulfur atom, with the sulfur and the oxygen atom, and only with the oxygen atom respectively.

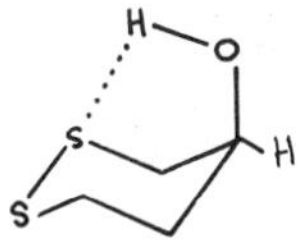

107 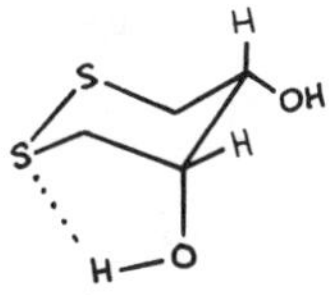 **108**

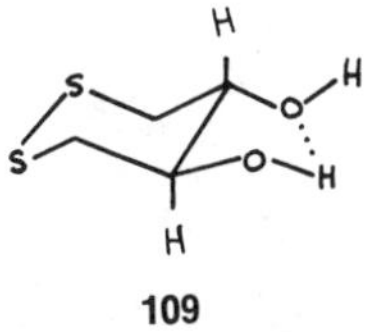

109 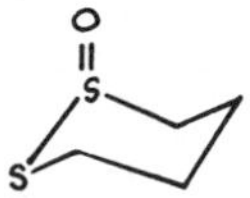 **110**

The axial preference of the oxygen atom (62%) in thiane 1-oxide, noted in Section VIII.B, is further enhanced by the presence of an adjacent sulfur atom in the ring. 1,2-Dithiane 1-oxide undoubtedly exists in the conformer with an axial oxygen atom (**110** $>$ 95%) and a conformational barrier of over 8.4 kJ mol^{-1} This is a dipolar effect analogous to the anomeric effect.[288,289]

B. 1,3-Dithianes

1,3-Dithianes are cyclic acetals or ketals of propane-1,3-dithiol. They are prepared from 1,3-dithiols and aldehydes or ketones, and the methods of preparation will be found in references cited below. 4*S*,6*S*(−)-4,6-Dimethyl-1,3-dithiane was formed, together with 4*S*,6*S*(−)-4,6-dimethyl-1,3-oxathiane, from 2*R*,4*R*(−)-2,4-bismesyloxypentane and potassium thiolacetate followed by 12 *N* $HCl/H_2C(OMe)_2/MeOH$. The optical activity was thus used as a probe in order to explain the mechanism which involves inversion of configuration at each of the carbon atoms.[290] The S–C–S chromophore does not contribute seriously to the ord and cd spectra[205,291] and makes optical activity a useful probe in this system.

As early as 1931, Mills and Saunders[292] resolved the hydrazone **111** (from 5-methyl-1,3-dithiane-2-one and *p*-carboxyphenylhydrazine) which owes its asymmetry to geometrical isomerism about the C=N bond.

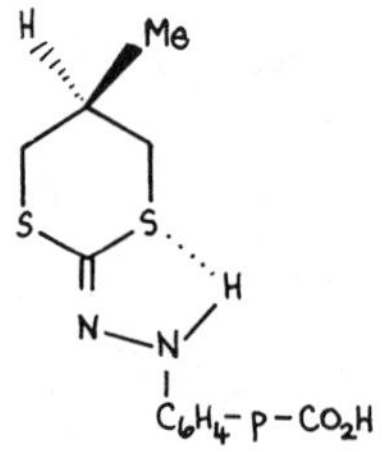

H Me S S H N N C_6H_4-p-CO_2H

111

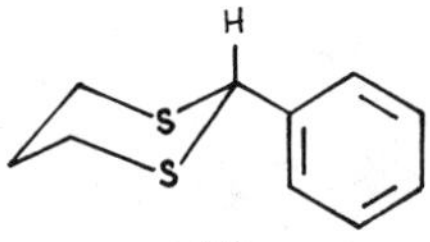

H S S

112

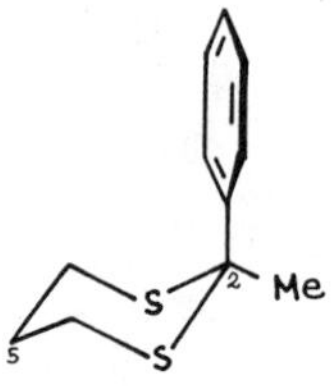

S S 2 Me 5

113

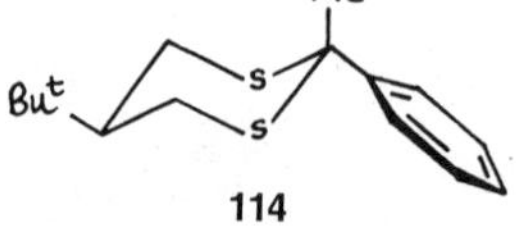

Me Bu^t S S

114

Kalff and Romers[293] have determined the X-ray structure of 2-phenyl-1,3-dithiane and found that the dithiane ring is in the chair conformation, and the phenyl ring is equatorial and is perpendicular to the plane of the dithiane ring (**112**). Kalff and Havinga[294] measured the ^1H nmr spectra and dipole moments of 1,3-dithiane, 2-methyl-, 2,2-dimethyl-, 2-phenyl-, 2-methyl-2-phenyl-, 2,2-diphenyl-, 2,2-diphenylene-, 2-*p*-chlorophenyl-, and 2-*p*-chlorophenyl-2-methyl-1,3-dithianes and showed that the geometry was consistent with the X-ray data. Whereas the aryl substituent in 2-phenyl-

1,3-dithiane is equatorial, the aryl group in 2-aryl-2-methyl-1,3-dithianes is axial but with its plane not along the C–2,C–5 axis (**113**). They also examined the two isomers of 5-*t*-butyl-2-methyl-2-phenyl-1,3-dithiane, and in each case the *t*-butyl group was equatorial (ananchomeric). In the isomer in which the phenyl group was also equatorial, the benzene ring was oriented at ring angles to the C–2,C–5 axis (**114**), as in the case of 2,2-diphenyl-1,3-dithiane, but unlike in 2,2-biphenylene-1,3-dithiane where the benzene rings are forced to be along the C–2,C–5 axis. Further similar data on 2-phenyl-1,3-dithianes were reported by Langer and Lehner.[295]

The 1,3-dithiane ring is flexible, and the inversion barriers for the parent substance, 2,2-dimethyl- and 5,5-dimethyl-1,3-dithianes, are $\Delta G^{\ddagger} = 39.3$, 41.0, and 43.1 kJ mol^{-1} respectively.[296] A careful study of coupling constants in the ^{1}H nmr spectra, by Gelan and Anteunis, revealed that the C_4–C_5–C_6 region of the heterocyclic ring in 1,3-dithianes (and 1,3-oxathianes) is less flattened than in 1,3-dioxanes, but the sulfur ring is more readily distorted by substituents.[297,298] Whereas *cis*-2,5-di-*t*-butyl-1,3-dioxane is in a chair conformation with an axial 5-*t*-butyl group,[299] *cis*-2,5-di-*t*-butyl-1,3-dithiane[300] adopts a twist-boat conformation (**115**), and *cis*-5-acetoxy-2-phenyl-1,3-dithiane may also be in this conformation.[301]

Eliel and Hutchins[300] examined the BF_3-catalyzed equilibration of several 1,3-dithianes and reported the ΔG values for alkyl substituents in various positions in the ring. They found that substituents at C–2 and C–4 have axial and equatorial preferences as in cyclohexane, but with substituents at C–5 the preference for the equatorial conformation is less marked than in cyclohexane. These results were interpreted in terms of weaker syn-axial interactions, the geometry of the dithiane ring, and known barriers of rotation about C–S bonds.[302] ^{1}H nmr and ir spectral studies of *cis*- and *trans*-5-hydroxy-2-phenyl-1,3-dithianes and their acetates also revealed that the energy differences between axial and equatorial 2-substituents in these and in the respective 1,3-dioxanes are much smaller than in the corresponding cyclohexanes.[301] In *trans*-4-methyl-6-alkyl-1,3-dithianes the methyl group has a high preference for the equatorial conformation and forces other alkyl groups (e.g., Et and iso-Pr) into the axial conformation.[303]

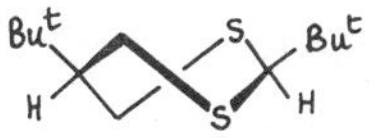

Bu^t S Bu^t H S H

115

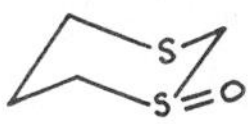

S S=O

116

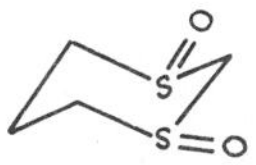

O S S=O

117

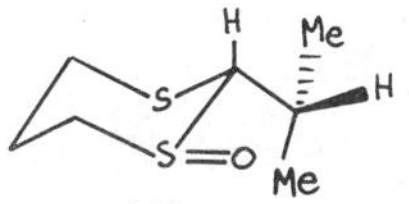

H Me S H S=O Me

118

In contrast with 1,2-dithiane 1-oxide, in which the oxygen atom is oriented axially (Section IX.A), the oxygen atom of 1,3-dithiane 1-oxide (**116**) has a preference for the equatorial conformation (84:16; $\Delta G^{\ddagger}=41$ kJ mol^{-1} at $-70.5°C$). The oxygen atoms in *cis*-1,3-dithiane 1,3-dioxide also assume a diequatorial conformation (**117**).[304,305,228,229] The presence of the equatorial oxygen atom in *trans*-2-isopropyl-1,3-dithiane 1-oxide has an effect on the rotameric behavior of the isopropyl group, and the ^{1}H nmr spectra indicated that the rotamer **118** was highly favored.[306] Variable temperature ^{1}H nmr spectra demonstrated a barrier for hindered rotation in 2-*t*-butyl-2-methyl-1,3-dithiane of $\Delta G^{\ddagger}=40.2$ kJ mol^{-1} at $-83.5°C$.[211]

Intramolecular hydrogen bonding between the hydroxyl group and the ring sulfur atom was demonstrated in 5-hydroxy-5-phenyl-[137] and 5-hydroxy-2-phenyl-1,3-dithianes (**119**).[301,307] *cis*- and *trans*-5-Hydroxy-2-phenyl-1,3-dithianes reacted with $POCl_3$ in pyridine stereospecifically and provided *cis*- and *trans*-4-chloromethyl-2-phenyl-1,3-dithiolanes.[307]

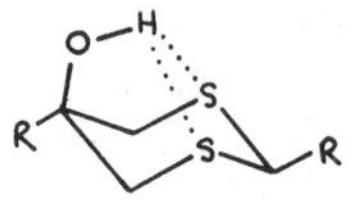

119

Direct alkylation of 1,3-dithianes with trialkyl oxonium borofluorides occurs on the sulfur atoms. Alkylation of 2,2-disubstituted 1,3-dithianes produces only one diastereomer, presumably the product of equatorial alkylation.[308] There is no evidence of inversion at the sulfur cation in *cis*- and *trans*-1,3-dimethyl-1,3-dithiane dications.[309] Remarkable stereospecificity was observed by Eliel and collaborators[310,311] in the protonation and methylation of conformationally biased 1,3-dithiane lithium derivatives. When the lithium salt of *cis*-4,6-dimethyl-1,3-dithiane was treated with DCl or MeI it exclusively gave the derivative with the deuterium and methyl group in the equatorial position, that is, cis to the two methyl groups (Equation 24). This is a thermodynamic effect because the lithium derivative (or ion pair) **120** is more stable than the diastereomer with the

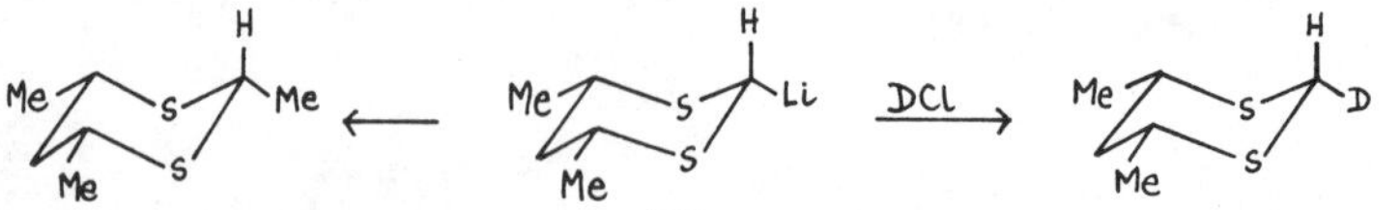

(24)

120

lithium atom in the axial orientation. Further studies on the lithiation of *r*-2-*trans*-4-*trans*-6-trimethyl-1,3-dithiane (axial 2-Me) followed by treatment with CO_2 gave the equatorial 2-carboxyl derivative. *trans*-4-*trans*-6-

Dimethyl-*r*-2-methoxycarbonyl-1,3-dithiane (axial 2-CO_2Me) and the all-cis isomer, after treatment with BuLi/MeI, gave 2,4,6-trimethyl-2-methoxycarbonyl-1,3-dithiane (all methyl groups equatorial) stereoselectively. These results, and deuterium isotope effects, demonstrate that the equatorial preference of the lithium derivative or carbanion is largely the result of thermodynamic control.[310–312]

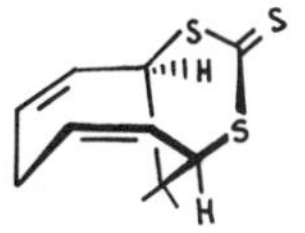

121

Arai and Ōki[312a] chlorinated 1,3-dithiane with *N*-chlorosuccinimide. The very reactive product, 2-chloro-1,3-dithiane, was examined in some detail by ^1H and ^{13}C nmr spectroscopy at various temperatures. There was no evidence for ring inversion, but the spectra in liquid SO_2 suggested that an equilibrium was present between one chair form of 2-chloro-1,3-dithiane with the chlorine axial, dissociated dithiane cation and Cl^- ions, and the other chair form of 2-chloro-1,3-dithiane, in which the chlorine atom had recombined but was also in the axial conformation.

Sodium trithiocarbonate reacted with *cis*-3,7-dibromocycloocta-1,5-diene, in aqueous MeCN, and gave 9-thiono-8,10-dithiabicyclo[5.3.1]undeca-2,5-diene (**121**). ^1H nmr spectra and X-ray crystallography showed that it possessed a rigid twist-boat structure.[313]

C. 1,4-Dithianes

1,4-Dithianes have been prepared by the dimerization of 2,2-diethoxyethylthiol,[314] acetonylthiol,[315,316] and by reaction of bis-(2-methylallyl)disulfide with SO_2Cl_2.[317] Mixtures of *cis*- and *trans*-2,5-disubstituted derivatives were formed and were separated. A convenient synthesis of 1,4-dithianes is by condensation of ethane-1,2-disulfinyl chloride with acetylene or but-2-yne. The products formed in these examples are *trans*-2,3-dichloro- and *cis*-2,3-dichloro-*cis*-2,3-dimethyl-1,4-dithianes.[318] For further examples of synthesis some of the references cited below should be consulted.

Electron diffraction[319] and X-ray crystallography[320–322] have conclusively shown that 1,4-dithiane (and 1,4-diselenane) exists in the preferred chair conformation, and some of these data were summarized by Romers and co-authors.[323] X-Ray,[324,325] ir, dipole moment, and ^1H nmr measurements[326] confirmed that the chair conformation for 2,3- and 2,5-dihalo-1,4-dithianes is the most stable, and that a strong anomeric effect is observed because both the halogen atoms assume the axial conformations **122** and **123** with X=Cl or Br. The free energy for the inversion process in 1,4-dithiane[327] is $\Delta G^{\ddagger}$ ~30.1 kJ mol^{-1}, and the value for 2,3-dichloro-1,4-dithiane (31.4 kJ mol^{-1}) is consistent with this result.[326] The value for 2,3,5,6-tetrachloro-1,4-dithiane is considerably higher (63.6 kJ mol^{-1}) and is a reflection of the nonbonded interactions involved.[326]

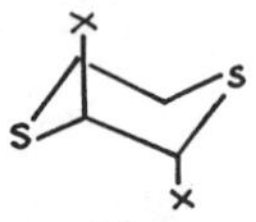

122

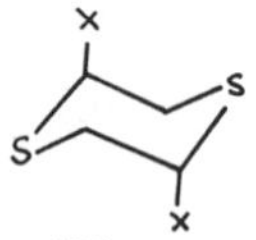

123

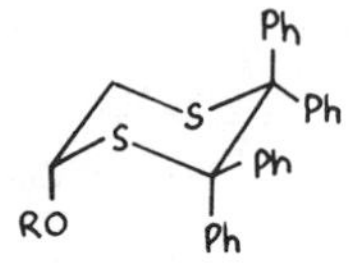

124

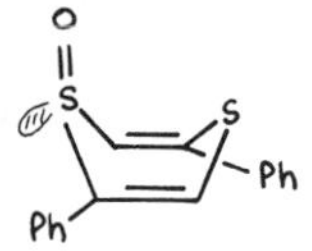

125

The substituent in 2-(6-chloropurin-9-yl)-1,4-dithiane has a preference for the axial conformation. This anomeric effect was attributed to hydrogen bonding between H–8 of the purine ring and the dithiane sulfur atom[229] (compare with the related thiane derivative in Section VIII.B). The anomeric effect is operating in 2-butoxy- and 2-ethoxy-5,5,6,6-tetraphenyl-1,4-dithianes (**124**) in which the conformers with axial alkoxy groups are favored. The 2-butylthio group in 2-butylthio-5,5,6,6-tetraphenyl-1,4-dithiane is obviously too large because it adopts the equatorial conformation.[57,327]

Electric polarization and nmr measurements demonstrated that 2,5-diphenyl-1,4-dithiin is in a boat conformation,[328] and X-ray analysis of the 1-oxide (**125**) confirmed this structure and showed further that the oxygen atom is oriented axially.[329] The strong axial preference of the oxygen atom in 1,4-dithianes was revealed in a study of *cis*- and *trans*-1,4-dithiane 1,4-dioxides. X-Ray data on *β-cis*-[330] (**126**) and *α-trans*-1,4-dithiane 1,4-

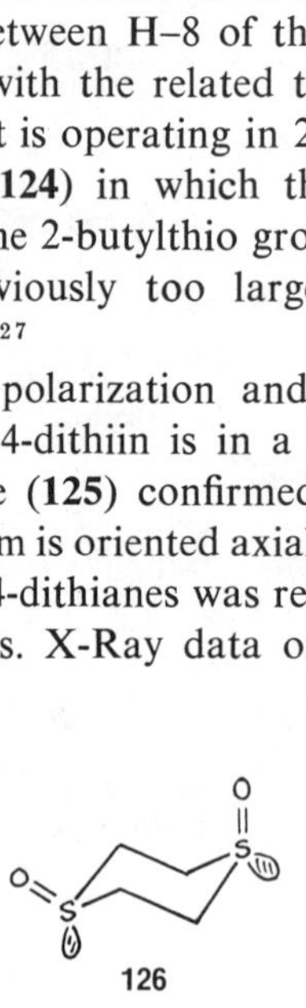

126

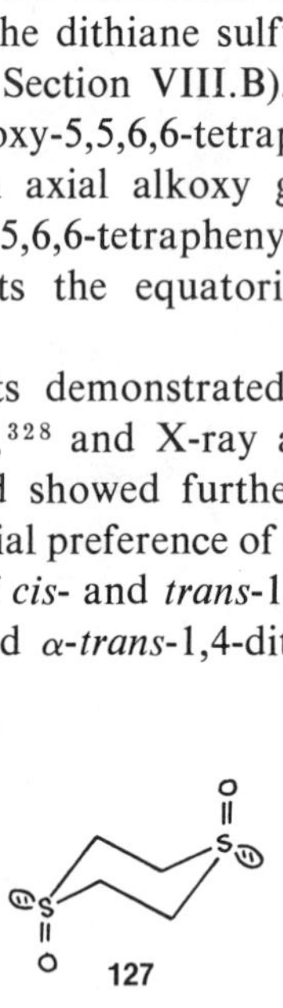

127

dioxides[331] (**127**) confirmed the relative structures.[332] The ^1H nmr spectrum of the β isomer had one line consistent with a flexible conformation, whereas the spectrum of the α isomer was complex because of the rigid structure with 1,4-diaxial oxygen atoms.[333] Ir and Raman spectra of these and 2,2-diphenyl-1,3-dithiane oxides are in agreement with the preferred diaxial oxygen atoms in the trans isomers and show that the dipolar effect is particularly strong in *trans*-1,4-dithiane 1,4-dioxide.[334]

The absolute configuration of 2*S*,3*S*(−)-*trans*-2,3-dicarboxy-1,4-dithiane was deduced by correlation with 2*S*,3*S*(+)-2,3-dimercaptosuccinic acid.[206]

X. TRITHIANES

Of the three trithianes the 1,2,3- and 1,3,5-trithianes have attracted some stereochemical interest. These are conformationally mobile, and, like cyclohexane, the chair conformations are the more stable. Lüttringhaus and co-workers[335] measured the free energy change for ring inversion in 1,2,3-trithiane and found it to be higher ($\Delta G^{\ddagger} = 55.2$ kJ mol^{-1}) than the values for 1,2-dithiane (48.5 kJ mol^{-1})[269] and 1,3-dithiane (39.3 kJ mol^{-1}),[296] and this is due to the energy necessary to rotate the S–S bonds. Anteunis and Goor[336] examined the ^1H nmr spectra of 5-methyl-1,2,3-trithiane (**128**), in which the C–4,C–5,C–6 region of the molecule is almost in the ideal chair arrangement, and found that the ΔG value for the methyl group is much smaller than in methylcyclohexane but similar to that in 5-methyl-1,3-dithiane. This clearly shows the smaller nonbonded interaction of the 5-methyl group with the syn-axial lone pair of electrons compared with the syn-axial hydrogen atom. They also examined the ring inversion barriers in 5,5-dimethyl-, diethyl-, and diisobutyl-1,2,3-trithianes and the 5,5-spiro derivatives all of which had, as expected, $\Delta G^{\ddagger}$ values larger than in unsubstituted 1,2,3-trithiane.[337]

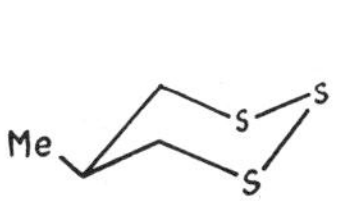

128 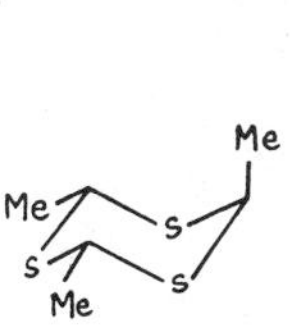 **129** 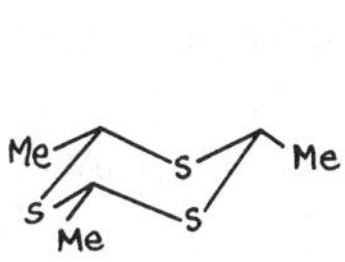 **130**

2,4,6-Trimethyl-1,3,5-trithiane is the trimer of thioacetaldehyde, which was isolated in three forms and was known for quite some time. One form was identified as the "trans" isomer (**129**) because it gave two monosulfones, and the other the "cis" isomer (**130**) because it gave only one monosulfone.[338] Both isomers, however, gave the same disulfone, but this was attributed to the ready isomerization of the product to the thermodynamically more stable disulfone.[339] Campaigne and collaborators[340] examined these compounds by ^1H nmr spectroscopy which demonstrated that these existed in interconverting chair conformations. They clearly showed two signals (2:1 ratio) for the three protons of the "trans" isomer but only one signal for the "cis" isomer. The third isomer was a eutectic mixture of both of these. *cis*- and *trans*-2,4,6-Triphenyl-[340] and tris-*o*-, *m*-, or *p*-fluorophenyl-1,3,5-trithianes were synthesized, and contrary to earlier claims it was found that the proportion of the β- or tri-equatorial (all-*cis*) 1,3,5-trithianes were always equal to or less than the α- or di-equatorial-mono-axial isomers. The ratios however varied considerably with the nature and position of the substituent in the original thiobenzaldehyde used.[341,342]

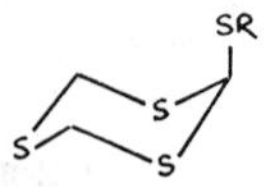

SR S S S

131

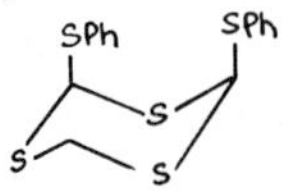

SPh SPh S S S

132

The anomeric effect in 1,3,5-trithianes was investigated through ^1H nmr spectroscopy by Ōki and co-workers.[343] The preference for the axial conformation of the substituents in 2-methylthio-1,3,5-trithiane in $CDCl_3$–CS_2 was stronger than in the 2-phenylthioderivative (**131,** R = Ph, 1.2 *ax*:1 *eq*) and increased with decrease of temperature. The proportion of diaxial conformer **132** in *cis*-2,4-bisphenylthio-1,3,5-trithiane was approximately 33% but decreased at lower temperatures. Equilibration experiments of this cis compound showed that *trans*-2,4-bisphenylthio-1,3,5-trithiane was thermodynamically more stable, and the barrier to inversion of the latter was 46 kJ mol^{-1}. The anomeric effect in these compounds is to be compared with that in 2-phenylthiothiane, in which the axial to equatorial ratio is 45:55 in $CDCl_3$ (see Section IV.B). Attack of benzoyloxy radicals in the reaction of *t*-butyl perbenzoate, in the presence of Cu_2Cl_2, with 1,3,5-trithianes takes place axially, and the conformer with the benzoyloxy substituent in the axial orientation is predominant.[226,344] 2-Benzoyloxy-1,3,5-trithiane yields *cis*-2,4-bisbenzoyloxy-1,3,5-trithiane, and the benzoyloxy group in *trans*-2-benzoyloxy-4-methyl-1,3,5-trithiane (similarly prepared from 2-methyl-1,3,5-trithiane) is displaced by an –SPh group with high retention of configuration, that is, to yield *trans*-2-methyl-4-phenylthio-1,3,5-dithiane.[344]

XI. 1,2,4,5-TETRATHIANES AND PENTATHIANES

The most extensively studied member of the tetrathiane series is 3,3,6,6-tetramethyl-1,2,4,5-tetrathiane (duplothioacetone), which was examined by Bushweller and co-workers.[345–347] ^{1}H nmr data revealed that it adopted a chair-twist conformation (**133**) in solution with a high barrier for conversion into the chair conformations. All the thermodynamic parameters were determined, and the free energy of conversion of chair to twist ($\Delta G^{\ddagger} = 65.3$ kJ mol^{-1}) was slightly different from the energy for conversion of the twist to the chair conformation (70 kJ mol^{-1}).[348,349] X-Ray crystallographic analysis of **133**[350] showed that the sulfur-containing ring in the crystal was in the boat conformation, which is of similar energy to the chair-twist conformation observed in solution. The nmr spectra of 3,3,6,6-bistetramethylene- and 3,3,6,6-bispentamethylene-1,2,4,5-tetrathiane and deuterated derivatives were examined by Bushweller and co-workers,[351,352] who found that there was a slow chair-to-twist and twist-to-twist stereomutation in the sulfur-containing ring. The thermodynamic parameters for these were determined. An X-ray structure determination of 3,3,6,6-bispentamethylene-1,2,4,5-tetrathiane, however, clearly showed that the tetrathiane ring in the crystal was in the chair conformation.[352,353]

133

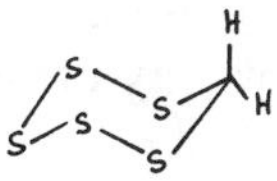

134

The ^{1}H nmr spectrum of pentathiane at 23°C consisted of two doublets indicating a coalescence of these signals above this temperature and, therefore, a very high barrier for inversion. It is therefore essentially locked in the chair conformation **134**.[354]

XII. SEVEN- AND LARGER-MEMBERED SULFUR-CONTAINING RINGS

A. Seven-Membered Rings

The cis and trans geometry and reactions of substituents on the saturated carbon atoms in the benzo-fused thiepins, such as 4,5-dichloro-2,3-benzo-4,5,6,7-tetrahydrothiepin,[355] 5,7-dihalo-2,3-benzo-4,5,6,7-tetrahydrothiepin-4-ones,[356] 4,5-dibromo-2,3-6,7-dibenzo-4,5-dihydrothiepin,[357] and the rearrangement of *cis*-5-bromo-4-hydroxy-2,3-benzo-4,5,6,7-tetrahydrothiepin into 2-(2-bromoethyl)thianaphthene[358] were described. Mock[359] had synthesized *cis*- and *trans*-2,7-dimethyl-2,7-dihydrothiepin 1,1-dioxides and studied their thermal decomposition. The cis and trans isomers eliminated SO_2 at 195 and 225°C in a trans-concerted, antarafacial (conrotatory) process and provided *trans,cis,cis*- and *trans,cis,trans*-2,4,6-octatrienes respectively and exclusively (e.g., Equation 25).

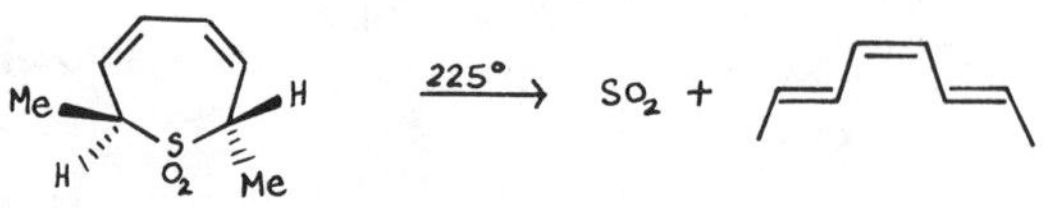

Me H S O2 H Me —225°→ SO_2 + (25)

The possibility of puckering in the ring of thiepin 1,1-dioxide was considered by Anet and co-workers.[360] They were unable to observe any evidence for this in the parent substance by variable temperature nmr, but found that the methyls of the isopropyl group in 3-isopropyl-6-methylthiepin 1,1-dioxide (**135**) were prochiral and appeared as a doublet at 250 MHz. The ring inversion barrier was estimated as 26.8 kJ mol^{-1}. By flanking the molecule with benzene rings, the inversion barrier can be raised considerably, to the extent that the invertomers became stable enough for optical resolution. 7-Carboxytribenzo[*b.d.f*]thiepin 1,1-dioxide (**136**) was resolved by way of the brucine salt. The optically active acids racemized slowly with an E_A value of 134 kJ mol^{-1}.[361] Overberger and Weise[362] prepared *R*(−)-4-methylthiepan-2-one (**137**), by cyclization of the respective 6-mercaptohexanoic acid, and studied the complex ord curves. The configuration of this compound and the isomeric *R*(+)-5-methylthiepan-2-one were deduced by a synthesis from *R*(+)-pulegone (**138**). The solvent dependent anomalous ord curves of these two thiolactones suggest that they have similar conformations in solution.

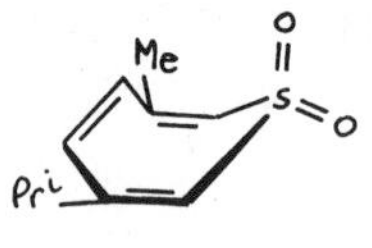

Me Pr^i S O O
135

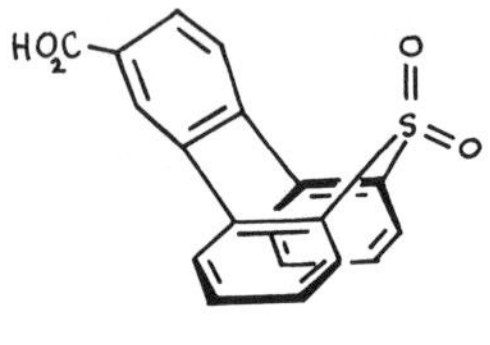

HO_2C S O O
136

Thermal decomposition of thiepane 1-oxide and *cis*-2-methylthiepane 1-oxide causes a stereospecific ring contraction with the formation of *cis*-2-methyl- and *cis*-2-ethylthiane 1-oxides respectively. *trans*-2-Methylthiepane 1-oxide decomposes more slowly but does not yield thiane deriva-

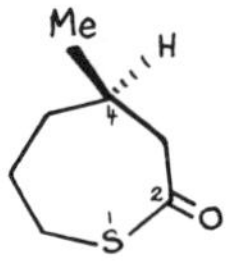

137

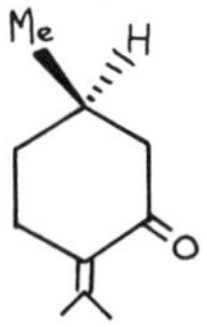

138

tives. 2-Methylthiane 1-oxide is inert under similar conditions because a special steric relationship between the S–O group and the β-hydrogen atom is necessary, and an olefinic sulfenic acid is probably involved in the rearrangement.[362a]

S(−)-4-Amino-1,2-dithiepane (**139**) was formed by oxidation of *S*-2-aminopentane-1,5-dithiol with iodine. The latter was prepared from *S*(+)-glutamic acid which establishes the absolute configuration. The optically active methanesulfonamido derivative was obtained from the corresponding 1,5-dithiocyanotopentane.[363] Schotte[364,365] similarly oxidized meso and racemic 2,6-dimercaptoheptane-1,7-dioic acid to *meso*(cis)- and *d,l*(trans)-3,7-dicarboxy-1,2-dithiepane (**140**). The racemic acid was best obtained by heating the meso isomer to 200°, and was partially resolved. The acid had a high optical rotation which was undoubtedly caused by the chiral S-S chromophore.

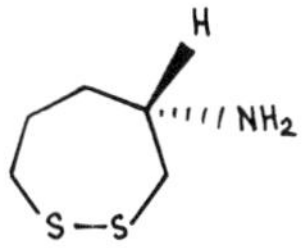

139

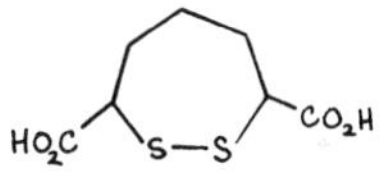

140

Data on the conformational mobility of seven-membered rings containing more than one sulfur atom is accumulating steadily, and the major contributors are Lüttringhaus, Friebolin, Kabuss, and Mecke. The flexibility of these molecules and the conformers in equilibrium are revealed by variable temperature nmr spectroscopy. The spectra of frozen conformations are essentially made up of sharp lines which broaden (coalesce) as the temperature is raised, and then sharpen up again (usually in a different pattern) to give the time-averaged spectra of the conformers in equilibrium. From the line shapes at the coalescence temperatures the thermodynamic parameters for the equilibria involved can be deduced. The flexibility of 2,2,7,7-d_4-5,5-dimethyl- and 4,4,6,6-tetramethyl-1,2-dithiepanes was investigated in this way and gave $\Delta G^{\ddagger}$ values of 37.7 and 46 kJ mol^{-1} for the free energy of ring inversion.[4,335] This is higher than in cyclohep-

141

142

143

144

tane[356a] and demonstrates how much more energy is required to rotate the S–S bond compared with the C–C bond. Three conformers are possible in 6,7-dihydro-3*H*-1,2-dithiepin: a rigid chair form (**141**), and the boat (**142**) and twist (**143**) forms which are flexible. These were studied in 6,7-dihydro-4,5-benzo-3*H*-1,2-dithiepin (**144,** R = H) from which the equilibria in Scheme 4 were demonstrated. The free energy for chair-chair interconversion in **144,** R = H) was 56.5 kJ mol^{-1}, with the free energy of the pseudorotation process, involving the two twist-boats and the two "boat" conformations, contributing up to 43.5 kJ mol^{-1}.[366] The $\Delta G^{\ddagger}$ value for the corresponding dimethyl derivative **144** (R = Me) is 50.6 kJ mol^{-1}.[4,335] (Scheme 4).

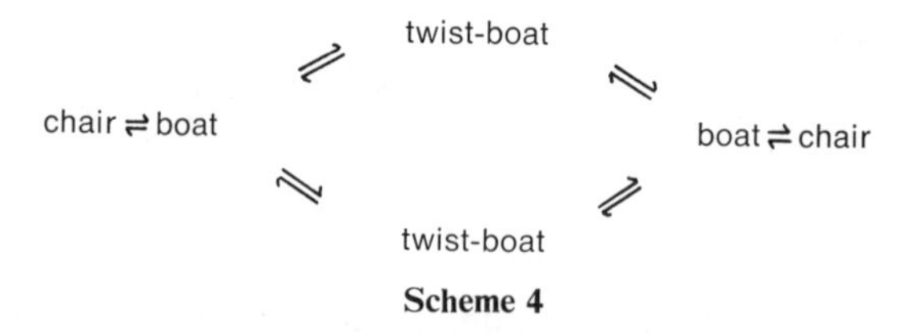

Scheme 4

The separation of the sulfur atoms by a CH_2 group lowers the inversion barrier, as do methyl groups to a lesser extent, but annelation raises it, as exemplified in 4,7-dihydro- (**145,** R = H), 2,2-dimethyl-4,7-dihydro- (**145,** R = Me), and 5,6-benzo-4,7-dihydro-2*H*-1,3-dithiepins which have $\Delta G^{\ddagger}$ values of 35.6, 34.3, and 42.7 kJ mol^{-1} respectively.[366a] A double bond in the ring and an increase in the number of sulfur atoms also raises the inversion barrier, as in 1,2,3-trithiepane (28.9 kJ mol^{-1}), 4,7-dihydro-1,2,3-trithiepin (37.2 kJ mol^{-1}), and 5,6-benzo-4,7-dihydro-1,2,3-trithiepin (**146,** R = H; 72.8 kJ mol^{-1}).[4,335,367] The last named is particularly rigid, and the free energy of interconversion increases with substitution in the ortho positions of the benzene ring; that is, the $\Delta G^{\ddagger}$ values for **146** (R = OMe), **146** (R = Me), and **146** (R = Ph) are 82.9, 83.7, and 87.9 kJ mol^{-1} respectively. These are almost stable enough to possess slight optical stability. A conformational study of tetra- and pentathiepanes was made by Moriarty and collaborators,[368] who confirmed that increasing the number of sulfur atoms, particularly adjacent sulfur atoms, in the ring increases the inversion barrier, and the flexibility of 1,2,3,5,6-pentathiepane (**147,** 54 kJ mol^{-1}) is to be compared with 1,2,3-trithiepane (28.9 kJ mol^{-1}) above.

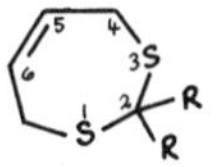

145 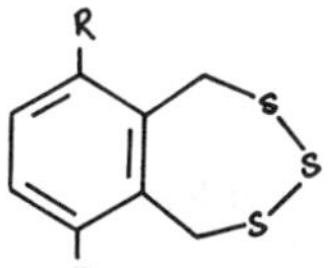 **146** 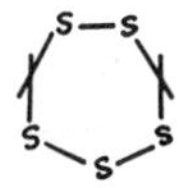 **147**

For further reading on seven-membered rings containing sulfur atoms and on conformational analysis of sulfur-containing rings, the excellent reviews of Field and Tulleen[369] and of Rahman and coauthors[4] should be consulted.

B. Eight- and Larger-Membered Rings

The facile rearrangement of *syn*-benzene dioxide (Chapter 2, formula **47,** R=H) into 1,4-dioxocin (Chapter 2, formula **48,** R=H) prompted a study of *syn*-benzene biepisulfide (**148**), by Vogel and co-workers.[370] They synthesized the bithiirane (**148**) from *r*-3-*cis*-6-diacetylthio-*trans*-4-*trans*-5-ditosyloxycyclohexene, and, although it was stable and characterized at −20°C, it decomposed at room temperature ($t_{1/2} \sim$ 30 min) into sulfur and benzene without giving the desired 10π-heterocycle 1,4-dithiocin. Kagabu and Prinzbach[370a] converted *syn*-benzene trioxide into *syn*-benzene triepisulfide (with complete inversion of configuration), but attempts to isomerize it to 1,4,7-trithionin lead to polymers, and treatment with trimethylphosphite gave benzene. The condensed 2,3-benzo-1,4-dithiocin (**149**), on the other hand, was obtained by thermal rearrangement of *cis*-3,4-benzo-2,5-dithiabicyclo[4.2.0]octa-3,7-diene (**150**), but only in 11% yield and under mild conditions (200°C). The parent 2,5-dithiabicyclo-[4.2.0]octa-3,7-diene and its 3,4-dicyano derivative gave benzene and a tar respectively, presumably through intermediates such as **148.** At higher temperatures (400°C) these bicyclo compounds gave naphthalene, benzene, and phthalonitrile respectively.[371]

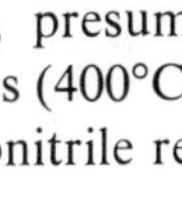

148 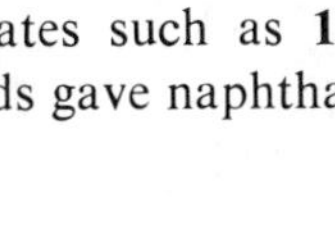 **149**

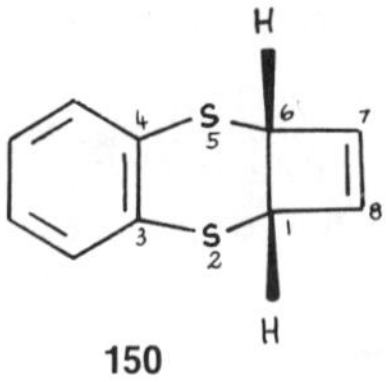

150

An X-ray diffraction analysis of 1-acetonylthionia-5-thiacyclooctane perchlorate (**151**) was made by Johnson, Maier, and Paul.[372] The eight-membered ring is in the chair-boat conformation. In solution it is rapidly inverting between two chair-boat conformations which suggests that the sulfonium cation is also inverting. Lüttringhaus and co-workers[373,374] examined the ^{1}H nmr spectra of 4,5,6,7-dibenzo-1,2-dithiacyclooctane [5*H*,-8*H*-dibenzo[*d.f*](1,2)dithiocin, **152**] and found that it had a very high inversion barrier (120.5 kJ mol^{-1}). They succeeded in obtaining it in optically active form by resolution on a "cellulose 2½-acetate" column. A single crystal X-ray diffraction investigation of this molecule indicated that the eight-membered ring was in the pseudo-chair conformation **152**,[375] rather than the expected pseudo-boat conformation, as predicted from nmr data and optical rotations in solution.[374]

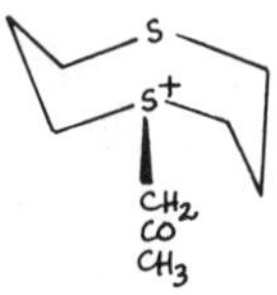

151

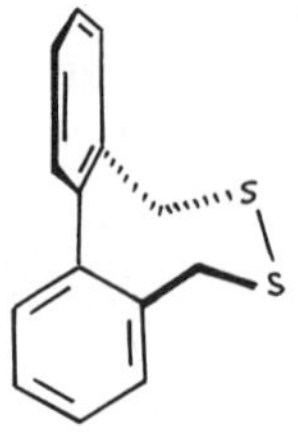

152

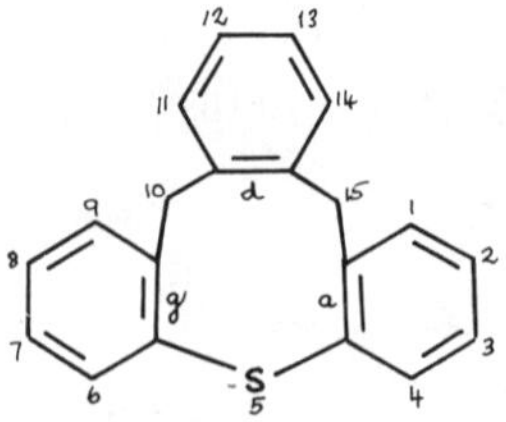

153

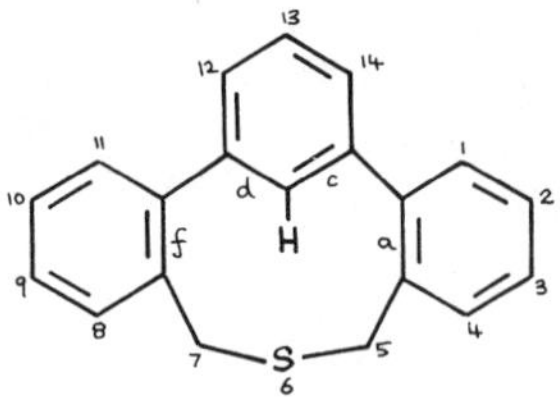

154

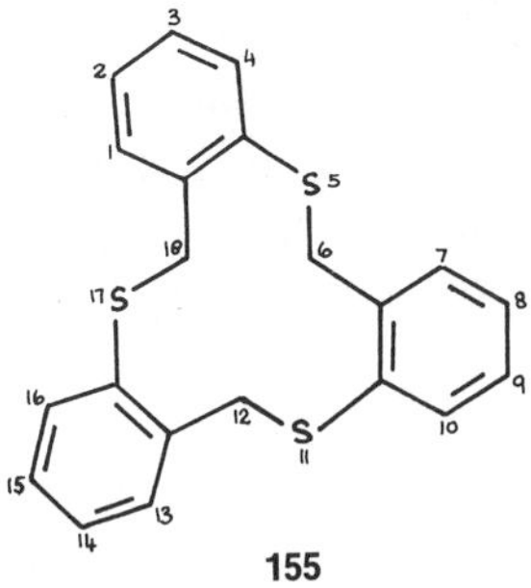

155

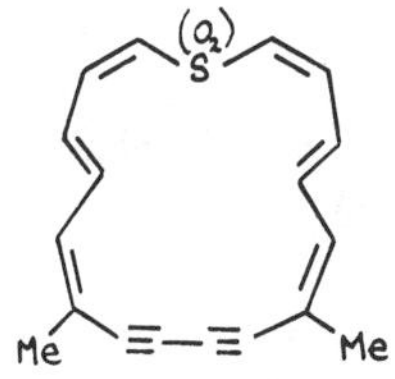

156

1,3,5,7-Tetrathiocane (tetrathiacyclooctane) was examined by X-ray crystal analysis and, in solution, by ^1H nmr spectroscopy. The data is inconsistent with even minor amounts of the crown form, and the chair-boat conformation must be predominant in solution as in the crystal.[376] This is in contrast with 1,3,5,7-tetraoxocane which may contain appreciable amounts of the crown conformation.[377]

The nine-membered rings in 10*H*,15*H*-5-thiatribenzo[*a.d.g*]cyclononatriene (**153**), and its 5,5-dioxide, are very rigid[378] compared with 5*H*,7*H*-6-thiatribenzo[*a.cd.f*]cyclononatriene (**154**), and its 6,6-dioxide, which exhibit ring inversion with $\Delta G^{\ddagger}$ values of 67 and 75.7 kJ mol^{-1} respectively.[379] The higher barrier for the latter dioxide is due to nonbonded interactions between the oxygen atoms and the hydrogen atom of the *cd*-benzene ring which is directed into the eight-membered ring (compare with **154**). As the size of the ring is increased the conformational mobility is increased, but the atoms still retain some preferred ordered arrangement. The 12-membered ring 6*H*,12*H*,18*H*-5,11,17-trithiatribenzo[*a.e.i*]cyclododecatriene (**155**) inverts between two preferred helical arrangements of atoms with a $\Delta G^{\ddagger}$ value of 38.9 kJ mol^{-1}.[380] Sondheimer and collaborators[381–384] prepared several thiaannulenes and found that *cis*-7,12-dimethylthia[17]annulene and its *S,S*-dioxide (**156**) did not give clear signals because of conformational mobility.[385] Macrocyclic polythioethers[386] analogous to the oxygen crown ethers are described in Chapter 2, Section XIV.D.

C. Transannular Interactions Involving the Sulfur Atom

Only a few examples are known in which the sulfur atom was shown to interact with atoms across a ring. Reactions similar to the ones noted here were also observed in the nitrogen (Part 1, Chapter 4, Section V.A.I) and oxygen (Chapter 2, Section XIV.B) heterocycles.

Irradiation of thiacyclohexan-4-one, and its 2- and 4-methyl derivatives, produces thietane 2-ones together with ethylene. This decomposition must involve the formation of an S–C_4 bond prior to degradation (Equation 26).[387] Participation of the sulfur atom occurs in the solvolysis of the *p*-nitrobenzoate ester of 3-hydroxythianes but not in 4-hydroxythianes. The former esters hydrolyze much faster than their 4 isomers and the interversion of the episulfonium ion (**157**) accounts for the data observed.[388] Paquette and collaborators,[389] however, showed that if the geometry of

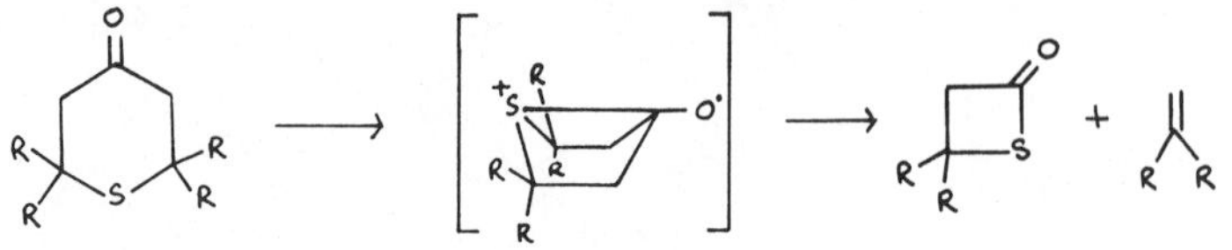

(26)

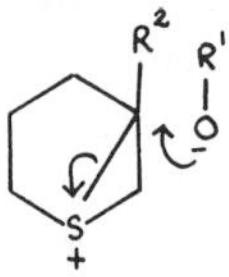

157

the molecule is carefully adjusted then sulfur participation in the solvolysis at the 4-position can be achieved. They synthesized the epimeric octahydro-6,2,5-ethanylylidene-2*H*-cyclobuta[*c.d*][2]benzothiophen-7-ol tosylates (**158,** R^1=H, R^2=OTs, and R^1=OTs R^2=H) as indicated in Equation 27.[390] The *endo*-tosylate (**158,** R^1=OTs, R^2=H) was found to solvolyze without participation of the sulfur atom because it occurred at about the same rate as the hydrocarbon **159** (R^1=OTs, R^2=H). The *exo*-tosylate (**158,** R^1=H, R^2=OTs), on the other hand, solvolyzed cleanly with retention of configuration and around 10^3 times faster than the corresponding hydrocarbon (**159,** R^1=OTs, R^2=H).[389] The interactions between the sulfur atom and C=O in molecules related to **158** (R^1, R^2=O) were examined by physical methods.[390]

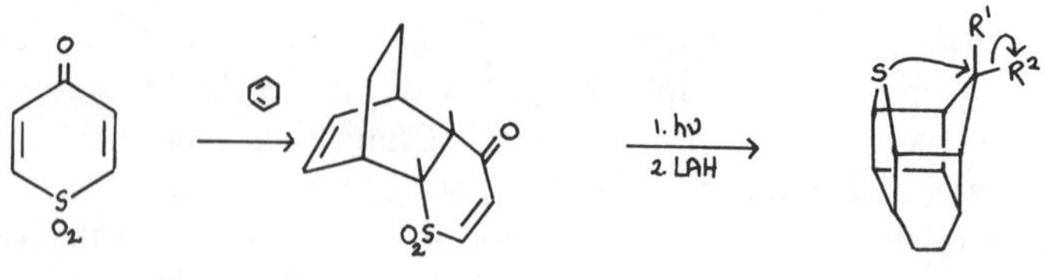

1. hν
2. LAH

(27)

158

159 160 161

In 1959 Overberger and Lusi[391] found that 5-hydroxythiacyclooctane could be converted into 1-thioniabicyclo[3.3.0]octane (**160**) which they isolated as the pictate in 57% yield. Leonard and collaborators[392] extended their illuminating studies of the transannular N.C_{co} (Part I, Chapter 4, Section V.A.1) and O.C_{co} (Chapter 2, Section XIV.B) interactions to the sulfur heterocycles. They found that the carbonyl stretching frequency in the ir spectrum of thiacyclooctan-5-one was lower compared with those of the corresponding sulfoxide and sulfone, and was consistent with differences in the uv and dipole moment data. This interaction, depicted in **161,** is weaker in thiacyclononan-5-one but was absent in thiacycloheptan-4-one and in open chain analogues.[393] These interactions are weaker than in the corresponding nitrogen analogues although much stronger than in the oxygen analogues. The oxygen atom on the ring sulfur atom of thiacyclooctan-5-one was shown to attack C–5, and in the presence of perchloric acid, 5-hydroxy-9-oxa-1-thioniabicyclo[3.3.1]nonane perchlorate (**162,** R= H) is formed. The hydroxy group can be methylated (**162,** R=Me) and acetylated (**162,** R=Ac). The S-^{18}O derivative, formed by oxidizing thiacyclooctan-5-one with $I^{18}O_4^-$, was shown to scramble the oxygen between the two oxygen atoms after hydrolysis of the intermediate (**162,** R=Ac) formed.[394] 5-Hydroxy (and methoxy) 8-oxa-1-thiabicyclo[3.2.1]octanes (**163,** R=H or Me) were similarly prepared from 1-thiacyclooctan-4-one,[395] and like the above thionia derivatives are very sensitive to hydrolysis and should be handled under strictly anhydrous conditions. These are examples of the rare sulfoxals which are the sulfur analogues of acetals, that is, re-

162 163

placing the alcohol C–OH by S^+–OH.

See the review by Field and Tuleen[369] for further examples of the involvement of the ring sulfur atom in transannular reactions.

D. Sulfur-Containing Cyclophanes

The present subsection contains mainly sulfur cyclophanes, also classified as ansa compounds (see Chapter 2, Section IX.C for definition). They were included in a recent review by Vögtle and Neumann,[396] who proposed an attractive general nomenclature for cyclophanes.[397] The compounds noted here should be compared with similar compounds which contain nitrogen (Part I, Chapter 4, Section V.A.2) and oxygen (Chapter 2, Section XIV.C) in place of sulfur atoms. The cyclophanes containing sulfur can be divided into two groups. The first group contains those cyclophanes in which the sulfur atom is in the aromatic ring (e.g., thiophenophanes) and the second group (for which there are more examples) have the sulfur atom in the chain that bridges the aromatic rings.

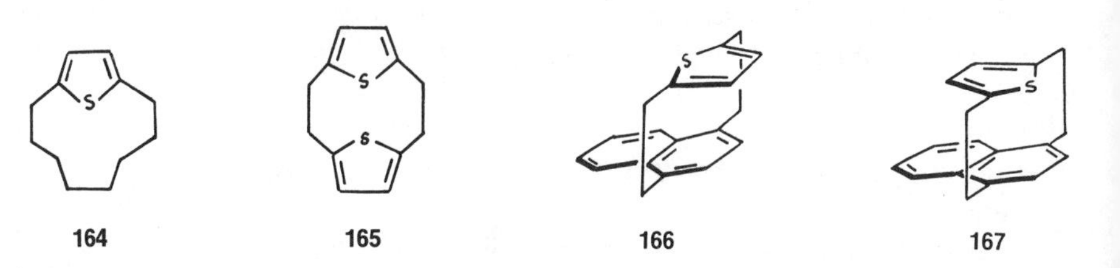

164 165 166 167

[8](2,5)-Thiophenophane (**164**) was prepared, by Nozaki and co-workers,[398,399] by a Paal-Knorr synthesis from cyclodecane-1,4-dione, and they also synthesized the nitrogen and oxygen analogues. ^{1}H nmr studies showed that these are conformationally mobile involving two more stable conformers which are interconverted by ring inversion (see formulae **64** and **65** in Part I, Chapter 4). Variable temperature experiments indicated that the sulfur analogue is the slowest to invert. Winberg and collaborators[400] made [2.2](2,5)thiophenothiophenophane (**165**) by pyrolysis of 2-methyl-5-trimethylammoniomethylthiophene hydroxide. An X-ray analysis of the 1:1 complex with benzotrifuroxan revealed that the two thiophene rings are not in the same plane, the sulfur atoms are facing each other, and the thiophen rings are distorted into boat shapes with the sulfur atoms slightly above and slightly below the planes of the thiophene rings.[401] Mizogami, Osaka, and Otsubo[402] used Winberg's method to condense a thiophene with a naphthalene ring and obtained a mixture of *anti*- (**166**) and *syn*-[2.2](1,4)naphthalene(2,5)thiophenophanes (**167**) in which the anti isomer (where π-π repulsive forces are least) predominated. A similar condensation with 1-methyl-4-trimethylammoniomethylanthracene gave only *anti*-[2.2]-(1,4)anthracene(2,5)thiophenophane. The "symmetrical" cyclophanes were of course also formed in these reactions and all were clearly identified by ^{1}H nmr spectroscopy.

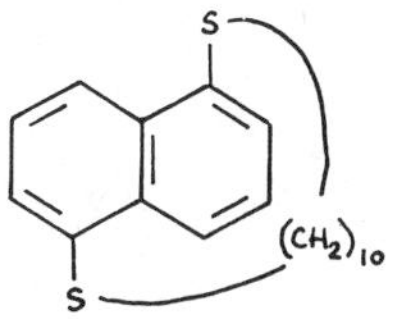

168

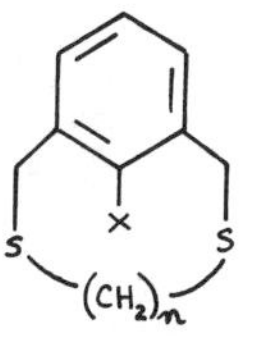

169

The earliest sulfur containing ansa compound which belongs to the second group is 1,5-(decamethylene-1,10-dithio)naphthalene [1,12-dithia-[12](1,5)naphthophane (**168**)]. It was prepared by Lüttringhaus and Gralheer,[403] in 1942, by intramolecular cyclization of 5-(10-bromodecamethylenethio)naphthalene 1-thiol in dilute solution. The yield of this compound (63%) was slightly higher than that of the corresponding oxygen analogue (see formula **321** in Chapter 2). Vögtle[404–407] investigated a large number of dithiacyclophanes with the general formula **169** in which X was H, F, Cl, or Br, and n was 3, 4, 5, or 6. These are capable of ring inversion like the thiophenophane **164**, and he examined the ^{1}H nmr spectra at various temperatures. The signals from the methylene protons adjacent to the ring were excellent probes because the sharp signals coalesced with change of temperature and provided a measure of the free energy change for the inversion process. The $\Delta G^{\ddagger}$ values generally increased as n became smaller and as the substituent X became larger. Thus when $n=2$ and X=H, $n=4$ and X=H, $n=4$ and H=F, $n=4$ and H=Cl or Br the $\Delta G^{\ddagger}$ values were 51.9, less than 40.2, 94.6, and greater than 97.9 kJ mol^{-1}, respectively, to give some idea of the energies involved.[404] The last value means that the coalescence temperatures are above 180°C. The system is not quite so simple because Mitchell and Boekelheide[408,409] found at least two coalescence temperatures for **169** ($n=3$, X=H), one for the benzylic CH_2, and one for the CH_2 on the chain between the sulfur atoms, indicating more than two relatively stable conformations. Vögtle and Risler[410] studied this system further and examined molecules similar to **169** with X=H, halogen, CN, Me, or OMe, to obtain another method for estimating the relative sizes of these substituents. The free energy required to move the carbon chain over the substituent can be obtained from ^{1}H nmr data, and they found that the energies were as generally predicted. They replaced the benzene ring of **169** with a pyridine ring, thus replacing C–X by N (lone pair), and demonstrated that the lone pair on the nitrogen atom was less space demanding than a hydrogen atom. The N-oxide with $n=6$, which undergoes the conformational changes **170** $\rightleftharpoons$ **171**, exhibits a free energy change for this equilibrium comparable with that for **169** ($n=5$, X=F).[410]

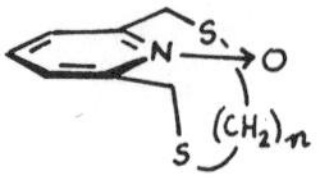

170 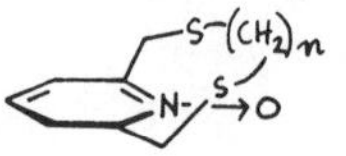 ⇌ 171

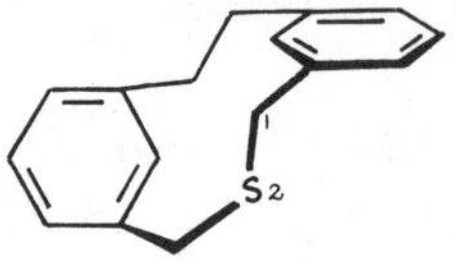

172

Sato and collaborators[411] studied 2-thia[3.2]- (**172**) and 2,10-dithia[3.3]-metacyclophanes (**173,** R = H) and found that the former was less mobile than the latter, and that its mobility decreased further if the sulfur atom was oxidized to the sulfoxide and to the sulfone. When methyl groups were introduced into the latter (**173,** R = Me) the $\Delta G^{\ddagger}$ value rose to above 99.6 kJ mol^{-1}. Similar metacyclophanes were investigated by Vögtle and coworkers, these contained benzene rings,[412–414] (1,7)-naphthalene rings,[415] and mixed benzene and pyridine rings.[410,416] ^{1}H nmr data[412,413] on metacyclophanes related to 17,18-dimethyl-2,10-dithia[3.3]metacyclophanes (**174**) suggest that the system is conformationally quite rigid; indeed, Mitchell and Boekelheide,[417] by reacting 2,6-bisbromomethyltoluene with Na_2S obtained the isomeric cyclophanes **174** and **175** in 14 and 2% yields respectively. X-Ray crystal analysis of the isomer **174** demonstrated that the two benzene rings are not parallel but are inclined towards each other.[418] Many of the conformationally more rigid cyclophanes should be obtainable in optically active forms.

Another interesting example is 1,10-dithia[2.2](1,3)(1,4)cyclophane (**176,** R = H) which inverts by rotation of the para substituted benzene ring ($\Delta G^{\ddagger}$ = 61.5 kJ mol^{-1}). In the 16-fluoro derivative **176** (R = F), on the other hand, the benzene ring finds it very difficult to rotate and the $\Delta G^{\ddagger}$ value is greater than 95.4 kJ mol^{-1}.[419] The sulfoxide **177** (X = SO) presents another case in which multiple, energy requiring, conformational changes

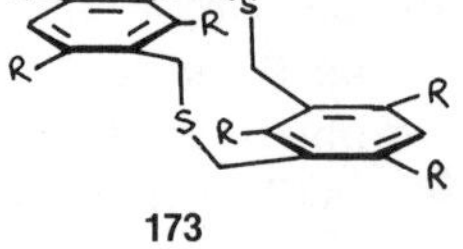

173

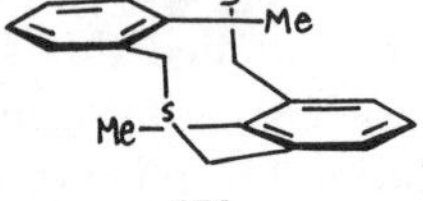

174

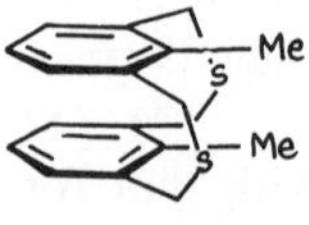

175

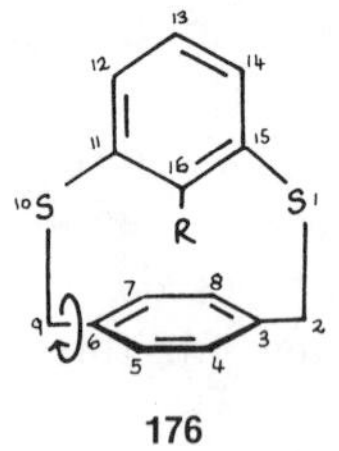

176

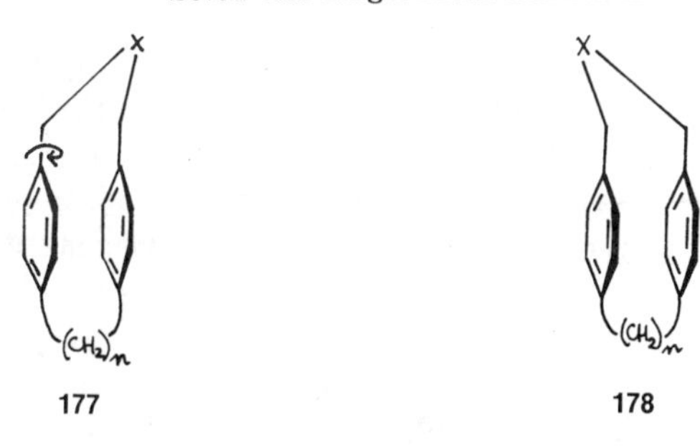

177 **178**

occur. Potter and Sutherland[420] have observed that the benzene rings can rotate about the para axis with $\Delta G^{\ddagger}$ values of greater than 104.6, 77.1, 58.2, and less than 9 kJ mol^{-1} for $n = 4$, 5, 6, and 7 respectively, and the energy for the bridge inversion process, **177** (X=S or SO_2) $\rightleftharpoons$ **178** (X=S or SO_2), is in the order 37.2–45.2 kJ mol^{-1}. The $(CH_2)_n$ bridge also exhibits conformational changes (for $n = 3$ and $n = 4$, $\Delta G^{\ddagger} \sim 48.1$ and 37.2 kJ mol^{-1}) which are observed only by the line-shape method.

Lehner[421] used the "isoconformational concept" to correlate the conformations of cyclophanes with the patterns of signs for torsional angles and an equal number of torsional degrees of freedom. He was able to deduce the conformations of some cyclophanes from the geometries of cyclohexanes.

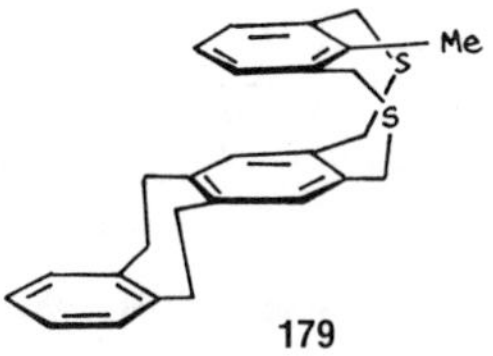

179 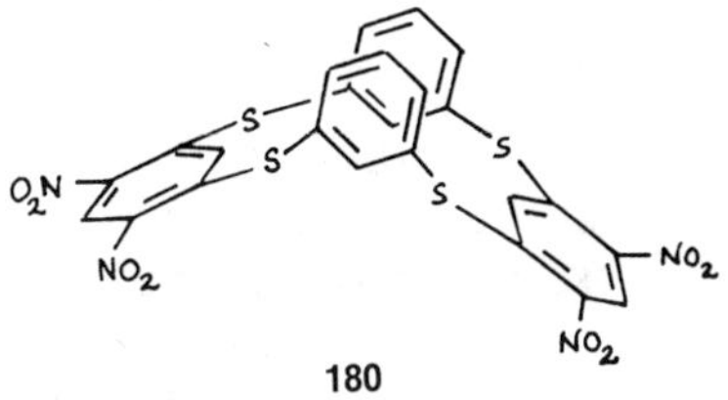 **180**

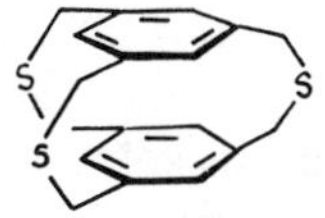

181

Other reports were on the preparation of multilayered cyclophanes such as **179,** in which isomers such as **174** and **175** were obtained,[422] and compounds containing four benzene rings, like **180,** which possess a macrocyclic ring.[423] The most interesting compounds came from Vögtle's laboratory,[424–426] and these had benzene rings joined by three bridges which form a cage molecule. Thus 1,3,5-trisbromomethylbenzene and 1,3,5-trismercaptomethylbenzene were condensed together using the dilution principle to form 2,11,20-trithia[3.3.3](1,3,5)cyclophane (**181**) which was oxidized to the trisulfone.[424] Macrocyclic cages, for example, 2,23,50-trithia-[3.3.3](4′,4″,4‴) *sym*-triphenylbenzeno<8>phane, were also formed from the condensation of 1,3,5-tris(4-bromomethylphenyl)benzene and 1,3,5-tris-(4-mercaptomethyl)benzene in Wurtz-type reactions in dilute solutions.[427] Cages such as these in which the benzene rings were joined by bridges having $-S-CH_2CH_2-S-$ units were also synthesized and were able to chelate with metal ions such as Cu, Hg, Au, *et cetera.*[425] Boekelheide and coworkers succeeded in extruding the sulfur atoms in cyclophanes similar to **174** by *S*-alkylation and rearrangement,[428,429] or by photolysis in the presence of Ph_3P,[430] and keeping the integrity of the cyclophane. By this means they succeeded in preparing the simplest cage [2.2.2.2](1,2,4,5)cyclophane (**182**).[430]

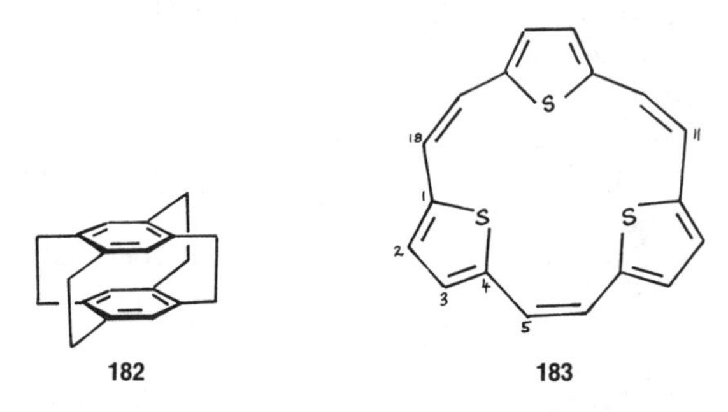

182 **183**

Badger, Elix, and Lewis[431] demonstrated that the sulfur atoms in the [18]annulene trisulfide (**183**) could not be readily accommodated in one plane in the center of the molecule. Two of the sulfur atoms must be above the plane of the molecule and one below it, or alternatively, three must be above and three must be below the plane of the molecule. The uv and ^1H nmr spectra were consistent with a nonplanar molecule. They showed that this system was capable of dissymmetry by preparing the brucine salt of the 5,11,18-tricarboxylic acid derivative which underwent marked, but rapid, mutarotation in solution. The free acid could not be obtained in its optically active forms, which means that the enantiomers are readily interconvertible.

XIII. THIABICYCLO[*x.y.z*]ALKANES

This section is a potpourri of thiabicycloalkanes, in which the three-dimensional structures of these compounds are pointed out. These should be known before a full understanding of their reactions and physical properties can be achieved. A few of these bridged systems have been discussed by Traynelis.[432]

The first example is 1-thioniabicyclo[2.2.1]heptane hydrochloride (**184**), which was prepared, by Prelog and Cerkovnikov[433] in 1939, by intramolecular cyclization of 4-chloromethylthiane. It readily formed a picrate and chloroplatinate. The isomeric 2-thiabicyclo[2.2.1]heptane[434] gave two sulfoxides, **185** (X=O, Y=lone pair) and **185** (X=lone pair, Y=O), which were assigned by ^1H nmr in conjunction with the shift reagent $Eu(dpm)_3$.[434a] These sulfoxides and their deuterated derivatives (in all positions but C–1 and C–3) have been very useful in the study of H-D exchange and alkylation at the carbon atoms α to the S-O group in relation to the sulfur lone pair of electrons. The rigidity of the system fixes the torsion angle between the α-CH bonds and the sulfur lone pair and provides more meaningful data for the H-D exchange. The base-catalyzed H-D exchange rates gave K_a/K_b (for H_a and H_b) values for **185** (X=O, Y=lone pair) and **185** (X=lone pair, Y=O) equal to 57:1 and 1:9. This indicates that the α-CH eclipsed by the oxygen atom is more acidic. There is also an endo-exo effect because in the sulfone **185** (X=Y=O), where

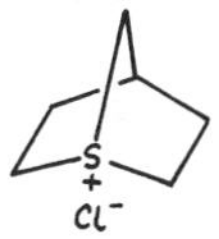

184 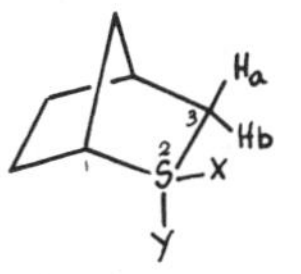 **185** 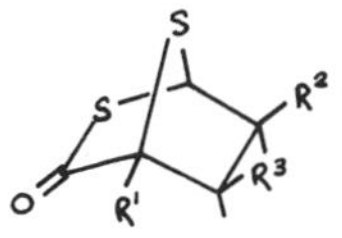 **186**

exchange is about 1000 times faster than in the sulfoxides, the K_a/K_b value is 8:1 demonstrating that the *exo*-hydrogen atom, H_a, is more readily exchanged than the sterically more hindered *endo*-hydrogen atom H_b.[435] The rigidity of this ring system has lead to interesting stereochemical observations in 7-thiabicyclo[2.2.1]heptanes. The acetolysis of the 2-*endo*-chloro derivative followed second order kinetics at a considerably faster rate (4.7×10^9) than the 2-*exo*-chloro isomer, which acetolyzed with first order kinetics. The *endo*-chloro compound gave the corresponding 2-*endo*-acetoxy derivative with retention of configuration and implies that a sulfonium ion assisted the reaction (Equation 28). The 2-*exo*-chloro-thiaheptane solvolyzed in a different manner which involved a skeletal rearrangement to give 3-*exo*-acetoxy-2-thiabicyclo[2.2.1]heptane (Equation 28a).[436] The parent, 7-thiabicyclo[2.2.1]heptane, was readily formed from alkaline treatment of 4-chloro-1-acetylthiocyclohex-1-ene,[437] and dehydrochlorination of bis-*endo*-2,5-dichloro-7-thiabicyclo[2.2.1]heptane gave the intermediate 7-thiabicyclo[2.2.1]hepta-2,5-diene which extruded sulfur readily and

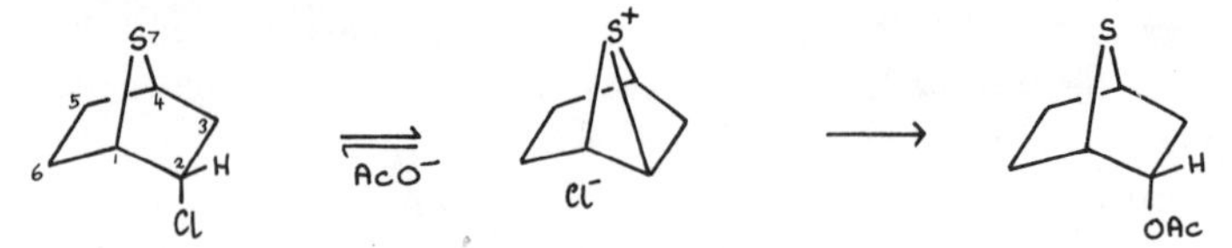

(28)

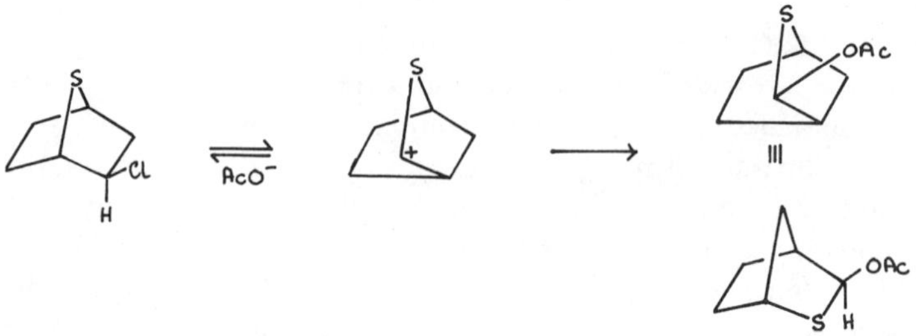

(28a)

produced benzene.[438] The mesoionic 1,3-dithiolium 4-oxides undergo 1,3-dipolar cycloaddition reactions with olefins (e.g., maleic ester)[439] or acetylenes[440] and yield 2,7-dithiabicyclo[2.2.1]heptan-3-ones (**186**) or hept-6-en-3-ones respectively.

Johnson and Billman[441] treated 1,2-dibromo-4-acetylthiomethylcyclohexane with alkali and obtained a 62% yield of *endo*-4-bromo-6-thiabicyclo-[3.2.1]octane, which was reduced to the parent compound with Ph_3SnH. Attempts to dehydrobrominate this compound in order to obtain a double bond between the bridgehead carbon atom and C–4 (i.e., violate the Bredt rule, see below) gave 6-thiabicyclo[3.2.1]oct-3-ene. A more direct cyclization of this type, reported by Surzur and co-workers,[442] is one in which 4-mercaptomethylcyclohex-1-enes were photolyzed. The unsubstituted compound **187** (R = H) cyclized exclusively to 2-thiabicyclo[2.2.2]octane (**188,** R = H) in 70% yield, whereas the methyl derivative (**187,** R = Me) gave 75% of 4-methyl-6-thiabicyclo[3.2.1]octane (**189,** R = Me) together with 25% of the isomeric 1-methyl-2-thia[2.2.2]octane (**188,** R = Me). The difference in the latter case was attributed to the greater stabilization of the radical **189a** (R = Me) by the methyl group compared with the radical **188a.** 2-Thiabicyclo[2.2.2]octane (**188,** R = H) was prepared in a completely different synthesis involving the 1,4-addition of thiophosgene to cyclohexa-1,3-diene to form 6,6-dichloro-5-thiabicyclo[2.2.2]oct-2-ene, followed by reduction first with LAH to remove the chlorine atoms and then $N_2H_4/Cu^{++}/O_2$ to reduce the double bond.[443] Plešek and collaborators[444] stated that a rearrangement had occurred when optically active 1,2,2-trimethyl-1,3-bistosyloxymethylcyclopentane (**190,** obtained from (+)-camphor) was treated with Na_2S, and that (+)-1,5,5-trimethyl-2-thiabicyclo-[2.2.2]octane (**191**) was formed. Dodson, Cahill, and Chollar[445] then

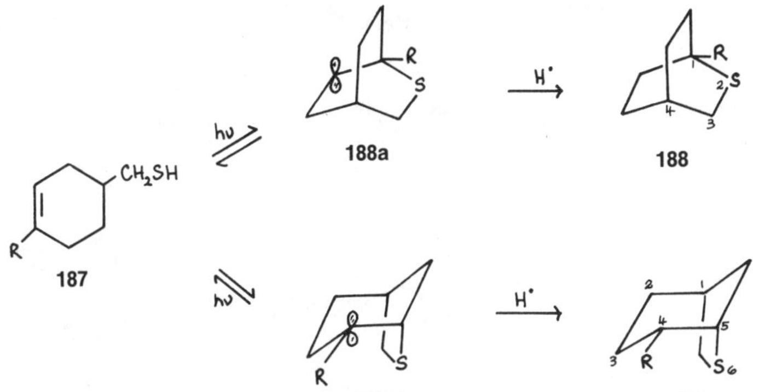

187

188a

188

189a

189

examined the chiroptical properties of this product and found that they were not consistent with the structure **191**, and that it should be (+)-1,8,8-trimethyl-3-thiabicyclo[3.2.1]octane (**193,** $R^1=R^2=$lone pair). Also 3-thiabicyclo[3.2.1]octane had been previously prepared in a similar way without apparent rearrangement.[446] They confirmed this by an unambiguous synthesis from the intermediate **192.** The oxidation of the bicyclic compound **193** ($R^1=R^2=$lone pair) is particularly interesting because with *m*-chloroperbenzoic acid a 95:5 ratio of the *exo*- (**193,** $R^1=O$, $R^2=$lone pair) and *endo-S*-oxides (**193,** $R^1=$lone pair, $R^2=O$) is formed, whereas with *t*-BuOCl a 4:96 ratio of *exo*- and *endo*-oxides is obtained. The absolute configurations for *exo-S*-oxide (*R*) and *endo-S*-oxide (*S*) follows from a knowledge of the absolute configuration of the (+)-thiabicyclic compound **193** ($R^1=R^2=$lone pair) which was derived from (+)-camphor without upsetting the asymmetric centers.[447] Binsch and Franzen[448] obtained an anisochronous nucleus by oxidizing 4-alkyl-2,6,7-trithiabicyclo[2.2.2]octanes to the 2,6,7-trioxides, and isolated an isomer in which the three *S*-oxide groups formed a chiral propellar (**194**). This was evidenced in the nmr spectra in which the methylene protons α to the sulfur atom had an AA′A″-BB′B″ pattern of signals.[448]

Entry into the bicyclic system 9-thiabicyclo[3.3.1]nonane is best achieved by the addition of SCl_2 to *cis,cis*-cycloocta-1,5-diene[449,450] and cyclooctatetraene[451] which provides *endo,endo*-2,6-dichloro-9-thiabicyclo[3.3.1]nonane and -nona-3,7-diene respectively. The former compound is also formed by the addition of chlorine to 5,6-epithio-*cis*-cyclooct-1-ene,[452] and the latter reaction was applied successfully to cycloheptatriene and 1,5,9-cyclododecatrienes, in which the sulfur atom added across two double bonds.[451] The sulfur atom in these systems has been oxidized to the sulfone, and the halogens can be reduced to yield the parent 9-thiabicyclo[3.3.1]-

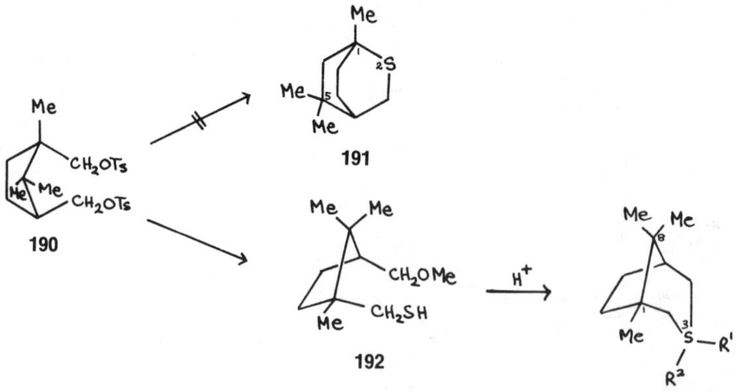

190

191

192

193

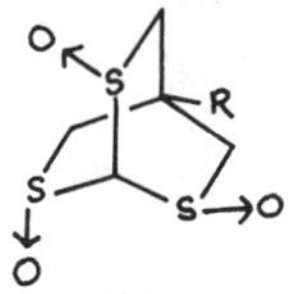

194

nonane.[450] The halogen atoms undergo substitution reactions with nucleophiles with high retention of configuration,[453–455] and a sulfonium ion intermediate was invoked to support the stereospecificity.[453] More recently with the use of ^{13}C nmr spectroscopy, Vincent and co-workers[456] found an explanation for the enhanced reactivity of the first halogen atom with respect to the second, and for the stereoselectivity. The ^{13}C nmr spectrum in FSO_3H/liq. SO_2 at $-60°C$ revealed that the intermediate sulfonium ion **196** is formed first and isomerizes to the ion **197,** which is stable even at room temperature. The 2,6-dibromo and diiodo derivatives behave in the same manner, but 2,6-dihydroxy-, 2,6-dimethoxy-, 2-chloro-6-hydroxy-, and 2-chloro-6-methoxy-9-thiabicyclo[3.3.1]nonanes give the stable sulfonium ion, similar to **197,** directly. These cations are also formed in SbF_5/$CDCl_3$/liq. SO_2 and $AlCl_3$/$CDCl_3$/liq. SO_2, and the energies involved in the isomerization have been discussed.

Direct chlorination and oxidation of 9-thiabicyclo[3.3.1]nonane did not give the α-chlorosulfone but rather the β-chlorosulfone *endo*-2-chloro-9-thiabicyclo[3.3.1]nonane 9,9-dioxide, presumably involving a ring cleavage recombination mechanism. The α anion of the sulfone, however, was formed when it was treated with alkyl lithium. The anion has the negative charge at the bridgehead carbon atom, and on reaction with SO_2Cl_2,[457] MeI, or Br_2[458] the respective 1-chlorosulfonyl-, 1-methyl-, or 1-bromo-9-thiabicyclo[3.3.1]nonane 9,9-dioxides are formed. The 1-chlorosulfonyl de-

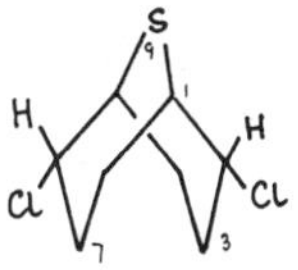

195

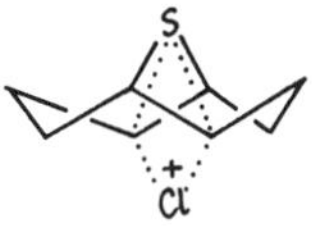

196

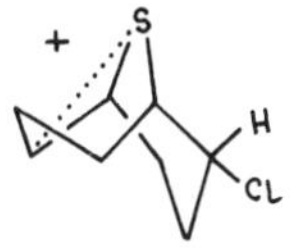

197

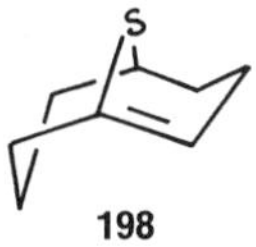

198

rivative loses SO_2 on heating and generates the 1-chloro derivative, which decomposes in the presence of base, with loss of HCl and SO_2, to form bicyclo[3.3.0]oct-1,5-ene.[457] The displacement of the halogen atom in *endo*-2-chloro-9-thiabicyclo[3.3.1]nonane 9,9-dioxide by *t*-butoxide occurs with retention of configuration. A sulfonium ion such as **197** cannot be involved here, and Paquette and Houser[457] suggested that the intermediate was the α,β-unsaturated sulfone (9-thiabicyclo[3.3.1]non-1-ene 9,9-dioxide). Quinn and Wiseman[459] showed that this was indeed possible although it violated the Bredt rule (see Part I, Chapter 4, Section B.2 and D.2]. By treating 1-methanesulfonyloxy-9-thiabicyclo[3.3.1]nonane with *t*-BuOK, they isolated 9-thiabicyclo[3.3.1]non-1-ene and provided evidence that it had the boat-chair conformation **198.** It slowly isomerized to the non-2-ene and formed the adduct **199** (R = lone pair) with 1,4-diphenylisobenzofuran. The structure of the dioxide of this adduct (**199,** R = O) was confirmed by an X-ray analysis.[458] The oxygen analogue 9-oxabicyclo[3.3.1]non-1-ene was also prepared in a similar manner.[460]

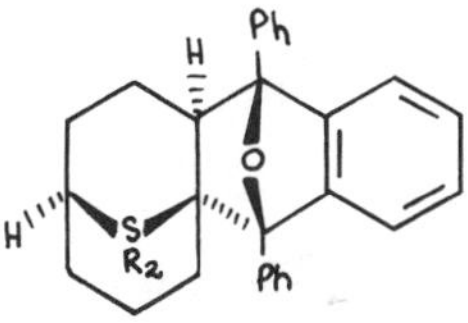

199

9-Thiabicyclo[3.3.1]non-6-en-2-one rearranges to 2-thiabicyclo[6.1.0]non-5-en-2-one upon uv irradiation, and the latter structure was confirmed by an X-ray analysis.[461] Irradiation of *endo*-2-hydroxy-9-thiabicyclo[3.3.1]-nonan-6-one 9-oxide undergoes stereomutation of the 9-oxygen atom; that is, the sulfoxide group is equilibrated and apparently involves energy transfer from the excited carbonyl to the sulfinyl group, but no skeletal rearrangement takes place.[462]

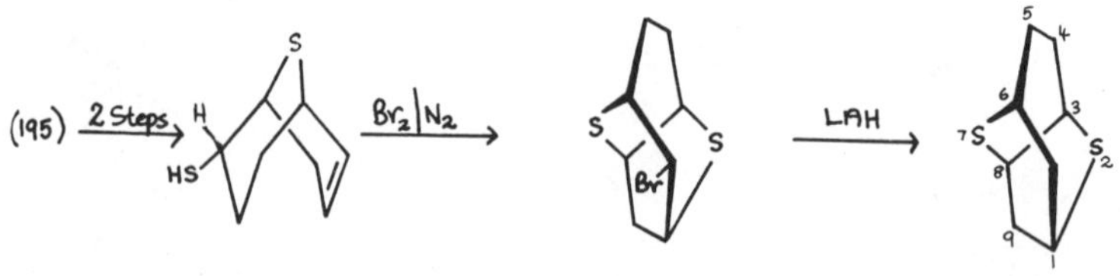

(29)

The dichloro compound **195** is a very useful intermediate for the preparations of 2,7-dithiatwistanes and isotwistanes, which were explored by Ganter and co-workers. One of the halogen atoms is eliminated as HCl, and the 6-chlorothianon-2-ene formed is then converted into the 6-thiol. When this thiol is brominated, transannular addition of sulfur occurs across the double bond, and 10-bromo-2,7-dithiaisotwistane [10-bromo-2,7-dithiatricyclo[4.3.1.$0^{3,8}$]decane] is formed (Equation 29).[463] 2-Oxa-7-thiatwistanes and their 2-oxides have been prepared by related reactions.[464–466] (compare with the aza- and oxatwistanes in Part I, Chapter 3, Section X.C, and Chapter 2, Section XI.D, respectively, and mixed heterotwistanes in Chapter 4, Section II.E and IV.D).

A new conformational effect was observed in some sulfur derivatives of bicyclo[3.3.1]nonanes by Zefirov and Rogozina. In the parent hydrocarbon, the 3,7-aza- and oxa-substituted derivatives, the physical data show that the system exists in the chair-chair conformation as in the 9-thiaderivative **195.** When sulfur atoms, however, are introduced at positions 3 and 7 considerable repulsion occurs between the sulfur atoms, and the ring system acquires the boat-chair conformation (**200,** X$=$$CH_2$ or S). Clearly there is strong repulsion between the 3*d* orbitals of the sulfur atoms in order to force the system into this conformation[467] (see Chapter 4, Section IV.C.3).

200 **201** **202**

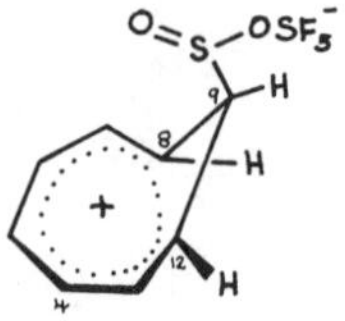

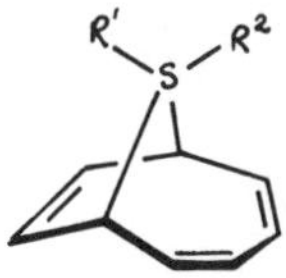

Several examples of electrophilic additions to cyclooctatetraene[468,469] are known, and with one equivalent of SbF_5 in liquid SO_2 the *endo*-8-substituted homotropilium ion **201** is formed. This ion rearranges into 9-thiabicyclo[4.2.1]nona-2,4,7-triene (**202,** $R^1 = R^2 = O$).[470] Paquette and co-workers reexamined this reaction and reported that, in addition to the triene (**202**), 9-thiabarbaralane 9,9-dioxide [9-thiatricyclo[3.3.1.$0^{2,8}$]nona-3,6-diene 9,9-dioxide, **203,** R=O] was formed in an unprecedented 1,5-cycloaddition.[471] 9-Thiabicyclo[4.2.1]nona-2,4,7-triene 9-oxide (**202,** R^1=lone pair and R^2=O; or R^1=O, R^2=lone pair) was produced by a cycloaddition reaction between cyclooctatetraene and thiirane 1-oxide in xylene at 140°C. The parent compound was obtained by reduction of this sulfoxide with LAH, and, on photolysis, gave 9-thiabicyclo[6.1.0]nona-2,4,6-triene (cyclooctatetraene monoepisulfide).[472] Irradiation of this thiirane in the presence of a sensitizer gave 9-thiabarbaralane (**203,** R=lone pair).[473] Anastassiou and collaborators[474] studied the fluxional behavior of 9-heterobarbaralanes and listed the free energies for the interconversions, which may go through a π-hybrid such as **204.** They listed the respective $\Delta G^{\ddagger}$ values of **204,** X=NH, NCN, S, and SO_2 as 30.1, 20.1, 40.6, and 39.8 kJ mol^{-1}. They also concluded that the bridging unit had a small influence on the Cope rearrangement and that the free energy of the rearrangement is mainly a function of the distance between the bridgehead carbon atoms C–1 and C–5.

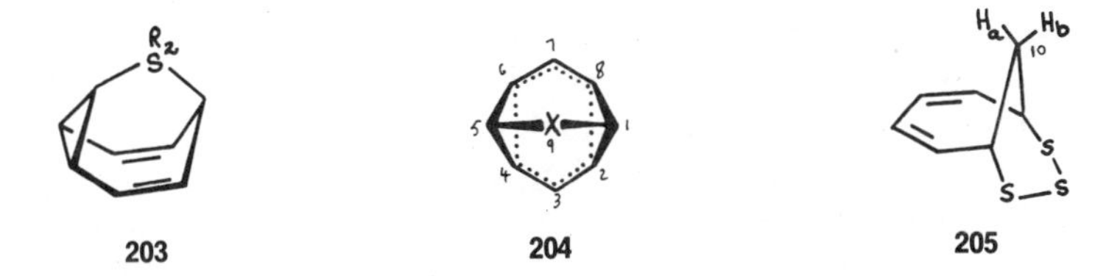

203 **204** **205**

Cycloheptatriene reacts with sulfur to yield 7,8,9-trithiabicyclo[4.3.1]deca-2,4-diene (**205**) in sulfolane at 70°C, and is catalyzed by pyridine.[475] It was analyzed by ^{13}C and ^{1}H nmr spectroscopy, and the geminal protons at C–10 differ by 0.7 ppm in chemical shift.

206

Much interest has been shown in sulfur derivatives of adamantane,[475–482] and theoretical studies using the 5,7-dicarbanion of 1,3-dimethyl-2,4,6,8-tetrathiaadamantane (**206,** R=H) as a rigid example revealed that the sulfur 3*d* orbitals are not necessarily involved in the structure of dimethyl sulfide dicarbanion.[483] The tetrathiaadamantane (**206,** R = H) is dissymmetric and was obtained in its optically active forms by Hedblom and Olsson.[484] They converted the racemic thiaadamantane into its lithium salt (**206,** R = Li) which gave (±)-1-carboxy-3,5-dimethyl-2,4,6,8-tetrathiaadamantane (**206,** R = CO_2H) on treatment with carbon dioxide. The acid was resolved by recrystallization of the cinchonidine salt to give the (+) acid, and the strychinine salt to give the (−) acid. Decarboxylation of these acids, in quinoline catalyzed by Cu, provided (+)- and (−)-1,3-dimethyl-2,4,6,8-tetrathiaadamantane without change in the sign of optical rotation.

XIV. MISCELLANEOUS

A. Thioxanthenes and Thianthrenes

Thioxanthene is not a planar molecule and the central ring is boat shaped. Molar Kerr constants and dipole moment measurements indicated that the molecule is folded to about 135°.[485] Ternay and Evans[486] studied the ^1H nmr spectra of several 9-alkyl-substituted derivatives and found that the 9-alkyl groups prefer a pseudo-axial conformation (**207**). The system is conformationally flexible, and in the case of the 9-*t*-butyl derivative the central ring is rather flattened. The conformational mobility of 9,9-dimethyl-10-phenylthioxanthylium perchlorate (**208**) was examined by variable temperature nmr, and the methyl signals were found to coalesce at about 200°C, indicating a $\Delta G^{\ddagger}$ for the pyramidal sulfonium cation inversion of approximately 106.3 kJ mol^{-1}. This inversion barrier is high enough for it to be resolvable into its optical antipodes.[487] 9-Substituted thioxanthenes have been oxidized to the 10-oxides in which the oxygen on the sulfur atoms can be cis or trans with respect to the 9-substituent. This substituent has a strong directive influence on the oxidizing agents used, and the orientation is also affected by the nature of the solvent. Thus oxidation of 9-hydroxythioxanthene with *m*-chloroperbenzoic acid gives a mixture of 10-oxides, whereas H_2O_2 in aqueous acetone produces only the

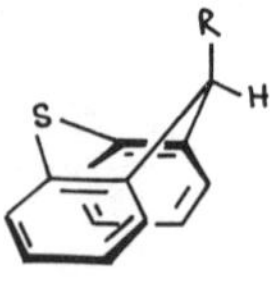

207 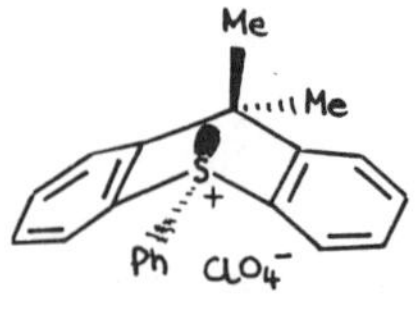 **208** 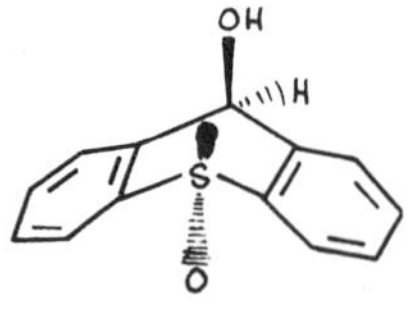 **209**

cis (α) isomer. 9-Chlorothioxanthene is oxidized with the former reagent in acetone to yield predominantly the trans (β) isomer.[488] Reduction of the 9-keto-10-oxide with excess $NaBH_4$, on the other hand, provides mainly *cis*-9-hydroxythioxanthene 10-oxide.[489] The structure of *trans*-9-hydroxythioxanthene 10-oxide (**209**) was confirmed by an X-ray crystallographic analysis, and the 10-oxygen atom is oriented pseudo-equatorially.[488] Several 9-hydroxy-, 9-trimethylsilyloxy-, and 9-acetoxythioxanthenes, including some 2-chloro derivatives of these, were oxidized, and their *cis*- and *trans*-10-oxides were separated and identified by uv and nmr spectroscopy. Generally, it was found that the 10-oxide group orients itself pseudo-equatorially.[490] The two racemates of 9-(*N*-methylpiperidin-3-yl)-thioxanthene 10-oxides were separated by solvent distribution between benzene–hexane, and aqueous buffers from which their *p*K*a* values could be predicted.[491]

Bennett and Glasstone,[492] in 1934, found that thianthrene has a dipole moment and, like thioxanthene above, is therefore not planar. Attempts to resolve 2-trimethylammoniothianthrene iodide, however, failed suggesting that the molecule may racemize very rapidly, and the free energy involved was calculated at approximately 29.3 kJ mol^{-1}.[493] LCAO-MO calculations by Chandra[494] predicted an energy barrier of 25–29 kJ mol^{-1} for the inversion process in thianthrene. Le Fèvre and co-workers[495] measured the dipole moments and molar Kerr constants of thianthrene, its 5,10-dioxide, and its 5,5,10,10-tetroxide in benzene solution and concluded that they existed as flexible folded molecules, but the angle of fold in each case was greater than reported from X-ray analyses of the crystals. Thianthrene 5,10-dioxide was obtained in two geometrical isomers, the cis and trans

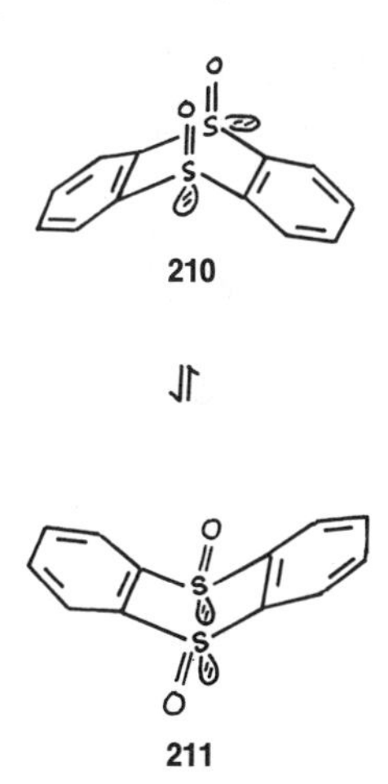

forms, which could be equilibrated in the presence of charcoal and give the same mixture if each was kept above its melting point for some time.[496] The dipole moments of the 5,10-dioxides have been measured and present one of the few cases in which the cis isomer has greater symmetry than the trans isomer, as shown by the smaller electric moment and higher melting point.[497] The cis isomer exists in two different conformations, **210** and **211,** whereas in the trans isomer the two conformations are identical unless one of the benzene rings bears a substituent. Purcell and Berschied[498] studied the temperature dependence of the ^1H nmr spectrum of *cis*-thianthrene 5,10-dioxide in $CHCl_3$ and found that the inversion rate was faster than 100 sec^{-1} at 40°C. This corresponds to an inversion and rotation barrier of around 4.2 kJ mol^{-1}. At 160°C the spectra indicated a 1 : 1 mixture of cis,syn (**211**) and cis,anti isomers (**210**).

Mislow and Ternay[499] prepared several stereoisomeric thianthrene 5,10-dioxides and assigned their configurations by uv and ir spectroscopy, which were diagnostic for the *cis*- and *trans*-5,10-dioxides. The dioxides undergo thermal stereomutation (isomerization), and they found that generally at elevated temperatures (200–300°C) all the trans isomers studied, for example, 2,7-dimethyl-, 2,7-dimethoxycarbonylmethyl-, 2-bromo-, and 2,7-dichlorothianthrene 5,10-dioxides, were converted into the cis isomers. Jancziwski and Charmas[500] had resolved *cis*- and *trans*-2-carboxythianthrene 5,10-dioxides into their optical antipodes and claimed that these were resistent to racemization, and that oxidation of the cis(+) and trans(+) isomers gave the same optically active (−)-2-carboxythianthrene 5,5,10,10-tetroxide. They stated that, similarly, the *cis*(−)- and *trans*(−)-dioxides gave the same (−)-tetroxide. Chikos and Mislow[501] repeated this work, and, although the optical resolutions proceeded as prescribed, they found that when the dioxides were oxidized only racemic tetroxides could be isolated, and they refuted the earlier statements. Also, on heating, the *cis*(−)-dioxide melted at 278–280°C and resolidified and then melted again at 296–298°. The latter melting point was that of the racemate, and the melted product was devoid of optical activity. The trans(+) isomer behaved similarly, and these results were consistent with the expected conformational mobility of the thianthrene system.

B. Chiral Thienyls

Gronowitz[502] discussed the optical activity of several thienyl compounds with a chiral center on the substituent, in a review on substituent effects in thiophenes. Absolute configurations were deduced by the Fredga quasi-racemate method[503] and by desulfurization and chemical correlation. The optical rotations of 2-thienyl-, 3-thienyl-, and phenylglycollic acids, succinic acids, and methyl succinic acids were of similar sign for the same absolute configuration. The most interesting resolution in this class of compounds involves the selective hydrolysis of an *N*-acyl group of a thienyl α-amino acid with the natural configuration by a beef pancreas acylase. Thus enzymatic hydrolysis of *R,S,N*-chloroacetyl(thien-2-yl)alanine gave a mixture of *S*-thien-2-ylalanine and *R,N*-chloroacetyl(thien-2-yl)alanine, which are readily separable.[504]

The study of molecular asymmetry of the biphenyl type was extended to thiophenes because in these examples the five-membered ring alters the bond distances between *o* and *o′* positions and gives valuable information regarding nonbonded intramolecular interactions between *o* and *o′* substituents. Rotation about the pivot bond in 2-(6-methyl-2-nitrophenyl)-3-carboxythiophene (**212**) was restricted enough for Owen and Nord[505] to resolve the acid by way of its brucine salt, but the optically active species are not very stable. The brucine salt exhibited mutarotation, and the isolated free acid racemized readily at room temperature. By placing substituents in all the ortho positions, as in 3-carboxy-2,5-dimethyl-4-(6-methyl-2-nitrophenyl)thiophene, Jean[506] obtained a relatively more stable acid which retained its optical activity at room temperature in MeOH and $CHCl_3$ for at least one week and in acetic acid for at least 24 hr, but at the boiling point of these solvents (65, 61, and 118° respectively) it racemized with $t_{1/2}$ values of 24.4, 29.6, and 2.9 min. respectively.

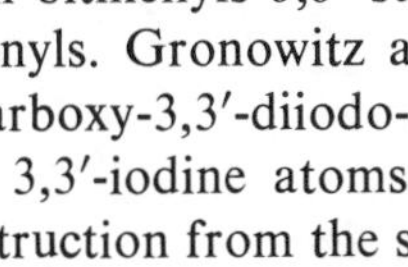

NO_2 CO_2H S Me
212

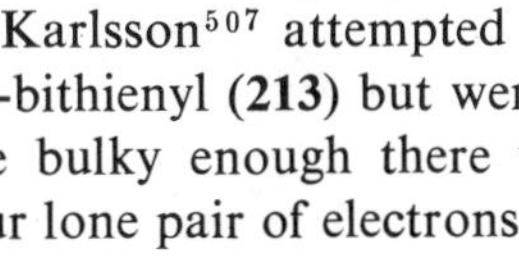

I I S S CO_2H
213

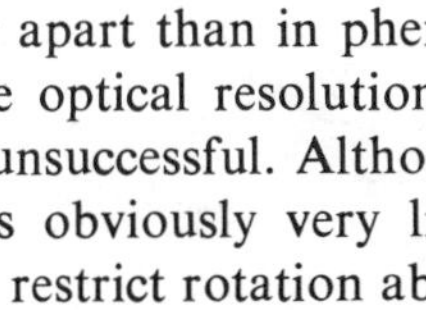

CO_2H HO_2C S S NO_2 O_2N
214

In bithienyls *o,o′* substituents are slightly further apart than in phenylthienyls. Gronowitz and Karlsson[507] attempted the optical resolution of 5-carboxy-3,3′-diiodo-2,2′-bithienyl (**213**) but were unsuccessful. Although the 3,3′-iodine atoms are bulky enough there was obviously very little obstruction from the sulfur lone pair of electrons to restrict rotation about the 2,2′ bond. These authors were unable to observe mutarotation in several salts of optically active amines and this racemic acid. Gronowitz and co-workers, however, obtained several 3,3′-bithienyls with large substituents in the 2,2′- and 4,4′-positions in optically active forms. These include 2,2′-dicarboxy-4,4′-dibromo-5,5′-dimethyl-,[508] 4,4′- dicarboxy-2,2′,-5,5′-tetramethyl-,[509,510] 2,2′-dicarboxy-4,4′-dinitro-,[511] 4,4′-dicarboxy-2,2′-dinitro-,[512] and 4,4′-dicarboxy-2,2′,5,5′-tetraethyl-3,3′-bithienyls.[510] Håkansson and Wiklund[513] resolved 2,2′,4,4′-tetrabromo-5,5′-dicarboxy-3,3′-bithienyl into its optically active forms and caused it to decarboxylate, with mercuric acetate in acetic acid, into 2,2′,4,4′-tetrabromo-3,3′-bithienyl without racemization. 2,2′-Dicarboxy-4,4′-dibromo-5,5′-dimethyl-3,3′-bithienyl was also optically stable, whereas 2,2′,5,5′-tetramethyl(and tetraethyl)-4,4′-dicarboxy-3,3′-bithienyls racemized above 100°C ($t_{1/2}$ for the tetramethyl compound is 19 min at 100°C) with a $\Delta G^{\ddagger}$ value of 116 (133.5) kJ mol^{-1}. This is to be compared with 5,5′,6,6′-tetramethyl-2,2′-dicarboxybiphenyl which could not be racemized under these conditions.[510] The absolute configuration *R*(+) for 4,4′-dicarboxy-2,2′-dinitro-3,3′-bithienyl (**214**) was deduced by the quasi-racemate method and ord comparison of the dimethyl ester with *R*(+)-2,2′-dinitro-6,6′-dimethoxycarbonyl-1,1′-biphenyl.[512] Similarly the absolute configuration *R*(+) for 2,2′-dicarboxy-4,4′-dinitro-3,3′-bithienyl was correlated with that of 2,2′-dinitro-6,6′-dicarboxy-1,1′-biphenyl and by comparison of the cd curves.[511] Caution, however, must be exercised in ord correlations because the spectra of bithienyl dicarboxylic acids can exhibit anomalous solvent effects that may be partly related to the ionization of the acidic groups.[510]

C. Chiral Sulfur Compounds Derived from 2,2′-Disubstituted Biphenyls

Two groups of biphenyl derivatives are discussed under this heading. The first group comprises diphenyls in which the sulfur atom is one carbon atom removed from the benzene ring (e.g., dibenzothiepins), and in the second group the sulfur atoms are directly bonded to the benzene rings (e.g., dibenzodithiins).

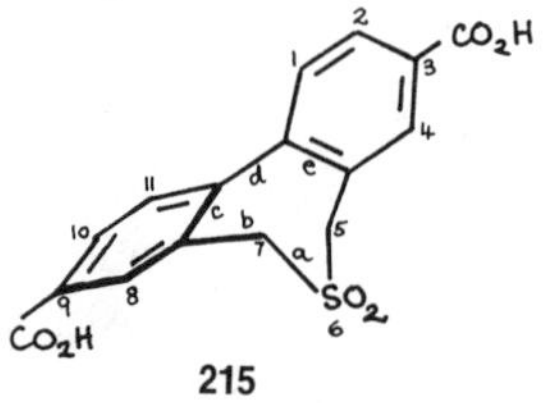

215 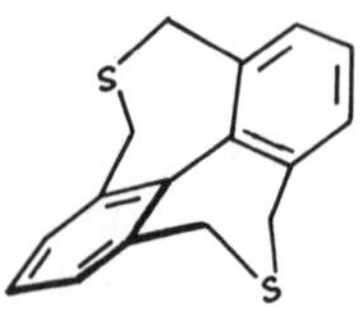 **216** 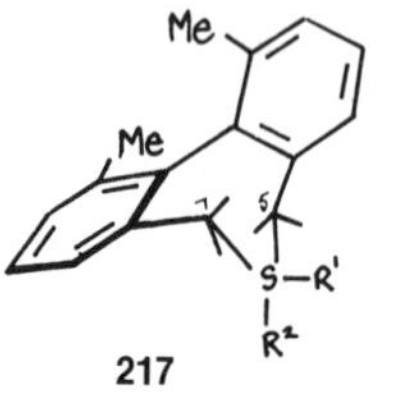 **217**

Truce and Emrick attempted to resolve the 6-methyl-5,7-dihydrodibenzo-[*c.e*]thiepinium cation without success, but realized the resolution of 3,9-dicarboxy-5,7-dihydrodibenzo[*c.e*]thiepin 6,6-dioxide (**215**) by recrystallization of the cinchonidine salt.[514] The conformational mobility of 5,7-dihydrodibenzo[*c.e*]thiepins was demonstrated by Sutherland and Ramsay[515] by using variable temperature ^{1}H nmr spectroscopy. They obtained signal coalescences for the parent compound and for its 6,6-dioxide, and calculated the free energy for the inversion process in pyridine at 64.9 (70 in $CDCl_3$) and 76.2 kJ mol^{-1} respectively. Mislow and his school[516] examined the chiroptical, optical, and nmr properties of several doubly bridged biphenyls including the dithiepin **216** that is an optically active example of the point group D_2. The nmr and uv data revealed that in the singly bridged compounds of the type **215** the benzene rings have the largest twist, whereas they have the smallest twist in the doubly bridged biphenyls such as **216.** The dibenzodithiepin [**216,** 10,12-dihydro-4*H*,6*H*-5,11-dithiadibenzo[*ef.kl*]heptalene] is particularly interesting because it is readily formed by disproportionation of two molecules of 1,11-bisbromomethyl-5,7-dihydrodibenzo[*c.e*]thiepin (bimolecular)—the other product being chiral 2,2′,6,6′-tetrakisbromomethylbiphenyl.

The rigid sulfones like 1,11-dimethyl-5,7-dihydrodibenzo[*c.e*]thiepin 6,6-dioxide (**217,** $R^1=R^2=O$) are useful in evaluating the relative H-D exchange rates of the protons α to the sulfonyl group. The protons on C–5 and C–7 in the sulfone (**217,** $R^1=R^2=H$) were assigned by ^1H nmr spectroscopy, and the rates of base-catalyzed exchange with deuterium were evaluated. These rates were found to be in qualitative agreement with MO theory, which predicts that the carbanion **218** is more stable than the

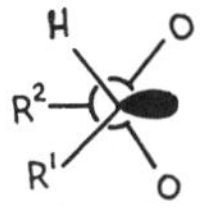

218 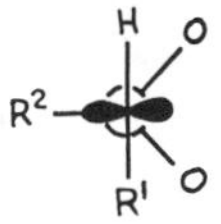 **219**

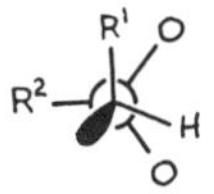

220 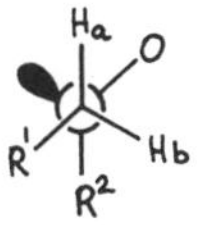 **221**

carbanions **219** (by 10.5 kJ mol^{-1}) and **220** (by 17.2 kJ mol^{-1}).[517] Similar studies were made with 1,11-dimethyl-5,7-dihydrodibenzo[*c.e*]thiepin 6-oxide (**217,** $R^1=O$, R^2=lone pair) in which theory predicts that the more stable carbanion will have the electron pair (formed by ionization of H_a in **221**) bisecting the projected angle between the lone pair and oxygen of the adjacent SO group. Relative H-D exchange rates of the order of 1:1250 were observed by Fraser and Schuber,[518] but solvent effects can obscure the issue.[519] $Eu(dpm)_3$ induced chemical shifts of the protons in the 6-oxide **217** ($R^1=O$, R^2=lone pair) were used to assign the configuration of the sulfoxide function.[520]

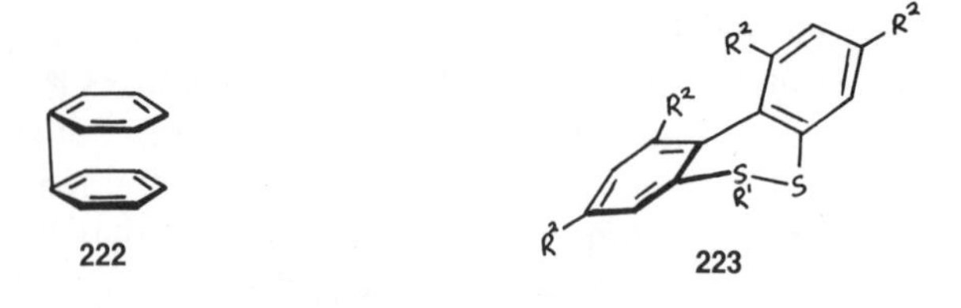

Among the important experimental evidence which refuted the Kaufler formula (**222**) for biphenyl is the study of the oxidation of 2,2′-, 3,3′-, and 4,4′-dimercaptobiphenyls by Barber and Smiles.[521] They found that whereas the 3,3′- and 4,4′-dimercapto compounds gave polymeric materials on oxidation, the 2,2′ isomer gave the monomeric dibenzodithiin (**223**, R^1=lone pairs, R^2=H) in high yields. This dithiin could not be obtained in optically active form because the material used for its preparation, biphenyl-2,2′-disulfonic acid, racemized readily at 100°C.[522] Armarego and Turner[523] resolved 4,4′,6,6′-tetramethylbiphenyl-2,2′-disulfonic acid into its very stable (optical) antipodes and converted these into their respective 2,2′-disulfonyl chlorides and 2,2′-dithiols. Oxidation of the dithiols gave a polymer instead of the expected dithiin (**223**, R^1=lone pairs, R^2=Me). The 2,2′-disulfonyl chloride, however, was smoothly reduced with sodium bisulfite into the cyclic thiolsulfonate, 1,2,8,10-tetramethyldibenzo[*c.e*]dithiin 5,5-dioxide (**223**, $R^1=O_2$, R^2=Me) in high yield. The inability of the 2,2′-dithiol to give the dithiin compared with the dithiin dioxide was attributed to the smaller size of the sulfur atom in –S– compared with $-SO_2-$. Hence the 6,6′-methyl groups were large enough in the least twisted molecule (i.e., almost touching each other) to keep the two SH groups far enough apart to cause them to react inter- rather than intramolecularly. The case is quite different in the formation of the slightly larger $-SO_2-S-$ linkage. The cyclic thiolsulfonate (**223**, $R^1=O_2$, R^2=Me) racemized with $t_{1/2}$ values of 24 and 291 min in boiling ethylbenzene and toluene respectively (E_a= 129.7 kJ mol^{-1}). The precursors to this compound could not be racemized without chemical destruction, and the relatively facile racemization of the cyclic thiolsulfonate must involve bending of the benzene rings towards each other followed by a twisting motion to allow the 2,2′- and 6,6′-substituents to cross over each other.[524] The reduction of 2,2′,4,4′-tetramethylbiphenyl-6,6′-disulfonyl chloride to the thiolsulfonate **223** (R^1= O_2, R^2=Me) is accompanied by a large change in optical rotation with inversion of sign. Mislow and his collaborators[525] applied their optical displacement rule (see Part I, Chapter 4, Section II.A) to this cyclic thiolsulfonate and deduced the *S*(+) absolute configuration for it and then the absolute configuration for the whole series.

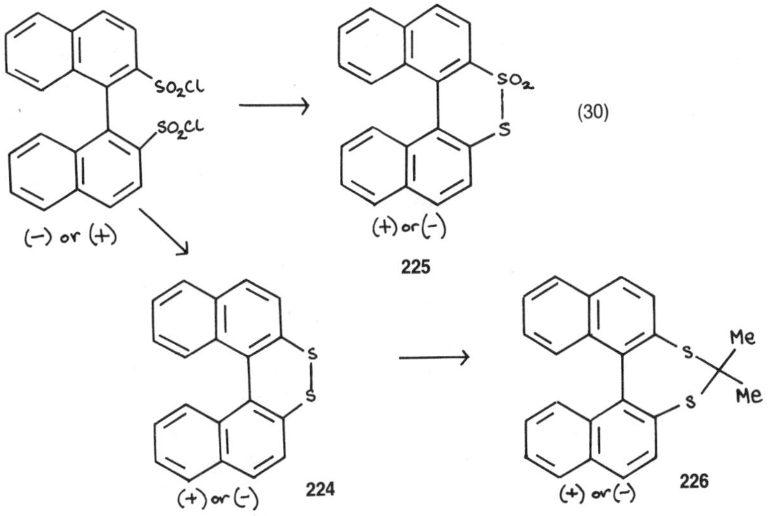

(30)

224 225 226

Armarego and Turner[526] also resolved 1,1′-dinaphthyl-2,2′-disulfonic acid and converted its optical antipodes into dinaphtho[2.1-*c*.1.′2′-*e*]dithiin (**224**), its 3,3-dioxide **225,** and the dinaphthodithiepin **226** (Equation 30). Mislow and co-workers[525] applied their optical displacement rule to these compounds also and deduced their absolute configurations (e.g., *R*(−) for **224** and **225**). The dinaphthodithiin **224** exhibited very high optical rotations at short wavelengths (anomalous dispersion with two maxima) as would be expected from the helical nature of the molecule. Djerassi and collaborators reported its ord spectrum and attributed the inflection at 400 nm in the uv spectrum to the chiral S–S bond.[527] The extrusion of one of the sulfur atoms from dibenzo[*c*.*e*]dithiins by copper bronze to form dibenzothiophenes occurs without rotation of the benzene rings. This was demonstrated in the desulfurization of 2,9-dimethyldibenzo[*c*.*e*]dithiin which gave 2,8-dimethyldibenzothiophene—not the isomeric 2,6-dimethyldibenzothiophene. The latter would have been formed by extrusion of one sulfur atom, rotation of one of the benzene rings, and cyclization of the remaining sulfur atom on to the second ortho position of this benzene ring.[528]

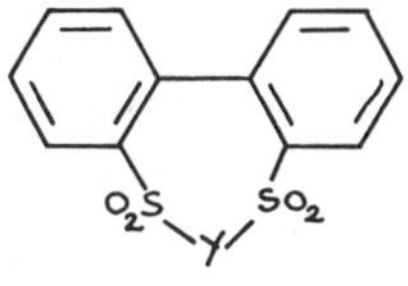

227

Braunton, Millar, and Tebby[529] examined the uv spectra of biphenyls related to the dithiepin tetroxides such as **227.** The spectra revealed that the dihedral angle between the benzene rings depends not only on the length of the 2,2′-bridge but also on the nature of the functional groups in it. They found that if the bridge –Y– contains four or five carbon atoms then the repulsive forces between the two sulfone groups causes them to assume a *transoid* conformation and to inhibit conjugation between the two benzene rings.

D. Heterohelicenes Derived from Thiophenes

Helicenes are polycyclic aromatic helical compounds that are chiral, and they were reviewed by Martin[530] in 1974. Wynberg and Groen[531] have synthesized several heterohelicenes containing thiophene and benzene by uv irradiation of thiophenes bridged by an ethylenic linkage (**228**), as in

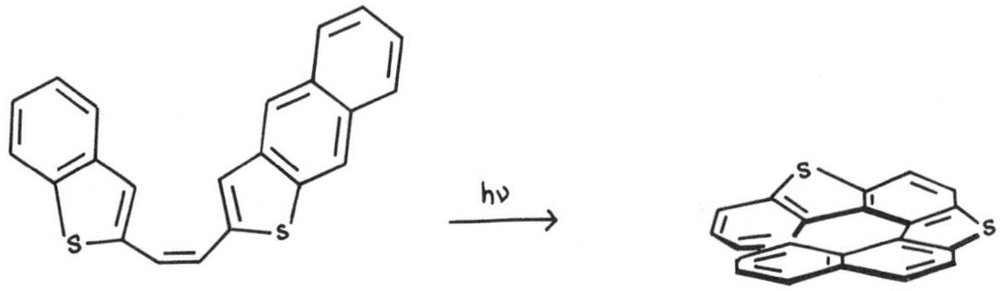

228 229

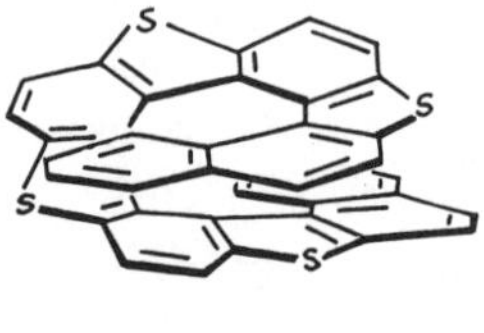

230 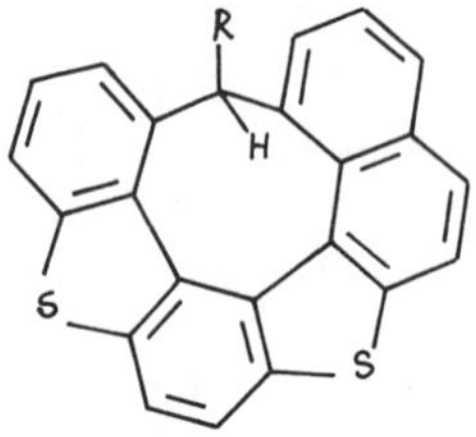 231

the formation of benzo[*d*]naphtho[1.2-*d*′]benzo[1.2-*b*:4.3-*b*′]dithiophene (**229**). Several of these were found to undergo spontaneous resolution, and, because of their helical and highly conjugated nature, they have extremely high optical rotations. The heterohexahelicene (six rings, **229**), has $[\alpha]^{20}_{546}+$ 2050°,[531] and its helical nature was confirmed by an X-ray crystal study.[532] By the flicking of the terminal aromatic rings over each other heterohexahelicenes are capable of racemization with $\Delta G^{\ddagger}$ values of the order of 100 kJ mol^{-1}.[533] Wynberg and Groen[534] have also succeeded in synthesizing the heteroundecahelicene **230** [bis(naphtho[2.1-*b*]thieno)[1.2-*e*:1′.2′-*e*]benzo-[1.2-*b*:4.3-*b*′]bis[1]benzothiophene] in optically active form, with $[\phi]_{578}^{25}$ 17,000, by a combination of a Wittig reaction followed by uv irradiation of the olefin in the presence of iodine. Methano-bridged heterohexahelicenes (**231**) were prepared by Numan and Wynberg and, no doubt, will be obtained in optically active forms because the molecules are not planar.[535]

E. Thiapropellanes

Aza- and oxapropellanes were described by Ginsburg,[536] who extended the work to include thiapropellanes (see Part I, Chapter 4, Section IX; and Chapter 2, Section XII.C.4). Lantos and Ginsburg[537] prepared the propellane **232** and its *S*,*S*-dimethyl dication, which underwent a Stevens rearrangement to give 1,4-bis-*exo*-methylene-2,3-bismethylthiocyclohexane.

S S

232

2 11 10 3 1 9 S12 4 6 8 5 7 13

233

S S $(CH_2)_n$

234

R R R R $(CH_2)_n$ R= CH_2OMes

235

Paquette and his school[538] reported the synthesis of 12-thia[4.4.3]popell-3-ene (**233**) from the reaction of *cis*-5,10-bismethanesulfonyloxymethyl-Δ^2-octalin and Na_2S, in HMPA, in very high yield. From this intermediate they prepared the corresponding 12-thia[4.4.3]propella-2,4-diene and 12-thia[4.4.3]propella-2,4,7,9-tetraene. Weinges and co-workers[539] prepared propellanes with two rings containing sulfur atoms, as in compounds 8,11-dithia[4.3.3]propellane (**234,** $n=4$) and 9,12-dithia[5.3.3]propellane (**234,** $n=5$) from the respective 1,1,2,2-tetrakismethanesulfonyloxymethyl-cycloalkanes, **235** ($n=4$) and **235** ($n=5$), by reaction with Na_2S/DMSO. They also reported the related oxathiapropellanes.

XV. REFERENCES

1. W. G. Salmond, *Quart. Rev.*, **22,** 253 (1968).
2. J. B. Lambert, *Topics Stereochem.*, **6,** 19 (1971).
3. K. Mislow, *Rec. Chem. Progr.*, **28,** 217 (1967).
4. R. Rahman, S. Safe, and A. Taylor, *Quart. Rev.*, **24,** 108 (1970).
5. W. L. F. Armarego in *Physical Methods in Heterocyclic Chemistry*, Vol. III, A. R. Katritzky, Academic, New York, 1971, Ch. 4, pp. 67–222.

6. F. G. Riddell, *Quart. Rev.*, **21,** 364 (1967).
7. S. Gronowitz, *Adv. Heterocyclic Chem.*, **1,** 1 (1963).
8. R. Connor in *Organic Chemistry*, Vol. I, H. Gilman, Ed., Wiley, New York, 1950, Ch. 10, pp. 835–943.
9. F. Challenger, *Aspects of the Organic Chemistry of Sulphur*, Butterworths, London, 1959.
10. E. L. Eliel, N. L. Allinger, S. J. Angyal, and G. A. Morrison, *Conformational Anaylsis*, Wiley, New York, 1965.
11. C. M. Suter, *Organic Chemistry of Sulphur*, Wiley, New York, 1944.
12. D. D. Reynolds and D. L. Fields, in *Heterocyclic Compounds with Three- and Four-Membered Rings, Part One*, A. Weissberger, Ed., Interscience, New York, 1964, Ch. III, p. 576.
13. N. Spassky and P. Sigwalt, *Tetrahedron Letters*, 3541 (1968).
14. M. G. Ettlinger, *J. Amer. Chem. Soc.*, **72,** 4792 (1950).
15. E. E. van Tamelen, *J. Amer. Chem. Soc.*, **73,** 3444 (1951).
16. C. C. Price and P. F. Kirk, *J. Amer. Chem. Soc.*, **75,** 2396 (1953).
16a. K. Jankowski and R. Harvey, *Canad. J. Chem.*, **50,** 3930 (1972).
17. J. S. Harding and L. N. Owen, *J. Chem. Soc.*, 1528 (1954).
18. R. M. Kellogg, S. Wassenaar, and J. Buter, *Tetrahedron Letters*, 4689 (1970).
19. P. Raynolds, S. Zonnebelt, S. Bakker, and R. M. Kellogg, *J. Amer. Chem. Soc.*, **96,** 3146 (1974).
20. K. S. Sidhu, E. M. Lown, O. P. Starusz, and H. E. Gunning, *J. Amer. Chem. Soc.*, **88,** 254 (1966).
21. E. Block and E. J. Corey, *J. Org. Chem.*, **34,** 896 (1969).
22. D. C. Dittmer and G. C. Levy, *J. Org. Chem.*, **30,** 636 (1965).
23. K. Kondo and A. Negishi, *Tetrahedron*, **27,** 4821 (1971).
24. B. F. Bonini and G. Maccagnani, *Tetrahedron Letters*, 3585 (1973).
25. L. Thijs, A. Wagenaar, E. M. M. van Rens, and B. Zwanenburg, *Tetrahedron Letters*, 3589 (1973).
26. F. G. Bordwell, J. M. Williams, Jr., E. B. Hoyt, Jr., and B. B. Jarvis, *J. Amer. Chem. Soc.*, **90,** 429 (1968).
27. N. Tokura, T. Nagai, and S. Matsumura, *J. Org. Chem.*, **31,** 349 (1966).
28. L. A. Carpino and L. V. McAdams III, *J. Amer. Chem. Soc.*, **87,** 5804 (1965).
29. L. A. Carpino, L. V. McAdams III, R. H. Rynbrandt, and J. W. Spiewak, *J. Amer. Chem. Soc.*, **93,** 476 (1971).
30. M-M. Rohmer and B. Roos, *J. Amer. Chem. Soc.*, **97,** 2025 (1975).
31. R. Hoffmann, H. Fujimoto, J. R. Swenson, and C-C. Wan, *J. Amer. Chem. Soc.*, **95,** 7644 (1974).
32. T. J. Batterham, *N.M.R. Spectra of Simple Heterocycles*, E. C. Taylor and A. Weissberger, Eds., Wiley-Interscience, 1973, pp. 423–461.
33. G. Gottarelli and B. Samori, *J.C.S. Chem. Comm.*, 398 (1974).
34. I. Moretti, G. Torre, and G. Gottarrelli, *Tetrahedron Letters*, 4301 (1971).
35. M. Ohtsuru, K. Tori, and M. Fukuyama, *Tetrahedron Letters*, 2877 (1970).
36. G. K. Helmkamp, D. J. Pettitt, J. R. Lowell, Jr., W. R. Mabey, and R. G. Wolcott, *J. Amer. Chem. Soc.*, **88,** 1030 (1966).

37. J. R. Lowell, Jr. and G. K. Helmkamp, *J. Amer. Chem. Soc.*, **88,** 768 (1966).
38. S. M. Iqbal and L. N. Owen, *J. Chem. Soc.*, 1030 (1960).
39. G. K. Helmkamp and D. J. Pettitt, *J. Org. Chem.*, **29,** 3258 (1964).
40. B. M. Trost and S. Ziman, *Chem. Comm.*, 181 (1969).
41. Y. Hata, M. Watanabe, S. Inoue, and S. Oae, *J. Amer. Chem. Soc.*, **97,** 2553 (1975).
42. N. P. Neureiter and F. G. Bordwell, *J. Amer. Chem. Soc.*, **81,** 578 (1959).
43. D. B. Denney and M. J. Boskin, *J. Amer. Chem. Soc.*, **82,** 4736 (1960).
44. C. C. J. Culvenor, W. Davies, and N. S. Heath, *J. Chem. Soc.*, 282 (1949).
45. K. Kondo, M. Matsumoto, and A. Negishi, *Tetrahedron Letters*, 2131 (1972).
46. G. E. Hartzell and J. N. Paige, *J. Org. Chem.*, **32,** 459 (1967).
47. R. B. Woodward and R. Hoffmann, *Angew. Chem. Internat. Edn.*, **8,** 781 (1969).
48. J. E. Baldwin, G. Höfle, and S. C. Choi, *J. Amer. Chem. Soc.*, **93,** 2810 (1971).
49. K. Kondo, A. Negishi, and I. Ojima, *J. Amer. Chem. Soc.*, **94,** 5786 (1972).
50. S. Matsumura, T. Nagai, and N. Tokura, *Bull. Soc. Chem., Japan*, **41,** 2672 (1968).
51. F. G. Bordwell, B. B. Jarvis, and P. W. R. Corfield, *J. Amer. Chem. Soc.*, **90,** 5298 (1968).
52. Y. Etienne, R. Soulas, and H. Lumbroso in *Heterocyclic Compounds with Three- and Four-Membered Rings, Part Two*, A. Weissberger, Ed., Interscience, New York, 1964, Ch. V. p. 647.
53. L. A. Paquette and J. P. Freeman, *J. Amer. Chem. Soc.*, **91,** 7548 (1969).
54. L. A. Paquette and J. P. Freeman, *J. Org. Chem.*, **35,** 2249 (1970).
55. R. M. Dodson, E. H. Jancis, and G. Klose, *J. Org. Chem.*, **35,** 2520 (1970).
56. B. M. Trost, W. L. Schinski, F. Chen, and I. B. Mantz, *J. Amer. Chem. Soc.*, **93,** 676 (1971).
57. A. Ohno, Y. Ohnishi, and G. Tsuchihashi, *J. Amer. Chem. Soc.*, **91,** 5038 (1969).
58. A. Padwa and R. Gruber, *J. Org. Chem.*, **35,** 1781 (1970).
59. W. Wucherpfennig, *Tetrahedron Letters*, 765 (1970).
60. J. H. Barlow, C. R. Hall, D. R. Russell, and D. J. H. Smith, *J.C.S. Chem. Comm.*, 133 (1975).
61. S. Allenmark, *Arkiv Kemi*, **26,** 73 (1967).
62. W. O. Siegl and C. R. Johnson, *Tetrahedron*, **27,** 341 (1971).
63. W. O. Siegl and C. R. Johnson, *J. Org. Chem.*, **35,** 3657 (1970).
64. G. Stork and I. J. Borowitz, *J. Amer. Chem. Soc.*, **84,** 313 (1962).
65. D. C. Dittmer and F. A. Davis, *J. Org. Chem.*, **29,** 3131 (1964), *J. Amer. Chem. Soc.*, **87,** 2065 (1965).
66. L. A. Paquette, *J. Org. Chem.*, **29,** 2851 (1964).
67. L. A. Paquette and M. Rosen, *J. Amer. Chem. Soc.*, **89,** 4102 (1967).
68. P. Del Buttero and S. Maiorana, *J.C.S. Perkin I*, 2540 (1973).
69. C. T. Goralski and T. E. Evans, *J. Org. Chem.*, **37,** 2080 (1972).
70. L. A. Paquette, *J. Org. Chem.*, **29,** 2854 (1964).
71. L. A. Paquette, J. P. Freeman, and S. Maiorana, *Tetrahedron*, **27,** 2599 (1971).
72. D. C. Dittmer and R. Glassman, *J. Org. Chem.*, **35,** 999 (1970).
73. L. A. Paquette, R. W. Houser, and M. Rosen, *J. Org. Chem.*, **35,** 905 (1970).
74. L. A. Paquette, *J. Org. Chem.*, **30,** 629 (1965).

75. L. A. Paquette, J. P. Freeman, and M. J. Wyvratt, *J. Amer. Chem. Soc.*, **93,** 3216 (1971).
76. D. O. Harris, H. W. Harrington, A. C. Luntz, and W. D. Gwinn, *J. Chem. Phys.*, **44,** 3467 (1966); H. W. Harrington, *J. Chem. Phys.*, **44,** 3481 (1966).
77. W. D. Keller, T. R. Lusebrink, and D. H. Sederholm, *J. Chem. Phys.*, **44,** 782 (1966).
78. A. d'Annibale, L. Lunazzi, G. Fronza, R. Mondelli, and S. Bradamante, *J.C.S. Perkin II*, 1908 (1973).
79. C. Guimon, D. Liotard, and G. Pfister-Guilouzo, *Canad. J. Chem.*, **53,** 1224 (1975).
80. S. Kumakura, T. Shimozawa, Y. Ohnishi, and A. Ohno, *Tetrahedron*, **27,** 767 (1971).
81. D. W. Scott, H. L. Finke, W. N. Hubbard, J. P. McCullough, C. Katz, M. E. Gross, J. F. Messerly, R. E. Pennington, and G. Waddington, *J. Amer. Chem. Soc.*, **75,** 2795 (1953).
82. S. Abrahamsson and G. Rehnberg, *Acta Chem. Scand.*, **26,** 494 (1972).
83. G. L. Hardgrove, Jr., J. S. Bartholdt, and M. M. Lein, *J. Org. Chem.*, **39,** 246 (1974).
84. P. Del Buttero, S. Maiorana, and M. Trautluft, *J.C.S. Perkin I*, 1411 (1974).
85. C. Cistaro, G. Fronza, R. Mondelli, S. Bradamante, and G. Pagani, *Tetrahedron Letters*, 189 (1973).
86. B. A. Arbuzov, O. N. Nuretdinova, and A. N. Vereshchagin, *Doklady Akad. Nauk S.S.S.R.* (Engl. Trans), **172,** 59 (1967).
87. H. J. Backer and K. J. Keuning, *Rec. Trav. chim.*, **53,** 798 (1934).
88. R. M. Dodson and J. Y. Fan, *J. Org. Chem.*, **36,** 2708 (1971).
89. R. M. Dodson, P. D. Hammen, and J. Y. Fan, *J. Org. Chem.*, **36,** 2703 (1971).
90. E. M. Dodson and P. D. Hammen, *Chem. Comm.*, 1294 (1968).
91. R. M. Dodson, P. D. Hammen, E. H. Jancis, and G. Klose, *J. Org. Chem.*, **36,** 2698 (1971).
92. R. M. Dodson, P. D. Hammen, and R. A. Davis, *J. Org. Chem.*, **36,** 2693 (1971).
93. S. Bradamante, P. Del Buttero, D. Landini, and S. Maiorana, *J.C.S. Perkin II*, 1676 (1974).
94. E. P. Adams, K. N. Ayad, E. P. Doyle, D. O. Holland, W. H. Hunter, J. H. C. Nayler, and A. Queen, *J. Chem. Soc.*, 2665 (1960).
95. M. Sander, *Monatsh.*, **96,** 896 (1965).
96. F. F. Blicke in *Heterocyclic Compounds*, Vol. 1, R. C. Elderfield, Ed., Wiley, New York, 1950, Ch. 5, p. 208.
97. H. D. Hartough, *Thiophene and Its Derivatives*, A. Weissberger, Ed., Interscience, New York, 1952.
98. G. Claeson and H-G. Jonsson, *Arkiv Kemi*, **31,** 83 (1970).
99. K. Mislow and W. C. Meluch, *J. Amer. Chem. Soc.*, **78,** 5920 (1956).
100. J. Trotter and J. A. Hamilton, *Biochemistry*, **5,** 713 (1966).
101. B. R. Baker, M. V. Querry, S. R. Safir, and S. Bernstein, *J. Org. Chem.*, **12,** 138 (1947).
102. B. R. Baker, M. V. Querry, and A. F. Kadish, *J. Org. Chem.*, **13,** 123, (1948); B. R. Baker, M. V. Querry, S. R. Safir, W. L. McEwen, and S. Bernstein, *J. Org. Chem.*, **12,** 174 (1947).
103. B. R. Baker, M. V. Queery, W. L. McEwen, S. Bernstein, S. R. Safir, L. Dorfman, and Y. Subbarow, *J. Org. Chem.*, **12,** 186 (1947).

104. B. R. Baker, W. L. McEwen, and W. N. Kinley, *J. Org. Chem.*, **12**, 322, (1947).

105. D. B. Melville, *J. Biol. Chem.*, **208**, 495 (1954).

106. L. D. Wright, E. L. Cresson, J. Valiant, D. E. Wolf, and K. Folkers, *J. Amer. Chem. Soc.*, **76**, 4163 (1954).

107. T. C. Bruice and S. J. Benkovic, *Bioorganic Mechanisms*, Vol. 2, Benjamin, New York, 1966, p. 380.

108. R. Bentley, *Molecular Asymmetry in Biology*, Academic, New York, 1970, p. 435.

109. I. T. Harrison, R. J. K. Taylor, and J. H. Fried, *Tetrahedron Letters*, 1165 (1975).

110. G. E. Wilson, Jr. and R. Albert, *J. Org. Chem.*, **38**, 2156 (1973).

111. R. F. Curtis, D. M. Jones, and W. A. Thomas, *J. Chem. Soc.* (*C*), 234 (1971).

112. J. Buter, S. Wassenaar, and R. M. Kellogg, *J. Org. Chem.*, **37**, 4045 (1972).

113. T. S. Cantrell, *J. Org. Chem.*, **39**, 2242 (1974).

114. G. O. Schenck, W. Hartmann, and R. Steinmetz, *Chem. Ber.*, **96**, 498 (1963).

115. D. N. Reinhoudt, H. C. Volger, and C. G. Kouwenhoven, *Tetrahedron Letters*, 5269 (1972).

116. D. N. Reinhoudt and C. G. Kouwenhoven, *J.C.S. Chem. Comm.*, 1233 (1972).

117. G. Stork and P. L. Stotter, *J. Amer. Chem. Soc.*, **91**, 7780 (1969).

118. E. Jonsson and S. Holmquist, *Arkiv Kemi*, **29**, 301 (1968).

119. J. J. Rigau, C. C. Bacon, and C. R. Johnson, *J. Org. Chem.*, **35**, 3655 (1970).

120. D. N. Jones and D. A. Lewton, *J.C.S. Chem. Comm.*, 457 (1974).

121. G-i. Tsuchihashi and S. Iriuchijima, *Bull. Chem. Soc., Japan*, **43**, 2271 (1970).

122. P. Chao and D. M. Lemal, *J. Amer. Chem. Soc.*, **95**, 921 (1973); D. M. Lemal and P. Chao, *J. Amer. Chem. Soc.*, **95**, 922 (1973).

123. W. R. Sorenson, *J. Org. Chem.*, **24**, 1796 (1959).

124. F. Ellis, P. G. Sammes, M. B. Hursthouse, and S. Neidle, *J.C.S. Perkin I*, 1560 (1972).

125. J. M. Landesberg and M. Siegel, *J. Org. Chem.*, **35**, 1674 (1970).

126. S. N. Lewis and W. D. Emmons, *J. Org. Chem.*, **31**, 3572 (1966).

127. G. Claeson and H-G. Jonsson, *Arkiv Kemi*, **26**, 247 (1967).

128. E. Jonsson, *Acta Chem. Scand.*, **19**, 2247 (1965).

129. H. Meguro, T. Konno, and K. Tuzimura, *Tetrahedron Letters*, 3165 (1972).

130. W. N. Hubbard, H. L. Finke, D. W. Scott, J. P. McCullough, C. Katz, M. E. Gross, J. F. Messerly, R. E. Pennington, and G. Waddington, *J. Amer. Chem. Soc.*, **74**, 6025 (1952); N. L. Allinger and M. J. Hickey, *J. Amer. Chem. Soc.*, **97**, 5167 (1975).

131. R. Lett and A. Marquet, *Tetrahedron*, **30**, 3365 (1974).

132. R. Lett and A. Marquet, *Tetrahedron*, **30**, 3379 (1974).

133. R. Lett and A. Marquet, *Tetrahedron*, **31**, 653 (1975).

134. O. Hofer and E. L. Eliel, *J. Amer. Chem. Soc.*, **95**, 8045 (1973).

135. A. Garbesi, G. Barbarella, and A. Fava, *J.C.S. Chem. Comm.*, 155 (1973).

136. R. Lett, S. Bory, B. Moreau, and A. Marquet, *Tetrahedron Letters*, 3255 (1971).

137. A. Lüttringhaus, S. Kabuss, H. Prinzbach, and F. Langenbucher, *Annalen*, **653**, 195 (1962).

138. C. H. Eugster and K. Allner, *Helv. Chim. Acta*, **45**, 1750 (1962).

139. S. Tamagaki, M. Mizuno, H. Yoshida, H. Hirota, and S. Oae, *Bull. Chem. Soc. Japan*, **44,** 2456 (1971).
140. L. Sagramora, A. Garbesi, and A. Fava, *Helv. Chim. Acta*, **55,** 675 (1972).
141. K. Ogura and G. Tsuchihashi, *Chem. Comm.*, 1689 (1970).
142. E. W. Truce and T. C. Klingler, *J. Org. Chem.*, **35,** 1834 (1970).
143. S. D. McGregor and D. M. Lemal, *J. Amer. Chem. Soc.*, **88,** 2858 (1966).
144. W. L. Mock, *J. Amer. Chem. Soc.*, **88,** 2857 (1966).
145. R. M. Kellogg and W. L. Prins, *J. Org. Chem.*, **39,** 2366 (1974).
146. W. L. Prins and R. M. Kellogg, *Tetrahedron Letters*, 2833, (1973).
147. H. D. Hartough and S. L. Meisel, *Compounds with Condensed Thiophene Rings*, A. Weissberger, Ed., Interscience, New York, 1954.
148. S. F. Birch, R. A. Dean, and E. V. Whitehead, *J. Org. Chem.*, **23,** 783 (1958).
149. S. F. Birch, R. A. Dean, N. J. Hunter, and E. V. Whitehead, *J. Org. Chem.*, **22,** 1590 (1957).
150. L. A. Paquette, R. H. Meisinger, and R. E. Wingard, Jr., *J. Amer. Chem. Soc.*, **95,** 2230 (1973).
151. L. N. Owen and A. G. Peto, *J. Chem. Soc.*, 2383 (1955).
152. K. Weinges, M. Weber, and K. Klessing, *Chem. Ber.*, **106,** 2305 (1973).
153. D. C. Neckers, J. H. Dopper, and H. Wynberg, *J. Org. Chem.*, **35,** 1582 (1970).
154. J. H. Dopper and D. C. Neckers, *J. Org. Chem.*, **36,** 3755 (1971).
155. D. N. Harpp and C. Heitner, *J. Org. Chem.*, **38,** 4184 (1973).
156. D. N. Reinhoudt and C. G. Kouwenhoven, *Tetrahedron Letters*, 2503 (1974).
157. C. G. Overberger and J. M. Whelan, *J. Org. Chem.*, **26,** 4328 (1961).
158. D. N. Harpp and C. Heitner, *J. Org. Chem.*, **35,** 3256 (1970).
159. W. L. Mock, *J. Amer. Chem. Soc.*, **92,** 6918 (1970).
160. S. F. Birch, R. A. Dean, N. J. Hunter, and E. V. Whitehead, *J. Org. Chem.*, **20,** 1178 (1955).
161. B. Sjöberg, *Arkiv Kemi*, **15,** 451 (1960).
162. S. Hagishita and K. Kuriyama, *J.C.S. Perkin II*, 686 (1974).
163. P. Lauer, J. Häuser, J. E. Gurst, and K. Mislow, *J. Org. Chem.*, **32,** 498 (1967).
164. N. M. Green, W. P. Mose, and P. M. Scopes, *J. Chem. Soc. (C)*, 1330 (1970).
165. E. Jonsson, *Arkiv Kemi*, **26,** 357 (1967).
166. B. J. Auret, D. R. Boyd, and H. B. Henbest, *J. Chem. Soc. (C)*, 2374 (1968).
167. K. K. Andersen, R. L. Caret, and I. Karup-Nielsen, *J. Amer. Chem. Soc.*, **96,** 8026 (1974).
168. G. Barbarella, A. Garbesi, A. Biocelli, and A. Fava, *J. Amer. Chem. Soc.*, **95,** 8051 (1973).
169. R. F. Gratz and P. Wilder, Jr., *Chem. Comm.*, 1449 (1970).
170. P. Wilder, Jr., and L. A. Feliu-Otero, *J. Org. Chem.*, **30,** 2560 (1965).
171. P. Wilder, Jr. and R. F. Gratz, *J. Org. Chem.*, **35,** 3295 (1970).
172. P. Wilder, Jr., L. A. Feliu-Otero, and G. A. Diegnan, *J. Org. Chem.*, **39,** 2153 (1974).
173. E. Casadevall and M. M. Bouisset, *Tetrahedron Letters*, 2975 (1973).
174. F. G. Yamagishi, D. R. Rayner, E. T. Zwicker, and D. J. Cram, *J. Amer. Chem. Soc.*, **95,** 1916 (1973).

175. J. N. Roitman and D. J. Cram, *J. Amer. Chem. Soc.*, **93,** 2225 (1971).
176. D. J. Cram and T. A. Whitney, *J. Amer. Chem. Soc.*, **89,** 4651 (1967).
177. T. A. Whitney and D. J. Cram, *J. Org. Chem.*, **35,** 3964 (1970).
178. T. A. Whitney and W. H. Pirkle, *Tetrahedron Letters*, 2299 (1974).
179. W. H. Pirkle, S. D. Beare, and R. L. Muntz, *Tetrahedron Letters*, 2295 (1974).
180. J. R. Du Manoir, J. F. King, and R. R. Fraser, *J.C.S. Chem. Comm.*, 541 (1972).
181. G. Claeson, G. Androes, and M. Calvin, *J. Amer. Chem. Soc.*, **83,** 4357 (1961).
182. L. J. Reed, *Adv. Enzymol.*, **18,** 319 (1957).
183. J. A. Baltrop, P. M. Hayes, and M. Calvin, *J. Amer. Chem. Soc.*, **76,** 4348 (1954).
184. M. Calvin and J. A. Baltrop, *J. Amer. Chem. Soc.*, **74,** 6153 (1952).
185. E. Walton, A. F. Wagner, F. W. Bachelor, L. H. Peterson, F. W. Holly, and K. Folkers, *J. Amer. Chem. Soc.*, **77,** 5144 (1955).
186. B. A. Lewis and R. A. Raphael, *J. Chem. Soc.*, 4263 (1962).
187. I. C. Gunsalus, L. S. Barton, and W. Gruber, *J. Amer. Chem. Soc.*, **78,** 1763 (1956).
188. D. S. Acker and W. J. Wayne, *J. Amer. Chem. Soc.*, **79,** 6483 (1957).
189. R. C. Thomas and L. J. Reed, *J. Amer. Chem. Soc.*, **78,** 6150 (1956).
190. K. Mislow and W. C. Meluch, *J. Amer. Chem. Soc.*, **78,** 2341 (1956).
191. G. Bergson and L. Schotte, *Arkiv Kemi*, **13,** 43 (1958).
192. C. Bergson and L. Schotte, *Acta Chem. Scand.*, **12,** 367 (1958); O. Foss and O. Tjomsland, *Acta. Chem. Scand.*, **11,** 1426 (1957).
193. M. Calvin, *Fed. Proc.*, **13,** 697 (1954).
194. G. Claeson, *Arkiv. Kemi*, **30,** 277 (1969).
195. H. Yanagawa, T. Kato, and Y. Kitahara, *Tetrahedron Letters*, 1073 (1973).
196. A. Kato and H. Numata, *Tetrahedron Letters*, 203 (1972).
197. H. Wolf, E. Bunnenberg, C. Djerassi, A. Lüttringhaus, and A. Stockhausen, *Annalen*, **674,** 62 (1964).
198. A. K. M. Anisuzzaman and L. N. Owen, *Chem. Comm.*, 16 (1966).
199. N. G. Kardouche and L. N. Owen, *J.C.S. Perkin I*, 754 (1975).
200. M. E. Ali, N. G. Kardouche, and L. N. Owen, *J.C.S. Perkin I*, 748 (1975).
201. M. Kyaw and L. N. Owen, *J. Chem. Soc.*, 1298 (1965).
202. D. J. Martin, *J. Org. Chem.*, **34,** 473 (1969).
203. L. W. C. Miles and L. N. Owen, *J. Chem. Soc.*, 2938 (1950).
204. M. Ebel and L. Legrand, *Bull. Soc. chim. France*, 176 (1971).
205. R. C. Cookson, G. H. Cooper, and J. Hudec, *J. Chem. Soc.*, (*B*) 1004 (1967).
206. M-O. Hedblom, *Arkiv Kemi*, **31,** 489 (1970).
207. W. Otting and F. A. Neugebauer, *Chem. Ber.*, **95,** 540 (1962).
208. L. A. Sternson, D. A. Coviello, and R. S. Egan, *J. Amer. Chem. Soc.*, **93,** 6529 (1971).
209. R. Keskinen, A. Nikkilä, and K. Pihlaja, *J.C.S. Perkin II*, 1376 (1973).
210. L. A. Sternson, L. C. Martinelli, and R. S. Egan, *J. Heterocyclic Chem.*, **11,** 1117 (1974).
211. C. H. Bushweller, G. U. Rao, W. G. Anderson, and P. E. Stevenson, *J. Amer. Chem. Soc.*, **94,** 4743 (1972).
211a. T. C. Shields and A. N. Kurtz, *J. Amer. Chem. Soc.*, **91,** 5415 (1969).

212. S. B. Tjan, J. C. Haakman, C. J. Teunis, and H. G. Peer, *Tetrahedron*, **28,** 3489 (1972).
213. M-F. Guimon, C. Guimon, and G. Pfister-Guillouzo, *Tetrahedron Letters*, 441 (1975).
214. C. A. R. Baxter and D. A. Whiting, *J. Chem. Soc.* (*C*), 1174 (1968).
215. B. R. Baker and F. Ablondi, *J. Org. Chem.*, **12,** 328 (1947).
216. D. M. Vyas and G. W. Hay, *J.C.S. Perkin I*, 180 (1975).
217, J. B. Lambert, R. G. Keske, and D. K. Weary, *J. Amer. Chem. Soc.*, **89,** 5921 (1967).
218. T. P. Forrest, *J. Amer. Chem. Soc.*, **97,** 2628 (1975).
219. J. B. Lambert and R. G. Keske, *J. Org. Chem.*, **31,** 3429 (1966).
220. J. B. Lambert, D. S. Bailey, and C. E. Mixan, *J. Org. Chem.*, **37,** 377 (1972).
221. M. J. Cook, H. Dorn, and A. R. Katritzky, *J. Chem. Soc.* (*B*), 1467 (1968).
222. E. L. Eliel, R. L. Willer, A. T. McPhail, and K. D. Onan, *J. Amer. Chem. Soc.*, **96,** 3021 (1974).
223. R. Gerdil, *Helv. Chim. Acta*, **57,** 489 (1974).
224. J. M. Cox and L. N. Owen, *J. Chem. Soc.* (*C*), 1130 (1967).
225. N. S. Zefirov, V. S. Blagoveshchenskii, I. V. Kazimirchik, and O. P. Yakovleva, *Zhur. org. Khim.*, **7,** 594 (1971).
226. N. S. Zefirov, V. S. Blagoveshchenskii, and I. V. Kazimirchik, *Zhur. org. Khim.*, **5,** 1150 (1969).
226a. T. Sugawara, H. Iwamura, and M. Ōki, *Bull. Chem. Soc. Japan*, **47,** 1496 (1974).
227. N. S. Zefirov and N. M. Shekhtman, *Doklady Akad. Nauk. S.S.S.R.*, **180,** 1363 (1968) and **177,** 842 (1967).
228. N. S. Zefirov and N. M. Shekhtman, *Russ. Chem. Rev.*, **40,** 315 (1971).
229. W. A. Szarek, D. M. Vyas, and B. Achmatowicz, *Tetrahedron Letters*, 1553 (1975).
230. J. B. Lambert, C. E. Mixan, and D. S. Bailey, *Chem. Comm.*, 316 (1971).
231. J. B. Lambert, C. E. Mxian, and D. S. Bailey, *J. Amer. Chem. Soc.*, **94,** 208 (1972).
232. G. Wood, C. C. Barker, and A. Kligerman, *Canad. J. Chem.*, **51,** 3329 (1973).
233. G. W. Buchanan and T. Durst, *Tetrahedron Letters*, 1683 (1975).
234. R. Curci, F. Di Furia, A. Levi, V. Lucchini, and G. Scorrano, *J.C.S. Perkin II*, 341 (1975).
235. R. Lett and A. Marquet, *Tetrahedron Letters*, 1579 (1975).
236. D. C. Appleton, D. C. Bull, J. McKenna, J. M. McKenna, and A. R. Walley, *J.C.S. Chem. Comm.*, 140 (1974).
237. C. R. Johnson and D. McCants, Jr., *J. Amer. Chem. Soc.*, **87,** 1109 (1965).
238. G. Barbieri, M. Ciquini, S. Colonna, and F. Montanari, *J. Chem. Soc.* (*C*), 659 (1968).
239. J. Klein and H. Stollar, *Tetrahedron*, **30,** 2541 (1974).
240. C. R. Johnson and D. McCants, Jr., *J. Amer. Chem. Soc.*, **87,** 5404 (1965).
241. C. R. Johnson, *J. Amer. Chem. Soc.*, **85,** 1020 (1963).
242. S. Bory, R. Lett, B. Moreau, and A. Marquet, *Tetrahedron Letters*, 4921 (1972).
243. R. Lett, S. Bory, B. Moreau, and A. Marquet, *Bull. Soc. chim. France*, 2851 (1973).
244. S. Iriuchijima and G-i. Tsuchihashi, *Bull. Chem. Soc. Japan*, **46,** 929 (1973).
245. M. Cinquini, S. Colonna, and F. Montanari, *J.C.S. Perkin I*, 1723 (1974).
246. J. Klein and H. Stollar, *J. Amer. Chem. Soc.*, **95,** 7437 (1973).
247. S. Iriuchijima, M. Ishibashi, and G-i. Tsuchihashi, *Bull. Chem. Soc. Japan*, **46,** 921 (1973).

248. H. Stollar and J. Klein, *J.C.S. Perkin I*, 1763 (1974).
249. C. R. Johnson and D. McCants, Jr., *J. Amer. Chem. Soc.*, **86,** 2935 (1964).
250. J. C. Martin and J. J. Uebel, *J. Amer. Chem. Soc.*, **86,** 2936 (1964).
251. U. Folli, D. Iarossi, and F. Taddei, *J.C.S. Perkin II*, 1658 (1974).
252. M. D. Brown, M. J. Cook, B. J. Hutchinson, and A. R. Katritzky, *Tetrahedron*, **27,** 593 (1971).
253. A. Rauk, S. Wolfe, and I. G. Csizmadia, *Canad. J. Chem.*, **47,** 113 (1969).
254. C. R. Johnson and J. J. Rigau, *J. Org. Chem.*, **33,** 4340 (1968).
255. C. R. Johnson, J. J. Rigau, M. Haake, D. McCants, Jr., J. E. Keiser, and A. Gertsema *Tetrahedron Letters*, 3719 (1968).
256. A. Chatterjee and B. Bandyopadhyay, *Indian J. Chem.*, **11,** 446 (1973).
257. W. D. Cotterill, J. J. Cottam, and R. Livingstone, *J. Chem. Soc. (C)*, 1006 (1970).
258. B. D. Tilak, R. B. Mitra, and Z. Muljiani, *Tetrahedron*, **25,** 1939 (1969).
259. E. E. Smissman, R. A. Schnettler, and P. S. Portoghese, *J. Org. Chem.*, **30,** 797 (1965).
260. P. T. Lansbury and D. J. Scharf, *J. Amer. Chem. Soc.*, **90,** 536 (1968).
261. J. Kattenberg, E. R. de Waard, and H. O. Huisman, *Tetrahedron*, **30,** 463 (1974).
262. T. E. Young and L. J. Heitz, *J. Org. Chem.*, **38,** 1562 (1973).
263. D. A. Pulman and D. A. Whiting, *J.C.S. Perkin I*, 410 (1973).
264. D. A. Pulman and D. A. Whiting, *Chem. Comm.*, 831 (1971).
265. C. C. Price, M. Siskin, and C. K. Miao, *J. Org. Chem.*, **36,** 794 (1971).
266. F. G. Holliman and F. G. Mann, *J. Chem. Soc.*, 37, (1945).
267. E. R. de Waard, W. J. Vloon, and H. O. Huisman, *Chem. Comm.*, 841 (1970).
268. A. van Bruijnsvoort, C. Kruk, E. R. de Waard, and H. O. Huisman, *Tetrahedron Letters*, 1737 (1972).
269. G. Claeson, G. M. Androes, and M. Calvin, *J. Amer. Chem. Soc.*, **82,** 4428 (1960).
270. G. Wood, R. M. Srivastava, and B. Adlam, *Canad. J. Chem.*, **51,** 1200 (1973).
271. H. T. Kalff and E. Havinga, *Rec. Trav. chim.*, **81,** 282 (1962).
272. O. Foss and L. Schotte, *Acta Chem. Scand.*, **11,** 1424 (1957).
273. N. Isenberg and H. F. Herbrandson, *Tetrahedron*, **21,** 1067 (1965).
274. R. M. Dodson and V. C. Nelson, *J. Org. Chem.*, **33,** 3966 (1968).
275. R. S. Cahn, C. Ingold, and V. Prelog, *Angew. Chem. Internat. Edn.*, **5,** 385 (1966).
276. A. F. Beecham, J. Fridrichsons, and Mc. L. Mathieson, *Tetrahedron Letters*, 3131 (1966).
277. D. B. Cosulich, N. R. Nelson, and J. H. van den Hende, *J. Amer. Chem. Soc.*, **90,** 6519 (1968).
278. J. E. McCormick and R. S. McElhinney, *J.C.S. Perkin I*, 2795 (1972).
279. M. Carmack and C. J. Kelley, *J. Org. Chem.*, **33,** 2171 (1968).
280. M. Carmack and L. A. Neubert, *J. Amer. Chem. Soc.*, **89,** 7134 (1967).
281. A. Lüttringhaus and A. Brechlin, *Chem. Ber.*, **92,** 2271 (1959).
282. J. P. Casey and R. B. Martin, *J. Amer. Chem. Soc.*, **94,** 6141 (1972).
283. A. Lüttringhaus, S. Kabuss, W. Maier, and H. Friebolin, *Z. Naturforsch.*, **16B,** 761 (1961).
284. G. Claeson, *Arkiv Kemi*, **30,** 511 (1969).
285. O. Foss, K. Johnsen, and T. Reistad, *Acta Chem. Scand.*, **18,** 2345 (1964).

286. J. P. Danehy and V. J. Elia, *J. Org. Chem.*, **36,** 1394 (1971).
287. G. Cignarella and G. Cordella, *Gazzetta*, **104,** 455 (1974).
287a. D. N. Harpp and J. G. Gleason, *J. Amer. Chem. Soc.*, **93,** 2437 (1971).
288. D. N. Harpp and J. G. Gleason, *J. Org. Chem.*, **36,** 1314 (1971).
289. L. Van Acker and M. Anteunis, *Tetrahedron Letters*, 225 (1974).
290. D. Danneels, M. Anteunis, L. Van Acker, and D. Tavernier, *Tetrahedron*, **31,** 327 (1975).
291. D. A. Lightner, C. Djerassi, K. Takeda, K. Kuriyama, and T. Komeno, *Tetrahedron*, **21,** 1581 (1965).
292. W. H. Mills and B. C. Saunders, *J. Chem. Soc.*, 537 (1931).
293. H. T. Kalff and C. Romers, *Acta Cryst*,. **20,** 490 (1966).
294. H. T. Kalff and E. Havinga, *Rec. Trav. chim.*, **85,** 467 (1966).
295. E. Langer and H. Lehner, *Monatsh.*, **106,** 175 (1975).
296. H. Friebolin, S. Kabuss, W. Maier, and A. Lüttringhaus, *Tetrahedron Letters*, 683 (1962).
297. J. Gelan and M. Anteunis, *Bull. Soc. chim. belges*, **77,** 423 (1968).
298. H. Gelan and M. Anteunis, *Bull. Soc. chim. belges*, **77,** 447 (1968); M. Anteunis, G. Swaelens, and J. Gelan, *Tetrahedron*, **27,** 1917 (1971).
299. F. G. Riddell and M. J. T. Robinson, *Tetrahedron*, **23,** 3417 (1967).
300. E. L. Eliel and R. O. Hutchins, *J. Amer. Chem. Soc.*, **91,** 2703 (1969).
301. R. J. Abraham and W. A. Thomas, *J. Chem. Soc.*, 335 (1965).
302. E. L. Eliel, *Bull. Soc. chim. France*, 517 (1970).
303. J. Gelan and N. Anteunis, *Bull. Soc. chim. belges*, **78,** 599 (1969).
304. S. A. Khan, J. B. Lambert, O. Hernandez, and F. A. Carey, *J. Amer. Chem. Soc.*, **97,** 1468 (1975).
305. M. J. Cook and A. P. Tonge, *J.C.S. Perkin II*, 767 (1974).
306. M. Anteunis, L. Van Acker, and D. Danneels, *Bull. Soc. chim. belges*, **83,** 301 (1974).
307. R. S. J. Beer, D. Harris, and D. J. Royall, *Tetrahedron Letters*, 1531 (1964).
308. I. Stahl and J. Gosselck, *Tetrahedron*, **29,** 2323 (1973).
309. I. Stahl and J. Gosselck, *Tetrahedron*, **30,** 3519 (1974).
310. A. A. Hartmann and E. L. Eliel, *J. Amer. Chem. Soc.*, **93,** 2572 (1971).
311. E. L. Eliel, A. Abatjoglou, and A. A. Hartmann, *J. Amer. Chem. Soc.*, **94,** 4786 (1972).
312. E. L. Eliel, A. A. Hartmann, and A. G. Abatjoglou, *J. Amer. Chem. Soc.*, **96,** 1807 (1974).
312a. K. Arai and M. Oki, *Tetrahedron Letters*, 2183 (1975).
313. E. Cuthbertson, D. D. MacNicol, and P. R. Mallinson, *Tetrahedron Letters*, 1345 (1975).
314. G. Hesse and I. Jörder, *Chem. Ber.*, **85,** 924 (1952).
315. O. Hromatka and E. Engel, *Monatsh.*, **78,** 38 (1948).
316. R. Haberl, F. Grass, O. Hromatka, K. Brauner, and A. Preisinger, *Monatsh.*, **86,** 551 (1955); R. Haberl and F. Grass, *Monatsh.*, **86,** 599 (1955).
317. W. A. Thaler and P. E. Butler, *J. Org. Chem.*, **34,** 3389 (1969).
318. W. H. Mueller and M. Dines, *J. Heterocyclic Chem.*, **6,** 627 (1969).

319. O. Hassel and H. Viervoll, *Acta Chem. Scand.*, **1,** 149 (1947).
320. G. Y. Chao and J. D. McCullough, *Acta Cryst.*, **13,** 727 (1960).
321. R. E. Marsh., *Acta Cryst.*, **8,** 91 (1955).
322. R. E. Marsh and J. D. McCullough, *J. Amer. Chem. Soc.*, **73,** 1106 (1951).
323. C. Romers, C. Altona, H. R. Buys, and E. Havinga, *Topics Stereochem.*, **4,** 39 (1969).
324. H. T. Kalff and C. Romers, *Acta Cryst.*, **18,** 164 (1965).
325. H. T. Kalff and C. Romers, *Rec. Trav. chim.*, **85,** 198 (1966).
326. H. T. Kalff and E. Havinga, *Rec. Trav. chim.*, **85,** 637 (1966).
327. G. Tsuchihashi, M. Yamauchi, and M. Fukyama, *Tetrahedron Letters*, 1971 (1967).
328. F. Lautenschlaeger and G. F. Wright, *Canad. J. Chem.*, **41,** 1972 (1963).
329. G. Bandoli, D. A. Clemente, C. Panattoni, A. Dondoni, and A. Mangini, *Chem. Comm.*, 1143 (1970).
330. H. Montgomery, *Acta Cryst.*, **13,** 381 (1960).
331. H. M. M. Shearer, *J. Chem. Soc.*, 1394 (1959).
332. E. V. Bell and G. M. Bennett, *J. Chem. Soc.*, 1798 (1927).
333. C-Y. Chen and R. J. W. Le Fèvre, *Austral. J. Chem.*, **16,** 917 (1963).
334. P. B. D. de la Mare, D. J. Millen, J. G. Tillett, and D. Watson, *J. Chem. Soc.*, 1619 (1963).
335. S. Kabuss, A. Lüttringhaus, H. Friebolin, and R. Mecke, *Z. Naturforsch.*, **21B,** 320 1966.
336. M. Anteunis and G. Goor, *Bull. Soc. chim. belges*, **83,** 463 (1974).
337. G. Goor, and M. Anteunis *Heterocycles*, **3,** 363 (1975), see also *Bull. Soc. chim. belges*, **84,** 337 (1975).
338. F. D. Chattaway and E. G. Kellett, *J. Chem. Soc.*, 1352 (1930).
339. R. M. Roberts and C-C. Cheng. *J. Org. Chem.*, **23,** 983 (1958).
340. E. Campaigne, N. F. Chamberlain, and B. E. Edwards, *J. Org. Chem.*, **27,** 135 (1962).
341. E. Campaigne and M. P. Georgiadis, *J. Heterocyclic Chem.*, **6,** 339 (1969).
342. E. Campaigne and M. Georgiades, *J. Org. Chem.*, **28,** 1044 (1963).
343. M. Ōki, T. Sugawara, and H. Iwamura, *Bull. Chem. Soc. Japan*, **47,** 2457 (1974).
344. T. Sugawara, H. Iwamura, and M. Ōki, *Tetrahedron Letters*, 879 (1975).
345. C. H. Bushweller, *J. Amer. Chem. Soc.*, **89,** 5978 (1967).
346. C. H. Bushweller, *J. Amer. Chem. Soc.*, **90,** 2450 (1968).
347. C. H. Bushweller, J. Golini, G. U. Rao, J. W. O'Neil, *Chem. Comm.*, 51 (1970).
348. C. H. Bushweller, *J. Amer. Chem. Soc.*, **91,** 6019 (1969).
349. C. H. Bushweller, J. Golini, G. U. Rao, and J. W. O'Neil, *J. Amer. Chem. Soc.*, **92,** 3055 (1970).
350. A. Fredga, *Acta Chem. Scand.*, **12,** 891 (1958).
351. C. H. Bushweller, G. U. Rao, and F. H. Bissett, *J. Amer. Chem. Soc.*, **93,** 3058 (1971).
352. C. H. Bushweller, *Tetrahedron Letters*, 2785 (1968).
353. C. H. Bushweller, G. Bhat, L. J. Letendre, J. A. Brunelle, H. S. Bilofsky, H. Ruben, D. H. Templeton, and A. Zalkin, *J. Amer. Chem. Soc.*, **97,** 65 (1975).
354. F. Fehér, B. Degen, and B. Söhngen, *Angew. Chem. Internat. Edn.*, **7,** 301 (1968).
355. V. J. Traynelis, Y. Yoshikawa, J. C. Sih, and L. J. Miller, *J. Org. Chem.*, **38,** 3978 (1973).

356. V. J. Traynelis, J. C. Sih, Y. Yoshikawa, R. F. Love, and D. M. Borgnaes, *J. Org. Chem.*, **38,** 2623 (1973).

357. M. Nógrádi, W. D. Ollis, and I. O. Sutherland, *J.C.S. Perkin I*, 621 (1974).

358. A. Chatterjee and B. K. Sen, *J.C.S. Chem. Comm.*, 626 (1974).

359. W. L. Mock, *J. Amer. Chem. Soc.*, **91,** 5682 (1969).

360. F. A. L. Anet, C. H. Bradley, M. A. Brown, W. L. Nock, and J. H. McCausland, *J. Amer. Chem. Soc.*, **91,** 7782 (1969).

361. W. Tochtermann, C. Franke, and D. Schäfer, *Chem. Ber.*, **101,** 3122 (1968).

362. C. G. Overberger and J. K. Weise, *J. Amer. Chem. Soc.*, **90,** 3525, 3533, and 3538 (1968).

362a. D. N. Jones, D. R. Hill, and D. A. Lewton, *Tetrahedron Letters*, 2235 (1975).

363. H. F. Herbrandson and R. H. Wood, *J. Medicin. Chem.* **12,** 617, 520 (1969).

364. L. Schotte, *Arkiv Kemi*, **9,** 441 (1956).

365. L. Schotte, *Arkiv Kemi*, **9,** 413 (1956).

365a. J. B. Hendrickson, *J. Amer. Chem. Soc.*, **83,** 4537 (1961); E. L. Eliel, N. L. Allinger, S. J. Angyal, and G. A. Morrison, *Conformational Analysis*, Wiley, New York, 1965, pp. 207–210.

366. S. Kabuss, A. Lüttringhaus, H. Friebolin, H. G. Schmid, and R. Mecke, *Tetrahedron Letters*, 719 (1966).

366a. H. Friebolin, R. Mecke, S. Kabuss, and A. Lüttringhaus, *Tetrahedron Letters*, 1929 (1964).

367. H. Friebolin and S. Kabuss, *Nucl. Magnetic Resonance Chem. Proc.* Symposium, Cagliara, Italy, 125 (1964), *Chem. Abs.*, **66,** 37265 (1967).

368. R. M. Moriarty, N. Ishibe, M. Kayser, K. C. Ramey, and H. J. Gisler, Jr., *Tetrahedron Letters*, 4883 (1969).

369. L. Field and D. L. Tuleen in *Seven Membered Heterocyclic Compounds Oxygen and Sulphur*, A. Rosowsky, Ed., Interscience, New York, 1972, Vol. 26, Ch. X, p. 573.

370. E. Vogel, E. Schmidbauer, and H. J. Altenbach, *Angew. Chem. Internat. Edn.*, **13,** 736 (1974).

370a. S. Kagabu and H. Prinzbach, *Angew. Chem. Internat. Edn.*, **14,** 252 (1975).

371. D. L. Coffen, Y. C. Poon, and M. L. Lee, *J. Amer. Chem. Soc.*, **93,** 4627 (1971).

372. S. M. Johnson, C. A. Maier, and I. C. Paul, *J. Chem. Soc.* (*B*), 1603 (1970).

373. A. Lüttringhaus, U. Hess, and H. J. Roesnbaum, *Z. Naturforsch.*, **22B,** 1296 (1967).

374. A. Lüttringhaus and H-J. Rosenbaum, *Monatsh.*, **98,** 1323 (1967).

375. G. H. Wahl, Jr., J. Bordner, D. N. Harpp, and J. G. Gleason, *J.C.S. Chem. Comm.*, 985 (1972).

376. G. W. Frank, P. J. Degen, and F. A. L. Anet, *J. Amer. Chem. Soc.*, **94,** 4792 (1972).

377. F. A. L. Anet and P. J. Degen, *J. Amer. Chem. Soc.*, **94,** 1390 (1972).

378. T. Sato and K. Uno, *J.C.S. Perkin I*, 895 (1973).

379. F. Vögtle and L. Schunder, *Annalen*, **721,** 129 (1969).

380. W. D. Ollis, J. F. Stoddart, and M. Nógrádi, *Angew. Chem. Internat. Edn.*, **14,** 168 (1975).

381. J. M. Brown and F. Sondheimer, *Angew. Chem. Internat. Edn.*, **13,** 339 (1974).

382. A. B. Holmes and F. Sondheimer, *Chem. Comm.*, 1434 (1971).

383. P. J. Garratt, A. B. Holmes, F. Sondheimer, and K. P. C. Vollhardt, *Chem. Comm.*, 947 (1971).
384. P. J. Garratt, A. B. Holmes, F .Sondheimer, and K. P. C. Vollhardt, *J. Amer. Chem. Soc.*, **92,** 4492 (1970).
385. R. L. Wife and F. Sondheimer, *Tetrahedron Letters*, 195 (1975).
386. L. A. Ochrymowycz, C-P. Mak, and J. D. Michna, *J. Org. Chem.*, **39,** 2079 (1974).
387. P. Y. Johnson and G. A. Berchtold, *J. Org. Chem.*, **35,** 584 (1970).
388. S. Ikegami, T. Asai, K. Tsuneoka, S. Matsuura, and S. Akaboshi, *Tetrahedron*, **30,** 2087 (1974).
389. L. A. Paquette, G. V. Meehan, and L. D. Wise, *J. Amer. Chem. Soc.*, **91,** 3231 (1969).
390. L. A. Paquette and L. D. Wise, *J. Amer. Chem. Soc.*, **89,** 6659 (1967).
391. C. G. Overberger and A. Lusi, *J. Amer. Chem. Soc.*, **81,** 506 (1959).
392. N. J. Leonard, T. L. Brown, and T. W. Milligan, *J. Amer. Chem. Soc.*, **81,** 504 (1959).
393. C. G. Overberger, P. Barkan, A. Lusi, and H. Ringsdorf, *J. Amer. Chem. Soc.*, **84,** 2814 (1962).
394. N. J. Leonard and C. R. Johnson, *J. Amer. Chem. Soc.*, **84,** 3701 (1962).
395. N. J. Leonard and W. L. Rippie, *J. Org. Chem.*, **28,** 1957 (1963).
396. F. Vögtle and P. Neumann, *Angew. Chem. Internat. Edn.*, **11,** 73 (1972).
397. F. Vögtle and P. Neumann, *Tetrahedron*, **26,** 5847 (1970).
398. H. Nozaki, T. Koyama, T. Mori, and R. Noyori, *Tetrahedron Letters*, 2181 (1968).
399. H. Nozaki, T. Koyama, and T. Mori, *Tetrahedron*, **25,** 5357 (1969).
400. H. E. Winberg, F. S. Fawcett, W. E. Mochel, and C. W. Theobald, *J. Amer. Chem. Soc.*, **82,** 1428 (1960).
401. B. Kamenar and C. K. Prout, *J. Chem. Soc.*, 4838 (1965).
402. S. Mizogami, N. Osaka, and T. Otsubo, *Tetrahedron Letters*, 799 (1974).
403. A. Lüttringhaus and H. Gralheer, *Annalen*, **550,** 67 (1942).
404. F. Vögtle, *Tetrahedron*, **25,** 3231 (1969).
405. F. Vögtle, *Tetrahedron Letters*, 5221 (1968).
406. F. Vöglte, *Chem. Ber.*, **102,** 1784 (1969).
407. F. Vögtle, *Tetrahedron Letters*, 3193 (1969).
408. R. H. Mitchell and V. Boekelheide, *J. Heterocyclic Chem.*, **6,** 981 (1969).
409. R. H. Mitchell and V. Boekelheide, *Tetrahedron Letters*, 2013 (1969).
410. F. Vögtle and H. Risler, *Angew. Chem. Internat. Edn.*, **11,** 727 (1972).
411. T. Sato, M. Wakabayashi, K. Hata, and M. Kainosho, *Tetrahedron*, **27,** 2737 (1971).
412. F. Vögtle and P. Neumann, *Tetrahedron*, **26,** 5299 (1970).
413. F. Vögtle and L. Schunder, *Chem. Ber.*, **102,** 2677 (1969).
414. F. Vögtle, W. Weider, and H. Förster, *Tetrahedron Letters*, 4361 (1974).
415. F. Vögtle, R. Schäfer, L. Schunder, and P. Neumann, *Annalen*, **734,** 102 (1970).
416. F. Vögtle, *Tetrahedron Letters*, 3623 (1968).
417. R. H. Mitchell and V. Boekelheide, *Tetrahedron Letters*, 1197 (1970).
418. B. R. Davis and I. Bernal, *J. Chem. Soc.* (*B*), 2307 (1971).
419. F. Vögtle, *Chem. Ber.*, **102,** 3077 (1969).
420. S. E. Potter and I. O. Sutherland, *J.C.S. Chem. Comm.*, 754 (1972).

421. H. Lehner, *Monatsh.*, **105,** 895 (1974).
422. T. Umemoto, T. Otsubo, and S. Misumi, *Tetrahedron Letters*, 1573 (1974).
423. G. Montaudo, F. Bottino, and E. Trivellone, *J. Org. Chem.*, **37,** 504 (1972).
424. F. Vögtle, *Annalen*, **735,** 193 (1970).
425. R. G. Lichtenhaler and F. Vögtle, *Chem. Ber.*, **106,** 1319 (1973).
426. F. Vögtle and R. G. Lichtenhaler, *Angew. Chem. Internat. Edn.*, **11,** 535 (1972).
427. F. Vögtle, G. Hohner, and E. Weber, *J.C.S. Chem. Comm.*, 366 (1973).
428. V. Boekelheide and C. H. Tsai, *J. Org. Chem.*, **38,** 3931 (1973).
429. V. Boekelheide and P. H. Anderson, *J. Org. Chem.*, **38,** 3928 (1973).
430. R. Gray and V. Boekelheide, *Angew. Chem. Internat. Edn.*, **14,** 107 (1975).
431. G. M. Badger, J. A. Elix, and G. E. Lewis, *Austral. J. Chem.*, **18,** 70 (1965).
432. V. J. Traynelis in *Seven-Membered Heterocyclic Compounds Containing Oxygen and Sulphur*, Vol. 26, A. Rosowsky, Ed., Interscience, New York, 1972, Ch. XI, p. 667.
433. V. Prelog and E. Cerkovnikov, *Annalen*, **537,** 214 (1939).
434. C. R. Johnson, J. E. Keiser, and J. C. Sharp, *J. Org. Chem.*, **34,** 860 (1969).
434a. R. R. Fraser and Y. Y. Wigfield, *Chem. Comm.*, 1471 (1970).
435. U. Folli, D. Iarossi, I. Moretti, F. Taddei, and G. Torre, *J.C.S. Perkin II*, 1655 (1974); U. Folli, D. Iarossi, and F. Taddei, *J.C.S. Perkin II*, 933 (1974).
436. I. Tabushi, Y. Tamura, Z-i. Yoshida, and T. Sugimoto, *J. Amer. Chem. Soc.*, **97,** 2886 (1975).
437. S. F. Birch, R. A. Dean, and N. J. Hunter, *J. Org. Chem.*, **23,** 1026 (1958).
438. T. J. Barton, M. D. Martz, and R. G. Zika, *J. Org. Chem.*, **37,** 552 (1972).
439. H. Gotthardt and B. Christl, *Tetrahedron Letters*, 4751 (1968).
440. H. Gotthardt and B. Christl, *Tetrahedron Letters*, 4747 (1968).
441. C. R. Johnson and F. L. Billman, *J. Org. Chem.*, **36,** 855 (1971).
442. J-M. Surzur, R. Nouguier, M-P. Crozet, and C. Dupuy, *Tetrahedron Letters*, 2035 (1971).
443. H. J. Reich, and J. E. Trend, *J. Org. Chem.*, **38,** 2637 (1973).
444. J. Plešek, S. Heřmánek, and B. Štíbr, *Coll. Czech. Chem. Comm.*, **33,** 2336 (1968).
445. R. M. Dodson, P. J. Cahill, and B. H. Chollar, *Chem. Comm.*, 310 (1969).
446. S. F. Birch and R. A. Dean, *Annalen*, **585,** 234 (1954).
447. R. Nagarajan, B. H. Chollar, and R. M. Dodson, *Chem. Comm.*, 550 (1967).
448. G. Binsch and G. R. Franzen, *J. Amer. Chem. Soc.*, **91,** 3999 (1969).
449. F. Lautenschlaeger, *Canad. J. Chem.*, **44,** 2813 (1966).
450. E. D. Weil, K. J. Smith, and R. J. Gruber, *J. Org. Chem.*, **31,** 1669 (1966).
451. F. Lautenschlaeger, *J. Org. Chem.*, **33,** 2627 (1968).
452. F. Lautenschlaeger, *J. Org. Chem.*, **33,** 2620 (1968).
453. E. J. Corey and E. Block, *J. Org. Chem.*, **31,** 1663 (1966).
454. E. Lautenschlaeger, *J. Org. Chem.*, **31,** 1679 (1966).
455. C. Ganter and J-F. Moser, *Helv. Chim. Acta*, **51,** 300 (1968).
456. J. A. J. M. Vincent, P. Schipper, Ae. de Groot, and H. M. Buck, *Tetrahderon Letters*, 1989 (1975).
457. L. A. Paquette and R. W. Houser, *J. Amer. Chem. Soc.*, **91,** 3870 (1969).

458. C. B. Quinn, J. R. Wiseman, and J. C. Calabrese, *J. Amer. Chem. Soc.*, **95,** 6121 (1973).
459. C. B. Quinn and J. R. Wiseman, *J. Amer. Chem. Soc.*, **95,** 6120 (1973).
460. C. B. Quinn and J. R. Wiseman, *J. Amer. Chem. Soc.*, **97,** 1342 (1973).
461. A. Padwa, A. Battisti, and E. Shefter, *J. Amer. Chem. Soc.*, **91,** 4000 (1969).
462. C. Ganter and J-F. Moser, *Helv. Chim. Acta*, **54,** 2228 (1971).
463. C. Ganter and N. Wigger, *Helv. Chim. Acta*, **55,** 481 (1972).
464. C. Ganter and J-F. Moser, *Helv. Chim. Acta*, **52,** 725 (1969).
465. N. Wigger, N. Stücheli, H. Szczepanski, and C. Ganter, *Helv. Chim. Acta*, **55,** 2791 (1972).
466. N. Wigger and C. Ganter, *Helv. Chim. Acta*, **55,** 2769 (1972).
467. N. S. Zefirov and S. V. Rogozina, *Tetrahedron*, **30,** 2345 (1974).
468. R. Huisgen, G. Boche, and H. Huber, *J. Amer. Chem. Soc.*, **89,** 3345 (1967).
469. R. Huisgen and J. Gasteiger, *Angew. Chem. Internat. Edn.*, **11,** 716, 1104 (1972); R. Huisgen and J. Gasteiger, *Tetrahedron Letters*, 3661 (1972).
470. J. Gasteiger and R. Huisgen, *J. Amer. Chem. Soc.*, **94,** 6541 (1972).
471. L. A. Paquette, U. Jacobsson, and M. Oku, *J.C.S. Chem. Comm.*, 115 (1975).
472. A. G. Anastassiou and B. Y-H. Chao, *Chem. Comm.*, 979 (1971).
473. A. G. Anastassiou and B. Y-H. Chao, *J.C.S. Chem. Comm.*, 277 (1972).
474. A. G. Anastassiou, E. Reichmanis, and J. C. Wetzel, *Tetrahedron Letters*, 1651 (1975).
475. H. Fritz and D. C. Weis, *Tetrahedron Letters*, 1659 (1974).
476. J. Janku and Landa, *Coll. Czech. Chem. Comm.*, **37,** 2269 (1972).
477. K. Olsson, *Arkiv Kemi*, **14,** 371 (1959).
478. K. Olsson, *Arkiv Kemi*, **26,** 435 (1967).
479. K. Olsson and C-O. Almqvist, *Arkiv Kemi*, **27,** 571 (1967).
480. H. Stetter, *Angew. Chem.*, **66,** 217 (1954).
481. M. Thiel, F. Asinger, K. Schmiedel, and H. Petschik, *Monatsh.*, **91,** 473 (1960).
482. K. Olsson, *Acta Chem. Scand.*, **12,** 366 (1958).
483. S. Wolfe, A. Rauk, L. M. Tel, and I. G. Csizmadia, *Chem. Comm.*, 96 (1970).
484. M-O. Hedblom and K. Olsson, *Arkiv Kemi*, **32,** 309 (1974).
485. M. J. Aroney, G. M. Hoskins. and R. J. W. Le Fèvre, *J. Chem. Soc.* (*B*), 980 (1969).
486. A. L. Ternay, Jr. and S. A. Evans, *J. Org. Chem.*, **39,** 2941 (1974).
487. K. K. Andersen, M. Cinquini, and N. E. Papanikolaou, *J. Org. Chem.*, **35,** 706 (1970).
488. A. L. Ternay, Jr., D. W. Chasar, and M. Sax, *J. Org. Chem.*, **32,** 2465 (1967).
489. A. L. Ternay, Jr. and D. W. Chasar, *J. Org. Chem.*, **32,** 3814 (1967).
490. A. L. Ternay, Jr. and D. W. Chasar, *J. Org. Chem.*, **33,** 2237 (1968).
491. W. Michaelis, O. Schindler, and R. Signer, *Helv. Chem. Acta*, **49,** 42 (1966).
492. G. M. Bennett and S. Glasstone, *J. Chem. Soc.*, 128 (1934).
493. G. H. Keats, *J. Chem. Soc.*, 1592 (1937).
494. A. K. Chandra, *Tetrahedron*, **19,** 471 (1963).
495. N. J. Aroney, R. J. W. Le Fèvre, and J. D. Saxby, *J. Chem. Soc.*, 571 (1965).
496. T. W. J. Taylor and W. C. J. Coughtrey, *J. Chem. Soc.*, 974 (1935).
497. T. W. J. Taylor, *J. Chem. Soc.*, 625 (1935).

498. K. F. Purcell and J. R. Berschied, Jr., *J. Amer. Chem. Soc.*, **89,** 1579 (1967).

499. K. Mislow, P. Schneider, and A. L. Ternay, Jr., *J. Amer. Chem. Soc.*, **86,** 2957 (1964).

500. M. Janczewski and W. Charmas, *Roczniki Chem.*, **40,** 1243 (1966); M. Janczewski, M. Dec, and W. Charmas, *Roczniki Chem.*, **40,** 1021 (1966).

501. J. Chikos and K. Mislow, *J. Amer. Chem. Soc.*, **89,** 4815 (1967).

502. S. Gronowitz, *Arkiv Kemi*, **13,** 295 (1958).

503. A. Fredga, *The Svedberg 1884, 30/4, 1944*, Uppsala 1944, p. 261 (taken from reference 502).

504. B. F. Crowe and F. F. Nord, *J. Org. Chem.*, **15,** 688 (1950).

505. L. J. Owen and F. F. Nord, *J. Org. Chem.*, **16,** 1864 (1951).

506. G. N. Jean, *J. Org. Chem.*, **21,** 419 (1956).

507. S. Gronowitz and L. Karlsson, *Arkiv Kemi*, **22,** 119 (1964).

508. S. Gronowitz and H. Frostling, *Acta Chem. Scand.*, **16,** 1127 (1962).

509. S. Gronowitz and R. Beselin, *Arkiv Kemi*, **21,** 349 (1964).

510. S. Gronowitz and J. E. Skramstad, *Arkiv Kemi*, **28,** 115 (1968).

511. S. Gronowtiz, *Arkiv Kemi*, **23,** 307 (1965).

512. S. Gronowitz and P. Gustafson, *Arkiv Kemi*, **20,** 289 (1963).

513. R. Håkansson and E. Wiklund, *Arkiv Kemi*, **31,** 101 (1970).

514. W. E. Truce and D. D. Emrick, *J. Amer. Chem. Soc.*, **78,** 6130 (1956).

515. I. O. Sutherland and M. V. J. Ramsay, *Tetrahedron*, **21,** 3401 (1965).

516. K. Mislow, M. A. W. Glass, Jr., H. B. Hopps, E. Simon, and G. H. Wahl, *J. Amer. Chem. Soc.*, **86,** 1710 (1964).

517. R. R. Fraser and F. J. Schuber, *Chem. Comm.*, 1474 (1969).

518. R. R. Fraser and F. J. Schuber, *Chem. Comm.*, 397 (1969).

519. R. R. Fraser, F. J. Schuber, and Y. Y. Wigfield, *J. Amer. Chem. Soc.*, **94,** 8795 (1972).

520. R. R. Fraser and Y. Y. Wigfield, *Chem. Comm.*, 1471 (1970).

521. H. J. Barber and S. Smiles, *J. Chem. Soc.*, 1141 (1928).

522. M. S. Lesslie and E. E. Turner, *J. Chem. Soc.*, 2394 (1932).

523. W. L. F. Armarego and E. E. Turner, *J. Chem. Soc.*, 3668 (1956).

524. C. K. Ingold, private communication 1956.

525. D. D. Fitts, M. Siegel, and K. Mislow, *J. Amer. Chem. Soc.*, **80,** 480 (1958).

526. W. L. F. Armarego and E. E. Turner, *J. Chem. Soc.*, 13 (1957).

527. C. Djerassi, A. Fredga, and B. Sjöberg, *Acta Chem. Scand.*, **15,** 417 (1961).

528. W. L. F. Armarego and E. E. Turner, *J. Chem. Soc.*, 1665 (1956).

529. P. N. Braunton, I. T. Millar, and J. C. Tebby, *J.C.S. Perkin II*, 138 (1972).

530. R. H. Martin, *Angew. Chem. Internat. Edn.*, **13,** 649 (1974).

531. H. Wynberg and M. B. Groen, *J. Amer. Chem. Soc.*, **90,** 5339 (1968).

532. S. Stulen and G. J. Visser, *Chem. Comm.*, 965 (1969).

533. H. Wynberg and M. B. Groen, *Chem. Comm.*, 964 (1969).

534. H. Wynberg and M. B. Groen, *J. Amer. Chem. Soc.*, **92,** 6664 (1970).

535. H. Numan and H. Wynberg, *Tetrahedron Letters*, 1097 (1975).

536. D. Ginsburg, *Accounts Chem. Res.*, **2,** 121 (1969).

537. I. Lantos and D. Ginsburg, *Tetrahedron*, **28,** 2507 (1972).

538. L. A. Paquette, R. E. Wingard, Jr., J. C. Philips, G. L. Thompson, L. K. Read, and J. Clardy, *J. Amer. Chem. Soc.*, **93,** 4508 (1971).

539. K. Weinges, K. Klessing, and R. Kolb, *Chem. Ber.*, **106,** 2298 (1973).

4 MIXED NITROGEN, OXYGEN AND SULFUR HETEROCYCLES

I. INTRODUCTION

Mixed nitrogen and oxygen, nitrogen and sulfur, oxygen and sulfur, and nitrogen, oxygen, and sulfur heterocycles are discussed in this chapter. The stereochemistry of these mixed heterocycles is predictable from a knowledge of the stereochemistry of the individual heterocyclic systems discussed in the preceeding chapters. The configurational and conformational properties in the mixed systems were studied in much the same way as in simpler systems, and the increase in the number of heteroatoms generally makes the stereochemistry simpler. This is partly because the heteroatom itself does not usually introduce an element of asymmetry into the molecule as does a carbon atom. A nitrogen atom inverts too rapidly to be asymmetric, the oxygen atom needs to be alkylated to an unstable oxonium ion to be asymmetric, and the sulfur atom is asymmetric in its monooxide or thionium derivatives. The stereochemistry of mixed heterocycles, however, has attracted interest. Some special features about their syntheses, configurations, conformations, and reactions have been demonstrated and are unraveled in the following sections.

II. NITROGEN AND OXYGEN HETEROCYCLES

A. Three-Membered Rings

The oxidation of aldimines and ketimines with perbenzoic acid is a satisfactory method for preparing oxaziridines.[1] In this way a variety of oxaziridines with substituents on the carbon and on the nitrogen atom can be obtained. When the substituents on the carbon atom are different the carbon atom is asymmetric. Previous experience (see Part I, Chapter 2, Section I.A.3) predicts that the nitrogen atom in oxaziridines should invert extremely slowly and that *N*-substituted derivatives should have a chiral nitrogen atom (Equation 1). Boyd and Graham[2] oxidized several aldimines and ketimines with (+)-peroxycamphoric acid and isolated optically active oxaziridines. The optically active oxaziridine from *N*-*t*-butyl formimine (i.e., *N*-*t*-butyloxaziridine, **1**, $R^1=R^2=H$, $R^3=t$-Bu) clearly has an asymmetric nitrogen atom. When C–3 and N–2 each have a substituent diastereomers are possible, but the oxidation is usually stereoselective. The highest optical rotation was obtained for 3-(4-nitrophenyl)-2-*t*-butyloxaziridine, $[\alpha]_{436}$ − 210°, and the optical purity was estimated at 66%. The stereoselectivity of this oxidation is dependent on the solvent and temperature used. Boyd and co-workers[3] were able to deduce the relative stereochemistry of the substituents on C–3 and N–2 from their spectral properties and found that the stereomers can be interconverted on irradiation with uv light. It is the configuration of the substituent on the nitrogen atom that is most probably inverted upon irradiation.

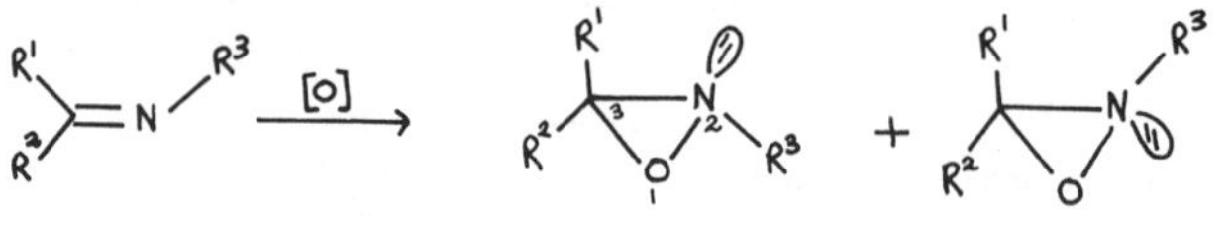

(1)

Bapat and Black[4] found that photolysis of 2,4-disubstituted 5,5-dimethyl-1-pyrroline 1-oxides gave a mixture of *cis*- and *trans*-3,5-disubstituted 2,2-dimethyl-6-oxa-1-azabicyclo[3.1.0]hexanes when the 2-substituent in the pyrroline ring was a methyl group (Equation 2). However, when the substituent at C–2 was an aryl group stereoselectivity was observed and only the trans isomers (**2**, R^2 = aryl) were formed. Oxidation of the substituted pyrrolines with perbenzoic acid, on the other hand, gave exclusively the *cis*-oxaziridine **3**. Isomerization to the *trans*-5-phenyl derivatives (**2**, R^2 = Ph) occurred upon irradiation of the *cis*-5-phenyl derivatives (**3**, R^2 = Ph), but the *cis*-5-methyl derivative was photostable. Oxaziridines were reviewed by

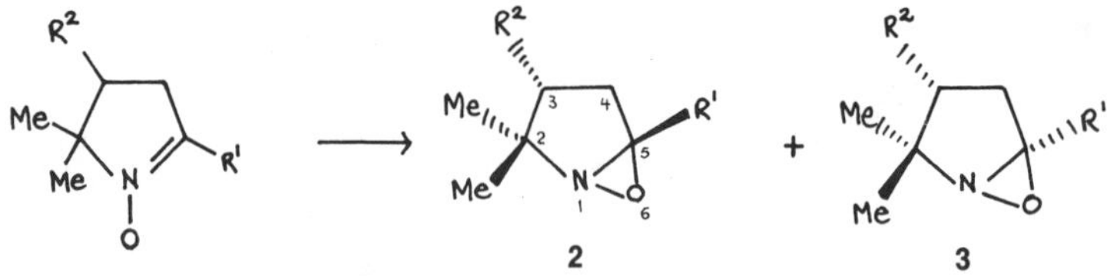

(2)

Schmitz in 1963.[5]

B. Four-Membered Rings

2-Chloro- and 2-fluoroperfluoro-1,2-oxazetidines were examined by ^{19}F nmr spectroscopy, which provided evidence for nonplanarity in these molecules. Nmr spectra over the range of 85 to −120°C revealed that the nonplanar conformers were in equilibrium (**4** ⇌ **4a**), and the free energy differences for ring inversion between them were of the order of 3.77 and 4.19 kJ mol^{-1} for the 2-fluoro and 2-chloro derivatives respectively.[6] Readio[7] also studied the ^{19}F nmr spectra of 4-chloro-2,3,3,4-tetrafluoro- and 2,4-dichloro-3,3,4-trifluoro-1,2-oxazetidines, and in addition to observing ring inversion they provided further evidence for the high barrier to pyramidal nitrogen inversion. The spectra consisted of two equilibrating (by ring inversion) species in each case, for example, *cis*- (**5** and **5a**) and *trans*-2,4-dichloro-3,3,4-trifluoro-1,2-oxazetidines (**6** and **6a**), without interconversion of the cis and trans isomers. The free energies for ring inversion were determined and found to be of the same order as in the above oxazetidines.

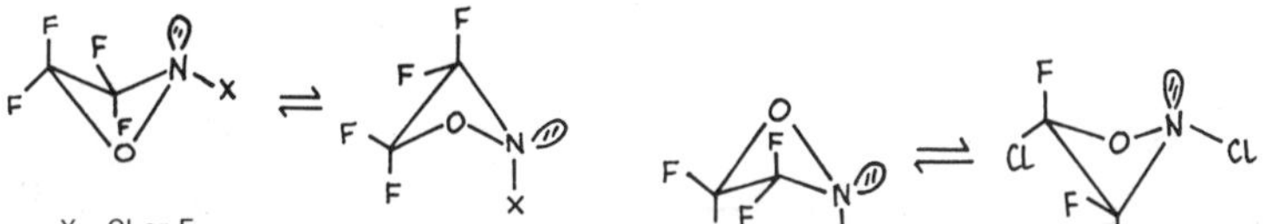

C. Five-Membered Rings

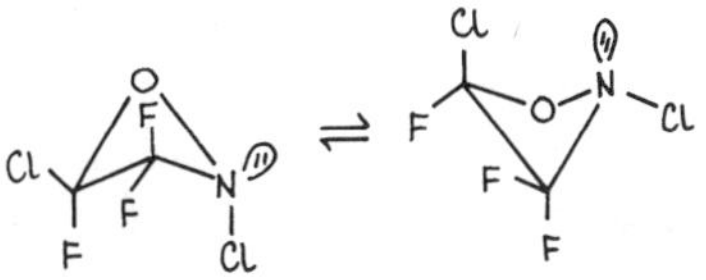

1. *Isoxazolines and Isoxazolidines*

Most of the stereochemical information about reduced isoxazoles comes from synthetic studies. The reduced isoxazole ring is readily formed by a [3 + 2] cycloaddition reaction which proceeds thermally. The simplest of these is the condensation of nitrile oxides (the most useful being arylnitrile oxides) with olefins to yield 2-isoxazolines (Equation 3). The addition is stereospecific with the configuration of the olefin being retained in the product; for example, condensation of benzonitrile oxide with maleic esters yields *cis*-4,5-dialkoxycarbonyl-3-phenyl-2-isoxazoline.[8–11] The polarity of the double bond in unsymmetrical olefins confers regiospecificity on the reaction.[12] Thus benzonitrile oxide reacts with *trans*-β-nitrostyrene,[13] citraconic anhydride,[14] cinnamic ester, cinnamic acid,[15] and cyclopentadiene[16] to yield *trans*-4-nitro-5-phenyl-, *cis*-4,5-dicarboxy-5-methyl-, *trans*-4-ethoxycarbonyl-5-phenyl-, *trans*-5-carboxy-4-phenyl-3-phenyl-2-isoxazolines, and 4-phenyl-*cis*-2-oxa-3-azabicyclo[3.3.0]octa-3,7-diene (**7**) respectively. The last named compound can react further with benzonitrile oxide at the 7,8 double bond in two orientations to give a 30:45 mixture of the *cis*- and *trans*-biisoxazolines (**8**), but these could also have the isoxazoline rings syn (e.g., **8**), or anti (e.g., **9**) with respect to each other. In a series of

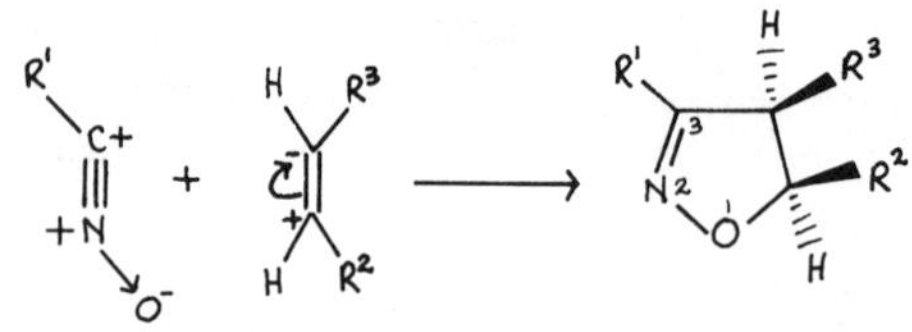

(3)

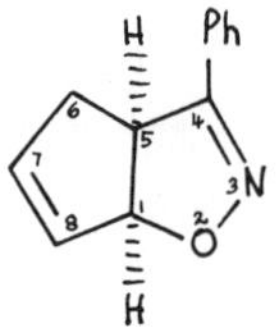

7

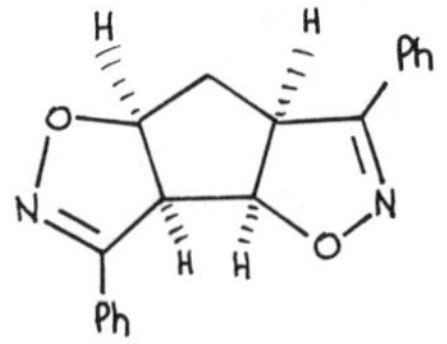

8

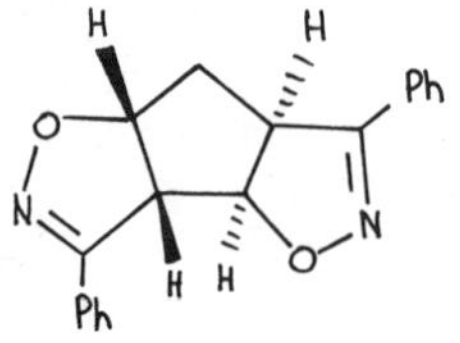

9

papers Huisgen and co-workers[17–20] examined these condensations in great detail and laid down the theoretical basis for these 1,3-dipolar cycloaddition reactions. By using fulminic acid ($HC{\equiv}N{\rightarrow}O \leftrightarrow H{-}C^{-}{=}N^{+}{=}O$) in place of a nitrile oxide they obtained 3-unsubstituted 2-isoxazolines.[21]

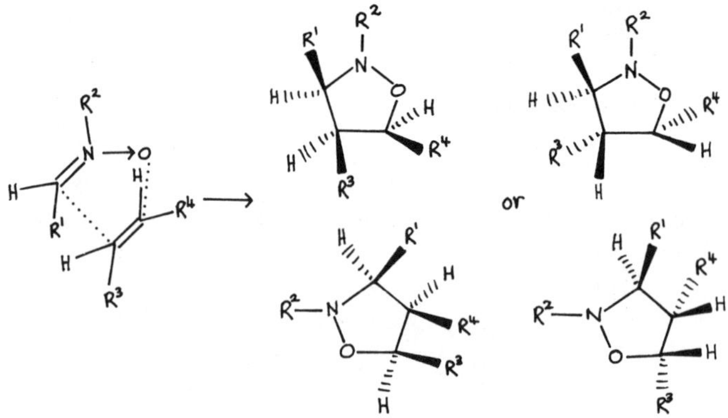

Scheme 1

Nitrones undergo 1,3-dipolar cycloaddition reactions with olefins in much the same way as the nitrile oxides above, but the system is more complicated because a new asymmetric center may be created at C–3 during the formation of isoxazolidines (Scheme 1).[22–27] Although the dipolar addition takes place with complete cis stereospecificity, the stereochemistry of the product or products is strongly dependent on the approach of the reagents and is therefore dictated by nonbonded and polar interactions in the transition states.[28] The number of isomers formed in these cyclizations is not as large as predicted by Scheme 1 because these nonbonded interactions strongly influence the direction of the reactions. Huisgen and co-workers[23] studied the kinetics of addition of many nitrones to ethyl crotonate in various solvents and interpreted the parameters in terms of dipolar effects. Cyclic nitrones, such as 1-pyrroline 1-oxide[29] and 3,4-dihydroisoquinoline 1-oxide,[23,30,31] similarly react with olefins to yield isoxazolidines with high stereospecificity.

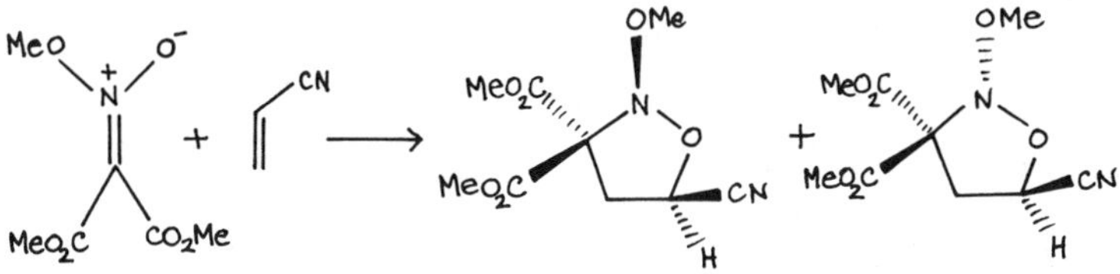

(4)

Nitronic esters react with olefins in the same way but yet another source of chirality is introduced. The substituent on the nitrogen atom of the isoxazolidine formed, usually an alkoxy group, and the adjacent oxygen of the ring make the barrier for pyramidal nitrogen inversion so high that syn and anti isomers are not readily interconverted. Müller and Eschenmoser[32] condensed the *O*-methyl nitronic ester of dimethyl α-nitromalonic ester with acrylonitrile and obtained two *N*-methoxy-5-cyano-3,3-bismethoxycarbonylisoxazolidines in about equal amounts, which they separated by tlc (Equation 4). The Arrhenius activation energy for the epimerization, that involves inversion at the nitrogen atom, was found to be 122.2 kJ mol^{-1} and is well in the region where the epimers could be isolated in optically active form. Grée and Carrié[33,34] investigated several 1,3-dipolar additions of nitronic esters to olefins and deduced the stereochemistry of the products by nmr methods. They found that the stereo- and regiospecificities are usually quite high. They rationalized their results in terms of the stereochemistry of the nitronic ester, in which case the *cis*-nitronic esters reacted faster than the trans esters, and with respect to the substituent on the olefin the "endo" approach **10** (not the "exo" approach **11**) is always observed.[35]

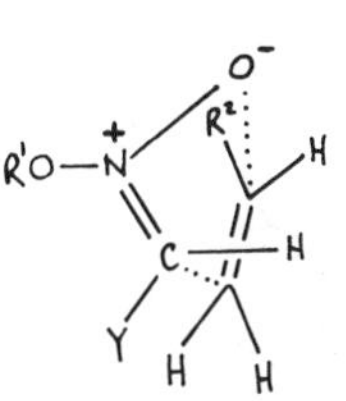

10 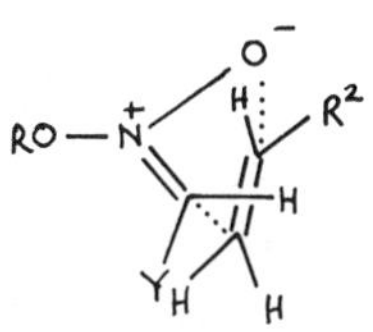 **11** 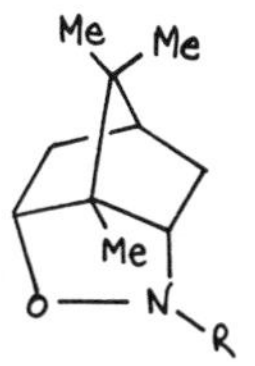 **12**

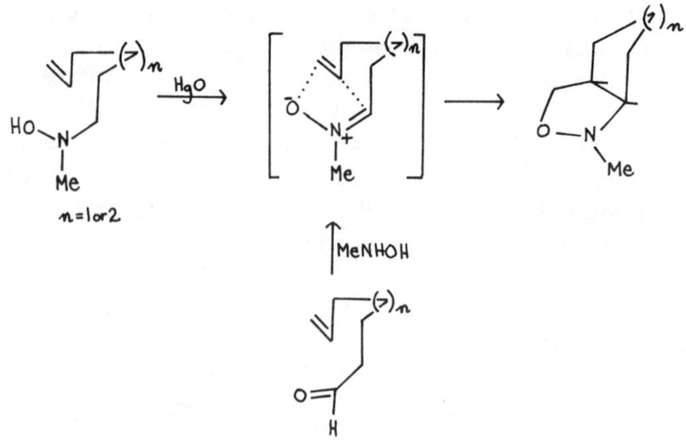

Scheme 2

LeBel and co-workers[36–38] used the condensation of nitrones with olefins in an intramolecular sense. Thus by generating a nitrone in the mercuric oxide oxidation of *N*-methyl-*N*-hex-5-enyl hydroxylamine they obtained 2-methyl-*cis*-3-oxa-2-azabicyclo[3.3.0]octane (Scheme 2, $n=1$). This reaction can also be carried out by condensation of hex-5-enal and methyl hydroxylamine without oxidation because of the higher oxidation state of the aldehyde compared with the hydroxylamine above. In this last reaction some trans isomer was formed. The synthesis was extended to a seven carbon chain (e.g., Scheme 2, $n=2$), and in this case the proportion of 7-methyl-*trans*-8-oxa-7-azabicyclo[4.3.0]nonane was much higher than of the cis isomer.[37] They adapted this intramolecular cyclization successfully to tetrasubstituted olefin derivatives[39] and to ω-nitroneallenes.[40] The nitrone of campholenic aldehyde and of cyclohex-3-ene aldehydes[41] (or cyclohex-3-enylacetonitrile)[42] reacted similarly with hydroxylamines to give tricyclic compounds such as **12.** The shorter chain length between the nitrone function and the olefinic double bond usually leads to polycyclic compounds, as in the cyclization of *N*-1-buten-3-yl-*anti*-benzaldehyde oxime (**13**) to *exo*-phenyl-7-oxa-1-azabicyclo[2.2.1]heptane (**14**) by slow distillation or by

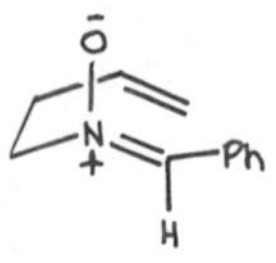

13

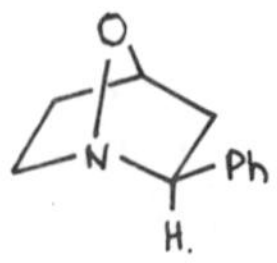

14

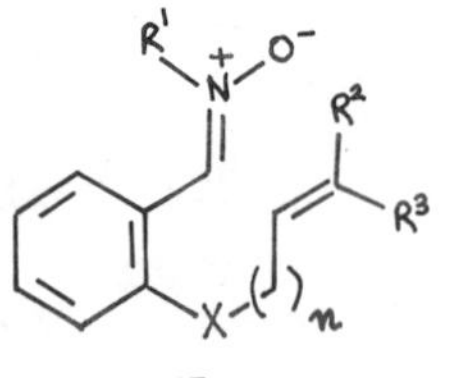

15

boiling in xylene.[43] Oppolzer and co-workers[44–46] adapted this type of reaction to the cyclization of nitrone olefins with the general formula **15** (where X is O or NR and *n* is 1 or 0), and the nature of some of the *cis*-cycloaddition products was confirmed by X-ray analyses.[45]

Benzo[*c*]isoxazole condenses with *N*-phenylmaleimide to give an adduct with a heterocyclic skeleton similar to **14**,[47] and uv irradiation of 3,4-dimethyl-1,3,5-oxadiazole must generate methylnitrile oxide because in the presence of cyclopentene it forms 4-methyl-*cis*-2-oxa-3-azabicyclo[3.3.0]oct-3-ene in the 23% yield.[48] Other miscellaneous syntheses include the base-catalyzed cyclization of 1,3-dinitro-1,2,3-triphenylpropane into *trans*-3,4,5-triphenyl-2-isoxazoline 2-oxide,[49] the Beckmann rearrangement of *cis*-bicyclo[4.2.0]octan-7-one to *cis*-2-oxa-3-azabicyclo[4.3.0]nonane,[50] the cycloaddition reactions of sydnones with acetylenes,[51,52] and the cycloaddition of nitrosobenzene to *N*-ethoxycarbonylazepine which yields the azaoxazolidine, 9-ethoxycarbonyl-8-phenyl-7-oxa-8,9-diazabicyclo[4.2.1]-nona-2,4-diene (**16**).[53] The latter reaction is most probably nonconcerted because thermally induced $[6 + 2]\pi$ cycloadditions of this type are not permissible according to the orbital symmetry rules.

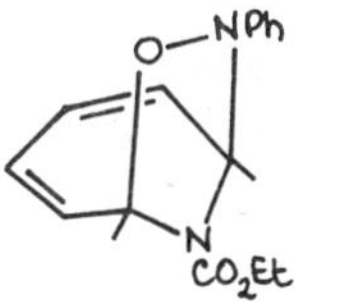

16 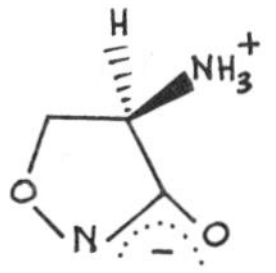 **17** 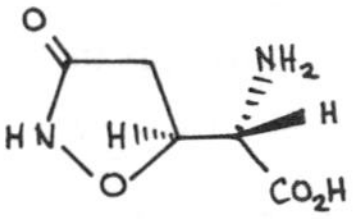 **18**

The broad spectrum antibiotic oxamycin is *R*-4-amino-3-isoxazolidinone (**17**) and is commonly known as D-cycloserine. Folkers and co-workers[54] synthesized the racemate from (±)-serine and resolved it by way of the (+)- and (−)-tartrate salts. Although it is stable as the zwitterion,[55] it racemizes in the presence of 5-chlorosalicylaldehyde because of the formation of a Schiff base which labilizes the α-hydrogen atom.[56] *S*-Erythro-α-amino-3-oxoisoxazolidin-5-ylacetic acid (**18**) is the tasty constituent of the mushroom *Tricholomic muscarium*, which has flycidal properties. It was isolated and identified by Kamiya and co-workers,[57–59] who synthesized it and its isomers.

The high barrier for nitrogen pyramidal inversion in *cis*- and *trans*-5-cyano-2-methoxy-3,3-bismethoxycarbonylisoxazolidines (see Equation 4) is mentioned earlier, and Müller[60] used three examples as models in the theoretical treatment of nitrogen inversion in *N*,*N*-dialkoxyalkylamines. LeBel and co-workers[61] studied the effects of solvent on the slow nitrogen inversion in 1,2-dimethyl-3-oxa-2-azabicyclo[3.3.0]octane, **19** ⇌ **19a**. They found that $\Delta G^{\ddagger}$ varied between 56.1 and 60.3 kJ mol^{-1}, depending on the solvent, and inversion in this compound was clearly faster than in the *N*-methoxy derivatives in Equation 4. The nitrogen inversion barrier in *N*-methylisoxazolidine in $CDCl_3$ has a $\Delta G^{\ddagger}$ value of 65.3 kJ mol^{-1}.[62]

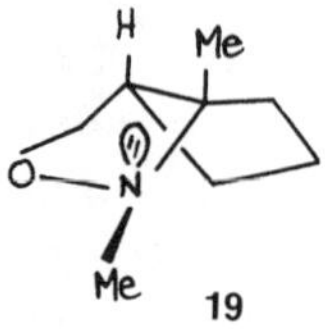

19 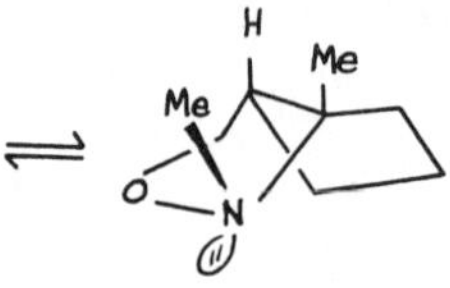 19a 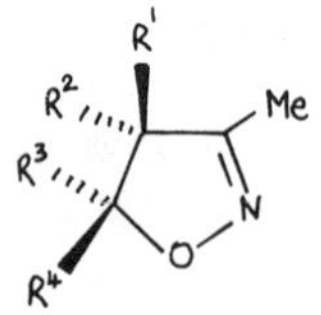 20

Two reactions of 2-isoxazolines need to be mentioned. The first is the facile epimerization that occurs in the 5-hydroxy derivatives (**20,** R^3 or R^4=OH) which were prepared by reacting 2-isoxazolin-5-ones with Grignard reagents. The epimerization between **20** (R^3=OH, R^4=H) and **20** (R^3=H, R^4=OH) must occur by ring opening between O–1 and C–5 and must be subject to the relative steric effects of the substituents R^1 and R^2.[63] The second reaction is the Lossen rearrangement of *cis*-3-phenyl-2-isoxazolin-*N*-benzenesulfonyloxy-4,5-dicarboximide, which gives a mixture consisting predominantly of *cis*-4-carboxy-3-phenyl-5-ureido-2-isoxazoline with retention of configuration at C–5.[64]

2. *Oxazolines and Oxazolidines*

The stereochemistry of formation of 2-oxazolines has received considerable attention because the compounds can be prepared in a variety of ways. These are basically cyclized *O*-acyl- or *N*-acyl-β-alkanolamines, and the mechanism of synthesis will dictate whether the carbon atom attached to the oxygen atom will undergo a Walden inversion or not. Sicher and collaborators[65–67] studied the formation of 2-oxazolines from β-hydroxyalkylamines and related compounds in several ways. For example *threo-N*-thiobenzoyl derivatives cyclized only to *trans*-4,5-disubstituted 2-oxazolines with retention of configuration (e.g., Equation 5), and the erythro isomers gave preferentially, but not entirely, the *cis*-4,5-disubstituted 2-oxazolines.[68]

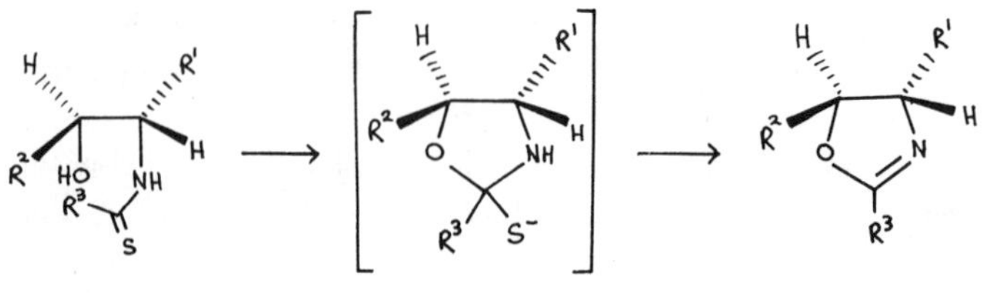

(5)

(6)

Treatment of *N*-acyl β-ethanolamines with thionyl chloride yields 2-oxazolines with complete inversion at the carbon atom bearing the oxygen atom (Equation 6).[69] The intramolecular cyclization of *O*-methanesulfonyl derivatives of *cis*- and *trans*-*N*-benzoyl-β-hydroxycycloalkylamines, by solvolysis in EtOH/KOAc, also occurs with a Walden inversion and loss of methanesulfonic acid which contains the oxygen atom from the alcohol. The cycloalkane ring in these examples contained more than eight carbon atoms, and the rates for the trans isomers were faster than for the cis isomers.[70] The cyclizations generally fall in the two classes described above, and much of the ideas of Sicher and co-workers were confirmed and amplified by others.[71–75] Pines and co-workers[76,77] used *N*-acyl and *N*-aroyl 3-phenyl- and 2-methyl-3-phenylserines, and while obtaining similar mechanisms showed that an S_N1 mechanism can intervene because the phenyl ring can stabilize a carbonium ion at C–3 after it loses the oxygen atom. They used the formation of a 2-oxazoline in order to convert *erythro*-3-aryl-2-methylserine esters into the threo isomers by inversion at C–3 (Equation 7).[78]

(7)

R=H, SOCl

Other stereospecific or stereoselective syntheses of 2-oxazolines involve the cyclization of β-ureido alcohols, which occur with inversion if $SOCl_2$ is used and yield the 2-amino derivatives.[79] When *trans*-2-aminocyclohexanol is condensed with the dimethyl ester of *N*-tosylamidodithiourethane [p-$CH_3C_6H_4$–$SO_2NHCON{=}(SMe)_2$] 2-tosylamino-*trans*-4,5-tetramethylene-2-oxazoline is formed with retention of configuration,[80] but cyclization of *trans*-2-*N*[1]-tosylureidocyclohexanol takes place with inversion of configuration, and 2-tosylamino-*cis*-4,5-tetramethylene-2-oxazoline is obtained.[81] *trans*-2-Amino-4,5-tetramethylene-2-oxazoline is formed from *trans*-2-aminocyclohexanol and cyanogen bromide with retention of configuration,[82] whereas the cis isomer is obtained by boiling *trans*-1-iodo-2-ureidocyclohexane in water.[83] The latter type of reaction can be enhanced by the addition of silver ions.[84] *cis*-4,5-Trimethylene-2-*p*-nitrophenyl-2-oxazoline is similarly formed from *trans*-1-iodo-2-*p*-nitrobenzamidocyclopentane.[85] The cyclization of *N*-benzoyl cyclohex-2-en-1-ylamine with bromine in methanol, followed by catalytic reduction, yields *cis*-4,5-tetramethylene-2-phenyl-2-oxazoline.[86]

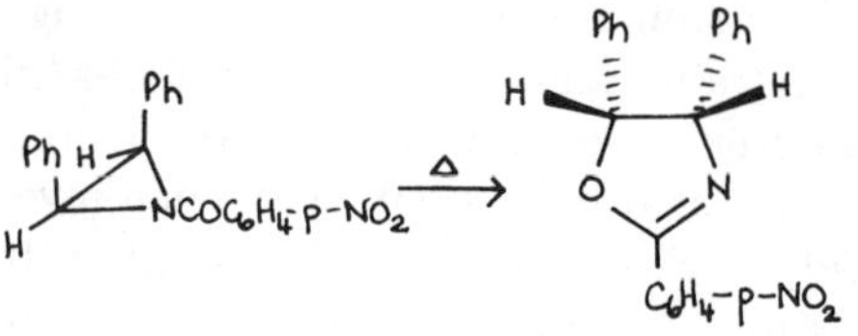

Ph, Ph H, H, $NCOC_6H_4$-p-NO_2 $\xrightarrow{\Delta}$ Ph, Ph, H, H, O, N, C_6H_4-p-NO_2 (8)

Thermolysis of *N*-*p*-nitrobenzoyl *cis*- and *trans*-2,3-diphenylaziridines causes them to rearrange into 2-*p*-nitrophenyl *cis*- and *trans*-4,5-diphenyl-2-oxazolines with retention of configuration (e.g., Equation 8). The nitro group appears to be essential because the corresponding *N*-benzoyl *cis*-2,3-diphenylaziridine does not rearrange.[87] The reaction of *S*-2-acetamido-2-(3,4-dimethoxybenzyl)propionitrile at a platinum anode furnishes a 3.5:1 mixture of *cis*-4-cyano-5-aryl(4*R*,5*S*)- and *trans*-4-cyano-5-aryl(4*R*,5*R*)-2,4-dimethyl-2-oxazolines. The stereospecificity of this reaction was attributed to asymmetric induction at the benzylic cation rather than to the heterogeneity of the reaction of the anode.[88]

Tosylmethyl isocyanide readily condenses with ketones in the presence of TlOEt and forms a mixture of *cis*- and *trans*-5,5-disubstituted 4-ethoxy-2-oxazolines in which the isomer that has the ethoxy substituent in a trans relationship with the bulkier C–5 substituent predominates. Originally the 4-tosyl derivative is formed, but it undergoes nucleophilic displacement by the ethoxide present in solution.[89] Photolysis of 2-substituted 3-phenyl-2*H*-azirines causes C–2,C–3 bond cleavage and produces a nitrile ylide. This ylide reacts with aldehydes to form *cis*- and *trans*-2,5-disubstituted 4-phenyl-3-oxazolines in which the substituent at C–5 is derived from the aldehyde. The cis isomer generally predominates in this reaction and a transition state such as **21** was suggested.[90]

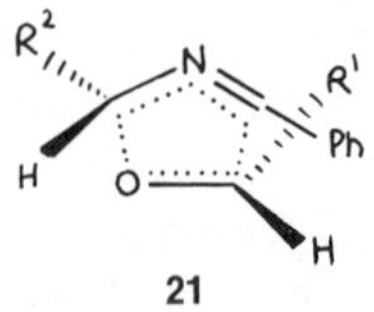

R^2, N, R^1, Ph, H, O, H

21

In the general synthesis of oxazolidin-2-ones from the cyclization of β-substituted alkylamines the same two types of reactions as above are observed, that is, one with retention and the other with inversion of configuration at one of the carbon centers involved. The thermal cyclization of *threo*- and *erythro*-β-iodocarbamates yields mainly the respective *cis*- and *trans*-oxazolidin-2-ones with inversion of configuration at the carbon atom which releases the halogen atom. The stereoselectivity for the erythro isomers is greater than for the threo isomers.[91] Fodor and Koczka[92] examined the stereochemical course of the conversion of 2-ureido alcohols into oxazolidin-2-ones and used examples from the ephedrine series. These are derivatives of β-*N*-methylaminoethanols and were found to cyclize with retention of configuration meaning that the urea NH_2 is lost in the process. β-Chloroethyl ammonium chlorides are transformed into oxazolidin-2-ones in one step by heating with Na_2CO_3 in dry dimethyl sulfoxide with inversion of configuration at the β-carbon atom.[93] As in the cyclization of β-iodourethanes, one would expect that β-hydroxyisocyanates should produce oxazolidin-2-ones. The isocyanate was prepared in situ in the Curtius reaction from 1*R*,2*S*(−)-*cis*-2-hydroxy-1-hydrazinocarbonylcyclohexane which gave 4*R*,5*S*(−)-*cis*-4,5-tetramethyleneoxazolidin-2-one directly with retention of configuration at the asymmetric carbon atoms (Equation 9).[94] *trans*-1-Iodo-2-isocyanatocyclohexane, however, yields 2-

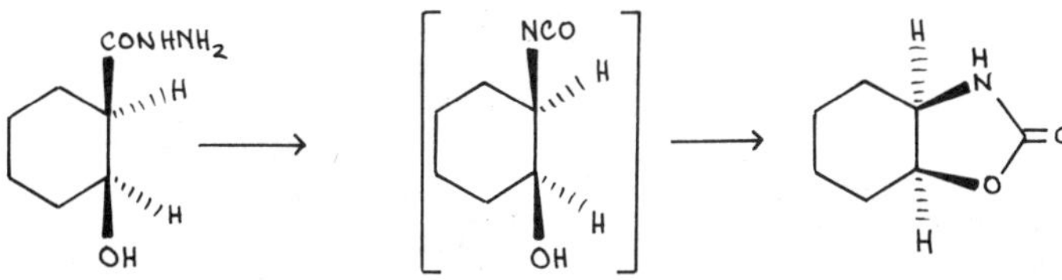

CONHNH2 H H OH → [NCO H H OH] → H H N O =O (9)

tosylamino-*cis*-4,5-tetramethylene-2-oxazoline with inversion of configuration at C–1.[95]

Intramolecular cyclization by attack of the amino group is also possible as in the formation of *cis*- and *trans*-4,5-disubstituted 3-aryloxazolidin-2-ones by treating the respective *threo*- and *erythro*-β-bromoethylurethanes with inversion of configuration at the β-carbon atom.[96] Similarly *trans*-2-chlorocyclohexylurethane gave *cis*-4,5-tetramethyleneoxazolidin-2-one.[97]

A very ingeneous synthesis of oxazolidin-2-ones, developed by Yamada and his school,[98–101] involves the intramolecular insertion of a nitrene. Photochemically or thermally generated nitrenes from optically active azidoformates that have a tertiary hydrogen atom (necessary for abstraction) four atoms removed from the nitrene nitrogen atom give optically active oxazolidin-2-ones (e.g., Equation 10). This is a general reaction, and the cyclization occurs with very high retention of configuration implying that a "singlet" nitrene is involved. The absolute configuration of the chiral oxazolidin-2-one was deduced by hydrolysis to the corresponding β-amino alcohol which was correlated with the respective α-amino acid. A stereospecific vicinal oxyamination of olefins by alkylimido osmium compounds

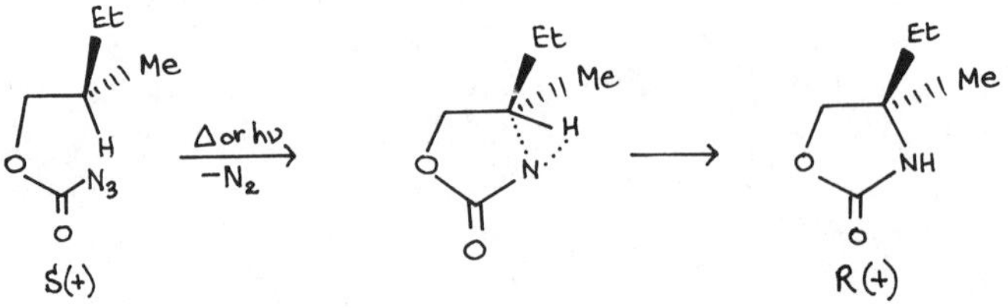

(10)

was observed by Sharpless and his co-workers, and this was applied to the synthesis of oxazolidin-2-ones. *trans*-1-Deuterodec-*trans*-1-ene was treated with *t*-butylimido osmium trioxide followed by reaction with LAH, OH^-, and carbonyl diimidazole and gave 3-*t*-butyl-*trans*-4-deutero-5-octyloxazolidin-2-one. If the decene were oxidized first to the oxirane with *m*-chloroperbenzoic acid followed by *t*-BuNHLi and then carbonyl diimidazole, isomeric 3-*t*-butyl-*cis*-4-deutero-5-octyloxazolidin-2-one could be formed (Equation 11). These reactions appear to exhibit remarkable regiospecificity.[102]

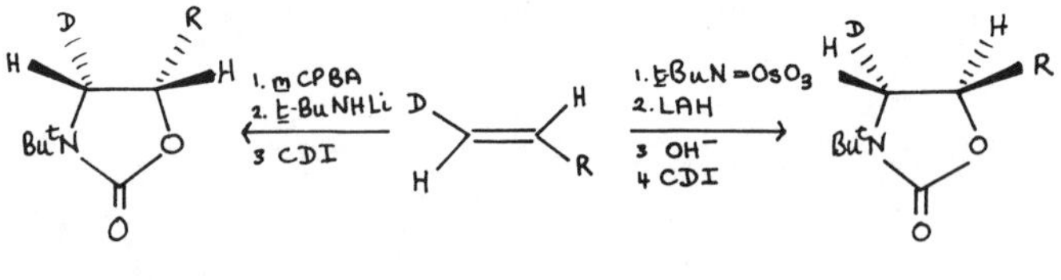

(11)

Methyl isothiocyanates which carry an ethoxycarbonyl, phenyl, or ethenyl group on the methyl group condense with aldehydes or ketones in the presence of *t*-BuOK or NaH to furnish mixtures of *cis*- and *trans*-4,5-disubstituted oxazolidin-2-thiones in high yields. The isomers in which the bulkier C–4 and C–5 groups are trans usually predominate.[103]

Optically active *S*-2-phenyl-4-benzyl-2-oxazolin-5-one was prepared from *N*-benzoyl *S*-phenylalanine with acetic anhydride in dioxane. Its racemization, and kinetics of ring cleavage, were studied in detail by Goodman and Levine[104] in connection with peptide synthesis. The use of *N*-carboxy-α-amino acid anhydrides (the 4-substituted oxazolidin-2,5-diones derived from α-amino acids) in polypeptide syntheses is well known.[105–108] Two geometrical isomers, one red and one yellow, of 2,4-dibenzylidene-3-

phenyloxazolidin-5-one were isolated, but their relative structures are not known.[105]

Oxazolidines are readily obtainable from β-amino alcohols and aldehydes or ketones.[109–112] Since the chiral centers are not involved in these condensations, then the original configuration in the β-amino alcohols is transferred to the oxazolidines.[113,114] Oxazolidines are isomeric with the Schiff bases of the parent β-amino alcohols. McCasland and Horswill[115] examined the effect of configuration on the formation of oxazolidines. *cis*- and *trans*-2-Aminocyclohexanols react with benzaldehyde to give the Schiff bases and not the oxazolidines. Benzoylation of these benzylidene derivatives causes them to cyclize to *cis*- and *trans*-3-benzoyl-4,5-tetramethylene oxazolidines. Cope and Hancock[116] have used molecular refractivities to distinguish between Schiff bases and oxazolidines, but uv, ir, and nmr spectroscopy show the differences much more clearly. 2-Aryl-1,1-dimethylaziridinium borofluorides undergo a highly stereoselective reaction with benzaldehyde and form *cis*-5-aryl-2-phenyl-3,3-dimethyloxazolidinium borofluorides. A transition state similar to **22** was proposed to account for the specificity.[117]

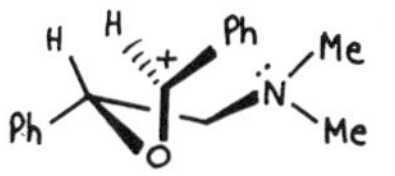

22

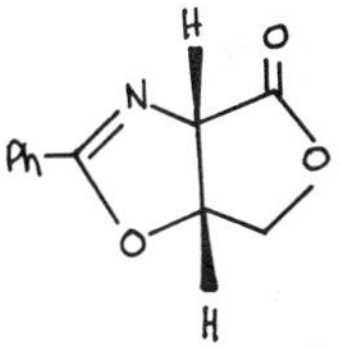

23

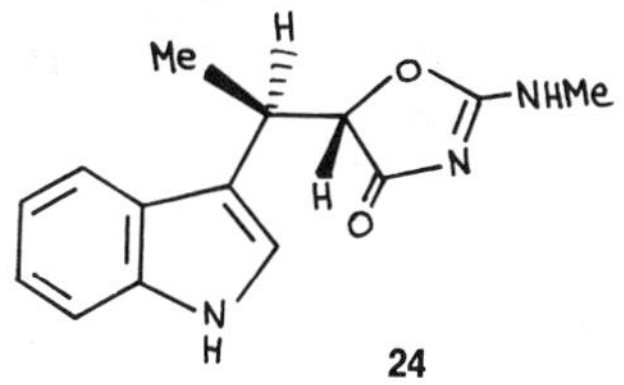

24

Optically active oxazolidin-2-ones derived from ephedrine and norephedrine, with known absolute configurations,[118] and also derived from 2,3-dihydroxypropylamines obtained from a synthesis which started with D-mannitol, have been reported.[119] The absolute configurations of 2-oxazolines derived from threonine follow from the reaction mechanisms,[120,121] and the resolution of *trans*-4-carboxy-5-methyl-2-phenyl-2-oxazoline by way of the brucine salt was accomplished and used to provide both enantiomers of threonine.[121] The cis-trans relationship of substituents at C–4 and C–5 in oxazolidin-2-ones can be obtained from their nmr spectra,[122] and for other nmr data on the oxazolines and oxazolidines the references cited in this section should be consulted. The absolute configuration of the diastereotopic protons in benzylamine were determined from an analysis of the ^{1}H nmr spectra of 4*R*(+)- and 4*S*(−)-3-benzyl-4-phenyloxazolidin-2-thione.[123] The mutarotation observed for the lactones of (+)- and (−)-*cis*-4-carboxy-5-hydroxymethyl-2-phenyl-2-oxazoline (**23**) in alkaline solution is attributed to opening of the lactone ring and not to alteration of the chiral centers.[124] The absolute configuration of the 2-oxazolin-4-one antibiotic indolomycin (**24**) was reported.[125]

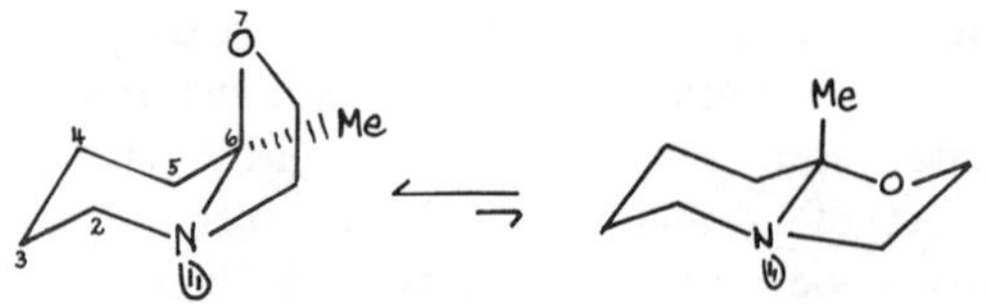

25 **26**

27 **28**

Leonard and Musker[126] realized the conformational properties of oxazolidines particularly when they were fused with six-membered rings and sharing the nitrogen atom, as in 6-methyl-7-oxa-1-azabicyclo[4.3.0]nonane. This compound is capable of existing in two conformations, a cis-fused (**25**) and a trans-fused (**26**) conformation which are interconvertible by pyramidal inversion of the bridgehead nitrogen atom. They concluded, from the absence of Bohlmann bands in the ir spectrum (see Part I, Chapter 4, Section VI.C), that this nonane and its 2-methyl derivative exist in the cis conformation **25.** Crabb and his school[127–132] have examined the conformational equilibria of a large number of substituted 8-oxa-1-azabicyclo[4.3.0]nonanes and related systems by ir ^1H and ^{13}C nmr spectroscopy and interpreted the preferred conformations and configurations in terms of nonbonded interactions. In the parent 8-oxa-1-azabicyclo[4.3.0]nonane, the equilibrium of cis **27** and trans **28** was shown by ^1H nmr[133] and ^{13}C nmr[132] spectroscopy to be in favor of the trans isomer to the extent of 76%. The free energy change is of the order of 10 kJ mol^{-1}.[134] A methyl group at C–3, but cis to the bridgehead hydrogen atom C–6 in this system, does not alter the situation. However, when the methyl group is at C–3, but trans to H–6, then the preferred conformation is cis, as in **27,** in order that the methyl group can assume the equatorial conformation.[133] The conformational properties of these and related bicyclic compounds were reviewed by Crabb, Newton, and Jackson.[135]

Katritzky and co-workers[136] examined and discussed the conformational equilibria of cyclohexanespiro-4′-2′-oxazolines and 2′-substituted derivatives, and found that in all cases the conformers in which the nitrogen atom assumed the equatorial conformation (**29**, R=H, Me, CF_3, Ph, SMe, or *t*-Bu) always predominated. This is contrary to the predicted syn-axial interactions of the CH_2 group compared with the ring nitrogen atom. They also determined conformational preferences and free energy changes in *N*-*t*-butylpiperidine-4-spiro-4′- and 5′-(2′,2′-dimethyloxazolidine) and the corresponding 2′-oxazolines by ^1H nmr and dipole moment methods.[137,138] In the spiro-4′-(2′,2′-dimethyloxazolidine) the conformer with the equatorial nitrogen atom is preferred, but in the spiro-5′-(2′,2′-dimethyloxazolidine) the conformer with the axial oxygen atom is predominant.

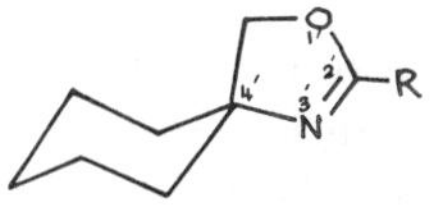

29 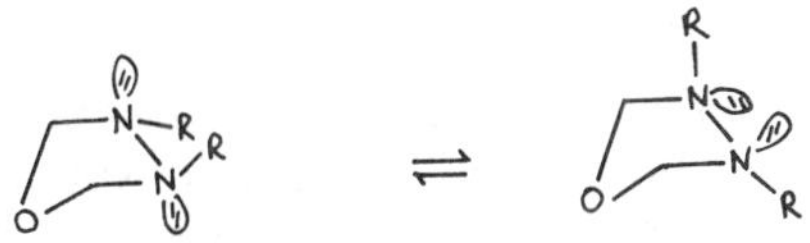 **30** ⇌ **31**

In the simultaneous nitrogen inversion in *N,N′*-disubstituted 1,3,4-oxadiazolidines (4-aza analogues of oxazolidines), Katritzky and co-workers have shown that there is strong steric interaction between the *N*-substituents in the transition state as inversion occurs. This is revealed in the free energy of inversion, as measured from coalescence temperatures, which are 44.3, 50.2, 79.4, and greater than 87.8 kJ mol^{-1}, for **30** ⇌ **31** with R=Me, R=Et, R=isoPr, and R=*t*-Bu respectively.[139]

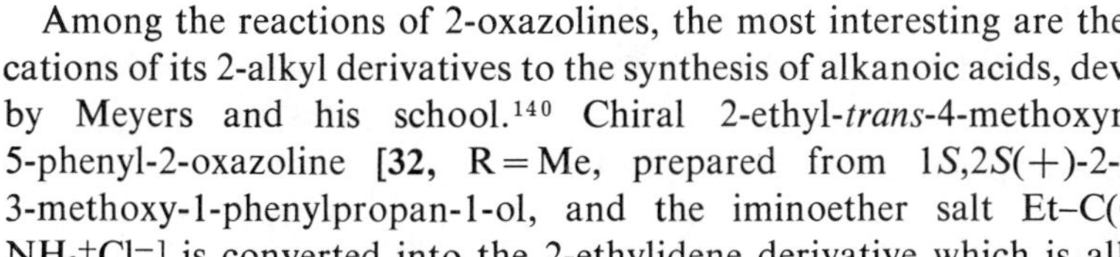

32 $\xrightarrow[-78]{LiN(i\text{-}Pr)_2}$ $\xrightarrow{R^2I}$ **33** (12)

Among the reactions of 2-oxazolines, the most interesting are the applications of its 2-alkyl derivatives to the synthesis of alkanoic acids, developed by Meyers and his school.[140] Chiral 2-ethyl-*trans*-4-methoxymethyl-5-phenyl-2-oxazoline [**32**, R=Me, prepared from 1*S*,2*S*(+)-2-amino-3-methoxy-1-phenylpropan-1-ol, and the iminoether salt Et–C(OEt)=$NH_2^+Cl^-$] is converted into the 2-ethylidene derivative which is alkylated stereoselectively. Hydrolysis of the 2-alkyl-2-oxazoline **33** provides an optically active 2-alkylalkanoic acid with an absolute configuration which can be deduced from a consideration of the transition state involved in the alkylation (Equation 12). The original chiral β-amino alcohol is recovered unchanged from this reaction and can be recycled. Similarly the 2-methyl derivative **32**, R = H) can be treated in this way, and several optically active alkanoic acids, with optical purities ranging from 20 to more than 70%, were prepared in this way.[141–143]

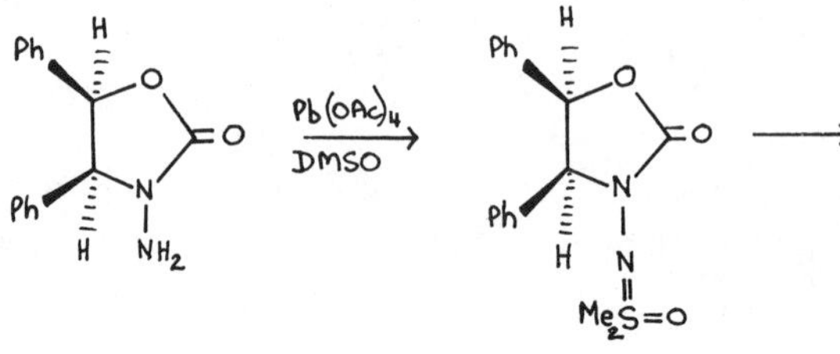

(13)

The stereospecific thermal fragmentation of sulfoximines, formed by reacting 4,5-disubstituted 3-aminooxazolidin-2-ones with $Pb(OAc)_4$, in the presence of dimethyl sulfoxide, provides the olefins with the configurations of substituents similar to the ones present in the original oxazolidin-2-ones (e.g., Equation 13). The stereospecificity and yield is better than 78% and affords a new synthesis for olefins.[144] Finally, mention must be made of the photolytic intramolecular cyclization of *N*-(ethoxycarbonyl)diazoacetyl-4,4-dimethyloxazolidine to the oxapenam, (*trans*-5*H*,6*H*)-6-ethoxycarbonyl-2,2-dimethyl-7-oxo-4-oxa-1-azabicyclo[3.2.0]heptane. The azetidinone ring, in this oxygen analogue of a penicillin, is readily cleaved, with benzylamine for example, to yield 2-(α-*N*-benzylcarbamoyl-α-ethoxycarbonylmethyl)-

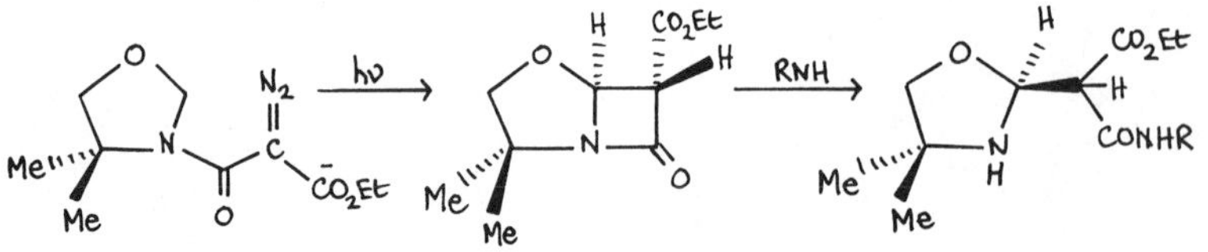

(14)

4,4-dimethyloxazolidine (Equation 14).[145] Several reviews can be consulted for the general chemistry of oxazoles and reduced oxazoles.[146–148]

D. Six-Membered Rings

1. *Reduced 1,2-Oxazines*

Reduced 1,2-oxazines are essentially cyclic hydroxylamines. A regio- and stereoselective synthesis of 3,6-dihydro-1,2-oxazines was investigated by Kresze and co-workers[149–151] and involves a Diels-Alder reaction. A $(4 + 2)\pi$ cycloaddition between aryl nitroso compounds and dienes occurs with cis stereospecificity, and if the butadiene is substituted then regioselectivity is also observed (Equation 15). Because these are cyclic hydroxylamines they can be readily reduced to hydroxyamines. α-Chloropropionaldonitrones are equally effective and are induced by silver ions. They react by a [4 + 2] cycloaddition in which only one C=C double bond is necessary, and tetrahydro-1,2-oxazines are formed in the presence of a nucleophile. The reaction proceeds with cis stereospecificity (Equation 16).[152–154] Noland and collaborators[155,156] noted that solvolysis occurred when the

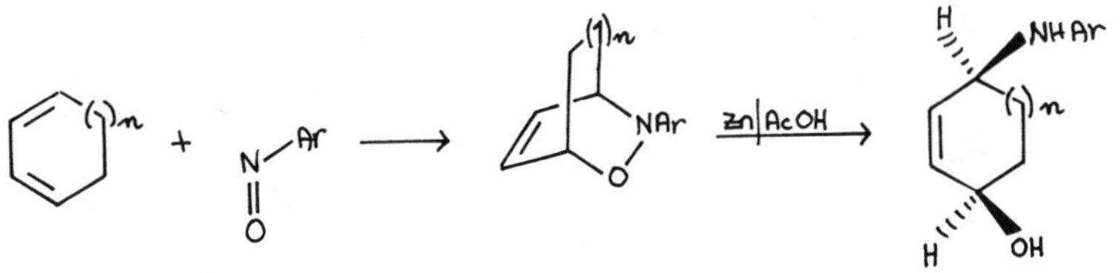

(15)

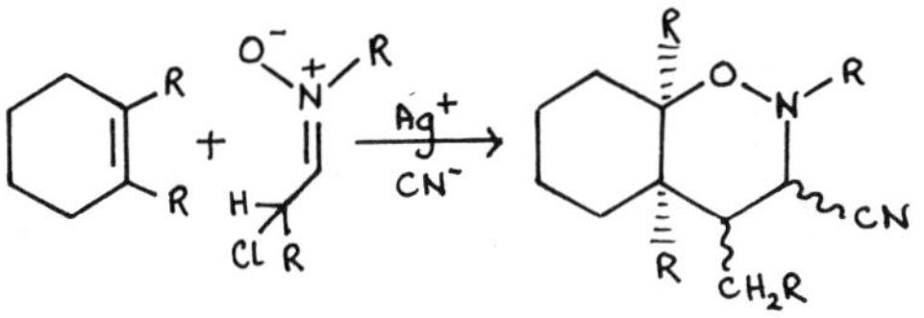

(16)

sodium salt of 5-nitronorbornenes, in ice-cold methanol, was acidified, and this resulted in a rearrangement that was assisted by the internal double bond. The product, 4-oxo-*cis*-2-oxa-3-azabicyclo[4.3.0]non-8-ene, was formed stereospecifically (Equation 17).

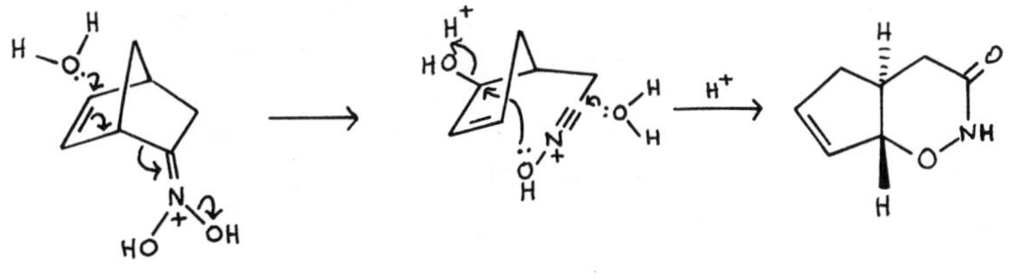

(17)

A very good source of the nitrogen and oxygen atoms for 3,6-dihydro-1,2-oxazines is 1-chloro-1-nitrosocyclohexane.[157] It undergoes a $[4 + 2]\pi$ cycloaddition reaction with isoprene and gives an 80:20 mixture of 5-methyl- and 4-methyl-3,6-dihydro-1,2-oxazines.[158] The cycloaddition must be cis and there is a clear regiospecificity. The *N*-ethoxycarbonyl derivative of the former product was reduced (Pd/C) to 2,5-dimethyltetrahydro-1,2-

oxazine. The double bond in 3,6-dihydro-1,2-oxazines can be oxidized with specificities which are characteristic of the reagent and the oxazine. Thus oxidation of 6-acetoxymethyl-2-benzoyl-3,6-dihydro-1,2-oxazine with OsO_4 gives *r*-6-acetoxymethyl-2-benzoyl-*cis*-4-*cis*-5-dihydroxytetrahydro-1,2-oxazine as a major product,[159] but a similar oxidation of 2-phenyl-6-(tetrahydropyran-2-yl)oxymethyl-3,6-dihydro-1,2-oxazine yields *trans*-4-*trans*-5-dimethyl-2-phenyl-*r*-6-(tetrahydropyran-2-yl)oxymethyltetrahydro-1,2-oxazine. The latter was converted into *cis*-2-(tetrahydropyran-2-yl)oxymethyl-4-phenyl-*cis*-3-oxa-4,7-diazabicyclo[4.1.0]heptane by tosylation, reaction with NaN_3 followed by LAH, and from the nmr analysis of its free alcohol, the half-chair conformation **34** was proposed.[160]

Other synthetic approaches to tetrahydro-1,2-oxazines were described by Riddell and Williams.[161]

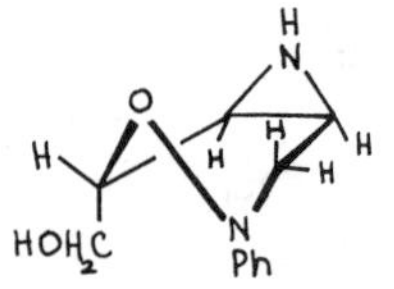

34 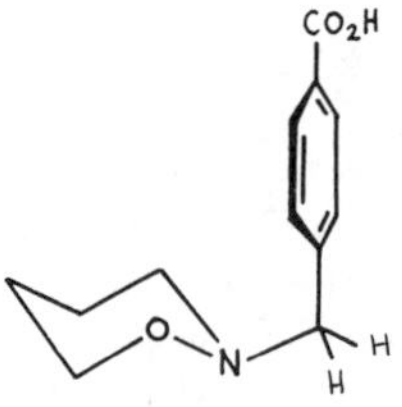 **35**

The six-membered ring of tetrahydro-1,2-oxazine is undoubtedly preferentially in the chair conformation. This was shown by Riddell and coworkers[162] from an X-ray investigation of *N*-(4-carboxybenzyl)-tetrahydro-1,2-oxazine in which the benzyl group was equatorial with the benzene ring almost at right angles to the oxazine ring (**35**). They also had evidence of significantly larger torsional angles around the nitrogen and oxygen atoms compared to those around the carbon atoms indicating a more puckered ring than in cyclohexane. Tetrahydro-1,2-oxazine is conformationally mobile and inverts between two chair conformations. The nitrogen atom also undergoes inversion. The ^1H nmr spectral changes of 2-methyltetrahydro-1,2-oxazine in the temperature range -45 to 55°C were attributed to nitrogen inversion because the free energy change, 57.3 kJ mol^{-1} in hexane, increased markedly to 62.8 kJ mol^{-1} in the hydrogen bonding solvent D_2O–CD_3OD.[62] Methyl groups on the carbon skeleton have a preference for the equatorial conformation, and Riddell and Williams[158,163] determined the free energy differences between axial and equatorial arrangements for the methyl groups on each of the carbon atoms. The $\Delta G^{\ddagger}$ values for a methyl group at C–3, C–4, C–5, and C–6 are

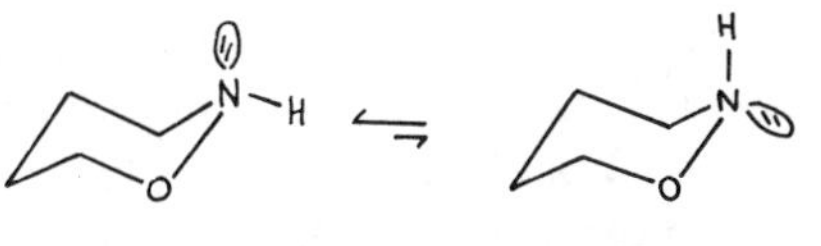

36 37

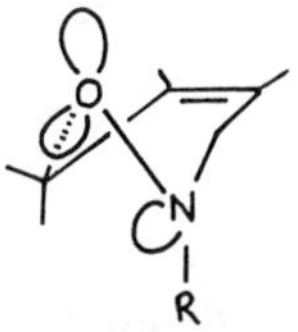

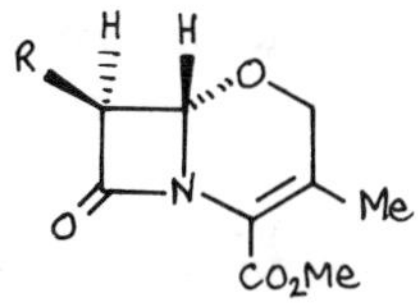

R=phthalimido

38 39

7.9, 7.1, 5.7, and 10.1 kJ mol^{-1} respectively.[158] Katritzky and his school[164,165] were more interested in the preferred orientation of the nitrogen lone pair of electrons in tetrahydro-1,2-oxazine. They used ir band shapes and dipole moment measurements to demonstrate that the conformer with the axial nitrogen lone pair (**36**) predominated over the one with the axial N–H (**37**). The equatorial preference for the *N*-methyl group in 2-methyltetrahydro-1,2-oxazine is of the order of 8.0 kJ mol^{-1} and is not to be confused with the free energy of pyramidal nitrogen inversion because ring inversion is mainly responsible for this value.[165] Jones, Katritzky, and Saba[166] studied nitrogen inversion in *N*-alkyl 3,6-dihydro- and 3,6-dihydrobenzo-[*d*]-1,2-oxazines by low temperature ^{1}H nmr spectroscopy and dipole moment measurements. They observed that the heterocyclic ring is in the half-chair conformation, and the *N*-alkyl groups had a very strong preference for the pseudo-equatorial conformation. This preference was partly caused by the unfavorable and large lone pair-lone pair repulsion in the conformers with *N*-alkyl groups oriented axially (**38**).

The general chemistry of oxazines was reviewed by Cromwell.[167]

2. *Reduced 1,3-Oxazines*

Reduced 1,3-oxazines can be synthesized by bridging the nitrogen and oxygen atoms of γ-amino alcohols by a one carbon unit. If the bridging carbon is supplied as an aldehyde or ketone a reduced 1,3-oxazine is formed in which the chiral centers present in the original γ-amino alcohol are preserved.[167,168] In the case of aldehydes and unsymmetrical ketones, however, a mixture of reduced 1,3-oxazines which are epimeric at C–2 is formed. The mixture can be identified by nmr spectroscopy[169] and can be used to deduce the configuration of the original γ-amino alcohol since the system is now more rigid and nmr data more meaningful. The ord curves of these derivatives, when compared with the original amino alcohols, can give some idea of the conformation of the amino alcohols in solution.[170] For the preparation of the 2-oxo derivatives, retention of configuration at the carbon bonded to the oxygen atom occurs when the ureido derivative of a γ-amino alcohol cyclizes to the reduced 1,3-oxazin-2-one.[171,172] The preparation of tetrahydro-1,3-oxazin-2-one is more readily conceived, without altering the chiral centers, from the γ-amino alcohol and carbonyl diimidazole.[173]

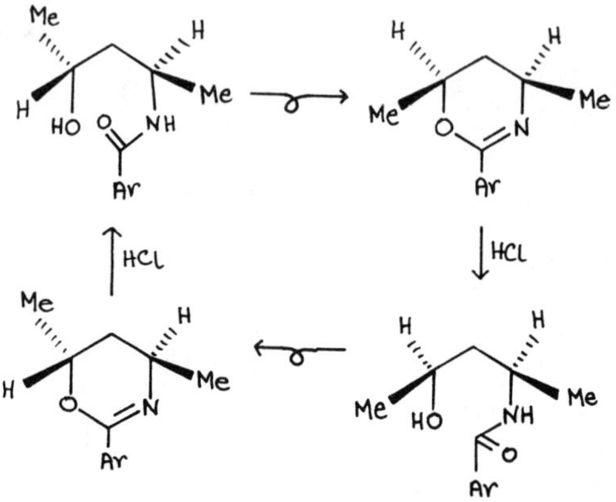

Me H H Me HO O NH Ar → H H Me Me O N Ar

↑ HCl ↓ HCl

Me H H Me O N Ar ← H H Me Me HO NH O Ar

Scheme 3

The cyclization of *N*-acyl and *N*-aroyl γ-amino alcohols, with thionyl chloride or by way of the methyl sulfonate ester, proceeds with a Walden inversion at the carbon atom bonded to the oxygen atom and forms the 2-substituted 5,6-dihydro-4*H*-1,3-oxazines.[171] Hydrolysis of the 5,6-dihydro-4*H*-1,3-oxazine with hydrochloric acid, on the other hand, yields the γ-amino alcohol with retention of configuration because it is C–2 that is involved in this reaction. The cycle in Scheme 3 was used by Sicher and co-workers[174] to prepare the two isomers of the γ-amino alcohols and of the oxazines. The configurations of C–4 and C–5 in the oxazine and C–2 and C–3 in the γ-amino alcohol remain unaltered on acid treatment. The intramolecular cyclization of methyl 2-(4′-chloro-3′-*S*-phthalimido-2′-oxo)-azetidin-1-yl-3-methyl-4-hydroxy-*trans*-2-butenoate occurs by attack of the 4-OH on C–2′ in the azetidinone with inversion of configuration and provides the oxacepham (**39**) or the epimer depending on the chloro compound used.[175]

A completely different synthesis of 5,6-dihydro-4*H*-1,3-oxazines requires a 1,4-dipolar cycloaddition of amidomethylenium ions with olefins which proceeds with cis stereospecificity and normally high regiospecificity. The regiospecificity becomes unimportant in symmetrically substituted olefins (Equation 18).[176] For a short summary of these reactions see Bradsher's review on cationic polar cycloadditions.[176a]

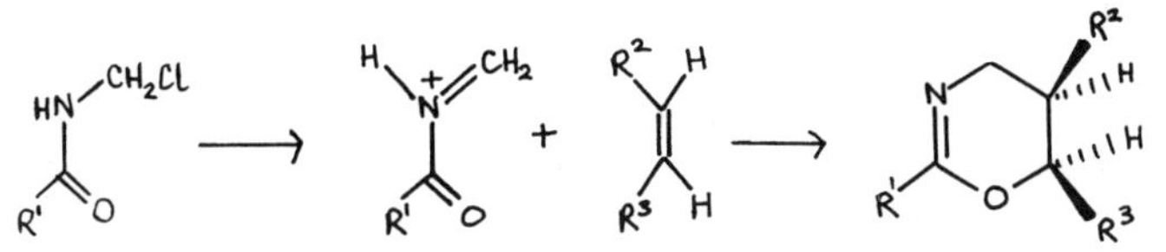

(18)

(+)-5-Ethyl-5-phenyl-5,6-dihydro-2,4-dioxo-1,3-oxazine was prepared by Testa and co-workers[177] from (−)-1-hydroxymethyl-1-phenylbutyric acid by way of the cyclization of the corresponding CH_2O–carbamoyl derivative with $SOCl_2$.

Most of the ^1H nmr and dipole moment data of 1,3-oxazane are consistent with it being in the preferred chair conformation. Katritzky and co-workers,[164,178] using this data, showed that of the two chair conformations the one with the axial NH and equatorial nitrogen lone pair (**40**) is the one that predominates. Booth and Lemieux[179] arrived at a similar conclusion and found that an *N*-methyl group had a partial axial character. They attributed this axial preference to the generalized anomeric effect (rabbit ear effect) (see Chapter 2, Section X.C). Katritzky and collaborators[180] measured the percentage of equatorial *N*-alkyl groups in 1,3-oxazanes and found that for Me, Et, isoPr, and *t*-Bu the percentages were 58, 68, 86, and about 100. They attributed the high proportion of axial NH in the parent compound and in the lower *N*-alkyl homologues to attractive forces between N–H and the substituent with the axial oxygen lone pair and the generalized anomeric effect (see above).[181] Geminal coupling constants on a methylene group adjacent to two heteroatoms, as in 1,3-oxazane, were discussed by Anteunis and co-workers[182] and Riddell and Lehn.[183] The latter authors concluded that although the chemical shifts in heterocyclic molecules are affected by the orientation of the lone pair on the heteroatoms and by other substituents, the J_{gem} values are mainly influenced by the orientation of the lone pair of electrons on the vicinal heteroatom. Katritzky and his school[184] made an extensive study of the chemical shifts and J_{gem} values of a number of hexahydropyrimidines, hexahydro-1,3,5-triazines, and 1,3-oxazanes and found no simple correlation between the chemical shifts and J_{gem} values with the orientation of the lone pair on the vicinal heteroatom.

H N O NO2 R N O NO2 R N O

40 **41** **42**

Urbański and collaborators[185,186] studied the conformational properties of 5-nitro-1,3-oxazanes by dipole moment and ^{1}H nmr measurements. The data were explained in terms of a chair conformation for the ring, and in all the examples studied the nitro group was axial (**41,** R=H). It appears that the nonbonded interactions between an equatorial 5-nitro group and the flanking equatorial H–4 and H–6 are too large to keep it in this conformation. Methyl, ethyl, *n*-propyl, *n*-butyl, and benzyl groups on N–3 in 5-nitro-1,3-oxazanes are predominantly axial, but a large group on N–3, such as a *t*-butyl and a cyclohexyl group, adopt the equatorial conformation (**42**). While confirming these results by more careful ^{1}H nmr analyses, Allingham, Cookson, Crabb, and Vary[187] also showed that the conformation of the *N*-isopropyl derivative is that of **42** (R=isoPr) and determined the conformation of 6-substituted 1,3-oxazanes. Alkylation of the nitro-oxazanes with the conformation **41** by alkyl halides proceeded with an equatorial attack of the reagent. Alkylation of derivatives with the conformation **42,** however, was unsuccessful most probably because of the large steric interference not only by the 5-NO_2 group but also because of the rather bulky group on N–3.[188]

Leonard and Musker[126] prepared 1,3-oxazanes with a tetramethylene bridge between C–2 and N–3. The nitrogen atom is at a bridgehead position and is capable of equilibrating between the cis (**43**) and trans (**44**) configurations, as in quinolizidines (see Part I, Chapter 4, Section VI.C). From the presence of Bohlmann bonds in the ir spectra they concluded that the trans structure **44** is the major component. Crabb and co-workers[129,135,189–193] studied a large number of similar and more complicated systems and were able to adequately explain the preferred configurations and conformations in terms of nonbonded interactions.

43 **44**

Dipole moment measurements, by Chylińska and Urbański,[194] of 3,4-dihydro-3-cyclohexyl-6-methyl-2*H*-1,3-benzoxazine and 2-methyl- and 2-*p*-

tolyl-2,3-dihydro-4*H*-naphtho[1.2-*e*]-1,3-oxazines revealed that the heterocyclic ring is in the half-chair conformation and that the substituent on the nitrogen atom is in the quasi-axial position.

3. *Reduced 1,4-Oxazines (Including Morpholines)*

Stereochemically, the most interesting reduced compounds in this subsection are the perhydro derivatives which are commonly known as morpholines. They have been prepared in various ways starting from β-amino alcohols. These include the cyclization of the *N*-chloroacetyl derivatives,[118] the *N*-2′-hydroxyethyl derivatives,[195–197] and the condensation of β-amino alcohols with dimethyl acetylene dicarboxylic ester.[198] Whether inversion

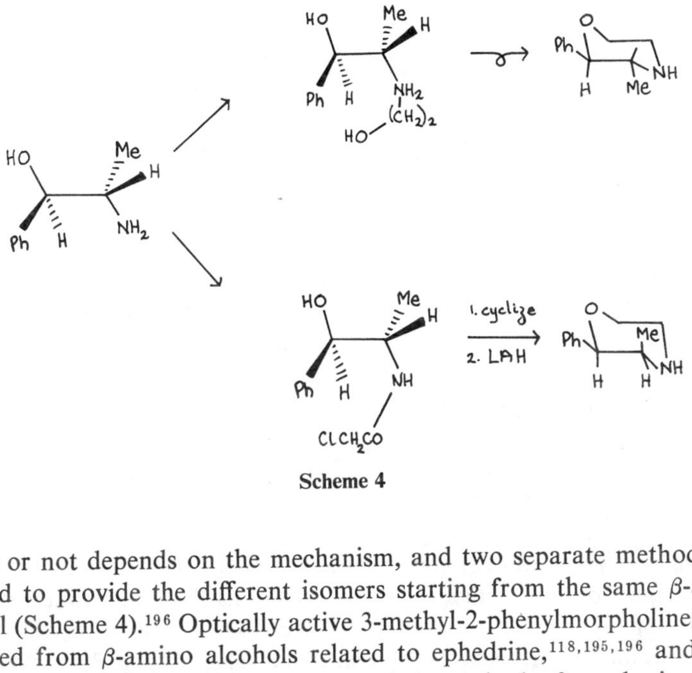

Scheme 4

occurs or not depends on the mechanism, and two separate methods can be used to provide the different isomers starting from the same β-amino alcohol (Scheme 4).[196] Optically active 3-methyl-2-phenylmorpholines were prepared from β-amino alcohols related to ephedrine,[118,195,196] and their absolute configurations followed from their method of synthesis. In an application of these compounds, Blaschke[199] made an optically active adsorbent by bonding polyacrylamide with 2*S*,3*S*-*trans*-3-methyl-2-phenylmorpholine through N–4, and used it to separate racemic mandelic acid into its optical antipodes. Gero and co-workers[200] obtained rigid morpholines, such as 5-tosyl-7-tosyloxy-2-oxa-5-azabicyclo[2.2.1]heptanes (**45**, R= Ts), by intramolecular cyclization of derivatives of polyhydroxyfurans, for example, *r*-2-tosylaminomethyl-*cis*-3-*trans*-4-bistosyloxytetrahydrofuran, with alkoxide. The related *N*-tosylamino derivative **45** (R=NHTs) was prepared from *r*-2-tosyloxymethyl-*cis*-3-*trans*-4-bistosyloxytetrahydrofuran and hydrazine followed by tosylation.[201] The same ring system can be obtained by intramolecular cyclization of *trans*-3-hydroxyproline derivatives.[202]

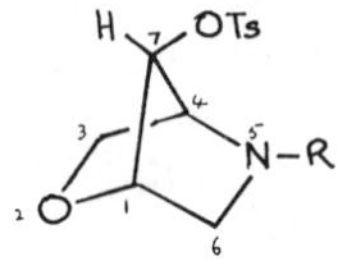

45

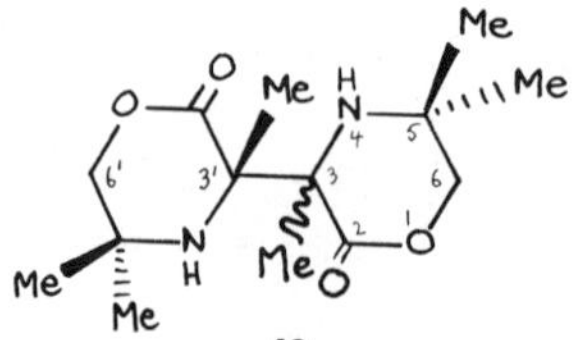

46

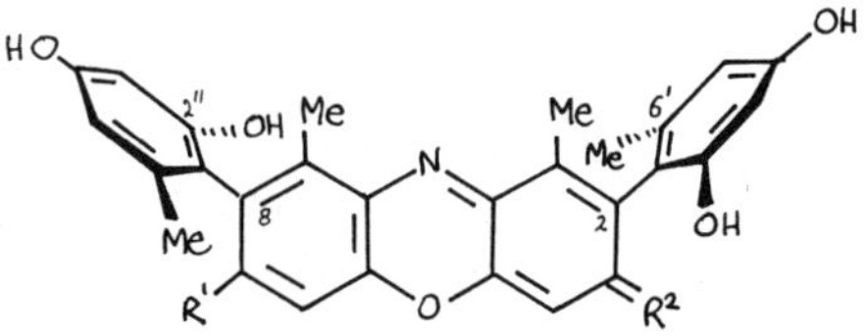

47

Irradiation of 5,6-dihydro-3,5,5-trimethyl-1,4-oxazin-2-one, at −15°C with a 450-W mercury lamp through a pyrex filter in propan-2-ol, gave a 3:2 mixture of *meso*- and *d,l*-bi-(5,6-dihydro-3,5,5-trimethyl-1,4-oxazin-2-on-3-yl) (**46**) which was separated by liquid chromatography. The stereochemistry was assigned by observing the changes in chemical shifts of the protons of these isomers on addition of the chiral shift reagent tris-[3-(trifluoromethylhydroxymethylene)-*d*-camphorato]europium (III). The 3,3′-methyl and *H*-6,6′ signals were separated in the *d,l* but not in the meso isomer on addition of the reagent.[203] The oxidation of *R*(+)-2,2′-diamino-4,4′-dimethoxy-6,6′-dimethylbiphenyl in the presence of ammonia gave the 2,8-diarylphenoxazine which was chiral and had the substituents in the 2- and 8-phenyl groups trans to each other (**47**, $R^1=R^2=O$). The ord of this compound and the related amines **47** ($R^1=NH_2$, $R^2=O$) and **47** ($R^1=NH_2$ and $R^2=NH$) were compared and discussed. A racemic mixture of the biphenyl, on the other hand, gave the meso isomer of **47** in which the substituents in the 2- and 8-phenyl rings (i.e., the 6′-Me and 2″-OH groups) are cis oriented with respect to each other.[204] For the general chemistry of 1,4-oxazines see the review by Cromwell.[167]

Dipole moment and molar Kerr measurements of morpholine and *N*-methylmorpholine were interpreted by Le Fèvre and co-workers[205,206] in terms of a chair conformation for the ring. From these data they deduced that the equatorial to axial ratios of the *N*H[37:63 (cyclohexane), 13:87 (benzene)] and *N*Me[57:43 (cyclohexane), 98:2 (benzene)] in morpholine varied with the solvent.[205] These values gave some indication of the relative space requirements of the nitrogen lone pair compared with a hydrogen atom and a methyl group. The morpholine ring is flexible and inverting between two chair conformations with the nitrogen atom also undergoing pyramidal inversion to accommodate the substituent on it, in its preferred conformation. This is similar to what is observed in piperidine (see Scheme 1 in Part I, Chapter 3). Harris and Spragg[207] determined the free energy for ring inversion in morpholine (41.4 kJ mol^{-1}) and *N*-methyl morpholine (48.1 kJ mol^{-1}), which were comparable with those obtained for related heterocycles. The *cis*- and *trans*-2,6-dimethyl morpholines were separated, and the ^{1}H nmr spectra of the former were explained by the presence of one conformer with both methyl groups equatorial, but the latter consisted of two equilibrating conformers each with an axial and an equatorial methyl group.[208] In order to obtain evidence for the preferred conformation of the *N*-methyl group in morpholines (and piperidines, see Part I, Chapter 3, Section II.C), Booth and Little[209] analyzed the ^{1}H nmr spectra of 2,4,6- and 3,4,5-trimethylmorpholines in CF_3CO_2H and assumed that protonation was very fast and that the cation did not dissociate with any appreciable speed. They found that the respective cations were predominantly, or almost exclusively, in the conformations **48** and **49** respectively. They inferred that in the free base the *N*-methyl group is preferentially in the equatorial conformation.

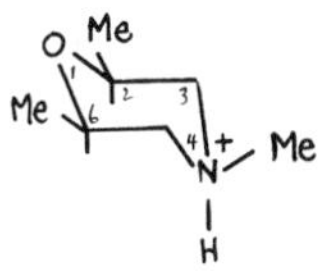

48 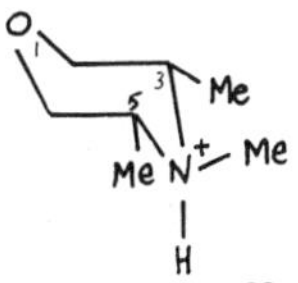 **49** 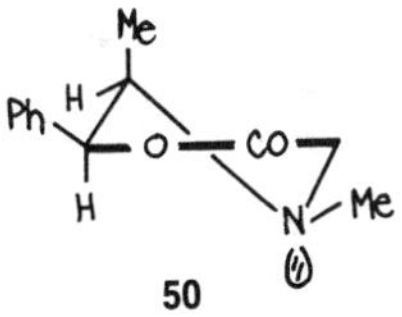 **50**

Cahill and Crabb[210] synthesized several morpholin-2-ones, which are essentially lactones, and analyzed their ^{1}H nmr spectra. The linear ester function caused these molecules to exist in preferred half-chair conformations, such as in *cis*-4,5-dimethyl-6-phenylmorpholin-2-one (**50**). Restricted rotation about the N–CO bond in benzoyl derivatives of the 2-oxa-5-azabicyclo[2.2.1]heptane and heptan-3-one (ring system similar to **45**) was observed by Portoghese and Turcotte.[202] They also estimated the free energies involved and found them to be 70.7 and 59.4 kJ mol^{-1} for the heptane and heptan-3-one respectively. This difference was explained in terms of the steric relationship of the carbonyl group and N–5.

4. *Miscellaneous*

A few dioxazines, oxadiazines, and dioxadiazines are described in this subsection, but stereochemical interest in them has not been extensive. Katritzky and co-workers[211,212] prepared 2-methyl- and 2,3,3-trimethyl-1,4,2-dioxazanes and examined their conformational mobility. Of the two conformers **51** and **52** the one with the equatorial methyl (**51**) is predominant, but this is dependent on the substituent R; when R=H, the ratio of **51** to **52** is 15:1 ($\Delta G^{\ddagger}=4.39$ kJ mol^{-1}), and when R=Me, the ratio is 5:1 ($\Delta G^{\ddagger}=2.55$ kJ mol^{-1}). The ratio for 2-ethyl-1,4,2-dioxazane falls between these two values.[212]

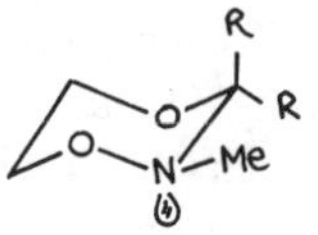

51 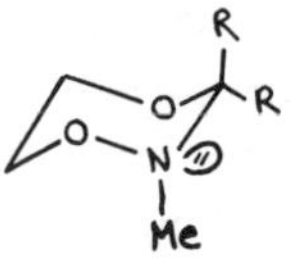 **52**

5,6-Polymethylene-2-phenyl-5,6-dihydro-4*H*-1,3,4-oxadiazines have attracted the most attention, and the authentic *cis*- and *trans*-5,6-tetramethylene derivatives were readily obtained from *cis*- and *trans*-2-aminocyclohexanols and ethyl phenylformimidoate hydrochloride [Ph–C(OEt)=NH_2Cl] with retention of configuration.[213] The related compounds with a cis configuration and a double bond in the polymethylene chain are formed by a sigmatropic rearrangement of 2,3-dibenzoyl-2,3-diazabicyclo[2.2.*n*]alk-5-

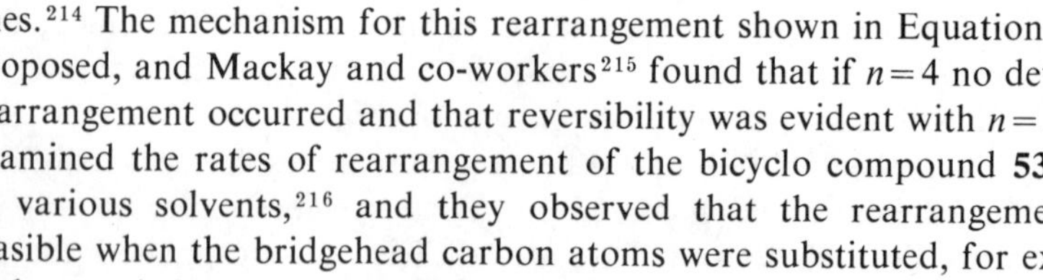

53 (19)

enes.[214] The mechanism for this rearrangement shown in Equation 19 was proposed, and Mackay and co-workers[215] found that if $n=4$ no detectable rearrangement occurred and that reversibility was evident with $n=3$. They examined the rates of rearrangement of the bicyclo compound **53** ($n=1$) in various solvents,[216] and they observed that the rearrangement was feasible when the bridgehead carbon atoms were substituted, for example, with a methyl group, when (CH_2): was replaced by C=O in **53**, and when other acyl and aroyl groups are used. The rates of the rearrangements were influenced by these changes and were rationalized.[217] In one example the rearrangement was performed in the presence of (+)-camphor-10-sulfonic acid, and a partial asymmetric synthesis of (+)-5-benzoyl-6,8-dimethyl-1,3,9-triphenyl-*cis*-2-oxa-4,5-diazabicyclo[4.3.0]nona-3,8-diene was achieved. By using a chiral shift reagent and chiral solvents, the enrichment in the isomerization was shown to be 4.21 ± 0.08%.[218] A cycloaddition occurred when diphenyl azidocarboxylate or dibenzoyldimide reacted with 2,3-dideuteronorbornene and the corresponding *cis,exo*-2,4,5,6-tetrasubstituted 5,6-dihydro-4*H*-1,3,4-oxadiazines **54** (R = OPh) and **54** (R = Ph) were formed stereospecifically.[219]

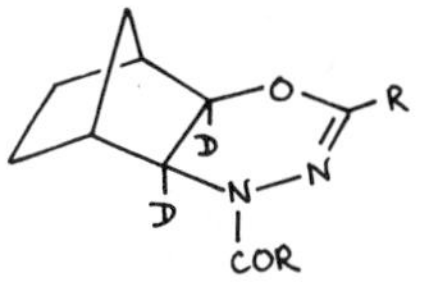

54

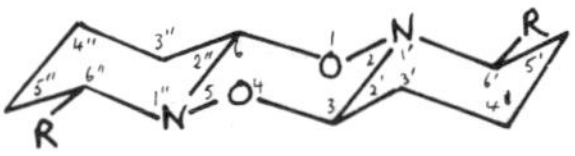

55

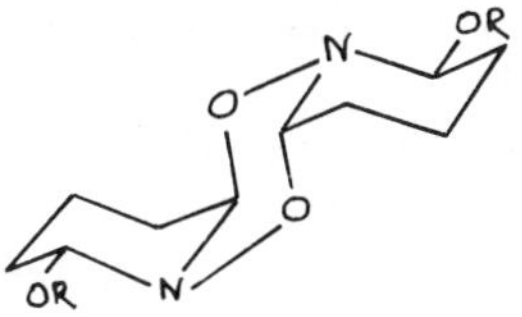

56

The oxidation of *N*-hydroxypiperidine with HgO or $K_3[Fe(CN)_6]$ yields the nitrone which dimerizes readily to dipiperidino[1.′2′:2.3:1.″2″:5.6]-perhydro-1,4,2,5-dioxadiazine (**55**, R = H) which is probably in the conformation and configuration shown.[220] The diphenyl derivative **55** (R = Ph) was similarly obtained.[221] The same ring system was formed when glutaraldehyde reacted with hydroxylamine hydrochloride in alcohols. The ^1H nmr data, however, suggested that the dioxadiazine existed in the conformation and cis configuration shown in **56** (R = Me or Et).[222]

E. Two or More Rings Containing N and O Heteroatoms in Each Ring

This subsection contains a variety of compounds which are considered in order of ring sizes.

For a comparison of effective sizes of a ring oxygen atom and a CH_2 group, Katritzky and collaborators[223] measured the dipole moment of 1-*t*-butylpiperidine-4-spiro-2′-oxirane. They found that the conformer with the axial oxygen atom **57** (X = O) was favored by 1.1 kJ mol^{-1}. The sulfur analogue **57** (X=S) also prefers the conformation with the axial sulfur atom, but to the extent of 1.8 kJ mol^{-1}.

Uv irradiation of carbonyl compounds such as benzophenone, benzoyl formamide, and methyl benzoyl formate in the presence of *N*-acyl- or *N*-aroylindole gave the *cis*-indolooxetanes **58** with high regiospecificity. The stereochemistry at the carbon atom adjacent to the oxygen atom in the latter two carbonyl compounds was not established.[224]

57 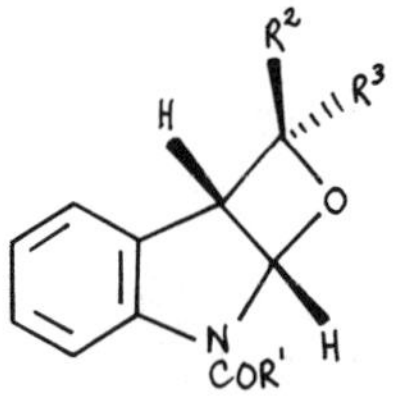 **58** 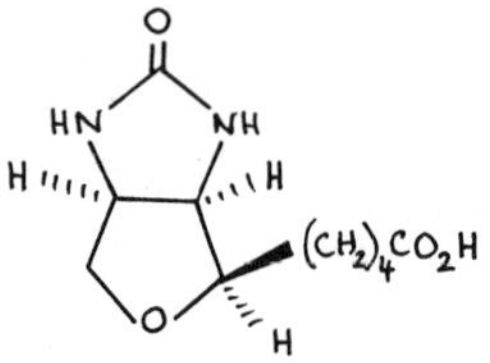 **59**

In the 5,5-fused ring systems the more interesting heterocycles are the ones that are related to biotin. Hofmann[225] synthesized *d,l*-oxybiotin (**59**) from 2-(5-hydroxypentyl)-3,4-dicarboxyfuran, and because the product had microbiological activity comparable to *d,l*-biotin, it was assumed that the relative configurations of the three chiral centers were similar. Several cis-fused reduced furo[3.4-*d*]imidazoles were synthesized[226] in racemic and optically active forms,[227] including some which were used in the total synthesis of biotin.[228] Ring-chain tautomerism is remarkably facile in 3,2′-hydroxyethyl-3-phenyl-2-methoxy-3*H*-indole which cyclizes to *cis*-8*a*-methoxy-3*a*-phenyl-3*a*,8*a*-dihydrofuro-[2.3-*b*]indole (**60**) stereospecifically.[229]

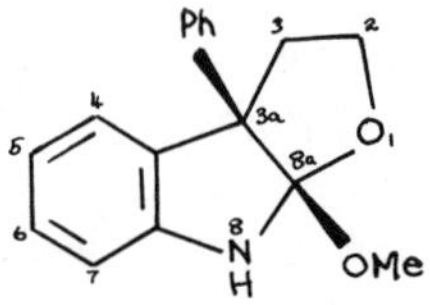

60

Two 1,4-cycloaddition reactions were developed for the synthesis of 5,6-fused ring systems. The first was for the preparation of reduced cis-fused furo[3.2-*c*]quinolines, which was studied mainly by Perricone, Elslager, and Worth.[230–232] It involved the condensation of benzylidene aniline derivatives with 5-methyl-2,3-dihydrofuran in the presence of BF_3. The reaction is cis stereospecific, but usually diastereomers are formed because the aryl group from the aldehyde portion of the Schiff base exhibits

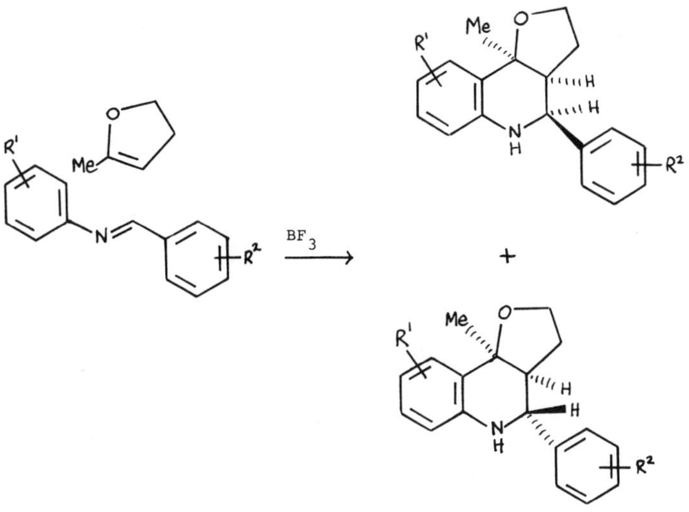

(20)

no specificity (Equation 20). The bisbenzylidene derivative from *p*-phenylenediamine reacted at the two azomethine linkages with excess of the dihydrofuran.[233] The second 1,4-cycloaddition reaction, developed by Desimoni and his collaborators,[234–237] required the addition of enol ethers to α arylidene pyrazolones and pyrano[3.2-*c*]pyrazoles were formed (Equation 21). The steric relationship of the substituents at C–2 and C–3 is dictated by the cis stereospecificity of the reaction and the configuration of the enol ether. Two products, epimeric at C–4, are usually obtained, but the

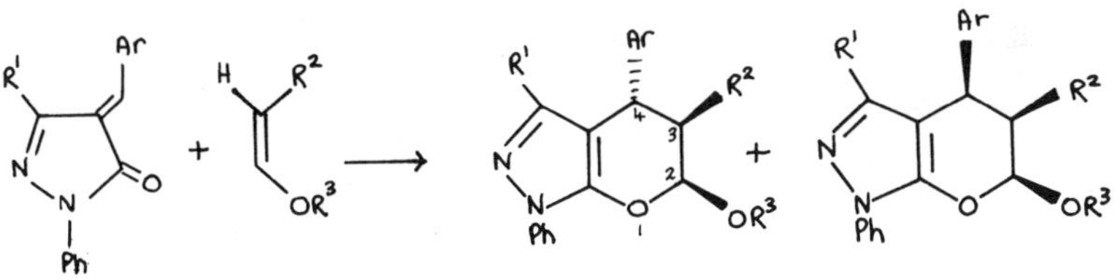

(21)

61 **62**

major isomer is the one in which the substituents at C–3 and C–4 are trans (**62**). The reaction is feasible also when N–Ph in **61** is replaced by a ring oxygen atom, that is, 4-arylidene isoxazol-5-ones,[238] and when the pyrazolone is replaced by *N*-acetyl 3-arylidene oxindoles.[239] The reaction is clearly regiospecific. The ^{1}H nmr spectra revealed that the dihydropyran ring is in the half-chair conformation with the 2-alkoxy group oriented pseudo-axially indicating that the anomeric effect is quite strong. In the examples where the substituents at C–2 and C–4 are cis, the aryl group is in the axial conformation also (e.g., **63**).[234,239] In connection with studies on the structure of retronecine and related bases, Adams and co-workers[240] obtained 1-cyano-*cis*-7-methyl *cis*-perhydropyrano[3.2-*b*]pyrrole in optically active form by treating retronecanol with cyanogen bromide.

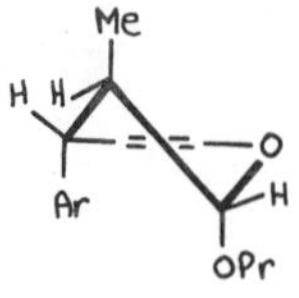

63 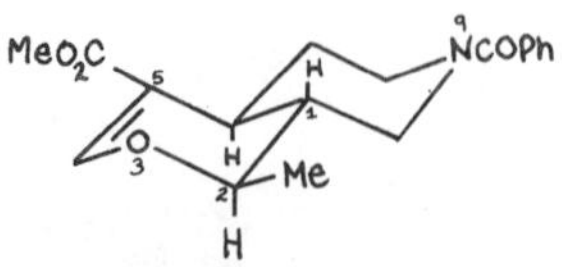 **64** 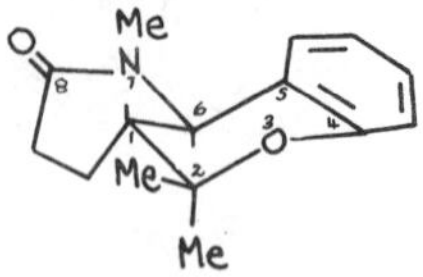 **65**

A number of 6,6-fused ring heterocycles with one oxygen and one nitrogen atom in each ring are known, but only two cases of stereochemical interest are mentioned here. One of the reactions among the elegant synthetic studies on indole and quinoline alkaloids, by Uskoković and his group,[241] is the intramolecular cyclization of *trans*-1-benzoyl-3-vinyl-4-(2-hydroxy-1-methoxycarbonylvinyl)piperidine. When carried out with Hg-$(OAc)_2$, followed by $NaBH_4$ reduction, a 2:1 mixture of 9-benzoyl-*cis*-2-methyl(equatorial)-5-methoxycarbonyl-*trans*-3-oxa-9-azabicyclo[4.4.0]dec-4-ene (**64**) and the *trans*-2-methyl (axial) epimer was formed. The second case was in the synthesis of nitrogen analogues of tetrahydrocannabinols by Cushman and Castagnoli Jr.[242] This example also involves an intramolecular cyclization whereby a mixture of olefins obtained by dehydrohalogenation of 5-bromo-6-*o*-hydroxyphenyl-5-isopropyl-1-methylpiperidin-2-ones is converted into 4,5-benzo-2,2,7-trimethyl-8-oxo-*trans*-3-oxa-7-azabicyclo[4.4.0]dec-4-ene (**65**) in trifluoroacetic acid.

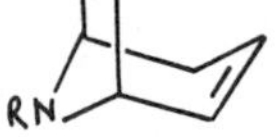

66 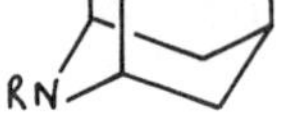 **67**

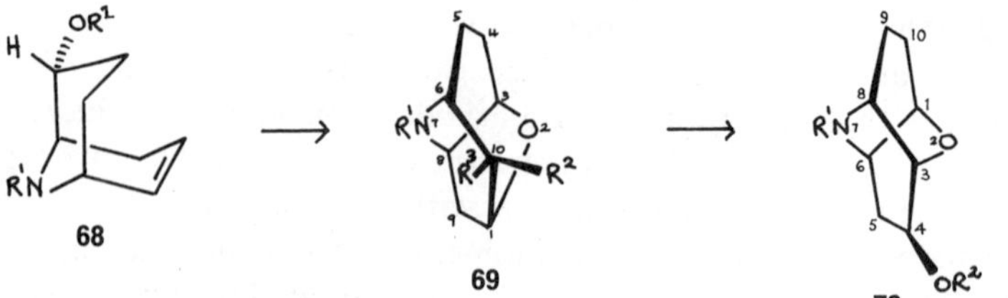

68 → **69** → **70** (22)

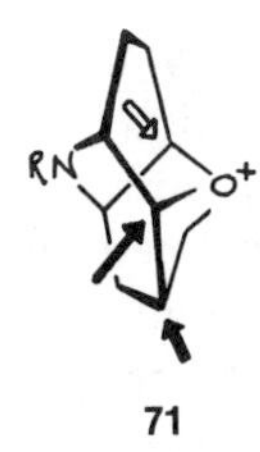

71

The investigations of Ganter and his co-workers on intramolecular cyclizations of heterocycles derived from cyclooctenes were extended to include mixed oxygen and nitrogen analogues of adamantane, twistane, and related compounds. 9-Formyl- or 9-tosyl-9-azabicyclo[3.3.1]nona-2,6-diene (**66**) undergo oxymercuration and afforded, after reductive removal of the mercury from the intermediate, 6-substituted 2-oxa-6-azaadamantanes (**67**).[243] The related acetoxynonene (**68,** R^1=Me, R^2=Ac), on the other hand, reacts with $Hg(OAc)_2$ to give the isotwistane mercuric acetate **69** (R^1=Me, R^2=HgOAc, R^3=H) which is converted into 7-methyl-10-iodo-2-oxa-7-azaisotwistane (**69,** R^1=Me, R^2=H, R^3=I) with potassium iodide. The latter skeleton rearranges in the presence of silver tosylate into 7-methyl-4-tosyloxy-2-oxa-7-azatwistane (**70,** R^1=Me, R^2=Ts) (Equation 22). The oxonium ion **71** is apparently involved, and the carbon atoms that are subject to nucleophilic attack are shown in the formula.[244] Aza- (Part I, Chapter 3, Section X.C), oxa- (Chapter 2, Section XI.D), thia- (Chapter 3, Section XIII), and oxathiatwistanes (Section IV.D) are reported in the sections cited.

F. Seven- and Larger-Membered Rings

The seven-membered azepines and oxepins are flexible molecules; the rings invert rather rapidly and rarely exceed an energy barrier of 42 kJ mol^{-1}.[245] Wasylishen, Rice, and Weiss[246] investigated the conformational mobility of hexahydro-3,3,7,7-tetramethyl-1,2-oxazepin-5-ones (**72**) by variable temperature ^1H nmr spectroscopy and found that the inversion process of this seven-membered ring was rather slow. The $\Delta G^‡$ values for **72** with R=H, COPh, and Me in d_6-dimethyl formamide were 79.4, 83.2, and 90.3 kJ mol^{-1} respectively.

Wong and Paudler[247] synthesized [2.2](2,5)furano(2,5)pyridinophane by a cross reaction between 2,5-dimethylene-2,5-dihydrofuran and 2,5-dimethylene-2,5-dihydropyridine. The preferred conformation of the cyclophane in solution has the furan ring directed towards the pyridine ring with the oxygen atom closer to C–3 and C–4 of the pyridine ring than to N–1 and C–6. No change in ^1H nmr spectrum was observed between −50 and 110°C suggesting that ring rotation is very slow. An X-ray study of this cyclophane revealed that the conformation in which the oxygen and nitrogen atoms are syn to each other (**73**) is present in the crystal, and is the first example in which the rings are inclined towards each other at an angle of 23°.[248] There appears to be little repulsion between the lone pairs of the heteroatoms in the solid state. The rings are not flat, and the uv spectra show that there is no conjugation between them.

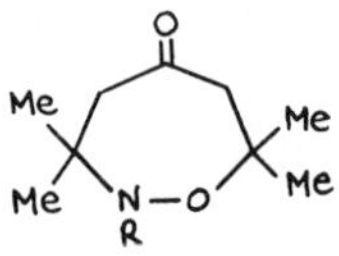

72 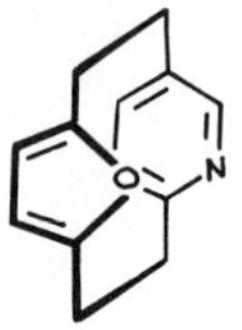 **73** 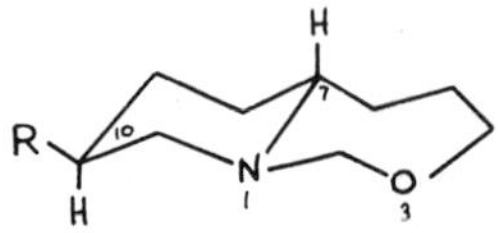 **74**

The ability of perhydrobicyclic systems possessing a bridgehead nitrogen atom to undergo cis-trans equilibria by virtue of an inverting nitrogen atom (e.g., **43** ⇌ **44** is described earlier for 5,5-, 5,6-, and 6,6-fused bicyclic systems (Section II.D.3 and Part I, Chapter 4, Sections VI.A, B, and C). Crabb and co-workers[249] extended the studies to 6,7-fused systems, the pyrido[1.2-*c*][1.3]oxazepines, and found that nonbonded interactions here as well have a controlling influence on the preferred conformation in the equilibria. Thus the *trans*-10-methyl and *trans*-10-ethyl (**74,** R = Me or Et) derivatives exist predominantly in the trans-fused configuration, but the *cis*-10-methyl derivative (**75**) prefers the cis-fused configuration making the equatorial preference of the alkyl group the dominating factor. The *trans*-10-alkyl derivatives, but not the cis isomers, dimerize in the liquid state, and an X-ray analysis of one of the crystalline dimers is consistent with the 14-membered macrocyclic ring structure (**76,** R = Et). Molecular weight determinations in solution suggest that reversible dissociation into monomers occurs.

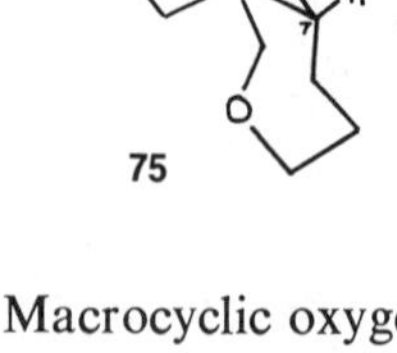

75 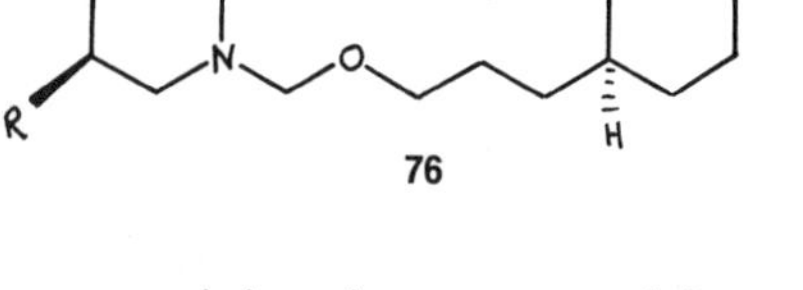 **76**

Macrocyclic oxygen- and nitrogen-containing rings, prepared from ethanolamine and ethylene glycol, form cryptates (formula **340,** Chapter 2) and bind selectively with alkali and alkaline earth metals (see Chapter 2, Section XIV.D).[250] These assist in solubilizing metals (e.g., potassium) in benzene or toluene and have possibilities as reducing agents in organic chemistry.[251] The structures of some of these metal complexes were established by X-ray analyses.[252] Crown ethers possessing a nitrogen atom in the ring derived from *o*-aminophenol or *o*-phenylene diamine and ethylene glycol,[253] and chiral macrocyclic ethers with nitrogen heteroatoms in the ring derived from 2,6-bishydroxymethylpyridine and chiral 2,2′-dihydroxy-1,1′-binaphthyl also complex with metal ions.[254] The selectivity of these amino ethers depends on the relative stereochemical arrangement of the oxygen and nitrogen atoms in the macrocycle, which has to coordinate (through the heteroatoms) with a metal by wrapping itself around the metal in such a way as to satisfy the coordination number and the stereochemistry of the bonds of the metal. (See Chapter 2, Section XIV.D).

III. NITROGEN AND SULFUR HETEROCYCLES

A. Four-Membered Rings

2-Thiaazetidine 2-oxides are formed by cycloaddition of *N*-sulfinyl *p*-toluenesulfonamide (TsN=S=O) onto vinyl ethers at about −30°C. A cis addition is involved, but the thiaazetidine oxides dissociate back to the starting materials at higher temperatures, and an equilibrium of diastereomeric oxides is attained (Equation 23). The dissociation was demonstrated by addition of 2,3-dimethylbutadiene which trapped TsN=S=O as 4,5-dimethyl-2-tosyl-3,6-dihydro-1,2-oxazine.[255] *N*-Sulfonyl benzamide behaves in the same manner but yields the 1-benzoyl-2-thiaazetidine 2,2-dioxides.[256] 2-Thiaazetidine 2,2-dioxides are also formed by the cycloaddition reaction of phenylsulfene and arylidene methylamine, but a mixture of *cis*- and *trans*-4-aryl-3-phenyl-1-methyl-2-thiaazetidine 2,2-dioxides is formed which can be separated (Equation 24). Thermolysis of the *cis*- and *trans*-3,4-diphenyl derivatives is not stereospecific because *trans*-stilbene and benzaldehyde are formed in each case.[257]

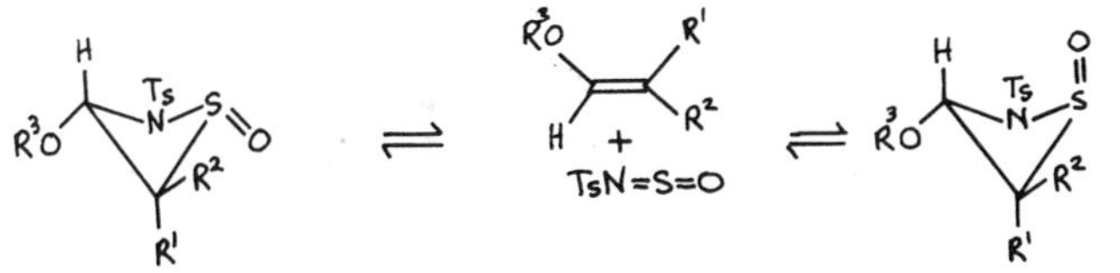

(23)

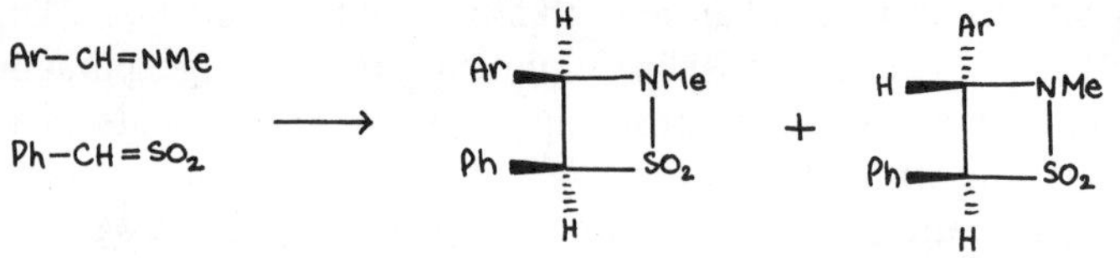

(24)

B. Five-Membered Rings

1. *Isothiazolidines*

Only two examples of isothiazolidines are noted in this section. The first is the intramolecular cyclization of *d,l-N,S*-diacetylerythro-3-phenylcysteine ethyl ester, which occurs on chlorination followed by treatment with ammonia, and which gave about a 1:1 mixture of *cis*- and *trans*-4-acetamido-5-phenylisothiazolidin-3-one 1,1-dioxides. These were separated and identified by ir, nmr, and *p*K*a* measurements. Both isomers decomposed into the same 4-benzylidene-2-methyl-2-oxazolin-5-one on storing in acetic anhydride–pyridine.[258] The second example is also a cyclic sultam which is obtained by the intramolecular cyclization of α- and β-2-anilino-*d*-camphor-10-sulfonic acid. When the α form is heated with $PhSO_2Cl$ or Ac_2O the isothiazolidine (**77**) is formed readily. The epimeric β form also cyclized in a similar manner but required slightly more drastic reaction conditions.[259] The absolute configurations of these two derivatives follows from that of the original camphor.

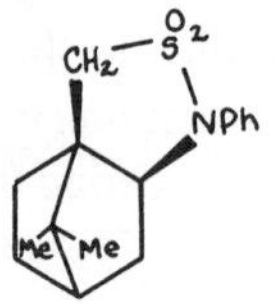

77

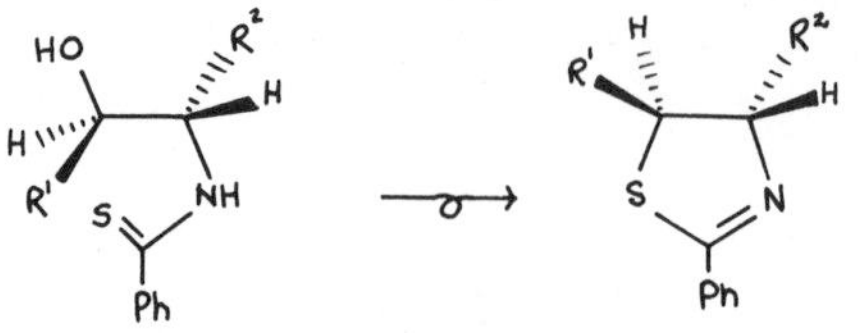

(25)

2. *Thiazolines and Thiazolidines*

Of the thiazolines, the stereochemistry of formation and properties of 2-thiazolines have attracted the most attention. Several syntheses were devised, and their stereochemical aspects were examined. Farkas and Sicher[68] studied the intramolecular cyclization of β-thiobenzamido alcohols in 10% ethanolic HCl at 100°C. The rates varied with the substituents, but the ring closure occurred with inversion of configuration at the α-carbon atom and extrusion of the OH group. Thus *erythro*-β-thiobenzamido alcohols provided *trans*-4,5-disubstituted 2-phenyl-2-thiazolines (Equation 25). Inversion also occurs if this reaction is carried out with P_2S_5. The latter reaction ensures that 2-oxazolines are not obtained because these could be formed as by-products in the former method. 2-Phenyl-*cis*-4,5-tetramethylene-2-oxazoline is converted into 2-phenyl-*trans*-4,5-tetramethylene-2-thiazoline with P_2S_5.[260] When β-aminothiols are cyclized to yield 2-thiazolines the process invariably takes place with retention of configuration. *Erythro*- and *threo*-*N*-benzamido-β-phenylcysteine ethyl esters give *cis*- and *trans*-4-ethoxycarbonyl-2,5-diphenyl-2-thiazolines, respectively, with ethanolic HCl. Prolonged treatment of the cis isomer with the reagent caused it to isomerize (by inversion at C–4) to the *trans*-thiazoline.[261]

The asymmetric centers, which are not involved in the cyclizations, generally remain unaffected unless the chiral center has an activated α-CH. Thus the cyclization of *N*-thiobenzoyl α-amino acids and 2-thiobenzamido-substituted ethyl alcohols respectively gave optically inactive 4-substituted 2-phenyl-2-thiazolin-5-ones and optically active 4-substituted 2-phenyl-2-thiazolines (with retention of configuration at C–4) on treatment with trifluoroacetic acid.[262] β-Chloroamines react with CS_2 to form 2-thiazolidine-2-thiones with inversion of configuration at the β-carbon atom, and they are readily converted into the respective 2-alkylthio-2-thiazolines without upsetting the chiral centers.[263] A synthesis which proceeds without racemization is the condensation of nitriles with β-aminothiols.[264] White, McCapra, and Field used this method to prepare firefly D-luciferine (**78**) from D-cysteine and 2-cyano-*p*-hydroxybenzothiazole without altering the chiral center.[265]

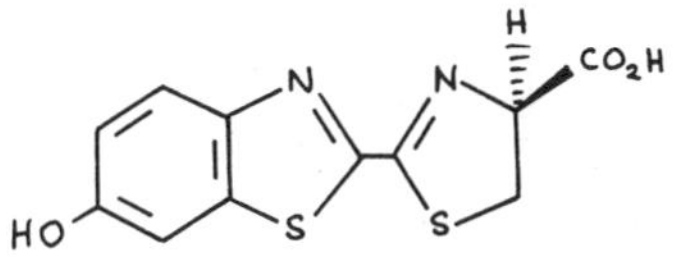

78

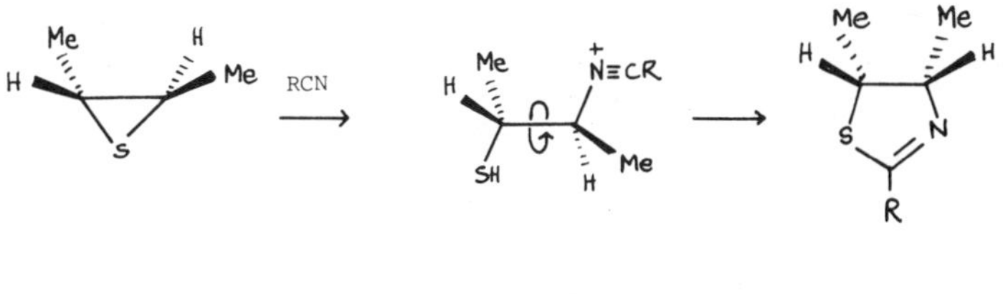

RCN $\longrightarrow$ $\longrightarrow$ (26)

Thiiranes are useful starting materials for 2-thiazolines. They react with nitriles at the more substituted carbon atom with inversion of configuration and ring opening, followed by ring closure to the 2-thiazoline. The relative configurations of the substituents at C–4 and C–5 in the 2-thiazoline are different from C–2 and C–3 substituents in the thiirane (Equation 26). A similar change in configuration occurs when aziridines are converted into 2-thiazolines with thioamides.[266] The acid-catalyzed rearrangement of optically active 2-methyl-1-*N′*-phenylcarbamylaziridine yields approximately a 1:1 mixture of 4- and 5-methyl-2-anilino-2-thiazolines. The retention of configuration at the chiral centers varies with the acid used. Complete retention was observed with *p*-toluenesulfonic acid in various solvents and with picric acid in dimethyl formamide, but complete racemization occurred with the latter acid in nitromethane.[75]

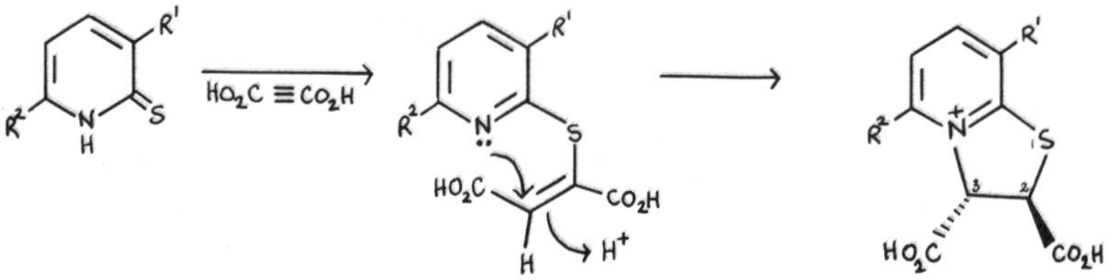

$HO_2C\equiv CO_2H$ (27)

Pyrid-2-thiones undergo Michael addition reactions with acetylene dicarboxylic[267] or α-bromocrotonic[268] acids, which are followed by intramolecular cyclization to 2,3-disubstituted 2,3-dihydrothiazolo[3.2-*a*]pyridinium derivatives. The 2- and 3-substituents invariably have the trans configuration (Equation 27). The related (−)-3-carboxy-7-hydroxy-4-methyl-2,3-dihydrothiazolo[2.3-*a*]pyridinium zwitterion (**79**) was isolated from bovine liver, and its structure was elucidated by Undheim and Nordal.[269] It gave the *cis*-*S*-oxide on oxidation with peracids, and on decarboxylation furnished the optically active 4-methyl-1-oxo-2,3-dihydrothiazolo[3.2-*a*]pyridinium-7-oxide with known absolute configuration.[270] The cd curves of these were determined.[271] Degradation of Penicillin G methyl ester by boiling in trifluoroacetic acid for 15 min gave optically active 4-methoxycarbonyl-5,5-dimethyl-2-thiazoline (**80**) in 50–60% yield.[272]

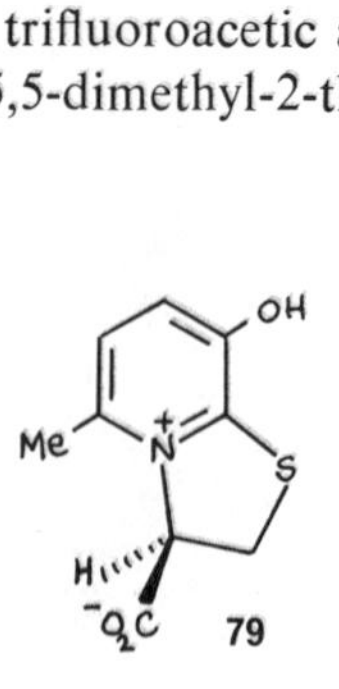

79 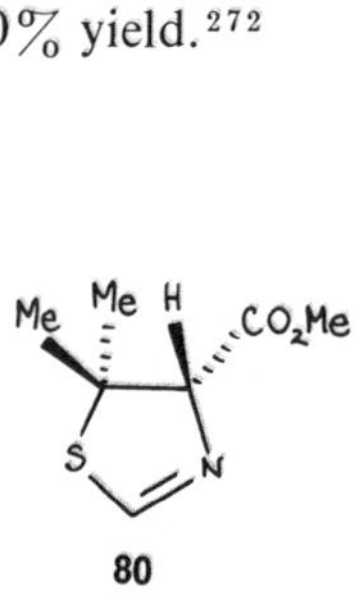 **80**

In the extensive dipole moment and variable temperature ^1H nmr investigations of the conformational preferences in 1-*t*-butylpiperidine-4-spiro heterocycles, Katritzky and co-workers[137] examined the spiro-4′- and spiro-5′-(2′-phenyl-2′-thiazolines). They found that the nitrogen atom in the former spiro compound was axial whereas the 5′-CH_2 was equatorial, and in the latter the sulfur heteroatom was also axial and the 5′-CH_2 equatorial. These results are in agreement with the smaller space requiring properties of the sp^2 nitrogen and the sulfur atoms compared with a CH_2 group.

(28)

Thiazolidines are readily obtained from β-aminothiols and aldehydes or ketones without involving the chiral centers of the aminothiol.[273–275] Unsymmetrical ketones and aldehydes introduce another chiral center at C–2. Sheehan and Yang[276] made the 2,2-dimethylthiazolidine derivatives of cysteine and cysteine peptides as a means for protecting the amino and thiol groups, and the thiazolidines retained the optical activity. 1-Aminopropane-2-thiol was resolved by the method of Taguchi and co-workers[275] by forming the respective diastereomic 2-substituted thiazolidines by condensation with D-glucose. Two diastereomers, of four possibilities, were obtained and separated, and when each was treated with benzaldehyde the enantiomeric (+)- and (−)-5-methyl-2-phenylthiazolidines were formed. There is some steric control in the last reaction also, because only the cis isomer, of the two possible isomers, is formed (Equation 28).[277] The antitubercular (*in vitro*) antibiotic (−)-2-(5-carboxypentyl)-thiazolidin-4-one (**81**) was synthesized by condensing methyl ω-aldehydopimelate with thiolacetamide and resolved into its optically active forms by way of the brucine salts. It gave an optically active sulfoxide, with hydrogen peroxide, which was unstable to alkali.[278] The optically active thiazolidine carboxylic acids derived from *R*- and *S*-cysteines and formaldehyde reacted with potassium cyanate to form the optically active bicyclic hydantoins 3-thia-1,7-diazabicyclo[3.3.0]octane-6,8-diones (**82**).[279] These mutarotated in dilute alkali probably because of cleavage in the thiazolidine ring. Clarke and Sykes[280,281] studied the reduction of 4- and 4,5-substituted *N*-alkylthiazolinium salts with $NaBH_4$ and $NaBD_4$. Although they demonstrated that the hydride ions are transferred to C–2 and C–4, they observed no special stereospecificity in the reduction.

Craig and co-workers[282] isolated a trithiaza heterocycle from the condensation of chloroacetaldehyde hydrate and ammonium dithiocarbamate. They proposed an adamantane structure for it which was shown to be incorrect.[283] The product is in fact hexahydro-1,3,5-trithia-6*b*-azacyclopenta[*c.d*]pentalene (**83**), and its structure is supported by its ^{1}H nmr spectrum and that of its cation. It is clearly nonplanar, and if the nitrogen atom does not invert freely, then the molecule should be capable of optical activity.[284]

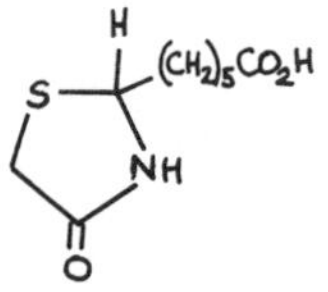

81 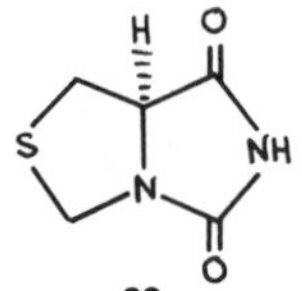 **82** 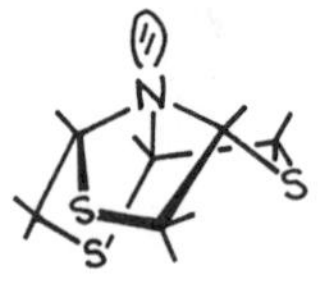 **83**

Bicyclic thiazolidin-2-ones were obtained by Potts and co-workers[285] from the cycloaddition reactions between the mesoionic compounds, such as anhydro-2-aryl-5-hydroxy-3-methylthiazolinium hydroxide and dimethylacetylene dicarboxylic ester or *N*-phenyl maleimide. The structure of the adducts, 4-aryl-5,6-disubstituted 7-methyl-2-thia-7-azabicyclo[2.2.1]hept-5-en-3-ones or heptan-3-ones (e.g., **84**) were deduced by ^{1}H nmr spectroscopy.

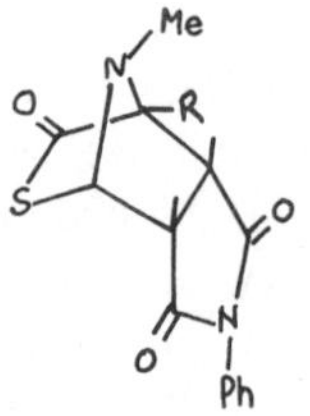

84 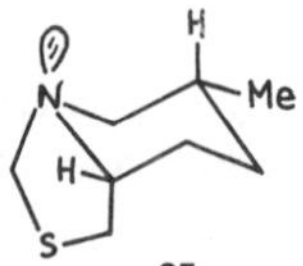 **85** 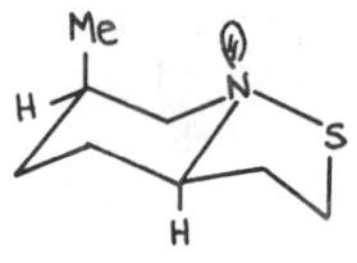 **86**

The conformational equilibria in methyl substituted 8-thia-1-azabicyclo-[4.3.0]nonanes were studied by Crabb and Newton[273] using the J_{gem} values for the N–CH_2–S in the ^{1}H nmr spectra and the Bohlmann bands at 2700–2850 cm^{-1} in the ir spectra. Like the oxygen analogues (Section II.C.2) the alkyl group has a strong influence on the position of equilibrium;

for example, *cis*-3-methyl-8-thia-1-azabicyclo[4.3.0]nonane prefers the conformation with the equatorial methyl group and cis-fused configuration **85** rather than **86.**

A considerable amount of stereochemically very interesting chemistry of thiazoline and its *S*-oxide was reported in the study of penicillin and related compounds. The structure of natural penicillins (**87**) was determined by an X-ray analysis, and the absolute configuration is *S* at C–3 and *R* at C–5 and C–6.[286] Renewed interest in the study of penicillins has been observed in the past five or six years and is probably because the earlier patent rights are beginning to expire. Many reviews on the chemistry of penicillin, and their chemical and stereochemical transformations, have been written and a repetition is unwarranted. Leading references and summaries of recent accounts have been reported by Barton[286a] and Stoodley[287] in 1973, and McGregor[288] in 1974, and should be consulted for reactions and stereochemical aspects.

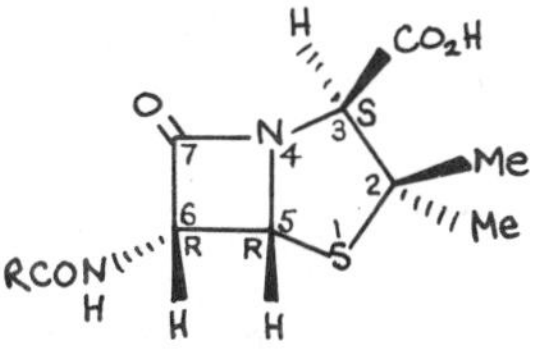

87

C. Six-Membered Rings

1. *Reduced 1,2-Thiazines*

Only a few examples of this class of compounds of stereochemical interest were reported. One case is the *cis*-5,6-disubstituted tetrahydro-3,4-benzo-1,2-thiazine 1-oxide (**88**). It is formed by the cycloaddition of *N*-sulfinylaniline (PhN=S=O) and norbornene, followed by a prototropic rearrangement. 3-Sulfinylaminopyridine and 1-amino-8-sulfinylaminonaphthalene behave similarly.[289]

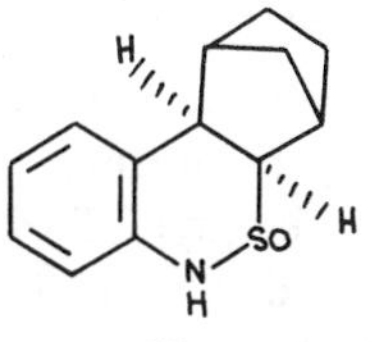

88 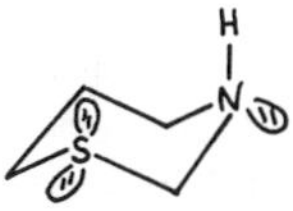 **89** 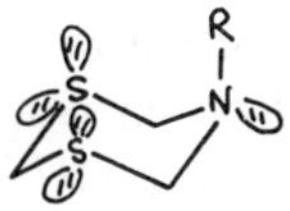 **90**

2. *Reduced 1,3-Thiazines*

Reduced 1,3-thiazines have assisted in providing further examples of the generalized anomeric effect (see Section II.D.2 and Chapter 2, Section X.C). Katritzky and co-workers[164,181] showed that in 1,3-thiazane the NH is predominantly axial (**89**) as in the corresponding oxazane. They extended this study to *N*-alkyl-1,3,5-dithiazanes and 3,5-dialkyl-1,3,5-thiadiazanes and found that the alkyl group of the former class of compounds was predominantly axial (**90**) even when it was a *t*-butyl group. In the second class of compounds they observed that the conformers with one or two axial alkyl groups can be predominant. The ring inversion barriers are between 38.9 to 50.6 kJ mol^{-1}, and increase in the size of the *N*-alkyl groups, or replacing N–R by S, decreases the inversion barrier. In these examples also, there was no correlation between J_{gem} values and the orientation of the adjacent nitrogen lone pair.[290] Crabb and Newton[291] examined the ir and ^{1}H nmr spectra of a series of 3-thia-1-azabicyclo[4.4.0]decanes. When they were compared with the corresponding 3-oxa-1-azadecanes it was found that the 1,3-thiazane ring showed marked deviations from the chair conformation.

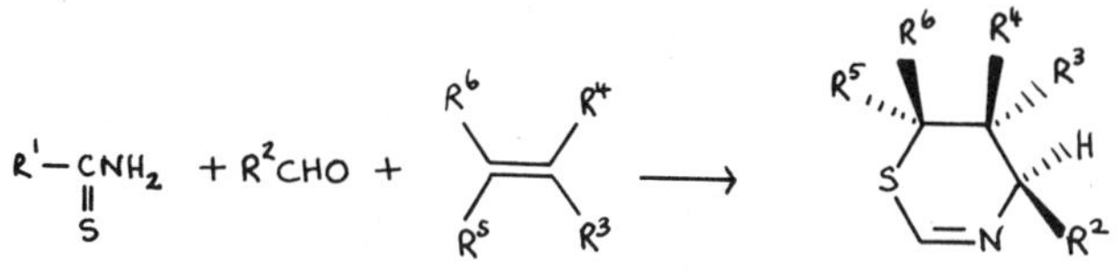

$$R^1-C(=S)NH_2 + R^2CHO + R^5R^6C{=}CR^4R^3 \longrightarrow \quad (29)$$

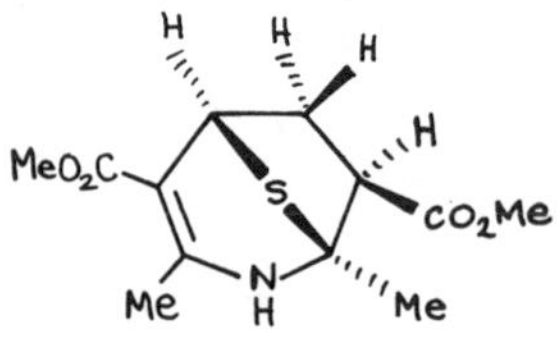

91

A new regio- and stereospecific synthesis of 5,6-dihydro-4*H*-1,3-thiazines involving a 1,4-dipolar cycloaddition was developed by Abis and Giordano.[292] These are formed by the reaction of a thioamide, an aldehyde, and an olefin and are essentially a cis addition of the Schiff base to the olefin (Equation 29). The configurations and preferred conformations of the products were deduced by ^{1}H nmr spectroscopy. An intramolecular cyclization and rearrangement occurs when 4-chloromethyl-3,5-bismethoxycarbonyl-2,6-dimethyl-1,4-dihydropyridine is treated with H_2S, and 4,7-bismethoxycarbonyl-1,3-dimethyl-8-thia-2-azabicyclo[3.2.1]oct-3-ene is formed stereospecifically.[293] The structure and stereochemistry shown in **91** was confirmed by X-ray diffraction studies.[294] 1,3-Thiazine-2,4-dione (1-thiauracil), like uracil, dimerizes upon uv irradiation, and the cis,syn photodimer **92** was isolated and identified.[295]

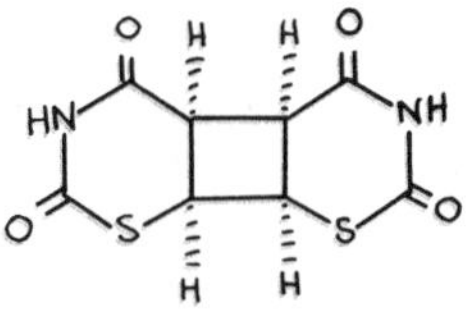

92 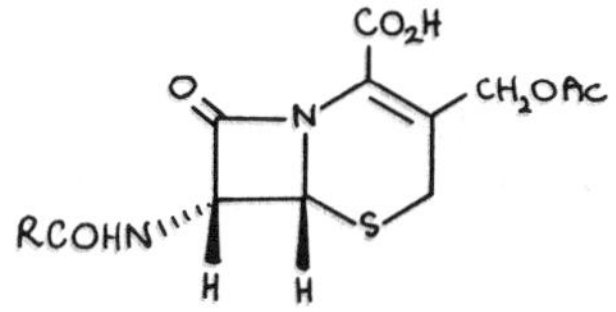 **93** 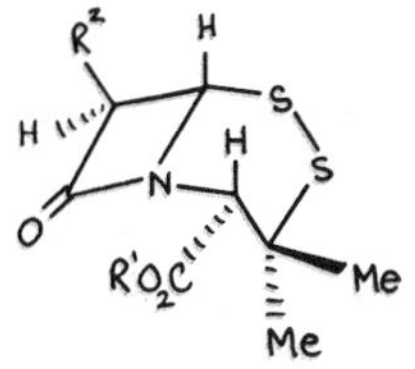 **94**

Much of the interest in the dihydro and tetrahydro 1,3-thiazines arose from attempts at understanding the chemistry of derivatives of cephalosporins (**93**). These are antibiotics related to penicillin which contain the same number of carbon atoms. The total synthesis of cephalosporin C [**93**, $R=(CH_2)_3CH(NH_3^+)CO_2^-$] was accomplished by Woodward[296] and his collaborators; also the conversion of a penam (penicillin ring system) into a cepham (cephalosporin ring system)[297] was achieved in the usual brilliant Woodward manner. The X-ray analysis of cephalosporins has been reported.[298] Other studies on the syntheses of the cepham nucleus,[299–301] the reactions of substituents, and the cd chromophoric properties of 3-cepham derivatives[302] together with the determination of the stereochemistry of 3-methylenecepham by lanthanide induced shifts[303] were described. The *P* or *M* helicity of penams and cephams containing another sulfur atom vicinal to the one already present (e.g., 2-alkoxycarbonyl-3,3-dimethyl-8-oxo-7-phthalimido-4,5-dithia-1-azabicyclo[4.2.0]octane, **94**) was elucidated by ^{1}H nmr (NOE) spectroscopy and confirmed by X-ray analysis.[304] Reviews which discuss the chemistry of cephams are available.[305]

3. *1,4-Thiazanes* (*Including Phenothiazines*)

One of the oldest examples of a stereospecific synthesis of a 2,3-disubstituted thiazane is the one described by Mousseron and Combes.[306] They heated *cis*-1-chloro-2-(2′-chloroethylamino)cyclohexane in a Na_2S solution at 140° and isolated *trans*-2,3-tetramethylene-1,4-thiazane with inversion of configuration at the carbon atom involved in the cyclization. Carson and Boggs[307] noted an intramolecular formation of 1,4-thiazanes by cyclizing the *S*-oxide of *S*-(*cis*-propen-2-yl)-L-cysteine in which the amino group adds stereospecifically on to the double bond. The (+)-diastereomeric propenyl-*S*-oxide gave optically active 1*S*,3*R*,5*S*-3-carboxy-5-methyl-1,4-thiazane 1-oxide (Equation 30), and the (−) isomer gave 1*R*,3*R*,5*R*-3-

carboxy-5-methyl-1,4-thiazane 1-oxide. These were reduced with HI to 3*R*,5*S*-*cis*- and 3*R*,5*R*-*trans*-3-carboxy-5-methyl-1,4-thiazanes respectively. The cyclization of *S*-vinyl-L-cysteine sulfone, similarly gave a mixture of *cis*- and *trans*-3-carboxy-5-alkyl(or phenyl)-1,4-thiazane 1,1-dioxides in which the *R* configuration in the cysteine was retained at C–3 in the thiazane. ^1H nmr spectra revealed that in the *trans*-1,4-thiazane 1,1-dioxide the 3-CO_2H group was axially disposed.[308] Some stereospecificity has also been observed in the oxidation of 5-alkyl(Et or Me)-3-carboxy-1,4-thiazanes. The trans isomer yields specifically the *S*-oxide on oxidation, but the cis isomers and 3-carboxy-1,4-thiazane yield a mixture of diastereomeric *R*- and *S*-1-oxides.[309] The structures and conformations of the isomers

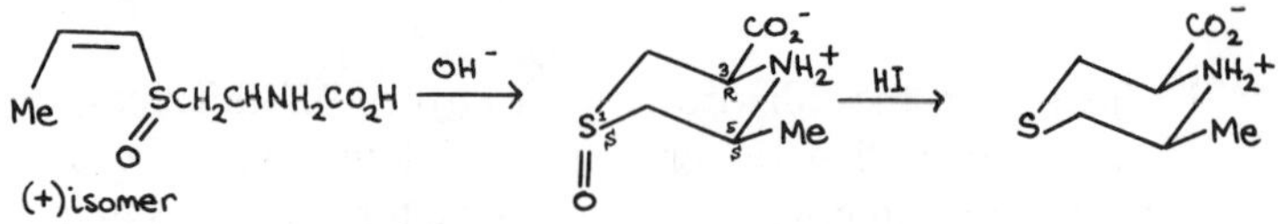

(+)isomer — OH^- → — HI → (30)

95 and **96**,[310] in the solid state, were confirmed by X-ray analyses. The conformations of these and of related compounds in solution were established by ^1H nmr, ir, and optical rotations. The diastereomeric *S*-oxides of *S*-(2-propynyl)-L-cysteine were also subjected to the cyclization conditions. The (+) isomer gave one 3*R*-3-carboxy-5-methyl-2,3-dihydro-1,4-thiazine 1-oxide (**97**), but the configuration at the sulfur atom is unknown. The (−) diastereomer, however, gave resins, but *S*-(2-propynyl)-L-cysteine *S*,*S*-dioxide cyclized satisfactorily. The ^1H nmr spectrum of **97** was consistent with a half-chair conformation.[311]

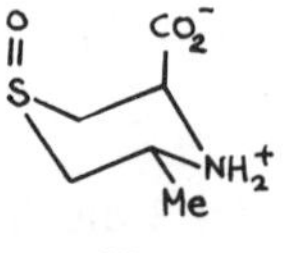

95 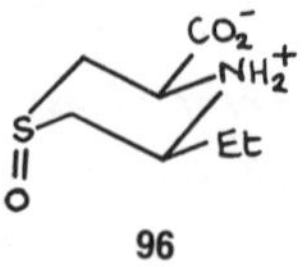 **96** 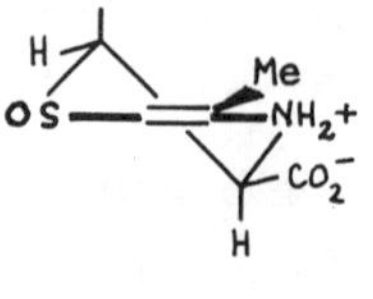 **97**

Stoodley and his collaborators studied the reactions and stereochemistry of a variety of substituted 6-alkoxycarbonyl-2,3-dihydro-1,4-thiazines and their 1-oxides,[312–318] and in particular their conformational mobility.[315,319,320] The two conformers **98** and **99** are involved, with the latter conformer usually predominating, but the ratio is sensitive to the solvent used. Portoghese and Telang prepared the rigid 1,4-thiazane (−)-*N*-tosyl-2-thia-5-azabicyclo[2.2.1]heptane (**100**), and its diastereomeric *S*-oxides of known absolute configuration, by intramolecular cyclization of *trans*-2-acetylthiomethyl-1-tosyl-4-tosyloxypyrrolidine, obtained from *trans*-4-hydroxy-*S*-proline.[321]

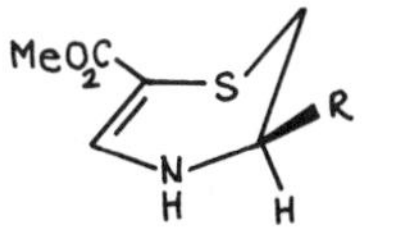

98 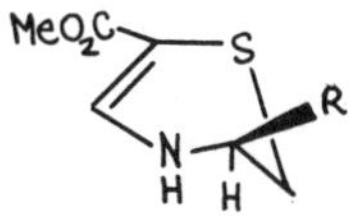 **99**  **100**

Lowe and co-workers[322–324] prepared 1,4-thiazanes for making nuclear analogues of cephams. The key reaction involved the photolytic intramolecular cyclization of diazomalonyl derivatives (Equation 31)—a method which has also been used in related systems (see Equation 14 in Section II.C.2 and Equation 28 in Part I, Chapter 2, Section II.A.1), and some stereoselectivity was observed because the cis isomer predominated.

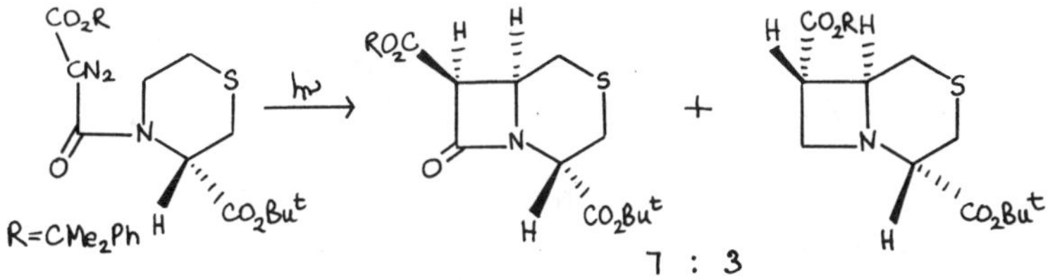

R=CMe$_2$Ph (31)

7 : 3

The folded structure of phenothiazine and its *N*-methyl and *N*-phenyl derivatives in solution was established by dipole moment and molar Kerr constants. Aroney, Hoskins, and Le Fèvre[325] interpreted the data in terms of structures in which the "angle of fold" was about 150°, and in which the substituents on the nitrogen atom were oriented in an equatorial conformation with respect to the heterocyclic ring (**101**). An X-ray analysis of the phenothiazine crystal indicated that although the molecule is folded, it is, however, slightly puckered. The dihedral angle at the sulfur atom is

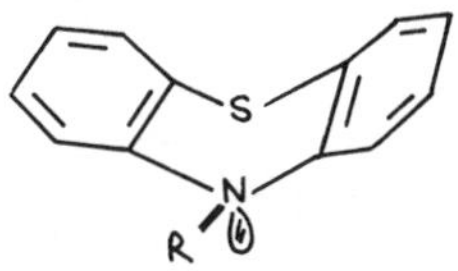

101 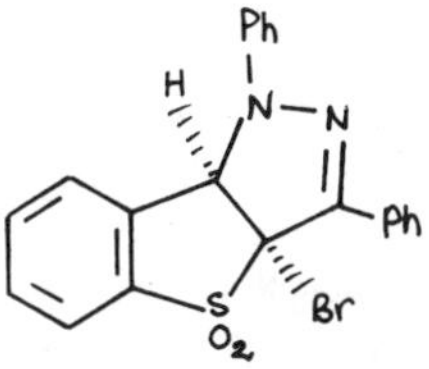 **102** 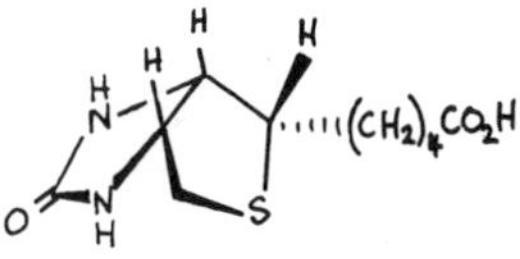 **103**

99.6° and at the nitrogen atom it is 121.5°, but all the other angles are 120°.[326] The cis and trans isomers of 1,2,3,4,4*a*,10*a*-hexahydrophenothiazines have been reported.[327]

D. Miscellaneous

cis- or *trans*-2-Benzoyl-3-(thien-2-yl)-1-cyclohexylaziridines react with dimethyl acetylene dicarboxylate by preliminary cleavage of the C–C bond of the aziridine ring. This is followed by a cycloaddition reaction and yields *cis*-2-benzoyl-1-cyclohexyl-3,4-bismethoxycarbonyl-5-(thien-2-yl)-2,3-dihydropyrrole while the thiophene ring remains unreacted.[328] Partial resolution of 2-(thien-2-yl)-pyrrolidine was apparently achieved and is attributed to the presence of an asymmetric carbon atom rather than restricted rotation about the bond joining the two nuclei.[329] A cycloaddition reaction onto a thiophene double bond does occur in thianaphthene 1,1-dioxides. The 2-bromo derivative condensed with benzaldehyde phenyl hydrazone to yield 1,3-diphenyl-3*a*-bromo-*cis*-3*a*,8*b*-dihydrobenzo[1]-thieno[3.2-*c*]pyrazole 4,4-dioxide (**102**).[330] Diazoalkanes, on the other hand, add in the opposite sense and yield *cis*-3*a*,8*b*-dihydrobenzo[1]thieno[2.3-*c*]pyrazoles.[331] Considerable effort has been expended in the study of the vitamin biotin (**103**) and related compounds. Its synthesis,[332,333] optical resolution,[334] stereochemical relationship with its stereoisomers,[335] its sulfoxides,[336,337] and synthesis of its α-dehydro derivative[338] have been described, and some aspects of it are reported in Chapter 3, Section IV. More recently Ellis and Sammes[339] have synthesized the ring system, *cis*-perhydrothieno[3.4-*d*]-imidazol-2-one *S*,*S*-dioxide (**104**), by base-catalyzed intramolecular cyclization of 3-ureido-2,3-dihydrothiophene 1,1-dioxide. Lett and Marquet[340,341] analyzed the nmr spectra of the sulfoxides and sulfones of the basic bicyclic ring system (**104**), and the related *cis*-3-thia-6,8-diazabicyclo[3.3.0]octane, and interpreted their data in terms of a pseudorotating thiophane ring.[342]

Leonard and co-workers[343,344] synthesized 3*R*,6*S*-3-methoxycarbonyl-2-2-dimethyl-5-oxo-6-phenylacetamido-1,4-thiazepane (**105**), and its 3*S*,6*S* diastereomer, by condensing α-phenylacetamidoacrylic acid and *S*-(D)-penicillamine methyl ester hydrochloride. This thiazepine is the monocyclic analogue of penicillin and was envisaged as an intermediate in its biosynthesis. All attempts, however, to cyclize it into the penam have been unsuccessful. The synthesis of the optically active *cis*-3-acetoxy-5-[2(dimethylamino)ethyl]-2,3-dihydro-2-*p*-methoxyphenyl-1,5-benzothiazepin-4(5*H*)-one hydrochlorides[345] and 3-methyl-3,5-dihydro-4,1-benzothiazepin-2(1*H*)-one *S*-oxide[346] were described, and their configurations were established by X-ray analyses. Szmant and Chow[347] reported two isomers of 2,3,6,7-dibenzo-1-thia-4,5-diazacyclohepta-2,4,6-triene 1,1-dioxide and provisionally assigned the chair conformation to one of them. Allinger and Youngdale[348] repeated the preparation and found that only one compound was produced (the other being only an impure form) and objected to the chair form because it could not be more stable than the boat conformation (**106**) which is the more reasonable structure for this compound. The ^{1}H nmr spectra of *cis*- and *trans*-2,7-dimethyl-3,6-diphenyl-2,7-dihydro-1,4,5-thiadiazepine revealed that the ring is puckered.[349]

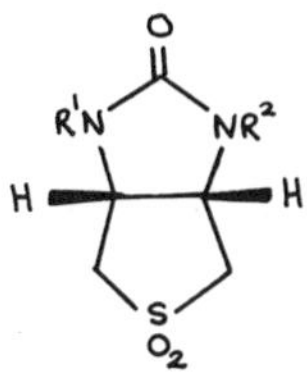

104 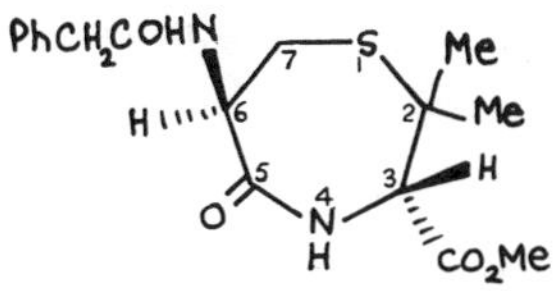 **105** 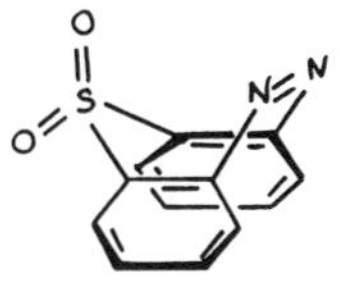 **106**

In the eight-membered sulfur and nitrogen heterocycles an interesting phenomenon was observed by Paul and Go.[350] They investigated the crystal structure of 5-methyl-1-thia-5-azacyclooctane 1-oxide perchlorate. The crystals of this molecule change when the temperature is lowered from 25 (β form) to 3°C (α form). Both forms are approximately in the boat-chair conformations and have the general structure **107** with a transannular N···H···O–S bridge. The transformation was explained by the flipping of the rings from **107** to **108**, followed by rotation of the molecule. The alternative explanation would require considerable translational movement of the cations within the crystal.

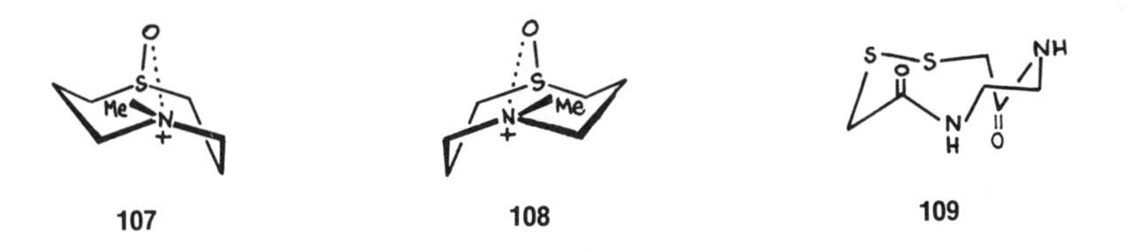

107 **108** **109**

Ring inversion has been observed in 6-substituted 6,7-dihydro-5*H*-dibenzo[*b.g*]-1,5-thiazocine.[351] The rate process of ring inversion in 1,5-dithia-3,7-diazacane was measured by variable temperature ^{1}H nmr spectroscopy, and the $\Delta G^\ddagger$ values for the 3,7-dimethyl, diethyl, diisopropyl, and phenyl derivatives were 61.9, 61.1, 59.0, and 56.1 kJ mol^{-1} and should be compared with the more flexible polydeuterocyclooctane ($\Delta G^\ddagger = 33.9$ kJ mol^{-1}). A pseudorotating system may be involved in these examples.[352] Large rings containing sulfur and nitrogen atoms derived from the oxidation of *N,N′*-bismercaptoacetylethylenediamine (to 1,2,5,8-dithiadiazadecane-4,9-dione, **109**)[353] and the chiral cyclic tetrapeptidebisdisulfide from L-cysteinyl-L-cysteine were reported.[354]

IV. SULFUR AND OXYGEN HETEROCYCLES

A. Four-Membered Rings

The addition of SO_3 across an olefinic double bond was known to form 1,2-oxathietane 2,2-dioxides (ethylene sultones).[355,356] But only recently Nagayama and co-workers[357] demonstrated that the addition reaction is highly cis stereospecific. *cis*- and *trans*-But-2-enes reacted with an equimolar amount of the 1:1 complex of SO_3–dioxane, or SO_3 alone at 0°C in $CHCl_3$, and gave *cis*- (**110**) and *trans*-3,4-dimethyl-1,2-oxathietane 2,2-dioxides respectively. Pyrolysis of *cis*-2,3-dibenzoyl-2,3-diphenylthiirane 1,1-dioxide produces SO_2 and diphenyl acetylene instead of an olefin. The decomposition was therefore assumed to proceed by a rearrangement of the sulfone into *cis*-3,4-dibenzoyl-3,4-diphenyl-1,2-oxathietane 2-oxide followed by elimination of benzil and the intermediate 3,4-diphenyl-1,2-oxathiete 2-oxide, which extruded SO_2 to form diphenyl acetylene.[358]

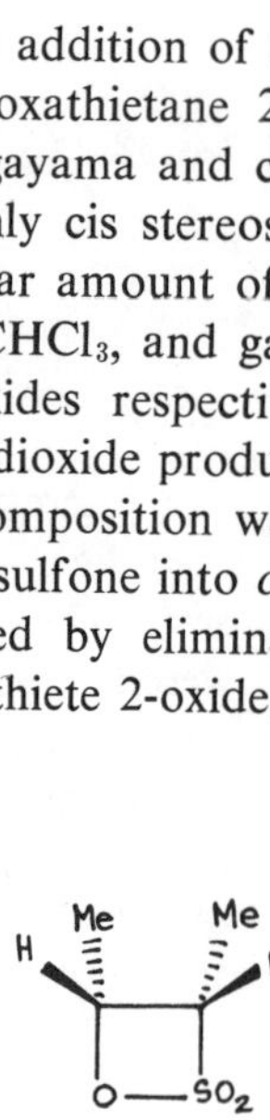

110

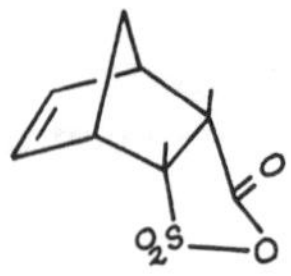

111

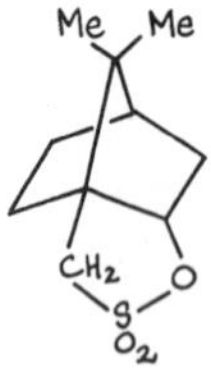

112

B. Five-Membered Rings

1. *1,2-Oxathiolanes*

1,2-Oxathiolane 2,2-dioxides are formed by the intramolecular cyclization of *β*-methanesulfonyloxyethyl bromide which is promoted by *n*-butyl lithium in THF. The methyl group is converted into its anion which displaces the *β*-bromine atom with inversion of configuration. *cis*-4,5-Tetramethylene and *cis*-4,5-hexamethylene 1,2-oxathiolane 2,2-dioxides are obtained in this way in 33 and 36% yields respectively (Equation 32).[359] They are basically cyclic sulfate esters and therefore reactive molecules. The 5-oxo derivatives which are cyclic carboxylicsulphonic anhydrides can be produced by Diels-Alder reactions between a diene (e.g., butadiene) and the cyclic β-sulfoacrylic anhydride. The bicyclic adducts, for example, **111** from cyclopentadiene, are cis-fused.[360] Dimmel and Fu[361,362] found that the 1,2-oxathiolane 2,2-dioxide (**112**), known as 10-isobornylene sultone, undergoes a variety of skeletal rearrangements, of the Wagner-Meerwein type, to *exo*-camphene sultone (**113**) while the heterocyclic ring remains intact during the transformations. The mechanism of the rearrangement was established by means of deuterated derivatives and was shown to involve an *exo*-3,2-methyl shift.

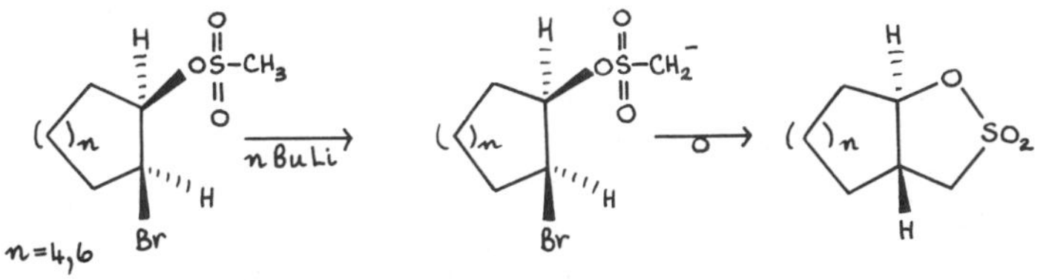

(32)

The dipole moments of 1,2-oxathiolane 2-oxide and 2,2-dioxide were measured and were consistent with gauche conformations for the ester function.[363]

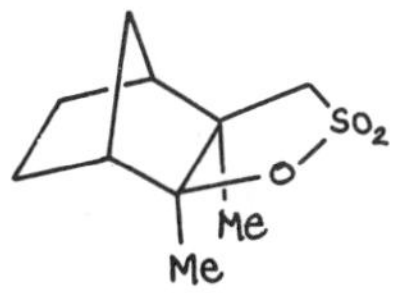

113 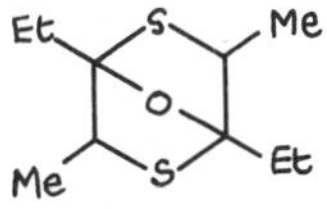 **114** 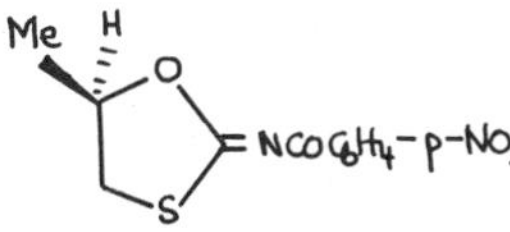 **115**

2. *1,3-Oxathiolanes*

1,3-Oxathiolanes are thioketals and are formed from carbonyl compounds and β-hydroxythiols by azeotropic removal of water.[364,365] The acid-catalyzed reaction between two molecules of pentan-3-on-2-thiol to form 3,6-diethyl-2,5-dimethyl-3,6-endoxo-1,4-dithiane (**114**) involves the removal of one molecule of water, but the stereochemistry of the product is not known.[366,367] Thiocarbonyl ylides generated by thermolysis of 2,5-disubstituted 1,3,4-thiadiazol-3-enes undergo 1,3-dipolar addition reactions with carbonyl compounds (e.g., ketenes) to give 1,3-oxathiolanes in which the configuration of the substituents at C–2 and C–5 is the same as in the original thiadiazolene (Equation 33)[368] (see Equation 2 in Chapter 3). 2-Imino-1,3-oxathiolane anions were postulated as intermediates in the conversion of oxiranes into thiiranes with KCNS (see Scheme 1, Chapter 3, Section II.A). This was confirmed experimentally by trapping the intermediate as the *p*-nitrobenzoyl derivative (e.g., **115**).[369] The sulfur atom in some 1,3-oxathiolanes can be oxidized to a sulfoxide,[368] and the diastereomeric *S*-oxides can be identified by nmr spectroscopy.[370]

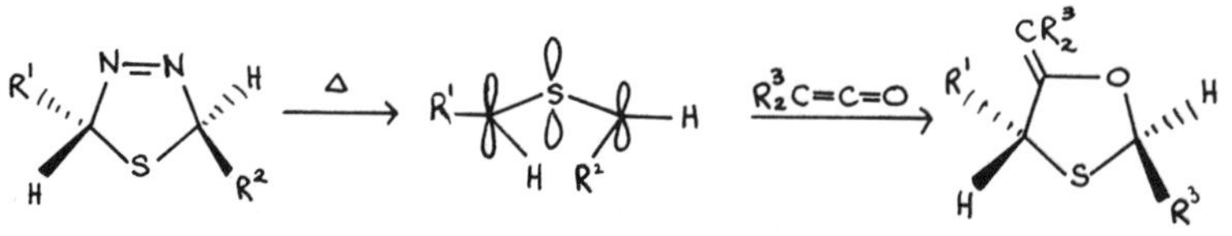

(33)

Djerassi and collaborators[371] examined the chiroptical properties of several 1,3-oxathiolanes including the 2-thione derivatives and 1,3-dithiolane 2-thiones. They derived empirical rules which relate the direction of twist of the chromophoric ring and the sign of rotation observed. Most of the examples are from the steroid field; that is, the heterocyclic ring is condensed with a steroid nucleus, and it should be interesting to see if the rules hold for the determination of absolute configuration of nonsteroidal 1,3-oxathiolanes.

The five-membered 1,3-oxathiolane ring is conformationally mobile and, like five membered rings generally, exhibits pseudorotation (see Scheme 5, Chapter 2). The ^1H nmr spectra of several 2-alkyl-5-methyl- and 2,2-dialkyl-5-methyl-1,3-oxathiolanes were examined together with the acid-catalyzed equilibration of cis and trans isomers. The data have confirmed that some pseudorotation is occurring, but have been used to assign relative configurations of substituents and to obtain thermodynamic parameters for the chemical equilibration of diastereomers.[364] Similar studies were made with 2-alkyl-4-methyl- and 2-alkyl-2,4-dimethyl-1,3-oxathiolanes by 100- and 220-MHz ^1H nmr spectroscopy, which showed that the envelope conformation was not very flexible.[365] Other studies revealed that some conformations are highly favored to the extent that pseudorotation has been negated in this ring system, and the free energy values for the conformational preferences for the 2-methyl, 2-ethyl, 2-isopropyl, and 2-phenyl groups were evaluated.[372] Wilson, Huang, and Bovey[373] obtained J values for a series of 2-mono- and 2,2-disubstituted 1,3-oxathiolanes and concluded that although some conformations are highly favored pseudorotation is taking place, and they confirmed this by examining the spectra of 2-ethyl-1,3-oxathiolane at low temperatures. The preferred envelope conformations **116** and **117** are consistent with the observed J values.

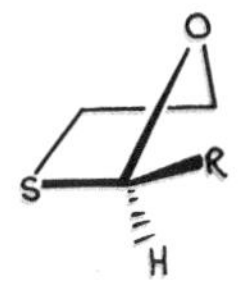

116

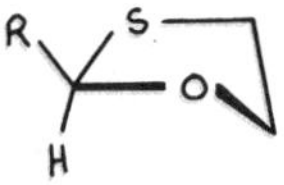

117

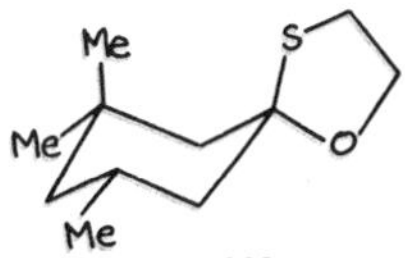

118

An assessment of the relative space requiring properties of the oxygen atom compared with the sulfur atom was made by finding the preferred conformation in cyclohexanespiro-2′-(1,3-oxathiolanes). Eliel and co-workers[374,375] found that in 1-*t*-butylcyclohexane-4-spiro-2′-(1,3-oxathiolane) the conformer with the axial sulfur atom was slightly (58%) preferred in the BF_3-catalyzed equilibration. This result was substantiated by Mertes, Lee, and Schowen,[376] who studied the kinetics of the BF_3-catalyzed equilibration of 1,1,3-trimethylcyclohexane-5-spiro-2′-(1,3-oxathiolane) and showed that the conformer with the axial sulfur atom (**118**) was favored by 3.2 kJ mol^{-1}. No complexing of the catalyst BF_3 with the oxygen or sulfur atoms was evidenced, which means that these conformational preferences are legitimate. This preference was referred to as the "lever effect" because of the longer C–S bond compared with the C–O bond.[377]

3. *1,3,2-Dioxathiolanes*

1,3,2-Dioxathiolane 2-oxide and 2,2-dioxide are the cyclic esters of ethylene sulfite and ethylene sulfate respectively. The 1-oxide is formed from ethylene glycol and $SOCl_2$. Pritchard and Lauterbur[378] separated two isomeric propylene sulfites which were 4-methyl-1,3,2-dioxathiolane *syn*- and *anti*-2-oxides. The chiral ring sulfur atom was also demonstrated in the cyclic thiono esters. When 1,2-diols are treated with sulfur monochloride (S_2Cl_2) a dioxadithiane is not formed, but instead 1,3,2-dioxathiolanes are obtained.[379] *meso*-Dihydrobenzoin provided *syn*- (**119**, R=Ph, X=S) and *anti,cis*-4,5-diphenyl-1,3,2-dioxathiolane 2-thiones (**120**, R=Ph, X=S), whereas racemic dihydrobenzoin gave only one *trans*-4,5-diphenyl-1,3,2-dioxathiolane 2-thione (**121**, R=Ph, X=S).[380] The oxygen analogues (**119–121**, X=O) also show similar isomerism.

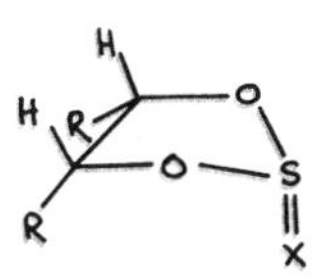

119

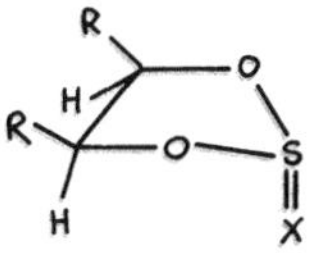

120

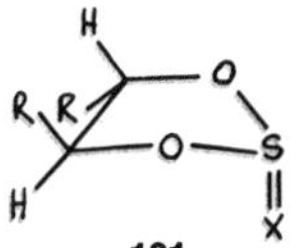

121

1,3,2-Dioxathiolane 2-oxides are cyclic sulfite esters and are therefore quite reactive. They react with benzaldehyde to form 2-phenyl-1,3-dioxalanes with loss of SO_2. Tracer experiments with ^{18}O indicated that the S–O and not the C–O bonds are cleaved and that this is confirmed by retention of configuration of substituents on the carbon atoms; that is, *cis*-4,5-dimethyl-1,3,2-dioxathiolane 2-oxide was converted into *cis*-4,5-dimethyl-2-phenyl-1,3-dioxalane.[381] Substituted 1,3,2-dioxathiolane *syn*- and *anti*-2-oxides have different ^1H nmr spectra, and they also behave differently on pyrolysis. When 4,5-diaryl derivatives are heated, a mixture of aldehydes ($R^1R^2CH–CHO$) and ketones ($R^1CH_2COR^2$) is formed. The proportion of ketone was found to be higher in the *anti*-2-oxide than in the *syn*-2-oxide and was attributed to the proximity of the oxygen atom on the sulfur atom to the α-hydrogen atom (Equation 34).[382] In the photolysis of *cis*- and *trans*-4,5-diphenyl-1,3,2-dioxathiolane 2-oxides, on the other hand, there was little difference in the distribution of products.[383]

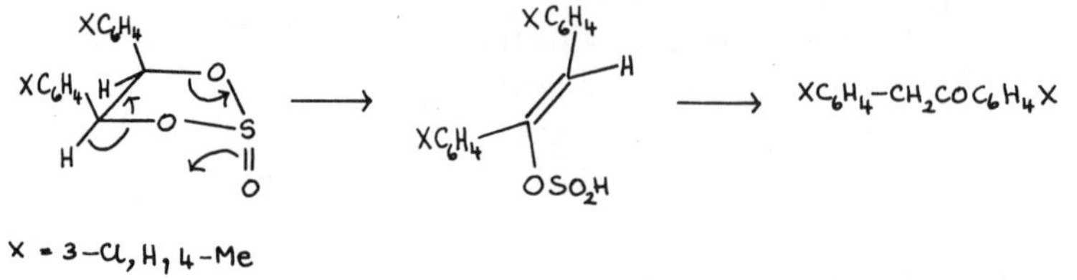

XC_6H_4, XC_6H_4, H, H, O, O, S, O ⟶ XC_6H_4, H, XC_6H_4, OSO_2H ⟶ $XC_6H_4–CH_2COC_6H_4X$ (34)

X = 3-Cl, H, 4-Me

The ^1H nmr and ir spectra of several substituted 1,3,2-dioxathiolane 2-oxides have been measured in order to establish the conformations of these compounds. The data suggest that certain conformers are more stable, and that the oxygen atom on the sulfur atom is preferentially in the pseudo-axial orientation in all cases.[384,385] They exist in twist-envelope conformations,[385] and the proportions of these can be estimated from J values. Green and Hellier[386] attempted to calculate these proportions for *syn*-(**122** and **122a**) and *anti*-4-methyl-1,3,2-dioxathiolane 2-oxides (**123** and **123a**) and demonstrated that the conformers **122** and **123** are less favored than **122a** and **123a** by about 5.4 and 0.84 kJ mol^{-1} respectively. The ΔG value between the syn and anti isomers from equilibration experiments is about 3.8 kJ mol^{-1}.

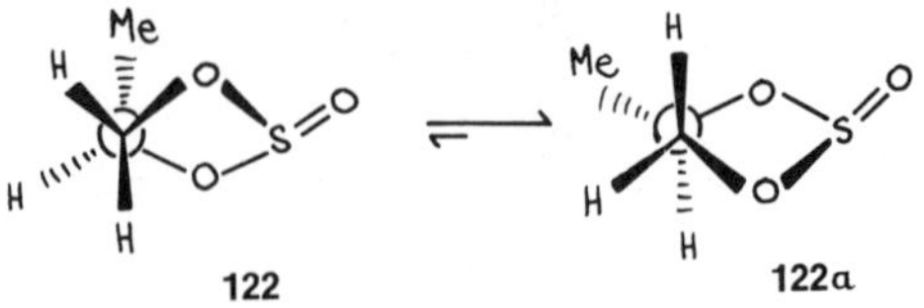

Me, H, H, H, O, O, S=O ⇌ H, Me, H, H, O, O, S=O

122 **122a**

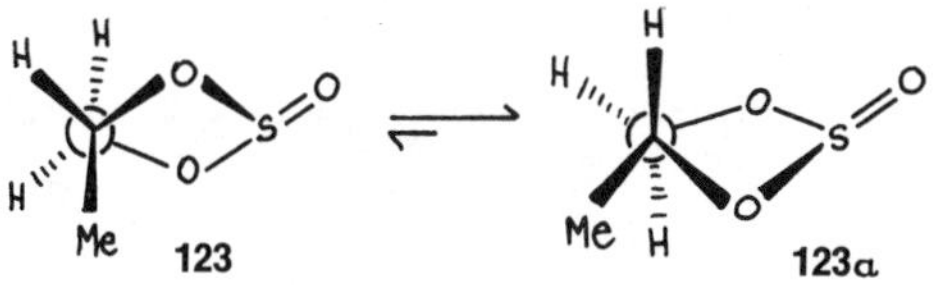

123 ⇌ 123a

C. Six-Membered Rings

1. *1,2-Oxathianes*

1,2-Oxathianes are rather unstable because they are esters of sulfenic acid. Their 2-oxide and 2,2-dioxides are, however, more stable since they are cyclic sulfites and sulfates respectively. Harpp and Gleason[387] prepared 1,2-oxathiane 2-oxide and 2,2-dioxide by oxidizing butane-1,4-dithiol with H_2O_2/AcOH to 1,2-dithiane 2,2-dioxide which rearranges with extrusion of sulfur, in the presence of tristriethylamino phosphine, to yield 1,2-oxathiane 2-oxide. Oxidation of the latter provided the 2,2-dioxide (Equation 35). Dipole moment measurements are consistent with a chair conformation for 1,2-oxathiane 1-oxide with the oxygen atom preferentially in the axial conformation **124.**[363] ^{1}H nmr data revealed only one conformation in which the exocyclic oxygen atom is axial, **124,** and conformationally less flexible than the thiane 1-oxide in which the ring oxygen atom is replaced by a CH_2 group (see Chapter 3, Section VIII.B). The stronger axial preference of the exocyclic oxygen atom in **124** is due to the strong dipolar effect which is essentially an anomeric effect.[387a]

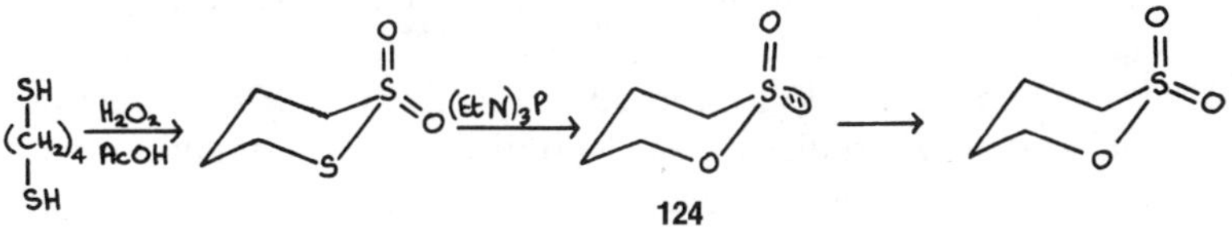

$HS(CH_2)_4SH$ —(H_2O_2, AcOH)→ … —($(EtN)_3P$)→ **124** → … (35)

Substitution at C–3 in 1,2-oxathiane 2,2-dioxide by lithium is stereospecific—the metal preferentially substitutes the equatorial hydrogen. A similar stereospecificity was previously noted in 1,3-dithianes (see Chapter 3, Section IX.B). The study was made by Durst,[388] who used the conformationally biased 5-*t*-butyl-1,2-oxathiane 2,2-dioxide. Lithiation of this compound was shown to occur at the equatorial C–3 atom because on alkylation or hydrolysis with D_2O it provided the *cis*-3-alkyl- or *cis*-3-deutero-5-*t*-butyl derivatives. Lithiation of *cis*-5-*t*-butyl-3-methyl-1,2-oxathiolane 2,2-dioxide occurs axially, but this epimerizes, and on hydrolysis yields the *trans*-5-*t*-butyl-3-methyl derivative (Scheme 5).

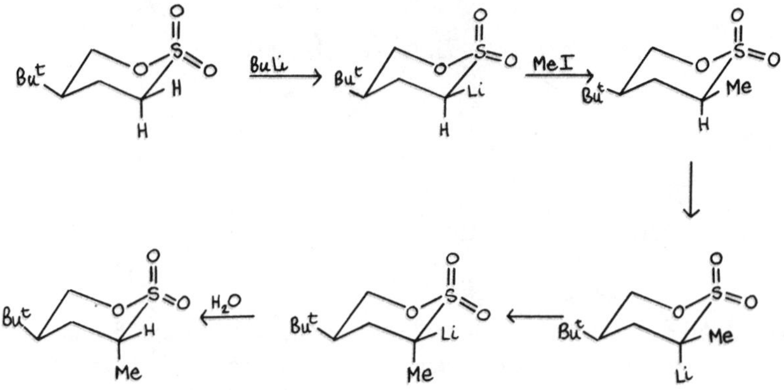

Scheme 5

2. *1,3-Oxathianes*

Stereochemical studies of 1,3-oxathianes were directed mainly to conformational studies of the ring and substituents. The ring is slightly, but observably, different from 1,3-dioxane as evidenced by ^{1}H nmr chemical shifts and *J* values,[389,390] and long-range *J* values indicate that the presence of a ring sulfur atom makes the ring more readily distorted.[391] This is revealed in the chair-twist conformational free energy differences (e.g., **125** ⇌ **126,** only one twist form is displayed) of 1,3-dioxane, 1,3-oxathiane, and 1,3-dithiane which are 33.5, 23.0, and 11.0 kJ mol^{-1} respectively.[392] 1,3-Oxathianes are distorted asymmetrically by comparison with 1,3-dioxanes, and 1,3-dithianes and the axial bonds at C–4 and C–6 are not parallel. An axial methyl group at C–4 (adjacent to S–3) is sterically less compressed than an axial methyl group at C–6 (adjacent to O–1),[393] and they have torsion angles of approximately 52 and 60° respectively.[394] This is demonstrated in conformational studies of *trans*-4,6-dialkyl-1,3-oxathianes where it is found that the conformers in which the alkyl group at C–4 is axial and C–6 is equatorial (**127**) predominate over the invertomers (**128**).[395–397]

125 ⇌ 126

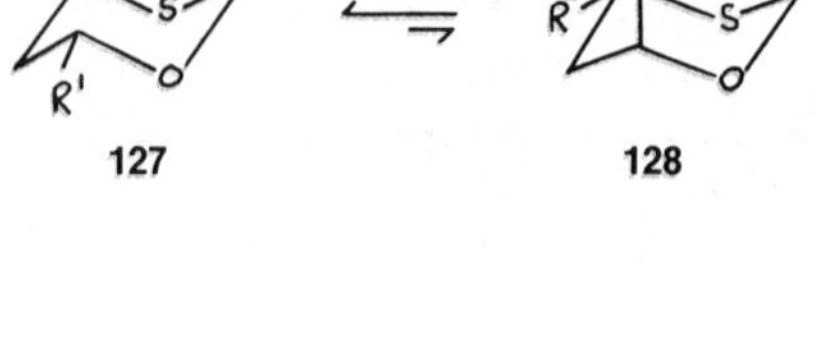

127 ⇌ 128

The conformation of the exocyclic oxygen atom in 1,3-oxathiane 3-oxide was also examined by ^{1}H nmr spectroscopy. The equilibrium is in favor of an axial oxygen atom **129** to the extent of 83.5%, but the introduction of *gem*-dimethyl groups (i.e., **130**) at C–5 alters the situation drastically, and only 10.5% of the molecules now have an axial exocyclic oxygen atom. These results are to be compared with 1,3-dithiane 1-oxide in which the conformer with an axial oxygen atom is favored only to the extent of about 15.4%, with less than 5% axial oxygen in 5,5-dimethyl-1,3-dithiane 1-oxide[398] (see Chapter 6, Section IX.B).

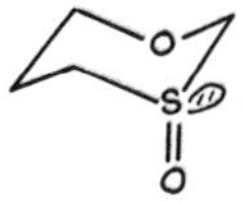

129 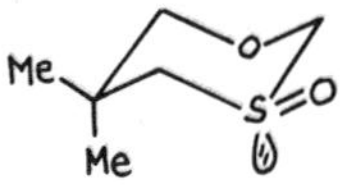 **130**

3. 1,4-Oxathianes

3,5-Bischloromethyl-1,4-oxathiane is formed when diallyl ether is treated with sulfur dichloride. The characteristic feature of this reaction is that the proportion of cis isomer is high, and the products are useful intermediates.[399] Diallyl sulfide reacts similarly to yield the corresponding *cis*- and *trans*-3,5-bischloromethyl-1,4-dithianes. *cis*- and *trans*-2,6-Bisiodomethyl-1,4-oxathianes are obtained by the reaction of diallyl sulfide with $Hg(OAc)_2$ followed by iodine. Similar treatment of diallyl sulfone, on the other hand, gave only *cis*-2,6-bisiodomethyl-1,4-oxathiane 4,4-dioxide. These are also very useful intermediates for preparing a variety of 1,4-oxathianes.[400]

Zefirov and collaborators[401] examined the conformational properties of 2-substituted 1,4-oxathianes with particular interest in the anomeric effect. They showed that the proportion of 2-axial substituent in 1,4-oxathianes is considerably less than in the corresponding 1,4-dioxanes and is probably due to the larger repulsion between the lone pairs, "hockey-sticks effect," in 1,4-oxathianes (e.g., **131**) than in 1,4-dioxanes. 1,4-Oxathiane-2-ones undergo ring inversion, and the free energy involved increases considerably in the presence of substituents. The $\Delta G^{\ddagger}$ values for 1,4-oxathiane 2-one and its 6-methyl, 5-phenyl, and 6-phenyl derivatives are 40.6, 72.8, 78.7, and 79.1 kJ mol^{-1} respectively.[402]

Foster and co-workers[403–405] studied the stereochemistry of several 1,4-oxathianes, related to monosaccharides, and their oxidation to the *S*-oxides. These were derivatives of 6-hydroxymethyl-1,4-oxathianes. They found that the configuration of the S–O group formed on oxidation was influenced by the presence and the configuration and conformation of the substituents already present. The configurations and conformations were deduced by ^1H nmr spectroscopy. The specificity is exemplified by the work of Foster[403] and of Watkins and their collaborators,[406,407] who showed, for example, that oxidation of 2*R*,6*S*-(*trans*)-2-hydroxymethyl-6-methoxy-1,4-oxathiane with HIO_4 gave mainly 2*R*,4*S*,6*S*-2-hydroxymethyl-6-methoxy-1,4-oxathiane 4-oxide (**132**), and the structure was confirmed by an X-ray analysis. The epimeric *cis*-2-hydroxymethyl-6-methoxy-1,4-oxathiane gave predominantly the 4-oxide **133** with HIO_4, but with ozone a 1:1 mixture of diastereomeric oxides was formed.[403]

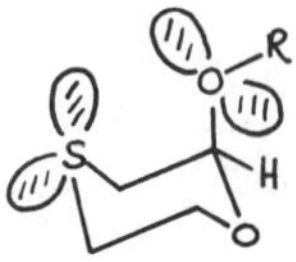

131 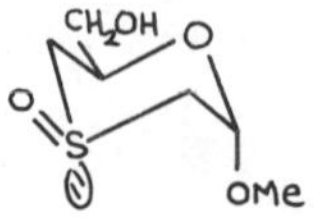 **132** 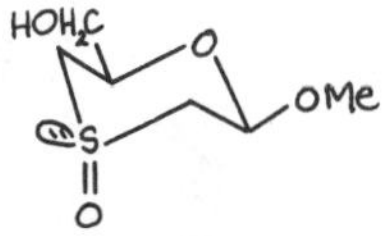 **133**

Bicyclo[3.3.1]nonanes generally exist in the chair-chair conformation even with nitrogen, and oxygen heteroatoms at the apical positions 3, 7, and 9 (Part I, Chapter 4, Section VII.D.1, 2 and 3) and in 3-oxa-7,9-dithia-bicyclo[3.3.1]nonane. In 3,7-dithia- (**134,** $X{=}CH_2$),[408] 9-oxa-3,7-dithia- (**134,** X=O),[409] and 3,7,9-trithia-bicyclo[3.3.1]nonanes,[408] however, the lone pair repulsion between the 3- and 7-sulfur atoms is strong enough for the molecule to prefer the boat-chair conformation (see Chapter 3, Section XIII). The boat-chair conformation in 9-oxa-3,7-dithiabicyclo-[3.3.1]nonane (**134,** X=O) was confirmed by an X-ray study.[409]

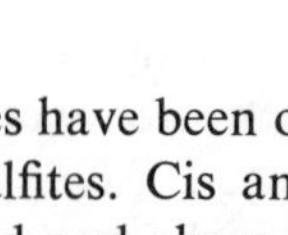

134

4. *1,3,2-Dioxathianes*

Most of the stereochemical studies have been on 1,3,2-dioxathiane 2-oxides which are cyclic trimethylene sulfites. Cis and trans isomers of the 4,6-dimethyl derivatives were isolated and characterized by ^1H nmr, and the exocyclic oxygen atom introduces yet another source of isomerism because it can be syn or anti to the substituents on the carbon atoms.[410–413] The conformational properties of 1,3,2-dioxathiane 2-oxides were studied by nmr, ir, dipole moment, and ultrasonic methods. The data for 1,3,2-

dioxathiane 2-oxide and some derivatives indicated that they existed as equilibrium mixtures of two chair conformations in which case one conformer, with the axial S=O, was highly favored,[414,415] but the data for several derivatives suggested that appreciable amounts of the non-chair form were also present.[416] ^{13}C nmr results were discussed in terms of chair and twist conformations, and the orientation of the S=O group has a large effect on the chemical shifts of substituents on C–4 and C–6. Only in some cases could conclusive conformational assignments be made.[417] The main difficulties encountered in this ring system are due to their flexibility and the relative stability of the twist conformation, and only in examples where this conformation is minimal could definite assignments be made. Whereas 4-methyl-1,3,2-dioxathiane 2-oxide exists predominantly in the conformation with an axial S=O group (**135**),[418] the 4-phenyl and 4,6,6-trimethyl derivatives exist as mixtures consisting of molecules in the chair conformation in equilibrium with a considerable proportion of molecules in the non-chair or twist conformation (e.g., **136**).[419] The X-ray analysis of 1,3,2-dioxathiane 2-oxide at −100°C revealed that it possessed a chair conformation with the S=O group in an axial direction in the crystalline state.[420]

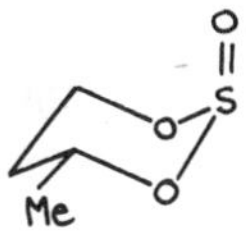

135 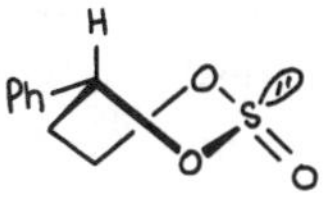 **136**

The conformational assignments of 4,6-dimethyl-1,3,2-dioxathiane 2-oxide were revised. The cis isomer has the S=O oriented axially, but the trans isomer was the first example in which the S=O group was in an equatorial conformation.[421] Several 5-substituted derivatives were also examined. The conformations of *cis*- and *trans*-5-chloro-1,3,2-dioxathiane 2-oxide were first deduced from ir data as having the S=O group equatorially disposed.[422] Later data on these, however, were consistent with preferred conformers with the axial S=O group,[423] and the ir assignments for the S=O group were revised. The conformational free energies of substituents for several 5-mono- and 5,5-disubstituted 1,3,2-dioxathiane 2-oxides were determined by the BF_3-catalyzed isomerization of cis and trans isomers.[424]

Some stereoselectivity was observed in the ester exchange between 2,4-dihydroxy-2-methylpentane and dimethyl sulfite, which provided a 2:3 mixture of *cis*- and *trans*-4,4,6-trimethyl-1,3,2-dioxathiane 2-oxides. Ir and nmr information showed that both isomers are in rigid chair conformations with the S=O axially oriented.[425] The conformational free energy for the S=O group was deduced in the rigid system *trans*-1,3,2-dioxathiadecalin-2-oxide by BF_3–HCl- and BF_3–$SOCl_2$-catalyzed equilibration experiments. The ratios of **137** to **138** at 40°C in CCl_4–BF_3–$SOCl_2$, C_6H_6–BF_3–HCl, $CHCl_3$–BF_3–HCl, and MeCN–BF_3–HCl are 25, 18.7, 13.9, and 12.9, which corresponded to $\Delta G^{\ddagger}$ values of 7.9, 7.5, 6.7, and 6.6 kJ mol^{-1} respectively.[426]

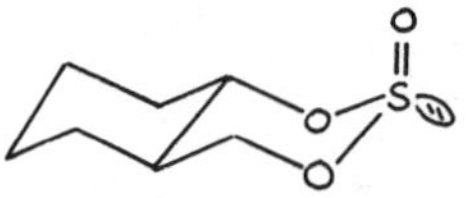

137 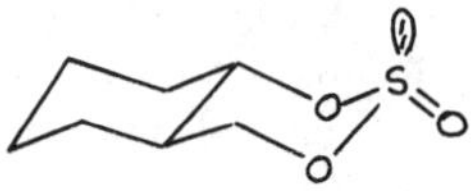 **138** 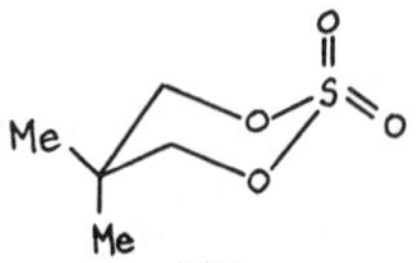 **139**

The conformational mobility of 5,5-dimethyl-1,3,2-dioxathiane 2,2-dioxide (**139**) was investigated by ^1H nmr at low temperatures. The $\Delta G^{\ddagger}$ value for ring inversion between "chair-chair" conformations (based on the methyl signals) was 35.1 kJ mol^{-1}, and preliminary data for the unsubstituted compound indicate a similar energy barrier. This is lower than that observed for cyclohexane (43.1 kJ mol^{-1}). The value of $\Delta H^{\ddagger}=25.9$ kJ mol^{-1}, previously obtained for 4-methyl-1,3,2-dioxathiane 2,2-dioxide by ultrasonic methods, is probably the energy barrier between a chair and a non-chair form.[427]

D. Miscellaneous

Prinzbach, Kaiser, and Fritz[428] extended their studies of benzene oxides to the oxothio derivatives. They prepared α,α-dioxo-β-thio- (**140,** X=O, Y=S) and α,α-dithio-β-oxobenzenes (**140,** X=S, Y=O) by judicious substitution reactions in *cis*-benzene trioxide. When these were treated with trimethyl phosphite they extruded sulfur at ambient temperatures, and the respective benzene *cis*-dioxide and benzene oxide were formed, but no ring enlargement was observed.

The studies of Ganter and his school on heterotwistanes (Section II.E; Part I, Chapter 3, Section X.C; Chapter 2, Section XI.D; and Chapter 3, Section XIII) also included 2-oxa-7-thiatwistanes (**141**) and isotwistanes. These were prepared by methods used for the other analogues starting from bicyclo[3.3.1]nonenes, and their properties were generally similar to those of their analogues.[429–431] The related starting material 9-oxabicyclo[3.3.1]-nona-2,6-diene reacted with $SOCl_2$ to form 3,8-dichloro-2-oxa-6-thiaadamantane which is a useful starting material for studies of this ring system.[432]

Spectral studies of [18]annulene-1,4-oxide-7,10:13,16-disulfide, like the related trisulfide (formula **183,** Chapter 3, Section XII.D), revealed that the ring system is nonaromatic and probably not planar.[433]

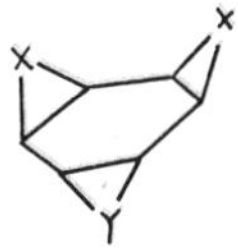

140

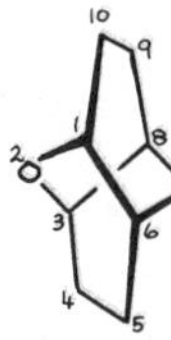

141

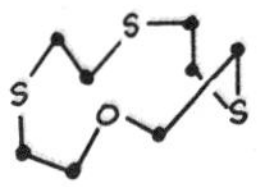

142

The cyclic disultone, 4,8-dimethyl-1,5-dioxa-2,6-dithiacyclooctane 2,2,6,6-tetraoxide was formed from propan-1-ol and sulfuryl chloride. Cis and trans isomers are possible in this case. ^{1}H nmr studies indicated a strong preference for one conformation in which the methine hydrogen atoms are anti to each other most of the time, but a definite conformation was not proposed.[434]

Several macrocyclic crown ethers containing ethylene 1,4-dithia, 1,4-oxathia, and 1,4-dioxa linkages in the same ring have been synthesized. Rings ranging from 12 atoms to 21 atoms were prepared, and X-ray analyses of two of these demonstrated that the arrangement of atoms in the crystals exhibit some nonplanar regularities, for example, the 1,4,7-trithia-[12 crown-4]ether (**142**).[435]

V. NITROGEN, OXYGEN, AND SULFUR HETEROCYCLES

Only a few examples of these mixed heterocycles of stereochemical interest are known, and they all belong to the 1,2,3-oxathiazolidine 2-oxide series. A variety of β-amino alcohols were treated with $SOCl_2$ in the presence of a base, usually Et_3N, and gave a mixture of isomeric 1,2,3-oxathiazolidine 2-oxides in very good yields.[436] When (−)-ephedrine reacted in a similar way it afforded a 28:72 mixture of *cis*-3,4-dimethyl-5-phenyl-1,2,3-oxathiazolidine *syn*-2-oxide and *anti*-2-oxide (Equation 36).[437] The syn isomer can

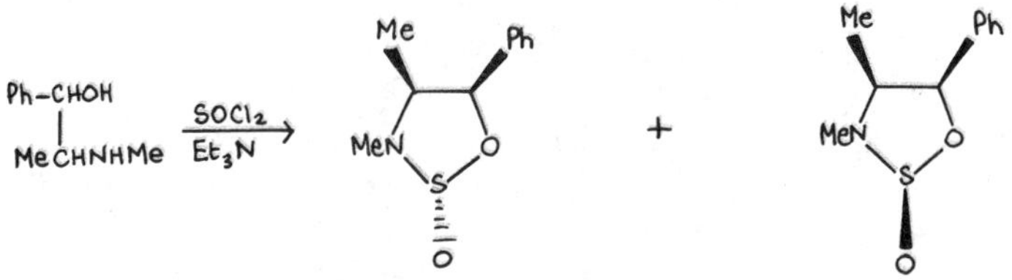

(36)

be converted into the more stable anti isomer by hydrochloric acid.[438] *S*-Serine *O*-sulfate, in DMF, cyclizes into 4-carboxy-1,2,3-oxathiazolidine 2,2-dioxide in the presence of dicyclohexyl carbodiimide, presumably with retention of configuration.[439]

VI. REFERENCES

1. W. D. Emmons, *J. Amer. Chem. Soc.*, **78,** 6208 (1956).
2. D. R. Boyd and R. Graham, *J. Chem. Soc.* (*C*), 2648 (1969).
3. D. R. Boyd, R. Spratt, and D. M. Jerina, *J. Chem. Soc.* (*C*), 2650 (1969).
4. J. B. Bapat and D. St. C. Black, *Chem. Comm.*, 73 (1967); J. B. Bapat and D. St. C. Black, *Austral. J. Chem.*, **21,** 2507 (1968).
5. E. Schmitz, *Adv. Heterocyclic Chem.*, **2,** 83 (1963).
6. J. D. Readio and R. A. Falk, *J. Org. Chem.*, **35,** 927 (1970).
7. J. D. Readio, *J. Org. Chem.*, **35,** 1607 (1970).
8. A. Quilico and P. Grünanger, *Gazzetta*, **85,** 1250 (1955).
9. A. Quilico and P. Grünanger, *Gazzetta*, **82,** 140 (1952).
10. A. Quilico, G. S. d'Alcontres, and P. Grünanger, *Gazzetta*, **80,** 479 (1950).
11. A. Quilico and P. Grünanger, *Gazzetta*, **85,** 1449 (1955).
12. P. Grünanger and I Grasso, *Gazzetta*, **85,** 1271 (1955).
13. P. Grünanger, *Gazzetta*, **84,** 359 (1954).
14. A. Quilico, P. Grünanger, and R. Mazzini, *Gazzetta*, **82,** 349 (1952).
15. F. Monforte, *Gazzetta*, **82,** 130 (1952).
16. G. Bailo, P. Caramella, G. Cellerino, A. G. Invernizzi, and P. Grünanger, *Gazzetta*, **103,** 47 (1973).
17. K. Bast, C. Christl, R. Huisgen, W. Mack, and R. Sustmann, *Chem. Ber.*, **106,** 3258 (1973).
18. M. Christl, R. Huisgen, and R. Sustmann, *Chem. Ber.*, **106,** 3275 (1973).
19. M. Christl and R. Huisgen, *Chem. Ber.*, **106,** 3345 (1973).
20. K. Bast, M. Christl, R. Huisgen, and W. Mack, *Chem. Ber.*, **106,** 3312 (1973).
21. R. Huisgen and M. Christl, *Chem. Ber.*, **106,** 3291 (1973).
22. R. Huisgen, H. Hauck, H. Seidl, and M. Burger, *Chem. Ber.*, **102,** 1117 (1969).
23. R. Huisgen, H. Seidl, and I. Brüning, *Chem. Brt.*, **102,** 1102 (1969).
24. R. Huisgen, R. Grashey, H. Hauck, and H. Seidl, *Chem. Ber.*, **101,** 2548 (1968).
25. R. Huisgen, H. Hauck, R. Grashey, and H. Seidl, *Chem. Ber.*, **101,** 2559, 2568 (1968).
26. M. Joucla, D. Grée, and J. Hamelin, *Tetrahedron*, **29,** 2315 (1973).
27. G. Bianchi, R. Gandolfi, and P. Grünanger, *Tetrahedron*, **26,** 5113 (1970),
28. M. Joucla, J. Hamelin, and R. Carrié, *Bull. Soc. chim. France*, 3116 (1973).
29. R. Bonnett, S. C. Ho, and J. A. Raleigh, *Canad. J. Chem.* **43,** 2717 (1965).
30. R. Huisgen, H. Hauck, R. Grashey, and H. Seidl, *Chem. Ber.*, **102,** 736 (1969).
31. G. Bianchi, A. Gamba, and R. Gandolfi, *Tetrahedron*, **28,** 1601 (1972).
32. K. Müller and A. Eschenmoser, *Helv. Chim. Acta*, **52,** 1823 (1969).

33. R. Grée and R. Carrié, *Tetrahedron Letters*, 2987 (1972).
34. R. Grée and R. Carrié, *Tetrahedron Letters*, 4117 (1971).
35. R. Grée, F. Tonnard, and R. Carrié, *Tetrahedron Letters*, 453 (1973).
36. N. A. LeBel and J. J. Whang, *J. Amer. Chem. Soc.*, **81,** 6334 (1959).
37. N. A. LeBel, M. E. Post, and J. J. Whang, *J. Amer. Chem. Soc.*, **86,** 3759 (1964).
38. A. C. Cope and N. A. LeBel, *J. Amer. Chem. Soc.*, **82,** 4656 (1960).
39. N. A. LeBel and E. G. Banucci, *J. Org. Chem.*, **36,** 2440 (1971).
40. N. A. LeBel and E. G. Banucci, *J. Amer. Chem. Soc.*, **92,** 5278 (1970).
41. N. A. LeBel, M. G. J. Slusarczuk, and L. A. Spurlock, *J. Amer. Chem. Soc.*, **84,** 4360 (1962).
42. N. A. LeBel, N. D. Ojha, J. R. Menke, and R. J. Newland, *J. Org. Chem.*, **37,** 2896 (1972).
43. W. C. Lumma, Jr., *J. Amer. Chem. Soc.*, **91,** 2820 (1969).
44. W. Oppolzer and K. Keller, *Tetrahedron Letters*, 1117 (1970).
45. W. Oppolzer and H. P. Weber, *Tetrahedron Letters*, 1121 (1970).
46. W. Oppolzer and K. Keller, *Tetrahedron Letters*, 4313 (1970).
47. E. C. Taylor, D. R. Eckroth, and J. Bartulin, *J. Org. Chem.*, **32,** 1899 (1967).
48. T. S. Cantrell and W. S. Haller, *Chem. Comm.*, 977 (1968).
49. A. T. Nielsen and T. G. Archibald, *J. Org. Chem.*, **34,** 984 (1969).
50. P. W. Jeffs and G. Molina, *J. C. S. Chem. Comm.*, 3 (1973).
51. K. T. Potts, S. Husain, and S. Husain, *Chem. Comm.*, 1360 (1970).
52. K. T. Potts and D. McKeough, *J. Amer. Chem. Soc.*, **96,** 4276 (1974).
53. W. S. Murphy and J. P. McCarthy, *Chem. Comm.*, 1155, (1968).
54. C. H. Stammer, A. N. Wilson, F. W. Holly, and K. Folkers, *J. Amer. Chem. Soc.*, **77,** 2346 (1955).
55. R. A. Payne and C. H. Stammer, *J. Org. Chem.*, **33,** 2421, (1968).
56. C. H. Stammer and J. D. McKinney, *J. Org. Chem.*, **30,** 3436 (1965).
57. H. Iwasaki, T. Kamiya, O. Oka, and J. Ueyanagi, *Chem. and Pharm. Bull.* (*Japan*), **17,** 866 (1969).
58. H. Iwasaki, T. Kamiya, C. Hatanaka, Y. Sunada, and J. Ueyanagi, *Chem. and Pharm. Bull.* (*Japan*), **17,** 873 (1969).
59. T. Kamiya, *Chem. and Pharm. Bull.* (*Japan*), **17,** 879, 886, 890, 895 (1969).
60. K. Müller, *Helv. Chim. Acta*, **53,** 1112, (1970).
61. M. Raban, F. B. Jones, Jr., E. H. Carlson, E. Banucci, and N. A. LaBel, *J. Org. Chem.*, **34,** 1496 (1970).
62. F. G. Riddell, J. M. Lehn, and J. Wagner, *Chem. Comm.*, 1403 (1968).
63. R. Jacquier, F. Petrus, J. Verducci, and Y. Vidal, *Tetrahedron Letters*, 387 (1974).
64. W. J. Tuman and L. Bauer, *J. Org. Chem.*, **37,** 2983 (1972).
65. M. Svoboda, J. Sicher, J. Farkaš, and M. Pánková, *Coll. Czech. Chem. Comm.*, **20,** 1426 (1955).
66. J. Sicher and M. Svoboda, *Coll. Czech. Chem. Comm.*, **23,** 1252 (1958).
67. J. Sicher, M. Tichý, F. Šipoš, and M. Pánková, *Coll. Czech. Chem. Comm.*, **26,** 2418 (1961).
68. J. Farkaš and J. Sicher, *Coll. Czech. Chem. Comm.*, **20,** 1391 (1955).

69. J. Sicher and M. Pánková, *Coll. Czech. Chem. Comm.*, **20,** 1409 (1955).
70. J. Sicher and M. Svoboda, *Coll. Czech. Chem. Comm.*, **23,** 2094 (1958).
71. E. E. van Tamelen and R. S. Wilson, *J. Amer. Chem. Soc.*, **74,** 6299 (1952).
72. T. Taguchi and M. Kojima, *J. Amer. Chem. Soc.*, **81,** 4318 (1959).
73. P. E. Fanta and E. N. Walsh, *J. Org. Chem.*, **31,** 59 (1966).
74. C. H. Stammer and R. G. Webb, *J. Org. Chem.*, **34,** 2306 (1969).
75. T. Nishiguchi, H. Tochio, A. Nabeya, and Y. Iwakura, *J. Amer. Chem. Soc.*, **91,** 5841 (1969).
76. S. H. Pines and M. A. Kozlowski, *J. Org. Chem.*, **37,** 292 (1972).
77. S. H. Pines, M. A. Kozlowski, and S. Karady, *J. Org. Chem.* **34,** 1621 (1969).
78. S. H. Pines, S. Karady, M. A. Kozlowski, and M. Sletzinger, *J. Org. Chem.*, **33,** 1762 (1968).
79. J. R. Carson, G. I. Poos, and H. R. Almond, Jr., *J. Org. Chem.*, **30,** 2225 (1965).
80. R. Gompper and W. Hägele, *Chem. Ber.*, **99,** 2885 (1966).
81. H. Egg, *Monatsh.*, **100,** 2125 (1969).
82. R. R. Wittekind, J. D. Rosenau, and G. I. Poos, *J. Org. Chem.*, **26,** 444 (1961).
83. L. Brickenbach and M. Linhard, *Ber.*, **64,** 1076 (1931).
84. F. Plantefeve and G. Descotes, *Bull. Soc. chim. France*, 2923 (1972).
85. P. E. Fanta, R. J. Smat, and J. R. Krikau, *J. Heterocyclic Chem.*, **5,** 419 (1968).
86. L. Goodman, S. Winstein, and R. Boschan, *J. Amer. Chem. Soc.*, **80,** 4312 (1958).
87. H. W. Heine and M. S. Kaplan, *J. Org. Chem.*, **32,** 3069 (1967).
88. S. H. Pines, *J. Org. Chem.*, **38,** 3854 (1973).
89. O. H. Oldenziel and A. M. van Leusen, *Tetrahedron Letters*, 163 (1974).
90. H. Giezendanner, H. Heimgartner, B. Jackson, T. Winkler, H-J. Hansen and H. Schmid, *Helv. Chim. Acta*, **56,** 2611 (1973).
91. T. A. Foglia and D. Swern, *J. Org. Chem.*, **43,** 1680 (1969).
92. G. Fodor and K. Koczka, *J. Chem. Soc.*, 850 (1952).
93. A. Hassner and S. S. Burke, *Tetrahedron*, **30,** 2613 (1974).
94. J. B. Kay and J. B. Robinson, *J. Chem. Soc.* (*C*), 248 (1969).
95. H. Bretschneider and H. Egg, *Monatsh.*, **103,** 1394 (1972).
96. R. A. Wohl, *J. Org. Chem.*, **38,** 3858 (1973).
97. H. Bretschneider and H. Egg, *Monatsh.*, **100,** 2131 (1969).
98. S. Terashima, M. Nara, and S-i. Yamada, *Chem. and Pharm. Bull.* (*Japan*), **18,** 1124 (1970); S-i. Yamada and S. Terashima, *Chem. Comm.*, 511 (1969).
99. S. Terashima and S-i. Yamada, *Chem. and Pharm. Bull.* (*Japan*), **16,** 1953 (1968).
100. S-i. Yamada, S. Terashima, and K. Achiwa, *Chem. and Pharm. Bull.* (*Japan*), **14,** 800 (1966).
101. S-i. Yamada, S. Terashima, and K. Achiwa, *Chem. and Pharm. Bull.* (*Japan*), **13,** 751 (1965).
102. K. B. Sharpless, D. W. Patrick, L. K. Truesdale, and S. A. Biller, *J. Amer. Chem. Soc.*, **97,** 2305 (1975).
103. D. Hoppe, *Angew. Chem. Internat. Edn.*, **11,** 933 (1972).
104. M. Goodman and L. Levine, *J. Amer. Chem. Soc.*, **86,** 2918 (1964).
105. H. N. Rydon, *Roy. Inst. Chem., Lecture Series*, No. 5 (1962).

106. D. G. H. Ballard and C. H. Bamford, *Chem. Soc. Special Publ.*, No. 2, 25 (1955).
107. J. P. Greenstein and M. Winitz, *Chemistry of the Amino Acids*, Vol. 2, Wiley, New York, 1961, p. 860.
108. S. Larsen and J. Bernstein, *J. Amer. Chem. Soc.*, **72,** 4447 (1950).
109. W. L. Nelson, *J. Heterocyclic Chem.*, **5,** 231 (1968).
110. E. Gil-Av, *J. Amer. Chem. Soc.*, **81,** 1602 (1959).
111. M. G. Ettlinger, *J. Amer. Chem. Soc.*, **72,** 4792 (1950).
112. T. Taguchi and Y. Kawozoe, *J. Org. Chem.*, **26,** 2699 (1961).
113. L. Neelakantan, *J. Org. Chem.*, **36,** 2256 (1971).
114. H. Pfanz and G. Kirchner, *Annalen*, **614,** 149 (1968).
115. G. E. McCasland and E. C. Horswill, *J. Amer. Chem. Soc.*, **73,** 3923 (1951).
116. A. C. Cope and E. M. Hancock, *J. Amer. Chem. Soc.*, **66,** 1453 (1944); A. C. Cope and E. M. Hancock, *J. Amer. Chem. Soc.*, **64,** 1503 (1942).
117. T. R. Keenan and N. J. Leonard, *J. Amer. Chem. Soc.*, **93,** 6567 (1971).
118. G. Fodor, J. Stefanovsky, and B. Kurtev, *Monatsh.*, **98,** 1027 (1967).
119. N. D. Harris, *J. Org. Chem.*, **28,** 745 (1963).
120. D. F. Elliott, *J. Chem. Soc.*, 589 (1949).
121. D. F. Elliott, *J. Chem. Soc.*, 62 (1950).
122. S. L. Spassov, J. N. Stefanovsky, B. J. Kurtev, and G. Fodor, *Chem. Ber.*, **105,** 2462 (1972).
123. H. Gerlach, *Helv. Chim. Acta*, **49,** 2481 (1966).
124. E. E. Hamel and E. P. Painter, *J. Amer. Chem. Soc.*, **76,** 919 (1954).
125. U. Hornemann, L. H. Hurley, M. K. Speedie, and H. G. Floss, *J. Amer. Chem. Soc.*, **93,** 3028 (1971).
126. N. J. Leonard and W. K. Musker, *J. Amer. Chem. Soc.*, **82,** 5148 (1960).
127. T. A. Crabb and M. J. Hall, *J. C. S. Perkin II*, 1419 (1974).
128. R. Cahill and T. A. Crabb, *J. Heterocyclic Chem.*, **9,** 875 (1972).
129. T. A. Crabb and M. J. Hall, *J. C. S. Perkin II*, 1379 (1973).
130. T. A. Crabb and R. F. Newton, *Tetrahedron*, **24,** 1997, (1968).
131. T. A. Crabb, M. J. Hall, and R. O. Williams, *Tetrahedron*, **29,** 3389 (1973).
132. Y. Takeuchi, P. J. Chivers, and T. A. Csabb, *J. C. S. Perkin II*, 51 (1975).
133. Y. Takeuchi, P. J. Chivers, and T. A. Crabb, *J.C.S. Chem. Comm.*, 210 (1974).
134. T. A. Crabb and R. F. Newton, *Tetrahedron Letters*, 1551 (1970).
135. T. A. Crabb, R. F. Newton, and D. Jackson, *Chem. Rev.*, **71,** 109 (1971).
136. R. A. Y. Jones, A. R. Katritzky, K. A. F. Record, R. Scattergood, and J. M. Sullivan, *J.C.S. Perkin II*, 402 (1974).
137. R. A. Y. Jones, A. R. Katritzky, P. G. Lehman, and B. B. Shapiro, *J. Chem. Soc. (B)*, 1308 (1971).
138. R. A. Y. Jones, A. R. Katritzky, and P. G. Lehman, *J. Chem. Soc. (B)*, 1316 (1971).
139. V. J. Baker, A. R. Katritzky, J-P. Majoral, S. F. Nelsen, and P. J. Hintz, *J.C.S. Chem. Comm.*, 823 (1974).
140. A. I. Meyers, G. Knaus, and K. Kamata, *J. Amer. Chem. Soc.*, **95,** 268 (1974).
141. A. I. Meyers and G. Knaus, *J. Amer. Chem. Soc.*, **96,** 6508 (1974).

142. A. I. Meyers and K. Kamata, *J. Org. Chem.*, **39,** 1603 (1974).
143. A. I. Meyers, E. D. Mihelich, and K. Kamata, *J. C. S. Chem. Comm.*, 768 (1974).
144. J. D. White and M-g. Kim, *Tetrahedron Letters*, 3361 (1974).
145. B. T. Golding and D. R. Hall, *J.C.S. Chem.Comm.*, 293, (1973).
146. J. W. Cornforth in *Heterocyclic Compounds*, Vol. 5, R. C. Elderfield, Ed., Wiley, New York, 1957, p. 298.
147. R. Lakhan and B. Ternai, *Adv. Heterocyclic Chem.*, **17,** 99 (1974).
148. R. Filler, *Adv. Heterocyclic Chem.*, **4,** 75 (1965).
149. G. Kresze and J. Firl, *Tetrahedron*, **24,** 1043 (1968).
150. G. Kresze and W. Kosbahn, *Tetrahedron* **27,** 1931 (1971).
151. G. Kresze and G. Schulz, *Tetrahedron*, **12,** 7 (1961).
152. M. Petrzilka, D. Felix, and A. Eschenmoser, *Helv. Chim. Acta*, **56,** 2950 (1973).
153. U. M. Kempe, T. K. Das Fupta, K. Blatt, P. Gygax, D. Felix, and A. Eschenmoser, *Helv. Chim, Acta*, **55,** 2187 (1972).
154. T. K. Das Gupa, D. Felix, U.M. Kempe, and A. Eschenmoser, *Helv. Chim. Acta*, **55,** 2198 (1972).
155. W. E. Noland, J. H. Cooley, and P. A. McVeigh, *J. Amer. Chem. Soc.*, **81,** 1209 (1959).
156. W. E. Noland, J. H. Cooley, and P. A. McVeigh, *J. Amer. Chem. Soc.*, **79,** 2976 (1957).
157. O. Wichterle and J. Novák, *Coll. Czech. Chem. Comm.*, **15,** 309 (1950); G. Kresze and J. Firl, *Fortschr. Chem. Forsch.*, **11,** 245 (1969).
158. F. G. Riddell, *Tetrahedron*, **31,** 523 (1975).
159. S. Oida and E. Ohki, *Chem. and Pharm. Bull.* (*Japan*), **17,** 1990 (1969).
160. S. Oida and E. Ohki, *Chem. and Pharm. Bull.* (*Japan*), **17,** 939 (1969).
161. F. G. Riddell and D. A. R. Williams, *Tetrahedron*, **30,** 1083 (1974).
162. F. G. Riddell, P. Murray-Rust, and J. Murray-Rust, *Tetrahedron*, **30,** 1087 (1974).
163. F. G. Riddell and D. A. R. Williams, *Tetrahedron*, **30,** 1097 (1974).
164. R. A. Y. Jones, A. R. Katritzky, A. C. Richards, S. Saba, A. J. Sparrow, and D. L. Trepanier, *J.C.S. Chem. Comm.* 673 (1972).
165. R. A. Y. Jones, A. R. Katritzky, S. Saba, and A. J. Sparrow, *J.C.S. Perkin II*, 1554 (1974).
166. R. A. Y. Jones, A. R. Katritzky, and S. Saba, *J.C.S. Perkin II*, 1737 (1974).
167. N. H. Cromwell in *Heterocyclic Compounds*, Vol. 6, R. C. Elderfield, Ed., Wiley, New York, 496 (1957).
168. Z. Eckstein and T. Urbański, *Adv. Heterocyclic Chem.*, **2,** 311 (1963).
169. G. A. Cooke and G. Fodor, *Canad. J. Chem.*, **46,** 1105 (1968).
170. G. Fodor, J. Stefanovsky, and B. Kurtev, *Chem. Ber.*, **100,** 3069 (1967).
171. M. Mousseron, F. Winternitz, and M. Mousseron-Canet, *Bull. Soc. chim. France*, 737 (1953).
172. G. Bernáth, G. Göndös, K. Kovács, and P. Sohár, *Tetrahedron*, **29,** 981 (1973).
173. J. Szmuszkovicz and L. L. Skaletzky, *J. Org. Chem.*, **32,** 3300 (1967).
174. J. Sicher, M. Pánková, J. Jonáš, and M. Svoboda, *Coll. Czech. Chem. Comm.*, **24,** 2727(1959).

175. S. Wolfe, J-B. Ducep, K-C. Tin, and S-L. Lee, *Canad. J. Chem.*, **52,** 3996 (1974).

176. R. R. Schmidt, *Chem. Ber.*, **103,** 3242 (1970).

176a. C. K. Bradsher, *Adv. Heterocyclic Chem.*, **16,** 312 (1974).

177. E. Testa, L. Fontanella, and G. E. Cristiani, *Annalen*, **626,** 121 (1959); E. Testa, L. Fontenella, G. Cristiani, and G. Gallo, *J. Org. Chem.*, **24,** 1928 (1959).

178. I. D. Blackburne, R. P. Duke, R. A. Y. Jones, A. R. Katritzky, and K. A. F. Record, *J.C.S. Perkin II*, 332 (1973).

179. H. Booth and R. U. Lemieux, *Canad. J. Chem.*, **49,** 777 (1971).

180. R. A. Y. Jones, A. R. Katritzky, and D. L. Trepanier, *J. Chem. Soc.* (*B*), 1300 (1971).

181. M. J. Cook, R. A. Y. Jones, A. R. Katritzky, M. M. Monas, A. C. Richards, A. J. Sparrow, and D. L. Trepanier, *J.C.S Perkin II*, 325 (1973).

182. M. A. Anteunis, G. Swaelens, and J. Gelan, *Tetrahedron*, **27,** 1917 (1971).

183. F. G. Riddell and J. M. Lehn, *J. Chem. Soc.* (*B*), 1224 (1968).

184. P. J. Halls, R. A. Y. Jones, A. R. Katritzky, M. Snarey, and D. L. Trepanier, *J. Chem. Soc.* (*B*), 1320 (1971).

185. D. Gürne and T. Urbański, *J. Chem. Soc.*, 1912 (1959).

186. D. Gürne, L. Stefaniak, T. Urbański, and M. Witanowski, *Tetrahedron*, suppl. **6,** 211 (1963).

187. Y. Allingham, R. C. Cookson, T. A. Crabb, and S. Vary, *Tetrahedron*, **24,** 4625 (1968).

188. D. Gürne, T. Urbański, M. Witanowski, B. Karniewski, and L, Stefaniak, *Tetrahedron*, **20,** 1173 (1964).

189. T. A. Crabb and R. F. Newton, *Tetrahedron*, **24,** 4423 (1968).

190. R. Cahill and T. A. Crabb, *J.C.S. Perkin II*, 1374 (1972).

191. T. A. Crabb and R. F. Newton, *Tetrahedron Letters*, 3361 (1971).

192. T. A. Crabb, P. J. Chivers, E. R. Jones, and R. F. Newton, *J. Heterocyclic Chem.*, **7,** 635 (1970).

193. T. A. Crabb and E. R. Jones, *Tetrahedron*, **26,** 1217 (1970).

194. J. B. Chylińska and T. Urbański, *J. Heterocyclic Chem.*, **1,** 93 (1964).

195. W. G. Otto, *Angew. Chem.*, **68,** 181 (1956).

196. F. H. Clarke, *J. Org. Chem.*, **27,** 3251 (1962).

197. V. A. Rao, P. C. Jain, and N. Anand, *Indian J. Chem.*, **10,** 1134 (1972).

198. J-P. Vigneron, H. Kagan, and A. Horeau, *Bull. Soc. chim. France*, 3836 (1972).

199. G. Blaschke, *Angew. Chem. Internat. Edn.*, **10,** 520 (1971).

200. J. Cleophax, J. Leboul, A-M. Sepulchre, and S. D. Géro, *Bull. Soc. chim. France*, 4412 (1970).

201. J. Cleophax, A. Gaudemer, A-M. Sepulchre, and S. D. Géro, *Bull. Soc. chim. France*, 4414 (1970).

202. P. S. Portoghese and J. G. Turcotte, *Tetrahedron*, **27,** 961 (1971).

203. T. H. Koch, J. A. O. Lesen, and J. DeNiro, *J. Org. Chem.*, **40,** 14 (1975).

204. H. Musso and W. Steckelberg, *Chem. Ber.*, **101,** 1510 (1968).

205. M. J. Aroney, C-Y. Chen, R. J. W. Le Fèvre, and J. D. Saxby, *J. Chem. Soc.*, 4269 (1964).

206. M. Aroney and R. J. W. Le Fèvre, *J. Chem. Soc.*, 3002 (1958).

207. R. K. Harris and R. A. Spragg, *J. Chem. Soc.* (*B*), 684 (1968).

208. H. Booth and G. C. Gidley, *Tetrahedron*, **21,** 3429 (1965).
209. H. Booth and J. H. Little, *J.C.S. Perkin II*, 1846 (1972).
210. R. Cahill and T. A. Crabb, *Tetrahedron*, **25,** 1513 (1969).
211. R. A. Y. Jones, A. R. Katritzky, A. R. Martin, and S. Saba, *J.C.S. Chem. Comm.*, 908 (1973).
212. R. A. Y. Jones, A. R. Katritzky, and A. R. Martin, *J.C.S. Perkin II*, 1561 (1974).
213. T. Taguchi, J. Ishibashi, T. Matsuo, and M. Kojima, *J. Org. Chem.*, **29,** 1097 (1964).
214. L. A. Capino and E. S. Rundberg, Jr., *Chem. Comm.*, 1431 (1968).
215. C. Y-J. Chung, D. McKay, and T. D. Sauer, *Canad. J. Chem.* **50,** 3315 (1972).
216. D. Mackay, J. A. Campbell, and C. P. R. Jennison, *Canad. J. Chem.*, **48,** 81 (1970).
217. J. A. Campbell, I. Harris, D. Mackay, and T. D. Sauer, *Canad. J. Chem.*, **53,** 535 (1975).
218. C. P. R. Jennison and D. Mackay, *Canad. J. Chem.*, **51,** 3726 (1973).
219. J. J. Tufariello, T. F. Mich, and P. S. Miller, *Tetrahedron Letters*, 2293 (1966).
220. J. Thesing and H. Mayer, *Chem. Ber.*, **89,** 2159 (1956).
221. J. Thesing and H. Mayer, *Annalen*, **609,** 46 (1957).
222. G. Eikelmann, W. Heimberger, G. Nonnenmacher, and W. M. Weigert, *Annalen*, **759,** 183 (1972).
223. R. A. Y. Jones, A. R. Katritzky, P. G. Lehman, A. C. Richards and R. Scattergood, *J.C.S. Perkin II*, 41 (1972).
224. D. R. Julian and G. D. Tringham, *J.C.S. Chem. Comm.*, 13 (1973).
225. K. Hofmann, *J. Amer. Chem. Soc.*, **67,** 1459 (1945).
226. K. Hofmann and A. Bridgewater, *J. Amer. Chem. Soc.*, **67,** 1165 (1945).
227. Y. Aoki, H. Suzuki, and H. Akiyam, *Heterocycles*, **3,** 67 (1975).
228. M. Gereck, J-P. Zimmermann, and W. Aschwanden, *Helv. Chim. Acta.* **53,** 991 (1970).
229. H. E. Zaugg and R. W. DeNet, *J. Amer. Chem. Soc.*, **84,** 4574 (1962).
230. S. C. Perricone, E. F. Elslager, and D. F. Worth, *J. Heterocyclic Chem.*, **7,** 135 (1970).
231. E. F. Elslager and D. F. Worth, *J. Heterocyclic Chem.*, **6,** 597 (1969).
232. D. F. Worth, S. C. Perricone, and E. Elslager, *J. Heterocyclic Chem.*, **7,** 1353 (1970).
233. S. C. Perricone, D. F. Worth, and E. F. Elslager, J. *Heterocyclic Chem.*, **7,** 537 (1970).
234. G. Desimoni, A. Gamba, P. R. Righetti, and G. Tacconi, *Gazzetta*, **101,** 899 (1971).
235. G. Desimoni, L. Astolfi, A. Cambieri, A. Gamba, and G. Tacconi, *Tetrahedron*, **29,** 2627 (1973).
236. G. Desimoni, G. Cellerino, A. Gamba, P. P. Righetti, and G. Tacconi, *Tetrahedron*, **29,** 2621 (1973).
237. G. Desimoni, G. Colombo, P. P. Righetti, and G. Tacconi, *Tetrahedron*, **29,** 2635 (1973).
238. G. Desimoni, G. Cellerino, G. Minoli, and G. Tacconi, *Tetrahedron*, **28,** 4003 (1972).
239. G. Tacconi, F. Marinone, and G. Desimoni, *Gazzetta*, **101,** 173 (1971).
240. R. Adams, M. Carmack, and J. E. Mahan, *J. Amer. Chem. Soc.*, **64,** 2593 (1942).
241. J. Gutzwiller, G. Pizzolato, and M. Uskoković, *J. Amer. Chem. Soc.*, **93,** 5907 (1971).
242. M. Cushman and N. Castagnoli, Jr., *J. Org. Chem.*, **38,** 440 (1973).
243. R. E. Portmann and C. Ganter, *Helv. Chim. Acta*, **56,** 1962 (1973).

244. R. E. Portmann and C. Ganter, *Helv. Chim. Acta*, **56,** 1991 (1973).
245. J. B. Lambert, *Topics in Heterocyclic Chem.*, **6,** 19 (1971); J. M. Lehn, *Fortschr. Chem. Forsch.*, **15,** 311 (1970); A. Rauk, L. C. Allen, and K. Mislow, *Angew. Chem. Internat. Edn.*, **9,** 400 (1970).
246. R. E. Wasylishen, K. C. Rice, and U. Weiss, *Canad. J. Chem.*, **53,** 414 (1975).
247. C. Wong and W. W. Paudler, *J. Org. Chem.*, **39,** 2570 (1974).
248. J. L. Atwood, W. E. Hunter, C. Wong, and W. W. Paudler, *J. Heterocyclic Chem.*, **12,** 433 (1975).
249. D. A. Whiting, R. Cahill, and T. A. Crabb, *J.C.S. Chem. Comm.*, 1307 (1972).
250. B. Dietrich, J. M. Lehn, and J. P. Sauvage, *J.C.S. Chem. Comm.*, 15 (1973).
251. B. Kaempf, S. Raynal, A. Collet, F. Schué, S. Boileau, and J. M. Lehn, *Agnew. Chem. Internat. Edn.*, **13,** 611 (1974).
252. D. Metz, D. Moras, and R. Weiss, *Chem. Comm.*, 217 (1970).
253. J. C. Lockhart, A. C. Robson, M. E. Thompson, D. Furtado, C. K. Kaura, and A. R. Allan, *J.C.S. Perkin I*, 577 (1973).
254. M. Newcomb, G. W. Gokel, and D. J. Cram, *J. Amer. Chem. Soc.*, **96,** 6810 (1974).
255. W. Wucherpfennig, *Tetrahedron Letters*, 1891 (1971).
256. G. M. Atkins, Jr. and E. M. Burgess, *J. Amer. Chem. Soc.*, **94,** 6135 (1972).
257. T. Hiraoka and T. Kobayashi, *Bull. Chem. Soc. Japan*, **48,** 480 (1975).
258. J. C. Howard, *J. Org. Chem.*, **36,** 1073 (1971).
259. R. S. Schreiber and R. L. Schriner, *J. Amer. Chem. Soc.*, **57,** 1896 (1935).
260. T. Taguchi and M. Kojima, *J. Amer. Chem. Soc.*, **78,** 1464 (1966).
261. J. Sicher, M. Svoboda, and J. Farkaš, *Coll. Czech. Chem. Comm.*, **20,** 1439 (1955).
262. G. C. Barrett and A. R. Khokhar, *J. Chem. Soc.* (*C*), 1117 (1969).
263. F. Winternitz, M. Mousseron, and R. Dennilauler, *Bull. Soc. chim. France*, 1228 (1956).
264. R. Kuhn and F. Drawert, *Annalen*, **590,** 55 (1954).
265. E. H. White, F. McCapra, and G. F. Field, *J. Amer. Chem. Soc.*, **85,** 337 (1963).
266. T. R. Lowell, Jr. and G. K. Helmkamp, *J. Amer. Chem. Soc.*, **88,** 678 (1966).
267. R. Lie and K. Undheim, *Acta Chem. Scand.*, **27,** 1756 (1973).
268. K. Undheim and R. Lie, *Acta Chem. Scand.*, **27,** 1749 (1973).
269. K. Undheim and V. Nordal, *Acta. Chem. Scand.*, **23,** 371 (1969).
270. K. Undheim and V. Nordal, *Acta Chem. Scand.*, **23,** 1966 (1969).
271. T. Greibrokk and K. Undheim, *Tetrahedron*, **28,** 1223 (1972).
272. M. R. Bell, J. A. Carlson, and R. Oesterlin, *J. Amer. Chem. Soc.*, **92,** 2177 (1970).
273. T. A. Crabb and R. F. Newton, *Tetrahedron*, **24,** 2485 (1968).
274. G. E. Woodward and E. F. Schroeder, *J. Amer. Chem. Soc.*, **59,** 1690 (1937); T. Taguchi, M. Kojima, and T. Muro, *J. Amer. Chem. Soc.*, **81,** 4322 (1959).
275. T. Taguchi, T. Takatori, and M. Kojima, *Chem. and Pharm. Bull.* (*Japan*), **10,** 245 (1962).
276. J. C. Sheehan and D-D. H. Yang, *J. Amer. Chem. Soc.*, **80,** 1158 (1958).
277. J. R. Piper and T. P. Johnston, *J. Org. Chem.*, **29,** 1657 (1964).
278. W. M. McLamore, W. D. Celmer, V. V. Bogert, F. C. Pennington, and I. A. Solomons, *J. Amer. Chem. Soc.*, **74,** 2946 (1952).

279. M. D. Armstrong, *J. Amer. Chem. Soc.*, **77,** 6049 (1955).

280. G. M. Clarke and P. Sykes, *Chem. Comm.*, **370** (7965).

281. G. M. Clarke and P. Sykes, *J. Chem. Soc.* (*C*), 1411 (1967).

282. D. Craig, J. J. Shipman, A. Hawthorne, and R. Fowler, *J. Amer. Chem. Soc.*, **77,** 1283 (1955).

283. M. Thiel, F. Asinger, K. Schmiedel, H. Petschik, R. Haberl, and O. Hromatka, *Monatsh.*, **91,** 473 (1960); M. Thiel, F. Asinger, and K. Schmiedel, *Annalen*, **611,** 121 (1958).

284. R. C. Fort, Jr. and W. L. Semon, *J. Org. Chem.*, **32,** 3685 (1967).

285. K. T. Potts, J. Baum, E. Houghton, D. N. Roy, and U. P. Singh, *J. Org. Chem.*, **39,** 3619 (1974).

286. G. J. Pitt, *Acta Cryst.*, **5,** 770 (1952).

286a. D. H. R. Barton, *Pure Appl. Chem.*, **33,** 1 (1973).

287. R. J. Stoodley, *Progr. Org. Chem.*, **8,** 102 (1973).

288. D. N. McGregor, *Fortschr. Chem. org. Naturstoffe*, **31,** 1 (1974).

289. H. Beecken, *Chem. Ber.*, **100,** 2159 (1967).

290. L. Angiolini, R. P. Duke, R. A. Y. Jones, and A. R. Katritzky, *J.C.S. Perkin II*, 674 (1972); L. Angiolini, R. A. R. Jones, and Al R. Katritzky, *Tetrahedron Letters*, 2209 (1971).

291. T. A. Crabb and R. F. Newton, *Tetrahedron*, **26,** 3941 (1970).

292. L. Abis and C. Giordano, *J.C.S. Perkin I*, 771 (1973).

293. J. Ashby and U. Eisner, *J. Chem. Soc.* (*C*), 1706 (1967).

294. U. Eisner, M. Z. Hag, J. Flippen, and I. Karle, *J.C.S. Perkin I*, 357 (1972).

295. J. B. Bremner, R. N. Warrener, E. Adman, and L. H. Jensen, *J. Amer. Chem. Soc.*, **93,** 4574 (1971).

296. R. B. Woodward, K. Heusler, J. Gosteli, P. Naegeli, W. Oppolzer, R. Ramage, S. Ranganathan, and H. Vorbrüggen, *J. Amer. Chem. Soc.*, **88,** 852 (1966).

297. R. Scartazzini, J. Gosteli, H. Bickel, and R. B. Woodward, *Helv. Chim. Acta*, **55,** 2567 (1972).

298. R. M. Sweet and L. F. Dahl, *J. Amer. Chem. Soc.*, **92,** 5489 (1970).

299. R. Lattrell and G. Lohaus, *Annalen*, 870 (1974).

300. R. W. Ratcliff and B. G. Christensen, *Tetrahedron Letters*, 4649 (1973).

301. K. Kühlein and H. Jensen, *Annalen*, 369 (1974); R. Nagarajan, L. D. Boeck, M. Gorman, R. L. Hamill, C. E. Higgens, M. M. Hoehn, W. M. Stark, and J. G. Whitney, *J. Amer. Chem. Soc.*, **93,** 2308 (1971).

302. R. Nagarajan and D. O. Spry, *J. Amer. Chem. Soc.*, **93,** 2310 (1971).

303. M. Ochiai, E. Mizuta, O. Aki, A. Morimoto and T. Okada. *Tetrahedron Letters*, 3245 (1972).

304. S. Kukolja, P. V. Demarco, N. D. Jones, M. O. Chaney, and J. W. Paschal, *J. Amer. Chem. Soc.*, **94,** 7592 (1972).

305. E. P. Abraham, *Quart. Rev.*, **21,** 231 (1967); E. H. Flynn, Ed., *Cephalosporins and Penicillins*, Academic, New York, 1972; M. S. Manhas and A. K. Bose, *Synthesis of Penicillin, Cephalosporin C. and Analogues*, Dekker, New York, 1969.

306. M. Mousseron and G. Combes, *Bull. Soc. chim. France*, 82 (1947).

307. J. F. Carson and L. O. Boggs, *J. Org. Chem.*, **31,** 2862 (1966).

308. J. F. Carson, L. E. Boggs, and R. Lundin, *J. Org. Chem.*, **33,** 3739 (1968).
309. J. F. Carson, L. M. Boggs, and R. E. Lundin, *J. Org. Chem.*, **35,** 1594 (1970).
310. K. J. Palmer and K. S. Lee, *Acta Cryst.*, **20,** 790 (1966).
311. J. F. Carson and L. E. Boggs, *J. Org. Chem.*, **36,** 611 (1971).
312. J. Kitchin and R. J. Stoodley, *J.C.S. Chem. Comm.*, 959 (1972).
312. A. G. W. Baxter, J. Kitchin, R. J. Stoodley, and R. B. Wilkins, *J.C.S. Chem. Comm.*, 285 (1973).
314. R. J. Stoodley and R. B. Wilkins, *J.C.S. Perkin I*, 1572 (1974).
315. A. R. Dunn and R. J. Stoodley, *Tetrahedron*, **28,** 3315 (1972).
316. J. Kitchin and R. J. Stoodley, *J.C.S. Perkin I*, 22 (1973).
317. J. Kitchin and R. J. Stoodley, *J.C.S. Perkin I*, 2464 (1973).
318. J. Kitchin and R. J. Stoodley, *J. Amer. Chem. Soc.*, **95,** 3439 (1973); A. R. Dunn and R. J. Stoodley, *Tetrahedron Letters*, 3367 (1969).
319. A. R. Dunn, I. McMillan, and R. J. Stoodley, *Tetrahedron*, **24,** 2985 (1968).
320. A. R. Dunn and R. J. Stoodley, *Tetrahedron Letters*, 2979 (1969).
321. P. S. Portoghese and V. G. Telang, *Tetrahedron*, **27,** 1823 (1971).
322. D. M. Brunwin and G. Lowe, *J.C.S. Chem. Comm.*, 589 (1972).
323. D. M. Brunwin and G. Lowe, *J.C.S. Perkin I*, 1321 (1973).
324. G. Lowe and M. V. J. Ramsay, *J.C.S. Perkin I*, 479 (1973).
325. M. J. Aroney, G. M. Hoskins, and R. J. W. Le Fèvre, *J. Chem. Soc.* (*B*), 1206 (1968).
326. J. D. Bell, J. F. Blount, O. V. Briscoe, and H. C. Freeman, *Chem. Comm.*, 1656 (1968).
327. O. Hromatka, J. Augl, and W. Grünsteidl, *Monatsh.*, **90,** 914 (1959).
328. J. M. Lown and K. Matsumoto, *Canad J. Chem.*, **48,** 3399 (1970).
329. J. G. Kirchner and I. B. Johns, *J. Amer. Chem. Soc.*, **62,** 2183 (1940).
330. F. Sauter, G. Büyük, and U. Jordis, *Monatsh.*, 869 (1974).
331. F. Sauter and G. Büyük, *Monatsh.*, **105**, 550 (1974).
332. S. A. Harris, D. E. Wolf, R. Mozingo, R. C. Anderson, G. E. Arth, N. R. Easton, D. Heyl, A. N. Wilson, and K. Folkers, *J. Amer. Chem. Soc.*, **66,** 1756 (1944); H. Ohrui and S. Emoto, *Tetrahedron Letters*, 2765 (1975).
333. B. R. Baker, M. V. Querry, W. L. McEwen, S. Bernstein, S. R. Safir, L. Dorfman, and Y. Subbarow, *J. Org. Chem.*, **12,** 186 (1947); B. R. Baker, W. L. McEwen, and W. N. Kinley, *J. Org. Chem.*, **12,** 322 (1947).
334. D. E. Wolf, R. Mozingo, S. A. Harris, R. C. Anderson, and K. Folkers, *J. Amer. Chem. Soc.*, **67,** 2100 (1945).
335. S. A. Harris, R. Mozingo, D. E. Wolf, A. N. Wilson, and K. Folkers, *J. Amer. Chem. Soc.*, **67,** 2102 (1945).
336. L. D. Wright and E. L. Cresson, *J. Amer. Chem. Soc.*, **76,** 4156 (1954).
337. H. Ruis, D. B. McCormick, and L. D. Wright, *J. Org. Chem.*, **32,** 2010 (1967).
338. G. F. Field, W. J. Zally, and L. H. Sternbach, *J. Amer. Chem. Soc.*, **92,** 3520 (1970).
339. F. Ellis and P. G. Sammes, *J.C.S. Perkins I*, 2866 (1972).
340. R. Lett and A. Marquet, *Tetrahedron Letters*, 2851, 2855 (1971).
341. R. Lett and A. Marquet, *Tetrahedron*, **30,** 3365, 3379 (1974).
342. R. Lett and A. Marquet, *Tetrahedron*, **31,** 653 (1975).
343. N. J. Leonard and G. E. Wilson, Jr., *J. Amer. Chem. Soc.*, **86,** 5307 (1964).

344. N. J. Leonard and R. Y. Ning, *J. Org. Chem.*, **31,** 3928 (1966).
345. H. Inoue, S. Takeo, M. Kawazu, and H. Kugita, *J. Pharm. Soc. Japan*, **93,** 729 (1973).
346. P. C. Thomas, I. C. Paul, T. Williams, G. Grethe, and M. Uskoković, *J. Org. Chem.*, **34,** 365 (1969).
347. H. H. Szmant and Y-L. Chow, *J. Amer. Chem. Soc.*, **79,** 4382, 5583 (1957).
348. N. L. Allinger and G. A. Youngdale, *J. Org. Chem.*, **24,** 2059 (1959).
349. I. Sataty, *J. Heterocyclic Chem.*, **7,** 431 (1970).
350. I. C. Paul and K. T. Go, *J. Chem. Soc.* (*B*), 33 (1969).
351. S. Tanaka, K. Hashimoto, H. Watanabe, and K. Sakaguchi, *Chem. and Pharm. Bull.* (*Japan*), **21,** 1683 (1973).
352. J. M. Lehn and F. G. Riddell, *Chem. Comm.*, 803 (1966).
353. T. C. Owen and J. M. Fayadh, *J. Org. Chem.*, **35,** 3198 (1970).
354. R. Wade, M. Winitz, and J. P. Greenstein, *J. Amer. Chem. Soc.*, **78,** 373 (1956).
355. F. G. Bordwell, R. D. Chapman, and C. E. Osborne, *J. Amer. Chem. Soc.*, **81,** 2002 (1959).
356. I. L. Khunyants and G. A. Sokolski, *Angew. Chem. Internat. Edn.*, **11,** 583 (1972).
357. M. Nagayama, O. Okumura, S. Noda, and A.Mori, *J.C.S. Chem. Comm.*, 841 (1973).
358. D. C. Dittmer, G. C. Levy, and G. E. Kuhlmann, *J. Amer. Chem. Soc.*, **91,** 2097 (1969).
359. T. Durst and K-C. Tin, *Canad. J. Chem.*, **48,** 845 (1970).
360. J. Hendrickson, *J. Amer. Chem. Soc.*, **84,** 653 (1962).
361. D. R. Dimmel and W. Y. Fu, *J. Org. Chem.*, **38,** 3778 (1973).
362. D. R. Dimmel and W. Y. Fu, *J. Org. Chem.*, **38,** 3782 (1973).
363. O. Exner, D. N. Harpp, and J. G. Gleason, *Canad. J. Chem.*, **50,** 548 (1972).
364. R. Keskinen, A. Nikkilä, and K. Pihlaja, *Tetrahedron*, **28,** 3943 (1972).
365. R. Keskinen, A. Nikkilä, J. Pihlaja, and F. G. Riddell, *J.C.S. Perkin II*, 466 (1974).
366. K. Rühlmann, D. Gramer, D. Heuchel, and U. Schräpler, *J. prakt. Chem.*, **10,** 316 (1960).
367. K. Rühlmann, U. Schräpler and D. Gramer, *J. prak. Chem.*, **10,** 325 (1960).
368. R. M. Kellogg, *J. Org. Chem.*, **38,** 844 (1973).
369. C. C. Price and P. F. Kirk, *J. Amer. Chem. Soc.*, **75,** 2396 (1953).
370. C. Th. Pedersen, *Acta Chem. Scand.*, **23,** 489 (1969).
371. D. A. Lightner, C. Djerassi, K. Takeda, K. Kuriyama, and T. Komeno, *Tetrahedron*, **21,** 1581 (1965).
372. D. J. Pasto, F. M. Klein, and T. W. Doyle, *J. Amer. Chem. Soc.*, **89,** 4368 (1967).
373. G. E. Wilson, Jr., M. G. Huang, and F. A. Bovey, *J. Amer. Chem. Soc.*, **92,** 5907 (1970).
374. E. L. Eliel, L. A. Pilato, and V. G. Badding, *J. Amer. Chem. Soc.*, **84,** 2377 (1962).
375. E. L. Eliel and L. A. Pilato, *Tetrahedron Letters*, 103 (1962).
376. M. P. Mertes, H-K. Lee, and R. L. Schowen, *J. Org. Chem.* **34,** 2080 (1969).
377. E. L. Eliel, E. W. Della, and M. Rogić, *J. Org. Chem.*, **30,** 855 (1965).
378. J. G. Pritchard and P. C. Lauterbur, *J. Amer. Chem. Soc.*, **83,** 2105 (1961).
379. Q. E. Thompson, M. M. Crutchfield, and M. W. Dietrich, *J. Amer. Chem. Soc.*, **86,** 3891 (1964).

380. Q. E. Thompson, M. M. Crutchfield, and M. W. Dietrich, *J. Org. Chem.*, **30,** 2696 (1965).
381. M. Kobayashi, A. Yabe, and R. Kiritani, *Bull. Chem. Soc. Japan*, **39,** 1785 (1966).
382. J. M. Coxon, M. P. Hartshorn, G. R. Little, and S. G. Maister, *Chem. Comm.*, 271 (1971).
383. R. L. Smith, A. Manmade and G. W. Griffin, *Tetrahedron Letters*, 663 (1970).
384. C. H. Green and D. G. Hellier, *J.C.S. Perkin II*, 243 (1973).
385. P. Albriktsen, *Acta Chem. Scand.*, **26,** 3671 (1972).
386. C. H. Green and D. G. Hellier, *J.C.S. Perkin II*, 190 (1975).
387. D. N. Harpp and J. G. Gleason, *Tetrahedron Letters*, 1447 (1969).
387a. D. N. Harpp and J. G. Glesaon, *J. Org. Chem.*, **36,** 1314 (1971).
388. T. Durst, *Tetrahedron Letters*, 4171 (1971).
389. N. de Wolf and H. R. Buys, *Tetrahedron Letters*, 551 (1970).
390. Y. Allingham, R. C. Cookson, and T. A. Crabb, *Tetrahedron*, **24,** 1989 (1968).
391. J. Gelan and M. Anteunis, *Bull. Soc. chim. belges*, **77,** 447 (1968).
392. K. Pihlaja and P. Pasanen, *J. Org. Chem.*, **39,** 1948 (1974).
393. J. Gelan and M. Anteunis, *Bull. Soc. chim. belges*, **77,** 423 (1968).
394. J. Gelan, G. Swaelens, and M. Anteunis, *Bull. Soc. chim. belges*, **79,** 321 (1970).
395. J. Gelan and M. Anteunis, *Bull. Soc. chim. belges*, **79,** 313 (1970).
396. P. Pasanen and K. Pihlaja, *Tetrahedron*, **28,** 2617 (1972).
397. P. Pasenen and K. Pihlaja, *Tetrahedron Letters*, 4515 (1971).
398. L. Van Acker and M. Anteunis, *Tetrahedron Letters*, 225 (1974).
399. F. Lautenschlaeger, *J. Org. Chem.*, **33,** 2620 (1968).
400. R. K. Summerbell and E. S. Poklacki, *J. Org. Chem.*, **27,** 2074 (1962).
401. N. S. Zefirov, V. S. Blagoveshchensky, I. V. Kazimirchik, and N. S. Surova, *Tetrahedron*, **27,** 3111 (1971).
402. K. Jankowski and R. Coulombe, *Tetrahedron Letters*, 991 (1971).
403. A. B. Foster, Q. H. Hasan, D. R. Hawkins, and J. M. Webber, *Chem. Comm.*, 1084 (1968).
404. A. B. Foster, T. D. Inch, M. H. Qadir, and J. M. Webber, *Chem. Comm.*, 1086 (1968).
405. K. W. Buck, A. B. Foster, W. D. Pardoe, M. H. Qadir, and J. M. Webber, *Chem. Comm.*, 759 (1966).
406. K. W. Buck, T. A. Hamor, and D. J. Watkin, *Chem. Comm.*, 759 (1955).
407. D. J. Watkin and T. A. Hamor, *J. Chem. Soc. (B)*, 1692 (1971).
408. N. S. Zefirov and S. V. Rogozina, *Tetrahedron*, **30,** 2345 (1974).
409. N. S. Zefirov, S. V. Rogozina, E. H. Kurkutova, A. V. Goncharov, and N. V. Belov, *J.C.S. Chem. Comm.*, 260 (1974).
410. L. Cazaux and P. Maroni, *Bull. Soc. chim. France*, 773 (1972).
411. L. Cazaux and P. Maroni, *Bull. Soc. chim. France*, 780 (1972).
412. G. Wood and M. H. Miskow, *Tetrahedron Letters*, 1775 (1970).
413. P. C. Lauterbur, J. G. Pritchard, and R. L. Vollmer, *J. Chem. Soc.*, 5307 (1963).
414. C. H. Green and D. G. Hellier, *J.C.S. Perkin II*, 458 (1972).
415. G. Wood, J. M. McIntosh, and M. H. Miskow, *Canad. J. Chem.* **49,** 1202 (1971).

416. G. Wood, G. W. Buchanan, and M. H. Miskow, *Canad. J. Chem.*, **50,** 521 (1972).
417. G. W. Buchanan, J. B. Stothers, and G. Wood, *Canad. J. Chem.*, **51,** 3746 (1973).
418. P. C. Hamblin, R. F. M. White, G. Eccleston, and E. Wyn-Jones, *Canad. J. Chem.*, **47,** 2731 (1969).
419. P. Albriktsen, *Acta. Chem. Scand.*, **26,** 3678 (1972).
420. C. Altona, H. J. Geise, and C. Romers, *Rec. Trav. Chim.*, **85,** 1197 (1966).
421. G. Wood and M. H. Miskow, *Tetrahedron Letters*, 1109 (1969); C. G. Overberger, T. Kurtz, and S. Yaroslavsky, *J. Org. Chem.*, **30,** 4363 (1965).
422. P. B. D. de la Mare, W. Klyne, D. J. Millen, J. G. Pritchard, and D. Watson, *J. Chem. Soc.*, 1813 (1956).
423. R. S. Edmunson, *Tetrahedron Letters*, 1649 (1965).
424. H. F. van Woerden, H. Cerfontain, C. H. Green, and R. J. Reijerkert, *Tetrahedron Letters*, 6107 (1968).
425. S. Sarel and V. Usieli, *Israel J. Chem.*, **6,** 885 (1968).
426. H. F. van Woerden and A. T. de Vries-Miedema, *Tetrahedron Letters*, 1687 (1971).
427. G. Wood, J. M. McIntosh, and H. M. Miskow, *Tetrahedron Letters*, 4895 (1970).
428. H. Prinzbach, C. Kaiser, and H. Fritz, *Angew. Chem. Internat. Edn.*, **14,** 253 (1975).
429. N. Wigger and C. Ganter, *Helv. Chim. Acta.*, **55,** 2769 (1972).
430. N. Wigger, N. Stücheli, H. Szczepanski, and C. Ganter, *Helv. Chim. Acta*, **55,** 2791 (1972).
431. C. Ganter and J-F. Moser, *Helv. Chim. Acta*, **52,** 725 (1969).
432. C. Ganter and K. Wicker, *Helv. Chim. Acta*, **51,** 1599 (1968).
433. G. M. Badger, G. E. Lewis, and U. P. Singh, *Austral. J. Chem.*, **19,** 257 (1966).
434. J. H. Markgraf and R. W. King, *J. Org. Chem.*, **29,** 3094 (1964).
435. J. S. Bradshaw, J. Y. Hui, Y. Chan, B. L. Haymore, R. M. Izatt, and J. J. Christensen, *J. Heterocyclic Chem.*, **11,** 45 (1974).
436. J. A. Deyrup and C. L. Moyer, *J. Org. Chem.*, **34,** 175 (1969).
437. F. Wudl and T. B. K. Lee, *J. Amer. Chem. Soc.*, **95,** 6349 (1973).
438. F. Wudl and T. B. K. Lee, *J.C.S.Chem. Comm.*, 61 (1972).
439. Y. Noda, *Bull. Chem. Soc. Japan*, **40,** 1554 (1967).

5 PHOSPHORUS HETEROCYCLES (INCLUDING SOME ARSENIC AND ANTIMONY HETEROCYCLES)

I. INTRODUCTION

It is nowadays commonplace and even trite to commence a survey of this type with a comment on the explosive growth of published work in the field in the preceding twenty years. The chemistry of phosphorus has certainly been no exception but merits comment because a disproportionately large part of the increase has been concerned with stereochemistry. The reasons for this are worth considering, Though as long ago as 1850 Thenard[1] began work on the organic chemistry of phosphorus, and Meisenheimer and Lichtenstadt[2] resolved a phosphine oxide in 1911, interest in the stereochemistry of this element (or indeed any element except carbon and nitrogen) was kept alive only by a few pioneers such as Mann and Campbell. As late as 1958, Mann could comfortably review the entire field of the stereochemistry of the group V elements, in all their valency states, in thirty pages.[3] Shortly thereafter Horner[4] obtained optically active phosphines, thus demonstrating unequivocally the nonplanarity and configurational stability of trivalent phosphorus and refuting the long-held belief in its facile inversion. At this time the ready resolvability of thiophosphoryl compounds[5] and their utility in stereochemical studies[6] (in marked contrast to their oxygen analogues) were also discovered. However, unquestionably the major impetus arose from an entirely different source, namely Westheimer's suggestion to account for the enormously enhanced rates of hydrolysis of five-membered cyclic phosphates when compared with their six-membered or acyclic analogues.[7] Though strain in the ring due to the greater length of the P–O bonds could account for the rate enhancement, it could not explain the formation of cyclic products in the acid hydrolysis (Equation 1). Westheimer proposed that the hydrolysis proceeded by way of a trigonal bipyramidal intermediate whose formation was facilitated by distortion of the ring angle at phosphorus, away from the normal tetrahedral, towards a right angle (Equation 2).

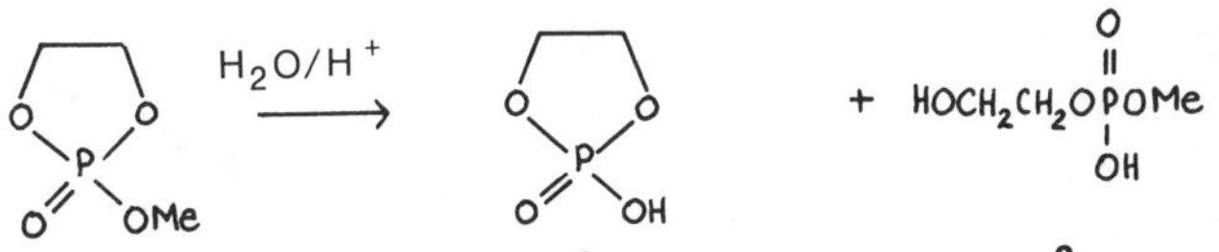

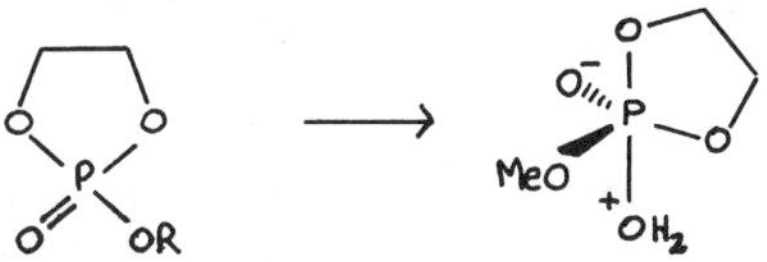

$$\mathbf{1} \xrightarrow{H_2O/H^+} \mathbf{2} + HOCH_2CH_2OP(O)(OH)OMe \; (\mathbf{3}) \quad (1)$$

$$\mathbf{4} \longrightarrow \mathbf{5} \quad (2)$$

A feature of the trigonal bipyramidal structure, first advanced by Berry[8] (at the suggestion of F. T. Smith) to account for the doublet ^{19}F spectrum of PF_5, is the ability of the substituents to undergo positional interchange. Westheimer elegantly applied this concept to further understanding of the hydrolytic behavior of cyclic organophosphorus compounds.[7] Until this time, though Wittig[9] had prepared pentaaryl phosphoranes and Ramirez[10] had convincingly demonstrated that pentaoxyphosphoranes were readily accessible, compounds of the type PR_5 were considered rather esoteric. Interestingly, the first suggestion of their occurrence as reaction intermediates had been made by Fenton and Ingold in 1929.[11] It was now apparent that they might well be extensively involved as reaction intermediates and this, coupled with the intrinsic interest in the dynamic stereochemistry of such structures, led to a flood of papers whose main theme was the novel stereochemistry of P(V) systems. Only very recently has this torrent shown signs of abating, and though other aspects of phosphorus stereochemistry have received much less attention, this field as a whole has probably been more active than any other. This imbalance is apparent in the relatively small number of chemical studies carried out on phosphoranes. Finally, it should be noted that the ease with which phosphoranes containing one or more five-membered rings can be prepared has resulted in the bulk of the published work dealing with such compounds, and hence falling within the scope of this review. As a result an exhaustive treatment is not possible here and considerable selection and condensation have been necessary. Excluded from any specific consideration are: phosphonitrilic compounds (recently reviewed by Allcock[12]), other compounds with no ring carbon, and metal complexes.

A number of general reviews of phosphorus stereochemistry have appeared,[3,13,14] but the older ones will have been overtaken by the weight of new results, and it is doubtful if the subject as a whole can be done justice to by a relatively short treatment. Hopefully, it is still possible-for cyclic compounds.

The basic source of data on specific compounds is Kosolapoff's and Maier's invaluable compendium[15] which covers the literature before 1970 with occasional later references. The literature since that date has been reviewed annually in the Chemical Society's excellent *Specialist Periodical Report* on phosphorus chemistry. These sources do not, of course, deal only with stereochemical data.

Phosphorus forms stable molecules with from two to six ligands and is the first element with this inherent complexity to be studied in any detail. The cyclic compounds are subdivided in terms of the number of ligands and the ring size, and for convenience, are written with the ligand number in roman numerals and the ring size in italic arabic numerals, for, example, P(IV) (*5*) and P(III) (*3*), et cetera. Polycyclic compounds are dealt with under the heading corresponding to the smallest ring containing the phosphorus atom.

II. NOMENCLATURE

There is no generally agreed nomenclature which covers all compounds of phosphorus. The most widespread system for cyclic compounds appears to be that of the *Ring Index* which is used by Mann in his monograph.[16] This is used extensively here to the extent that four-membered rings are named as derivatives of phosphetane (**6**), five-membered are phospholanes (**7**), and six-membered are phosphorinanes (**8**), and this usage is followed here regardless of the other atoms in the ring or the coordination number of phosphorus. Figures are used as much as possible. The terminology of cis and trans is used, the groups referred to being the two highest in the normal sequencing of substituents by the Cahn-Ingold-Prelog system.

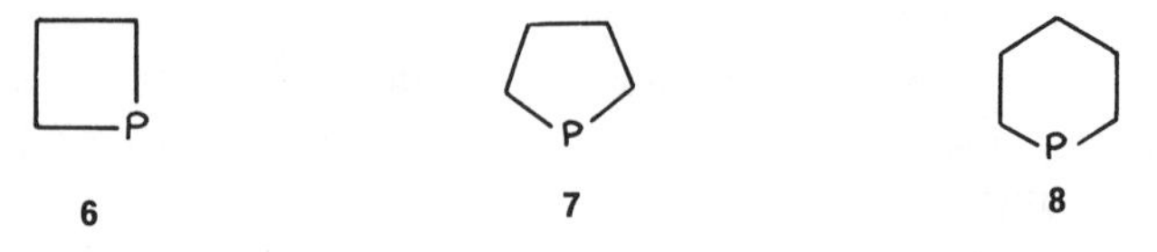

The writer feels it unwise to compound the existing confusion with suggested changes in nomenclature, but there are two minor points of some importance which are followed here that seem to be worthy of general adoption.

(*a*) The term phosphorane is used only where the coordination number is five; substances formed by proton abstraction (formally or directly) from atoms adjacent to P(IV) are best termed ylids or carbanions.

(*b*) The terms apical, *a*, and radial, *r*, to describe the two positions on a trigonal bipyramid should be used in place of the more common apical and equatorial or, even worse, axial and equatorial, which only invite confusion with the terminology of conformational analysis in carbocyclic systems. The latter point is important when one remembers that the bulk of trigonal bipyramidal molecules contain at least one ring.

The remaining Group V elements have attracted little attention no doubt largely because they are not observable by nmr techniques and hence lend themselves less to convenient analysis. Some work has appeared, however, and is treated along with the phosphorus analogues.

III. SPECTROSCOPY

A. Introduction

In common with all branches of chemistry, spectroscopic studies of organophosphorus heterocycles have increased enormously in the last decade corresponding with the growth of nmr investigations. The principal reason for the intense interest in cyclic as opposed to acyclic compounds (for which only a handful of studies has been described) is stereochemical, or more specifically conformational, analysis. This is facilitated by the ease of preparation in many instances and (in the case of phosphorus compounds) the presence of a second spin active nucleus (^{31}P; 100% abundance; $I = 1/2$) which greatly assists analysis. In addition to nmr spectroscopy a considerable number of X-ray and electron diffraction structures of cyclic compounds have been published and recently reviewed.[17] Table I summarizes bond angles at phosphorus for cyclic compounds. Bond lengths do not differ substantially from acyclic compounds, but some representative data are given in Table II. Corbridge's comprehensive survey[18] of structural features gives detailed coverage of the literature until 1972.

By comparison, uv and ir spectra have been almost completely neglected except for a few specialized areas which are described in the appropriate sections. An attempt to correlate $\nu P{=}O$ with ring strain[31] has yielded ambiguous results. In the following paragraphs the *general* conclusions arising from nmr studies are considered. More detailed discussion is given where it seems justifiable, in the relevant sections. A survey of studies on five- and six-membered ring systems is given in Tables III and IV. These have been selected for their stereochemical content; a more general review, covering the literature until 1971 can be found in Batterham's book.[19]

Mavel[22] has recently published an exhaustive review of the literature dealing with nmr spectra of phosphorus compounds for the period 1965–1969 inclusive. All nuclei of interest are covered, and many cyclic compounds are shown with spectral data, which is very convenient for quick reference. Sufficient work has been published in this field to justify a substantial survey in itself and must necessarily be condensed here.

B. Chemical Shift Data

1. ^{31}*P Chemical Shifts*

All chemical shifts quoted are in parts per million (ppm) with respect to 85% $H_3PO_4 = 0$; upfield shifts are *negative* and downfield shifts are *positive*.

Though large numbers of compounds have been examined and summarized particularly by the Monsanto group[20] and more recently by Fluck[60] and Mavel,[22] almost no attention has been paid to systematic stereochemical studies, and the major use of chemical shifts has been to identify coordination number and to classify measurements in terms of number and type of substituents attached to phosphorus. Part of the reason for this undoubtedly lies in the low sensitivity of ^{31}P ($\sim$6% that of ^{1}H) and the broad signals obtained as a consequence of multiple PH couplings when organic groups are attached. The widespread use of pulsed nmr techniques together with noise decoupling of protons should go far towards resolving both drawbacks and high resolution ^{31}P spectra of good sensitivity should become very common in the next few years. In the instances where pairs of disastereoisomers have been compared a substantial $\Delta\delta$ is found (1–3 ppm), and when sufficient data accumulate this will undoubtedly prove the quickest and simplest method for assigning stereochemistry as well as determining stereochemical purity. For example, Bentrude has noted[48] that for six-membered cyclic phosphites and phosphonites the more upfield δ ^{31}P is associated with the cis isomer, whereas the reverse is true for the five-membered analogues.[49] Bulky groups can alter this trend, but the correlation is a useful one.

Substantial differences are observed in δ^{31}P for P (IV) (*5*) and P (IV) (*6*) when compared[64] with each other or with their acyclic analogues, and though these are probably steric in origin, the situation is still unclear. Very recently Gorenstein described an empirical correlation between δ^{31}P and O$\hat{P}$O in phosphates[65] based on the considerable amount of X-ray data now available. The curve passes through a minimum close to the normal tetrahedral angle and seems to hold for cyclic and acyclic compounds. Examination of other classes of phosphorus compounds would be of great interest.

Chemical shift values can be used[54] to determine ring size in cyclic polyphosphines ($[RP]_n$; $n = 4$, $\delta = +50$ to $+70$; $n = 5$, $\delta = -20$ to $+10$);[70] this is presumably a steric effect.

Chemical shifts in P(III) have been the subject of a reasonably successful theoretical treatment, and in particular they are related to the percentage *s* character in the lone pair and hence to the bond angles of phosphorus.[20] Though it would be very unwise to rely on δ values as a criterion for geometry at P(III), nevertheless, when making a comparisons between closely related compounds, for example, heterotriptycenes (**9**),[67] (**10**)[68] and (**11**),[69]

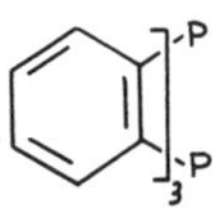

$\delta^{31}P - 43$ ppm

9

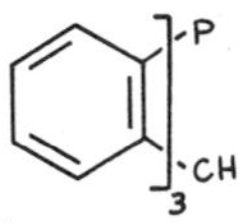

$\delta^{31}P - 64.8$ ppm

10

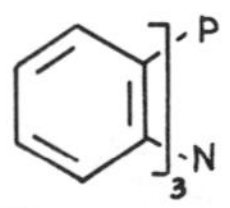

$\delta^{31}P - 80$ ppm

11

it can be seen that increased *s* character in the lone pair results in a marked shift to higher field. In this instance the variation is brought about by the diminishing bond lengths CN<CC<CP progressively closing the angle at phosphorus. The large differences in $\delta^{31}P$ values found[64,65] within a given class of compounds as the ring size varies are presumably a consequence of steric effects, though no simple relationship is obvious (e.g., **12** **13**, and **14**. Quin has examined[26] a number of P(III) (*5*) structures, both saturated and unsaturated, and found chemical shift values predictable in terms of effects found in acyclic systems with marked differences between stereoisomers. Anomalies are observed with unsaturated phospholenes, however.

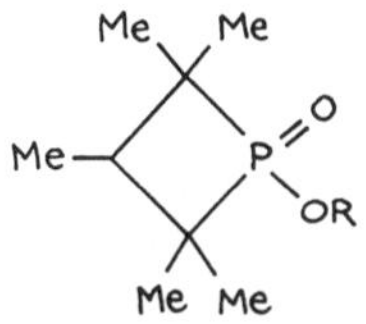

$\delta^{31}P + 52$ ppm

12

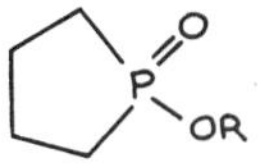

$\delta^{31}P + 70$ ppm

13

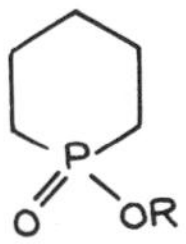

$\delta^{31}P + 45$ ppm

14

2. ^{13}C *Chemical Shifts*

The complexity of ^{1}H spectra when phosphorus is the only heteroatom in the ring led Quin and his collaborators to examine ^{13}C δ values in some detail for P(III) (*5*),[27] P(III) (*6*),[28,29] and P(IV) (*6*).[29,30] Axial groups on phosphorus in six-membered rings absorb at slightly higher field than equatorial, and the 3,5-carbons of six-membered rings are sensitive to compressive effects by substituents at phosphorus. These facts, together with the angular dependance of $^{2}J_{PC}$ discussed below, allow confident conformational assignments to be made in the P(III) series. On the other hand, there is little difference between the ^{13}C spectra of the 4-hydroxy-1-methylphosphorinane 1-sulfide (i.e., P(IV) (*6*)) conformers,[29,30] presumably reflecting a rapid interconversion and roughly 50:50 mixture. This is supported by the striking observation of both chair conformers of **15** in an X-ray crystal study.[30] Similar results have been obtained by Reisse and co-workers with the P(III) (*6*) 2-methoxy-1,3,2-dioxaphosphorinanes (**16**).[51]

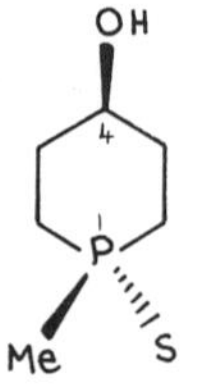

15 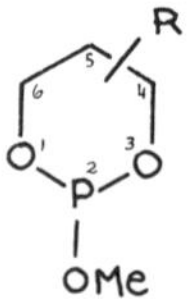 **16**

Both the ^{31}P and the 4,6 ^{13}C nuclei are shifted to higher field by about 3–4 ppm when the methoxyl group is axial as compared with the equatorial conformer.

3. *Other Nuclei*

Extensive compilations of ^{1}H and ^{19}F data have been prepared, particularly by Mavel[21,22] and Schmutzler,[66] but it is only occasionally that a general statement of use in determination of stereochemistry at phosphorus can be made, and these are described when discussing the various ring sizes.

C. Coupling Constant Data

1. $^{1}J_{PX}$

The only stereochemically useful information here is in the magnitude of this normally large constant. For P(III) compounds the value of $^{1}J_{PH}$ reflects the *s* character of the lone pair and hence bond angles at phosphorus. These effects are quite large[23–25] (e.g. **17–20**), although they are of limited value because of the rarity of this type of compound. With increasing coordination number $^{1}J_{PH}$ increases rapidly ($^{1}J_{P\,(IV)H}$ 480–730

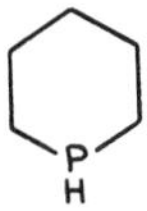

$^1J_{PH}$ 200Hz
17

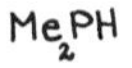

Me_2PH
$^1J_{PH}$ 190Hz
18

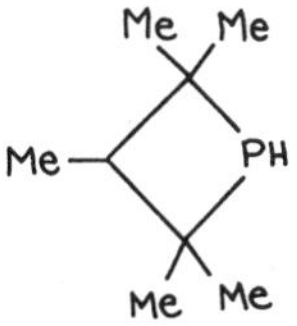

$^1J_{PH}$ 170Hz
19

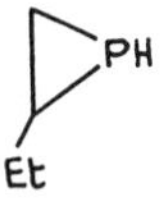

$^1J_{PH}$ 158Hz
20

Hz, $^1J_{P(V)H}$ 700–800 Hz $^1J_{P}$ (VI)H 800–1000 Hz) but affords little direct stereochemical information. $^1J_{PF}$ values, on the other hand, have proved valuable stereochemical probes in the study of P(V) molecules.[66] In a trigonal bipyramid (e.g., **21**) there would be two different environments for ligands as shown. If *a* and *r* are fluorine then in general *rF* will absorb at higher field (about 20–90 ppm) and have a larger coupling constant (100–200 Hz greater) than *a*F, $^1J_{PF}$ is large in both instances but is normally less than 900 Hz for *a*F and greater than 900 Hz for *r*F. A similar relationship seems to hold for P(VI) though fewer examples are available. Although differences are noted in $^1J_{PC_{ax}}$ and $^1J_{PC_{eq}}$ no obvious trend is apparent.[28–30] Gray and Cremer have described $^1J_{PC}$ couplings of opposite sign between an endocyclic and an exocyclic carbon in the same P(III) (*4*) molecule.[37]

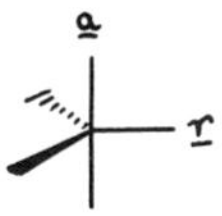

21

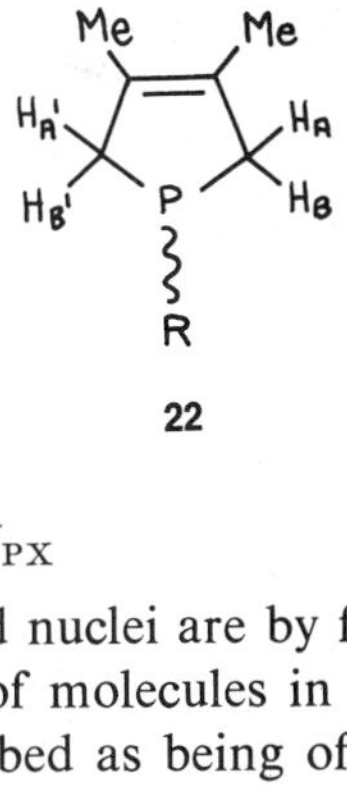

22

2. *Angular Dependence of $^2J_{PX}$ and $^3J_{PX}$*

Angular relationships between coupled nuclei are by far the most widespread method of assessing the shapes of molecules in solution. Most of these relationships can be loosely described as being of the type first advanced by Karplus for $^3J_{HH}$, and it has been a subject of intensive study ever since. Such relationships are too well known to be discussed here in any detail, and the reader is referred elsewhere[32] for the utility of, and pitfalls inherent in, such treatments. Coupling in organophosphorus heterocycles may be interpreted in similar ways and has the added advantage that the heteroatom is spin active. With the advent of routine ^{13}C nmr spectra much stereochemical information can now be obtained by combining ^1H and ^{13}C studies.

3. $^2J_{PX}$

Gagnaire and his collaborators first observed[33] that in heterocycles such as the 3-phospholenes (**22**), the CH_2 protons appeared as an AA′BB′X spectrum (X is P), and J_{AX} and J_{BX} differed markedly in both sign and magnitude. Further examination revealed a relationship between $^2J_{PCH}$ and dihedral angle of the Karplus type. The angle is defined as that formed by two planes defined by: (a) the directed lone pair of electrons on phosphorus, phosphorus, and carbon; and (b) phosphorus, carbon, and hydrogen. The curve shows two maxima at 0° (+25 Hz) and 180° (0 Hz) with a minimum at about 110° (−6 Hz). Similar relationships have been observed in five- six- and seven-membered rings. Not a large number of compounds have been examined, since when the ring is carbocyclic except for the phosphorus atom the ^1H spectra observed are often too complex for analysis. In such instances the proton decoupled ^{13}C spectra are likely to be more useful, and stereospecific $^2J_{PC}$ have been observed in P(III) (*4*),[34,36] P(IV) (*4*),[35] P(III) (*5*),[27,34] P(IV) (*5*),[31,34] P(III) (*6*),[28,29] and P(IV) (*6*).[29,31,34] Though for P(III) heterocycles the variation of $^2J_{PC}$ can be correlated with the orientation of the lone pair on phosphorus, more complex situations arise in P(IV) heterocycles, as the extensive data of Gray and Cremer on phosphetane derivatives demonstrate.[34,35] A similar angular relationship of $^2J_{POC}$ has been described for P(III) (*6*) compounds.[51]

In a study of a number of phosphoranes containing CF_3 groups, Cavell and co-workers[43a] observed a marked difference in $^2J_{PF}$ for the apical (57–72 Hz) and radial (108–157 Hz) positions. The absence of overlap should make this a useful, if limited, correlation.

4. $^3J_{PXCH}$

Much of our knowledge of the conformation of five- and six-membered phosphorus heterocycles in solution rests upon data derived from the angular dependence of couplings over three bonds. This is largely a consequence of the simple preparative methods available (Scheme 1).

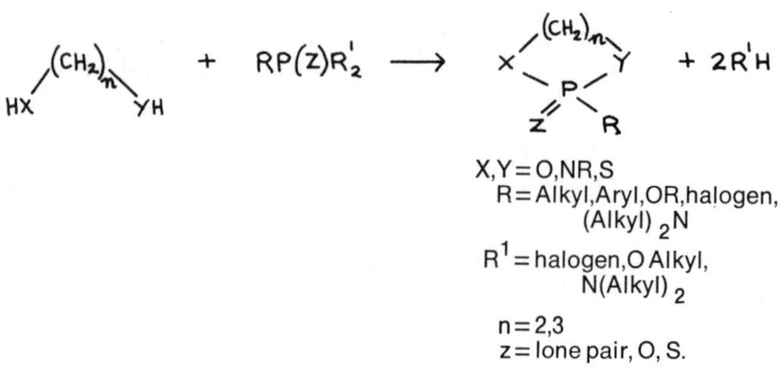

HX–$(CH_2)_n$–YH + $RP(Z)R'_2$ ⟶ cyclic X–$(CH_2)_n$–Y–P(=Z)R + $2R'H$

X,Y = O,NR,S
R = Alkyl,Aryl,OR,halogen, $(Alkyl)_2N$
R^1 = halogen,O Alkyl, $N(Alkyl)_2$
n = 2,3
z = lone pair, O, S.

Scheme 1

The most extensive examination of these relationships has been for six-membered rings (X = Y = O) both P(III)[38–42] and P(IV)[41,43–47] (Scheme 1). For P(III) both $J_{POCH_{ax}}$ and $J_{POCH_{eq}}$ are positive with the axial coupling substantially smaller (by about 8–10 Hz) than the equatorial, and a similar relationship is found for P(IV), although in this case $^3J_{PH_{eq}} - {^3J_{PH_{ax}}}$ is a little larger (about 10–20 Hz). Such results are expected for a Karplus-type relationship. For P(III) rings, as in the $^2J_{PX}$ case, the orientation of the lone pair on phosphorus exerts a considerable influence on the magnitude of the couplings which are otherwise rather insensitive to the nature of the exocyclic group at that atom, with the exception of amino and sulfur substituents.[38,39] In marked contrast $^3J_{PNCH}$ in P(III) (*6*) molecules is affected by substituents at phosphorus.[50]

Two groups have published complete analyses of the P(III) (*5*) compounds,[71,72] such as the substituted 1,3,2-dioxaphospholanes (**23**), and

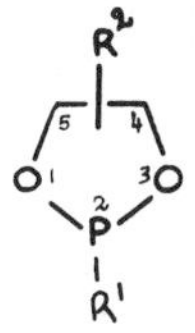

23

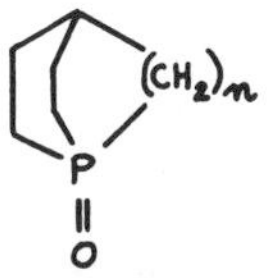

n=1,2

24

have observed two sets of different $^3J_{POCH}$, the larger value (about 9 Hz) being associated with the higher field protons while the lower field $^3J_{POCH}$ is much smaller (about 1 Hz). Many P (IV) (*5*) systems show similar behavior for $^3J_{POCH}$, and $^3J_{PNCH}$, and the original literature should be consulted for details (Table III).

Few studies of the angular dependence of $^3J_{PCCC}$ have appeared, but Wetzel and Kenyon in work[31] on the bridgehead P(IV) molecules, 1-phosphabicyclo [2.2.1] heptane 1-oxide (**24**, $n = 1$) and 1-phosphabicyclo-[2.2.2]octane 1-oxide (**24**), $n=2$), suggested the likelihood of a similar relationship. Here the coupling between phosphorus and the bridgehead carbon ($n=1$, $J=35$ Hz; $n=2$, $J=47$ Hz) is the largest reported. Significantly, the smaller value is found for the bicycloheptane where distortion from the theoretical dihedral angle (0°) is minimized by the one carbon bridge.

D. Shift Reagents

As might be expected from the known properties of phosphoryl compounds in metal extraction, molecules containing the P=O group complex selectively and strongly at this functionality with lanthanide shift reagents.[53,54]

Though this site specificity is valuable in interpretation of the shifted spectra, care should be used in conformational analysis since the equilibria may be markedly affected.[54] Good results have been obtained in determining the static stereochemistry of four-,[55] five,[52,53,56,61,63] six,[56–58,61] and seven-membered[59] rings containing a phosphoryl group. It is interesting that the technique is less successful with acyclic compounds as considerable signal broadening is observed. The major interest is in ^1H spectra, and although the ^{31}P nucleus will, of course, experience substantial shifts, these are normally considerable in the first place. It should be valuable in the ^{31}P spectra of phosphates whose chemical shift values are normally not very different.

Trivalent phosphorus complexes very poorly with the lanthanides, and the shift reagent will normally bind at oxygen or nitrogen where these atoms are available. Quin, however, found substantial benzene induced solvent shifts in his studies of P(III) phosphorinanes which greatly assisted isomer analysis.

E. Conclusion

Overall, angular relationships of the Karplus type are now established in the conformational analysis of organophosphorus heterocycles and may be expected to hold sufficiently well to allow disastereoisomer and conformer differentiation within a given class. It must be stressed, however, that they are not precise; a spread of results is always found even in molecules where the angles are fixed by ring constraints and other factors. Different curves are obtained for P(III), P(IV)=O, and P(IV)=S in the case of $^3J_{POCH}$,[41] and presumably the same will be true for other relationships. Sign changes as found in $^2J_{PCH}$[33] require that the magnitude of a coupling constant be interpreted with care, though a value of greater than 9 Hz is strongly indicative of a small dihedral angle in the system being considered. Comparisons should only be made with published data when the criteria of coordination number, substituent at phosphorus, and ring size are met. Exocyclic σ-bonded P-substituents normally do not greatly influence nmr parameters, but exceptions are noted when the substituent is N or S, particularly the former. Finally, with P(III) rings having an exocyclic halogen the possibility of an intermolecular exchange in solution resulting in inversion at phosphorus must always be considered.[61,73]

Tables III and IV give details of stereochemical studies of monocyclic P(*5*) and P(*6*) systems. Polycyclic systems are discussed under the appropriate headings.

IV. TABLES OF SPECTROSCOPIC AND STRUCTURAL DATA

A. Bond Angles

Table I. Bond Angles in Organophosphorus Heterocycles (in Degrees)

Co-ordination Number	Ring Size 3.	4.	5.	6.	Higher
P (II)			NP=C 89–92	CPC 102–103	
P (III)	47[a] 60[b]	PPP 85	CPC 97–98 NPN 89 OPN 94–95 OPO 96–99 PPP 97–101 SPS 100	OPO 103 PPP 95 CPC 102	
P (IV)		CPC 82–86	CPC 95–97 OPO 98 SPS 101	CPC 104–105 OPO 104–107	(8) NPN 108 (7) CPN 104
P (V)		CPC 76–78 OPC 71, 75	OPO 85–92 OPC 87 OPN 84–90	POC 93 OPO 106, 97–98[d]	
P(VI)			OPO 91		

[a] phosphirane [b] Pw [c] diradial [d] apical-radial.

B. Bond Lengths

Table II. Bond Lengths in Phosphorus Heterocycles (in Nanometers)

Other Atom	Coordination Number P (II)	P (III)	P (IV)	P (V)[a]	P (VI)
C	174–175[b]	178–187	183–188 172–175[b]	183–196	
N	166–170	168–170	164	167–170	
O	—	160–165	155–158 138–158	162–179	171–172
S	—	210–212	205–214 191–198[b]	215–221	

[a] No distinction is made between apical and radial bonds.

[b] Formal double bond.

C. Five-Memebered Rings

24a

Table III. Structural and Stereochemical Studies of Phospholane Derivatives[a, b, c]

X	Y	Z	R_1	R_2	R_3	R_4	Method	References
O	O		H	H	H	H	^1H, E	71, 72, 75, 85, 97 (R = Cl)
O	O		H	H	H	CH_3	^1H, ^{13}C, ^{31}P	96, 49
O	O		H	CH_3	H	CH_3	^1H	72, 95
O	O		CH_3	CH_3	H	H	^1H, ^{31}P	94, 95
O	O		CH_3	CH_3	CH_3	CH_3	^1H, E	73, 97
O	O		H	H	Ph	H	^1H, ^{31}P	94
O	O		H	Ph	H	Ph	X	86 (R = OCH_3)
O	NR′		H	H	H	H	^1H, E	76 (R = CH_3, Ph, C–C_6H_{11}) 100 (R = CH_3, Cl)
O	S		H	H	H	H	^1H	78, 79
NCH_3	NCH_3		H	H	H	H	^1H, E	80, 99 (X = Cl)
S	S		H	H	H	H	^1H, ^{31}P, X, E	81, 82, 96 [R = S (CH_2) S–] 98 (R = Cl)
S	S		H	H	H	CH_3	^1H	196
CH_2	CH_2			←$CH_3C{=}CCH_3$→			^1H	33

O	O			$1,2—C_6H_4$			E	101 (X = Cl)
O	O	O	H	H	H	H	X	89 (R = OCH_3)
O	O	O	H	Ph	H	Ph	^1H, X	74, 86 (R = OCH_3)
O	O	O	CH_3	CH_3	H	H	^1H, ^{31}P	94, 95
O	O	O	CH_3	CH_3	CH_3	CH_3	X	88 (R = OCH_3)
O	O	O	←$CH_3C{=}CCH_3$→				X	87 (R = OCH_3)
O	O	O		←o-C_6H_4→			X	90 (R = OH)
O	O	O	CH_3CO	CH_3	C=O		X	93 (R = OCH_3)
O	O	S	H	H	H	H	^1H	85
O	NR′	O	H	H	H	H	^1H	76 (R = CH_3, Ph, $C—C_6H_{11}$)
O	NR′	S	H	H	H	H	^1H	76 (R = CH_3, Ph, $C—C_6H_{11}$)
O	NCH_3	O	H	H	H	H	^1H	76, 77
O	NCH_3	O	Ph	H	CH_3	H	^1H, ^{31}P	83, 84
O	NCH_3	S	Ph	H	CH_3	H	^1H, ^{31}P	83
S	S	O	H	H	H	H	^1H, ^{31}P	82
S	S	S	H	H	H	H	^1H, ^{31}P, X	82, 91 (R = Cl), 92 (R = Ph)

[a] Usually nmr (nucleus given); X = X-ray; μ = dipole moment; ir = infrared; E = electron diffraction.

[b] Where a reference number is followed by identification of the R group this was the only compound studied. Most workers have studied a range of substituents, R, which most commonly is alkyl, aryl, alkoxy, aryloxy, halogen, or alkylamino. Thiolo substituents are much less commonly found.

[c] Diastereoisomers are not indicated; they are normally observed and data is recorded for each, but occasionally the method of preparation gives rise to only one isomer.

D. Six-Membered Rings

24b

Table IV. Structural and Stereochemical Studies of Phosphorinane Derivatives

X	Y	Z	R_1	R_2	R_3	R_4	R_5	R_6	Method	Reference
O	O		H	H	H	H	H	H	E	142 (R = Cl)
O	O		H	H	H	Bu^t	H	H	^1H, ^{13}C, ^{31}P	39, 48, 102
O	O		H	CH_3	H	H	H	H	^1H, ^{31}P	61, 103, 104, 42, 127, 304 (R = H)
O	O		H	CH_3	H	H	H	CH_3	^1H, ^{31}P	61, 103, 104
O	O		H	H	CH_2Cl	CH_3	H	H	^1H, ^{31}P	61, 118
O	O		H	H	CH_3	CH_3	H	H	^1H, ^{31}P	61, 103, 104, 43, 118
O	O		H	H	CH_3	$CH(CH_3)(OCH_3)$	H	H	^1H, ^{31}P	61
O	O		H	H	CH_3	H	H	H	^1H	103, 104, 106
O	O				←1,2-cyclohexyl→		H	H	^1H, ^{13}C	104
O	O	O	H	CH_3	H	H	H	CH_3	μ, ^1H, ^{31}P, ir	103, 105, 43–45 293
O	O	O	H	H	CH_3	H	H	H	^1H	103

O	O	O	H	H	CH_3	CH_3	H	H	^1H, ^{13}C, μ, ir, ^{31}P	103, 119, 120, 123, 125, 126, 294, 44
O	O	O	H	CH_3	H	H	H	H	^1H, ^{13}C,μ	103, 122, 123, 125, 126
O	O	O	H	CH_3	H	H	H	CH_3	X	106 (R = Ph_3C)
O	O	O	H	H	CH_3	CH_2Cl	H	H	X	107 (R = N⬭)
O	O	O	H	H	CH_3	CH_2Cl	H	H	X	108 (R = OPh)
O	O	O	H	H	H	H	H	H	X	109, 110 (R = OH), 111 (R = OPh), 112 (R = Br), 113 (R = CH_3), 114 (R = Ph)
O	O	O	H	H	CH_3	CH_2Br	H	H		115 (R = Br)
O	O	O	H	H	H	H	H	CH_3	X, ^1H, ^{31}P	116 (R = H), 125, 126
O	O	O	H	H	H	H	CH_3	CH_3	^1H	46
O	O	O	H	H	H	Bu^t	H	H	^1H	117 (R = Bu^t), 118, 58, 43
O	O	O	CH_3	H	CH_3	H	CH_3	H	^1H, ^{31}P, ir	294
O	O	O	H	Ph	H	H	H	H	^1H, μ, ir	74, 121
O	O	O	H	CH_3	H	H	CH_3	CH_3	^1H, ^{31}P, μ, ir	123, 124, 293
O	O	S	H	CH_3	H	H	H	H	^1H, ^{31}P	42, 125, 126, 128
O	O	S	H	H	CH_3	CH_3	H	H	^1H, ^{31}P, X	125, 126, 129 (R = CH_3), 47
O	O	S	H	H	H	H	H	H	^1H, ^{31}P	125, 126
O	O	S	H	H	H	Bu^t	H	H	^1H	43

(Continued)

Table IV (*continued*)

X	Y	Z	R_1	R_2	R_3	R_4	R_5	R_6	Method	Reference
O	O	BH_3	H	CH_3	H	H		CH_3	^1H, X	130, 224 (R = OMe)
O	NCH_3	O	H	H	CH_3	CH_3	H	H	^1H	132
O	NCH_3	S	H	H	CH_3	CH_3	H	H	^1H	132
O	NH	O	H	H	H	H	H	H	X	133, 134
O	NH	O	H	H	H	H	C=O		X	135 [R = $N(CH_2CH_2Cl)_2$]
NCH_3	NCH_3		H	H	H	CH_3	H	H	^1H	50
S	S		H	H	H	Bu^t	H	H	^1H	131
S	S		H	H	CH_3	CH_3	H	H	^1H	131
CH_2	CH_2		H	H	H	H	H	H	^1H, ^{13}C, ^{31}P	25, 29, 200 (R = CH_3)
CH_2	CH_2		H	H	OH	R^1	H	H	^1H, ^{13}C, ^{31}P	28, 137
CH_2	CH_2		H	H	C=O		H	H	^1H, X	33, 46, 138 (R = Ph)
CH_2	CH_2		H	H	OCH_3	OCH_3	H	H	X	139 (R = Ph)
CH_2	CH_2	S	H	H	H	OH	H	H	^{13}C, X	30 (R = CH_3)
CH_2	CH_2	S	H	H	CH_3	OH	H	H	^{13}C, X	30 (R = CH_3)
CH_2	CH_2	S	H	H	H	H	H	H	^{13}C	29
CH_2	CH_2	S	H	H	NCH_3		H	H	X	136 (R = Ph)
CH_2	CH_2	S	H	H	H	H	H	H	^1H	25 (R = H)
CH_2	CH_2	CH_3	H	H	H	H	H	H	^1H	25 (R = H)
CH_2	CH_2	O	H	H	C=O		H	H	^1H, ^{31}P	140
CH_2	CH_2	Ph	H	H	$PhPCH_2Ph$		H	H	X	141 (R = CH_2Ph)

V. P(II) COMPOUNDS

This class of compounds has increased rapidly in number since the initial[143] preparations by Dimroth and Hoffmann, but their stereochemical interest is rather limited. The P(II) (*6*) compounds and related derivatives have been reviewed recently.[144]

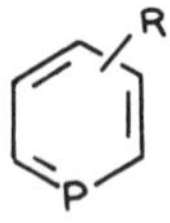

25 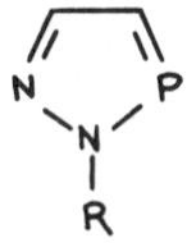 **26** 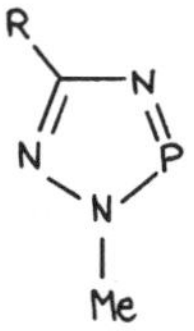 **27**

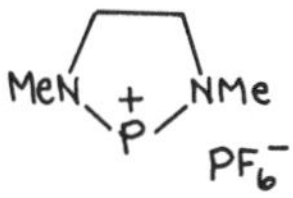

28 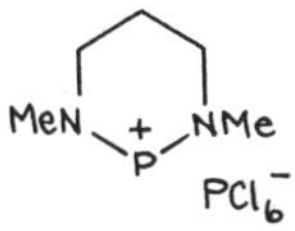 **29** 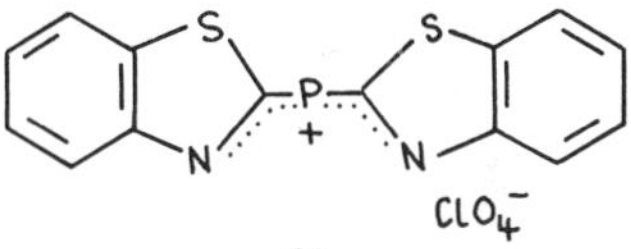 **30**

The compounds (**25–29**) exhibit ^{31}P chemical shifts (δ, ppm) at very low field (about +170 to +270) while **30** absorbs much higher (+50). This is apparently unrelated to steric effects since the CPC angles in **30** (105°)[145] and **25** (103°)[146] are very similar, whereas for **26**[147,148] the NPC angle is smaller but unexceptional for a five-membered ring. Structural data on **27**[149] are not yet available. Compounds **28**[150] and **29**[151] are formed only if the phosphorus atom is in a ring. It has been suggested[151] that only then is it possible for effective overlap to occur between the lone pairs on nitrogen and the vacant *p* orbital on phosphorus.

Arsenic analogues of **25**, **26**, and **30** are also known.

VI. P(III) COMPOUNDS

A. Introduction

All available evidence is in agreement with a pyramidal geometry for tricoordinated phosphorus. It was earlier thought that the pyramid would rapidly invert, by analogy with ammonia and its derivatives. The reasons for this thinking were largely based on the inability to resolve quaternary salts, and to account for this it was proposed that an equilibrium existed of the type:

$$R_4M^+X^- \rightleftharpoons R_3M + RX$$

In 1959, however, Kumli, McEwen, and Vander Werf[152] showed that the (−)-dibenzoyl hydrogen tartrate anion was generally useful in resolving phosphonium salts, and that once resolved they were configurationally stable. Subsequently, Horner[153] reduced resolved salts electrolytically, obtained resolved phosphines, and showed[154] that, far from being unstable, the activation energy for thermal racemization was in the region of 130 kJ mol^{-1}. In the interim Campbell had prepared what was until recently the only chiral resolved heterocyclic phosphine, 6-(*p*-dimethylaminophenyl)-5,6-dihydrodibenzo[*c.e*]azaphosphorine (**31**),[155] by reduction of the corresponding oxide with LAH. This reaction must be considered suspect since similar reduction of acyclic phospine oxides proceeds with racemization,[14] and indeed a subsequent attempt to reduce optically active 3-carboxy-5-phenyldibenzophosphole 5-oxide (**31a**) gave only a racemic product.[155a] Recent studies have shown that the P–N bond is liable to cleavage under the conditions used for reducing the oxide of **31**.[155b]

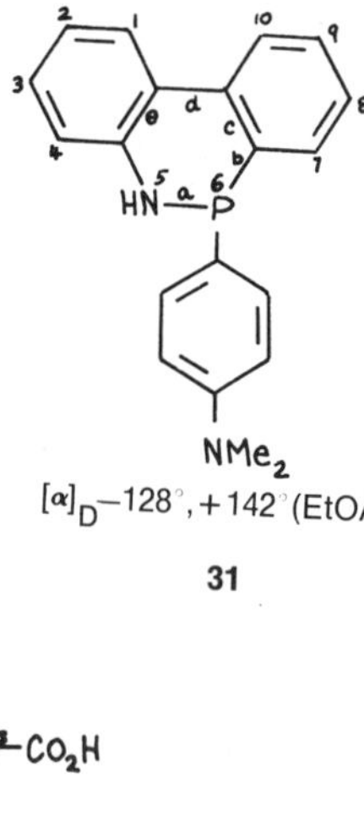

$[\alpha]_D -128°, +142°$ (EtOAc)

31

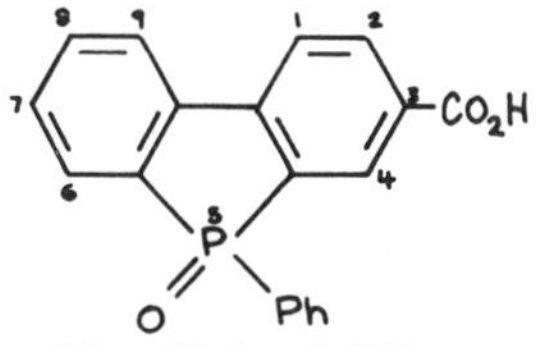

$[\alpha]_D \pm 126$ (aq.NaOH)

31a

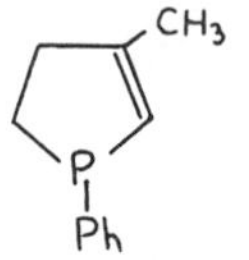

$[\alpha]_D^{25}$ −32.1° to +14.6°

31b

This example, **31**, is also ambiguous since the molecule is a bridged biphenyl and could exhibit chirality even if the P(III) group were planar. The same criticism is not applicable to the arsenic and antimony heterocycles **32** to **35**, also prepared and resolved by Campbell and her collaborators.[156–160] Very recently the chiral phosphole **31b** has been obtained by asymmetric reduction of the corresponding oxide with an amino-alane prepared using optically active α-phenylethylamine. The sign and magnitude of the rotation of **31b** vary markedly with reaction conditions.[443]

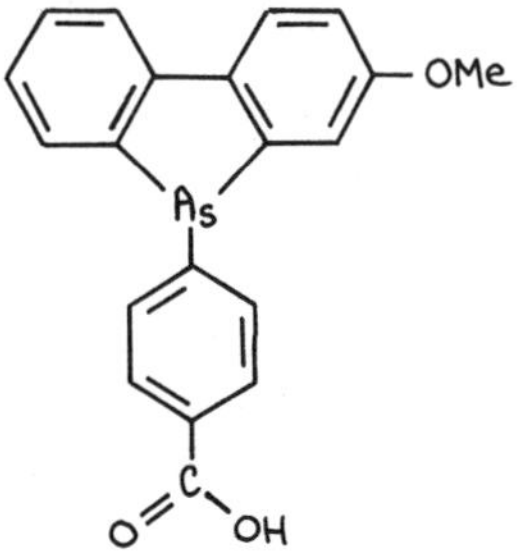

$[\alpha]_D \pm 160°$ (Pyridine)

32

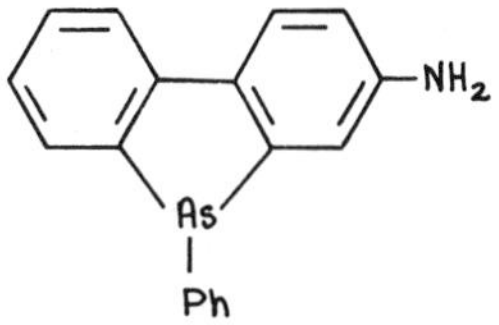

$[\alpha]_D \pm 255°$ (EtOH)

33

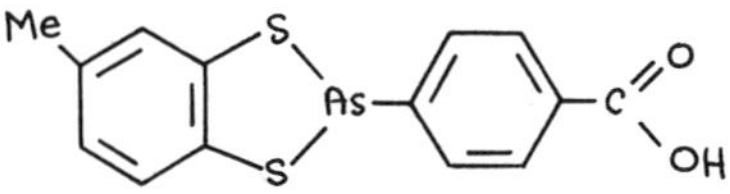

$[\alpha]_D -8.7°, +8.9°$ ($CHCl_3$)

34

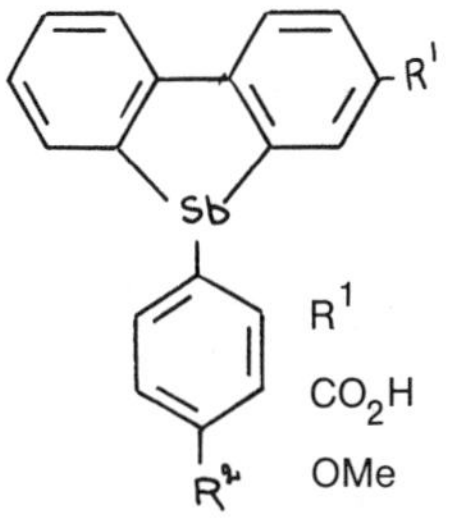

R^1	R^2	$[\alpha]_D$	Solvent
CO_2H	CH_3	+79°	$CHCl_3$
OMe	CO_2H	+153°	$CHCl_3$
OCH_2CO_2H	CH_3	±112°	Pyridine
NH_2	CH_3	±250°	PhH
CO_2H	CH_3	±250°	Pyridine

35

The first compounds containing resolved group IV B elements in the tricoordinate state were the heterocycles **36**.[161–163] Since these molecules are folded about the axis defined by the two heteroatoms, they could be chiral even if the atom M was not, provided that flipping about this axis has a high energy barrier. This does not appear to be the case,[164,165] and the resolvability of these systems is almost certainly due to the pyramidal stability of the triligated atom M.

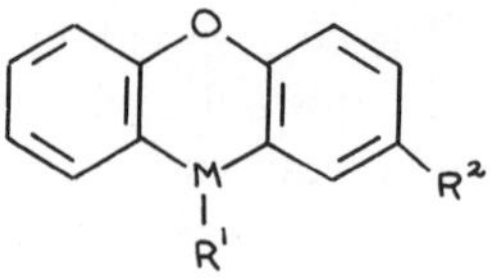

M=As; R^1=Me,Et,Ph; R^2=CO_2H
M=Sb; R^1=p-$HO_2CC_6H_4$;
R^2=CH_3

36

All the above compounds were resolved by standard methods. Despite their relative abundance and accessibility, no further work has appeared on chiral resolved heterocycles whose optical activity is due to P, As, or Sb in the tricoordinate state.

Pyramidal stability for phosphorus was predicted by Kincaid and Henriques[167] 35 years ago, and recently calculations on phosphine itself indicated a high barrier to inversion.[168,169] This was experimentally confirmed by the observation[170] that the PH protons of 1-phenylethylphospine **37** are anisochronous; separation of signals is still apparent at 150°C,

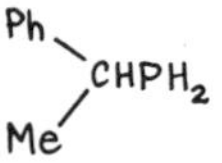

Ph(Me)$CHPH_2$

37

indicating an inversion barrier of not less than 110 kJ mol^{-1}. Mislow and co-workers[166] have recently calculated inversion barriers for a range of cyclic and acyclic arsines that agree well with experimental values and show that they range from about 80–220 kJ mol^{-1}.

Inversion of P(III) or As(III) cannot be assumed facile regardless of substituents. Indeed, if it is to be proposed at all, convincing supporting evidence must be presented.

Assignments of absolute configuration to P(III) rest upon the assumption that either quaternization, oxidation, or sulfuration proceeds with retention at phosphorus. There is no reason to suppose this is not the case, but an absolute proof is lacking. Recently, however, Newton and co-workers[86] have determined by X-ray analysis the crystal structures of *r*-2-methoxy-*trans*-4-*trans*-5-diphenyl-1,3,2-dioxaphospholane and its 2-oxide (**38**) the corresponding phosphate. The latter was prepared from the former by oxidation with ozone, and no evidence for the inverted isomer was obtained in the product of oxidation.

Mosbo and Verkade have arrived at similar conclusions for the N_2O_4 oxidation of six-membered cyclic phosphites, such as 2-methoxy-*cis*-4,6-

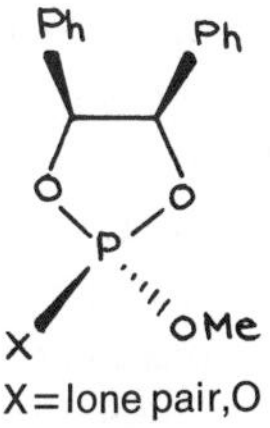

X = lone pair, O

38

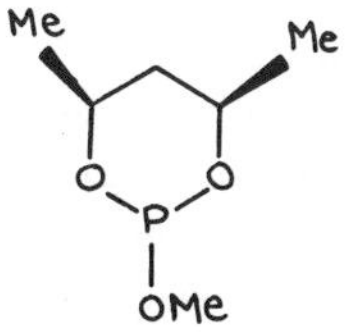

39

dimethyl-1,3,2-dioxaphosphorinane (**39**), on the basis of dipole moment and nmr measurements. The compound **39** was selected for study since the ring methyls should effectively lock the molecule into one chair conformation in which both are equatorial.[105]

B. P(III) Three-Membered Ring Compounds (Phosphiranes)

Only a handful of such compounds are known, the phosphiranes **40**, prepared by Goldwhite and colleagues,[23,171] and the more complex example **41**, due to Katz and coworkers.[173] The structures of phosphiranes (**40**,

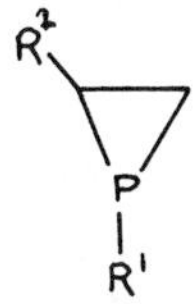

R^1=	H	H	H	Ph	Me
R^2=	H	Me	Et	H	H

40

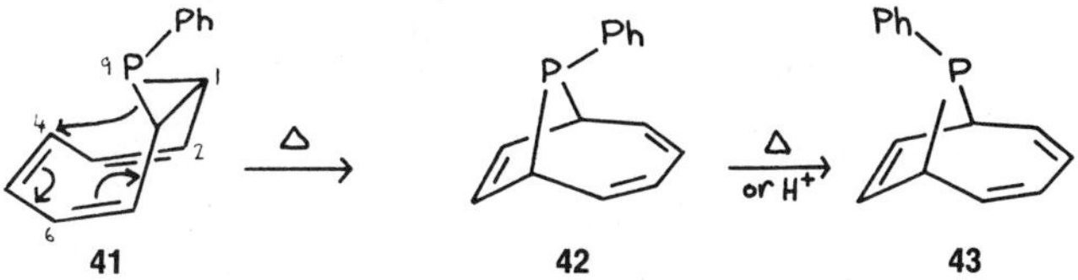

41 —Δ→ 42 —Δ or H^+→ 43 (3)

$R^1 = CH_3$) have been established by microwave spectra,[174,175] and mass, ir, and nmr studies have been described. Unfortunately, detailed ^1H nmr studies have not been reported. Stereoisomerism is evident for the phosphiranes **40** ($R^1 = H$, $R^2 = Me$, Et) and no interconversion is apparent. Despite the apparent generality of the preparation of **40** (essentially a dehydrohalogenation of a 2-haloethylphosphine generated in situ), no further stereochemical or chemical studies have been described. Of particular interest is the large difference (28 ppm) in the chemical shift of the two diastereoisomers of phosphiranes **40** ($R^1 = H$, $R^2 = Me$). As might be expected from the highly strained environment of the P nucleus ($C\hat{P}C = 47°$) the chemical shift is characteristically high (−341 ppm for the parent compound). Substitution at carbon in the ring, however, produces a profound downfield shift (50–100 ppm). In this context it is interesting to note that the chemical shift of P_4 is much higher (−451) though the endocyclic angle at P is larger (60°), a factor which Van Wazer and Letcher attribute to *d*-orbital involvement in P–P bonding.[20]

9-Phenyl-9-phosphabicyclo[6.1.0]nona-2,4,6-triene (**41**) was obtained, by Katz and co-workers,[173] by reaction between the dianion of cyclooctatetraene and phenylphosphenous dichloride. The stereochemistry shown is assigned largely on the basis that it is the most reasonable in view of the facile thermal rearrangement which it undergoes (Equation 3). The ^{31}P chemical shift of **41** is high (> −181) in agreement with the three-membered ring structure, but is again considerably downfield compared with less substituted derivatives. The observance of only one peak supports the contention that a single stereoisomer is formed in the preparation.

C. P(III) Four-Membered Ring Compounds (Phosphetanes)

Structures of this type are readily accessible by way of the corresponding P(IV) oxides (see section VII.C) from which they can be obtained by reduction with trichlorosilane–triethylamine[176] or hexachlorodisilane.[177] Both these processes proceed with retention of configuration (Equation 4).

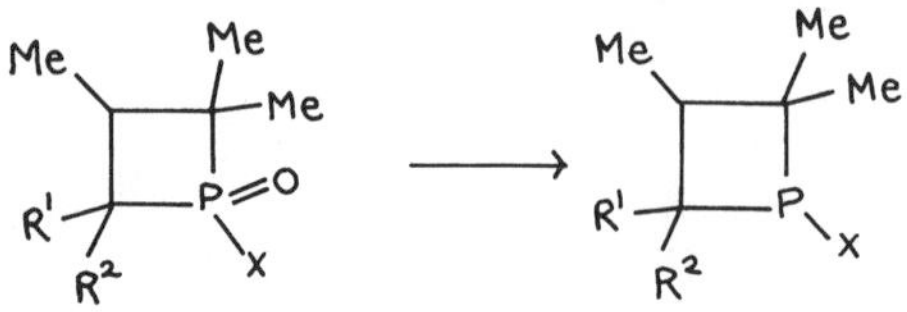

Me Me Me P=O X R¹ R² → Me Me Me P X R¹ R² (4)

$R^1 = R^2 = $ Me or H

44 **45**

Although the stereochemistry has been assigned unambiguously in the P(IV) compounds[178] by X-ray studies, its extension to the P(III) series rests on the assumption that either quaternization with alkyl halides or oxidation with hydrogen peroxide proceeds with retention of configuration.

Phosphetanes have been the subject of much active study, particularly from the stereochemical viewpoint, largely because of the constraint placed on the CPC angle by the four-membered ring. A rather lopsided view emerges, however, because of the very limited degree of variation possible in the carbocyclic portion of the ring. This is exemplified in the study of the thermal inversion at phosphorus. In view of the constraint imposed by the ring, one would expect[179] an increased energy of activation for the inversion process, and indeed 1,2,2,3,4,4-hexamethylphosphetane **45**, $R^1=R^2=X=Me$) inverts only very slowly at 162°C. However 1-*t*-butyl- (**45**, $R^1=Me$, $X=t$-Bu) and 1-phenyl-2,2,3,4,4-pentamethylphosphetanes (**45**, $R^1=R^2=Me$, $X=Ph$) have $\Delta H^{\ddagger}$ values of 118.1 and 124.7 kJ mol^{-1} respectively, which are little different from their acyclic analogues.[180] This is presumably a consequence of nonbonded interactions between substituents at P and on the ring, but other interpretations are possible for **45** (X=Ph).[180]

Most stereochemical studies of phosphetanes have been concerned with the P(IV) system. Trippett's group have shown, however, that nucleophilic substitution at P (e.g., of Cl by RNH_2) proceeds with inversion.[24] This may simply be due to a "normal" S_N2 process or may involve the formation of a R_4P^- system, a pseudophosphorane, in which one of the ligands is a lone pair. Such structures are discussed in more detail elsewhere (Section VIII.D.1).

Spectroscopic studies of P(III) phosphetanes are limited to nmr. Proton and ^{31}P spectra[35,180] allow a ready distinction between cis and trans isomers in the phosphetanes (**45**, $R^1=R^2=Me$, X=Cl, Ph, Me, *t*-Bu) since the chemical shift of the ring H in the trans isomer is 0.4–0.7 ppm downfield with respect to the same atom in the cis isomer, and with the ^{31}P nucleus the difference is much greater (20–30 ppm) with the cis isomer again being to higher field than the trans. In the ^{13}C spectra a strong dihedral dependence of $^2J_{PC}$ allows easy determination of the stereochemistry at phosphorus. When the α-methyl is cis to the lone pair $^2J_{PC}$ is large (about 27–37 Hz) and when trans it is much smaller (0–5 Hz).[35]

The chemical shift of the 2,4 ring carbons is 2–4 ppm to higher field in the cis isomer than in the trans isomer.[35]

These correlations will undoubtedly prove very useful when the development of synthetic methods allows greater variation in the ring than is presently possible.

D. P(III) Five-Membered Ring Compounds (Phospholanes)

This group, together with its six-membered analogues, comprises the most readily accessible and hence most studied classes of P(III) heterocycles. The P(III) (5) are invariably prepared by the general reaction in Scheme 2. The selection of Z is normally a matter of experimental convenience; when it is Cl or Br the reaction is carried out in the presence of a tertiary amine to remove the acid. When Z is not a halogen ZH is normally distilled off to

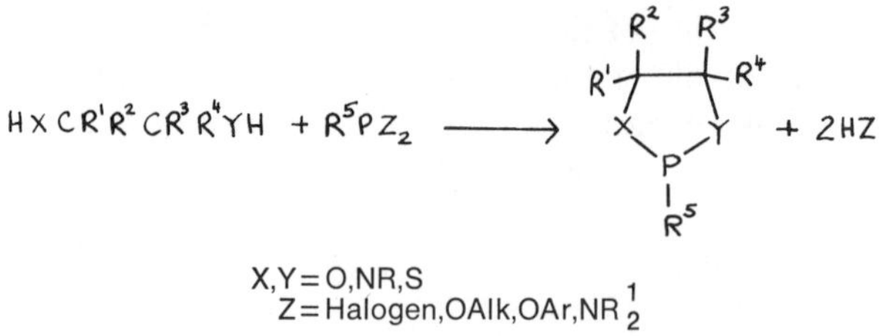

$$HXCR^1R^2CR^3R^4YH + R^5PZ_2 \longrightarrow \text{(ring)} + 2HZ$$

X,Y = O,NR,S
Z = Halogen,OAlk,OAr,NR^1_2

Scheme 2

drive the reaction to completion. The latter method has been used for the synthesis of the interesting bicyclic compounds 2,6,7-trioxa-1-phosphabicyclo[2.2.1]heptane (**46**)[181] and 2,7,8-trioxa-1-phosphabicyclo[3.2.1]octane (**47**).[182]

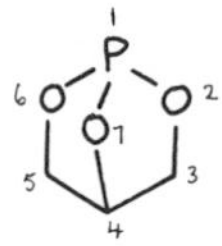

46 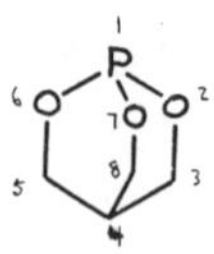 **47**

Recently, other general methods leading to the preparation of heterocycles containing two rather than three heteroatoms have been developed by Issleib[183–185] (phosphorus) and Tzschach[186–187] (arsenic), but stereochemical data are not yet available (Scheme 3). Carbocyclic rings with a

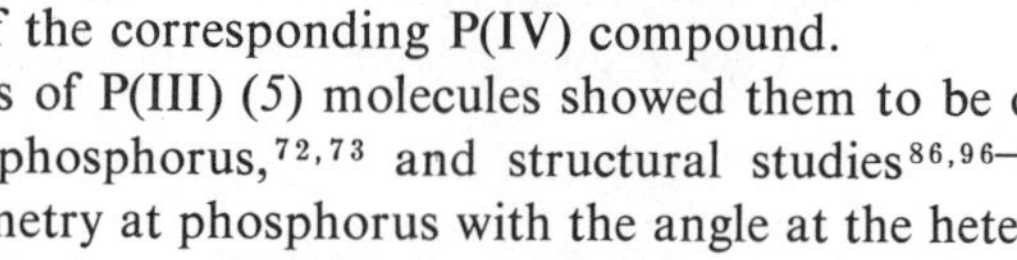

$$R^1PHCH_2CH_2XH + R^2R^3C{=}O \xrightarrow{H^+} \text{(ring)}$$

Scheme 3

phosphorus heteroatom, other than phospholes, are invariably obtained by reduction of the corresponding P(IV) compound.

Initial studies of P(III) (*5*) molecules showed them to be configurationally stable at phosphorus,[72,73] and structural studies[86,96–101] confirmed pyramidal geometry at phosphorus with the angle at the heteroatom in the ring 90–100°. Barriers to inversion are similar to those in acyclic systems, though when the ring is a phosphole (e.g., **48**) the barrier is substantially

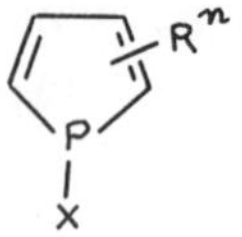

48

reduced. Mislow[188] attributed this fact to reduction in energy of the planar transition state by $(2p-3p)$ π electron delocalization. It was recently found, using photoelectron spectroscopy, that no such delocalization occurs in the ground state.[189] Similar results are obtained [191] with the arsenic analogues.

The energy barrier to inversion is too high to be observable at room temperature in 1,3,2-dioxaphospholanes[71–73] and arsolanes,[194] but inversion occurs readily when the exocyclic substituent in these compounds is chlorine, apparently by way of an intermolecular reaction.[73] This behavior cannot be generally assumed, however, because in at least one instance[196] (a 1,3,2-dithiaphospholane) it is not observed. Preparations by way of the general method normally afford mixtures of stereoisomers when the ring carbons are substituted,[190] and these are thought to be the thermodynamic equilibrium ratios, since different methods of preparation yield the same mixtures. Nothing is known about how the equilibria are established. Katz and co-workers invoked[173] the intermediacy of a configurationally mobile phosphorane (see Section VIII) to explain the epimerization of **42** by HCl. The formation of a phosphorane in this way is unexpected, but it could be favored here by the strain present in the ring system. Less understandable is the racemization of the Sb(III) 5-(*p*-tolyl)-3-carboxydibenzostilbole (**49**) by aqueous HCl, but not generally by acids. Whatever the mechanism, it presumably accounts for the second order asymmetric transformation which accompanies resolution.[158]

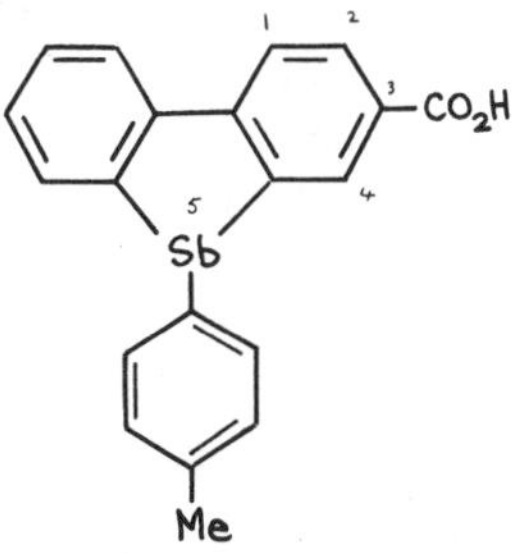

49

Denney,[190] and more recently Bentrude,[49] have examined a total of eleven stereoisomeric pairs of 1,3,2-dioxaphospholanes. Increasing substitution in the ring alters the relative proportions to favor the (presumably) more stable isomer to which both workers assign the trans structure.[190,49] The chemical shift of the phosphorus atom in the less stable isomer of **50** (X and R cis) is always to lower field (3–15 ppm) than the more abundant form. Both ^{1}H[190] and ^{13}C[49] spectra show marked differences, and assignment of geometry for these compounds is simple and straightforward. A useful general guide in the ^{1}H spectra is that ring substituents cis to the substituent at phosphorus, **50**, or arsenic, **51**, are deshielded in comparison with the trans isomer.

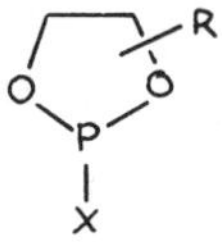

R=Me,t-Bu
X=Alk,Ar,OAlk,NR^{1}_{2}

50

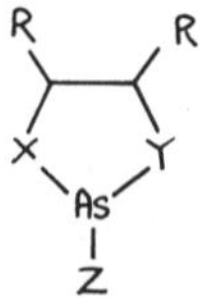

R=H or Me
X,Y=O,S
Z=Ph,Cl,OMe,SMe

51

Optically active compounds of both phosphorus and arsenic have been prepared by using resolved amino alcohols (ephedrine or ψ-ephedrine) in the preparation.[192] In some instances only a single isomer is obtained. It is worth noting that although a large number of P(III) (*5*) compounds [and a few As(III) (*5*) compounds] have been prepared and examined, mixtures of stereoisomers are almost invariably obtained, and little attention has been paid to their separation in pure form. This is another example of the imbalance between spectroscopic and chemical studies.

Proton spectra of both P(III) (*5*) (see Table III) and As(III) (*5*) compounds[193–195] have been examined in considerable detail. They are normally second order, but full analysis leads to the assignment of a twist-envelope conformation (**52a**) as the one most likely in solution, which is in substantial agreement with electron diffraction and X-ray studies. More

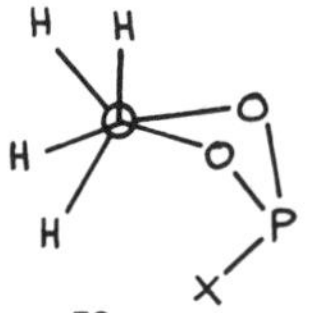

52a 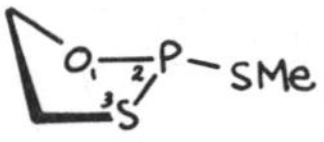 **52b**

complex still are the spectra obtained when the exocyclic substituent on phosphorus is fluorine, but the 1,3-dioxa (aza, thia) compounds have been analyzed completely.[197] Pseudorotation is probably normal in these systems though preferred conformers, for example, with the phosphorus in the "flap"[71] have been suggested, and in at least one case (2-methylthio-1,3,2-oxathiaphospholane, **52b** the conformation is reportedly largely fixed as shown, though the $POCH_3$ analogue is rapidly equilibrating.[78]

Compounds of the P(III) (*5*) type with phosphorus as the only heteroatom (phospholanes and phospholenes) have been studied much less, e.g., **53–56**. Their proton spectra are normally too complex to be useful, and although the spectra of the monounsaturated rings (phospholenes) are simpler, only a few studies have been described. Angular dependancy of the coupling $^2J_{PCH}$ is well established,[33] and the $^2J_{PCC}$ dependancy observed for phosphetanes[35,36] is also to be found for the phospholenes.[27]

Quin has examined the case of the 1-methylphospholane-3-ols (**53**) and has assigned the stereochemistry on the basis of the shielding of the ^1H resonance of the methyl group when the hydroxyl is cis to it.[199] Marsi has obtained the two diastereoisomers of 1,3-dimethylphospholane (**54**) by

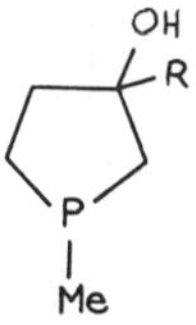

OH
R
P
Me
R=H,Me
53

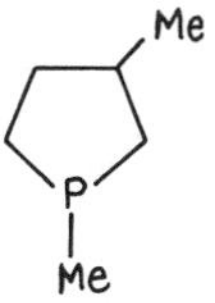

Me
P
Me
54

reduction of the corresponding oxide;[198] one isomer was obtained by fractionating the mixture obtained by reduction of the corresponding oxide mixture with $Et_3N–HSiCl_3$, and the other was obtained directly by stereospecific reduction of an isomerically pure oxide using phenylsilane. The latter reaction, which proceeds with retention of configuration,[198] seems the most promising method to date for the preparation of isomerically pure P(III) heterocycles. Trichlorosilane–triethylamine may cause racemization[422] (e.g., **129**). Unexpectedly, the two isomers of **54** have the same $\delta^{31}P$ (−33.8 ppm). Quin has described $\delta^{31}P$ values for a number of diastereoisomeric pairs and observed considerable differences.[26] ^{31}P chemical shift data for phospholanes as a whole are consistent and allow general predictions, but not for individual diastereoisomers.

Most preparations of P(III) (*5*) compounds afford mixtures of stereoisomers whose separation is laborious and rarely undertaken. Occasionally, reactions will yield largely a single stereoisomer whose geometry has been unequivocally established, for example, *r*-2-methoxy-*trans*-4-*trans*-5-diphenyl-1,3,2-dioxaphospholane and its 2-oxide **57**[86] and *r*-1-*trans*-2-*cis*-3-triphenyl-2,3-dihydro-1,2,3-triphosphaindane (**58**),[172] but there is no doubt that development of means for the separation of pure stereoisomers would greatly assist studies in this area.

1. *Reaction Stereochemistry of P(III) Five-Membered Ring Compounds*

The longer bond lengths to phosphorus compared with bonds between first row elements introduce an element of strain into cyclic systems con-

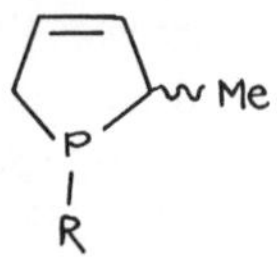

R=Ph,Me

55

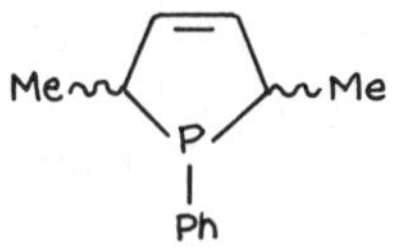

56

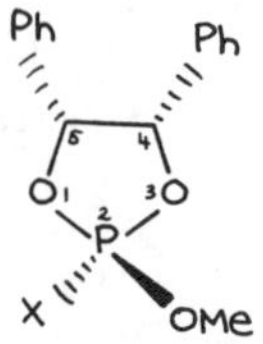

X=lone pair,O

57

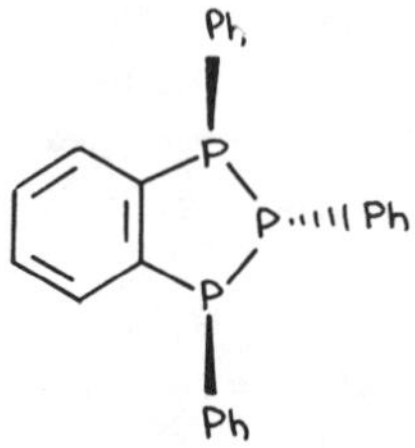

58

taining phosphorus. For example, the P(*5*) structure can no longer approximate a regular pentagon. With P(III) (*5*) systems this strain is fairly minimal since the calculated angle at phosphorus does not differ greatly from the normal angle at P(III) (90–100°; see Table II). For P(IV) (*5*) structures where the expected angle at phosphorus will be close to the normal tetrahedral angle (about 109°) considerable strain will be present, and this feature of these molecules has a profound effect on their chemistry.[7] For P(III) (*5*) compounds Hudson has extended this thinking, so successful for P(IV) (*5*) compounds, to account for the considerable variations in reactivity observed between cyclic and acyclic P(III) compounds.[201] Essentially, the hypothesis is that a reaction involving a P(III) center as a nucleophile, that is, leading to the formation of a P(IV) transition state or intermediate:

$$\gtrdot P: + X \rightarrow \gtrdot P^{+}{-}X^{-}$$

will have an additional energy barrier to overcome if phosphorus is in a five-membered ring (or smaller ring of course, but kinetic data are not available). This is borne out by experimental observation[202,203] on the Arbuzov reaction of cyclic versus acyclic (or rings larger than five) P(III) compounds. Conversely, if P(III) functions as an electrophile and accepts an electron to give what is formally a P(V) species, the energy barrier will be lowered by the favorable geometry at phosphorus in the starting material when it is constrained in a ring (four- or five-membered). The evidence for the latter is less compelling, partly because of uncertainty concerning the geometry of the P(V) intermediate (or transition state). There is a paucity of kinetic data in this area, but what seems certain is that many quite characteristic reactions of P(III) compounds (e.g., deoxygenation) are better understood in terms of phosphorus acting as an electrophile rather than as a nucleophile.

It can be anticipated that when sufficient kinetic and thermodynamic data become available, the Hudson hypothesis will provide a valuable and practicable probe of the nature of the transition state or intermediate for many reactions for which this knowledge is now purely speculative.

E. P(III) Six-Membered Ring Compounds (Phosphorinanes)

The vast amount of information available from carbocyclic, carbohydrate, and related systems makes the stereochemistry of this class of P(III) compounds more extensively studied than any other. A wide range of heterocycles is readily available by the same general method that is widely used for P(III) (*5*) compounds, namely, reaction of a disubstituted alkane derivative with a bifunctional P(III) species (Scheme 4).

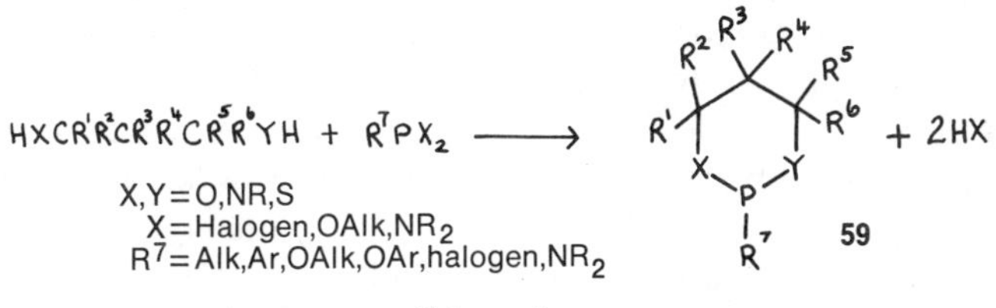

$$HXCR^1R^2CR^3R^4CR^5R^6YH + R^7PX_2 \longrightarrow \mathbf{59} + 2HX$$

X,Y = O,NR,S
X = Halogen,OAlk,NR_2
R^7 = Alk,Ar,OAlk,OAr,halogen,NR_2

59

Scheme 4

Rings in which all the atoms, other than phosphorus, are carbon atoms must be more laboriously assembled, and our knowledge of the stereochemistry of such systems is in large part due to Quin and his co-workers. However, it was Davis and Mann who first demonstrated diastereoisomerism in P(III) compounds by separating the 5,10-dihydrodiphosphanthrene (**60** R = Et) into two forms melting points 52–53° and 96–97° but it was not possible then to assign geometry. Only very small amounts of the higher melting form were obtained, and a tentative assignment of a cis geometry was made to the most abundant form on chemical grounds.[204] Interestingly, neither the diphenyl analogue 5,10-diphenyl-5,10-dihydrophosphanthrene (**60**, R = Ph)[204] nor the arsenic compounds (e.g., 5,10-dimethyl-5,10-dihydroarsanthrene, **61**)[205] showed any evidence of diastereoisomerism. The X-ray crystal structure of **61** has been reported[206] and is cis as shown, with the molecule folded about the As-As axis and the methyl group within the dihedral formed by the two rings.

There have been a large number of conformational studies on P(III) (*6*) compounds, particularly on the general type, **59** (X, Y are O, S, NR); the

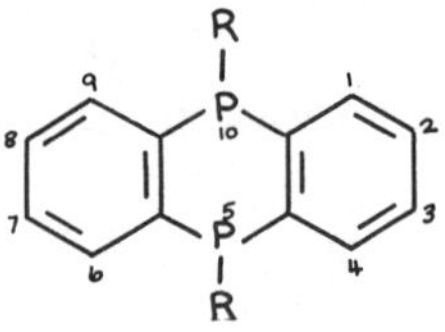

60

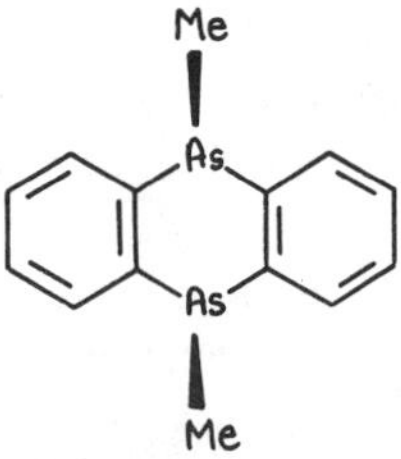

61

investigations are summarized in Table IV. Assignments have largely stemmed from analogies with other systems that have been exhaustively studied. There is general agreement that the ring is in the chair conformation, but somewhat flattened at the phosphorus end as a consequence of the greater length of the bonds to phosphorus. This is fully borne out by X-ray and electron diffraction studies. Of the two possible chair conformers, one normally predominates and this is most commonly the form with the substituent at phosphorus in an axial position, a preference that persists in the solid state. A similar preference has been observed by Aksnes and Vikane[207] for the chloroarsine, 2-chloro-1,3,2-dioxarsorinane, **62**. No really satisfactory explanation has been advanced for this phenomenon, and exceptions are observed when the substituent on phosphorus is alkylamino.[48,105] The influence of nitrogen is very probably a stereoelectronic effect associated with partial π character of the P(III)–N bond. This is a common phenomenon seen, for example, in the restricted rotation about the P–N bond in $(CH_3)_2NPX_2$ derivatives.[208] It may, however, be unnecessary to assume the operation of some special effect to account for the axial preference of substituents other than nitrogen. Wertz and Allinger[209] have recently advanced the hypothesis that gauche H-H interactions are of fundamental importance in determining the energy relationships of cyclohexane derivatives and that what is normally considered an equatorial preference by a group is in fact an axial preference by hydrogen. In the case of the great bulk of conformational studies on P(III) (6) compounds; the triad composed of the phosphorus and the two adjacent ring atoms have no hydrogen. Hence the conformational preference may be determined by gauche interactions between the P–R bond and the adjacent lone pairs leading to a thermodynamic preference of the substituent for the axial position since this minimizes such interactions. Axial orientation in these systems will be less affected by unfavorable 1,3-repulsions because of the ring flattening arising from the bond length and bond angle effects due to the presence of phosphorus (or another second or higher row atom). Significantly, the preference is less marked when phosphorus is the only heteroatom,[210] though in the case of phosphorinane itself (**17**), the hydrogen is predominantly axial.[25] Similar reasoning also applies to other hetero-

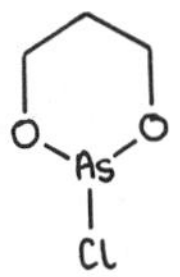

62

cycles, and similar axial preferences have been observed for sulfur[211] and selenium[212] analogues (**63**). Mosbo and Verkade[105] have described an alternative gauche effect explanation, and lone pair-lone pair repulsions lead to the same conclusion for 1,3,2-dioxaphosphorinanes.[105a].

It is interesting to note that the X-ray crystal structure of $(PhP)_6$ shows it to be a ring of six phosphorus atoms in the chair conformation with all the phenyl groups equatorial.[214] In this molecule there can be no relief of 1,3-diaxial interactions such as occurs in molecules where the longer bonds to a heteroatom result in a flattening of the molecule. It is possible that only when some mechanism for relief of 1,3-interactions is available can other influences become dominant.

The axial preference of the exocyclic phosphorus substituent would be expected to lead to the cis isomer being the more stable when the substituents at phosphorus and carbon are in a 1,4 relationship, and this, in fact, has been observed.[48] The ^{31}P spectra of isomer pairs of P(III) (*6*) phosphorinanes have been reported by Bentrude,[48] and show that the cis, isomer resonates at higher field than the trans isomer, a correlation supported by Quin and his collaborators[213] who have determined the X-ray crystal structures of the two isomers of 4-*t*-butyl-1-methyl-4-hydroxyphosphorinane (**64**). If this generalization holds true, determination of geometry in such systems should be a relatively trivial matter, at least in those cases where both isomers are available. If this fails use of the angular dependence of $^3J_{PH}$ should provide a fairly unambiguous answer. Differences have also been noted in the $^2J_{PCH}$ for the exocyclic substituent, but these are small, for example, 2.1 and 4.0 Hz for the equatorial and axial P–Me, respectively, of **64.**

The simple distinction possible between trans and cis isomers also corresponds to the exocyclic substituent at phosphorus being either equatorial or axial.[48,213] Furthermore, Quin[200] used this fact in the conformational analysis of **64**, provided further data to support the correlation.

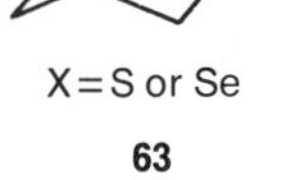

X=S or Se

63

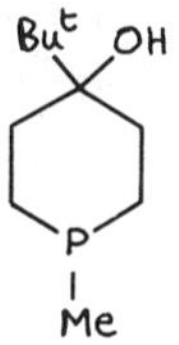

64

This would seem to be the method of choice in the conformational analysis of P(III) (*6*) phosphorinanes, particularly when there are no other heteroatoms present to simplify the proton spectra. Thus, the P-signals of the two conformers of 1-methylphosphorinane (**65**) are separated by 3.1 ppm at $-80°C$, and simple integration gave an axial to equatorial ratio of 2:1. Interestingly, the axial conformer predominantes at room temperature.[200] Variable temperature studies of conformational equilibria are rare, and this seems to be the most detailed example described. Gagnaire and co-workers[38] considered that **66** (R = Me, PhO, Cl, F) were conformationally rigid over the temperature range -40 to $+155°C$ since no variations in nmr spectra were observed. In contrast, White[219] found evidence for considerable changes in equilibria for 2-methyl- (**66**, R = Me) and 2-phenyl-5,5-dimethyl-1,2,3-dioxaphosphorinanes (**66**, R = Ph) between about 30 and 160°C. White has suggested that this unexpected difference may be explained if one conformer is dominant throughout and that equilibrium is very rapidly established. Certainly, the Featherman and Quin figure for $\Delta G^{\ddagger}$ for ring inversion (38.4 kJ mol^{-1} at $-75°C$) is close to that observed for cyclohexane itself.[200] Presumably, Gagnaire and others saw no conformational freezing because they did not go to a low enough temperature.

In an interesting study, Quin and Somers[210] reduced 1-methylphosphorinan-4-one (**67**) with no fewer than seven different systems and always

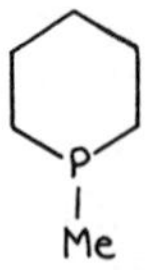

Me Me

P Me R O O P Me

65 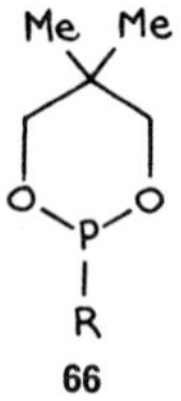 66 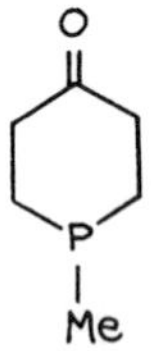 67

obtained the same (55 cis:45 trans) mixture of isomeric alcohols, a strong indication of the lack of configurational preference of the P–Me group. This is true even when the reducing agent [e.g., LiH $(OBu^t)_3$] is known to be highly selective in carbocyclic systems, though the latter's lack of specificity is probably due to the considerable flattening effect of both P(III) and C = O on the ring as a whole. Such an effect would, of course, minimize steric differentiation by an approaching reagent.

Following the initial separation of P(III) (*6*) disastereoisomers, it has become commonplace to observe them, though only occasionally can they be conveniently separated. In some cases this is due to the ease with which equilibrium is reestablished. Hydrogen chloride[39,42,215] or alcohols[103] rapidly equilibrate samples of dioxaphosphorinanes (**68**) which have been

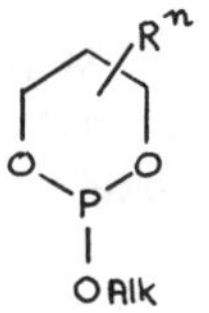

68 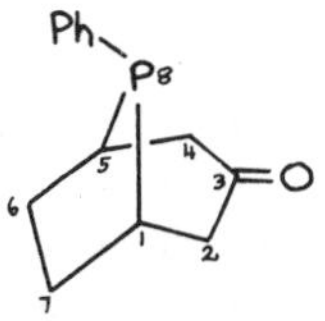 69 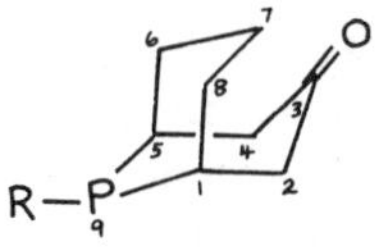 70

enriched in one isomer by distillation[103] or alternative methods of preparation.[39,40,103,215]

With this high chemical reactivity, it is not surprising that most methods of preparation afford very similar mixtures. This does not apply to the carbocyclic compounds containing phosphorus heteroatoms only, which Quin and co-workers succeeded in separating in a number of instances.[137,213] Compounds such as 2-alkoxy-1,3,2-dioxaphosphorinanes (**68**) are normally analyzed by oxidation or sulfurization to the corresponding P(IV) (*6*) compounds. By using an appropriate reagent[103,190,216] these reactions give good yields, are stereospecific, and proceed with retention of configuration.[107] The phosphoryl and thiono compounds give much more reproducible results by normal analytical techniques.

Recently Kashman and his colleagues obtained the stereochemically pure P(III) (*6*) compounds 8-phenyl-8-phosphabicyclo-[3.2.1]octan-3-one (**69**) and 9-phenyl-9-phosphabicyclo[3.3.1]nonan-3-one (**70**) by addition of primary phosphines to conjugated cyclohepta-[217] and cyclooctadienones.[218] Stereochemistry was assigned largely on the basis of ^1H nmr analysis of the corresponding oxides and phosphonium salts.

1. *Reaction Stereochemistry of P(III) Six-Membered Ring Compounds*

Relatively few studies of the chemistry of these systems have appeared, undoubtedly because of their high reactivity towards either oxygen or water and the difficulty of obtaining stereochemically homogeneous samples without a great deal of labor. Added to this, in the case of cyclic phosphites, is the uncertainty due to the unpredictable influence of traces of acids, bases, or alchols on stereochemical integrity. Thus Bodkin and Simpson[220] observed a marked lack of stereospecificity in the Arbuzov reaction of purified isomers of the monocyclic phosphite **71**, whereas the cage phos-

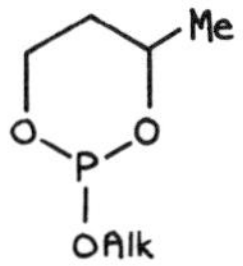

71 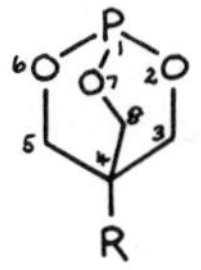 72 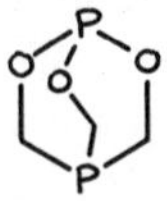 73

phites 4-substituted 2,6,7-trioxa-1-phosphabicyclo[2.2.2]octane (**72**) and 2,6,7-trioxa-1,4-diphosphabicyclo[2.2.2]octane (**73**) opened stereospecifically. Subsequently, Denney and co-workers[221] found that when the reaction of (**71**) was carried out in the presence of acid and base scavengers (Et_2NSiMe_3 and charcoal) stereospecificity was restored. Bodkin and Simpson[220] have described the preparation of both isomers of 2-alkoxy-4-methyl-1,3,2-dioxaphosphorinanes (**71**, Alk=Me, Et, Pr^i) using the method discovered by Aksnes, and others[215] (Scheme 5). Aksnes also noted that changing the reagent in what seems a relatively minor way results in the direct formation of the more stable isomer from the chloridite **74** (Scheme 5). Similar results were obtained by Bentrude and Hargis[39] for the 2-alkoxy-5-*t*-butyl-1,3,2-dioxaphosphorinanes (77), and it would seem that

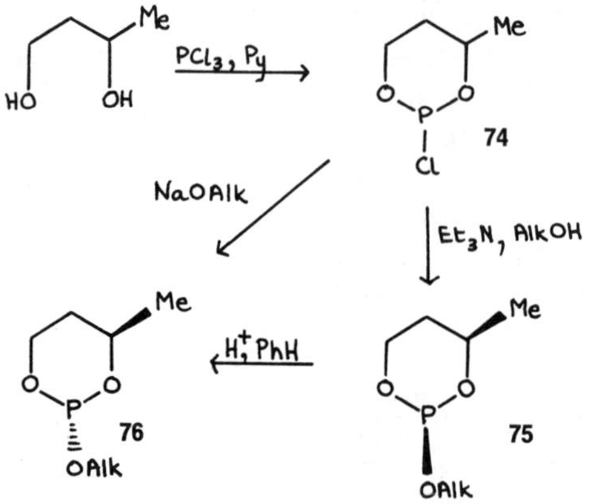

Me
HO OH
PCl_3, Py
Me
O O
P
74
Cl
NaOAlk
Et_3N, AlkOH
Me
H^+, PhH
Me
O O
P
76
OAlk
O O
P
75
OAlk

Scheme 5

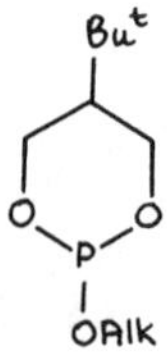

Bu^t
O O
P
OAlk
77

both isomers should be fairly readily accessible for 1,3,2-dioxaphosphorinanes in general, but the field is fraught with experimental difficulties of an unpredictable kind. Thus, where *cis*-2-alkoxy-4-methyl-1,3,2-dioxaphosphorinane (75) is stable at room temperature, 2-alkoxy-5-*t*-butyl-1,3,2-dioxaphosphorinane (77) is reported to isomerize slowly. Traces of acids or alcohols may bring about very rapid (even exothermic[42]) isomerization, and until some coherent picture emerges about the chemical reactions of these substances, no assumptions of stereochemistry should be made on the basis of preparative methods alone.

Oxidation and sulfurization of P(III) (*6*) compounds proceed stereospecifically and very probably with retention of configuration at phosphorus.[13,14] Widely used oxidizing agents include O_3, N_2O_4, and *t*-BuOOH. It cannot be assumed that oxidation in general will follow the same stereochemical path (e.g., HgO,[105]) and Finley and Denney[222] have even observed an example (neopentylhypochlorite on **75** and **76**) where the oxidation is stereospecific for one isomer and nonstereospecfic for the other. On the other hand, the Arbuzov reaction appears to be generally stereospecific even with such halides as PhSCl.[223] This stereospecificity has been used to assign the geometry at phosphorus in the all-*cis*-2-alkoxy-4,6-dimethyl-1,3,2-dioxaphosphorinane **78** indirectly by way of X-ray structure determination of the product phosphate obtained by reaction with Ph_3CCl (Scheme 6).[106] The P epimer **79**, was identified by conversion to its borane

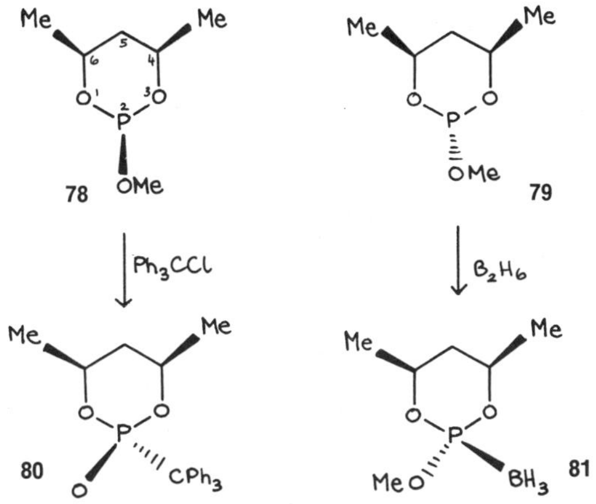

Scheme 6

adduct for crystal structure determination.[224] These two cyclic phosphites constitute the only pair of such compounds whose stereochemistry rests on a solid foundation, if one accepts that the 1,3-dimethyl groups lock the two isomers in that conformer with these two groups equatorial. This does not automatically follow, although it is supported by the observation that the less stable isomer **75** is conformationally mobile.[42] However, very rapid equilibration can produce the same result as fixed conformations in nmr analyses, on which most deductions are based, and to this extent conclusions from such data will remain uncertain.

The flattening of the ring in P(III) (*6*) due to the bond lengths to phosphorus would be expected to reduce the effect of steric factors on reaction rates. An example of this has been described in the quaternization reactions of the stereo isomers of 4-*t*-butyl-4-hydroxy-1-methylphosphorinane (**64**) which proceed at essentially the same rate.[210]

2. *Bicyclic Structures with a Bridgehead Heteroatom*

Molecules of this type are of general stereochemical interest since their geometry is fixed, facilitating spectroscopic examination, while their reactions can provide valuable information concerning the nature of transition states and so on. Two main classes have been examined incorporating Group V heteroatoms: bicyclic phosphites and aminophosphines (**82**), and heterosubstituted triptycenes (**83**).

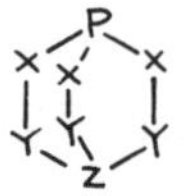

X=O,R,CH,$_2$(R=Alkyl)
Y=CH$_2$,NR, (R=Alkyl)
Z=As,P,CR, (R=Alkyl or Aryl)

82

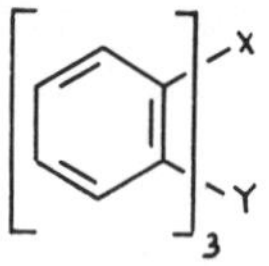

X=P,As
Y=CH,N,P,As

83

The bicyclic phosphites **82** (X=O, Z=P or CR) have been examined principally by Verkade and his collaborators.[225,226] Their stereospecific ring opening was used by Wadsworth to prepare P(IV) (*6*) compounds of known geometry,[227,228] for example, (**84**)→**86** in Scheme 7.

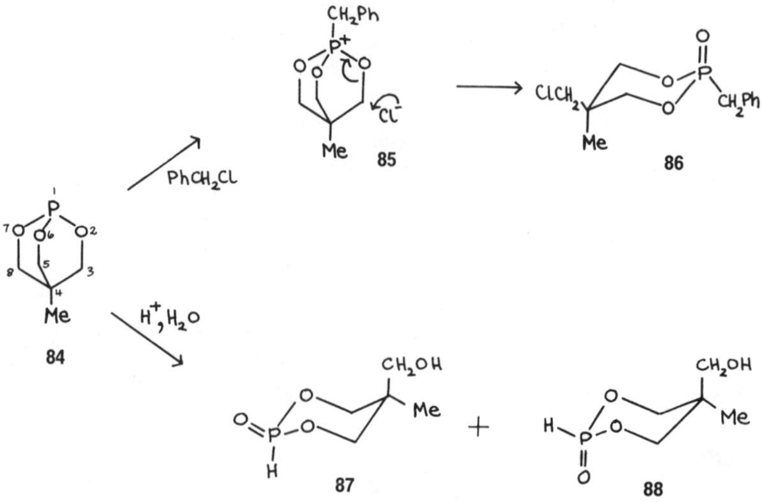

PhCH$_2$Cl

H$^+$, H$_2$O

84 **85** **86** **87** **88**

Scheme 7

It is reasonable to suppose that ring opening reactions of these rigid molecules always proceed stereospecifically, but this is not the case. Verkade and co-workers reported[229] the formation of both isomers in the acid-catalyzed hydrolysis of 4-methyl-2,6,7-trioxa-1-phosphabicyclo[2.2.2]-octane (**84**, Scheme 7). The suggested rationale for this is initial protonation at phosphorus followed by formation of a P(V) intermediate which may then open either radially or apically to give the observed products. However, protonation at oxygen would seem a more likely first step, followed by nucleophilic attack by water to give a ring opened product, formally a P(III)OH compound, capable of equilibration under the conditions of the reaction.

The other major class of bridgehead molecules, that is, heterotriptycenes (**83**), exhibits some interesting chemical properties, presumably as a consequence of their rigidity. They are generally unreactive, forming quaternary salts with some difficulty, and they do not form bis salts (where X and Y are As or P). The azarsatriptycene (**83**, X=As, Y=N) does not form a salt at all,[230] though with methyl fluorosulfonate an ether-insoluble solid is obtained which regenerates the triptycene on attempted dissolution in polar solvents.[231] This marked lack of reactivity was explained in terms of resistance of the molecule to distortion, particularly as a consequence of the unequal bond lengths joining the rings to the two heteroatoms. However, it was recently reported[232] that arsatriptycene (**83**, X=As, Y=CH) does form a quaternary salt, and all these molecules may be oxidized relatively easily—a reaction which would be expected to bring about much the same angular changes as quaternization. An extreme example of steric retardation of reactivity is 1,4,7-trimethyl-2,5,8-trioxa-10-phosphatricyclo-[5.2.1.0^{4,10}]deca-3,6,9-trione (**89**), obtained by the action of phosphine on pyruvic acid, which shows none of the reactions characteristic of a trialkyl phosphine.[233]

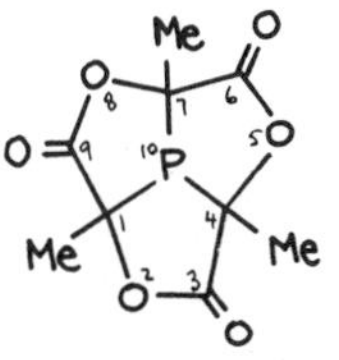

89

90

Aliphatic (and hence less rigid) counterparts of structure **83** should show a more varied chemistry, but though the diphosphorus compound 1,4-diphosphabicyclo[2.2.2]octane (**90**) has long been known, it is obtainable only in low yield after a laborious synthesis.[234] A convenient route to such a compound would provide a useful stereochemical probe.

A point which should always be borne in mind when working with these types of compounds is possible toxicity; the bicyclic phosphine **84** and related compounds are extremely toxic, the LD_{50} being much the same as the well known nerve gases, though the mode of action is different.[235]

F. P(III) Seven-Membered Ring and Larger Ring Compounds

No systematic work on larger rings has appeared, and there are only isolated examples of stereochemical interest. Marsi obtained both isomers of 4-methyl-1-phenylphosphepane (**91**) by reduction of the corresponding oxides, but the stereochemistry has not been assigned.[236]

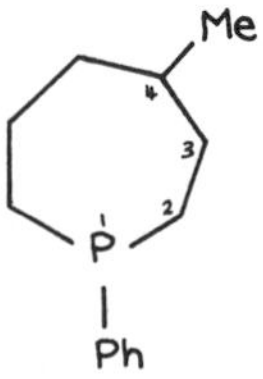

Very recently a method for the synthesis of macrocyclic diphosphines was developed (Scheme 8). The stereochemistry of the intermediate phosphine oxides was assigned on the basis of molecular dipole moments, and the oxides were reduced to the corresponding phosphines using trichlorosilane.[237] These are the only examples of P(III) (>*6*) heterocycles whose stereochemistry is known with any certainty.

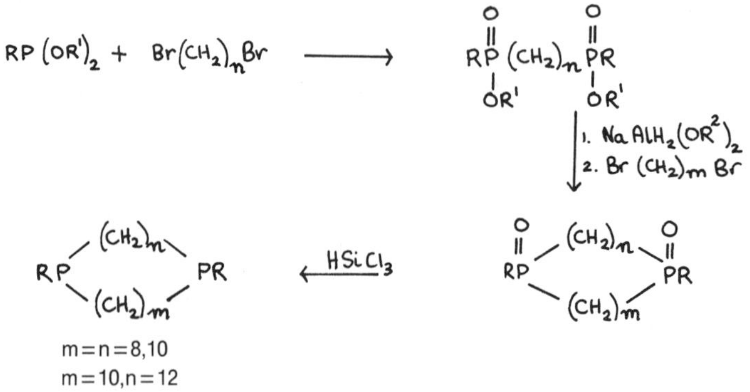

m=n=8,10
m=10,n=12

92

Scheme 8

Of great interest are the macrocycles 1,6-disubstituted 2,5,7,10-tetraoxa-1,6-diphosphacyclodecane (**93**), 1,4,4,7,10,10-hexamethyl-2,6,8,12-tetraoxa-1,7-diphosphacyclododecane (**94**), and 1,4,4,7,10,10,13,16,16-nonamethyl-2,6,8,12,14,18-hexaoxa-1,7,13-triphosphacyclooctadecane (**95**), which have been observed spectroscopically, by Robert and his colleagues, as products of the long known polymerization of 1,3,2-dioxaphospholanes[238] and

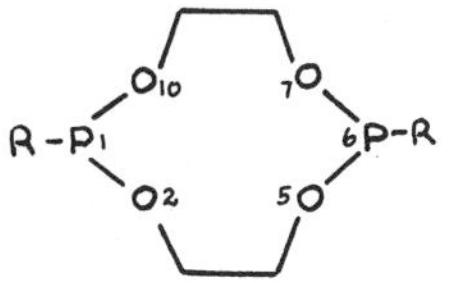

R=Ph,Me

93

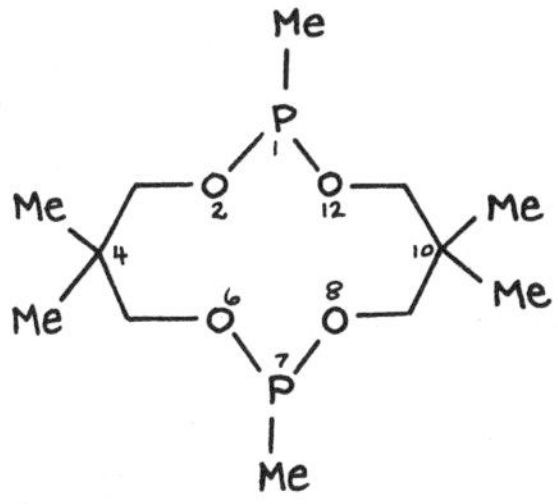

94

-phosphorinanes.[239] Isomer separation has not been achieved and though separation was accomplished after sulfurization for **93** and **94**, stereochemical assignments have not been made.

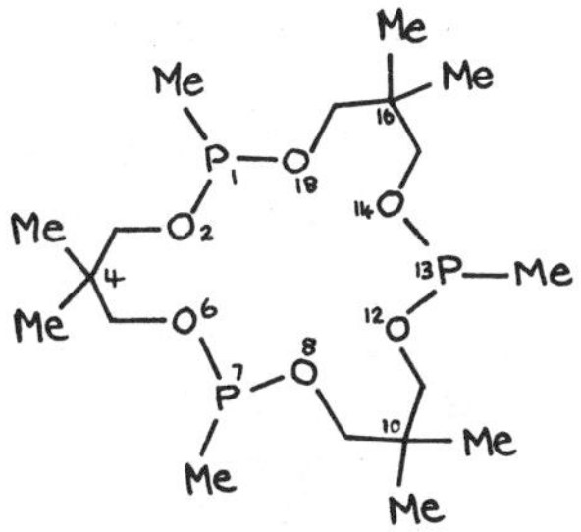

95

It is fascinating to contemplate the mode of formation of these rings since it represents a type of reaction hitherto unrecognized. It is probably related mechanistically to the scrambling reactions which are not uncommon in P(III) chemistry, though the intermediates involved are obscure. Equally as interesting is the virtually quantitative reversal of the reaction by distillation. It seems likely[238] that adventitious catalysts may be involved, but the reverse reaction, together with the apparently greater ease of reaction of P(III) (*5*) as opposed to P(III) (*6*) compounds, make it likely that the phosphorus is behaving as an electrophile perhaps as in Equation 5. If the apparent catalysis involved traces of acid, as in the inversion of dioxaphosphorinanes,[42,103,216] then this could facilitate the reaction by protonating oxygen and making P(III) more electrophilic. However, the nature of the catalyst (or even if it is always required) is not known, and the 1,6-dimethyl derivatives of 2,5,7,10-tetraoxa-1,6-diphosphacyclodecane (**93**, R = Me) is much more readily formed than the 1,6-diphenyl derivative (**93**, R = Ph).

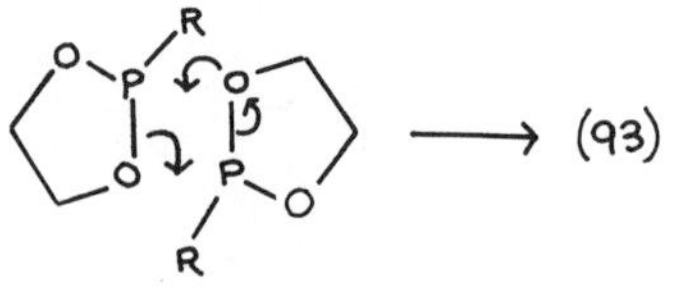

→ (93) (5)

96

VII. P(IV) COMPOUNDS

A. Introduction

This is by far the largest single group of phosphorus compounds. Such molecules are configurationally stable, much easier to handle than P(III) or P(V) compounds in general, and hence more amenable to study. In addition, the ubiquitous occurrence of phosphates, both cyclic and acyclic, in living systems has made this class more intensively studied than any other. By and large, this interest has also extended to stereochemical investigations beginning with the resolution of phosphine oxides by Meisenheimer and his collaborators early in this century.[2] Even so, nearly forty years passed before another type of P(IV) compound was resolved, namely, 2-phenyl-2-*p*-hydroxyphenyl-1,2,3,4-tetrahydroisophosphinolinium bromide (**97**),[240] which was followed not long after by the spirocycle, *P*-spiro-bis-1,2,3,4-tetrahydrophosphinolinium iodide (**98**).[241] The resolution of the latter provided strong evidence for a tetrahedral phosphorus atom (as distinct from nonplanar), and all subsequent investigations have supported this geometry for P(IV) compounds.

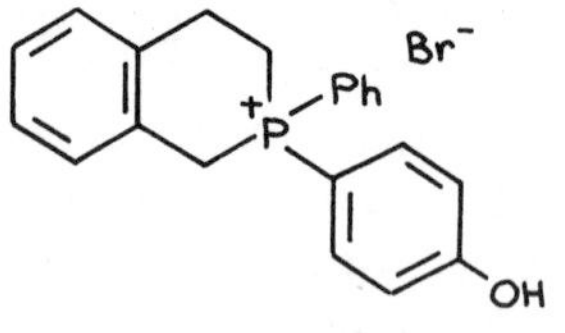

$[M]_D$ +32.9° (EtOH)

97

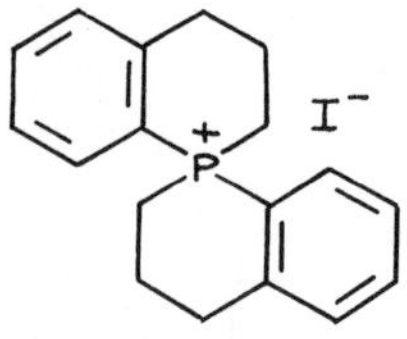

$[M]_D$ +66°, −65° ($CHCl_3$)

98

Much of the interest in P(IV) compounds has, naturally enough, centered around the chemistry of the biologically important phosphates, particularly on aspects of phosphorylation and the synthesis of complex molecules. The culmination of these is undoubtedly the synthesis of a gene by Khorana and his colleagues. Since phosphates in living systems are invariably substantially ionized, stereoisomerism arising as a consequence of the presence of this atom is never encountered, and consequently stereochemical studies have made up a relatively small proportion of the total. Stereochemical studies of cyclic As(IV) compounds are virtually unknown, though the salts **99**[242], **100**[243], and **101**[244] were resolved by Mann and his collaborators and, more recently, bis-2,2′-biphenylenearsonium iodide (**102**) by Wittig and Hellwinkel.[245] 5,7-Dimethyl-5,7-diarsa-1,2:3,4-dibenzocyclo-

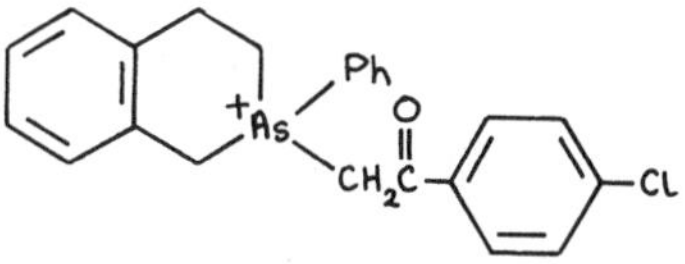

$[M]_D$ +457°, −450° ($CHCl_3$)

99

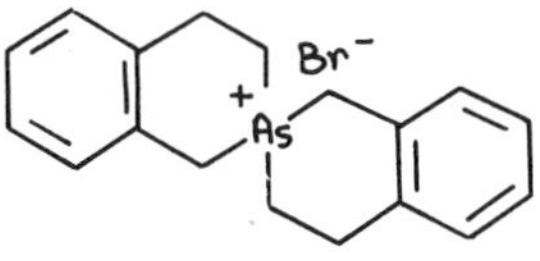

$[M]_D$ +342°, −344° ($CHCl_3$)

100

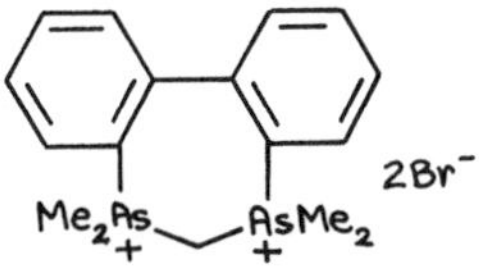

$[M]_D$ +174°, −179° (H_2O)

101

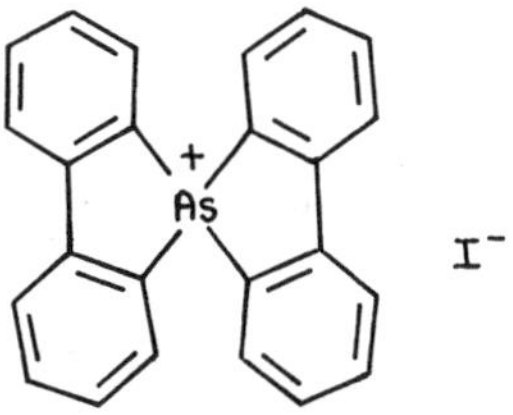

$[M]_D$ +78°, −77°

102

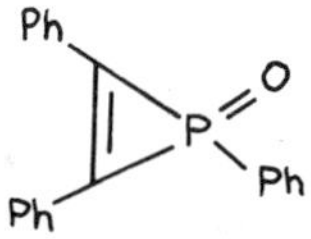

$\delta\ ^{31}P$ −30ppm

103

heptadiene 5,7-bismethyl bromide (**101**) is, of course, a bridged biphenyl, and optical activity stems from this source and not from chiral arsenic atoms.

Much of the chemistry of P(IV) systems is bound up with the formation of P(V) structures as reaction intermediates and, as such, is better discussed under that heading (Section VIII.D). A negligible number of studies has been reported on ring systems other than four-, five-, and six-membered. In particular, only a solitary example of a reasonably well authenticated P(IV) (*3*) compound is known,[246] 1,2,3-triphenyl-Δ^2-phosphirene 1-oxide (**103**), and even this material was not obtained pure. Its chemical shift is a useful pointer to further studies in this rather surprisingly neglected area.

B. P(IV) Four-Membered Ring Compounds (Phosphetanes)

Compounds of this type (phosphetanes) were first obtained by McBride and co-workers, in 1962, by the action of PCl_3 on 2,4,4-trimethylpent-2-ene in the presence of $AlCl_3$.[247] The mechanism shown in Scheme 9 was advanced. Subsequently, Cremer and Chorvat[176] took the reaction up and

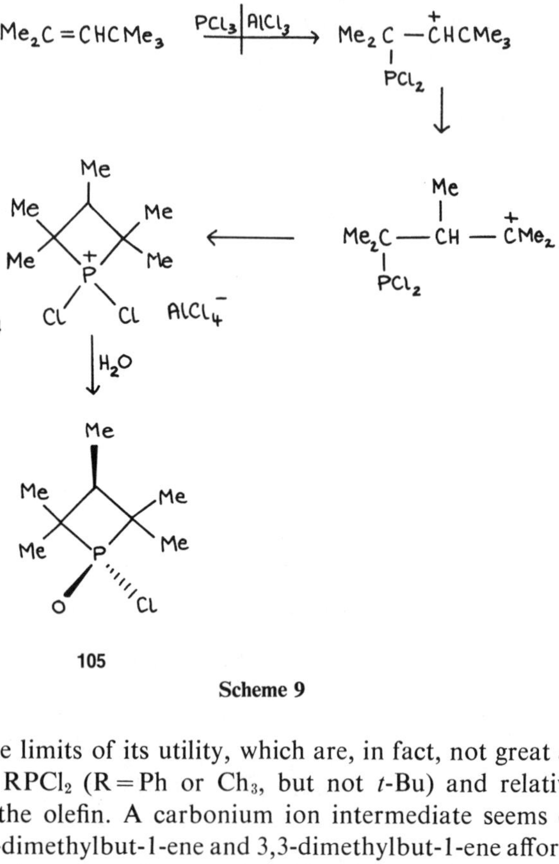

104

105

Scheme 9

determined the limits of its utility, which are, in fact, not great and extend only to using $RPCl_2$ (R = Ph or Ch_3, but not *t*-Bu) and relatively minor variations in the olefin. A carbonium ion intermediate seems established since both 2,3-dimethylbut-1-ene and 3,3-dimethylbut-1-ene afford 2,2,3,4,4-tetramethyl-1-phenylphosphetane 1-oxide (**106**) when reaction with phenylphosphonous dichloride. The two ring types, *r*-1-chloro-*trans*-

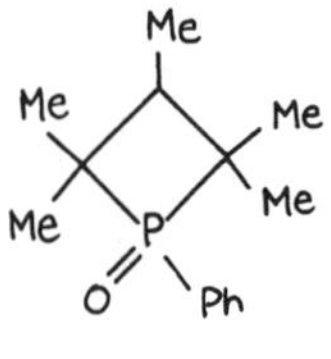

106

3,2,2,4,4-pentamethylphosphetane 1-oxide (**105**) and **106**, have been used almost exclusively in chemical and stereochemical studies, and this is in some respects unfortunate since the generally crowded nature of the molecule may give rise to an atypical result (for example, the very sluggish reactivity of the anhydride obtainable by hydrolysis of **105**[248]).

Cremer and his co-workers examined the McBride reaction in some detail[249] and confirmed the intermediacy of chlorophosphetanium salts such as **104**. A mixture of diastereoisomers is obtained when $RPCl_2$ is used in place of PCl_3, and this can be hydrolyzed to the corresponding mixture of oxides without inversion at phosphorus. The stereochemistry can be easily assigned from spectral data (Section III). Interestingly, hydrolysis of **104** gives virtually a single diastereoisomer, **105**.

Two other types of P(IV) (*4*) compounds were prepared by the Russian group Shermergorn and collaborators. The 3-substituted 3-thiaphosphetanes **107** (X=OH, OEt, or Cl) are readily accessible (Scheme 10).[250–252] In the vapor phase 3-chloro-1,3-thiaphosphetane 3-oxide (**107**, R=Cl) is almost flat.[253] The synthesis (Equation 6) and some reactions of 3-ethoxycarbonyl-1-phenylphosphetane 1-oxide (**108**) have been reported,[254] but

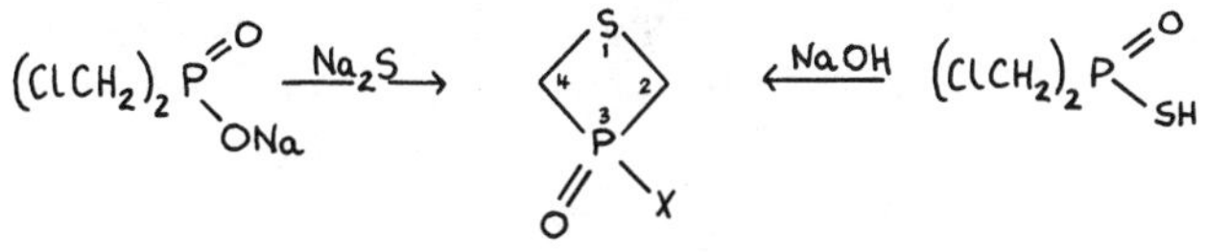

$(ClCH_2)_2P(=O)ONa \xrightarrow{Na_2S}$ **107** $\xleftarrow{NaOH} (ClCH_2)_2P(=O)SH$

107

Scheme 10

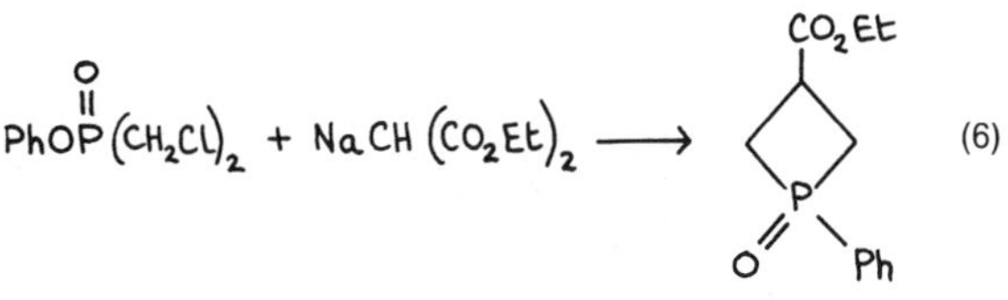

$$PhOP(O)(CH_2Cl)_2 + NaCH(CO_2Et)_2 \longrightarrow \textbf{108} \quad (6)$$

108

the great difficulty of isolation has presumably precluded further studies on this interesting heterocycle. The parent compound, 1-hydroxyphosphetane 1-oxide (**109**), was reported by Kosolapoff and Struck in such minute yield (about 0.01%) as to deter any further studies of it.[255]

X-ray crystal structures of an acid[256] (**110**), a phosphonium salt[257] (**111**), the trans[258] (m.p. 117–118°) and cis[259] (m.p. 126–127°) isomers of 2,2,3,4,4-pentamethyl-1-phenylphosphetane 1-oxide (**106**), and the acid chloride **105**[260] establish that the four-membered ring is puckered with the angle between the two planes about 25°. The heteroatom is the center of a distorted tetrahedron with the smallest angle being the endocyclic one (about

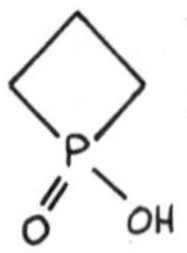

109 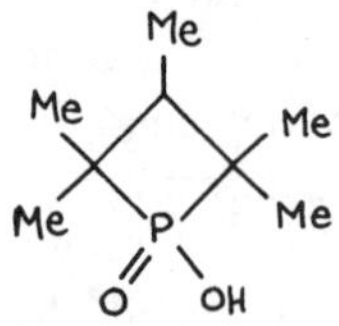 **110** 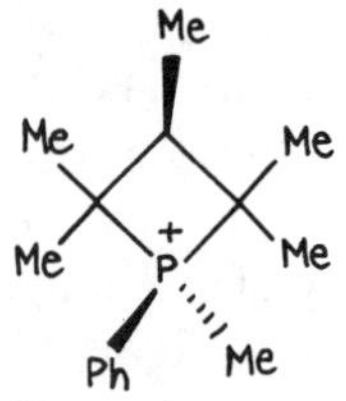 **111**

85–88°). Spectroscopic evidence favors a fixed conformation,[35] and in P(IV) (*4*), as in P(III) (*4*) compounds, the δ value for the ring proton at position 3 allows ready distinction between diastereoisomers, particularly in the presence of $Eu(dpm)_3$.[54]

1. *Reaction Stereochemistry of P(IV) Four-Membered Ring Compounds*

Numerous workers have examined the action of nucleophiles on P(IV) (*4*) compounds and have interpreted their results on the basis of the intermediate involvement of P(V) species and the constraints imposed on this by the presence of the small ring. These interpretations are discussed in the section on P(V) systems.

Nucleophiles attack P(IV) (*4*) compounds either nonstereospecifically or with retention of configuration. Thus 2,2,3,4,4-pentamethyl-1-trideuteromethoxyphosphetane 1-oxide **112** ($R = CD_3$ cis or trans) reacts with MeONa–MeOH to afford the corresponding OMe ester (**112**, cis or trans, R = Me) with no evidence for isomerism.[261]

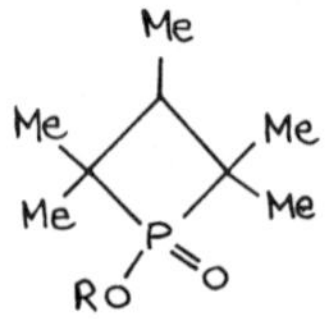

112 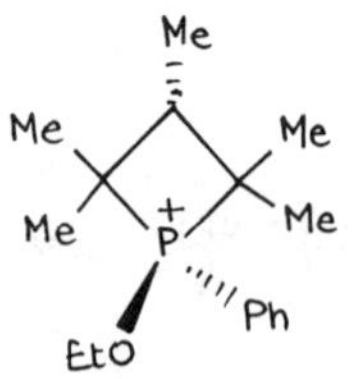 **113** 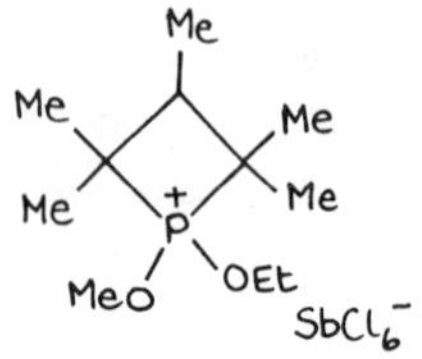 **114**

The alkyoxyphosphonium salt **113**, when treated with alkali, regenerates the phosphine oxide from which it was prepared by alkylation,[262] whereas either diastereoisomer of the salt 2,2,3,4,4-pentamethyl-1-ethoxy-1-methoxyphosphetanium hexachloroantimonate (**114**) is converted to the same mixture of esters (**112**, trans or cis, R = OMe or OEt) by aqueous alkaline dioxane.[263] Racemization is also observed in the alkaline hydrolysis of the simple tetramethylphosphetanium salts (**115**).[264,265]. This behavior may be traced directly to the formation of a P(V) intermediate; indeed, the trans and cis isomers of **115** (R^1 or R^2 = Ph, Me) are rapidly equilibrated by

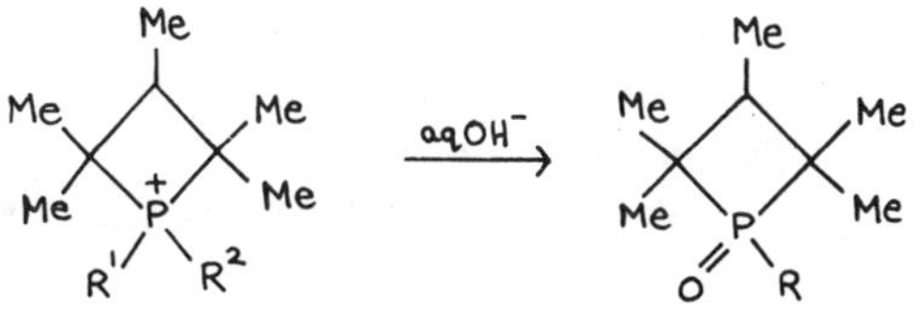

115 → (aq OH^-) → 116

catalytic amounts of aqueous sodium hydroxide.[265] On the other hand, the oxides such as **116** are stereochemically unaffected by an acid or base, though they may be racemized with $SiCl_4$. This latter observation is important since $SiCl_4$ is a by-product in the reduction of P(IV) oxides to P(III) compounds by Si_2Cl_6,[262] and P(III) compounds may also be racemized in this manner. Other reducing agents such as Et_3N–$HSiCl_3$ and $PhSiH_3$ also effect reduction to P(III) (*4*) compounds with retention of configuration. No example of predominant inversion at phosphorus in the reactions of P(IV) (*4*) series has been reported.

Appreciable differences have been noted in the rates of reaction of diastereoisomers of phosphetanes which differ in geometry only at phosphorus. It is reasonable to suppose this is a simple steric effect arising from the more difficult access of a nucleophile to phosphorus when the approach must be made from the same side as the 3-methyl. Hawes and Trippett have noted that the rate of hydrolysis of the ester **117** is extremely rapid in contrast with that of **112** as a result of reduction in the hindrance due to the

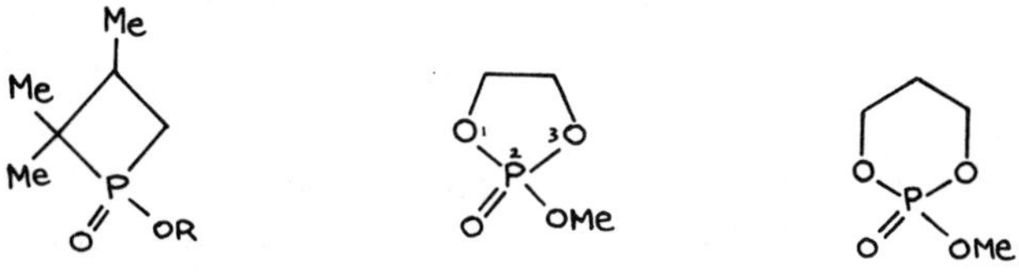

117 118 119

four methyl groups on the α-carbons in **112**. Normally, nucleophilic displacements at phosphorus with two *t*-butyl type substituents are very slow.[266]

Strikingly, the base-catalyzed decomposition of α-hydroxyimino-*p*-nitrobenzyl esters proceeds at substantially the same rate for both **112** and **117**. Cadogan and Eastlick interpreted this to be indicative of reduced steric requirements for intramolecular attack[267] and observed the same effect in acyclic systems.

Proof of the stereochemical consequences of nucleophilic attack in P(IV) (*4*) compounds is normally fairly straightforward and utilizes the assumptions of retention of configuration in oxidation, and quaternization, or the Wittig reaction. For example, reduction of an oxide to a phosphine gives, on reoxidation, the same oxide, or by quaternization affords the phosphonium salt from which the oxide may be obtained by alkaline hydrolysis (Scheme 11). Using simple cycles of this type, the Leicester group established the generality of retention of substitution of P(IV) (*4*) oxides and sulfides for a range of nucleophiles (RO^-, RNH_2,RS^-).[24]

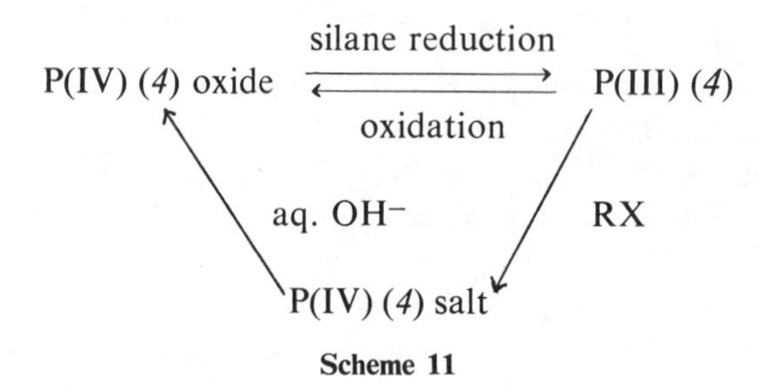

Scheme 11

The X-ray structure determinations of **106** and **111** substantiate many of the assignments in this field. Note, however, a reversal of assignments due to the better crystalline shape of the minor (5%) component of a mixture of phosphonium salts.[178,257]

Retention of configuration has also been observed in the reaction of P(IV) (*4*) oxides with isocyanates[268] and the cleavage of the P–N bond of P(IV) (*4*) amides with hydrogen chloride.[269]

C. P(IV) Five-Membered Ring Compounds (Phospholanes)

This very large class of compounds has attracted a great deal of attention, both chemical and spectroscopic. In Table III are recorded the structural and spectroscopic data available for monocyclic compounds. The long bonds to phosphorus introduce strain in these molecules which is taken up by angular distortion, the endocyclic angle at phosphorus being close to 98°. This strain is believed both to facilitate the transformation P(IV)→P(V) and to exert a profound effect on the chemistry of these systems. The strain may be observed by comparing the heats of hydrolysis of cyclic phosphates with acyclic analogues when the difference observed is 21–26 kJ mol^{-1} for the phosphate, 2-methoxy-1,3,2-dioxaphospholane 2-oxide (**118**),[270] whereas no such difference is observed for the P(IV) (*6*) homologue, 2-methoxy-1,3,2-dioxaphosphorinane 2-oxide (**119**).[176]

However, the generality of the concept of strain relief in a transition state or intermediate formed in this way should be accepted only with some reserve since this strain effect is inappreciable in the cyclic phosphate of cytosine,[272] a matter of considerable importance in view of the fact that such cyclic phosphates occur widely in nature. Nevertheless, all solid state structures so far determined confirm the considerable distortion of the endocyclic angle at phosphorus from the normal tetrahedral arrangement, and there can be no reasonable doubt that P(IV) (*5*) systems pass readily to P(V) (*5*), particularly under the conditions of nucleophilic attack. Increasing the strain in the ring leads to an enhancement in reactivity. Thus, the enediol phosphate, 2-methoxy-1,3,2-dioxaphospholene 2-oxide (**120**), hydrolyzes more readily than its saturated analogues,[273,274] and the ketophospholane (**121**) reacts so readily with alcohols that it has been claimed to be the most active phosphorylating agent known.[275] A similar

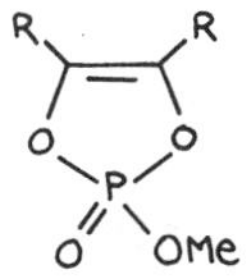

120 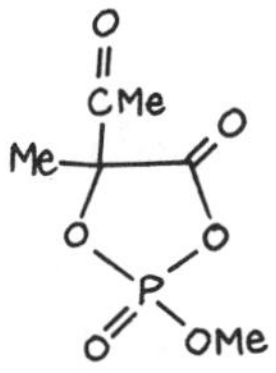 **121**

interpretation accounts for the fact that in the molecule 3,10-diethoxy-3,10-diphosphatricyclo[5.2.1.0^{2,6}]deca-4,8-diene 3,10-dioxide (**122**) one phosphorus atom, P*–10, undergoes hydrolysis much faster than the other, and a similar, though less marked, effect is observed for the saturated compound.[276]

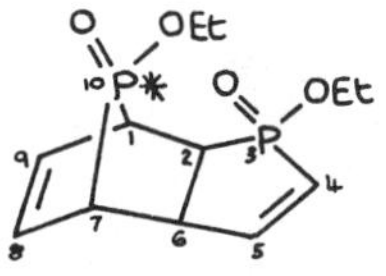

122 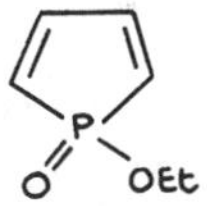 **123**

Interestingly, **122** is obtained by the spontaneous dimerization of 1-ethoxyphosphole 1-oxide (**123**), which presumably derives some of its reactivity from strain due to the presence of P(IV) since the corresponding P(III) (*5*) compounds show no tendency to dimerize.

Phospholane derivatives, where the two ring atoms attached to phosphorus are not carbon, are invariably prepared by the action of a P(IV) halide on the appropriate glycol, amino-alcohol or analogous compound in the presence of an amine, pyridine, or triethylamine usually, to remove the hydrohalide formed. The most general method of preparation of the carbocyclic analogues, is the McCormack reaction (Scheme 12).[277]

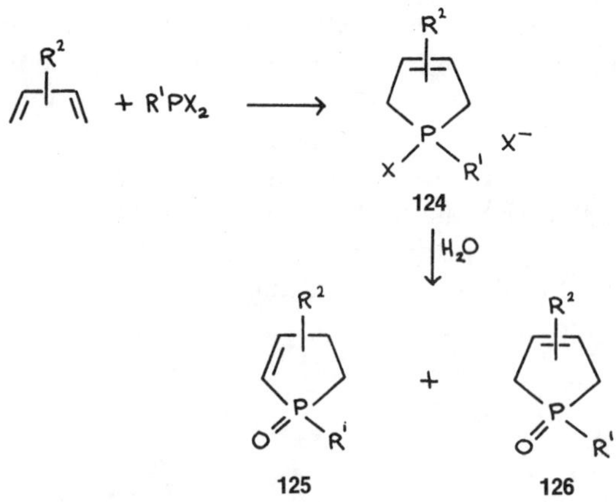

124

125

126

Scheme 12

The formation of 3-phospholenium salts (**124**) was shown to proceed stereospecifically in at least one instance,[278] but, in general, preparations of P(IV) (*5*) compounds are not stereospecific. Where stereoisomers are expected, they are normally found and may be separated by standard chemical means. The ^{1}H nmr spectra of the isomers generally allow a less clear-cut distinction than for P(III) (*5*) compounds since the presence of two exocyclic substituents at phosphorus tends to reduce the chemical shift differences. Similarly, the substantial difference in ^{31}P chemical shifts between P(III) (*5*) diastereoisomers is much reduced in P(IV) (*5*) analogues. However, the $\delta^{31}P$ for P(IV) (*5*) compounds occur at substantially lower field (about 20–30 ppm) than their acyclic or larger-membered ring analogues. This is presumably, at least in part, a reflection of the angular distortion.

Overall, spectroscopic and structural data indicate no surprises in the conformation of P(IV) (*5*) compounds and suggest an envelope or twist-envelope shape. In the majority of cases conformer interconversion is a low energy process often giving averaged nmr values at room temperature.

Of particular interest are the 1,3,2-oxazaphospholanes, **127**, since these have been prepared from the optically active amino alcohols, ephedrine[83,84,279,a] and ψ-ephedrine,[83,279,a] and the isomers obtained are epimeric only at phosphorus.

In this fashion, Feldman and Berlin[279,a] were the first to resolve a heterocyclic compound of phosphorus which was not also carbocyclic. They were unable to make firm assignments of configuration at phosphorus but noted stereoscpecficity in the displacement of Cl by RNH[279,a] an observation subsequently substantiated by Inch and his collaborators.[84] The Russians have used meso- and d,l-stilbene derivatives to obtain **128**.[279b]

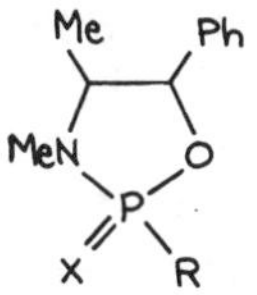

X=O,S
R=Cl,OAlk,OAr,Alk,Ar,NR_2

127

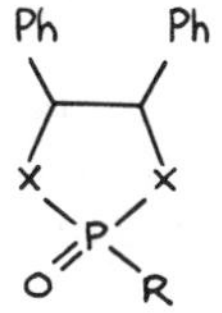

X=O,NH

128

Determination of configuration at phosphorus in this series rests on nmr interpretation, and there is a difference of opinion here between the French[83] and British[84] groups, at least in regard to the (−)-ephedrine series. Of the two studies, that of Inch and his group perhaps has a stronger basis since they were able to obtain both isomers in most cases, whereas Devillers and Navech and co-workers were reasoning by analogy with the ψ-ephedrine series. A single X-ray structural determination would resolve this problem and make available the only readily accessible and quite large group of P(IV) (*5*) compounds (or indeed any other organophosphorus compounds) of known absolute configuration at phosphorus. Inch based his nmr assignments on the deshielding influence of the P=O group on a ring hydrogen atom cis to it. There is precedent for this,[39] and if confirmed, it would be a most convenient probe. A similar conclusion was reached by Baccolini and Todesco in assigning configuration to the 1,2,3-diazaphospholenes (**129**),[280] but these authors relied on the more familiar ground of deshielding

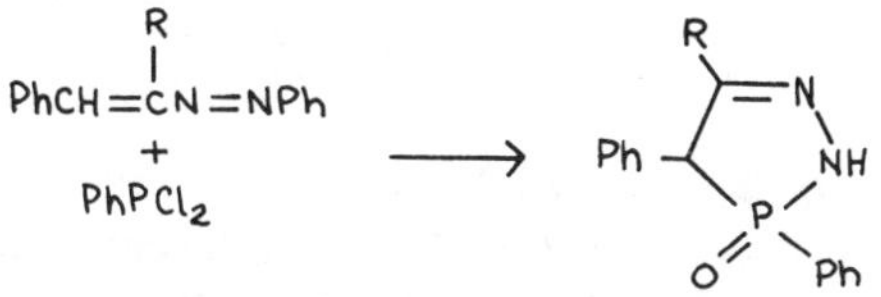

129

by the phenyl group on phosphorus. Marsi obtained both pure isomers of the 1-benzyl-1,3-dimethylphospholanium cation (**130**), but the stereochemistry was not assigned. These two salts have the same chemical shifts δ ^{31}P (+51.8), an unusual coincidence.[189]

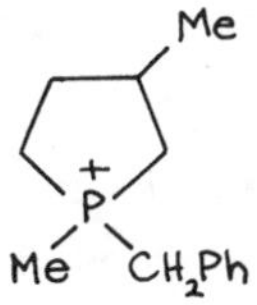

130

The great bulk of studies of P(IV) (*5*) compounds has been concerned with monocyclic derivatives, but recently examples of polycyclic compounds have been reported mostly by Kashman and his collaborators. Thus McCormack addition to the cycloheptadienes (**131**) affords the 8-phosphabicyclo[3.2.1]oct-6-ene **132** (Equation 7). The addition appears to be

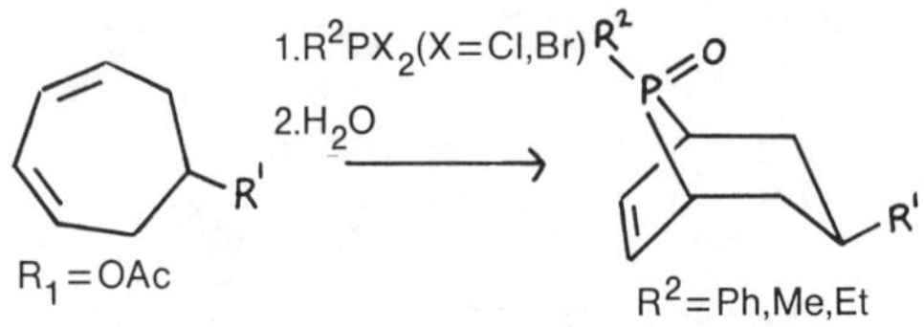

1. R^2PX_2 (X = Cl, Br) 2. H_2O (7)

R_1 = OAc

131

R^2 = Ph, Me, Et

132

stereospecific, but when R^2 = Me two products are obtained after hydrolysis. The configuration of the principal isomer is shown and was deduced by ^1H nmr shift reagent studies.[62] This series of compounds has been correlated chemically with those obtained by addition of primary phosphines to cycloheptadienone.[217] Stereoisomers have been obtained, but not assigned, for the interesting tricyclic products, **133**, of the variant of the McCormack reaction shown in Equation 8.[281]

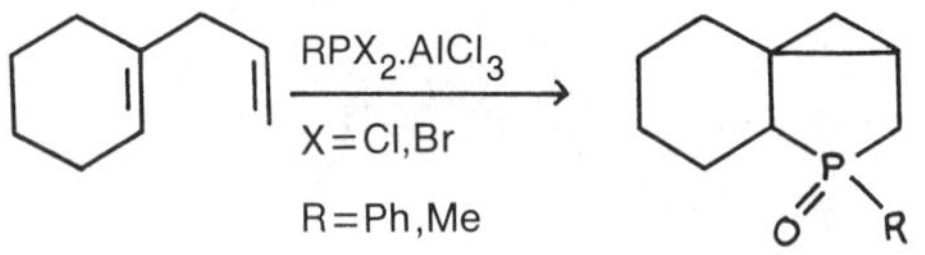

$RPX_2 \cdot AlCl_3$, X = Cl, Br, R = Ph, Me (8)

133

The very interesting bridgehead bicyclic compounds 1-phosphabicyclo[2.2.1]heptane 1-oxide (**134**) and 1-phosphabicyclo[2.2.2]octane 1-oxide (**135**) were prepared by Wetzel and Kenyon[31,282] from $POCl_3$ and the appropriate Grignard reagents.

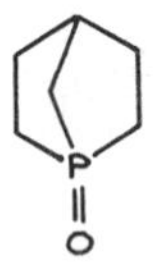

134 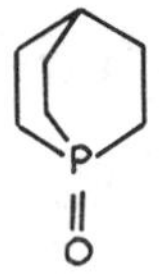 **135**

The Diels-Alder reactions of 1-substituted 3-phospholene 1-oxides (**136**),[283] 1-phenyl-1-thionophosphole (**137**),[284] and 1-phenyl-2-phospholene 1-oxide (**138**)[285] have been reported. Where the diene component is cyclic,

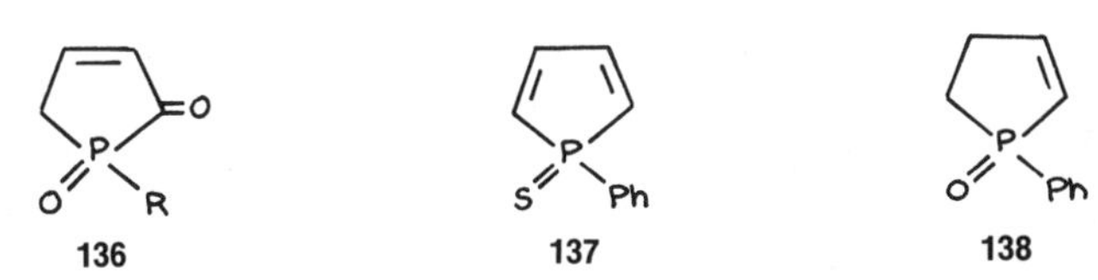

the reactions give largely one P epimer, presumably as a consequence of steric hindrance in the transition state; with similar dienes, however, both epimers are obtained.[285] The cycloaddition reactions themselves are cis stereospecific as expected.

The chemistry of rings containing more than one phosphorus atom has been relatively neglected, although the compounds $(RP)_n$ have attracted some structural interest. Horner and his group have, however, synthesized the 1,3-diphospholanium salt **140** by bisquaternization of the *meso*-diphosphine (**139**,[286] Equation 9). This phosphine, and both enantiomers of

Me, Ph, P, CH_2CH_2, P, Me, Ph $\xrightarrow{BrCH_2CH_2Br}$ Ph, Me, P+, +P, Ph, Ph, $2Br^-$ (9)

139 **140**

the *d,l* form also, is relatively readily accessible, and it is surprising that they have not attracted any further attention to the synthesis of diphosphorus heterocycles of known stereochemistry.

1. *Reaction Stereochemistry of P(IV) Five-Membered Ring Compounds*

Only results are summarized in this section; their explanation is deferred to the section on P(V) compounds.

As with P(IV) (*4*) compounds the stereochemical consequences of reaction at phosphorus in P(IV) (*5*) compounds are either retention or racemization, and there appears to be only one report of a reaction proceeding with predominant inversion.[405] Thus, the alkaline hydrolysis of phospholanium salts,[189] the reduction of phospholane oxides with phenylsilane,[189] and nucleophilic displacement in 1,3,2-oxazaphospholanes[84] have all been shown to proceed with retention of configuration. The methods used are chemical and involve a Walden-type cycle of reactions, utilizing oxidations and/or quaternizations which are accepted as proceeding with retention at phosphorus. Racemization, on the other hand, is self-evident and normally reflects the reversible formation of a configurationally mobile P(V) species. This is discussed in more detail in Section VIII.D.

D. P(IV) Six-Membered Ring Compounds (Phosphorinanes)

This is the largest single class of organophosphorus heterocycles. The extensive conformational and structural studies are summarized in Table IV. The commonest mode of synthesis is by the reaction of a 1,3-disubstituted alkane with an appropriate phosphoryl derivative. Carbocyclic rings are usually formed by ring closure of a suitably substituted acyclic compound or by addition of phosphines to unsaturated systems. Molecules with only one heteroatom other than phosphorus are much rarer; Bergesen[287] has separated isomers and assigned stereochemistry in the 1,2-oxaphosphorinane systems, **141** and **142**. The assignments are based on the modified Auwers-Skita rule.[288]

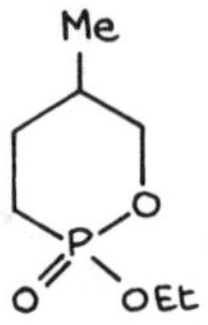

141 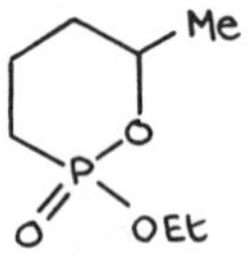 **142**

No other group of organophosphorus compounds is so well served with basic stereochemical data as the P(IV) (*6*) system. Two stereoisomeric pairs have been the subject of X-ray crystal analysis (e.g., 5-chloromethyl-5-methyl-2-phenoxy-1,3,2-dioxaphosphorinane 2-oxide, **143**,[107,108] and one isomer of a pair has been described for compounds **146–149**. It is thus

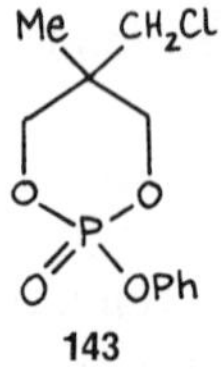

143 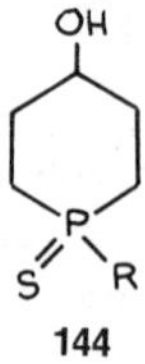 **144** 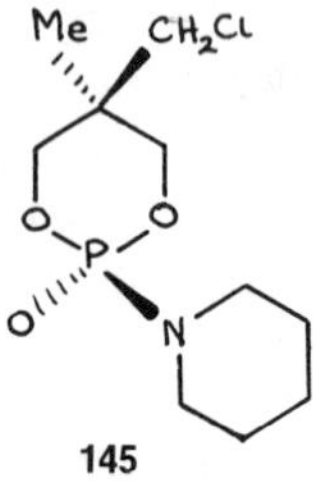 **145**

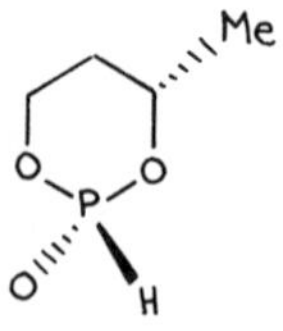

146 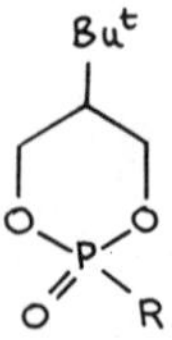 (R=OMe,Me) **147** **148**

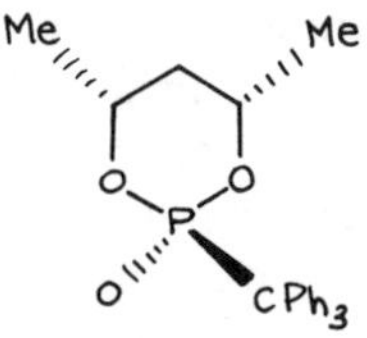

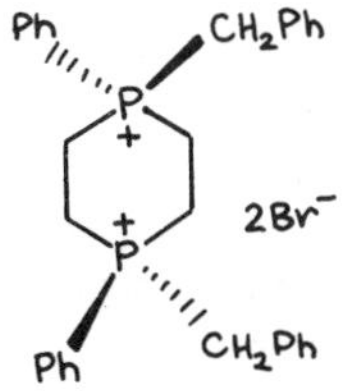

149

possible to make a wide range of direct correlations without any assumptions. Crystal structures all show much the same features. The ring is in the chair form somewhat flattened at the phosphorus end with the P=O group in the equatorial position; the angles at phosphorus are essentially tetrahedral. The same overall picture has been obtained for rings where stereoisomerism is not possible. There are, inevitably, exceptions. The phosphoramidate **145** and cyclophosphamide **150** in the solid state have the P=O group axial. This is possibly an electronic effect due to the nitrogen atom, and departures from the norm are also observed in the nmr spectra of P(IV)(*6*) compounds with an exocyclic amino substituent (Section III). Similarly, P=S in 2-substituted 4,4-dimethyl-1,3,2-dioxaphosphorinane 2-sulfides (**151**) is axial in the crystal, and in solution the percentage of P=S which is equatorial or axial varies with substituent and solvent.[129]

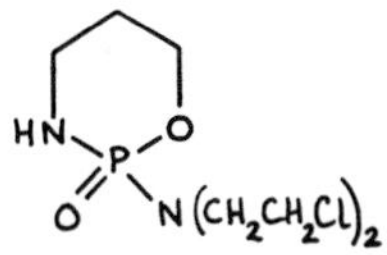

150

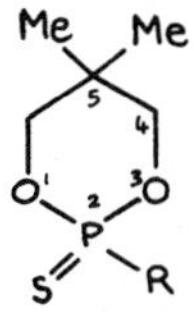

(R = Me, Ph, OPh, Cl)
151

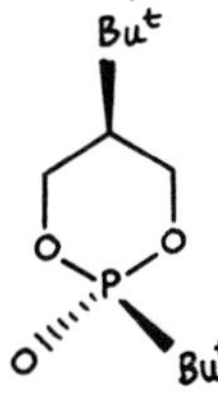

152

Not surprisingly, the presence of very bulky groups results in different conformational preferences. 1,5-*cis*-Di-*t*-butyl-1,3,2-dioxaphosphorinane 2-oxide (**152**) adopts a boat shape in solution, on the basis of nmr studies, and *trans*-4-*trans*-6-dimethyl-*r*-2-trityl-1,3,2-dioxaphosphorinane 2-oxide (**148**) has a half-boat structure in the crystal.

If the reasonable assumption is made that solid state structures are an acceptable basis for assessing the preferred conformation in solution, it would be expected that in the absence of severe distorting influences, P(IV) (*6*) molecules in solution will adopt a chair conformation with the P=O group equatorial (when present). The large number of nmr studies substantially support the hypothesis that P(IV) (*6*) molecules are present largely in one conformation and are generally conformationally mobile.

Note, however, that the claim to have isolated conformers of the *P* epimer of **145**[289] has been refuted.[290] For thiophosphoryl compounds the situation is less clear-cut, and it seems likely that there is no great energy difference between equatorial and axial positioning of this group. As pointed out before (Section III), two conformations have been observed in the same crystal. Conformational free energies for a number of P(IV) 1,3,2-dioxaphospholanes have been determined by Majoral and his collaborators.[299] Apart from solid state and solution nmr studies, other spectroscopic means for assigning stereochemistry are rare, though dipole moment studies support the general overall picture.[119] Nmr spectroscopy remains the most convenient method, particularly with the use of shift reagents,[54,58,291] but it is both necessary to be careful here that equilibria are not affected by the reagent and to choose the optimum ratio (usually about two substrate:one shift reagent in proton studies). Distinction between stereoisomers is always simpler when both isomers are available, and Majoral and Navech[293,294] have shown that for 2-oxo-1,3,2-dioxaphosphorinanes the P═O stretching frequency is higher (about 5–20 cm^{-1}) when the phosphoryl oxygen is equatorial than when it is axial. Similarly, the ^{31}P chemical shift is to higher field for the equatorial isomer (3–5 ppm). Other workers have obtained similar results.[48,221,295,431]

A number of stereospecific syntheses have been described which provide compounds of known configuration at phosphorus. The first of these was due to Wadsworth and coworkers and starts from a bicyclic compound, for example, 4-methyl-2,6,7-trioxa-1-phosphabicyclo[2.2.2]octane (**72**), and proceeds by way of the intermediate **153**, as shown in Scheme 13, to afford

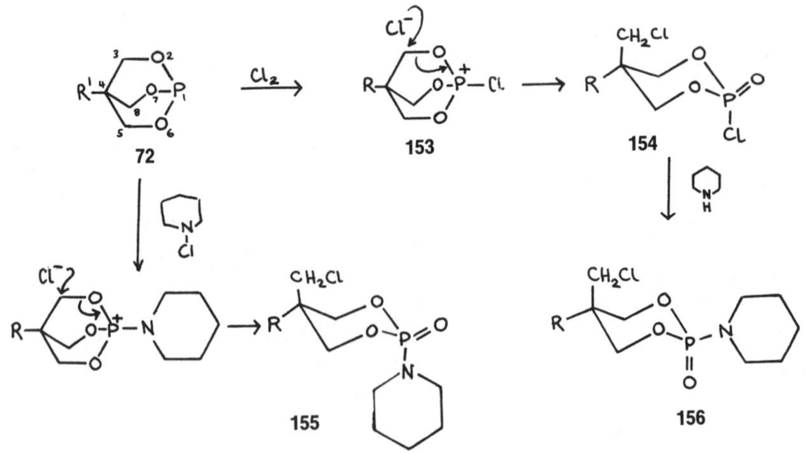

72 153 154 155 156

Scheme 13

the two phosphoramidates, **155** and **156**, unambiguously.[228] The sequence also provides strong evidence for inversion at phosphorus in the reaction of the phosphorochloridate with piperidine. Subsequent X-ray studies have established the stereochemistry of the halogenation product *r*-1-bromo-*trans*-5-chloromethyl-*cis*-5-methyl-1,3,2-dioxaphosphorinane 2-oxide (**154**, R = Me, Br in place of Cl)[115] and the phosphoramidate *r*-1-morpholino-*cis*-5-chloromethyl-*trans*-5-methyl-1,3,2-dioxaphosphorinane 2-oxide (**156**, R = Me).[109]

A different approach has been used by Inch and Lewis;[295,6] they prepared cyclic phosphates using methyl 2,3-di-*O*-methyl-α-D-glucopyranoside (**157**) as the diol and isolated crystalline compounds epimeric at

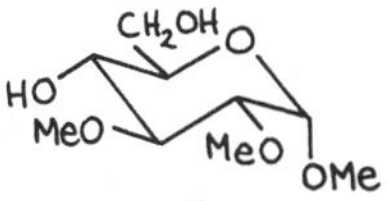

CH₂OH, HO, MeO, MeO, OMe

157

R², P, R¹, O, O, O, MeO, MeO, OMe

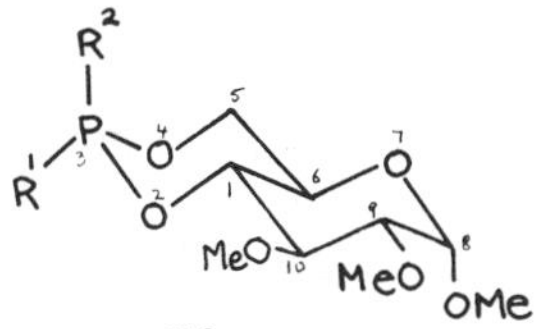

158

phosphorus. The absolute configuration was elegantly assigned by treating 1*S*, 2*R*, 6*R*, 8*R*, 10*R*-*r*-2-methyl(axial)-*cis*-8-*trans*-9-*cis*-10-trimethoxy-1*H*(axial)-2,4,7-trioxa-3-phospha-*trans*-bicyclo[2.2.0]decane 3-oxide (**158**, R^1 = Me; R^2 = O) sequentially with PhMgBr and EtMgBr thus converting it stereospecifically to the known (−) PhMeEtP=O whose absolute configuration (s) is well established.[297] (Note that the assignments of absolute configurations should be revised in many of the papers from this group due to an incorrect usage of the sequence rule.)[297a] This work is also of great interest because it records the first preparation of an optically active phosphate (**158**, R^1 = OEt, R^2 = O), but almost simultaneously Dudman and Zerner prepared an acyclic example by a much more laborious procedure.[298] The use of the relatively readily accessible diol **157** as a starting point for stereochemical studies of phosphorus, seems very promising and capable of extension to P(III) (*6*) compounds, but his has not been reported. The availability of an optically active P(III) 1,3,2-dioxaphospholane would be of exceptional interest in phosphorus stereochemical studies both for mechanistic and synthetic reasons.

Mikolajczyk, Stec, and their collaborators have reported syntheses of a number of useful isomerically pure derivatives of **159** (R = H, Cl, Br, NHR; X = O, S, Se)[300–302,304,305] starting from the P(III) (*6*) compounds,[300,306] and they have assigned the stereochemistry largely on spectroscopic grounds and by the use of stereospecific methods (e.g., halogenation, sufurization). ^{31}P chemical shift data for 13 cis-trans pairs have been obtained.[302]

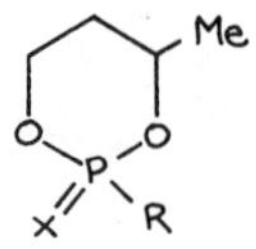

159 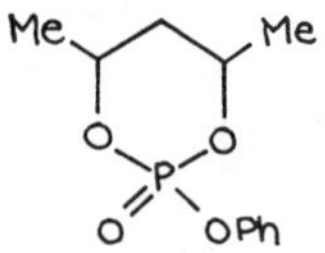 160 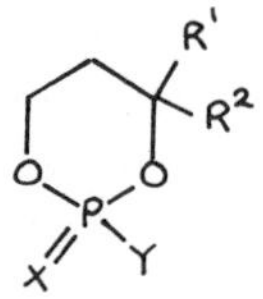 161

Hall and Malcolm[44] have described the separation of all three possible isomers of 4,6-dimethyl-2-phenoxy-1,3,2-dioxaphorinane 2-oxide (**150**) in addition to a detailed analysis of their ^{1}H spectra.

In the midst of the welter of instrumental studies, it is interesting to note that the photoelectron technique fails to distinguish between stereoisomers of **161** (X=O, Y=NHPh; X=S, Y=OH).[303]

The bis, salt 1,4-dibenzyl-1,4-diphenyl-1,4-diphosphorinanium dibromide (**162**) may be separated readily into its stereoisomers,[308–308a] but the original stereochemical assignment has been reversed by X-ray analysis.[141]

1. *Reaction Stereochemistry of P(IV) Six-Membered Ring Compounds*

Whereas reactions at phosphorus in four- and five-membered rings are normally stereospecific or at least stereoselective, no such general pattern emerges for P(IV) (*6*) compounds. This is probably a reflection of a lessening of steric restraints in the transition states and /or intermediates as a consequence of the larger ring. Thus, hydrolysis of cyclic six-membered ring phosphates shows no abnormalities when compared with their acyclic analogues,[271] in sharp contrast with P(IV) (*5*) compounds.

Reduction of P(IV) (*6*) oxides to P(III) (*6*) compounds with phenylsilane proceeds with retention of configuration,[307] whereas reaction of phosphorochloridates with amines occurs with inversion.[289,292] On the other hand, P(IV) (*6*) phosphonium salts react nonstereospecifically with aqueous alkali to give mixtures of the corresponding oxides,[308,309] but not necessarily the same mixture from each isomer. This is undoubtedly a consequence of the formation of a P(V) intermediate of sufficient lifetime to undergo extensive stereomutation. Nonstereospecificity has also been observed in the conversion of 4-methyl-2*H*-1,3,2-dioxaphosphorinane 2-oxide (**163**) into its 2-chloro derivative, **164**, with triethylamine–carbon tetrachloride.

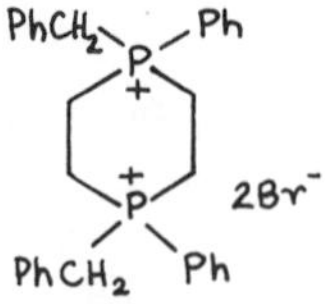

162 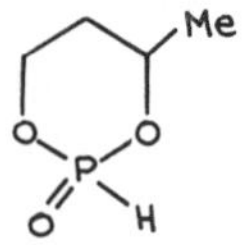 163 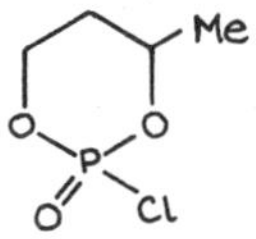 164

The thermodynamically less stable trans- isomer gives rise to a mixture of products (**164**), whereas the cis compound reacts stereospecifically. It seems probable that the intermediate in these reactions, possibly the anion, strongly prefers one geometry. Nifantev and Borisenko have shown that E_A for acid- or base-catalyzed trans → cis isomerization of **163** is about 95–100 kJ mol^{-1}.[306a] This is in the range observed for inversion of P(III), and the anion should possess some of the characteristics of P(III) by virtue of the resonance:

$$>P^{-}{=}O \leftrightarrow >P{-}O^{-}$$

Substitution reactions at phosphorus in P(IV) (*6*) compounds of halogen by amines proceeds with inversion of configuration, but the same is not true with other nucleophiles. Inch and Lewis report that methyl magnesium halides react with **158** ($R^1{=}Cl$, $R^2{=}O$) with 80% retention, possibly as a consequence of complexation.[296] Wadsworth has found a much more complex picture in the reactions of sodium phenoxides with the chloridate **154** (R = Me).[292] Very pronounced solvent and salt effects are observed; in particular inversion of isomer distribution occurs (86% retention → 84% inversion) on going from heterogeneous to homogeneous reaction mixtures, and addition of $Me_4N^+Cl^-$ produces a similar profound change (from 37 to 85% inversion). The latter salt also allows isomerization of the *p*-nitrophenyl ester of **154** (*p*-nitrophenoxy replaces Cl)—no reaction occurring in its absence. Wadsworth suggests that retention of configuration may arise from the formation of P(V) intermediates (assisted by cations), whereas inversion may be similar to an S_N2 process, that is, P(V) transition state. These conclusions are supported by Inch and his collaborators who found, for carbohydrate derivatives, inversion with weak nucleophiles and good leaving groups, but retention with good nucleophiles and less good leaving groups.[432] These workers[431] also assigned the configuration at phosphorus on the basis of δ ^{31}P, in support of Bentrude and Tan,[48] and described[430] one of the few studies of steric effects on the formation of P(IV) (*6*) compounds.

The stereochemistry of ring opening reactions in dioxaphosphorinanes is complex, varying with the reagent and substituents at phosphorus,[433] and displacements at the P=S group do not necessarily follow the pattern for the P=O group.[433–435]

E. P(IV) Seven-Membered Ring and Larger Ring Compounds

Very little work on these systems has appeared and even less of stereochemical interest. X-ray structures have been reported for 6-chloro-5,6,7,12-tetrahydro-5,7-dimethyldibenz[*d.g*]-1,3,2-diazaphosphocin 6-oxide(**165**)[310] and 4,5-dimethyl-2,7-diphenyl-2,3-dihydro-1*H*-1,2-azaphosphepin 2-oxide (**166**).[311] The latter compound and its oxygen analogues are obtained

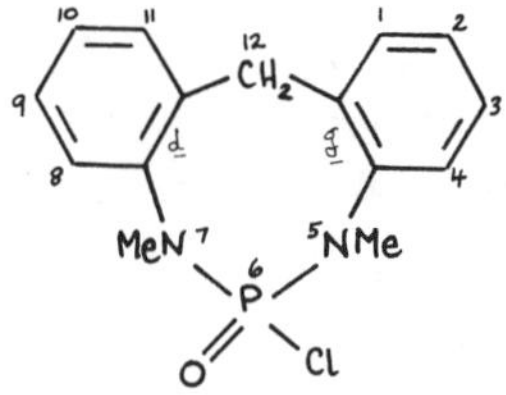

165

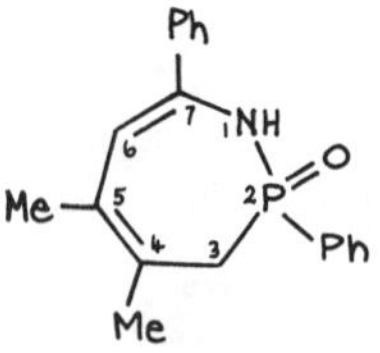

166

by ring-expansion reactions starting from phospholes and this route comprises the only potentially general entry into this field. Regrettably, attempts to reduce them to P(III) (7) compounds result in ring contraction to P(III) (6) derivatives.

The only conformational studies available are those of Sato and Goto,[59] for 2-chloro- 2-methoxy-, and 2-phenoxy-5,6-benzo-4,7-dihydro-1,3,2-dioxaphosphepin 2-oxides (**167**), who interpret their results to indicate the mobile conformational equilibria **167**$\rightleftharpoons$**168**. The predominant isomer is

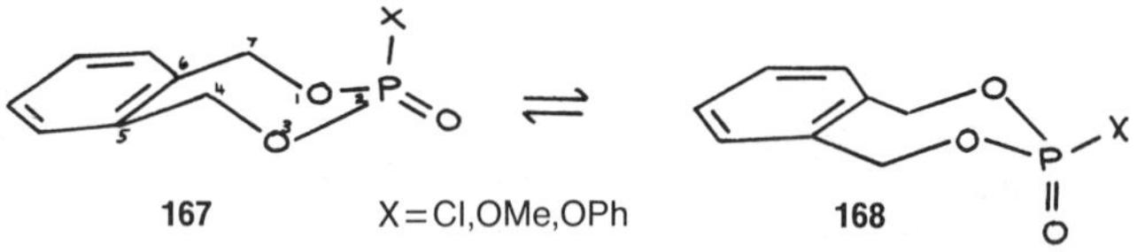

167 X=Cl,OMe,OPh **168**

R=Me,Ph

169

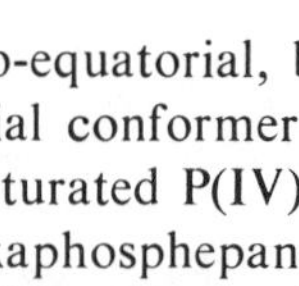

170

the one with the phosphoryl oxygen pseudo-equatorial, but this equilibrium is displaced in favor of the pseudo-axial conformer by added shift reagents. These authors also mention that saturated P(IV) (7) compounds (e.g., 2-methoxy- and 2-phenoxy-1,3,2-dioxaphosphepane 2-oxides,**169**) have equilibrating twist-boat conformations. Marsi[236] has isolated the two isomeric phosphine oxides, 3-methyl-1-phenylphosphepane 1-oxide (**170**), but the stereochemistry has not been assigned.

VIII. P(V) COMPOUNDS

A. Introduction

Before 1960 any discussion of phosphorus chemistry would have dismissed this topic in a few lines, but such has been the enormous interest in the last decade that it is now the single most studied facet of the subject. The reasons for this have been briefly outlined in the Introduction, but some further comment is necessary here. The introduction of phosphoranes as intermediates in the chemistry of phosphorus compounds coincided with an increasing interest in the stable members of the family, mainly fluorinated derivatives. This interest was due to convenient instrumentation (^{19}F nmr spectroscopy) becoming available and developed rapidly largely due to the Du Pont group of Muetterties and his collaborators.[312,313] As a consequence of the vast amount of work, principally spectroscopic, which has appeared, the literature has been reviewed several times,[7,10,314] and in particular the stereochemistry has been the subject of a book[315] which covers the literature thoroughly up until the middle of 1972. Hence an exhaustive or even comprehensive survey is not warranted here. In this section an attempt is made, however, to outline the principles with references to leading articles incorporating new material which have appeared until the end of 1974. It should be realized that virtually all the studies reported have dealt with cyclic compounds for reasons that will become apparent. This section is not subdivided with respect to ring size since the comparative effects of different sized rings are essential to the discussion.

B. Permutational Isomerization Processes (π-Processes)

The major reason for the intensive study of all aspects of P(V) stereochemistry is that it is a completely new field. Though chemists were accustomed to thinking of planar, tetrahedral, and octahedral molecules, these had, so far as was known, the simplifying feature of structural integrity. Indeed, it was and is considered a fundamental tenet of tetrahedral molecules that enantiomers will not interconvert without bond breaking. When Berry examined the ^{19}F nmr spectrum of PF_5, he therefore expected to observe two doublets since a fully symmetrical (i.e., all positions equivalent) pentaligated molecule is not possible.[8] The fact that only one doublet was observed called for an explanation in terms of rapid equilibration of isomers without bond breaking (since J_{PF} is not lost).

Calculations[320] indicate that the most likely structures for P(V) are either a trigonal bipyramid (TBP, **171**) or a square pyramid (SP, **172**), with the former being slightly preferred energetically. Many square pyramids are possible; the one normally taken as a basis for discussion, and which approximates well with the available X-ray data, has a basal-apical angle of about 105°. Berry adopted the TBP as a working model and to account for the equilibration suggested the following:

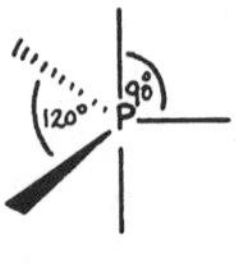

D_{3h}
171

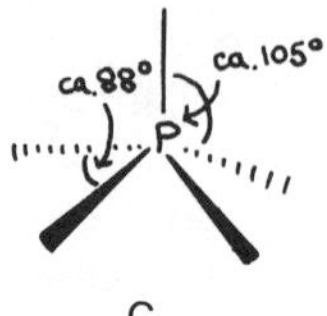

C_{4v}
172

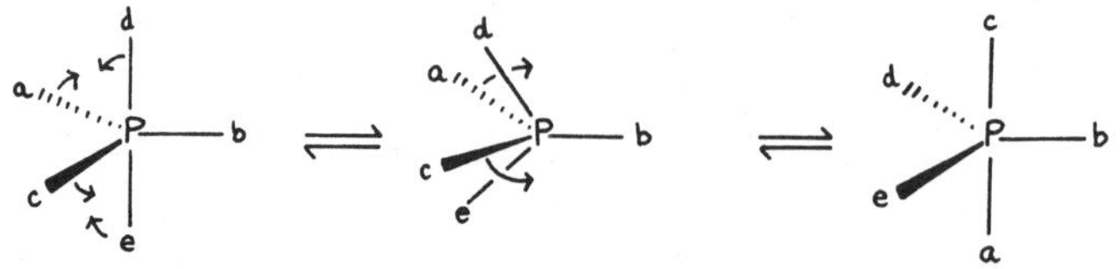

173 ⇌ **174** ⇌ **175**

Scheme 14

Normal thermal vibrations allow the angle aPc to open and the angle dPe to close converting the TBP, **173**, to an SP, **174**, in which the four substitutents a, c, d, and e lie in or close to the same plane. Such a process involving relatively minor angular distortions would be expected to have a low energy requirement, and, in fact, all subsequent evidence has supported this contention. The reverse process may now occur by opening one of the angles aPc or dPe and closing the other pair to afford either **175** or **173** respectively. The overall result (Scheme 14) is that the two apical substituents have become radial and two radial, apical. Whitesides and coworkers have presented convincing spectroscopic evidence that the exchange does occur in pairwise fashion.[318] One radial substituent (b in the scheme) is unaffected and is normally called the pivot.[10] Though the process appears complicated, it is simple to envisage with a little practice if one imagines taking any two radial substituents and straightening them out to give a new diapical pair (e.g., Scheme 15). This concept, with extensions and elaborations, remains the most widely used explanation of the equilibration process. Apparent symmetry arises when these conversions are rapid enough to allow the observation of only an averaged structure.

More recently the observation of rapid interconversions in molecules where this process could not apply, for example, 4′, 4′, 5′, 5′-tetra(trifluoro-

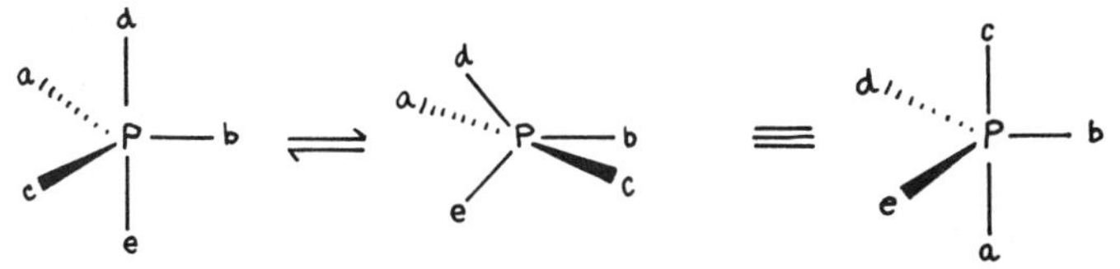

Scheme 15

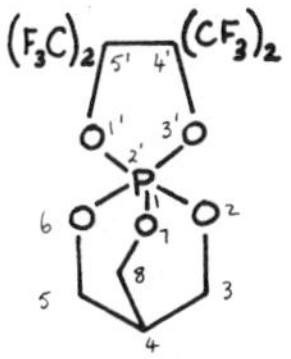

176

methyl)-spiro-[2,6,7-trioxa-1-phosphabicyclo[2.2.2]octane]-1, 2′-(1′, 3′, 2′-dioxaphospholane) (**176**), has led to the advancement of another mechanism by Ramirez, Ugi, and their collaborators,[316a,b,c,d,317] which they have termed "Turnstile Rotation" (TR). The ^{19}F nmr spectrum of **176** reveals a single signal above about −80°, a result incompatible with the rigid structure implied in the diagram. However, rotation of the CF_3-substituted ring with respect to the remainder of the molecule would bring about equivalence. If, by thermal vibrations, it were possible to arrange the substituents into a pair and a trio this process could be envisaged as general for TBPs. A net rotation of 60° of the two groups with respect to one another followed by reformation of the TBP achieves the same overall result as the Berry process (Scheme 16) and is equally compatible theoretically, though admittedly less symmetrical. Multiple rotations giving net rotations of 120°, 180°, and so on, are also possible where energetically desirable. A sixfold rotation, of course, regenerates the original TBP. Its great virtue lies in its ability to account for the behavior of strained molecules such as **176**, where the Berry path cannot be followed because of constraints

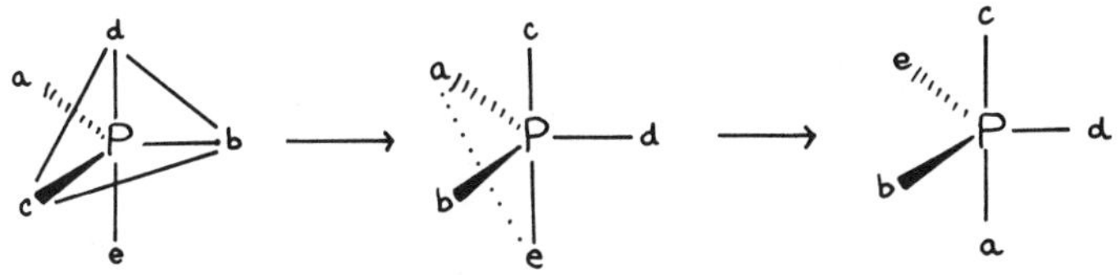

Rotate *bcd* by 120 — **177**

Rotate *ae* by 180 — **178**

179≡175

Scheme 16

on the movement of substituents relative to one another imposed by the structure of the molecule. To envisage the TR process, it is easier to consider it as occurring in the original TBP structure and as occurring in two stages. Thus in **177** the trio bcd rotates clockwise through 120° giving **178** the pair ae now rotates through 180° to give **179**. Of course, the whole process is synchronous and occurs in the molecule after the distortion from idealized TBP geometry. There are four possibilities of trio-pair combination, and each leads to the same result as a single Berry process.

It is most important to remember that the *net* result of both processes is the same, that is, one group (the pivot) is uninvolved, overall, and the remaining four groups have exchanged positions from apical to radial or vice versa. The essential difference from the point of view of the overall stereochemistry is that the TR process, since it involves rotation of groups, allows isomer interconversion without passing through TBPs, which are strained for steric reasons, for example, those in which a four-or five-membered ring lies in the radial plate. The major objection to the widespread application of the TR process is negative: if the process avoids strained TBPs why does it not occur all the time? As is discussed below, the presence or a four-or five-membered ring in a P(V) species does place an observable barrier between interconversion of some isomers. One might also ask why the TR process must have a dual-rotation character; there would seem no objection to the occurrence of only one such (formal) operation resulting in the interchange of only two substituents, (e.g., **177**→**178**, or **178**→**179**). The studies of Whitesides and his colleagues[318] indicate that this is not the case, and such an exchange is impossible by the Berry process. On the other hand, Berry processes cannot explain the very rapid isomer equilibration observed in cage phosphoranes such as **176**, whereas TR processes can. Further, the availability of four processes to achieve by TR what is done by one BP makes the entropy of the former more favorable. It is conceivable, however, that some element of strain, or other factor favoring C_2 symmetry,[320] must be present before the TR intermediate can form readily. Thus the more strained adamantanoid P(V) compound, **180**, is observed to undergo more rapid equilibration than the less strained analogues, **176** or **181**.

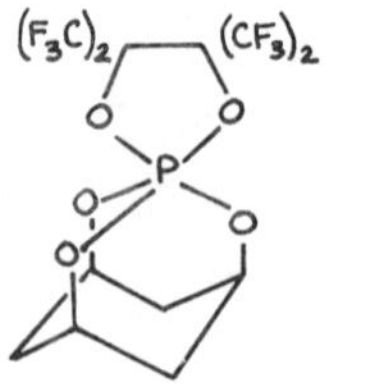

$(F_3C)_2$ $(CF_3)_2$ O O P O O O

180

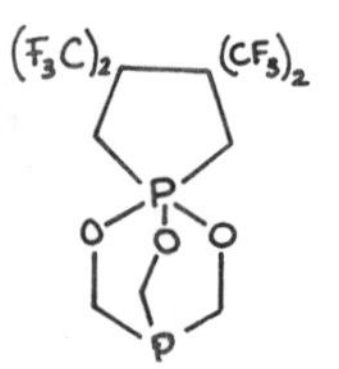

$(F_3C)_2$ $(CF_3)_2$ P O O O P

181

Clearly, much more work remains to be done in this area, but for now BP seems to have more supporting evidence than TR, *as a general process*, and it is used as the basis for discussion.

It should not be imagined that the two processes so far described are the only ones possible. A large number of other processes may be envisaged,[312,313,315,319] but they have been rejected since they are incompatible with the findings, of Whitesides and his collaborators, that exchange of substituents must occur in pairwise fashion. The method used was variable temperature line width analysis in the proton decoupled ^{31}P spectra of XPF_4 (X = Me_2N, Cl, Me).[318a,b] In addition, it has been found that the isomerization is independent of concentration, and thus is intramolecular, and that the activation energy for the process is low (about 17–38 kJ mol^{-1}). Interestingly, the rate for Me_2NPF_4 is accelerated by basic unhindered ethers probably as a consequence of reversible complexation (**182**), as in

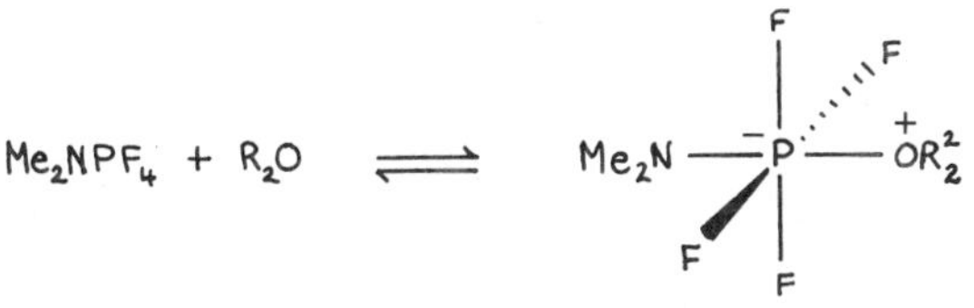

$$Me_2NPF_4 + R_2O \rightleftharpoons Me_2N\overset{-}{P}F_4\overset{+}{O}R_2^2 \qquad (10)$$

182

Equation 10 which can be considered as solvent assistance in the Berry process. Whitesides and his co-workers have also attempted to detect and obtain information about the intermediate between two TBP isomers.[318,b,c] Again using nmr methods, they concluded that a square pyramidal intermediate best accounts for the equivalence of the isopropyl methyls in **183**⇌**183a**, a process occurring simultaneously with the isomerization.

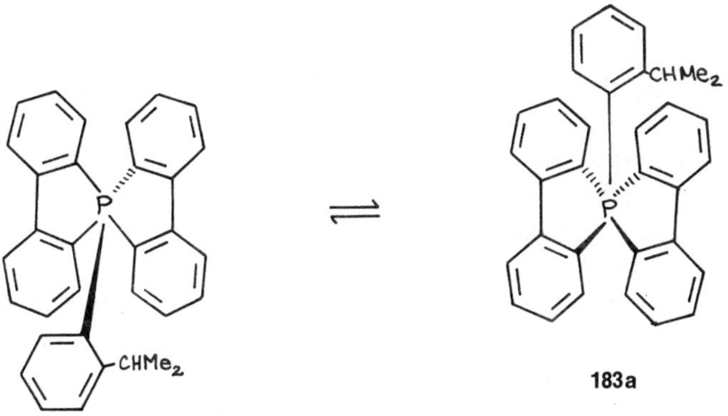

CHMe₂

CHMe₂

183

183a

Overall, permutational isomerizations in P(V) systems can be reasonably considered to involve TBP ground states and to proceed through a square pyramidal intermediate. It must be stressed that the process TBP⇌SP only involves relatively minor angular deformations, and whereas most solid state structure determinations support the TBP hypothesis, some SP structures have been found.[321–323] Given that the differences between the two structures, in energetic terms, are small and that small differences between solid state and solution structure are to be expected, serious consideration must be given to the possibility that there may be whole families of P(V) structures which are best considered as SPs interconverting by way of TBPs. Holmes recently cogently argued this case[324] and pointed out that what are called distorted TBPs are sometimes, in fact, less distorted SPs. It would not be surprising to find a whole range of structures between TBP and SP, but spectroscopic methods are not yet sufficiently precise to allow useful distinction in solutions, and solid state structures are always open to the criticism that deformations may be due to crystal forces. Further, as Holmes points out, interpreting structural data as SP rather than TBP gives a best fit only with spirocyclic oxyphosphoranes where the basal oxygen atoms are part of rings of roughly the same electronegativity. Very recent data have provided exceptions to even this in the compounds **184** which are essentially SP.[325]

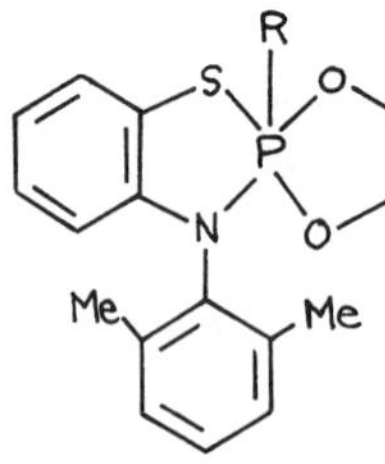

R
S
O
P
N
O
Me
Me

184

Nevertheless, theoretical considerations[320] indicate that electronegative substituents should preferentially occupy the basal plane of an SP, and it is reasonable to suppose that an accumulation of electronegative substituents might favor the SP structure, though other factors must be considered. For the moment, however, TBP constitutes a better general starting point for interpretation and is used here.

It is, of course, possible to isomerize P(V) heterocycles by more common thermal or acid- or base-catalyzed ring rupture and reclosure.[7,344] Ironically, such processes are termed "irregular" as distinct from the "regular" permutational isomerization processes described above. Although rare, they must be checked for before assuming their absence.

1. *Nomenclature of Permutational Isomerization Processes*

The most widespread name given to these intramolecular reorganizations of substituents is pseudorotation. The term is a bad one since it is already used in the conformational analysis of five-membered rings. Hence the letters PI (Permutational Isomerization) or even simpler, the homonym π, are preferable, and it is certainly easier to say π-process than the mouth-filling full name.

Molecules capable of undergoing π-processes have been described by Muetterties as stereochemically nonrigid. This casts a rather wide net, and the writer prefers the term "configurationally mobile," referring to any ligand reorganization process involving (formally at least) only angular distortion. The term has the added advantage of connoting what is, to organic chemists, the most unusual property of these processes.

2. *Factors Influencing Positional Isomerism in Phosphoranes*

Since their initial use as intermediates to explain aspects of the chemistry and stereochemistry of phosphorus, a number of features have been recognized which assist in determining which is the most likely structure to be found. Some of these are self evident; for example, when two substituents are the termini of a chain then they must occupy adjacent positions either diapical or apical-radial. Since the apical-radial angle is 90°, four and five-membered rings will preferentially occupy such a position, whereas six-membered rings being more flexible do not have nearly such a strong preference. Further, there is a positive energy barrier to either four- or five membered rings being placed diradial. The reason for this is that the angular strain will be much less when the ring is in the apical-radial position. Indeed, the relative stability of P(V) (*5*) systems has led to their being by far the most abundant single class of P(V) compounds and the first such compound to be prepared was 2*H*-2,2'-spirobi(1,3,2-benzodioxaphosphole) (**185**) obtained by Anschutz and Broeker in 1928 (Equation 11).[326]

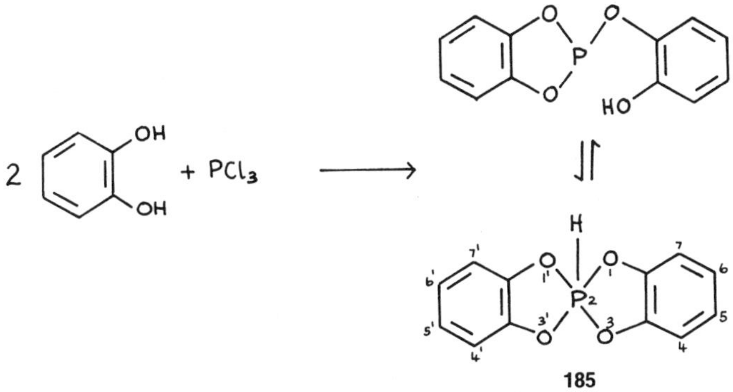

$$2\ \text{C}_6\text{H}_4(\text{OH})_2 + \text{PCl}_3 \longrightarrow \quad (11)$$

185

Not surprisingly, these workers assumed a P(III) (*5*) structure, and the true nature of the compound was not recognized until much later. In fact, spirocyclic P(V) (*5*) structures formed in this way now constitute the largest and most diverse class of phosphoranes. They are formed readily from the following types of 1,2-disubstituted compounds and a P(III) derivative: α-diols,[327,328,338,339] α-amino alcohols,[329,338,339] α-amino acids,[330] α-hydroxyacids,[331] hydroxamic acids,[332] amidoximes,[333] hydrazides,[334,335] and amidrazones.[336] Nothing is known of the mechanism of the final ring closure step, but some studies have been done on the influence of substituents on the chain-ring P(III)⇌P(V) equilibria which exist in solution.[329,337,340,341,343] In general, substitution at the ring carbon atoms increases the percentage of the P(V) (*5*) compounds, whereas the P(III) (*5*) compound is favored by increased temperature and basic solvent. This last effect is particularly noticeable when the hydrogens of the ring-forming group are notably acidic, for example *o*-diphenols or α-hydroxyacids.

This enhanced stability and ease of formation of P(V) (*5*) have been attributed[344] to reduction of steric crowding in cyclic as compared with acyclic P(V) compounds. In view of the great ease with which these rings form, however, even at the expense of a phosphoryl linkage,[345,346,438,439] this explanation is probably an oversimplification, and other steric and electronic effects may be operating. Electronic effects seem particularly probable in the case of *o*-diphenols and related aromatic substrates. Both hydroxyl and carboxyl groups are known to cyclize onto P=O to form P(V) (*5*). Thus, the phosphorane (**185**) is formed easily from catechol and anhydrous phosphonic acid in the presence of dicyclohexylcarbodiimide, and simply boiling phosphorus oxychloride with catechol or substituted catechols converts it to pentaoxyphosphoranes.[346] Similar results are obtained with *o*-aminophenols,[345] and the phosphorane 1,4-dihydro-2,3-dioxa-*P*,*P*-diphenyl-2*a*-phospha(V)cyclopent[*c.d*]indene (**185b**) is formed easily from its precursor diol (**185a**) by heating.[438] The P-substituted 1,6-dioxa-2,7-dioxo-5-phospha(V)spiro[*b.f*]dibenz[4.4] nonanes (**185d**) are formed so readily from **185c** that their precursor dicarboxylic acids are unknown. These P(V) (*5*) lactones display a quite extraordinary chemical and thermal stability, being unaffected by boiling aqueous alkali and temperatures as high as 340°. On the other hand, although As-substituetd 1,6-dioxa-2,7-dioxo-5-arsa(V)spiro-[4.4]nonanes (**185f**) are formed spontaneously from the dicarboxylic acids[440] (**185e**), an analogous phosphorus compound does not cyclize.[441]

It is clear that the formation of a five-membered ring containing two or more sp^2 carbon atoms constitutes a strong driving force for the formation of both mono- and spirocyclic phosphoranes.[314] It is not yet clear why this is so. Such rings are nearly flat, and although not very strained, they

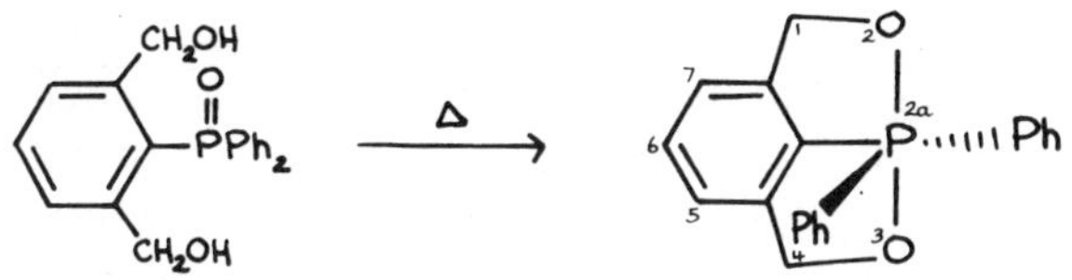

185a 185b

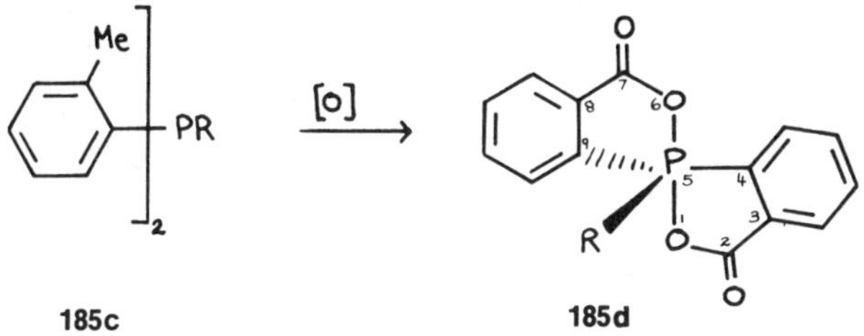

185c 185d

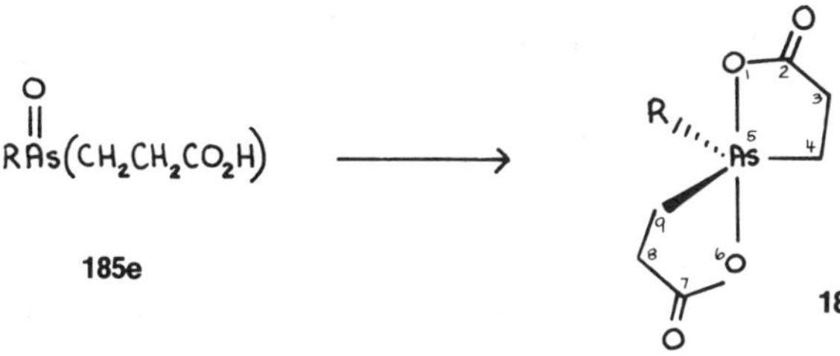

185e 185f

are certainly not strain-free; so it is far from obvious why their formation should be so favored.

The second major effect apparent in P(V) compounds is the electronegativity of the substituents, first suggested by Muetterties and his collaborators[347,348] on the basis of spectroscopic studies. Electronegative substituents prefer the apical position, in a TBP, and electropositive substituents bind better in the radial position. Theoretical calculations support the experimental observations.[320] While this distinction is quite clear for elements of widely differing electronegativity, (e.g., fluorine and carbon) it is much less so for pairs of elements where the difference is not so large.[349] In order to measure this preference quantitatively, or at least place various atoms in some order, spectroscopic means have been devised. Essentially the same methods can be used to determine the energy required to place a ring diradial, and the two are described together below.

Examples of the competitive operation of these effects are numerous. At room temperature acyclic R_2PF_3 and cyclic analogues with six-membered rings show two types of nuclei in a ratio of 2:1 in their ^{19}F spectra. This is compatible with the structure **186** where two fluorines occupy the apical positions and the ring is diradial. The P(V) (*5*) analogue, however, shows only a single type of fluorine. This is easily explained since if the ring occupied an apical-radial position (**187**) it would be destabilized by having two radial fluorines and one apical carbon. The increase in energy makes the isomer comparable in energy with the isomer having a diradial ring (**188**), and hence π-processes occur readily resulting in apparent equivalence of the fluorine atoms.

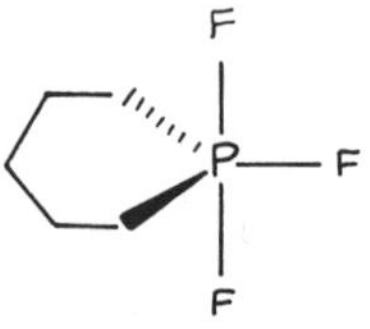

186 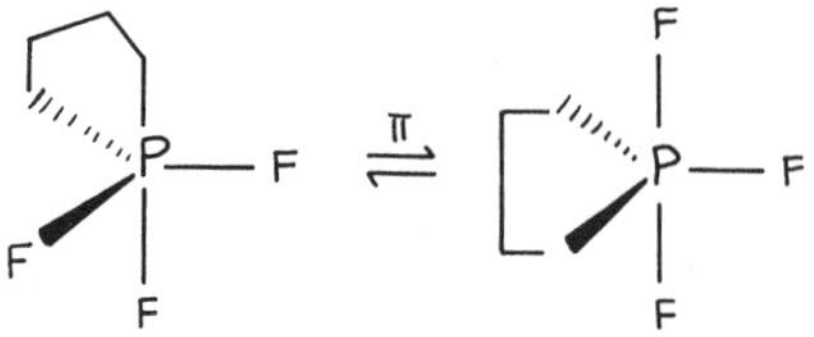 **187** **188**

On the other hand, the steric effect outweighs the electronegative effect in 2-fluoro-2,2′-spirobi(1,3,2-benzodioxaphosphole) (**189**) where the presence of the two five-membered rings forces the fluorine apical.[350] A combination

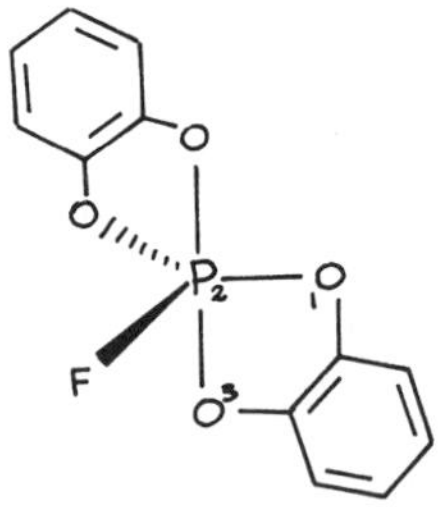

189 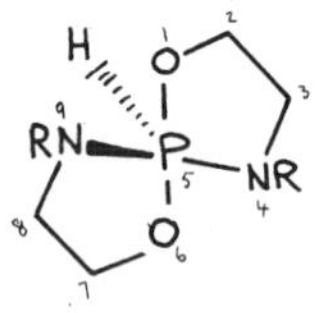 **190**

of steric and electronic effects has been applied by Wolf, Burgada, and their collaborators, who used β-amino alcohols to prepare the 4,9-disubstituted 1,6-dioxa-4,9-diaza-5-phospha(V)spiro[4.4]nonanes (**190**) whose structure is virtually fixed.

Steric effects arising from the bulk of substituents have been studied little, and few gross influences have been noted for P(V) heterocycles whether as discrete substances, reaction intermediates, or transition states. Substantial effects have been observed in acyclic systems, and it is suggested that the *t*-butyl group preferentially goes apical to relieve strain in the P(V) intermediate in the alkaline hydrolysis of P(IV) salts containing this group.[351] It has been found that the *t*-butyl group thermodynamically prefers the radial position in monocyclic P(V) compounds.[351a]

Two further and somewhat more subtle effects need to be considered: substituents with lone pairs of electrons (π-donors) when radially placed will lie with the lone pair in the radial plane, and substituents with an acceptor orbital (π-acceptors) will lie with this orbital normal to the radial plane. An alternative statement is that π-acceptors prefer apical sites and π-donors prefer radial sites.[320] Since π-donors are normally electronegative, this effect tends to offset the influence of electronegativity. Experimentally, the effect may be observed in the barrier to rotation about radial bonds to sulfur[352] and nitrogen.[353–356] Where this occurs care must be taken to distinguish between such restricted rotation and π-processes.[356] The π-donor or π-acceptor effect is apparently much weaker than the electronegative effect; hence fluorine preferentially occupies apical positions, but it is likely to be more important for substituents where electronegativity is less but not the π-donor effect, (e.g., phenyl).[357–359]

3. *Energy Barriers in π-Processes*

It would, of course, be of great value in predicting and understanding P(V) structures to have some idea of the energy required to bring about the isomerization processes, for example, the energy necessary to make a four- or five-membered ring adopt a diradial position. It would also be of great interest to have an ordering of common substituents in terms of their preference for an apical (or radial) position. Because of the interplay of conflicting effects neither of these desirable goals has been achieved. There is, however, a considerable body of data which allows some general trends to be perceived. Nearly all this information has been acquired in the same way, namely by variable temperature nmr studies with activation energies being calculated from signal coalescence temperatures. The approach is described generally in Scheme 17. Consider a monocyclic P(V) compound, **191**, bearing substituents A and B on the ring. π-Processes which do not involve a diradial ring (e.g., **191**$\rightleftharpoons$**192**) effectively interchange only the apical and radial environments and do not alter geometrical relationships (e.g., of A and B with R^2). However, when an isomerization occurs giving an isomer in which the ring does occupy a diradial position (**193**) a π-process is available which does alter geometrical relationships (**193** $\rightleftharpoons$ **194**). The ring is effectively symmetrized, with respect to the remaining substituents.

If A and B are detectable by nmr spectroscopy, the two processes should be distinguishable if they proceed at a rate which is slow on the nmr timescale. Thus, there are two environments for A (apical **191** and radial **192**), and determination of the coalescence temperature of the two peaks is a measure of the energy barrier **191**$\rightleftharpoons$**192**. For simplicity A and B are normally identical, and when these two signals coalesce it is possible to estimate the second energy barrier. The first of these barriers is much lower than the second and is not often observed unless the two atoms which terminate the ring are of widely different electronegativity,[360–362] for

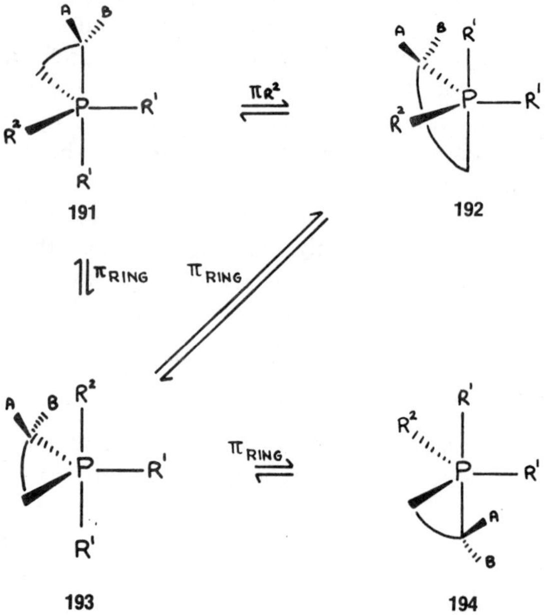

191 192 193 194

Scheme 17

example, carbon and oxygen. Where they are the same, the processes is difficult to freeze out even at low temperatures. Estimates of the barrier when there is no great difference in electronegativity vary, but a range of 20–30 kJ mol^{-1} is probably reasonable.[363] With one terminus carbon and the other oxygen, a figure of 40–70 kJ mol^{-1} was found[361] for P(V) (*5*) compounds.

The energy for the second process, placing the ring diradial, depends on ring size, but for four-[364,358] and five-membered[361,365–368] rings it is normally considerable and falls in the range 40–100 kJ mol^{-1}. It may seem unusual that there is not a substantially greater energy requirement for four- as opposed to five-membered rings, but it must be remembered that the range of P(V) (*4*) compounds available is very small and restricted virtually to carbocyclic compounds which prefer the diradial position on electronegativity grounds. Six-membered rings have a much lower barrier to diradial placement.[369]

This spectroscopic method may be extended to the determination of the relative preference of substituents for an apical (or radial) position. Thus, Trippett and co-workers studied the effect of varying R in **195** on the barrier to placing the four- or five-membered ring diradial, since such

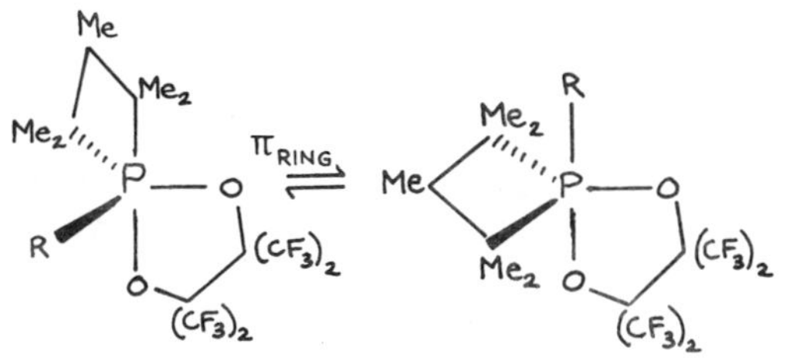

195 ⇌ (π_{RING}) 196

a process must also place R apical. This[358] and other [357,359,370] studies have provided the approximate apicophilicity series:

$$\text{H, OMe, }(CF_3)_2\text{CHO} > \text{PhO, PhS} > \text{NMe}_2 > \text{Alkyl} > \text{Ph, Cl}$$

There are a number of surprises here, namely, the high standing of hydrogen, the similarity of oxygen and sulfur, and the low position of chlorine. Though these anomalies can be rationalized, the position of a substituent in the series is sensitive to the other groups present, and the order shown can only be taken as a rough guide.

C. Stereoisomerism and Optical Activity

An acyclic TBP phosphorane with five nonidentical ligands will have 20 isomers and 10 pairs of enantiomers. With a SP the situation is worse with 30 isomers and 15 enantiomeric pairs. There has been no experimental confirmation of this awesome complexity presumably because barriers to π-processes are very low. Though the problem becomes simpler if rings are present and some substituents are the same, it still remains bewilderingly complex.[315] With hindsight, it is interesting to consider that this complexity may have discouraged workers from studying the stereochemistry of these molecules. The introduction of a ring reduces the number to 18, but the problem of considering the possibilities remains formidable. The introduction of a second ring reduces the number of diastereoisomers to two (if the rings are the same) or three (if they are different), but many studies have been carried out on monocyclic compounds, and it is desirable to have some method to cope with the complexity. This was solved by a number of workers who used three-dimensional graphs;[371–374] of which the easiest to remember and to use is due to Mislow.[373] It is shown (**197**) in the form used for monocyclic compounds.

Each apex represents an isomer which is identified by its two apical substituents as one of a pair of enantiomers; for example, 23 and $\overline{23}$ are

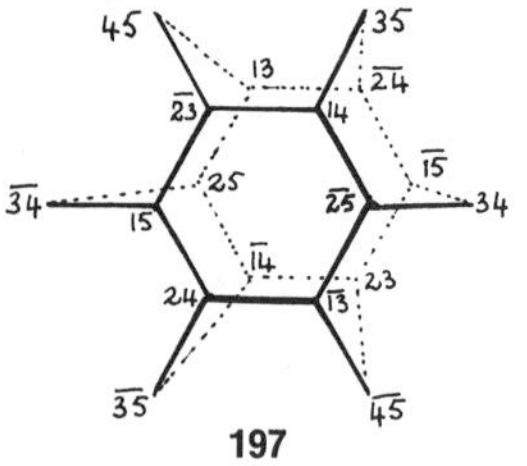

197 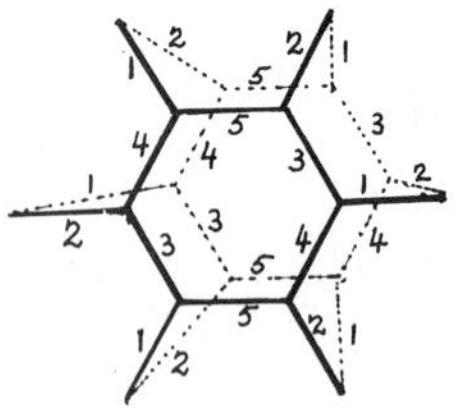 **198**

such a pair. By convention the termini of the ring are usually numbered 1 and 2, but since they cannot both be apical they do not appear this way on this graph. Each isomer is connected to its enantiomer by the five paths, each utilizing a different pivot, necessary to bring about inversion by π-processes.[14a] By adopting a suitable convention chirality may also be specified. Each line represents a π-process which may be identified by its pivot, and the simple pattern of these numbers (**198**) makes it relatively simple to construct the graph.

The graph is not merely a representation of isomers in a convenient form; it is possible to trace paths whereby π-processes interconvert isomers without actually drawing out the TBPs—a process prone to error. Remembering that 1 and 2 are the ring termini, it can be seen that the front and back hexagons represent isomers in which the ring is *ar*, whereas the bridges represent isomers with *rr* rings, that is, high energy forms if the ring is four- or five-membered. Since inversion requires going from one hexagon to another it must proceed by an *rr*-ring form and hence will face a substantial energy barrier. Mislow's original paper should be consulted for a more detailed discussion.[373]

From the above it should be clear that the construction of an optically active phosphorane, whose optical activity is due solely to phosphorus, will present formidable problems, but these have been resolved successfully by two approaches. The first was due to Hellwinkel[375] who prepared and resolved the P(VI) anion, **199**, which on treatment with acid afforded a mixture of spirophosphoranes, **200–202** of which only **202** proved to be optically active (Scheme 18). Since all the substituents have the same electronegativity, **202** is certainly a mixture of the two siomers shown, and variable temperature polarimetry should clarify this point. Both bis-2,2′-biphenylene derivatives, **200** and **201**, are chiral, but they racemize very easily because only a relatively minor angular enylene distortion converts them to a symmetrical SP. This is not possible for the bis-2,2′-bisphenylene **202**;

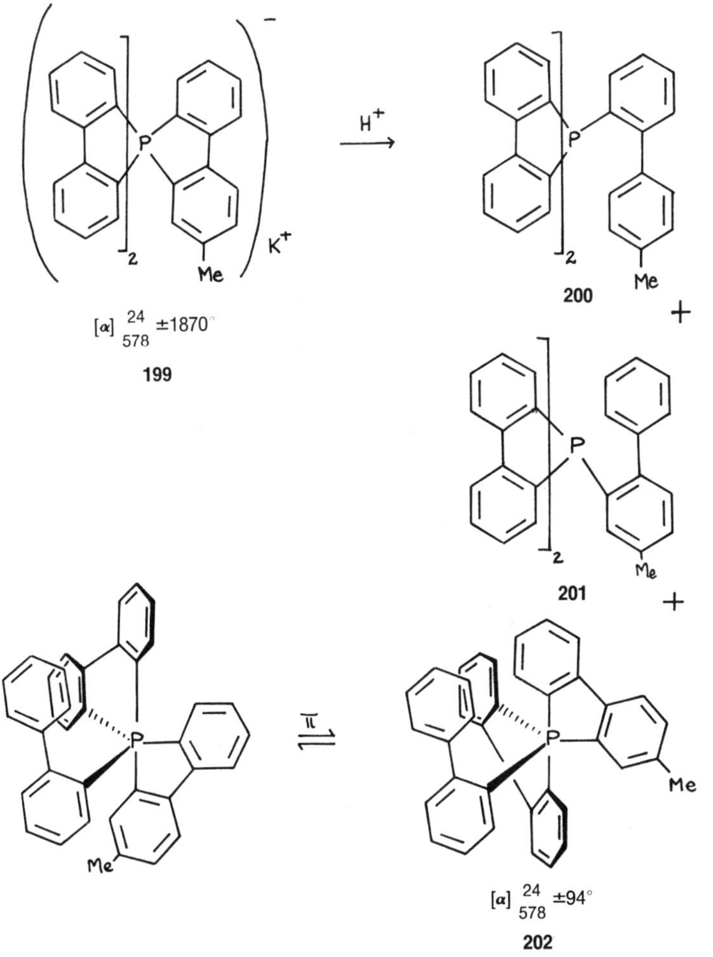

$[\alpha]^{24}_{578}$ ±1870°

199

200

201

$[\alpha]^{24}_{578}$ ±94°

202

Scheme 18

since the SP intermediate between the two forms is still chiral and inversion requires a π-process involving an *rr*-ring.

An entirely different approach has been used by the French groups of Wolf and Burgada and their collaborators. The spirophosphoranes, 1,6-dioxa-4,9-diaza-5-phospha(V)spiro[4.4]nonanes (**203**), prepared from amino alcohols have a fixed structure[342] with the oxygens apical and the nitrogens radial because of a combination of electronegativity and π-donor effects. Hence the isomerization **203**→**204**, so facile when both ring termini are the same, is blocked in this case (Equation 12). The prediction is confirmed by nmr[342] and X-ray studies.[378] By commencing with an optically

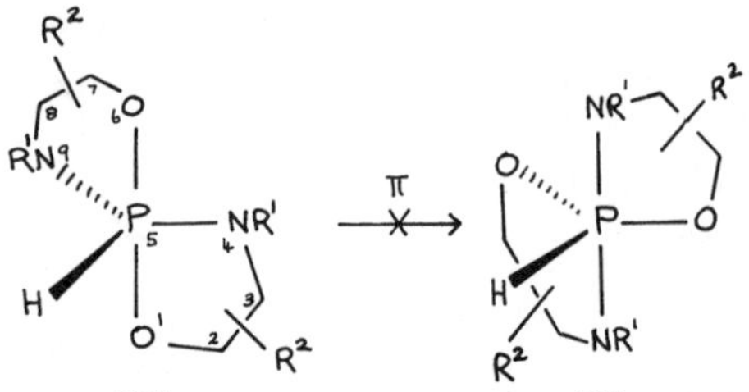

active amino alcohol (ephedrine and related compounds), Wolf and his collaborators were able to prepare a mixture of diastereoisomeric P(V) (5) compounds which differed in configuration only at phosphorus.[377–379] Slow crystallization from a solvent converted this mixture into a single pure epimer,[378] another example of second order asymmetric transformation which arises here as a consequence of slow epimerization of the P(V) compound in solution. Redissolution of a pure epimer results in reequilibration of diastereoisomers, which is apparent from the mutarotation of the solution. This process has been studied in some detail,[367,380] and mutarotation was observed in the first synthesis of such a phosphorane,[381] presumably as a consequence of stereoselectivity in the synthetic process. The activation energy for the epimerization is close to the values obtained by other means for placing a five-membered ring *rr*. However, in all these studies the possibility of a very rapid P(V)$\rightleftharpoons$P(III) equilibrium cannot be excluded (though no P(III) compound has been detected), and the energy involved in either an irregular or regular process is not greatly different.[361]

It is not, of course, necessary for both rings in a spiro-P(V) compound to be optically active, and indeed Bernard and Burgada have found[382] the initial preparation of the P(III) (*5*) oxazaphospholane to be stereospecific for **206** (R^1 = NMe_2; X = O, CHPh) and stereoselective for **206** (R^1 = OCH_3; X = O, CHPh), thus allowing stereoselective conversion to P(V) **207** by the general synthetic method developed by Ramirez (Equation 13).[10]

Epimerization and equilibration have been demonstrated for these systems also.[382] No example of the resolution of a spiro-P(V) compound in which at least one ring is not itself optically active is known, although chirality has been demonstrated by use of an optically active shift reagent.[383] ^1H nmr spectra of spiro-P(V) compounds are simplified by the use of shift reagents, but they may influence P(V)$\rightleftharpoons$P(III) equilibria.[384]

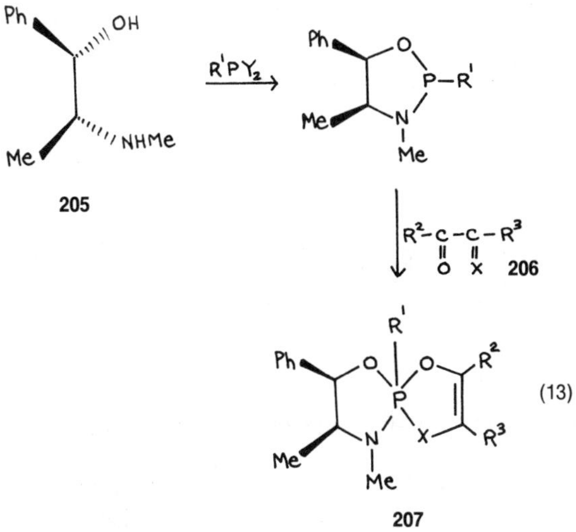

205

R^1PY_2

206

207

(13)

1. *Absolute Configuration*

The only work in this field is that due to Wolf and his colleagues.[377] They consider their spirocyclic P(V) compounds to be right- and left- handed helices (*P* and *M* respectively in the Cahn-Ingold-Prelog nomenclature[385]) (**208** and **209**). The view point selected is along the H–P bond in the direc-

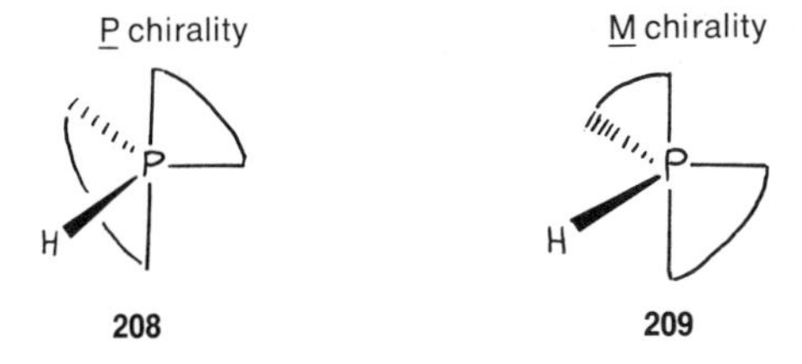

208 **209**

tion H→P. The X-ray structure of one pure epimer is known,[376] and assignments have been made to a number of others[378,379] on the basis of their rotations.

The preparation and resolution of P(V) compounds chiral only at phosphorus and of known absolute configuration is a challenging problem full of experimental difficulties yet to be overcome.

D. Diastereoisomerism

Apart from chirality, diastereoisomerism would be expected in P(V) compounds wherever there is a substantial barrier to the π-processes which would interconvert them (e.g., **210** and **211**). This was first observed by Westheimer and Gorenstein[360] and by Ramirez and co-workers,[362] and numerous other examples can be found in Hellwinkel's review.[314] Stereo-

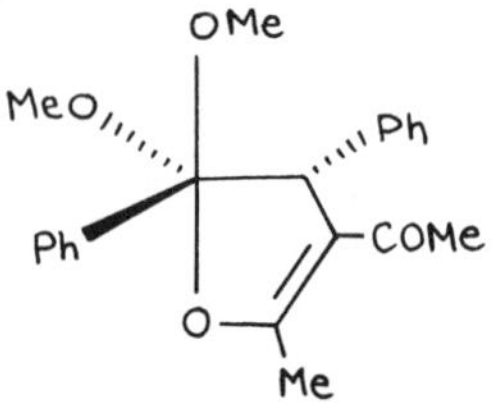

210 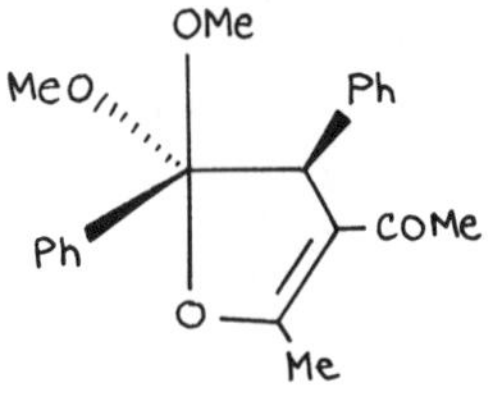 **211**

chemical assignments are usually tentative and based on nmr studies where both isomers are available.

The facile *a,r* interchange, which normally occurs with spirocyclic carbophosphoranes such as **200**, can be slowed by increasing the bulk of the noncyclic substituent;[386] when the substituent is 1-naphthyl, the *a,r* exchange does not become rapid until 54°C,[386] and when it is 8-dimethylamino-1-naphthyl, freezing of both the TBP and rotation about nitrogen apparently occurs.[387]

Stereoisomerism can also arise as a sole consequence of the ring substituents, for example, the 2:1 aldehyde-phosphite adducts, such as **212**, prepared by Ramirez and his colleagues (Equation 14).[390] Here rapid interchange of the OMe groups makes them effectively equivalent, and the only isomerism detectable is that due to the ring substituents. It is usually possible to asign this stereochemistry on the basis of J_{POCH}.

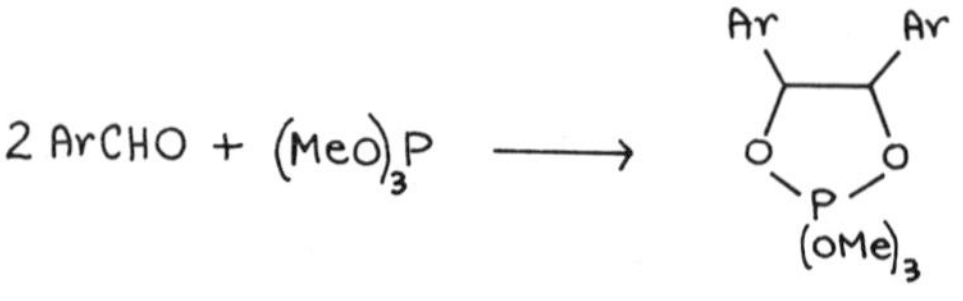

$$2\,ArCHO + (MeO)_3P \longrightarrow \mathbf{212} \quad (14)$$

212

E. Reaction Stereochemistry of P(V) Compounds

The involvement of P(V) compounds as intermediates or transition states has attracted great interest because of its success in explaining many features of the reactions of P(IV) compounds. Consequently the subject has been reviewed twice in recent years,[344,388] and the most recent review is very detailed.[344] Only a survey of the main classes is given here.

The reaction stereochemistry of phosphoranes themselves is very limited when the studies of intramolecular processes are excluded. The factors important in determining structure in P(V) compounds may be used to explain why some reactions proceed when others fail. For example, trialkyl phosphines react with α,β-unsaturated carbonyl compounds to give zwitterions, whereas if one or more of the groups on phosphorus is alkoxy, a P(V) (*5*) compound is formed[10] (Scheme 19). This is understandable since

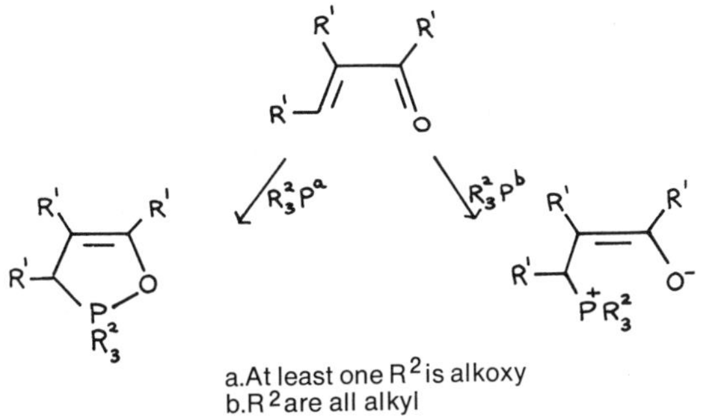

Scheme 19 (*a*) At least one R^2 is alkoxy; (*b*) all R^2 are alkyl.

with a trialkyl phosphine the product phosphorane must have an apical alkyl group. Similarly, one may understand the P(IV)$\rightleftharpoons$P(V) equilibrium of the 1:1 adduct of $P(NMe_2)_3$ and benzil.

Reactions of one phosphorane to give another are uncommon, probably the best known being the exchange reaction of $(RO)_5P$ with a 1,2-diol to give a five-membered cyclic P(V) compound and ROH.[369,389] These reactions occur readily, sometimes exothermally, and have been attributed to reduced steric crowding in the cyclic P(V) compound. A more striking example[391] is the preparation of **212b**, which occurs with loss of hydrogen (Equation 15). This reaction proceeds by way of the P(VI) intermediate, **212a**.[391b]

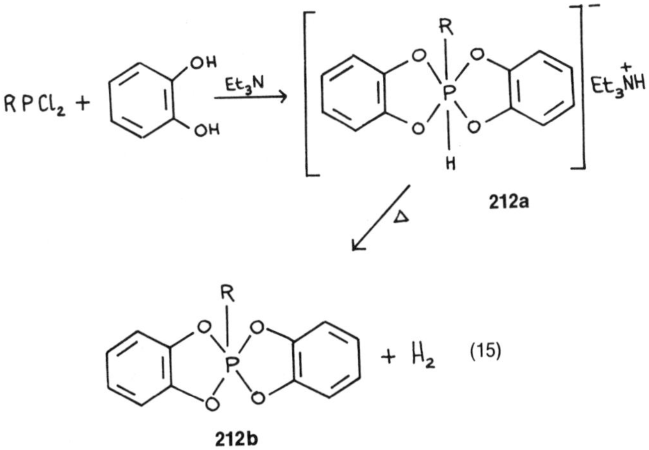

1. *P(V) Intermediates and Transition States*

These normally arise in two types of reactions; the first and less common of these is the reaction of P(III) compounds with electrophiles. An example is the reaction of phosphites and peroxides to give P(V) compounds which vary in stability and which are normally more stable when five-membered rings are present. The cyclic peroxide **213** reacts much more readily with Ph_2POMe than with Ph_3P; a reasonable explanation is that in the intermediate phosphorane from the phosphinite, two oxygen atoms can occupy the apical positions, whereas this is not possible with the phosphine.[396]

The P(V) compound **214**, obtained from the tricyclic phosphite and ozone, is much more stable than acyclic analogues, although the reason

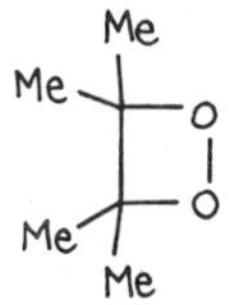

213

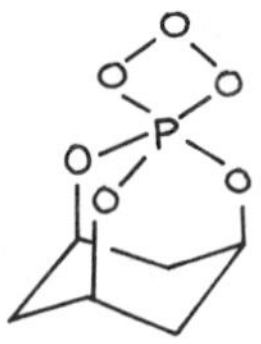

214

for this stability is unclear.[392] It may be that the decomposition (to give singlet oxygen) is a concerted process. If this is true the O_3 ring should be *rr*,[320] a difficult process for **214**, and much simpler for acyclic analogues such as the triphenyl phosphite–ozone adduct. Indeed, the unusual stability of **214** may well stem from the energy barrier to placing the four-membered ring *rr*.

The sequence of reactions shown in Scheme 20 was recently reported by Hall, Bramblett, and Lin[393] and provides a thought provoking example of how the application of stereochemical principles could be put to good synthetic use.

Theoretical considerations indicate[320] that concerted processes of the Woodward-Hoffmann type will occur only for groups situated *aa* and *rr*. Since the former is impossible for all save the largest rings, most examples will be found in the *rr* series. Thus conversion of 1,2,5-trimethyl-3-phospholene (**215**) to the P(V) compound 1,1-diethoxy-1,2,3-trimethyl-3-phospholene (**216**) will either force the five-membered ring *rr* or substantially reduce the energy necessary to place it there. Strain is relieved by the electrocyclic process forming the diene and the phosphonate. The same is true for **217**, which is destabilized because any isomer must have at least one carbon apical, and hence the energy difference between an *ar* ring and an *rr* ring is again reduced.

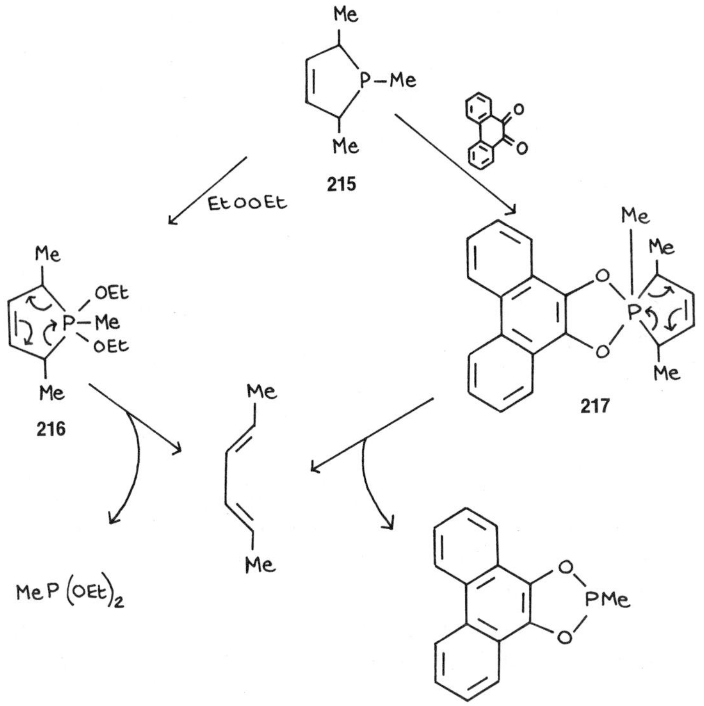

Scheme 20

The logical extension of this approach to both the synthesis of phosphoranes and synthetic organic chemistry should prove a fertile field.

Attack of a free radical on a P(III) compound gives rise to a phosphoranyl radical which is probably TBP, but available evidence[394] indicates that subsequent reactions proceed faster than π-processes. Little work has been done with cyclic compounds, though Hellwinkel obtained a relatively stable radical (**219**) by the thermal decomposition of the spirophosphorane 5*H*-5,5′-spirobi(dibenzophosphole) (**218**).[397] Roberts and co-workers have, however, made some interesting observations based on ESR spectra of P(V) radicals formed by hydride abstraction from the corresponding P-H phosphoranes. They find that for a monocyclic P(V) (5)[397a] in which one ligand is a half-filled orbital a low energy π-process (*ca.* 2.9 kJ mol^{-1}) is occurring which they assign to the *ar* interchange of the odd electron. On the other hand the spirocyclic analogues[397b] show little spectral change up to 130°C. These workers attribute this to the absence of π-processes but it appears much more likely that the process utilizing the half-filled orbital as pivot is occurring very rapidly at all temperatures used. However, this

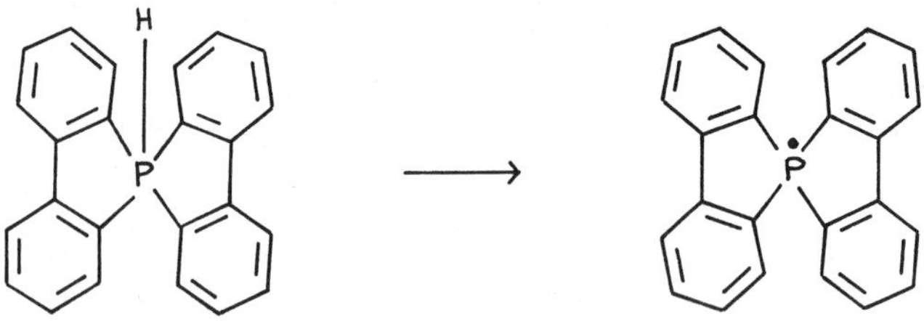

218 219

still suggests a remarkably high energy barrier to placing the odd electron apical.

The second and far more widely studied group of P(V) intermediates and/or transition states is that formed by the attack of nucleophiles on electrophilic phosphorus, usually, but not always, P(IV) compounds. Attack at electrophilic P(III) is a commonplace occurrence (e.g., reaction of PCl_3 with alcohols, amines, etc.), but nothing is known of the intermediates involved. Displacement of chloride from 1-chloro-2,2,3,4,4-pentamethylphosphetane (**220**) proceeds with inversion,[24] For this to happen by way of a P(V) intermediate involves at least three π-processes or placing the ring *rr*, an energetically expensive step with a barrier that may be reduced by the essentially unknown properties of the lone pair ligand. Alternatively, no discrete intermediate may be formed—only an

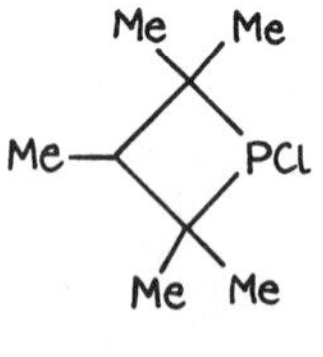

220 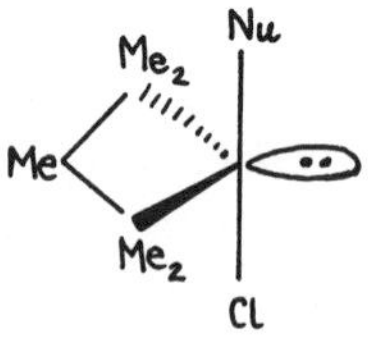 221

S_N2 type transition state. Since a stable adduct Me_3N-PCl_3[395] and an anion PBr_4^- [395a] whose δ ^{31}P (+187 and +150 ppm respectively) are in the range expected for P(V) compounds of their type, this latter possibility seems less likely. 1-Phenyl phosphetanes are epimerized by phenyl lithium,[24] and this may be an example of a multiple π-process occurring in a P(V) intermediate similar to **221**. On the other hand, there can be no doubt that nucleophilic attack on P(IV) molecules gives rise to P(V) species, and unless there is strong evidence to the contrary these must be considered as intermediates with sufficient lifetime to undergo π-processes. Phosphoranes have, in fact, been observed by ^{31}P spectroscopy in two well-known types of reactions: the action of base on P(IV) salts[398] and the Wittig reaction.[399]

Attack by a nucleophile on the face of P(IV) tetrahedron, the most likely process, will give rise to P (V) species in which the incoming group is apical. Since apical bonds are longer and weaker than radial, they are normally accepted as the preferred positions for the entering and leaving groups by analogy with the familiar transition state for an S_N2 displacement at carbon. With cyclic compounds this desirable arrangement is not always possible, but it is generally accepted that the departing group does so from the apical position. Hence, factors determining the stability of the intermediate and the energy barriers to the π-processes, which may be necessary to place the leaving group apical, assume a dominant influence on the rate and the stereochemical consequences of the reaction.

Historically, the first type of reaction to be considered was the hydrolysis of cyclic phosphates formed from 1,2-diols. This subject has been thoroughly reviewed[7,315] and only an outline is presented here. Phosphates such as 2-methoxy-1,3,2-dioxaphospholane 2-oxide (**222**) undergo hydrolysis much more rapidly than acyclic or six-membered analogues. This rate enhancement is due to a lowering of the energy barrier to formation of a P(V) intermediate, whose existence may also be inferred from isotope

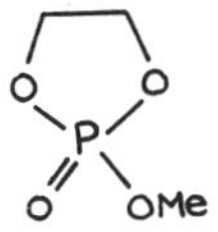

222 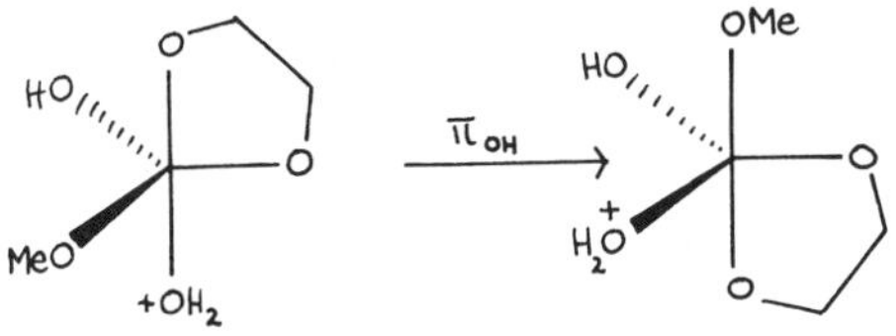 **223** **224**

experiments[7] and from Denney's observation that epimerization at phosphorus occurred together with hydrolysis.[190] Although normally assumed to be a consequence of the release of steric strain, this is not the only possible explanation.[201] Nevertheless, increased strain in the ring, such as unsaturation, increases the rate of hydrolysis. Relief of strain, however, does not explain why both ring opening and loss of exocyclic substituent proceed at an enhanced rate, but in terms of the intermediates it is quite straightforward. The P(V) intermediate **223** is initially formed and may decompose with ring opening, or by a low-energy π-process, to give **224** with the exocyclic substituent in the preferred leaving position. Replacement of one of the ring oxygens with carbon blocks a π-process similar to **223**→**224**, and thus hydrolysis occurs exclusively with ring opening. If both oxygens are replaced by carbon atoms, hydrolysis is no longer rapid since energetically unfavorable TBPs will be formed initially, thereby offsetting the strain factor. The direct formation of a P(V) intermediate with the leaving group apical would similarly be disfavored by the *rr* disposition of the five-membered ring (e.g., **225**→**226**). Rapid hydrolysis of phosphinates

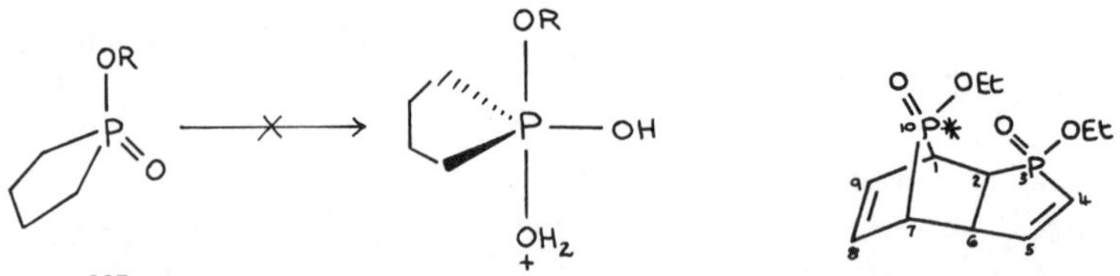

225 **226** **122**

does occur, however, if the ring is sufficiently strained (e.g., 1-ethoxyphosphole 1-oxide dimer, **122**, Section VII).

A related reaction is the exchange of phosphoryl oxygen and epimerization at phosphorus of cyclic phosphine oxides, but here the picture is less clear. Acyclic phosphine oxides are stable to aqueous alkali and epimerize or exchange oxygen relatively sluggishly under acidic conditions. Marsi[198] reported the equilibration of 1,3-dimethyl phospholane 1-oxides (**227**) in hot aqueous acid, and oxygen exchange was noted for 6-methyl-6-phosphatricyclo[3.1.1.1.^{3,7}]octane 6-oxide (**228**, cf. *Ring Index* No, *2847*)[401] and 2,2,3-trimethyl-1-phenylphosphetane 1-oxide (**229**).[400] The last also epimerizes under acid and alkaline conditions, but ,strikingly, 2,2,3,4,4-pentamethyl-1-phenylphosphetane 1-oxide (**106**) is inert.[262,264,400] In general, exchange is more rapid than epimerization. The exchange reaction is best

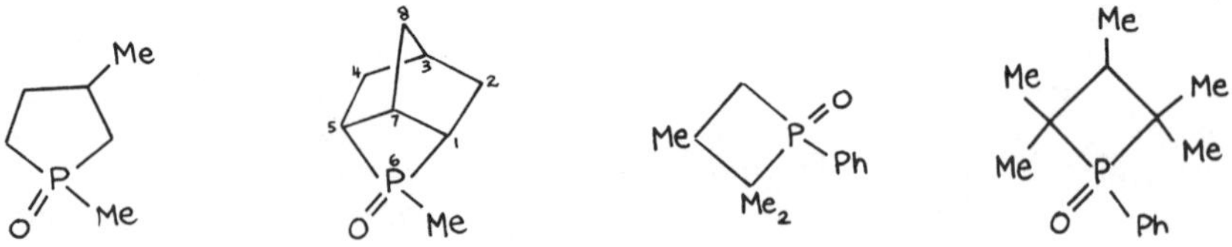

227 **228** **229** **106**

viewed as the formation of a dihydroxy phosphorane followed by a single low energy π-process using the exocyclic substituent as a pivot and followed by loss of apical OH. Epimerization requires an additional π-process to place the exocyclic substituent apical and the hydroxyl groups radial, and it will be of higher energy for this reason (Scheme 21).

The isomerization of **228** can similarly be accounted for, but if this is the case then the unreactivity of the phosphetane (**229**) is difficult to understand. Simple steric hindrance due to the extra α-methyl groups would seem insufficient. It is possible however, that the P(V) (*5*) compound epimerizes by way of an intermediate with the ring *rr* (**230**). It is known that placement of a four-membered ring *rr* requires about 85 kJ mol^{-1} and although a similar figure has been found for five-membered rings, these have

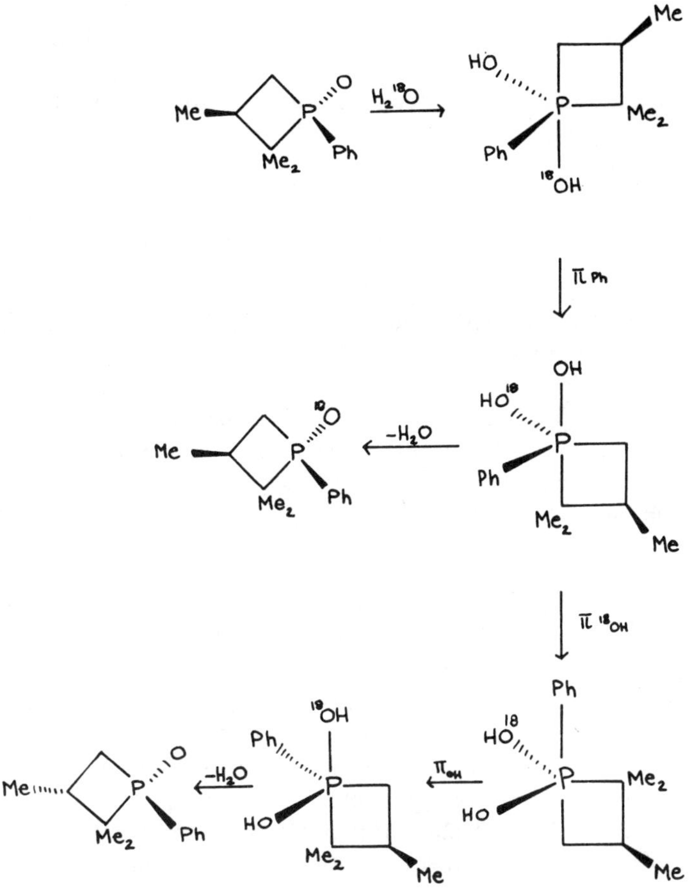

Scheme 21

had electronegative substituents at the ring termini and for a carbocyclic five-membered ring a lower value would be expected. Recent comparative exchange studies for a series of oxides of different ring sizes support this view.[282]

Phosphoryl compounds seem to require some type of strain assistance to allow easy P(V) formation; this normally means a four-or five-membered

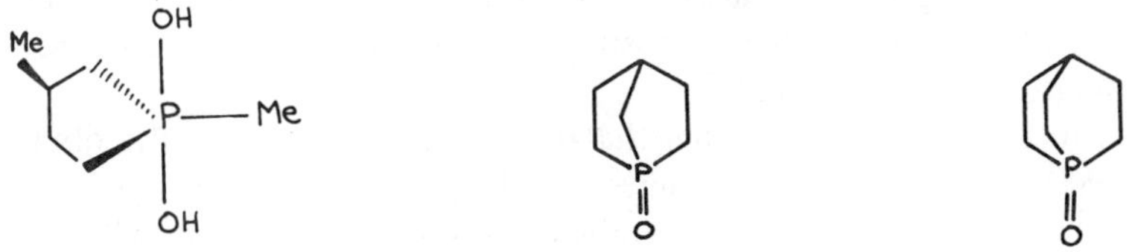

230

134

135

ring. Particularly striking examples are the oxides **134** and **135**; only the molecule containing the five-membered ring can be induced to exchange the oxygen.[282]

Another important class of reactions which has attracted much attention is the attack of nucleophiles, specifically hydroxide ion, on phosphonium compounds, $R_4P^+X^-$. The overall reaction is the loss of an R group and the formation of a phosphine oxide, and it is believed to proceed as shown:

$$R_4P^+ + OH^- \rightleftharpoons \underset{\mathbf{231}}{R_4POH} \rightleftharpoons \underset{\mathbf{232}}{R_4PO^-} \rightarrow R_3PO + R^-$$

Two intermediate phosphoranes are involved in one of which the incoming group is electronegative (**231**), but in the other (**232**) has been transformed into an electropositive one. The stereochemical results are: four-membered rings give retention[265,402] or racemization,[263,265] as do five-membered rings;[198,403] six-membered rings give racemization,[308,309] and seven-membered rings give inversion.[236] This is reasonably accounted for as shown in Scheme 22.

The simple process of shifting the leaving group (R_L) to the apical position for departure requires only a single π-process, which should be of low energy if it is the anionic P(V) (**234**; O^- instead of OH), which rearranges, but is more difficult if it is the hydroxy P(V) (**234**) since this would necessitate placing an electronegative substituent radial. Loss of R_L gives the oxide with retention of configuration (**237**), whereas for inversion to occur three further π-processes are necessary, **235**→**238**, **238**→**239**, and **239**→**(240)**, before decomposition can occur to give the inverted oxide (−) **237**. Note that equilibration of isomers is possible after three π-processes if the reaction is reversible **239**→(−) **233**, but that **239** is not the enantiomer of **234**. Cremer has observed such an equilibration for phosphetanium salts.[265]

It is clearly important from the viewpoint of π-processes to know which phosphorane is present, either the hydroxy or the O^- form. There is no evidence on this point, but since nucleophiles attack phosphoryl and thiophosphoryl P(IV) (*4*)[24] and P(IV) (*5*) compounds with retention of configuration, it is reasonable to assume that decomposition of anionic phosphoranes proceeds rapidly and does not allow more than one π-process. It seems reasonable to have at least one in order to remove what is now an electropositive group from the apical position. The increasing amount of inversion as ring size increases presumably reflects the increasing ability of the rings to occupy the *rr* position and to allow both incoming and outgoing groups to be apical.

It would be naive to imagine that the whole range of results obtained from P(IV) species and nucleophiles could be explained in terms of π-processes alone. The nature of the substituents on phosphorus will clearly affect the lifetime of any intermediate and the ease with which it is formed

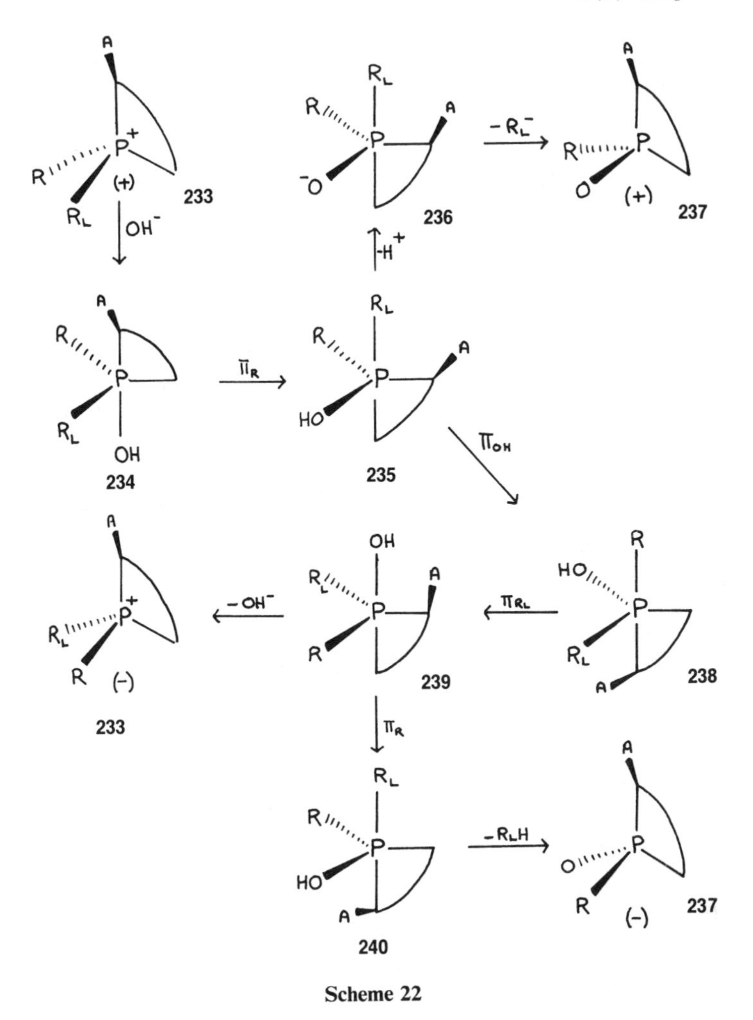

Scheme 22

and decomposed. For example, Marsi found retention of configuration in the action of aqueous alkali on phospholanium salts when the substituent is a good leaving group (e.g., benzyl).[198] but epimerization occurred when this was replaced by a phenyl group.[403] Phosphetanium salts may undergo ring opening[404] or ring expansion[436,437] when a good leaving group is not available, and structural variations in P(V) (*5*) compounds, such as unsaturation, may alter the course of the reactions to ring opening. Brophy and Gallagher[308a] have observed two different ring-opening pathways for decomposition of the P(IV) (*6*) bis salt, 1,1,4,4-tetraphenyl-1,4-diphosphorinanium dibromide (Scheme 23) depending on whether the salt or the alkali is present in excess, whereas when an exocyclic substituent is the better leav-

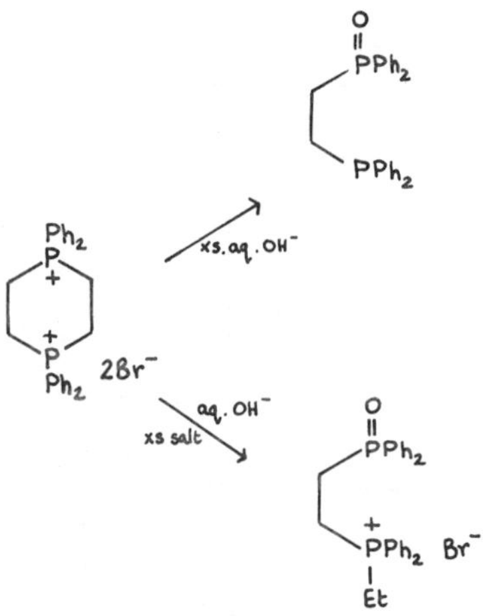

Scheme 23

ing group benzyl, no ring opening is observed.[308] The greatly enhanced rate of decomposition of these bis salts when compared with acyclic or cyclic monosalts cannot be explained in terms of substituent effects and has been attributed to relief of strain in the P(V) intermediate formed.[308a]

In other reactions believed to proceed by way of P(V) intermediates, varying the reagent may affect the stereochemistry; for example, phenylsilane reduces phospholane oxides with retention of configuration, whereas hexachlorodisilane gives predominant inversion (Equation 16),[405] presumably as a consequence of the high electronegativity of the silicon substituents in the supposed intermediate, **242**.

2. *Sterically Controlled Reactions*

The strain inherent in P(IV) (*4*) and P(IV) (*5*) compounds that facilitates transformation to P(V) intermediates can give rise to reactions not otherwise observed or expected. For example, the Wittig reaction proceeds by way of a 1,2-oxaphosphetane (**243**)[399] which readily eliminates phosphine oxide to give an olefin (Equation 17). If the phosphoryl group is part of a five-membered ring the reverse reaction can be observed, as in the catalytic conversion of isocyanates to carbodiimides by phospholene oxides (Scheme 24).[406,407] In a recent study Hall and Smith[408] found that p-toluenesulfonyl isocyanate reacted with phosphetane 1-oxides, and 1-sulfides, to give the imine analogous to **244** with retention of configuration, whereas acyclic phosphine oxides reacted three orders of magnitude more slowly and racemized much more rapidly than imine formation proceeded. In the absence of the driving force of relief of strain, the intermediate in the acyclic case is much more reluctant to form. Racemization

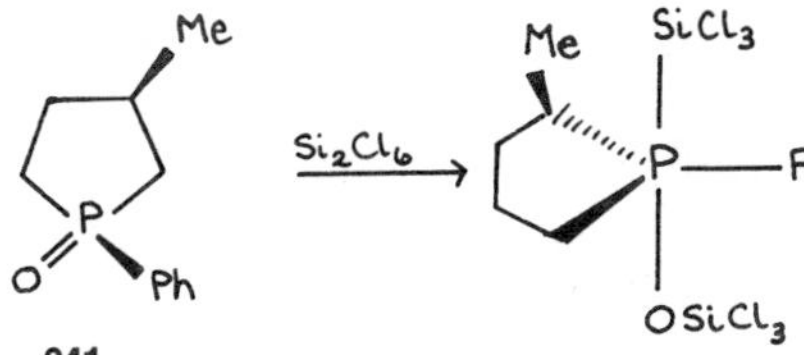
Me
Si_2Cl_6
Me
$SiCl_3$
P
Ph
$OSiCl_3$

241 242 (16)

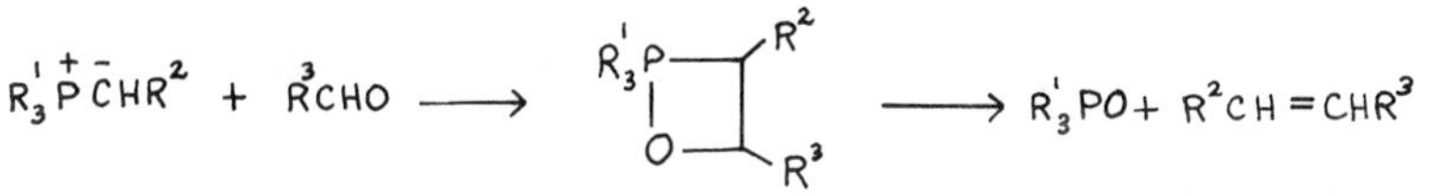
$R^1_3\overset{+}{P}\overset{-}{C}HR^2$ + R^3CHO
R^1_3P
R^2
O
R^3
R^1_3PO + $R^2CH{=}CHR^3$

(17)

243

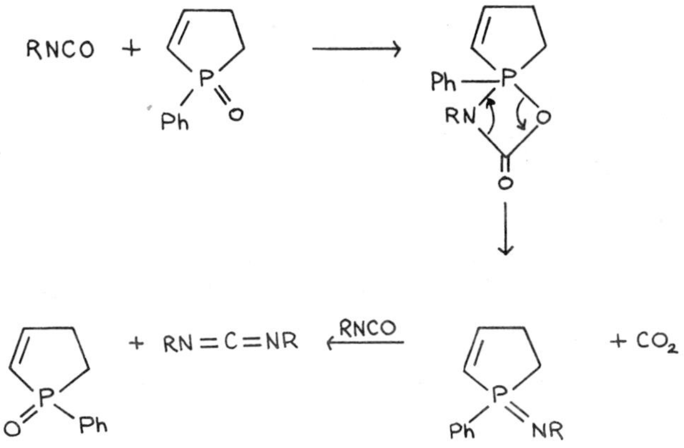
RNCO +
P
Ph
O
Ph
P
RN
O
O
+ RN=C=NR
RNCO
Ph
NR
+ CO_2

244

Scheme 24

is probably a consequence of more complex reaction pathways not requiring the formation of a four-membered ring.

Undoubtedly, the most elegant examples of this steric control are the studies of Katz and Turnblom on the novel phosphahomocubyl systems, **245**.[409,410] The phosphorus atom is so strained that organo lithium compounds simply add to give P(V) intermediates rather than, as is otherwise observed, abstracting a proton.[409] This has allowed the synthesis of the only pentaalkyl phosphorane[410] (**246**, $R^1{=}R^2{=}R^3{=}$Me). In another reaction of a type otherwise unknown, addition of phenyl lithium to the oxide **245** ($R^1{=}$O; $R^2{=}$Ph) followed by HBr gave the salt **245** ($R^1{=}R^2{=}$Ph). This reaction is known in arsenic chemistry.

In a further demonstration of steric control over reaction path, the strained P(IV) 9,9-diphenyl-9-phosphatricyclo[4.2.1.0^{2,5}]nonane cation **247** reacts with phenyl lithium to give a P(V) intermediate, whereas the less strained 9,9-diphenyl-9-phosphabicyclo[4.2.1]nonane cation undergoes

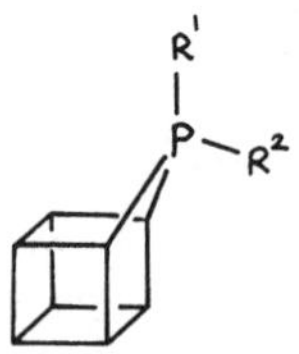

245

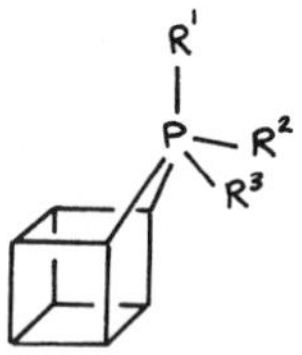

246

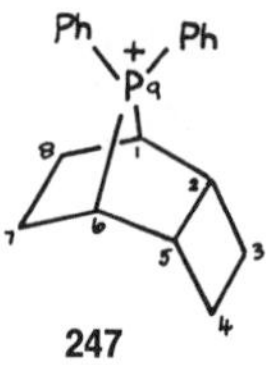

247

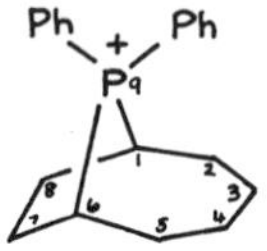

248

proton abstraction to give an ylid.[410a] Interestingly, **247** is itself made from the corresponding oxide.

F. As(V) Compounds

This field has been much less studied largely because it is less amenable to nmr studies. This is unfortunate because As(V) compounds seem to be easy to prepare, and in great variety, although many of the structures proposed need reexamination by modern techniques. The available evidence supports the occurrence of rapid π-processes here also. Hellwinkel,[411] Goldwhite,[412] Mislow,[413] and their collaborators have examined the spirocyclic arsoranes 5-substituted 5,5'-spirobi(dibenzarsole) (**249**) and 5-substituted 1,4,6,9-tetraoxa-5-arsaspiro[4.4]nonanes (**250**) respectively. Energy barriers associated with π-processes seem much the same as for the P(V) analogues; the suggestion that they are much faster in some instances[412] is apparently due to fortuitous isochrony of the ring substituents[413] in the ^{1}Hnmr spectra. Nmr evidence for π-processes has also been obtained for the antimony analogue of **249**.[414]

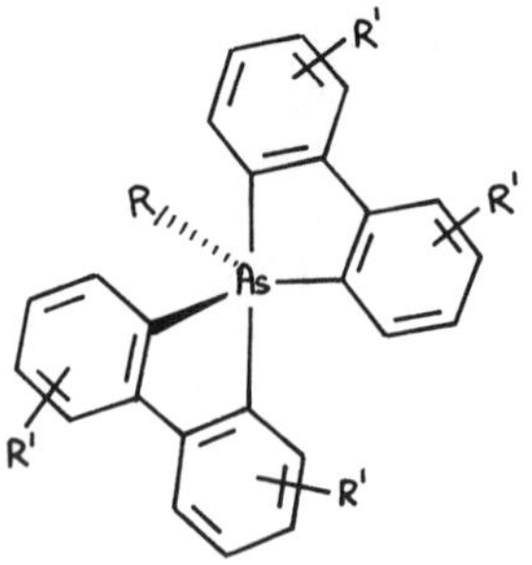

249

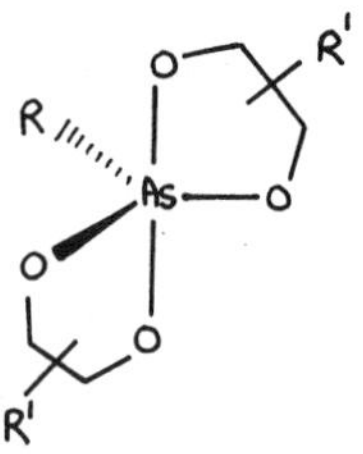

250

IX. P(VI) COMPOUNDS

By comparison with other aspects of phosphorus stereochemistry, this field has received very little attention. It will probably develop rapidly, however, as the chemistry of P(V) compounds is explored in more detail, particularly those reactions involving attack at P(V) by nucleophiles. Hellwinkel prepared[415] and resolved tris-(2,2′-biphenylene)phosphate anion (**251**) by classical means, thereby demonstrating an essentially octahedral geometry about phosphorus.[375] This was subsequently confirmed by the X-ray struc-

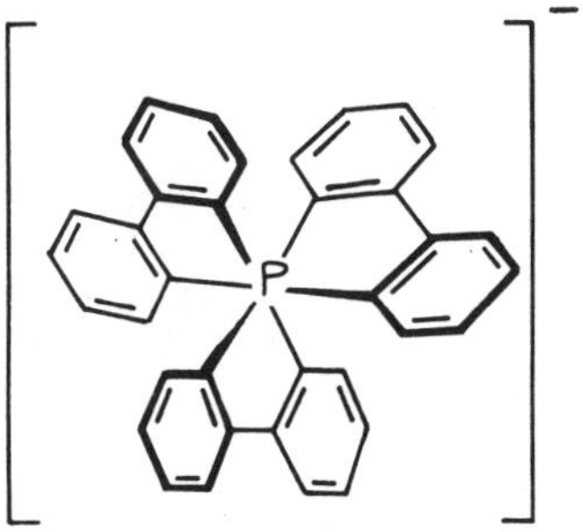

251

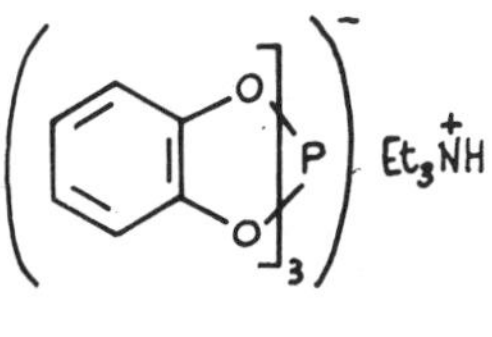

252

ture of triethylammonium tris-(*o*-phenylenedioxy)phosphate (**252**).[416] An entirely different approach has been taken by Wolf and his colleagues,[417,419] Reaction of the spirocyclic P(V) compound **253** with (+)-mandelic acid affords the optically active P(VI) anion **254** (Equation 18). As in the P(V)

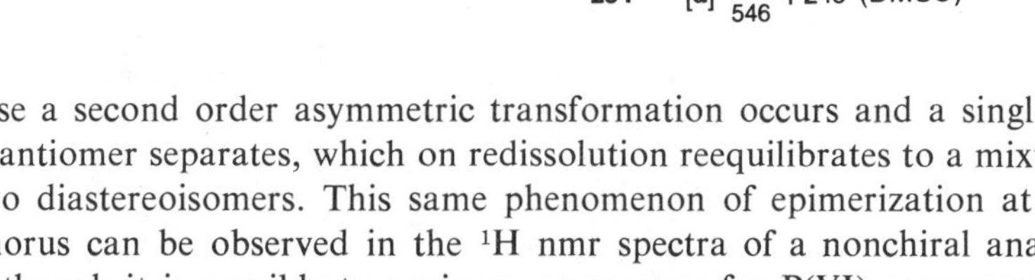

253 **254** $[\alpha]^{18}_{546}$ +245° (DMSO) (18)

case a second order asymmetric transformation occurs and a single pure enantiomer separates, which on redissolution reequilibrates to a mixture of two diastereoisomers. This same phenomenon of epimerization at phosphorus can be observed in the ^{1}H nmr spectra of a nonchiral analogue. Although it is possible to envisage π-processes for P(VI) compounds, the kinetic studies of these workers[419] led them to conclude that the most likely process was ring opening to P(V) intermediates which then under-

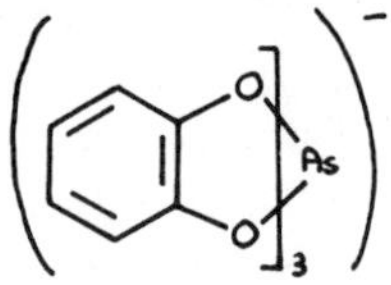

255

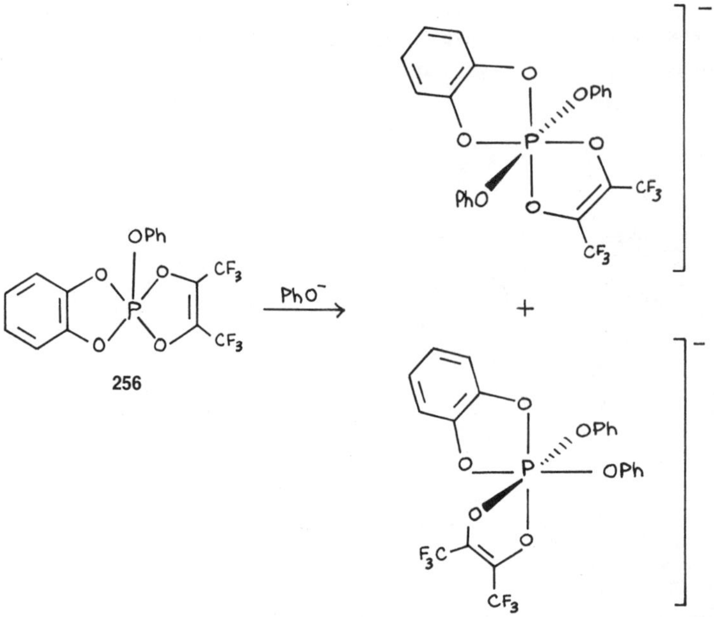

256

257

(19)

went stereomutation followed by ring closure. No intermediate was detected however, and the rate was concentration dependent, possibly due to autocatalysis.

The same group of workers has developed methods for the synthesis of P(VI) compounds with three different cyclic ligands,[417,418] and a cycloaliphatic P(VI) compound has been obtained by Denney and co-workers.[420]

Curiously, for a field so little studied the absolute configurations of two compounds have been determined: (−)-tris-2,2′-biphenylenephosphate (**251**) by circular dichroism[421] and its long known As (VI) analogue tris-*o*-phenylenedioxyarsenate anion (**255**) by X-ray.[422,423] Recently, Ramirez and collaborators in a study of the action of phenoxide ions and pyridines on the P(V) **256**, have detected spectroscopic evidence for the existence of geometrical isomerism in the P(VI) compounds **257**, presumably as shown in Equation 19. In general, however, they found the P(V)$\rightleftharpoons$ P(VI) equilibrium very rapid,[424] but they were able to isolate a stable crystalline P(VI) compound by reaction with pyridine; predictably the more hindered collidines gave less stable adducts. The incorporation of six ligands about a central atom would lead to considerably greater sensitivity to bulk than in the case of five ligands. Similarly, π-processes such as occur readily for P(V) and As(V) compounds are much more energetically expensive, and as a corollary, the irregular processes of bond breaking are favored by virtue of relief of crowding. Examples of this are found in the increased P–Cl bond length in PCl_6^-[425] and the preferential displacement at P(VI) in the reaction:[426]

$$PCl_4^+PCl_6^- + 2AsF_3 \rightarrow PCl_4^+PF_6^- + 2AsCl_3$$

Spectroscopically, P(VI) compounds are detected by the large upfield shift in δ ^{31}P.[20] Although data are still sparse, ring effects are to be expected, for example, as in **258** (-202 ppm)[427] [compare with **254** (-90 ppm) and PCl_6^- (-300 ppm).] A very high $^1J_{PH}$ value (about 800 Hz) is indicative

$$(MeNH)_2C{=}O + PCl_5 \longrightarrow \text{MeN}\overset{+}{\underset{Cl}{C}}\text{NMe} \; / \; \overset{-}{P}Cl_4 \text{ (four-membered ring)}$$

258

of a P(VI) species containing this bond,[428,429] but, from the few examples available, it seems that this is very sensitive to substituent effects.[319b]

π-Processes in six-coordinate iron have been observed, and Muetterties has discussed the possibilities for hexa and higher ligated atoms.[312,313] They would seem likely to be rare, however, and unlikely to play the dominant role in P(VI) compounds that they do for P(V) compounds.

X. REFERENCES

1. A. Thenard, *Compt. rend.*, **21,** 144 (1845).
2. J. Meisenheimer and L. Lichtenstadt, *Chem. Ber.*, **44,** 356 (1911).
3. F. G. Mann, *Progr. Stereochem.*, **2,** 196 (1957).
4. L. Horner, H. Winkler, A. Rapp, A. Mentrup, H. Hoffmann, and P. Beck, *Tetrahedron Letters*, 161 (1961).

5. H. S. Aaron, J. Braun, T. M. Shryne, H. F. Frack, G. E. Smith, R. T. Uyeda, and J. I. Miller, *J. Amer. Chem. Soc.*, **82,** 596 (1960).
6. J. Michalski and M. Mikolajczyk, *Tetrahedron*, **22,** 3055 (1966).
7. F. H. Westheimer, *Accounts Chem. Res.*, **1,** 70 (1968).
8. R. S. Berry, *J. Chem. Phys.*, **32,** 933 (1960); see footnote 19, p. 937.
9. G. Wittig and M. Rieber, *Naturwiss.*, **35,** 345 (1948).
10. F. Ramirez, *Accounts Chem. Res.*, **1,** 168 (1968).
11. G. W. Fenton and C. K. Ingold, *J. Chem. Soc.*, 2342 (1929).
12. H. R. Allcock, *Phosphorus-Nitrogen Compounds*, Academic, New York, 1972.
13. W. E. McEwen, *Topics Phosphorus Chem.*, **2,** 1 (1965).
14. M. J. Gallagher and I. D. Jenkins, *Topics Stereochem.*, **3,** 1 (1968).

14a. H. Christol and H. J. Cristau, *Ann. Chim. (France)*, **6,** 191 (1971).

15. G. M. Kosolapoff and L. Maier, Eds., *Organic Phosphorus Compounds*, Vols. 1–7, Wiley-Interscience, New York, 1972–1974.
16. F. G. Mann, *The Heterocyclic Derivatives of Phosphorus, Arsenic, Antimony and Bismuth*, Wiley-Interscience, New York, 1970.
17. L. S. Khaikin and L. V. Vilkov, *Russ. Chem. Rev.*, **41,** 1060 (1972).
18. D. E. C. Corbridge, (a) *Topics in Phosphorus Chem.*, **3,** 57 (1966), (b) *The Structural of Phosphorus*, Elsevier, Amsterdam (1974).
19. T. J. Batterham, *N. M. R. Spectra of Simple Heterocycles*, E. C. Taylor and A. Weissberger, Eds., Wiley-Interscience, New York, 1973, p. 467.
20. M. M. Crutchfield, C. H. Dungan, J. H. Letcher, V. Mark, and J. R. Van Wazer, *Topics Phosphorus Chem.*, **5,** whole volume (1967).
21. G. Mavel, *Progr. N. M. R. Spectroscopy*, **1,** 251 (1966).
22. G. Mavel, *Ann. Reports on N. M. R. Spectroscopy*, E. F. Mooney, Ed., Academic, New York, 1973, Vol. 5B, whole volume.
23. S. Chan, H. Goldwhite, H. Keyzer, G. G. Rowsell, and R. Tang, *Tetrahedron*, **25,** 1097 (1969).
24. J. R. Corfield, R. K. Oram, D. J. H. Smith, and S. Trippett, *J. C. S. Perkin I*, 713 (1972).
25. J. B. Lambert and W. L. Oliver, *Tetrahedron*, **27,** 4245 (1971).
26. J. J. Breen, J. F. Engel, D. K. Meyers, and L. D. Quin, *Phosphorus*, **2,** 55 (1972).
27. J. J. Breen, S. I. Featherman, L. D. Quin, and R. D. Stocks, *J. C. S. Chem. Comm.*, 657 (1972).
28. S. I. Featherman and L. D. Quin, *Tetrahedron Letters*, 1955 (1973).
29. S. I. Featherman, S. O. Lee, and L. D. Quinn, *J. Org. Chem.*, **39,** 2899 (1974).
30. L. D. Quin, A. T. McPhail, S. O. Lee, and K. D. Onan, *Tetrahedron Letters*, 3473 (1974).
31. R. B. Wetzel and G. L. Kenyon, *J. Amer. Chem. Soc.*, **96,** 5189 (1974).
32. L. M. Jackman and S. Sternhell, *Applications of NMR Spectroscopy in Organic Chemistry*, Pergamon, New York, 1969.
33. J. P. Albrand, D. Gagnaire, J. Martin, and J. B. Robert, *Bull. Soc. chim. France*, 40 (1969).
34. G. A. Gray and S. E. Cremer, *J. Org. Chem.*, **37,** 3470 (1972).

35. G. A. Gray and S. E. Cremer, *J. Org. Chem.*, **37,** 3458 (1972).
36. G. A. Gray and S. E. Cremer, *J. C. S. Chem. Comm.*, 367 (1972).
37. G. A. Gray and S. E. Cremer, *J. C. S. Chem. Comm.*, 451 (1974).
38. D. Gagnaire, J. B. Robert, and J. Verrier, *Bull. Soc. chim. France*, 2392 (1968).
39. W. G. Bentrude and J. H. Hargis, *J. Amer. Chem. Soc.*, **92,** 7136 (1970).
40. E. J. Boros, K. J. Coskran, R. W. King, and J. G. Verkade, *J. Amer. Chem. Soc.*, **88,** 1140 (1966).
41. D. W. White and J. G. Verkade, *J. Magn. Resonance*, **3,** 111 (1970).
42. C. L. Bodkin and P. Simpson, *J. Chem. Soc.* (*B*), 1136 (1971).
43. W. G. Bentrude and J. H. Hargis, *Chem. Comm.*, 1113 (1969).
43a. R. G. Cavell, D. D. Poulin, K. I. The, and A. J. Tomlinson, *J.C.S. Chem. Comm.*, 19 (1974).
44. L. D. Hall and R. B. Malcolm, *Canad. J. Chem.*, **50,** 2092 (1972).
45. L. D. Hall and R. B. Malcolm, *Canad. J. Chem.*, **50,** 2102 (1972).
46. J. P. Majoral, J. Navech, and K. Pihlaja, *Phosphorus*, **2,** 111 (1972).
47. J. P. Dutasta, A. Grand, J. B. Robert, and M. Taieb, *Tetrahedron Letters*, 2659 (1974).
48. W. G. Bentrude and H. W. Tan, *J. Amer. Chem. Soc.*, **95,** 4666 (1973).
49. H. W. Tan and W. G. Bentrude, *Tetrahedron Letters*, 619 (1975).
50. R. O. Hutchins, B. E. Maryanoff, J. P. Albrand, A. Cogne, D. Gagnaire, and J. B. Robert, *J. Amer. Chem. Soc.*, **94,** 9151 (1972).
51. M. Haemers, R. Ottinger, D. Zimmermann, and J. Reisse, *Tetrahedron*, **29,** 3539 (1973).
52. Y. Kashman and O. Awerbouch, *Tetrahedron*, **27,** 5593 (1971).
53. B. D. Cuddy, K. Treon, and B. J. Walker, *Tetrahedron Letters*, 4433 (1971).
54. W. G. Bentrude, H. W. Tan, and K. C. Yee, *J. Amer. Chem. Soc.*, **94,** 3264 (1972).
55. J. R. Corfield and S. Trippett, *Chem. Comm.*, 721 (1971).
56. Y. Kashman and S. Cherkez, *Tetrahedron*, **28,** 155 (1972).
57. Y. Kashman and E. Benary, *Tetrahedron*, **28,** 4091 (1972).
58. K. C. Yee and W. G. Bentrude, *Tetrahedron Letters*, 2775 (1971).
59. T. Sato and K. Goto, *J.C.S. Chem. Comm.*, 494 (1973).
60. E. Fluck, *Chem.-Ztg.*, **96,** 517 (1972).
61. D. W. White, R. D. Bertrand, G. K. McEwen, and J. G. Verkade, *J. Amer. Chem. Soc.*, **92,** 7125 (1970).
62. O. Awerbouch and Y. Kashman, *Tetrahedron*, **31,** 33 (1975).
63. Y. Kashman and O. Awerbouch, *Tetrahedron*, **31,** 53 (1975).
64. G. M. Blackburn, J. S. Cohen, and I. Weatherall, *Tetrahedron*, **27,** 2903 (1971).
65. D. Gorenstein, *J. Amer. Chem. Soc.*, **97,** 898 (1975).
66. R. Schmutzler, *Angew. Chem.*, **77,** 530 (1965).
67. K. G. Weinberg and E. B. Whipple, *J. Amer. Chem. Soc.*, **93,** 1801 (1971).
68. C. Jongsma, J. P. de Kleijn, and F. Bickelhaupt, *Tetrahedron*, **30,** 3465 (1974).
69. D. Hellwinkel and W. Schenk, *Angew. Chem.*, **81,** 1049, *Internat. Edn.*, **8,** 987 (1969).
70. L. R. Smith and J. K. Mills, *J.C.S. Chem. Comm.*, 808 (1974).
71. P. Haake, J. P. McNeal, and E. J. Goldsmith, *J. Amer. Chem. Soc.*, **90,** 715 (1968).

72. D. Gagnaire, J. B. Robert, J. Verrier, and R. Wolf, *Bull. Soc. chim. France*, 3719 (1966).
73. B. Fontal and H. Goldwhite, *Tetrahedron*, **22,** 3275 (1966).
74. M. Tsuboi, F. Kuriyagawa, K. Matsuo, and Y. Koygoku, *Bull. Chem. Soc. Japan*, **40,** 1813 (1967).
75. K. Bergesen and T. Vikane, *Acta Chem. Scand.*, **26,** 2153 (1972).
76. J. Devillers, J. Roussel, J. Navech, and R. Burgada, *Org. Magn. Resonance*, **5,** 511 (1973).
77. J. Devillers, J. Navech, and J. P. Albrand, *Org. Magn. Resonance*, **3,** 177 (1971).
78. K. Bergesen and M. Bjory, *Acta. Chem. Scand.*, **27,** 357 (1973).
79. K. Bergesen, M. Bjory, and T. Gramstad, *Acta. Chem. Scand.*, **26,** 2532 (1972).
80. J. P. Albrand, A. Cogne, D. Gagnaire, and J. B. Robert, *Tetrahedron*, **28,** 819 (1972).
81. J. P. Albrand, D. Gagnaire, J. Martin, and J. B. Robert, *Org. Magn. Resonance*, **5,** 33 (1973).
82. S. C. Peake, M. Fild, R. Schmutzler, R. K. Harris, J. M. Nichols, and R. G. Rees, *J.C.S. Chem. Comm.*, 380 (1972).
83. J. Devillers, F. Mathis, and J. Navech, *Compt. rend.*, **267,** 849 (1968).
83a. J. Devillers, L. T. Tran, and J. Navech, *Bull. Soc. chim. France*, 182 (1970).
83b. J. Devillers and J. Navech, *Bull. Soc. chim. France*, 4341 (1970).
84. D. B. Cooper, J. M. Harrison, and D. T. Inch, *Tetrahedron Letters*, 2697 (1974).
85. M. Revel, M. Bon, and J. Navech, *Compt. rend.*, **274C,** 430 (1972).
86. M. G. Newton and B. S. Campbell, *J. Amer. Chem. Soc.*, **96,** 7790 (1974).
87. D. Swank, C. N. Caughlan, F. Ramirez, O. P. Madan, and C. P. Smith, *J. Amer. Chem. Soc.*, **89,** 6503 (1967).
88. M. G. Newton, J. R. Cox, and J. A. Bertrand, *J. Amer. Chem. Soc.*, **88,** 1503 (1966).
89. T. Steitz and W. N. Lipscomb, *J. Amer. Chem. Soc.*, **89,** 2488 (1965).
90. F. P. Boer, *Acta Cryst.*, **28B,** 1201 (1972).
91. J. D. Lee and W. Goodacre, *Acta Cryst.*, **27B,** 1055 (1971).
92. J. D. Lee and W. Goodacre, *Acta Cryst.*, **27B,** 1841 (1971).
93. G. D. Smith, C. N. Caughlan, F. Ramirez, S. L. Glaser, and P. Stern, *J. Amer. Chem. Soc.*, **96,** 2698 (1974).
94. R. H. Cox and M. G. Newton, *J. Amer. Chem. Soc.*, **94,** 4212 (1972).
95. R. H. Cox, B. S. Campbell, and M. G. Newton, *J. Org. Chem.*, **37,** 1557 (1972).
96. M. G. Newton, H. C. Brown, C. J. Finder, J. B. Robert, J. Martin, and D. Tranqui, *J.C.S. Chem. Comm.*, 455 (1974).
97. V. A. Naumov and N. M. Zairipov, *Zhur. strukt. Khim.*, **11,** 1108 (1970).
98. G. Y. Schultz, I. Hargittai, J. Martin, and J. B. Robert, *Tetrahedron*, **30,** 2365 (1974).
99. V. A. Naumov, N. A. Gulyaeva, and M. A. Pudovik, *Doklady Akad. Nauk S.S.S.R.*, **203,** 590 (1972).
100. V. A. Naumov and M. A. Pudovik, *Doklady Akad. Nauk S.S.S.R.*, **203,** 351 (1972).
101. B. A. Arbuzov, V. A. Naumov, S. A. Shaidullin, and E. G. Mukmenev, *Doklady Akad. Nauk S.S.S.R.*, **204,** 859 (1972).
102. W. G. Bentrude, K. C. Yee, R. D. Bertrand, and D. M. Grant, *J. Amer. Chem. Soc.*, **93,** 797 (1971).

103. D. Z. Denney and D. B. Denney, *J. Amer. Chem. Soc.*, **88,** 1830 (1966).
104. M. Haemers, R. Ottinger, D. Zimmermann, and J. Reisse, *Tetrahedron Letters*, 461 (1971), 2241(1973).
105. J. A. Mosbo and J. G. Verkade, *J. Amer. Chem. Soc.*, **95,** 4659 (1973).
105a. R. F. Hudson and J. G. Verkade, *Tetrahedron Letters*, 3231 (1975).
106. M. G. B. Drew, J. Rodgers, D. W. White, and J. G. Verkade, *Chem. Comm.*, 227 (1971).
107. R. E. Wagner, W. Jensen, W. Wadsworth, and Q. Johnson, *Acta Cryst.*, **29B,** 2160 (1973).
108. R. E. Wagner, W. Jensen, and W. Wadsworth, *Cryst. Struct. Comm.*, **2,** 507 (1973).
109. W. Murayama and M. Kainosho, *Bull. Chem. Soc. Japan*, **42,** 1819 (1969).
110. Mazhar-ul-Haque, C. N. Caughlan, and W. L. Moats, *J. Org. Chem.*, **35,** 1446 (1970).
111. H. J. Geise, *Rec. Trav. chim.*, **86,** 362 (1967).
112. T. A. Beineke, *Acta Cryst.*, **25,** 413 (1969).
113. Mazhar-ul-Haque, C. N. Caughlan, J. H. Hargis, and W. G. Bentrude, *J. Chem. Soc.(A)*, 1786 (1970).
114. R. G. G. Killean, J. L. Lawrence, and I. M. Magennis, *Acta Cryst.*, **27B**, 189 (1971).
115. T. A. Bieneke, *Chem. Comm.*, 860 (1966).
116. W. Sanger and M. Mikolajczyk, *Chem. Ber.*, **106,** 3519 (1973).
117. W. G. Bentrude and K. C. Yee, *Chem. Comm.*, 169 (1972).
118. K. D. Bartle, R. S. Edmundson, and D. W. Jones, *Tetrahedron*, **23,** 1701 (1967).
119. M. Kainosho and T. Shimozawa, *Tetrahedron Letters*, 865 (1969).
120. M. Kainosho, T. Morofushi, and A. Nakamura, *Bull. Chem. Soc. Japan*, **42,** 845 (1969).
121. M. Kainosho, A. Nakumura, and M. Tsuboi, *Bull. Chem. Soc. Japan*, **42,** 1713 (1969).
122. C. L. Bodkin and P. Simpson, *J. C. S. Perkin II*, 676 (1973).
123. A. A. Borisenko, N. M. Sergeyev, E. Y. Nifantev, and Y. A. Ustynuk, *Chem. Comm.*, 406 (1972).
124. E. Y. Nifantev, A. A. Borisenko, I. S. Nosonovsky, and E. I. Matrosov, *Dokaldy Akad. Nauk S.S.S.R.*, **196,** 121 (1971).
125. J. F. Brault, J. P. Majoral, P. Savignac, and J. Navech, *Bull. Soc. chim. France*, 3149 (1973).
126. J. P. Majoral, C. Bergounhou, J. Navech, P. C. Maria, L. Elegant, and M. Azzaro, *Bull. Soc. chim. France*, 3142 (1973).
127. W. Stec, B. Uznanski, and J. Michalski, *Phosphorus*, **2,** 237 (1973).
128. W. Stec, B. Uznanski, and J. Michalski, *Phosphorus*, **2,** 235 (1973).
129. J. P. Dutasta, A. Grand, and J. B. Robert, *Tetrahedron Letters*, 2655 (1974).
130. D. W. White and J. G. Verkade, *Phosphorus*, **3,** 15 (1973).
131. R. O. Hutchins and B. W. Marynhoff, *J. Amer. Chem. Soc.*, **94,** 3266 (1972).
132. J. Durrie, R. Kraemer, and J Navech, *Org. Magn. Resonance*, **4,** 709 (1972).
133. J. C. Clardy, J. A. Mosbo, and J. G. Verkade, *Chem. Comm.*, 1163 (1972).
134. S. Garcia-Blanco and A. Perales, *Acta Cryst.*, **28B,** 2647 (1972).
135. N. Camerman and A. Camerman, *J. Amer. Chem. Soc.*, **95,** 5038 (1973).
136. B. M. Gatehouse and B. K. Miskin, *Acta Cryst.*, **30B,** 2112 (1974).

137. H. E. Shook and L. D. Quin, *J. Amer. Chem. Soc.*, **89,** 1841 (1967).
138. A. T. McPhail, J. J. Breen, and L. D. Quin, *J. Amer. Chem. Soc.*, **93,** 2574 (1971).
139. A. T. McPhail, J. J. Breen, J. H. Somers, J. C. H. Steele, and L. D. Quin, *Chem. Comm.*, 1020 (1971).
140. M. J. Gallagher and J. Sussman, *Phosphorus*, **5,** 91 (1975).
141. J. P. Beale, M. J. Gallagher, and I. D. Rae, *Acta Cryst.*, **B,** submitted for publication 1975.
142. V. A. Naumov and N. M. Zaripov, *Zhur. struct. Khim.*, **13,** 768 (1972).
143. K. Dimroth and P. Hoffmann, *Chem. Ber.*, **99,** 1325 (1966).
144. K. Dimroth, *Topics Current Chem.*, **38,** 1 (1973).
145. R. Allmann, *Chem. Ber.*, **99,** 1332 (1966).
146. J. C. J. Bart and J. J. Daly, *Angew. Chem.*, **80,** 843 (1968); *Angew. Chem. Internat. Edn.*, **7,** 811 (1968).
147. L. V. Vilkov, L. S. Khaikin, A. F. Vasilev, N. P. Ignatova, N. N. Melnikov, V. V. Negrebetskii, and N. I. Shvetsov-Shilovskii, *Doklady Alad. Nauk S.S.S.R.*, **197,** 1081 (1971).
148. A. F. Vasilev, L. V. Vilkov, N. P. Ignatova, N. N. Melnikov, U. N. Negretskii, N. I. Shvestsov-Shilovskii, and L. S. Khaikin, *Doklady Akad. Nauk S.S.S.R.*, **183,** 95 (1968).
149. Y. Charbonnel and J. Barrans, *Compt. rend.*, **278C,** 355 (1974).
150. S. Fleming, M. K. Upton, and K. Jekot, *Inorg. Chem.*, **11,** 2534 (1972).
151. B. E. Maryanoff and R. O. Hutchins, *J. Org. Chem.*, **37,** 3475 (1972).
152. K. F. Kumli, W. E. McEwen, and C. A. Wander Werf, *J. Amer. Chem. Soc.*, **81,** 248 (1959).
153. L. Horner and A. Mentrup, *Annalen*, **646,** 65 (1961).
154. L. Horner and H. Winkler, *Tetrahedron Letters*, 461 (1964).
155. I. G. M. Campbell and J. K. Way, *J. Chem. Soc.*, 5034 (1960).
155a I. G. M. Campbell and J. K. Way, *J. Chem. Soc.*, 2133 (1961).
155b. P. D. Henson, S. B. Ockrymiek, and R. E. Markham, *J. Org. Chem.*, **39,** 2296 (1974).
156. I. G. M. Campbell and R. C. Poller, *J. Chem. Soc.*, 1195 (1956).
157. I. G. M. Campbell, *J. Chem. Soc.*, 1976 (1956).
158. I. G. M. Campbell, *J. Chem. Soc.*, 3109 (1950).
159. I. G. M. Campbell, *J. Chem. Soc.*, 4448 (1952).
160. I. G. M. Campbell and D. J. Morrill, *J. Chem. Soc.*, 1662 (1955).
161. M. S. Lesslie and E. E. Turner, *J. Chem. Soc.*, 1170 (1934).
162. M. S. Lesslie and E. E. Turner, *J. Chem. Soc.*, 1051, 1268 (1935).
163. I. G. M. Campbell, *J. Chem. Soc.*, 4 (1947).
164. K. Mislow, A. Zimmerman, and J. T. Melillo, *J. Amer. Chem. Soc.*, **85,** 594 (1963).
165. F. G. Mann, *J. Chem. Soc.*, 4266 (1963).
166. J. D. Adose, A. Rauk, and K. Mislow, *J. Amer. Chem. Soc.*, **96,** 6904 (1974).
167. J. F. Kincaid and F. C. Henriques, *J. Amer. Chem. Soc.*, **62,** 1474 (1940).
168. G. W. Koeppl, D. S. Sagatys, G. S. Krishnaumurthy, and S. I. Miller, *J. Amer. Chem. Soc.*, **89,** 3396 (1967).
169. J. M. Lehn and B. Munsch, *Chem. Comm.*, 1327 (1969).

170. D. Gagnaire and M. St. Jacques, *J. Phys. Chem.*, **73,** 1678 (1969).
171. I. R. Wagner, L. D. Freeman, H. Goldwhite, and D. G. Rowsell, *J. Amer. Chem. Soc.*, **89,** 1102 (1967).
172. F. G. Mann and M. J. Pragnell, *J. Chem. Soc.* (*C*), 916 (1966).
173. T. J. Katz, C. R. Nicholson, and C. A. Reilly, *J. Amer. Chem. Soc.*, **88,** 3832 (1966).
174. M. T. Bowers, R. A. Beaudet, H. Goldwhite, and R. Tang, *J. Amer. Chem. Soc.*, **91,** 17 (1969).
175. M. T. Bowers, R. A. Beaudet, H. Goldwhite, and S. Chan, *J. Chem. Phys.*, **52,** 2831 (1970).
176. S. E. Cremer and R. J. Chorvat, *J. Org. Chem.*, **32,** 4066 (1967).
177. K. E. DeBruin, G. Zon, K. Naumann, and K. Mislow, *J. Amer. Chem. Soc.*, **91,** 7027 (1969).
178. S. E. Cremer, *Chem. Comm.*, 616 (1970).
179. A. Rauk, L. C. Allen, and K. Mislow, *Angew. Chem.*, **82,** 453, *Angew. Chem. Internat. Edn.*, **9,** 400 (1970).
180. S. E. Cremer, R. J. Chorvat, C. H. Chang, and D. W. Davis, *Tetrahedron Letters*, 5799 (1968).
181. D. B. Denney and S. L. Varga, *Phosphorus*, **2,** 245 (1973).
182. R. S. Edmundson and E. W. Mitchell, *Chem. Comm.*, 482 (1966).
183. H. Oehme, K. Issleib, and E. Leissring, *Tetrahedron*, **28,** 2587 (1972).
184. K. Issleib and H. Oehme, *Tetrahedron Letters*, 1489 (1967).
185. K. Issleib and H. J. Hanning, *Phosphorus*, **3,** 113 (1973).
186. A. Tzschach and J. Heinicke, *J. prakt. Chem.*, **315,** 65 (1973).
187. A. Tzschach, D. Drohne, and J. Heinicke, *J. Organometallic Chem.*, **60,** 95 (1973).
188. W. Egan, R. Tang, G. Zon, and K. Mislow, *J. Amer. Chem. Soc.*, **93,** 6205 (1971).
189. W. Schäfer, A. Schweig, G. Märkl, H. Hauptmann, and F. Mathey, *Angew. Chem.*, **85,** 140, *Angew. Chem. Internat. Edn.*, **12,** 145 (1973).
190. D. Z. Denney, G. Y. Chen, and D. B. Denney, *J. Amer. Chem. Soc.*, **91,** 6838 (1969).
191. K. Mislow and R. H. Bowman, *J. Amer. Chem. Soc.*, **94,** 2861 (1972).
192. P. Maroni, Y. Madaule, and J. G. Wolf, *Bull. Soc. chim. France*, 668 (1973).
193. D. W. Aksnes and O. Vikane, *Acta Chem. Scand.*, **27,** 1337 (1973).
194. D. W. Aksnes and O. Vikane, *Acta Chem. Scand.*, **26,** 2532 (1972).
195. D. W. Aksnes and O. Vikane, *Acta Chem. Scand.*, **26,** 835 (1972).
196. K. Bergesen and M. Bjory, *Acta Chem. Scand.*, **27,** 1103 (1973).
197. J. P. Albrand, A. Cogne, D. Gagnaire, J. Martin, J. B. Robert, and H. Verrier, *Org. Magn. Resonance*, **3,** 75 (1971).
198. K. L. Marsi, J. *Amer. Chem. Soc.*, **91,** 4724 (1969).
199. L. D. Quin and R. C. Stocks, *J. Org. Chem.*, **39,** 1339 (1974).
200. S. I. Featherman and L. D. Quin, *J. Amer. Chem. Soc.*, **95,** 1699 (1973).
201. R. F. Hudson and C. Brown, *Accounts Chem. Res.*, **5,** 204 (1972).
202. G. Aksnes and R. Eriksen, *Acta Chem. Scand.*, **20,** 2463 (1966).
203. R. Greenhalgh and R. F. Hudson, *Chem. Comm.*, 1300 (1968).
204. M. Davis and F. G. Mann, *J. Chem. Soc.*, 3770 (1964).

205. E. R. H. Jones and F. G. Mann, *J. Chem. Soc.*, 411 (1955).
206. O. Kennard, F. G. Mann, D. G. Watson, J. K. Fawcett, and K. A. Kerr, *Chem. Comm.*, 269 (1968).
207. D. W. Aksnes and O. Vikane, *Acta Chem. Scand.*, **26,** 4170 (1972).
208. M. P. Simonnin, C. Charrier, and R. Burgada, *Org. Magn. Resonance*, **4,** 113 (1972).
209. D. H. Wertz and N. L. Allinger, *Tetrahedron*, **30,** 1579 (1974).
210. L. D. Quin and J. H. Somers, *J. Org. Chem.*, **37,** 1217 (1972).
211. J. B. Lambert, D. S. Bailey, and C. E. Mixan, *J. Org. Chem.*, **37,** 377 (1972).
212. J. B. Lambert, C. E. Mixan, and D. H. Johnson, *Tetrahedron Letters*, 4335 (1972).
213. A. T. McPhail, P. A. Luhan, S. I. Featherman, and L. D. Quin, *J. Amer. Chem. Soc.*, **94,** 2126 (1972).
214. J. J. Daly, *J. Chem. Soc.*, 4789 (1965).
215. G. Aksnes, R. Eriksen, and K. Melligen, *Acta Chem. Scand.*, **21,** 1028 (1967).
216. J. Michalski, A. Okruszek, and W. Stec, *Chem. Comm.*, 1495 (1970).
217. Y. Kashman and O. Awerbouch, *Tetrahedron*, **26,** 4213 (1970).
218. Y. Kashman and E. Benary, *Tetrahedron*, **28,** 4091 (1972).
219. D. W. White, *Phosphorus*, **1,** 33 (1971).
220. C. L. Bodkin and P. Simpson, *J.C.S. Perkin II*, 2049 (1972).
221. R. D. Adamick, L. L. Chang, and D. B. Denney, *J.C.S. Chem. Comm.*, 987 (1974).
222. J. H. Finley and D. B. Denney, *J. Amer. Chem. Soc.*, **92,** 362 (1970).
223. D. B. Denney and M. A. Moskal, *Phosphorus*, **4,** 77 (1974).
224. J. Rodgers, D. W. White, and J. G. Verkade, *J. Chem. Soc. (A)*, 77 (1971).
225. J. G. Verkade, T. J. Huttermann, M. K. Fung, and R. W. King., *Inorg. Chem.*, **4,** 83 (1965).
226. J. W. Rathke, J. W. Guyer, and J. G. Verkade, *J. Org. Chem.*, **35,** 2310 (1970).
227. W. S. Wadsworth and W. D. Emmons, *J. Amer. Chem. Soc.*, **84,** 610 (1962).
228. W. S. Wadsworth, *J. Org. Chem.*, **32,** 1603 (1967).
229. R. D. Bertrand, H. J. Berwin, G. K. McEwen, and J. G. Verkade, *Phosphorus*, **4,** 81 (1974).
230. R. A. Earley and M. J. Gallagher, *J. Chem. Soc. (C)*, 158 (1970).
231. M. J. Gallagher and J. Gondosowito, unpublished results (1975).
232. H. Veermeer and P. C. J. Kevenaar, *Annalen*, **763,** 155 (1972).
233. S. A. Buckler, *J. Amer. Chem. Soc.*, **82,** 4215 (1960).
234. R. C. Hinton and F. G. Mann, *J. Chem. Soc.*, 2835 (1959).
235. E. M. Bellet and J. E. Casida, *Science*, **182,** 1135 (1973).
236. K. L. Marsi, *J. Amer. Chem. Soc.*, **93,** 6341 (1971).
237. T. H. Chan and B. S. Ong, *J. Org. Chem.*, **39,** 1749 (1974).
238. J. P. Dutasta, A. C. Guimaraes, J. Martin, and J. B. Robert, *Tetrahedron Letters*, 1519 (1975).
239. J. P. Albrand, J. P. Dutasta, and J. B. Robert, *J. Amer. Chem. Soc.*, **96,** 4584 (1974).
240. F. G. Holliman and F. G. Mann, *J. Chem. Soc.*, 1634 (1947).
241. F. A. Hart and F. G. Mann, *J. Chem. Soc.*, 4107 (1955).
242. F. G. Holliman and F. G. Mann, *J. Chem. Soc.*, 550 (1943).

243. F. G. Holliman and F. G. Mann, *J. Chem. Soc.*, 45 (1945).
244. M. H. Forbes, D. M. Heinekey, F. G. Mann, and I. T. Millar, *J. Chem. Soc.*, 2762 (1961).
245. G. Wittig and D. Hellwinkel, *Chem. Ber.*, **97,** 769 (1964).
246. E. W. Koos, J. P. Vander Kooi, E. E. Green, and K. J. Stille, *J.C.S. Chem. Comm.*, 1085 (1972).
247. J. J. McBride, E. Jungermann, J. V. Killheffer, and R. J. Clutter, *J. Org. Chem.*, **27,** 1833 (1962).
248. J. Emsley, T. B. Middleton, J. K. Williams, and M. F. Crook, *Phosphorus*, **3,** 45 (1973).
249. S. E. Cremer, F. I. Weitl, F. R. Farr, P. W. Kremer, G. A. Gray, and H. Hwang, *J. Org. Chem.*, **38,** 3199 (1973).
250. N. V. Ivasyuk and I. M. Shermergorn, *Izvest. Akad. Nauk S.S.S.R., Ser. khim.*, 481 (1969), *Chem. Abs.*, **71,** 13174 (1969).
251. M. M. Gilyazov, T. A. Zyablikova, E. K. Mukhametzyanov, and I. M. Shermergorn. *Izvest. Akad. Nauk S.S.S.R., Ser. khim.*, 1177 (1970), *Chem. Abs.*, **73,** 77339 (1970).
252. R. P. Arshinova, T. A. Zyabhkova, E. K. Mukhmetzyanova, and I. M. Shermergorn, *Doklady Akad. Nauk S.S.S.R.*, **204,** 1118 (1972), *Chem. Abs.*, **77,** 126751 (1972).
253. V. A. Naumov and V. N. Semashko, *Doklady Akad. Nauk S.S.S.R.*, **200,** 882 (1971), *Chem. Abs.*, **76,** 18986 (1972).
254. A. R. Pantleeva and I. M. Shermergorn, *Izvest. Akad. Nauk S.S.S.R., Ser. khim.*, 1652 (1968), *Chem. Abs.*, **69,** 87106 (1968).
255. G. M. Kosolapoff and R. F. Struck, *J. Chem. Soc.*, 3739 (1957).
256. D. D. Swank and C. N. Caughlan, *Chem. Comm.*, 1051 (1968).
257. C. Moret and L. M. Trefonas, *J. Amer. Chem. Soc.*, **91,** 2255 (1969).
258. Mazhar-ul-Haque, *J. Chem. Soc. (B)*, 117 (1971).
259. Mazhar-ul-Haque, *J. Chem. Soc. (B)*, 938 (1970).
260. Mazhar-ul-Haque, *J. Chem. Soc. (B)*, 934 (1970).
261. S. E. Cremer and B. C. Trivedi, *J. Amer. Chem. Soc.*, **91,** 7200 (1969).
262. K. E. DeBruin, G. Zon, K. Naumann, and K. Mislow, *J. Amer. Chem. Soc.*, **91,** 7027 (1969).
263. K. E. DeBruin and M. J. Jacobs, *Chem. Comm.*, 59 (1971).
264. W. Hawes and S. Trippett, *Chem. Comm.*, 295 (1968).
265. S. E. Cremer, R. J. Chorvat, and B. C. Trivedi, *Chem. Comm.*, 769 (1969).
266. W. Hawes and S. Trippett, *Chem. Comm.*, 577 (1968).
267. J. I. G. Cadogan and D. T. Eastlick, *Chem. Comm.*, 238 (1973).
268. C. R. Hall and D. J. H. Smith, *Tetrahedron Letters*, 1693 (1974).
269. K. Ellis, D. J. H. Smith, and S. Trippett, *J.C.S. Perkin I*, 1184 (1972).
270. E. T. Kaiser, M. Panar, and F. H. Westheimer, *J. Amer. Chem. Soc.*, **85,** 602 (1963).
271. H. G. Khorana, G. M. Tener, R. S. Wright, and J. G. Moffatt, *J. Amer. Chem. Soc.*, **79,** 430 (1957).
272. J. T. Bahr, R. E. Cathon, and G. G. Hammes, *J. Biol. Chem.*, **240,** 3372 (1965).
273. V. E. Belskii and I. P. Gozman, *Zhur. obschei Khim.*, **37,** 2730 (1967), *Chem. Abs.*, **69,** 58646 (1968).

274. P. Haake, R. D. Cook, W. Schwarz, and D. R. McCoy, *Tetrahedron Letters*, 5251 (1968).
275. F. Ramirez, S. L. Glaser, P. Stern, I. Ugi, and P. Lemmen, *Tetrahedron*, **29,** 3741 (1973).
276. R. Kluger, F. Kerst, D. G. Lee, E. A. Dennis, and F. H. Westheimer, *J. Amer. Chem. Soc.*, **89,** 3918 (1967).
277. L. D. Quin in 1,4-*Cycloaddition Reactions*, J. Hamer, Ed., Academic, New York, 1967, Ch. 3, pp. 47–94.
278. A. Bond, M. Green, and S. C. Pearson, *J. Chem. Soc.*, (*B*), 929 (1968).
279. I. K. Feldman and A. I. Berlin, *Zhur. obshchei Khim.*, **32,** 575 (1962), *Chem. Abs.*, **58,** 6820 (1963).
279a. I. K. Feldman and A. I. Berlin, *Zhur. Obschei Khim.*, **32,** 1604 (1962), *Chem. Abs.*, **58,** 12563 (1963).
279b. I. K. Feldman and A. I. Berlin, *Zhur. obshchei Khim.*, **32,** 3379 (1962); *Chem. Abs.*, **58,** 12563 (1963).
280. G. Baccolini and P. E. Todesco, *J. Org. Chem.*, **39,** 2650 (1974).
281. Y. Kashman, Y. Menachem, and E. Benary, *Tetrahedron*, **29,** 4279 (1973).
282. R. B. Wetzel and G. K. Kenyon, *J. Amer. Chem. Soc.*, **96,** 5199 (1974).
283. Y. Kashman and O. Awerbouch, *Tetrahedron Letters*, 3217 (1973).
284. Y. Kashman and O. Awerbouch, *Tetrahedron*, **29,** 191 (1973).
285. D. L. Morris and K. D. Berlin, *Phosphorus*, **4,** 69 (1974).
286. L. Horner, J. P. Bercz, and C. V. Bercz, *Tetrahedron Letters*, 5783 (1966).
287. K. Bergesen, *Acta Chem. Scand.*, **21,** 578 (1967).
288. E. L. Eliel, *Stereochemistry of Carbon Compounds*, McGraw-Hill, New York, 1962, p. 216.
289. W. S. Wadsworth and H. Horton, *J. Amer. Chem. Soc.*, **92,** 3785 (1970).
290. E. Duff and S. Trippett, *Phosphorus*, **1,** 291 (1972).
291. J. A. Mosbo and J. G. Verkade, *J. Amer. Chem. Soc.*, **95,** 204 (1973).
292. W. S. Wadsworth, *J. Org. Chem.*, **38,** 2921 (1973).
293. J. P. Majoral and J. Navech, *Bull. Soc. chim. France*, 1331 (1971).
294. J. P. Majoral and J. Navech, *Bull. Soc. chim. France*, 2609 (1971).
295. T. D. Inch and G. J. Lewis, *Chem. Comm.*, 310 (1973).
296. T. D. Inch and G. J. Lewis, *Tetrahedron Letters*, 2187 (1973).
297. O. Korpiun, R. A. Lewis, J. Chickos, and K. Mislow, *J. Amer. Chem. Soc.*, **90,** 4842 (1968).
297a. C. R. Hall, T. D. Inch, G. J. Lewis, and R. A. Chittenden, *J.C.S. Chem. Comm.*, 119 (1976) and D. B. Cooper, C. R. Hall, and T. D. Inch, *J.C.S. Chem. Comm.*, 120 (1976).
298. N. P. B. Dudman and B. Zerner, *J. Amer. Chem. Soc.*, **95,** 3019 (1973).
299. J. P. Majoral, C. Bergounhou, and J. Navech, *Bull. Soc. chim. France*, 3146 (1973).
300. M. Mikolajaczyk and J. Luczak, *Tetrahedron*, **28,** 5411 (1972).
301. W. Stec and M. Mikolajczyk, *Tetrahedron*, **29,** 539 (1973).
302. W. Stec and A. Lopusinski, *Tetrahedron*, **29,** 547 (1973).
303. W. Stec, W. E. Morgan, J. R. Van Wazer, and W. G. Proctor, *J. Inorg. Nuclear Chem.*, **34,** 1100 (1972).

304. W. Stec, B. Uznanski, and J. Michalski, *Phosphorus*, **2,** 237 (1973).
305. W. Stec, B. Uznanski, and J. Michalski, *Phosphorus*, **2,** 235 (1973).
306. A. E. Nifantev and J. S. Nasonoski, *Doklady Akad. Nauk S.S.S.R.*, **203,** 841 (1972).
306a. E. Y. Nifantev and H. H. Borisenko, *Tetrahedron Letters*, 309 (1972),
307. K. L. Marsi, *J. Org. Chem.*, **39,** 265 (1974).
308. G. E. Driver and M. J. Gallagher, *Chem. Comm.*, 150 (1970).
308a. J. J. Brophy and M. J. Gallagher, *Austral. J. Chem.*, **22,** 1385 (1969).
309. K. L. Marsi and R. T. Clark, *J. Amer. Chem. Soc.*, **92,** 3791 (1970).
310. C. Y. Cheng, R. E. Shaw, T. S. Cameron, and C. K. Prout, *Chem. Comm.*, 616 (1968).
311. J. P. Lampin, F. Mathey, and W. S. Sheldrick, *Acta Cryst.*, **30B,** 1626 (1974).
312. E. L. Muetterties, *Accounts Chem. Res.*, **3,** 266 (1970).
313. E. L. Muetterties, *M.T.P. International Review of Science*, Vol. 9, M. L. Tobe, Ed., Butterworths, London, 1972, p. 37.
314. D. Hellwinkel in *Organic Phosphorus Compounds*, G. M. Kosolapoff and L. Maier, Eds., Vol. 3, Wiley-Interscience, New York, 1972, p. 185.
315. D. Luckenbach, *Dynamic Stereochemistry of Pentacoordinated Phosphorus*, G. Thieme, Stuttgart (1973).
316. I. Ugi, F. Ramirez, D. Marquarding, H. Klusacek, and P. Gillespie, *Accounts Chem. Res.*, **4,** 288 (1971).
316a. P. Gillespie, P. Hoffman, H. Klusacek, D. Marquarding, S. Pfohl, F. Ramirez, E. A. Tsolis, and I. Ugi, *Angew. Chem.*, **83,** 691, *Internat. Edn.*, **10,** 687 (1971).
316b. F. Ramirez and I. Ugi, *Progr. Phys. Org. Chem.*, **9,** 25 (1971).
316c. I. Ugi and F. Ramierez, *Chem. in Britain*, 198 (1972).
317. F. Ramirez, S. Pfohl, E. A. Tsolis, J. F. Pilot, C. P. Smith, I. Ugi, D. Marquarding, P. Gillespie, and P. Hoffman, *Phosphorus*, **1,** 1 (1971).
318. G. M. Whitesides and W. M. Bunting, *J. Amer. Chem. Soc.*, **89,** 6801 (1967).
318a. G. M. Whitesides and H. L. Mitchell, *J. Amer. Chem. Soc.*, **91,** 5384 (1969).
318b. M. Eisenhut, H. L. Mitchell, D. D. Traficante, R. J. Kaufman, J. M. Deutch, and G. M. Whitesides, *J. Amer. Chem. Soc.*, **96,** 5385 (1974).
318c. G. M. Whitesides, M. Eisenhut, and W. M. Bunting, *J. Amer. Chem. Soc.*, **96,** 5398 (1974).
319. W. G. Klemperer, *J. Amer. Chem. Soc.*, **94,** 6940 (1972).
320. R. Hoffmann, J. M. Howell, and E. L. Muetterties, *J. Amer. Chem. Soc.*, **94,** 3047 (1972).
321. J. A. Howard, D. R. Russell, and S. Trippett, *J.C.S. Chem. Comm.*, 856 (1973).
322. H. Wunderlich, D. Mootz, R. Schmutzler, and M. Wieber, *Z. Naturforsch.*, **29,** 32 (1974).
323. H. Wunderlich, *Acta Cryst.*, **30B,** 939 (1974).
324. R. R. Holmes, *J. Amer. Chem. Soc.*, **96,** 4143 (1974).
325. J. I. G. Cadogan, R. O. Gould, and N. Tweddle, *J.C.S. Chem. Comm.*, 773 (1975).
326. L. Anschutz and W. Broeker, *Ber.*, **61,** 1264 (1928).
327. R. Burgada, D. Houalla, and R. Wolf, *Compt. rend.*, **264C,** 356 (1967).
328. H. Germa, M. Sanchez, R. Burgada, and R. Wolf, *Bull. Soc. chim. France*, 612 (1970).
329. M. Sanchez, J. F. Brazier, D. Houalla, and R. Wolf, *Bull. Soc. chim. France*, 3930 (1967).

330. A. Munoz, M. Koenig, B. Garrigues, and R. Wolf, *Compt. rend.*, **274C,** 1413 (1972).
331. M. Koenig, A. Munoz, and R. Wolf, *Bull. Soc. chim. France*, 4185 (1971).
322. A. Munoz, M. Koenig, R. Wolf, and F. Mathis, *Compt. rend.*, **277C,** 277 (1973).
333. L. Lopez and J. Barrans, *Compt. rend.*, **273C,** 1540 (1971).
334. A. Schmidpeter and J. Luber, *Angew. Chem.*, **84,** 349, *Internat. Edn.*, **11,** 306 (1972).
335. R. Wolf, M. Sanchez, D. Houalla, and A. Schmidpeter, *Compt. rend.*, **275C,** 151 (1972).
336. Y. Charbonnel and J. Barrans, *Compt. rend.*, **274C**, 2209 (1972).
337. M. Sanchez, R. Wolf, R. Burgada, and F. Mathis, *Bull. Soc. chim. France*, 773 (1968).
338. N. P. Gretchkin, R. R. Shagidullin, and G. D. Goulanova, *Izvest. Akad. Nauk S.S.S.R., Ser. khim.*, 1797 (1968), *Chem. Abs.*, **70,** 11641 (1969).
339. D. Bernard, C. Laurenco, and R. Burgada, *J. Organometallic Chem.*, **47,** 113 (1973).
340. R. Burgada and C. Laurenco, *J. Organometallic Chem.*, **66,** 255 (1974).
341. M. Koenig, A. Munoz, R. Wolf, and D. Houalla, *Bull. Soc. chim. France*, 1413 (1972).
342. D. Houalla, M. Sanchez, L. Beslier, and R. Wolf, *Org. Magn. Resonance*, **3,** 45 (1971).
343. A. Munoz, M. Sanchez, M. Koenig, and R. Wolf, *Bull. Soc. chim. France*, 2193 (1974).
344. P. Gillespie, F. Ramirez, I. Ugi, and D. Marquarding, *Angew. Chem. Internat. Edn.*, **12,** 91 (1973).
345. T. Koizumi, Y. Watanabe, Y. Yoshida, and E. Yoshii, *Tetrahedron Letters*, 1075 (1974).
346. M. J. Gallagher, A. Munoz, G. Gence, and M. Koenig, *J.C.S. Chem. Comm.*, 321 (1976).
347. E. L. Muetterties, W. Mahler, and R. Schmutzler, *Inorg. Chem.*, **2,** 613 (1963).
348. E. L. Muetterties, W. Mahler, K. J. Packer, and R. Schmutzler, *Inorg. Chem.*, **3,** 1298 (1964).
349. W. Mahler and E. L. Muetterties, *Inorg. Chem.*, **4,** 1520 (1965).
350. G. O. Doak and R. Schmutzler, *Chem. Comm.*, 476 (1970).
351. J. R. Corfield, N. J. De'ath, and S. Trippett, *J. Chem. Soc.* (*C*), 1930 (1971).
351a. A. P. Stewart and S. Trippett, *Chem. Comm.*, 1279 (1970).
352. S. C. Peake and R. Schmutzler, *J. Chem. Soc.* (*A*), 1049 (1970).
353. M. A. Solkalskii, G. I. Drozd, M. A. Landau, and S. S. Dubov, *Zhur. strukt. Khim.*, **10,** 1113 (1969).
354. J. J. Harris and B. Rudner, *J. Org. Chem.*, **33,** 1392 (1968).
355. J. S. Harman and D. W. A. Sharp, *Inorg. Chem.*, **10,** 1538 (1971).
356. S. Trippett and P. J. Whittle, *J.C.S. Chem. Comm.*, 2302 (1972).
357. K. E. DeBruin, A. G. Padilla, and M. T. Campbell, *J. Amer. Chem. Soc.*, **95,** 4681 (1973).
358. R. K. Oram and S. Trippett, *J.C.S. Chem. Comm.*, 554 (1972).
359. J. I. Dickstein and S. Trippett, *Tetrahedron Letters*, 2203 (1973).
360. D. Gorenstein and F. H. Westheimer, *J. Amer. Chem. Soc.*, **92,** 634 (1970).
361. D. Gorenstein, *J. Amer. Chem. Soc.*, **92,** 644 (1970).
362. F. Ramirez, J. F. Pilot, O. P. Madan, and C. P. Smith, *J. Amer. Chem. Soc.*, **90,** 1275 (1968).

362a. F. Ramirez, M. Nagabushanam, and C. P. Smith, *Tetrahedron*, **24,** 1785 (1968).

363. C. H. Bushweller, H. S. Bilofsky, E. W. Turnblom, and T. J. Katz, *Tetrahedron Letters*, 2401 (1972).

364. R. E. Duff, R. K. Oram, and S. Trippett, *Chem. Comm.*, 1011 (1971).

365. D. Houalla, R. Wolf, D. Gagnaire, and J. B. Robert, *Chem. Comm.*, 443 (1969).

366. L. Beslier, M. Sanchez, D. Houalla, and R. Wolf, *Bull. Soc. chim. France*, 2563 (1971).

367. A. Klaebe, J. F. Brazier, F. Mathis, and R. Wolf, *Tetrahedron Letters*, 4367 (1972).

368. S. A. Bone, S. Trippett, M. W. White, and P. J. Whittle, *Tetrahedron Letters*, 1795 (1974).

369. B. C. Chang, W. E. Conrad, D. B. Denney, D. Z. Denney, R. Edelman, R. L. Powell, and D. W. White, *J. Amer. Chem. Soc.*, **93,** 4004 (1971).

370. S. Bone, S. Trippett, and P. J. Whittle, *J.C.S. Perkin I*, 2125 (1974).

371. J. D. Dunitz and V. Prelog, *Angew. Chem.*, **80,** 700, *Internat. Edn.*, **7,** 725 (1968).

372. P. C. Lauterbur and F. Ramirez, *J. Amer. Chem. Soc.*, **90,** 6722 (1968).

373. K. E. DeBruin, K. Naumann, G. Zon, and K. Mislow, *J. Amer. Chem. Soc.*, **91,** 7031 (1969).

374. D. J. Cram, J. Day, D. R. Rayner, D. M. von Schriltz, D. J. Duchamp, and D. C. Garwood, *J. Amer. Chem. Soc.*, **92,** 7369 (1970).

375. D. Hellwinkel, *Chem. Ber.*, **99,** 3642 (1966).

376. M. G. Newton, J. E. Collier, and R. Wolf, *J. Amer. Chem. Soc.*, **96,** 6888 (1974).

377. R. Contreras, J. F. Brazier, A. Klaebe, and R. Wolfe, *Phosphorus*, **2,** 67 (1972).

378. J. F. Brazier, A. C. Carrelhas, A. Klaebe, and R. Wolf, *Compt. rend.*, **277C**, 183 (1973).

379. A. Klaebe, A. C. Cachapuz, J. F. Brazier, and R. Wolf, *J.C.S. Perkin II*, 1668 (1974).

380. A. Klaebe, A. C. Carrelhas, J. F. Brazier, M. R. Marre, and R. Wolf, *Tetrahedron Letters*, 3971 (1974).

381. R. Burgada, M. Bon, and F. Mathis, *Compt. rend.*, **265C,** 1499 (1967).

382. D. Bernard and R. Burgada, *Phosphorus*, **3,** 187 (1974).

383. D. Houalla, M. Sanchez, and R. Wolf, *Org. Magn. Resonance*, **5,** 451 (1973).

384. D. Houalla, M. Sanchez, R. Wolf, M. Bois, D. Gagnaire, and J. B. Robert, *Org. Magn. Resonance*, **6,** 340 (1974).

385. R. S. Cahn, C. K. Ingold, and V. Prelog, *Angew. Chem.*, **78,** 413, *Internat. Edn.*, **5,** 385 (1966).

386. D. Hellwinkel, *Chimia (Aarau)*, **22,** 488 (1968).

387. D. Hellwinkel and H. J. Wilfinger, *Tetrahedron Letters*, 3423 (1969).

388. K. Mislow, *Accounts Chem. Res.*, **3,** 321 (1970).

389. F. Ramirez, K. Tasaka, and R. Hershberg, *Phosphorus*, **2,** 41 (1972).

390. F. Ramirez, S. B. Bhatia, and C. P. Smith, *Tetrahedron*, **23,** 2067 (1967).

391. M. Wieber and W. R. Hoos, *Tetrahedron Letters*, 5333 (1968).

391a. M. Wieber and K. Foroughi, *Angew. Chem.*, **85,** 444, *Internat. Edn.*, **12,** 419 (1973).

392. A. P. Schapp, K. Kees, and A. L. T. Layer, *J. Org. Chem.*, **40,** 1185 (1975).

393. C. D. Hall, J. D. Bramblett, and F. F. S. Lin, *J. Amer. Chem. Soc.*, **94,** 9264 (1972).

394. W. G. Bentrude in *Free Radicals*, J. K. Kochi, Ed., Wiley-Interscience, New York, 1973.

395. R. R. Holmes, *J. Phys. Chem.*, **64,** 1295 (1960).

395a. K. B. Dillon and T. C. Waddington, *Chem. Comm.*, 1317 (1969).

396. P. D. Bartlett, A. L. Baumstark, M. E. Landis, and C. L. Lerman, *J. Amer. Chem. Soc.*, **96,** 5267 (1974).

397. D. Hellwinkel, *Angew. Chem.*, **78,** 985, *Internat. Edn.*, **5,** 968 (1966).

397a. R. W. Dennis and B. P. Roberts, *J. Organomet. Chem.*, **47,** C8 (1973).

397b. D. Griller and B. P. Roberts, *J. Chem. Soc. Perkin II*, 1416 (1973).

398. D. W. Allen, B. G. Hutley, and M. T. J. Mellor, *J.C.S. Perkin II*, 63 (1972).

399. E. Vedejs and K. A. J. Snoble, *J. Amer. Chem. Soc.*, **95,** 5778 (1973).

400. D. G. Gorenstein, *J. Amer. Chem. Soc.*, **94,** 2808 (1972).

401. D. Samuel and B. L. Silver, *Adv. Phys. Org. Chem.*, **3,** 123 (1965).

402. J. R. Corfield, J. R. Shutt, and S. Trippett, *Chem. Comm.*, 789 (1969).

403. K. L. Marsi, F. B. Burns, and R. T. Clark, *J. Org. Chem.*, **37,** 238 (1972).

404. S. E. Fishwick and J. A. Flint, *Chem. Comm.*, 182 (1968).

405. W. Egan, G. Chauvière, K. Mislow, R. T. Clark, and K. L. Marsi, *Chem. Comm.*, 733 (1970).

406. J. J. Monagle, T. W. Campbell, and H. F. McShane, *J. Amer. Chem. Soc.*, **84,** 4288 (1962).

407. H. Hoffmann, H. Forster, and G. Tor-oghossian, *Monatsh.*, **180,** 311 (1969).

408. C. R. Hall and D. J. H. Smith, *Tetrahedron Letters*, 1693 (1974).

409. T. J. Katz and E. W. Turnblom, *J. Amer. Chem. Soc.*, **92**, 6701 (1970).

410. E. W. Turnblom and T. J. Katz, *J. Amer. Chem. Soc.*, **93,** 4065 (1971).

410a. E. W. Turnblom and T. J. Katz, *J.C.S. Chem. Comm.*, 1270 (1972).

411. D. Hellwinkel and B. Knabe, *Phosphorus*, **2,** 129 (1972).

412. H. Goldwhite, *Chem. Comm.*, 651 (1970).

413. J. P. Casey and K. Mislow, *Chem. Comm.*, 1410 (1970).

414. D. Hellwinkel and M. Bach, *Naturwiss.*, **56,** 214 (1969).

415. D. Hellwinkel, *Chem. Ber.*, **98,** 576 (1965).

416. H. R. Allcock and E. C. Bissell, *J. Amer. Chem. Soc.*, **95,** 3154 (1973).

417. A. Munoz, G. Gence, M. Koenig, and R. Wolf, *Bull. Soc. chim. France*, 1433 (1975).

418. M. Koenig, A. Munoz, D. Houalla, and R. Wolf, *J.C.S. Chem. Comm.*, 182 (1974).

419. M. Koenig, A. Klaebe, A. Munoz, and R. Wolf, in the press (1975).

420. B. C. Chang, D. B. Denney, R. L. Powell, and D. W. White, *Chem. Comm.*, 1070 (1971).

421. D. Hellwinkel and S. F. Mason, *J. Chem. Soc.* (*B*), 640 (1970).

422. G. E. Ryschkewitsch and J. M. Garrett, *J. Amer. Chem. Soc.*, **90,** 7234 (1968).

423. T. Ito, A. Kobayashi, F. Marumo, and Y. Saito, *Inorg., Nuclear Chem. Letters*, **7,** 1097 (1971.)

424. F. Ramirez, V. A. V. Prasad, and J. F. Marecek, *J. Amer. Chem. Soc.*, **96,** 7269 (1974).

425. D. Clark, H. M. Powell, and H. F. Wells, *J. Chem. Soc.*, 642 (1942).

426. L. Kolditz, *Z. Anorg. Allgem. Chem.*, **284,** 144 (1956).

427. H. P. Latscha and P. B. Hormuth, *Angew. Chem.*, **80,** 281; *Internat. Edn.*, **7,** 299 (1968).

428. L. Lopez, M. T. Boisdon, and J. Barrans, *Comp. rend.*, **275C,** 295 (1972).
429. R. Burgada, D. Bernard, and C. Laurenco, *Compt. rend.*, **276C,** 297 (1973).
430. D. B. Cooper, T. D. Inch, and G. J. Lewis, *J.C.S. Chem. Comm.*, 1043 (1974).
431. D. B. Cooper, J. M. Harrison, T. D. Inch, and G. J. Lewis, *J.C.S. Perkin I*, 1049 (1974).
432. J. M. Harrison, T. D. Inch, and G. J. Lewis, *J.C.S. Perkin I*, 1053 (1974).
433. D. B. Cooper, J. M. Harrison, T. D. Inch, and G. J. Lewis, *J.C.S. Perkin I*, 1058 (1974).
434. T. D. Inch, G. J. Lewis, R. G. Wilkinson, and P. Watts, *J.C.S. Chem. Comm.*, 500 (1975).
435. W. S. Wadsworth and Y. G. Tsay, *J. Org. Chem.* **39**, 984 (1974).
436. S. E. Fishwick, J. A. Flint, W. Hawes, and S. Trippett, *Chem. Comm.*, 1113 (1967).
437. S. E. Cremer, *Chem. Comm.*, 1132 (1968).
438. D. Hellwinkel and B. Krapp, *Angew. Chem.*, **86,** 524; *Internat. Edn.*, **13,** 542 (1974).
439. Y. Segall, I. Granoth, A. Kalir, and E. D. Bergmann, *J.C.S. Chem. Comm.*, 399 (1975).
440. J. T. Braunholtz and F. G. Mann, *J. Chem. Soc.*, 3285 (1957).
441. M. J. Gallagher and F. G. Mann, *J. Chem. Soc.*, 5110 (1962).
442. G. Baccolini and P. E. Todesco, *J. Org. Chem.*, **40,** 2318 (1975).
443. E. Cernia, G. M. Giongo, F. Marcati, and N. Palladino, *Inorg. Chim. Acta*, **11,** 195, (1974).

INDEX

This index contains most of the names of heterocyclic and other compounds of stereochemical interest, but not all the compounds in the text. Authors' names mentioned in the text are indexed.